U0902638

中国高等植物

彩色图鉴

Higher Plants of China in Colour

《中国高等植物彩色图鉴》编委会　主编

Edited by
Editorial Committee of
Higher Plants of China in Colour

科学出版社
北京

中国高等植物
彩色图鉴
Higher Plants of China in Colour

《中国高等植物彩色图鉴》编委会　主编
Edited by Editorial Committee of Higher Plants of China in Colour

第3卷　被子植物　木麻黄科－莲叶桐科
Volume Ⅲ　Angiosperms　Casuarinaceae—Hernandiaceae

卷编辑　王文采 刘 冰
Edited by Wentsai WANG
Bing LIU

内 容 简 介

本套图鉴精选中国境内野生高等植物和重要栽培植物1万余种，配以图片近2万张，每一物种以中英文形式简要介绍植物的中文名、拉丁学名、形态特征、花果期、生境和分布。图鉴共分为9卷，收载苔藓植物100科、蕨类植物40科、裸子植物11科、被子植物232科，共计383科，且除苔藓植物之外，已收全所有科。本套图鉴是继《中国高等植物图鉴》、《中国植物志》、Flora of China之后，又一部大型植物分类学巨著。本卷为第3卷。

本书适合植物学领域的科研人员、管理人员及爱好植物学的普通大众阅读和收藏。

This set of pictorial books contains nearly 20 thousand photographs, presenting the cream of wild higher plants and important cultivated plants in China, the species of which number more than 10 thousand. Each of the species is concisely introduced in both Chinese and English from such aspects as Chinese name, Latin name, morphological features, flowering and fruiting season, habitat and distribution. Divided into nine volumes, this work includes 100 bryophyte families, 40 pteridophyte families, 11 gymnosperm families and 232 angiosperm families, 383 families altogether; the inclusion of all the said families is complete except for the bryophytes. The set of pictorial books is another monumental work on plant taxonomy, after *Iconographia Cormophytorum Sinicorum*, *Flora Reipublicae Popularis Sinicae*, and *Flora of China*. This is volume Ⅲ of the series.

This work is intended for scientific researchers and administrators in the field of botany and also for botany enthusiasts. As well as for reading, the work can be a classic collection.

图书在版编目(CIP)数据

中国高等植物彩色图鉴＝Higher Plants of China in Colour. 第3卷，被子植物. 木麻黄科—莲叶桐科：汉英 /《中国高等植物彩色图鉴》编委会主编；王文采，刘冰分册主编. —北京：科学出版社，2016.1

ISBN 978-7-03-047061-4

Ⅰ. ①中… Ⅱ. ①中… ②王… ③刘… Ⅲ. ①高等植物-中国-图集 ②木麻黄科-中国-图集 ③莲叶桐科-中国-图集 Ⅳ. ①Q949.4-64

中国版本图书馆CIP数据核字（2016）第013524号

责任编辑：王 静 付 聪 马 俊 / 责任校对：李 影

责任印制：肖 兴 / 书籍设计：北京美光设计制版有限公司

科 学 出 版 社 出版

北京东黄城根北街16号

邮政编码：100717

http://www.sciencep.com

北京汇瑞嘉合文化发展有限公司 印刷

科学出版社发行 各地新华书店经销

*

2016年1月第 一 版 开本：787×1092 1/16

2016年1月第一次印刷 印张：39

字数：1 779 000

定价：580.00元

（如有印装质量问题，我社负责调换）

《中国高等植物彩色图鉴》编委会

Editorial Committee of Higher Plants of China in Colour

丛书图片主要拍摄者

(按姓氏汉语拼音排序)

阿不都拉·阿巴斯　白鹭　白重炎　毕延超　邴艳红　曹同　车晋滇　陈彬　陈高
陈丽　陈庆　陈鑫　陈炳华　陈世品　陈贤兴　陈又生　陈志雄　成晓　程文达
迟敏杰　戴攀峰　邓涛　邓云飞　丁炳扬　丁学欣　董仕勇　杜诚　杜巍　杜玉芬
段长虹　段士民　方振东　方振兴　费勇　冯君茹　傅连中　甘啟良　高贤明　高信芬
高云东　葛斌杰　耿玉英　古训铭　顾余兴　管开云　郭世伟　韩国营　郝加琛　郝云庆
何海　何理　何春梅　何国生　和兆荣　侯元同　胡光万　胡国雄　华国军　黄健
黄江华　黄圣卓　黄向旭　黄俞淞　惠肇祥　季定乾　贾渝　姜林　蒋宏　蒋蕾
蒋日红　金伟涛　金孝锋　金效华　康世昌　赖阳均　郎楷永　黎斌　黎兴江　李东
李恒　李凯　李敏　李攀　李不言　李策宏　李东辉　李家湘　李建民　李剑武
李良千　李文军　李先源　李晓东　李小杰　李新华　李新伟　李学东　李泽贤　李振宇
李志奇　李智选　李中阳　梁同军　廖明林　廖云标　林敏　林祁　林维　林广旋
林建勇　林俊杰　林茂祥　林秦文　林哲丽　刘冰　刘博　刘静　刘军　刘坤
刘夙　刘翔　刘鑫　刘演　刘莹　刘大伟　刘光裕　刘海桑　刘红梅　刘伦辉
刘全儒　刘晟源　刘怡涛　刘正宇　刘宗才　柳永红　卢刚　卢元　罗柳青　骆适
吕碧凤　吕志学　马林　马炜梁　马文章　马欣堂　莫水松　牟善杰　沐先运　慕泽泾
南程慧　倪静波　倪素碧　农东新　潘勃　潘建斌　彭博　彭镜毅　彭日成　乔明明
秦卫华　秦祥堃　覃海宁　邱志敬　仁琛　任飞　任丽华　任明波　任昭杰　尚策
邵剑文　沈阳肇　施忠辉　石硕　寿海洋　税玉民　宋纬文　宋柱秋　买买提明·苏来曼
苏丽飞　苏享修　孙航　孙苗　孙观灵　孙明洲　孙卫邦　孙小美　谭运洪　陶国达
田乾福　田新民　童毅　童毅华　汪远　王辰　王东　王泓　王晖　王健
王进　王强　王颖　王耘　王喆　王长荣　王钧杰　王慷林　王清隆　王秋美
王文卿　王雅琼　王亚玲　王英伟　王玉兵　王正元　王祝年　韦宏金　韦毅刚　韦玉梅
卫然　魏来　温韩东　温九良　翁茂伦　吴丰　吴磊　吴双　吴棣飞　吴凤琴
吴光弟　吴国晞　吴林芳　吴声华　吴望辉　吴问舫　吴永红　吴增源　伍凯　武全安
武素功　武玉东　夏念和　肖翠　肖亮　肖艳　肖红菊　谢磊　辛夷　辛晓伟
辛益群　辛宇明　熊源新　徐徭　徐锦泉　徐克学　徐连升　徐申健　徐文斌　徐晔春
徐永福　许为斌　寻路路　严新富　严岳鸿　阳文静　杨浩　杨永　杨成梓　杨建昆
杨金财　杨科明　杨青山　杨世雄　杨奕维　杨增宏　姚永飚　叶德平　叶建飞　叶喜阳
叶幸儿　易思荣　殷建涛　尹志坚　于胜祥　郁文彬　喻勋林　袁彩霞　曾孝濂　曾云保
张力　张良　张强　张伟　张莹　张勇　张彩飞　张重岭　张代贵　张凤秋
张海华　张宏伟　张金龙　张金政　张守君　张淑梅　张树仁　张维柱　张宪春　张霄林
张志翔　赵宏　赵伟　赵大昌　郑宝江　郑希龙　郑小明　钟智明　周繇　周重建
周海成　周家宝　周兰平　周喜乐　周小林　周浙昆　朱弘　朱大海　朱仁斌　朱淑霞
朱维明　朱鑫鑫　David E. Boufford　Dmitry Sokoloff　Jan Thomas Johansson
Jozef Lemmens　Kirill Tkachenko　Pavel Novák　Ralf Knapp　Richard Ree
Susan Kelley

香港植物标本室(免费提供)

Major Photographers of the Series

(in the order of Chinese pinyin)

Abdulla ABASI	Lu BAI	Chongyan BAI	Yanchao BI	Yanhong BING	Tong CAO
Jindian CHE	Bin CHEN	Gao CHEN	Li CHEN	Qing CHEN	Xin CHEN
Binghua CHEN	Shipin CHEN	Xianxing CHEN	Yousheng CHEN	Chihhsiung CHEN	Xiao CHENG
Wenda CHENG	Minjie CHI	Panfeng DAI	Tao DENG	Yunfei DENG	Bingyang DING
Xuexin DING	Shiyong DONG	Cheng DU	Wei DU	Yufen DU	Changhong DUAN
Shimin DUAN	Zhendong FANG	Zhenxing FANG	Yong FEI	Junru FENG	Lianzhong FU
Qiliang GAN	Xianming GAO	Xinfen GAO	Yundong GAO	Binjie GE	Yuying GENG
Xunming GU	Yuxing GU	Kaiyun GUAN	Shiwei GUO	Guoying HAN	Jiachen HAO
Yunqing HAO	Hai HE	Li HE	Chunmei HE	Guosheng HE	Zhaorong HE
Yuantong HOU	Guangwan HU	Guoxiong HU	Guojun HUA	Jian HUANG	Jianghua HUANG
Shengzhuo HUANG	Xiangxu HUANG	Yusong HUANG	Zhaoxiang HUI	Dingqian JI	Yu JIA
Lin JIANG	Hong JIANG	Lei JIANG	Rihong JIANG	Weitao JIN	Xiaofeng JIN
Xiaohua JIN	Shihchang KANG	Yangjun LAI	Kaiyong LANG	Bin LI	Xingjiang LI
Dong LI	Heng LI	Kai LI	Min LI	Pan LI	Buyan LI
Cehong LI	Donghui LI	Jiaxiang LI	Jianmin LI	Jianwu LI	Liangqian LI
Wenjun LI	Xianyuan LI	Xiaodong LI	Xiaojie LI	Xinhua LI	Xinwei LI
Xuedong LI	Zexian LI	Zhenyu LI	Zhiqi LI	Zhixuan LI	Zhongyang LI
Tongjun LIANG	Minglin LIAO	Yunbiao LIAO	Min LIN	Qi LIN	Wei LIN
Guangxuan LIN	Jianyong LIN	Junjie LIN	Maoxiang LIN	Qinwen LIN	Zheli LIN
Bing LIU	Bo LIU	Jing LIU	Jun LIU	Kun LIU	Su LIU
Xiang LIU	Xin LIU	Yan LIU	Ying LIU	Dawei LIU	Guangyu LIU
Haisang LIU	Hongmei LIU	Lunhui LIU	Quanru LIU	Shengyuan LIU	Yitao LIU
Zhengyu LIU	Zongcai LIU	Yonghong LIU	Gang LU	Yuan LU	Liuqing LUO
Shi LUO	Pifong LU	Zhixue LÜ	Lin MA	Weiliang MA	Wenzhang MA
Xintang MA	Shuisong MO	Shannjye MOORE	Xianyun MU	Zejing MU	Chenghui NAN
Jingbo NI	Subi NI	Dongxin NONG	Bo PAN	Jianbin PAN	Bo PENG
Ching-I PENG	Richeng PENG	Mingming QIAO	Weihua QIN	Xiangkun QIN	Haining QIN
Zhijing QIU	Chen REN	Fei REN	Lihua REN	Mingbo REN	Zhaojie REN
Ce SHANG	Jianwen SHAO	Yangzhao SHEN	Zhonghui SHI	Shuo SHI	Haiyang SHOU

Yumin SHUI	Weiwen SONG	Zhuqiu SONG	Mamtimin SULAYMAN	Lifei SU	Xiangxiu SU
Hang SUN	Miao SUN	Guanling SUN	Mingzhou SUN	Weibang SUN	Xiaomei SUN
Yunhong TAN	Guoda TAO	Qianfu TIAN	Xinmin TIAN	Yi TONG	Yihua TONG
Yuan WANG	Chen WANG	Dong WANG	Hong WANG	Hui WANG	Jian WANG
Jin WANG	Qiang WANG	Ying WANG	Yun WANG	Zhe WANG	Changrong WANG
Junjie WANG	Kanglin WANG	Qinglong WANG	Chiumei WANG	Wenqing WANG	Yaqiong WANG
Yaling WANG	Yingwei WANG	Yubing WANG	Zhengyuan WANG	Zhunian WANG	Hongjin WEI
Yigang WEI	Yumei WEI	Ran WEI	Lai WEI	Handong WEN	Jiuliang WEN
Maolun WENG	Feng WU	Lei WU	Shuang WU	Difei WU	Fengqin WU
Guangdi WU	Guoxi WU	Linfang WU	Shenghua WU	Wanghui WU	Wenfang WU
Yonghong WU	Zengyuan WU	Kai WU	Quan'an WU	Sugong WU	Yudong WU
Nianhe XIA	Cui XIAO	Liang XIAO	Yan XIAO	Hongju XIAO	Lei XIE
Yi XIN	Xiaowei XIN	Yiqun XIN	Yuming XIN	Yuanxin XIONG	Yao XU
Jinquan XU	Kexue XU	Liansheng XU	Shenjian XU	Wenbin XU	Yechun XU
Yongfu XU	Weibin XU	Lulu XUN	Hsinfu YEN	Yuehong YAN	Wenjing YANG
Hao YANG	Yong YANG	Chengzi YANG	Jiankun YANG	Jincai YANG	Keming YANG
Qingshan YANG	Shixiong YANG	Yifei YANG	Zenghong YANG	Yongbiao YAO	Deping YE
Jianfei YE	Xiyang YE	Xing'er YE	Sirong YI	Jiantao YIN	Zhijian YIN
Shengxiang YU	Wenbin YU	Xunlin YU	Caixia YUAN	Xiaolian ZENG	Yunbao ZENG
Li ZHANG	Liang ZHANG	Qiang ZHANG	Wei ZHANG	Ying ZHANG	Yong ZHANG
Caifei ZHANG	Chongling ZHANG	Daigui ZHANG	Fengqiu ZHANG	Haihua ZHANG	Hongwei ZHANG
Jinlong ZHANG	Jinzheng ZHANG	Shoujun ZHANG	Shumei ZHANG	Shuren ZHANG	Weizhu ZHANG
Xianchun ZHANG	Xiaolin ZHANG	Zhixiang ZHANG	Hong ZHAO	Wei ZHAO	Dachang ZHAO
Baojiang ZHENG	Xilong ZHENG	Xiaoming ZHENG	Zhiming ZHONG	You ZHOU	Chongjian ZHOU
Haicheng ZHOU	Jiabao ZHOU	Lanping ZHOU	Xile ZHOU	Xiaolin ZHOU	Zhekun ZHOU
Hong ZHU	Dahai ZHU	Renbin ZHU	Shuxia ZHU	Weiming ZHU	Xinxin ZHU
David E. Boufford	Dmitry Sokoloff	Jan Thomas Johansson		Jozef Lemmens	Kirill Tkachenko
Pavel Novák	Ralf Knapp	Richard Ree	Susan Kelley		

Hong Kong Herbarium (free of charge)

丛书文字主要编写者

(按姓氏汉语拼音排序)

陈世龙　陈又生　陈之端　成　晓　崔逸群　邓云飞　杜　宁　段士民　樊　杰　方瑞征
费　勇　傅立国　高　凡　高　乞　谷粹芝　郭　慧　韩　宇　郝加琛　何　理　侯学良
侯元同　胡光万　黄向旭　黄俞淞　蒋　宏　金孝锋　金效华　赖阳均　雷立公　黎　斌
李　恒　李　嵘　李秉滔　李宏哲　李剑武　李梦华　李文军　李锡文　李晓贤　李新华
李新伟　李章海　李振宇　李中阳　廖文波　林秦文　刘　冰　刘　博　刘　演　刘大伟
刘海桑　刘全儒　刘衍男　卢金梅　马欣堂　潘　勃　覃海宁　邱志敬　萨　仁　尚　策
税玉民　孙　苗　孙久琼　万　涛　王　东　王　晖　王　健　王　强　王德艺　王文采
王英伟　卫　然　魏　来　吴　磊　吴鹏程　吴声华　向巧萍　向小果　谢　磊　徐松芝
徐晓婷　薛大伟　闫瑞亚　严岳鸿　杨世雄　叶建飞　游旨价　于胜祥　袁　慷　张　力
张　梅　张　强　张重岭　张钢民　张红瑞　张树仁　张宪春　张志翔　赵　宏　赵存峰
周兰平　左　勤

Major Textwriters of the Series

(in the order of Chinese pinyin)

Shilong CHEN	Yousheng CHEN	Zhiduan CHEN	Xiao CHENG	Yiqun CUI
Yunfei DENG	Ning DU	Shimin DUAN	Jie FAN	Ruizheng FANG
Yong FEI	Liguo FU	Fan GAO	Qi GAO	Cuizhi GU
Hui GUO	Yu HAN	Jiachen HAO	Li HE	Xueliang HOU
Yuantong HOU	Guangwan HU	Xiangxu HUANG	Yusong HUANG	Hong JIANG
Xiaofeng JIN	Xiaohua JIN	Yangjun LAI	Ligong LEI	Bin LI
Heng LI	Rong LI	Bingtao LI	Hongzhe LI	Jianwu LI
Menghua LI	Wenjun LI	Xiwen LI	Xiaoxian LI	Xinhua LI
Xinwei LI	Zhanghai LI	Zhenyu LI	Zhongyang LI	Wenbo LIAO
Qinwen LIN	Bing LIU	Bo LIU	Yan LIU	Dawei LIU
Haisang LIU	Quanru LIU	Yannan LIU	Jinmei LU	Xintang MA
Bo PAN	Haining QIN	Zhijing QIU	Ren SA	Ce SHANG
Yumin SHUI	Miao SUN	Jiuqiong SUN	Tao WAN	Dong WANG
Hui WANG	Jian WANG	Qiang WANG	Deyi WANG	Wentsai WANG
Yingwei WANG	Ran WEI	Lai WEI	Lei WU	Pengcheng WU
Shenghua WU	Qiaoping XIANG	Xiaoguo XIANG	Lei XIE	Songzhi XU
Xiaoting XU	Dawei Xue	Ruiya YAN	Yuehong YAN	Shixiong YANG
Jianfei YE	Zhijia YOU	Shengxiang YU	Qian YUAN	Li ZHANG
Mei ZHANG	Qiang ZHANG	Chongling ZHANG	Gangmin ZHANG	Hongrui ZHANG
Shuren ZHANG	Xianchun ZHANG	Zhixiang ZHANG	Hong ZHAO	Cunfeng ZHAO
Lanping ZHOU	Qin ZUO			

丛书前言

中国是世界上植物最丰富的国家之一，已知有三万五千多种野生和重要栽培的高等植物，其中特有种达一万五千多种，形成复杂而独具特色的植物区系。中国的先人们创造了古老而辉煌的农业文明，选育出水稻、大豆、茶、枣、桃、柿等重要作物，其中水稻的栽培历史可追溯到约七千年前新石器时期的河姆渡文化，如今稻米已成为世界上近一半人口的粮食。丰富的植物资源和灿烂的历史文化，使中国成为“花园之母”和世界农作物七大起源中心之一。

中国植物学家为了系统地展示中国植物的多样性，历经艰辛，相继编研了《中国高等植物图鉴》和《中国植物志》，并与外国专家合作出版Flora of China等大型志书，这些著作在国内外应用广泛、影响巨大，客观地展现了不同时期的植物分类学研究和植物资源调查的成果，成为植物分类学领域最重要的大型经典著作。但是，它们都有一个共同的缺憾，即仅有黑白线条图，难以充分表达植物各器官的质地和颜色等自然状态下的外貌特征，其效果难以满足部分读者鉴赏植物的需要。

大多数发达国家都有自己的植物彩色图鉴，这些图鉴不仅展示了本国的生物多样性，还兼备工具书功能和富有感染力的艺术效果，具有很高的应用和收藏价值。迄今为止，国内出版的植物彩色图书多为地区性的，或局限于某一类植物的，如观赏植物、栽培作物和药用植物。作为世界生物多样性大国，中国应当拥有一套全面体现本国野生植物多样性的的大型鉴赏类彩色图册。

将灿烂的瞬间变为永恒是广大植物爱好者和摄影爱好者的追求。为了填补上述空白，台湾吴声华研究员策划并启动了这项工作。在海峡两岸学者的共同努力下，本书的规模在不断扩大，从最初的云南植物写真集扩展到全国性大型彩色植物图鉴。中国科学院植物研究所王文采院士出任丛书编委会主任，吴声华研究员和中国科学院植物研究所李振宇研究员任副主任。编委会遴选国内从事植物分类学研究的专家担任各卷卷编辑，邀请中国大陆、台湾和香港近200位植物学家承担各科的编写和审稿工作，卷编辑在专家审稿的基础上，再次对本卷内容进行核查。近400位摄影作者提供了大量精美的植物彩色照片。丛书还采用了著名的动植物科学画大师曾孝濂先生绘制的20余幅优雅而灵动的彩色图片。

本丛书划分为九卷，共收录中国高等植物1万余种，种类以野生植物为主，同时收载重要的栽培植物，精选图片近2万张。本丛书中科的系统排列如下：苔藓植物主要参考《中国苔藓志》中的系统；蕨类植物按张宪春2015年在《石松类和蕨类名词及名称》提出的系统；裸子植物和被子植物的系统排列按第尔斯(L. Diels, 1936)于A. Engler's Syllabus der Pflanzenfamilien中采用的系统。仅第三卷将毛茛科分为星叶草科、毛茛科和芍药科，将木兰科分为木兰科、八角科、五味子科和水青树科。全书收载中国高等植物383科，其中苔藓植物100科，占全国苔藓科总数的大多数；其余是蕨类植物40科，裸子植物11科，被子植物232科，分别代表了国产三大门类所有的科。本丛书收载的植物中有一些是Flora of China出版后发表的国产新种，如香港鹅耳枥(*Carpinus insularis*)、球柱楼梯草(*Elatostema globosostigmatum*)和西藏小囊兰(*Micropera tibetica*)，以及中国分布新记录，如轮叶三棱栎(*Trigonobalanus verticillata*)和格力兜兰(*Paphiopedilum gratrixianum*)。

为了方便更多的读者阅读，本丛书的文字采用中英文，简要介绍各种植物的中文名、拉丁学名、形态特征、花果期、生境和分布。

本书在编写过程中，承中国科学院植物研究所中国植物图像库和中国自然标本馆提供了许多方便和帮助，在此向他们表示衷心的感谢。

感谢国家出版基金和科学出版社对本丛书出版的大力支持。

由于编著者的业务水平有限、错漏之处，欢迎批评指正。

《中国高等植物彩色图鉴》编委会

2015年10月31日

Preface to the Series

As one of the countries with the richest diversity of plant species in the world, China has more than 35 000 known species of wild and important cultivated higher plants, among which there are over 15 000 endemic species, forming a complex and unique flora. The ancestors of the Chinese people created an ancient and splendid agricultural civilization. They selected and cultivated significant crops like rice, soya bean, tea, jujube, peach, and persimmon. Among these crops, the cultivation history of rice can be traced back to the Hemudu culture of the Neolithic Period around 7000 years ago. Nowadays, rice has become the staple food for nearly half of the world's population. With abundant plant resources and a long history and great culture, China is renowned as 'the mother of gardens' and is one of the seven important centers of origin for crops in the world.

In order to present the diversity of China's plants systematically, botanists from China have made pain-taking efforts to compile a series of large-volume floras including *Iconographia Cormophytorum Sinicorum*, *Flora Reipublicae Popularis Sinicae* (Chinese version) by themselves, and *Flora of China* (English version) with the collaboration of international specialists. These books are well known both in China and abroad and have been used extensively for studying Chinese plants and plants from adjacent areas, these works present the results of plant taxonomic study and study of plant resources in China at different periods, and constitute some of the most important large classic volumes in the field of plant taxonomy. However, in all these works the plants are only partly illustrated by black and white line-drawings, unable to present the texture and colour of flowers and leaves fully in their natural state, and they hardly reveal the spectacular beauty and fascination of the wealth of plant species.

Most developed countries have colour pictorial books of their plants, which form greatly desirable works, because they are not only a presentation of the plant diversity of the countries, but are also an attractive record of the beauty of the nature. So far, most of the Chinese colour pictorial books of plants are regional treatments, or concentrate on particular groups, such as ornamental plants, cultivated crops and medicinal plants or certain taxonomic groups. As a country with a high level of biodiversity, China merits a large-scale colour pictorial book with high appreciation value featuring the wild plants that occur within its territory.

It is the goal of every lover of plants and plant photography to capture the essence of plant beauty and make it permanent. In order to fill the above-mentioned gap, Professor Shenghua WU from Taiwan, planned and launched the present project. With the joint effort of specialists from all over China, the scale of the book has expanded from the initial pictorial book of plants of Yunnan to a many-volume colour pictorial book of plants of the whole country. Academician Wentsai WANG of the Institute of Botany, Chinese Academy of Sciences, took up the post of the chairman of the editorial committee, and the positions of vice chairmen of the editorial committee were assumed by Prof. Shenghua WU, Taiwan, and Prof. Zhenyu LI of the Institute of Botany, Chinese Academy of Sciences. The editorial committee then selected experienced plant taxonomists as volume editors for each volume, and invited nearly 200 botanists from mainland China, Taiwan and Hong Kong to undertake the compilation and reviewing work for each plant family by volumes. Nearly 400 photographers provided numerous beautiful full colour plant photos. The well known zoological and botanical artist, Xiaolian ZENG, kindly allowed the use of more than twenty of his elegant and vivid plant portraits in this series.

This whole work is divided into nine volumes, depicting more than 10 thousand species of higher plants from China, dealing mainly with wild plants, but also including some important cultivated plants, and has involved the careful selection of nearly 20 thousand photographs. The system arrangement for plant families are as follows: bryophytes are mainly arranged according to the system used in *Flora Bryophytorum Sinicorum*; pteridophytes are arranged according to the system proposed by Professor Xianchun ZHANG in *A Glossary of Terms and Names of Lycopods and Ferns* (2015); gymnosperms and angiosperms are arranged according to the system used in *A. Engler's Syllabus der Pflanzenfamilien* (L. Diels, 1936), with the difference that in Volume III, Ranunculaceae is divided into Circaesteraceae, Ranunculaceae, and Paeoniaceae, and Magnoliaceae is divided into Magnoliaceae, Illiciaceae, Schisandraceae, and Tetracentraceae. The higher plants of China included in this work comprise 383 families; with 100 families of bryophytes, which represent the majority of the bryophyte families in China; the others are 40 pteridophyte families, 11 gymnosperm families and 232 angiosperm families, which represent all the families distributed in China respectively. The work includes some new additions of species published since *Flora of China*, such as *Carpinus insularis*, *Elatostema globosostigmatum*, *Micropera tibetica*, and new distribution records for China, such as *Trigonobalanus verticillata* and *Paphiopedilum gratrixianum*.

To facilitate and attract readers from both China and abroad, the text of this book series is bilingual in Chinese and English, providing the Chinese name, Latin name, morphological features, flowering and fruiting season, habitat and distribution.

In the process of compiling this work, Plant Photo Bank of China (PPBC) and Chinese Field Herbarium, both of which are under the Institute of Botany of Chinese Academy of Sciences, provided great help with the selection of photographs, for which we express our gratitude.

We also thank National Publication Foundation and Science Press, Beijing, for their great support for the publication of this book series.

It will be appreciated if mistakes and omissions are brought to our attention.

Editorial Committee of *Higher Plants of China in Colour*
31 October, 2015

关于本图鉴

1995年夏天，我参加由中国科学院昆明植物研究所臧穆教授带领的云南野外工作，同行的还有国际真菌学会理事长德籍的Franz Oberwinkler教授与法国的学者。臧教授爽朗好客，外国人都喜欢他的热情。那年去丽江，再去南部的西双版纳。西双版纳热带植物园的陶国达先生带领我们的野外工作，他是当地植物鉴定首席专家，知道好的树林在何处。一天，在傣族传统农家的木架房子吃中饭。臧教授建议陶先生既然喜欢摄影，何不出一本版纳植物图鉴，问我能不能帮忙在台湾找出版。我答应回去问问。

先问自然科学博物馆的李家维馆长，他对植物研究及保育充满热诚，对这项工作有兴趣。但未久他感觉这项工作所需时间过久，博物馆经费也不足以出版。我又问其他出版公司，没有得到响应。我想应该先有成果再问出版吧，就请陶先生持续植物拍摄。臧教授和夫人黎兴江教授推荐了费勇帮忙这项工作。1997年夏天，我在昆明机场与臧教授和费勇会合，一同飞去版纳。费勇年纪与我相当，长得瘦黑，话不太多。陶先生带领我们野外工作。回程时费勇说他想找几个同事一起负责滇西北的植物拍摄工作，与陶先生滇南的工作结合成为云南植物图鉴。回台后看陶先生给我的幻灯片，感觉质量不是太好，问他才知道所用的相机是正牌，镜头却是小厂牌。我汇钱请他购买一套相机，以利拍摄质量。

1998年，我到昆明植物所，和陶国达、费勇及孙航，讨论植物图鉴工作。费勇对此工作充满兴趣，人缘也好，决定由他征集昆明植物所人员拍摄的植物照片，并且中、英文字也由他撰写。翌年臧穆夫妇介绍昆明植物所的著名画家曾孝濂先生。曾先生长期考察云南山野的植物与动物，画作结合了科学性与艺术美感，是中国写实花鸟画得最好的。

2000年，费勇在日本富山县中央植物园半年，其间拍摄植物园栽培的中国植物。那年秋天费勇带我去大理点苍山和楚雄紫溪山。一天，我们在大理古城一间白族旧庭院吃风味晚餐，他兴致好，畅所欲言。费勇起初给我的印象是有些木讷，几次往来后就把我当熟人。几次的讨论，感觉他满心想做好这件事，并不在意条件。大理巷弄中有摊贩卖当地特产乳扇，他说闺女爱吃，买了两大张带走。

2001年年初一个早上，臧教授发来邮件，通知我费勇前一日在丽江不幸去世。一个年轻健康生命的突然离去，令人难以承受。出席完上午的会议后即打电话到昆明。黎教授说费勇到丽江出差，半夜室友听到声响，见他口吐白沫急送医院。地方医院初以为是癫痫，到清晨就不治了。

几个月后我有事联络曾孝濂先生，他告诉我费勇太太想与我联系。费勇太太姓向，我们称小向。她电话中希望植物图鉴工作能够继续，而且费勇的几个同事愿意帮忙。当年夏天在昆明的一个晚上，小向同昆明植物所的成晓、孙航、周浙昆一起和我见面，商讨后续的工作。成晓说他与费勇是同学，同时毕业，同时上班，他一定会帮忙。他确实尽力后续工作的联系与推动。2002年在昆明，几个朋友见面，小向带初中的女儿同来。女儿乖巧懂事，我说长得像费勇，她眼眶微微红了。成晓研究蕨类，他的岳父武素功先生及岳母方瑞征女士也是昆明植物所学者，两位在图片提供及文稿修订均提供协助。昆明植物所李锡文教授对植物分类的造诣比较全面，负责图片和文稿审查。昆明植物所还有多位专家对本书工作做出贡献，不在此逐一罗列。

2001年，曾孝濂先生介绍昆明植物所的画家刘怡涛先生。刘先生在版纳热带植物园待过，建立独特的版纳风光绘画风格，也喜好摄影，带过我几次野外工作。他建议我把植物图鉴工作扩大到全中国。艺术家天生具有美感，曾、刘两位画家拍摄的植物图，构图与取景皆有独到之处。2003年我到河北与吉林进行野外工作，2004年到新疆与吉林时决定把植物图鉴范围扩大到全中国。我和小向说明书的分量和质量要到位，才能彰显费勇的努力精神。费勇原本即有中国植物图鉴的梦想，干脆一次到位。

2004年，中国科学院植物研究所覃海宁博士来台，我们是1994年在英国邱园认识的。中国植物图像库在海宁领导下建立得有声有色。海宁总是满脸笑容，热诚谦虚，听我说植物图鉴的事，立即寻思找人帮忙。他人面广，介绍不少人，拍摄较好的有福建的何国生、四川的吴光弟、广西的刘演、广东的李泽贤。刘演的图片色彩饱满令人赞叹。我去爱丁堡皇家植物园时知道David Chamberlain博士是杜鹃花科专家，他同意审查杜鹃花科及小檗科图片。彭镜毅介绍哈佛大学David E. Boufford博士，他的图片是从中国西南的横断山脉植物调查工作所拍摄。David又推荐Susan Kelley及Richard Ree提供植物图片。

中科院植物所吴鹏程教授是苔藓专家，1990年我在芬兰赫尔辛基大学即将取得博士学位时他在赫大待了几个月。吴教授介绍几位中科院植物所的专家帮忙图片审查及文字撰写。台湾真菌学前辈吕理燊博士介绍昆明市农业局副局长惠肇祥先生提供杂草图片，惠先生又介绍北京的车晋滇先生提供华北的杂草图片。台湾赖明洲教授介绍上海自然博物馆的秦祥堃先生提供华东植物图片，又介绍中国科学院沈阳应用生态研究所的赵大昌先生提供长白山植物图片。我2004年到乌鲁木齐开会，组织会议的新疆大学阿不都拉教授拍了不少新疆植物图片，也提供给我。

大学同学康世昌是植物及计算机高手，拍摄的植物图片也提供给我。他早预想到网络世界的影响力，不推荐大部头实体书的出版构想。多年前他写个网址要我去看，那是我不知道的“Google”，可以查询信息。网络上图片的数量越来越多，趋势是如此。我在芬兰的指导教授Tuomo Niemelä出版过大型真菌的小书，亲自编排，图片与文字搭配得美感十足，我每翻阅总是心情愉悦。我向Tuomo请教对这套植物实体书的意见。他说网络的数据有时会消失，且许多没经过审查。我想这套书终要完成，无法顾及趋势与新世代人类的想法。

2007年年初，我在网络发现中科院植物所的中国植物图像库有影像部分。负责的是李敏，我问他图片提供者，他推荐几位拍摄较好的。多数是中科院植物所的年轻人，有刘冰、林秦文、于胜祥、李敏、高贤明、郦艳红，还有陕西的王耘。我当时已收集中国植物5000种的图片，工作超过10年理应收尾。然不加入这批有许多北方植物的图片实在不舍。刘冰是植物分类奇葩，这么年轻就拍到数量惊人的植物图片。刘冰和刘怡涛是给这项工作提供图片最多的两位。刘博帮忙不少文字撰写及图片审查，工作积极。当年年底，图片收集到6500种以上，接着准备文字、图片审查等出书的各项工作。

2010年在台湾“中央研究院”召开一项研讨会，覃海宁和李敏也来了，他们的报告显示中国植物图像库已收到数十万张图片。我如果再搜寻一次图片，能收到更好及更多的图片，但面对许多人殷盼这套书问世，时间的延长，压力更大。终究，我相信费勇会支持这最后一批图片的征求。湖南喻勋林及张代贵两位教授寄来许多华中植物图片，浙江张宏伟先生及安徽施忠辉先生也送来图片。吉林通化的周繇教授寄来他辛苦拍摄的长白山植物图片。近三年送来较多图片的还有朱鑫鑫、陈又生、徐晔春、陈彬、陈世品、何海、周喜乐，以及蕨类的张宪春和兰科的金效华。好友张力负责苔藓部分。

“中央研究院”彭镜毅博士提供了许多秋海棠科图片，也修订这科的文稿。牟善杰是台湾的蕨类学家，提供一些蕨类图片给本书，也审查过蕨类图片及文字。我在台湾大学念博士时，善杰是大学生，见他圆圆的笑脸，成天在标本馆研究。2010年11月，44岁的他突然中风走了，令人感慨！吕碧凤小姐是台湾优秀的业余蕨类专家，提供一些好的蕨类图片。还要感谢提供及审查图片的几位同事：王秋美、陈志雄、胡维新、黄俊霖、严新富和邱少婷。

早期收到的是幻灯片及少数印好的照片，2005年以后送来的是数码影像。数码图片干净，缺点是饱和度、清晰度和锐利度表现稍差，绿色部分有时偏黄。图效调整可改善这些问题。幻灯片的影像则会受到底片、冲洗、保存、扫描等质量的影响，好质量的并不多。图片须裁切出重点部位，再调整影像效果，这些工作大多是我处理。商请到一批人分别撰写文字。虽然有范本给撰写人参考，但各人的写法与仔细程度难免不一，有疏漏或小错误的情形普遍存在。起初我自己参考文献逐一查核，修订了约两千种的文字，但工作量太大，无法继续亲为。文字工作贡献较多的有费勇、刘博、萨仁、谷粹芝、成晓、杜宁、李锡文、徐晓婷和方瑞征等。

吴鹏程教授与科学出版社生物分社社长王静女士提及这项工作，王静有兴趣了解出版的可能，我们2010年在北京见面。自己过于深入这项工作，甚至如排版形式、字形等都亲自研究。像是自己养大的小孩，不放心交给他人处理，而且书稿已经在台湾找设计公司开始排版了。王静有毅力，持续两年逐渐消减我的疑虑。二十年来两岸的社会经济形势改变，使得这套书在大陆出版成为自然。德高望重的王文采院士及植物分类权威李振宇教授鼎力相助、组织动员，国家出版基金给予资助，促使整体工作能顺利完成。

吴声华

2015年10月20日

About the Pictorial Series

In the summer of 1995, I took part in the Yunnan fieldwork led by Prof. Mu ZANG from Kunming Institute of Botany, Chinese Academy of Sciences. Joining us were German professor Franz Oberwinkler, director general of International Mycological Association, and some French scholars. Prof. ZANG was candid, cordialand hospitable, which impressed everyone, especially the foreign guests. We first went to Lijiang and then Sipsongpanna in the south. During this fieldwork, we were guided by Mr. Guoda TAO from Xishuangbanna Tropical Botanical Garden, Chinese Academy of Sciences. He was the chief expert of plant identification in the area, knowing which areas of the woods were worth this field inspection of ours. One day, when we were having lunch together in a traditional wooden house of an ethnic Dai family, Prof. ZANG proposed to Mr. TAO: "Since you are so fond of photography, why not compile a pictorial book of Banna's plants?" Prof. ZANG then turned to me, asking whether I could give help in getting the book published in Taiwan, and I promised to give it a try after returning to Taiwan.

I first contacted Dr. Chiawei LI, director of Museum of Natural Science, who was passionate about plant research and conservation and interested in the project. But before long, his passion faded due to his sense that the project was likely to take too long a time, and the Museum did not have sufficient fund to support the publishing. I then inquired of other publishing companies, but none of them gave a positive response. These setbacks sent me thinking that perhaps we should make some tangible achievements first before our work could be accepted for publication. So I asked Mr. TAO to proceed with shooting plants. Prof. ZANG and his wife Prof. Xingjiang LI recommended Yong FEI to provide assistance to the work. In the summer of 1997, I met Prof. ZANG and Yong FEI at Kunming Airport, and we flew to Sipsongpanna together for the fieldwork led by Mr. TAO. Of the same age as mine, Yong FEI was a thin and swarthy man, not very talkative. On our way back, Yong FEI said he was considering asking several of his colleagues to join him in shooting plants of the northwest of Yunnan so that the pictures taken in the two areas (the SouthYunnan and the Northwest Yunnan) could be combined to make a single pictorial book that could be called "Plants of Yunnan". After returning to Taiwan, I browsed the slides given by Mr. TAO, feeling that their quality was not ideal. Having asked Mr. TAO about this, I learned that it had been caused by his camera whose main body was of good brand and quality but whose lens was made by a mediocre producer. I remitted money to him for purchasing a new camera set, hoping that the quality of photos could be ensured by a high-quality camera.

In 1998, I visited Kunming Institute of Botany to discuss with Guoda TAO, Yong FEI and Hang SUN about the work of pictorial book for plants. Yong FEI was full of enthusiasm on the work, and had good relations with people, so we decided to commission him to collect plant photos taken by staff from the Kunming Institute, and to compose text both in English and Chinese. The next year, Prof. Mu ZANG and Prof. LI introduced to me Mr. Xiaolian ZENG, who had been engaging in the investigation of plants and animals in the wilds of Yunnan Province for a long time and was also a famous painter from the Kunming Institute of Botany. His paintings are the best realistic bird-and-flower works in China, blending scientificity with artistic beauty.

In 2000, Yong FEI spent half a year in Botanic Gardens of Toyama (Japan), taking photos of Chinese plants grown in the Gardens. In autumn of the same year, with Yong FEI as my guide, we went to Diancang Mountain in Dali and Zixi Mountain in Chuxiong. One day, when we were having local delicacies for super in an old courtyard of Bai nationality, Yong FEI got into high spirit and chatted with me without restraint. My first impression of Yong FEI was that he was a bit unapproachable, but after several rounds of conversations, he regarded me as his close friend. After several discussions with him, I felt that he very much concentrated on doing the work well, paying no attention to remuneration. In a lane of Dali, we found a vendor selling milk fan cake, a kind of local specialty, and he bought two big pieces, saying that they were for his daughter who liked such food.

One early morning in the early 2001, Prof. ZANG sent me an email, saying sadly that Yong FEI passed away in Lijiang the day before. It was really unbearable to hear of the sudden passing of such a young life. As soon as the meeting in that morning ended, I called to Kunming. The call was answered by Prof. LI who said that Yong FEI had been on a working trip at the time. At midnight, his roommates were awakened by some noises and found him foaming at the mouth. He was rushed to a local hospital and initially diagnosed as only having a fit of epilepsy, but no amount of treatment took effect on him; he passed away just as dawn came.

Several months later, when contacting Mr. Xiaolian ZENG, I was told that Yong FEI's wife was looking for me. The family name of Yong FEI's wife was XIANG, so we called her Little XIANG, a traditional way of Chinese people addressing their acquaintances who were younger than themselves. Little XIANG expressed her wish over phone that the project of the pictorial book should go on as usual and she also said that several of Yong FEI's colleagues were willing to help. One summer evening of the same year in Kunming, Little XIANG, together with Xiao CHENG, Hang SUN and Zhekun ZHOU all from the Kunming Institute of Botany, had a meeting with me to discuss about the remaining work of the project. Xiao CHENG said he and Yong FEI were classmates, graduating and first getting employed at the same time, so he would definitely offer his help. And in fact he did try his best to facilitate the progress of the work through networking. In 2002, we had a gathering in Kunming. Little

XIANG brought her daughter there, who was then a junior-secondary-school student. The girl was both clever and well-behaved, and when I said to her that she looked like her father, her eyes moistened slightly. Xiao CHENG was a fern researcher. His father-in-law Mr. Sugong WU and mother-in-law Mrs. Ruizheng FANG were also scholars of the Kunming Institute of Botany, both of whom offered their assistance in providing photos and editing texts. Prof. Xiwen LI also from the Kunming Institute of Botany, who had comprehensive attainments in plant taxonomy, was responsible for examining photos and texts. There were many other experts from the Kunming Institute of Botany who made contributions to this book, but due to space constraint, their names are not listed here one by one.

In 2001, Mr. Xiaolian ZENG introduced to me Mr. Yitao LIU, another painter from Kunming Institute of Botany. Mr. LIU used to stay in Xishuangbanna Tropical Botanical Garden, where he developed his distinctive painting style with which to depict typical Sipsongpanna's landscape. He was also a lover of photography, and used to be my fieldwork guide for several times. Mr. LIU suggested that I expand the pictorial plant book project to cover the whole territory of China. Due to the innate aesthetic sense of artist, the plant photos taken by the two painters - Mr. ZENG and Mr. LIU - had unique characteristics both in picture composing and view finding. My fieldwork in Hebei and Jilin in 2003 and then my travelling in Xinjiang and Jilin in 2004 prompted my final decision to expand the pictorial plant book project to the whole country. I explained to Little XIANG that only when the book was comprehensive enough and of high quality, could Yong FEI's hardworking spirit and aspiration in this regard be fully manifested. And only in this way could Yong FEI's cherished dream of compiling a pictorial book on plants of China be realized without unnecessary pre-steps.

2004 saw Dr. Haining QIN's visit to Taiwan. Dr. QIN was from Institute of Botany, Chinese Academy of Sciences, and we got to know each other at British Kew Gardens in 1994. Under the leadership of Haining, the construction of Plant Photo Bank of China was making marvelous progress. Haining was a cordial and modest man, with his face always shining with smile. Upon knowing that I was conducting the project of pictorial plant book, he offered to give help. Taking advantage of his wide network, he brought in many talents, among whom Guosheng HE from Fujian, Guangdi WU from Sichuan, Yan LIU from Guangxi, and Zexian LI from Guangdong were all good at photography. In terms of color, Yan LIU's photos were particularly good, which was admirable. In addition, Dr. David Chamberlain, an expert in Ericacea, whom I got to know when I visited Royal Botanic Garden Edinburgh, agreed to review the photos of Ericaceae and Berberidaceae. Besides, Dr. Ching-I PENG introduced Dr. David E. Boufford from Harvard University who provided photos taken when he was investigating the plants of the Hengduan Mountains in the southwest of China. And David also recommended Susan Kelley and Richard Ree who both offered their plant photos.

Prof. Pengcheng WU from Institute of Botany, Chinese Academy of Sciences. was an expert in bryophytes. In 1990, he stayed in University of Helsinki, Finland for a few months when I was about to obtain my doctorate awarded by the University. Prof. WU introduced several experts from Institute of Botany to help review photos and write text. Dr. Liisin LEU, a Taiwan veteran in mycology, recommended Mr. Zhaoxiang HUI, deputy director of Kunming Municipal Bureau of Agriculture, to provide photos of weeds. And Mr. HUI invited Mr. Jindian CHE from Beijing to provide photos of weeds in Northern China. Prof. Mingjou LAI from Taiwan involved Mr. Xiangkun QIN from Shanghai Natural History Museum in contributing photos of plants in Eastern China, and then recommended Mr. Dachang ZHAO from Shenyang Institute of Applied Ecology, Chinese Academy of Sciences to offer photos of plants in Changbai Mountains. In 2004, I went to Urumqi to attend a conference whose organizer, Prof. Abdulla from Xinjiang University, gave me many photos of Xinjiang plants taken by himself.

My college classmate Shihchang KANG, an expert in plants and computer, also sent me plant photos taken by himself. Having long foreseen the power of internet, he did not quite agree with the idea of publishing a bulky physical book. Years ago, he wrote down a website address and asked me to visit it. The website, which I had never heard of before, was 'Google', a 'search engine' enabling us to search for information easily. And it turned out that this became a strong upward trend, with more and more photos being uploaded onto internet for people to view or download. However, Prof. Tuomo Niemelä, my Finnish adviser, had a different view on this phenomenon. He had published a handbook about large fungi, whose formatting was done by himself. The photos and text were arranged so well that a full sense of beauty permeated the entire book, and this always made me in a good mood each time I read it. When being consulted about the idea of publishing a physical plant book like this one, he encouraged me to continue doing so, saying that sometimes online data and materials would vanish for no reason and many online materials could not be said to be authentic because they had not undergone necessary review and approval. With this encouragement, I decided to carry out this project through to the end, paying no attention to the trends and fashionable ideas of new generations.

At the beginning of 2007, I found on internet that the Plant Photo Bank of China owned by Institute of Botany contained image data being managed by Min LI. So I asked him for sources of these photos.

Min LI recommended several persons whose photos in the Database were regarded as excellent. Most of these photo-takers were young people from the Institute of Botany. They were Bing LIU, Qinwen LIN, Shengxiang YU, Min LI, Xianming GAO, Yanhong BING. Besides, Yun WANG from Shaanxi was also added to the list of recommendation. By that time, I had already collected photos of 5000 species of plants in China through over 10 years of my hardwork which could have very well wound up. However, it would have been regrettable if I had not added so many fine photos of plants in Northern China to this important book. Bing LIU was a wonder in plant taxonomy - so young as he was, he had taken astonishingly large number of plant photos. It was Bing LIU and Yitao LIU who provided the largest number of photos for this work. Bo LIU, who was very active in work, helped a lot in writing text and reviewing photos. By the end of the same year, we had collected photos of more than 6500 species, paving the way for doing other publication-related preparatory work such as text writing and photo reviewing.

In 2010, "Academia Sinica" held a seminar in Taiwan, at which Haining QIN and Min LI delivered their reports which revealed that the Plant Photo Bank of China had collected hundreds of thousands of photos. In this circumstance, one more round of photo searching and collecting would certainly make more and better photos available for this upcoming book. Only, it would take more time. With so many people looking forward to the publication of the book, the longer time we took in publishing, the heavier pressure we would face. But in final analysis, I believed that Yong FEI, if he were still alive, would support this last round of photo searching and collecting. Prof. Xunlin YU and Prof. Daigui ZHANG from Hunan sent me many photos of plants of Central China. Mr. Hongwei ZHANG from Zhejiang and Mr. Zhonghui SHI from Anhui also sent photos to me. Prof. You ZHOU from Tonghua of Jilin contributed the photos of plants of Changbai Mountain that he took with great efforts. In the recent three years, a lot of photos were also provided by Xinxin ZHU, Yousheng CHEN, Yechun XU, Bin CHEN, Shipin CHEN, Hai HE, and Xile ZHOU. Xianchun ZHANG offered many photos of ferns. Xiaohua JIN submitted many photos of Orchidaceae plants. My good friend Li ZHANG was responsible for bryophytes.

Dr. Ching-I PENG from "Academia Sinica" provided a lot of pictures of Begoniaceae and edited the draft for this family. Shannjye MOORE, an expert of ferns from Taiwan, contributed some pictures of ferns to this book, and reviewed the pictures and text for the fern part. When I studied for doctorate in Taiwan University, Shannjye was still an undergraduate of the University. With a lovely round face often with smile, he was always seen studying in herbarium. In November 2010, however, he suddenly died of a stroke at the age of 44, making us very sad and regretful. Miss Pifong LU, an excellent amateur expert of ferns from Taiwan, contributed some good pictures of ferns. I also would like to express my thanks to the following colleagues who provided and reviewed pictures for me: Chiumei WANG, Chihhsiung CHEN, Weihsin HU, Chunlin HUANG, Hsinfu YEN and Shauting CHIU.

What we received in earlier stages were slides and a small number of prints,and after 2005, all contributions were in the form of digital image. Digital photos are clean, but their saturation, definition and sharpness are not very ideal, with green parts tending to turn slightly yellowish. Fortunately, these problems could be solved through photo-effect modification. As to the slides, high-quality ones were not many, as the quality of such images hinged on such factors as: quality of the film, developing process, storage condition and scanning, etc. The photos first needed some trimming so as to highlight their essential parts and then required modification to the image effect; most of the work was done by myself. In the meantime, we engaged a group of people to do text writing. Although templates were provided to text writers, inconsistency still appeared in some places due to different writing styles of different writers. There were also not a few oversights or slips caused by some writers who were not conscientious enough. At first, I myself did the correction and revision one by one against reference literature, finishing the work on about two thousand species, but as the amount of this kind of work was so big that I could not continue to do it all by myself. Here, I would like to list those who made greater contributions to the text. They are: Yong FEI, Bo LIU, Ren SA, Cuizhi GU, Xiao CHENG, Ning DU, Xiwen LI, Xiaoting XU and Ruizheng FANG, etc.

Prof. Pengcheng WU mentioned this work to Ms. Jing WANG, director of Biological Division of Science Press, who was interested in exploring the possibility of publishing the work, so we met each other in Beijing in 2010. Before this, I had devoted myself to the work so deeply that even small details like typesetting and font were studied and arranged by myself. Therefore, the work was like a child brought up by myself, so I would feel uneasy if I put it in the care of someone else, and moreover, we had already commissioned a design company in Taiwan to start typesetting the draft. However, my concern and worry were gradually dispelled by Jing WANG's sincerity and her perseverance in persuasion and explanation over two successive years. And the changes in social and economic situations across the Straits also made it natural for the book to be published in the Mainland. Also worthy of mentioning are: the generous support, organization and mobilization given or conducted by both Wentsai WANG, a renowned academician and Prof. Zhenyu LI, an expert on plant taxonomy, as well as the funding by National Publication Foundation. All this facilitated the smooth completion of the entire work.

Shenghua WU

20 October, 2015

第3卷编审者分工

木麻黄科－三白草科	刘　冰
胡椒科	刘　演
金粟兰科	张　强　孔宏智
杨柳科	何　理　张志翔　刘　博
杨梅科	刘　冰
胡桃科	刘　演　路安民
桦木科	刘　冰　陈之端　彭　华
壳斗科	刘　博　李新伟　刘　冰　周浙昆
榆科	刘　冰　萨　仁
马尾树科	刘　冰
桑科	刘　博　曹子余
荨麻科	王文采　陈家瑞
川苔草科	李振宇
山龙眼科	刘　冰　彭　华
铁青树科－檀香科	刘　冰
桑寄生科	刘　冰　邱少婷
马兜铃科	黄星凡　马金双
大花草科	金效华
蛇菰科	刘　博　赖阳均　彭　华
蓼科	侯元同　古训铭
藜科	黎　斌　李安仁
苋科	李振宇　李安仁
紫茉莉科	徐松芝　李振宇
商陆科	李振宇
番杏科	徐松芝　李振宇
马齿苋科	刘　冰　彭　华
落葵科	李振宇
石竹科	黎　斌　张代贵
睡莲科－连香树科	刘　冰
星叶草科－毛茛科	王文采
芍药科	刘　冰
木通科	覃海宁
小檗科	李新华　游旨价　David Chamberlain
防己科	刘　冰　刘丽霞
木兰科－水青树科	邓云飞　夏念和
蜡梅科	刘　冰
番荔枝科	侯学良　李秉滔
肉豆蔻科	李秉滔
樟科	李锡文　刘　冰
莲叶桐科	刘　冰

Authors and Reviewers of Volume Ⅲ

Casuarinaceae—Saururaceae	Bing LIU			
Piperaceae	Yan LIU			
Chloranthaceae	Qiang ZHANG	Hongzhi KONG		
Salicaceae	Li HE	Zhixiang ZHANG	Bo LIU	
Myricaceae	Bing LIU			
Juglandaceae	Yan LIU	Anmin LU		
Betulaceae	Bing LIU	Zhiduan CHEN	Hua PENG	
Fagaceae	Bo LIU	Xinwei LI	Bing LIU	Zhekun ZHOU
Ulmaceae	Bing LIU	Ren SA		
Rhoipteleaceae	Bing LIU			
Moraceae	Bo LIU	Ziyu CAO		
Urticaceae	Wentsai WANG	Jiarui CHEN		
Podostemonaceae	Zhenyu LI			
Proteaceae	Bing LIU	Hua PENG		
Olacaceae—Santalaceae	Bing LIU			
Loranthaceae	Bing LIU	Shauting CHIU		
Aristolochiaceae	Xingfan HUANG	Jinshuang MA		
Rafflesiaceae	Xiaohua JIN			
Balanophoraceae	Bo LIU	Yangjun LAI	Hua PENG	
Polygonaceae	Yuantong HOU	Xunming GU		
Chenopodiaceae	Bin LI	Anren LI		
Amaranthaceae	Zhenyu LI	Anren LI		
Nyctaginaceae	Songzhi XU	Zhenyu LI		
Phytolaccaceae	Zhenyu LI			
Aizoaceae	Songzhi XU	Zhenyu LI		
Portulacaceae	Bing LIU	Hua PENG		
Basellaceae	Zhenyu LI			
Caryophyllaceae	Bin LI	Daigui ZHANG		
Nymphaeaceae—Cercidiphyllaceae	Bing LIU			
Circaeasteraceae—Ranunculaceae	Wentsai WANG			
Paeoniaceae	Bing LIU			
Lardizabalaceae	Haining QIN			
Berberidaceae	Xinhua LI	Zhijia YOU	David Chamberlain	
Menispermaceae	Bing LIU	Lixia LIU		
Magnoliaceae—Tetracentraceae	Yunfei DENG	Nianhe XIA		
Calycanthaceae	Bing LIU			
Annonaceae	Xueliang HOU	Bingtao LI		
Myristicaceae	Bingtao LI			
Lauraceae	Xiwen LI	Bing LIU		
Hernandiaceae	Bing LIU			

目录 | Contents

第3卷
Volume Ⅲ
被子植物
木麻黄科－莲叶桐科
Angiosperms
Casuarinaceae－Hernandiaceae

木麻黄科 Casuarinaceae

木麻黄

Casuarina equisetifolia L.

乔木，雌雄同株，高可达35米，大树根部无萌蘖，树干通直。叶直立，贴生在小枝上，每轮(6-)7(-8)，披针形或三角形。雄花序长1-4厘米。球果椭圆形，翅果连翅长5-8毫米。花期4-5月，果期7-10月。栽培云南、广西、广东、台湾和福建。原产南亚、东南亚和大洋洲。

Trees, monoecious, up to 35 m tall, not suckering from roots, trunk straight. Leaves erect and appressed to branchlets (6-)7(-8) per whorl, lanceolate or triangular. Male inflorescences 1-4 cm long. Cones ellipsoid, samaras 5-8 mm long including wings. Fl. Apr-May. Fr. Jul-Oct. Cultivated in Yunnan, Guangxi, Guangdong, Taiwan and Fujian. Native to S and SE Asia, and Oceania.

木麻黄 *Casuarina equisetifolia*

三白草科 Saururaceae

三白草

Saururus chinensis (Lour.) Baill.

草本。根状茎匍匐，白色，粗厚。叶卵形至卵状披针形，纸质，密具腺点，无毛；顶端叶较小，2或3枚生茎顶，常花瓣状，花期白色。花序为一伸长的总状，腋生或顶生；雌蕊由4个基部合生的心皮组成。果实近球形。花期4-6月，果期6-7月。生海拔1700米以下的林下潮湿处、草地、河崖、灌丛、田边或水边。产中国除东北以外大部分地区。南亚、朝鲜半岛和日本南部亦有。

Herbs. Rhizomes creeping, white, thick. Leaves ovate to ovate-lanceolate, papery, densely glandular, glabrous; apical leaves smaller, 2 or 3 at stem apex, usually petal-like, white at anthesis. Inflorescences an elongated, axillary or terminal raceme; pistils with 4 carpels, base connate. Fruits subglobose. Fl. Apr-Jun. Fr. Jun-Jul. Wet places in forests, meadows, riverbanks, thickets, field edges or by waters below 1700 m. Distributed in most parts of China, except NE China. Also in S Asia, Korean Peninsula and S Japan.

三白草 *Saururus chinensis*

裸蒴

Gymnotheca chinensis Decne.

无毛草本。叶肾状心形，基部耳状，先端阔锐尖或圆形；托叶鞘长1.5-2厘米。花序长2-7.5厘米；每花下苞片倒披针形；雌蕊由4个合生心皮组成，子房下位，长倒卵球形。花期4-11月。生海拔(100-)600-2000米的水边或山谷。产中国西南、华南和华中。越南北部亦有。

Herbs, glabrous. Leaves reniform-cordate, base auriculate, apex broadly acute or rounded; stipular sheath 1.5-2 cm long. Inflorescences 2-7.5 cm long; bracts beneath each flower oblanceolate; pistils with 4 connate carpels, ovary inferior, long obovoid. Fl. Apr-Nov. Streamsides or valleys at (100-)600-2000 m. Distri-

裸蒴 *Gymnotheca chinensis*

buted in SW, S and C China. Also in N Vietnam.

白苞裸蒴

Gymnotheca involucrata C. Pei

草本。茎多匍匐。叶纸质，心形或肾形，无腺；叶鞘为叶柄的1/4-1/3。花序单生，常与叶对生于茎中部，基部有3-4片白色、叶状苞片；每花下苞片倒卵状长圆形或倒披针形；子房倒圆锥状。花期2-6月。生海拔700-1000米的路边或林中湿地。产四川南部。

Herbs. Stems stoloniferous. Leaves papery, cordate or reniform, without glands; leaf sheath 1/4-1/3 as long as petiole. Inflorescences solitary, usually leaf-opposed at middle of stems, with 3-4, white and foliaceous bracts at base; bracts beneath each flower obovate-oblong or oblanceolate; ovary obconical. Fl. Feb-Jun. Roadsides or moist places of forests at 700-1000 m. Distributed in S Sichuan.

蕺菜

Houttuynia cordata Thunb.

草本。叶阔卵形或卵状心形，薄纸质，密具腺点，下面常淡紫色，基部心形。总苞长圆形或倒卵形；每花下的苞片条形，圆柱形，不明显；雌蕊由3个下部合生的心皮组成，子房上位。花期4-9月，果期6-10月。生海拔2500米以下的水边、灌丛、路边、山坡或林中。产中国除东北以外大部分地区。南亚、朝鲜半岛和日本南部亦有。

Herbs. Leaves broadly ovate or ovate-cordate, thinly papery, densely glandular, usually purplish abaxially, base cordate. Involucral bracts oblong or obovate; bracts beneath each flower linear, terete, inconspicuous; pistils with 3 carpels, underneath connate, ovary superior. Fl. Apr-Sep. Fr. Jun-Oct. Streamsides, thickets, roadsides, slopes or forests below 2500 m. Distributed in most parts of China, except NE China. Also in S Asia, Korean Peninsula and S Japan.

白苞裸蒴 *Gymnotheca involucrata*

蕺菜 *Houttuynia cordata*

胡椒科 Piperaceae

齐头绒

Zippelia begoniifolia Blume ex Schultes et Schultes. f.

草本。茎于基部节间生根。叶卵状长圆形或卵形，膜质，密具透明腺点，基部斜心形，5-7条基出脉。花排列成疏松的总状花序。果密被锚状刺毛。花期5-7月。生海拔600-700米的林中。产云南、广西和海南。老挝、越南、马来西亚、印度尼西亚和菲律宾亦有。

Herbs. Stems rooting at basal nodes. Leaves ovate-oblong or ovate, membranous, densely pellucid dotted, base obliquely cordate, veins 5-7, all basal. Flowers in a laxly raceme. Fruits densely glochidiate. Fl. May-Jul. Forests at 600-700 m. Distributed in Yunnan, Guangxi and Hainan. Also in Laos, Vietnam, Malaysia, Indonesia and the Philippines.

短蒟

Piper mullesua D. Don

木质攀援藤本。叶纸质或薄革质，椭圆形或狭椭圆形或卵状披针形，无腺点，基部楔形，5-7条脉。花两性；花序短，近球形，在果期延长增大；子房倒卵球形，柱头3-4。核果倒卵球形。花期5-7月。生海拔800-2100米的山坡、山谷林中或溪边，攀援树上。产云南、四川南部、西藏南部和海南。印度、尼泊尔和不丹亦有。

短蒟 *Piper mullesua*

Vines climbing. Leaves papery or thin-coriaceous, elliptic or narrowly elliptic or ovate-lanceolate, without glands, base cuneate, with 5-7 veins. Flowers bisexual; inflorescences short, subglobose, elongated when fruiting; ovary obovoid; stigmas 3-4. Drupes obovoid. Fl. May-Jul. Slopes, forests in valleys, or by streams at 800-2100 m, climbing on trees. Distributed in Yunnan, S Sichuan, S Xizang and Hainan. Also in India, Nepal and Bhutan.

胡椒

Piper nigrum L.

木质藤本，雌雄同株。叶卵形至卵状长圆形，厚，近革质，

齐头绒 *Zippelia begoniifolia*

胡椒 *Piper nigrum*

基部圆形，两面无毛，叶脉每边5-7条。穗状花序与叶对生，与叶等长。核果球形，无柄，直径3-4毫米，成熟时红色，干后黑色。花期6-10月。栽培于中国西南和华南。广植于世界热带地区；原产东南亚。

Climbers woody, monoecious. Leaves ovate to ovate-oblong, thick, ± leathery, base rounded, both surfaces glabrous, veins 5-7 per side. Spikes leaf-opposed, to as long as leaves. Drupes globose, sessile, 3-4 mm diam, red when ripe, later black. Fl. Jun-Oct. Cultivated in SW and S China. Also cultivated in tropical regions of the world; native to SE Asia.

大叶蒟

Piper laetispicum C. DC.

木质攀援藤本，雌雄异株。叶革质，有腺点，叶脉羽状。花单生；雄花序长约10厘米；苞片叶状，有缘毛；雄蕊2；子房卵形；柱头4，顶端短尖。核果近球形。花期8-12月。生海拔100-600米的密林中，攀援于树上或石上。产广东南部和海南。

Climbers woody, dioecious. Leaves coriaceous, glandular, veins pinnate. Flowers solitary; male inflorescences ca. 10 cm long; bracts leaf-like, ciliate; stamens 2; ovary ovoid; stigmas 4, acute at apex. Drupes subglobose. Fl. Aug-Dec. Dense forests at 100-600 m, climbing on trees or rocks. Distributed in S Guangdong and Hainan.

毛蒟

Piper hongkongense C. DC.

攀援藤本，雌雄异株，长达数米。叶心状椭圆形或卵形，基部近心形，硬纸质，5-7条脉。花单性；花序与叶对生；子房近球形；柱头4。核果球形，直径约2毫米。花期3-5月。生海拔100-1300米的疏林或密林中，攀援树上或石上。产广西、广东和海南。

毛蒟 *Piper hongkongense*

Vines climbing, dioecious, up to several meters long. Leaves ovate-lanceolate or ovate, base ± cordate, hard-papery, with 5-7 veins. Flowers unisexual; spikes leaf-opposed; ovary subglobose; stigmas 4. Drupes globose, ca. 2 mm diam. Fl. Mar-May. Sparse or dense forests at 100-1300 m, climbing on trees or rocks. Distributed in Guangxi, Guangdong and Hainan.

缘毛胡椒

Piper semiimmersum C. DC.

攀援藤本。枝、叶被粗毛。叶长圆状卵形或卵状披针形，有细腺点，基部斜心形，纸质。穗状花序与叶对生；苞片近圆形，盾状，密具缘毛。核果球形。花期1-5月。生海拔200-900米的林中或潮湿处。产云南、贵州和广西。越南北部亦有。

Vines climbing. Branches and leaves hirsute. Leaves oblong-ovate or ovate-lanceolate, finely glandular, base obliquely cordate, papery. Spikes leaf-opposed; bracts suborbicular, peltate, densely ciliate. Drupes globose. Fl. Jan-May. Forests or damp places at 200-900 m. Distributed in Yunnan, Guizhou and Guangxi. Also in N Vietnam.

大叶蒟 *Piper laetispicum*

缘毛胡椒 *Piper semiimmersum*

荜拔 *Piper longum*

荜拔
Piper longum L.

藤本，达数米长，雌雄异株。叶纸质，两面脉上均被极细的短柔毛，叶脉7条，密具腺点，基部心形。穗状花序与叶对生，反折；苞片近圆形，有时稍楔形。核果球形。花期7-10月。生海拔580-700米的林中。产云南南部；栽培于广西、广东、海南和福建。东亚和东南亚亦有。

Climbers, up to several meters long, dioecious. Leaves papery, both surfaces thin-pubescent on veins, veins 7, densely glandular, base cordate. Spikes leaf-opposed, recurved; bracts suborbicular, sometimes slightly cuneate. Drupes globose. Fl. Jul-Oct. Forests at 580-700 m. Distributed in S Yunnan; cultivated in Guangxi, Guangdong, Hainan and Fujian. Also in E and SE Asia.

假蒟
Piper sarmentosum Roxb.

匍匐草本，长过10米。叶近膜质，卵形或卵状披针形，具细腺点，基部心形至圆形，叶脉7条。穗状花序与叶对生；苞片横向椭圆形。核果近球形，具4棱。花期4-11月。生海拔1000米以下的林中或潮湿处。产中国西南、华南和东南。南亚和东南亚亦有。

Herbs creeping, longer than 10 m. Leaves submembranous, ovate or ovate-lanceolate, finely glandular, base cordate to rounded, veins 7. Spikes leaf-opposed; bracts transversely elliptic. Drupes subglobose, 4-angled. Fl. Apr-Nov. Forests or damp places below 1000 m. Distributed in SW, S and SE China. Also in S and SE Asia.

假蒟 *Piper sarmentosum*

蒌叶 *Piper betle*

蒌叶
Piper betle L.

藤本，雌雄异株。叶卵形至卵状长圆形，纸质至近革质，下面叶脉密具腺点及细粉末状柔毛，基部心形。穗状花序与叶对生。核果合生成圆柱状，肉质，淡红色，复合果。花期5-7月。栽培于中国西南、华南、华中和华东。印度、斯里兰卡、越南、马来西亚、印度尼西亚、菲律宾和马达加斯加亦有。

Climbers, dioecious. Leaves ovate to ovate-oblong, papery to ± leathery, abaxially densely glandular with very finely powdery pubescent veins, base cordate. Spikes leaf-opposed. Drupes fused to form terete, fleshy, pale red, compound fruit. Fl. May-Jul. Cultivated in SW, S, C and E China. Also in India, Sri Lanka, Vietnam, Malaysia, Indonesia, the Philippines and Madagascar.

蒟子

Piper yunnanense Y. C. Tseng

直立亚灌木，雌雄异株。叶常卵形，纸质，下面有细腺点，叶脉9条。花单生；雄蕊3；柱头3。核果球形，成熟时红色，具疣，部分与轴合生。花期4-6月。生海拔1100-2000米的林边或湿润地。产云南西南部至西北部。

Subshrubs erect, dioecious. Leaves usually ovate, papery, abaxially densely minutely glandular, veins 9. Flowers solitary; stamens 3; stigmas 3. Drupes globose, red when ripe, tuberculate, partly connate to rachis. Fl. Apr-Jun. Forest edges or moist places at 1100-2000 m. Distributed in SW to NW Yunnan.

蒟子 *Piper yunnanense*

苎叶蒟

Piper boehmeriifolium (Miq.) C. DC.

近直立亚灌木。叶腹面和总花梗无毛，纸质至薄革质，密具细腺点。穗状花序多数与叶对生；苞片具柄。核果密簇生，近球形。花期12月至翌年7月。生海拔500-2200米的林中。产云南、贵州、广西和广东。南亚和东南亚亦有。

Subshrubs erect. Leaves adaxially and peduncles glabrous, papery to thinly papery, densely finely glandular. Spikes mostly leaf-opposed; bracts stipitate. Drupes densely clustered, subglobose. Fl. Dec to next Jul. Forests at 500-2200 m. Distributed in Yunnan, Guizhou, Guangxi and Guangdong. Also in S and SE Asia.

苎叶蒟 *Piper boehmeriifolium*

黄花胡椒

Piper flaviflorum C. DC.

攀援藤本，雌雄异株。叶椭圆形或卵状长圆形，纸质，具细腺点。穗状花序与叶对生，黄色；苞片倒卵形，具宽约0.5毫米或几与苞片等大的粗柄。核果球状，黄色。花期11月至翌年4月。生海拔500-1800米的林下或村旁树上。产云南中部和南部。

Vines climbing, dioecious. Leaves elliptic or ovate-oblong, papery, finely glandular. Spikes leaf-opposed, yellow; bracts obovate, with ca. 0.5 mm wide thick stipes and nearly equal to bracts in width. Drupes globose, yellow. Fl. Nov to next Apr. Forests or by villages on trees at 500-1800 m. Distributed in C and S Yunnan.

黄花胡椒 *Piper flaviflorum*

山蒟 *Piper hancei*

山蒟

Piper hancei Maxim.

攀援植物，雌雄异株。叶卵状披针形或椭圆形，脉5(-7)条。穗状花序与叶对生；苞片近圆形；柱头(3或)4。核果球状，黄色。花期3-8月。生海拔1700米以下的林中树上或岩石上。产中国西南、华南和华中。

Climbing plants, dioecious. Leaves ovate-lanceolate or elliptic, veins 5(-7). Spikes opposite to leaves; bracts suborbicular; stigmas (3 or)4. Drupes globose, yellow. Fl. Mar-Aug. Forests on trees or rocks below 1700 m. Distributed in SW, S and C China.

豆瓣绿

Peperomia tetraphylla (G. Forst.) Hook. et Arn.

多年生草本，肉质，成丛。茎、叶无毛或稀疏被毛。叶密生，阔椭圆形或近圆形，肉质，具透明腺点。穗状花序顶生及腋生，单生；花序长2-4.5厘米。小坚果近卵球形。花期2-4月，果期9-12月。生海拔600-3100米的林下，树上或潮湿岩石上。产中国西南、华南、东南和华西。世界热带和亚热带地区广布。

Perennial herbs, fleshy, forming clumps. Stems and leaves glabrous or sparsely hairy. Leaves dense, broadly elliptic or suborbicular, fleshy, pellucid dotted. Spikes terminal and axillary, solitary; inflorescences 2-4.5 cm long. Nutlets subovoid. Fl. Feb-Apr. Fr. Sep-Dec. Forests, on trees or damp rocks at 600-3100 m. Distributed in SW, S, SE and W China. Widespread in tropical and subtropical regions of the world.

蒙自草胡椒

Peperomia heyneana Miq.

多年生草本，成丛生。茎具分枝。叶倒卵状长圆形或倒卵状楔形，膜质，无毛，有腺点。穗状花序顶生或腋生。小坚果卵球形至卵球状长圆形。花期4-10月。生海拔800-2000米的密林下、沟边或湿润岩石上。产云南、四川、西藏、贵州和广西。印度、尼泊尔、不丹和缅甸亦有。

Perennial herbs, forming clumps. Stems branched. Leaves obovate-oblong or obovate-cuneate, membranous, glabrous, punctate. Spikes terminal or axillary. Nutlets ovoid to ovoid-oblong. Fl. Apr-Oct. Dense forests, by streams or damp rocks at 800-2000 m. Distributed in Yunnan, Sichuan, Xizang, Guizhou and Guangxi. Also in India, Nepal, Bhutan and Myanmar.

豆瓣绿 *Peperomia tetraphylla*

蒙自草胡椒 *Peperomia heyneana*

金粟兰科 Chloranthaceae

草珊瑚

Sarcandra glabra (Thunb.) Nakai

常绿亚灌木。叶对生，革质或纸质，边缘锯齿具腺状短尖，无毛。穗状花序顶生，通常有分枝；花黄绿色；雄蕊两倍长于花粉囊。核果球形或卵球形，成熟时亮红色。花期6月，果期8-12月。生海拔2000米以下的林中、灌丛、山谷、山坡、路边或溪边。产中国西南、华南、东南、华中和华东。南亚、东南亚、朝鲜半岛和日本亦有。

Subshrubs, evergreen. Leaves opposite, coriaceous or papery, glandular mucronate on marginal teeth, glabrous. Spikes terminal, usually branched; flowers yellowish green; stamens overall more than 2 × as long as thecae. Drupes globose or ovoid, shiny red at maturity. Fl. Jun. Fr. Aug-Dec. Forests, thickets, valleys, slopes, roadsides or streamsides below 2000 m. Distributed in SW, S, SE, C and E China. Also in S and SE Asia, Korean Peninsula and Japan.

海南草珊瑚 *Sarcandra glabra* subsp. *brachystachys*

海南草珊瑚

Sarcandra glabra (Thunb.) Nakai subsp. **brachystachys** (Blume) Verdc.

常绿亚灌木。叶纸质，无毛，边缘除基部外有钝锯齿。花序顶生，常分枝，稍成穗状；花黄绿色；雄蕊略长于花粉囊。核果球形或卵球形，成熟时橙红色。花期10月至翌年5月，果期翌年3-8月。生海拔400-1600米的山坡湿地或路边。产云南、海南、广西和广东。老挝、泰国北部和越南亦有。

Subshrubs, evergreen. Leaves papery, glabrous, margin dully serrate except base. Inflorescences terminal, usually branched, ± spiciform; flowers yellowish green; stamens only slightly longer than thecae. Drupes globose or ovoid, orange-red at maturity. Fl. Oct to next May. Fr. next Mar-Aug. Wet places on slopes or roadsides at 400-1600 m. Distributed in Yunnan, Hainan, Guangxi and Guangdong. Also in Laos, N Thailand and Vietnam.

金粟兰

Chloranthus spicatus (Thunb.) Makino

亚灌木，茎斜伸或稍匍匐。叶椭圆形或倒卵状椭圆形，5-11×2.5-5.5厘米，下面浅黄绿色，边缘具圆齿状锯齿，先端锐尖或钝。穗状花序排列为圆锥状，常顶生；花黄绿色，极芳香；雄蕊(药隔)顶端不规则3浅裂，中央裂片稍大，有时再3浅裂。花期4-7月，果期8-9月。生海拔150-1000米的林中。产云南、四川、贵州、广东、福建和河北。泰国和日本亦有。

Subshrubs, erect or slightly prostrate. Leaves elliptic or obovate-elliptic, 5-11 × 2.5-5.5 cm, pale yellowish green abaxially, margin crenate-serrate, apex acute or obtuse. Spikes arranged in panicles, usually terminal; flowers yellowish green, very fragrant; apical part of stamens (connectives) irregularly 3-lobed, central lobe larger, sometimes apex shallowly 3-lobed again. Fl. Apr-Jul. Fr. Aug-Sep. Forests at 150-1000 m. Distributed in Yunnan, Sichuan, Guizhou, Guangdong, Fujian and Hebei. Also in Thailand and Japan.

草珊瑚 *Sarcandra glabra*

金粟兰 *Chloranthus spicatus*

鱼子兰 *Chloranthus erectus*

鱼子兰

Chloranthus erectus (Buch.-Ham.) Verdc.

直立亚灌木。叶对生，宽椭圆形、倒卵形至长倒卵形，10-20 × 4-8厘米，纸质，边缘具腺头锯齿，先端渐尖。穗状花序常顶生，4-13，在总花梗顶部排列为圆锥状；花白色，芳香；雄蕊卵形，不伸长，顶端3浅裂。果幼时绿色，成熟时白色，倒卵形。花期4-7月，果期7-9月。生海拔100-2000米的林下、山谷中。产云南、四川、广西、贵州和西藏。南亚和东南亚亦有。

Subshrubs, erect. Leaves opposite, leaf blade broadly elliptic, obovate to long obovate, 10-20 × 4-8 cm, papery, pale green abaxially, glandular mucronate on margin serrate, apex acuminate. Spikes, 4-13, arranged in panicles, usually terminal; flowers white, fragrant; stamens ovoid, not elongated, 3-lobed apically. Drupes green when young, white at maturity, obovoid. Fl. Apr-Jul. Fr. Jul-Sep. Forests, valleys at 100-2000 m. Distributed in Yunnan, Sichuan, Guangxi, Guizhou and Xizang. Also in S and SE Asia.

狭叶金粟兰

Chloranthus angustifolius Oliver

多年生草本。叶对生，常8-12，长矛形，纸质，边缘具腺头锯齿，顶端渐尖。单个穗状花序顶生；花白色；药隔较长，水平伸展或向上斜伸，3深裂，线形。核果绿色，近球形或倒卵球形。花期4-6月，果期5-8月。生海拔650-1200米的林下、灌丛、路边和阴湿处。产重庆和湖北的部分地区。

Perennial herbs. Leaves opposite, usually 8-12, lanceolate, papery, glandular mucronate, margin serrate, apex acuminate. Spike solitary, terminal; flowers white; connectives elongated, horizontally spreading or ascending, 3-lobed, linear. Drupes green, subglobose or obovoid. Fl. Apr-Jun. Fr. May-Aug. Forests, thickets, roadsides and wet places at 650-1200 m. Distributed in partial areas of Chongqing and Hubei.

银线草

Chloranthus japonicus Sieb.

多年生草本。叶对生，常4片生于茎顶且近轮生，阔椭圆形或倒卵形，纸质，具腺点，边缘具锯齿或牙齿状锯齿，先端急尖。单个穗状花序顶生；花白色；药隔长而伸出，3深裂，线形，仅基部联合；中央药隔上通常无花粉囊，两侧药隔各具1个花粉囊。核果绿色，近球形或倒卵球形。花期4-5月，果期5-7月。生海拔100-2300米的林下、溪边和阴湿处。产华中、华北、西北部分地区和东北。日本、朝鲜半岛和俄罗斯亦有。

Herbs perennial. Leaves opposite, usually 4 clustered at stem apex and false-whorled, leaf blade broadly elliptic or obovate, papery, margin sharply serrate or dentate serrate, glandular mucronate, blade apex acute. Spikes solitary, terminal; flowers white; stamens 3-lobed; connectives

狭叶金粟兰 *Chloranthus angustifolius*

银线草 *Chloranthus japonicus*

elongated, exserted, connate at base; central connective usually without anther, lateral connectives with a 1-loculed anther for each. Drupes green, subglobose or obovoid. Fl. Apr-May. Fr. May-Jul. Forests, streamsides, shaded and wet places at 100-2300 m. Distributed in C, N, partial NW and NE China. Also in Japan, Korean Peninsula and Russia.

丝穗金粟兰

Chloranthus fortunei (A. Gray) Solms-Laub.

多年生草本。叶对生，4-6，常4枚生于茎顶或交互对生于茎上部，阔椭圆形、长椭圆形或倒卵形，先端急尖。穗状花序单生，自茎顶伸出；苞片倒卵形，常2-3齿裂；花白色，有香气；雄蕊3深裂；药隔伸长，线形，仅基部联合。核果球形，淡黄绿色。花期3-5月，果期4-6月。生海拔200-300米的山坡、林下阴湿处或沟边草丛中。产中国西南、华南、华中和华东。朝鲜半岛和日本亦有。

Perennial herbs. Leaves opposite, 4-6, usually 4-aggregated at stem apex or upper part, leaf blade broadly elliptic, long elliptic or obovate, apex acute. Spike solitary, arising from stem apex; bracts obovate, usually 2-3-dentate; flowers white, fragrant; stamens 3-lobed; connective elongated, linear, connate at base. Drupes globose, pale yellowish green. Fl. Mar-May. Fr. Apr-Jun. Slopes, wet places under forests or grasslands by streams at 200-300 m. Distributed in SW, S, C and E China. Also in Korean Peninsula and Japan.

全缘金粟兰

Chloranthus holostegius (Hand.-Mazz.) Pei et Shan

多年生草本。叶对生，常4个生于茎顶且轮生；叶阔椭圆形或倒卵形，坚纸质，具腺点，边缘具锯齿或牙齿状锯齿。穗状花序顶生及腋生，常1-5个成一束；花白色；药隔长而伸出，3深裂，线形。核果绿色，近球形或倒卵球形。花期5-6月，果期7-8月。生海拔400-2800米的灌丛或林中。产云南、四川、贵州和广西。

Perennial herbs. Leaves opposite, usually 4 at stem apex and whorled; leaves broadly elliptic or obovate, rigidly papery, glandular, margin serrate or dentate-serrate. Spikes terminal or terminal and axillary, usually 1-5 in a fascicle; flowers white; connectives elongated and exserted, 3-lobed, linear. Drupes green, subglobose or obovoid. Fl. May-Jun. Fr. Jul-Aug. Thickets or forests at 400-2800 m. Distributed in Yunnan, Sichuan, Guizhou and Guangxi.

全缘金粟兰 *Chloranthus holostegius*

及己

Chloranthus serratus (Thunb.) Roem. et Schult.

多年生草本。叶对生，4-6，常4片生于茎上部或茎顶，长椭圆形或长倒卵状椭圆形，5-15 × 2.5-6厘米，纸质，边缘具圆齿状锯齿，先端渐尖。穗状花序单一或2-4分叉，常顶生，有时具腋生花序；花白色；3裂雄蕊(药隔)合生至中部，内弯，2-3毫米；花粉囊着生于药隔中上部。核果绿色，球形或梨形。花期3-5(-7)月，果期(4-)6-8月。生海拔100-1800米林下、灌丛、峡谷、沼泽和溪边等阴湿处。产中国西南、华南、华中、东南和华东。日本和俄罗斯亦有。

Herbs perennial. Leaves opposite, 4-6, usually 4 at stem apex or upper part, leaf blade elliptic or oblong, 5-15 × 2.5-6 cm, papery, glandular, margin serrate, apex acuminate. Spikes 1-4-branched, terminal, sometimes axillary; flowers white; stamens 3, connectives oblong, connivent-connate up to apical part, introrse, 2-3 mm; thecae at middle or apical part of connectives. Drupes green, globose or pyriform. Fl. Mar-May(-Jul). Fr. (Apr-)Jun-Aug. Wet places in forests, thickets, valleys, swamps and streamsides at 100-1800 m. Distributed in SW, S, C, SE and E China. Also in Japan and Russia.

丝穗金粟兰 *Chloranthus fortunei*

及己 *Chloranthus serratus*

宽叶金粟兰 *Chloranthus henryi*

宽叶金粟兰
Chloranthus henryi Hemsl.

多年生草本。叶对生，常4片生于茎顶，宽椭圆形或倒卵形或倒卵状椭圆形，9-20 × 5-11厘米，纸质，边缘具圆齿状锯齿，先端渐尖。穗状花序1-4(-6)，着生在总花梗顶端，有时具腋生花序；花白色；雄蕊(药隔)3裂至基部，仅内侧基部联合，外展；药隔长圆形；花粉囊位于药隔基部。花期4-7月，果期7-8月。生海拔200-2100米的林下或灌丛等阴湿处。产中国西南、华南、华中、东南和华东。

Herbs perennial. Leaves opposite, usually 4 at apex, leaf blade broadly elliptic or obovate or obovate-elliptic, 9-20 × 5-11 cm, papery, margin serrate or crenate, apex acuminate. Spikes 1-4(-6)-branched, terminal or terminal and axillary; flowers white; stamens 3, base nearly free, only insides connected; connectives oblong; thecae at base of connectives. Fl. Apr-Jul. Fr. Jul-Aug. Shaded and wet places in forests or thickets at 200-2100 m. Distributed in SW, S, C, SE and E China.

华南金粟兰
Chloranthus sessilifolius K. F. Wu

多年生草本。叶对生，近无柄，4枚聚生茎顶呈轮生状，倒卵形、菱形或椭圆形，纸质，顶部渐尖。穗状花序顶生，具2-4(-6)个下垂的分枝；花白色；雄蕊3，仅基部内侧联合；药隔长圆形。核果棕褐色，近球形。花期3-4月，果期5-7月。生海拔500-1200米林中湿地或灌丛。产四川、贵州、广西、广东、福建和江西。

Perennial herbs. Leaves opposite, nearly sessile, 4 at stem apex and whorled, leaf blade obovate, rhombic or elliptic, papery, apex acuminate. Spikes terminal, with 2-4(-6) pendulous branches; flowers white; stamens 3, base nearly free; connectives oblong. Drupes brown, subglobose. Fl. Mar-Apr. Fr. May-Jul. Wet places in forests or thickets at 500-1200 m. Distributed in Sichuan, Guizhou, Guangxi, Guangdong, Fujian and Jiangxi.

雪香兰
Hedyosmum orientale Merr. et Chun

亚灌木，雌雄异株。叶对生，长矛形，纸质，边缘具密锯齿，先端渐狭，尾状。雄花序3-5，聚生于枝的顶端；雄蕊1，花丝无；药隔顶端具一突出、急尖附属物，0.7-1毫米长；雌花序顶生或在叶腋处着生，少分枝，1.5-5厘米长；苞片较长，0.8-1.2厘米。核果绿色，近椭圆形或三棱形，约4毫米；苞片上部与果实紧密黏合，伸长成长喙状。花期12月至翌年3月，果期翌年2-6月。生海拔500-2365米林下、灌丛、斜坡和峡谷等阴湿处。产广西、广东和海南。越南、马来西亚和印度尼西亚亦有。

Subshrubs, dioecious. Leaves opposite, lanceolate, papery, margin densely serrulate, apex gradually angustate becoming caudate. Staminate spikes 3-5, clustered at apex of branches; stamen 1; filament absent; connectives with a projected, acute appendage at apex, 0.7-1 mm; pistillate inflorescences terminal or axillary, branches few, 1.5-5 cm long; bracts long, 0.8-1.2 cm. Drupes green, subellipsoid-trigonous, ca. 4 mm; apical part of bracts tightly adnate to fruits, elongated into a long beak. Fl. Dec to next Mar. Fr. next Feb-Jun. Shaded and wet places in forests, thickets, slopes and ravines at 500-2365 m. Distributed in Guangxi, Guangdong and Hainan. Also in Vietnam, Malaysia and Indonesia.

华南金粟兰 *Chloranthus sessilifolius*

雪香兰 *Hedyosmum orientale*

杨柳科
Salicaceae

银白杨 *Populus alba*

银白杨
Populus alba L.

乔木。树皮通常灰白色，光滑。芽及小枝密被白色绒毛。叶卵圆形或椭圆状卵形，长枝上叶3-5掌裂，短枝上叶和叶柄下面密具白色绒毛；叶柄侧扁。蒴果圆锥形，2瓣裂。花期4-5月，果期5月。野生新疆；其他省有栽培。西北亚、欧洲和北非亦有。

Trees. Bark usually gray-white, smooth. Buds and branchlets densely covered with white tomentum. Leaves ovate-orbicular or elliptic-ovate, leaves of long shoots 3-5-palmately lobed, leaves of short branchlets and petioles abaxially densely white tomentose; petioles flattened. Capsules coniform, 2-valved. Fl. Apr-May. Fr. May. Native to Xinjiang; planted in other provinces. Also in NW Asia, Europe and N Africa.

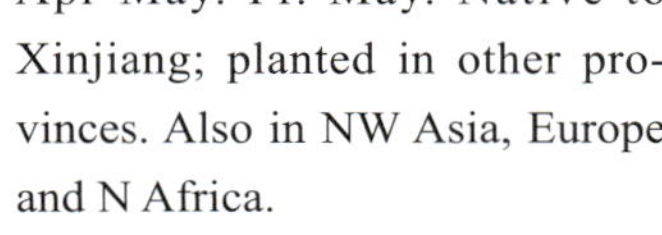

山杨
Populus davidiana Dode

乔木。树皮、小枝、叶柄和叶片光滑无毛。叶三角状卵形至圆形或近圆形，幼时淡红色，下面具毛，基部圆形、截形或浅心形，边缘密具波状齿，顶端锐尖或短渐尖；叶柄侧扁。蒴果卵球状圆锥形，2瓣裂。花期3-4月，果期4-5月。生海拔100-3800米的山区。产中国西南、华中、华北、西北和东北。俄罗斯、蒙古和朝鲜半岛亦有。

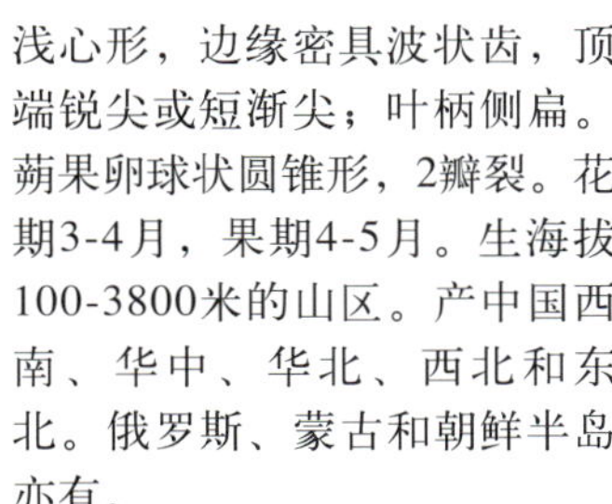

Trees. Bark, branchlets, petioles and leaves smooth and glabrous. Leaves deltoid-ovate to orbicular or suborbicular, pale red when very young, abaxially pilose, base rounded, truncate, or shallowly cordate, margin with dense, sinuolate teeth, apex acute or shortly acuminate; petioles flattened. Capsules ovoid-coniform, 2-valved. Fl. Mar-Apr. Fr. Apr-May. Mountains at 100-3800 m. Distributed in SW, C, N, NW and NE China. Also in Russia, Mongolia and Korean Peninsula.

山杨 *Populus davidiana*

河北杨

Populus × hopeiensis Hu et Chow

乔木。小枝灰褐色。叶卵状圆形，下面淡绿色，未展开时具绒毛，上面暗绿色，起初多毛，边缘具糙锯齿，齿锐尖、内弯，有时波状；叶柄侧扁。雄花序长约5厘米；雌花序长3-5厘米；花序轴被长毛。蒴果2瓣裂，具柄。花期4月，果期5-6月。生海拔700-1600米的河岸、山谷或冲积地。产河北、内蒙古、陕西和甘肃；各地常栽培。

Trees. Branchlets grayish brown. Leaves ovate-orbicular, abaxially pale green, tomentose when leaves unfold, adaxially dull green, pilose at first, margin coarsely serrate, teeth acute, incurved, sometimes sinuous; petioles flattened. Staminate catkin ca. 5 cm long; pistillate catkin 3-5 cm; rachis pubescent. Capsules 2-valved, shortly stipitate. Fl. Apr. Fr. May-Jun. Along rivers, valleys or alluvial deposits at 700-1600 m. Distributed in Hebei, Neimenggu, Shaanxi and Gansu; often planted elsewhere.

圆叶杨 *Populus rotundifolia*

圆叶杨

Populus rotundifolia Griff.

乔木。小枝褐色或暗褐色。叶三角状圆形，基部浅至深心形或截形，边缘具波状钝锯齿；叶柄侧扁。雌花序常长于10厘米；子房长卵球形，无毛；花柱短或近无柄；柱头2裂。蒴果长卵球形，2瓣裂。生海拔约2800米的山坡。产中国西南和华西。

Trees. Branchlets brown or dull brown. Leaves deltoid-orbicular, base shallowly to deeply cordate or truncate, margin sinuously longer serrate; petioles flattened. Pistillate catkins usually longer than 10 cm; ovary long ovoid, glabrous; styles short or subsessile; stigmas 2-parted. Capsules long ovoid, 2-valved. Mountain slopes at ca. 2800 m. Distributed in SW and W China.

响叶杨

Populus adenopoda Maxim.

乔木。叶卵状圆形或卵形，先端具长渐尖或尾尖，基部具2个凸起的腺体，边缘反卷，具腺状浅圆齿或粗齿；叶柄侧扁。雄花序长6-10厘米；苞片掌状分裂。蒴果长卵球状椭圆体形，2瓣裂。花期3-4月，果期4-5月。生海拔300-2500米的山坡。产中国西南、东南、华中、华北、华西和华东。

Trees. Leaves ovate-orbicular or ovate, long acuminate or caudate at apex, base with 2 raised glands, margin incurved, glandular crenate-serrate or dentate; petioles flattened. Staminate catkins 6-10 cm long; bracts palmatiparted. Capsules long ovoid-ellipsoid, 2-valved. Fl. Mar-Apr. Fr. Apr-May. Mountain slopes at 300-2500 m. Distributed in SW, SE, C, N, W and E China.

河北杨 *Populus × hopeiensis*

响叶杨 *Populus adenopoda*

毛白杨

Populus tomentosa Carr.

乔木。树冠圆锥形、卵球形或球形。叶卵形或三角状卵形，叶未开放时下面具绒毛，渐无毛，上面深绿色，光亮，边缘具波状锯齿，先端渐尖；叶柄侧扁。蒴果圆锥状或长卵球形，2裂。花期3月，果期4-5月。生平原至海拔1500米山地。产华西、华北、华中和华东。

Trees. Crown conical, ovoid or globose. Leaves ovate or deltoid-ovate, abaxially tomentose when leaves unfold, glabrescent, adaxially dark green, shiny, margin sinuate-dentate, apex acuminate; petioles flattened. Capsules conical or long ovoid, 2-valved. Fl. Mar. Fr. Apr-May. Mountain at plains to 1500 m. Distributed in W, N, C and E China.

大叶杨

Populus lasiocarpa Oliv.

乔木。树皮深灰色，纵裂。叶卵形，大，15-30 × 10-15厘米，基部深心形，常耳状，具2腺体，边缘具腺状圆锯齿，反卷。雄葇荑花序长9-12厘米。蒴果卵球形，3瓣裂，长1-1.7厘米。花期4-5月，果期5-6月。生海拔1300-3500米的山坡或路边林中。产云南、四川、贵州、湖北和陕西。

Trees. Bark dark gray, furrowed. Leaves ovate, large, 15-30 × 10-15 cm, base deeply cordate, often auriculate, with 2 glands, margin glandular crenate-serrate, revolute. Staminate catkins 9-12 cm long. Capsules ovoid, 3-valved, 1-1.7 cm long. Fl. Apr-May. Fr. May-Jun. Mountain slopes or riverside woods at 1300-3500 m. Distributed in Yunnan, Sichuan, Guizhou, Hubei and Shaanxi.

大叶杨 *Populus lasiocarpa*

小叶杨

Populus simonii Carr.

乔木。萌枝有明显棱脊。叶菱状卵形、菱状椭圆形或菱状倒卵形，3-12 × 2-8厘米，最宽部分在中部，下面灰绿色或银白色，边缘具锯齿，无毛。雄花序长2-7厘米；雌花序长2.5-6(-15)厘米。蒴果小，无毛或被柔毛，2(-3)瓣裂。花期3-5月，果期4-6月。生海拔2500米以下山地、平原、冲积地或河谷。产中国西南、华北、华西和东北。蒙古亦有。

Trees. Branchlets of young trees usually angulate. Leaves rhombic-ovate, rhombic-elliptic, or rhombic-obovate, 3-12 × 2-8 cm, broadest above middle, abaxially grayish green or slightly white, margin serrulate, glabrous. Staminate catkin 2-7 cm long; pistillate inflorescences 2.5-6 (-15) cm long. Capsules small, glabrous or sparsely pubescent, 2 (-3)-valved. Fl. Mar-May. Fr. Apr-Jun. Mountains, plains, alluvial deposits or valleys below 2500 m. Distributed in SW, N, W and NE China. Also in Mongolia.

毛白杨 *Populus tomentosa*

小叶杨 *Populus simonii*

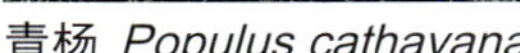
青杨 *Populus cathayana*

冬瓜杨 *Populus purdomii*

青杨

Populus cathayana Rehder

乔木。树冠宽卵状。小枝黄绿色或灰黄色。叶片卵形或椭圆形，先端短尖或渐尖，叶缘有腺点，圆锯齿。雄花序长5-6厘米；雌花序长4-5厘米；雄蕊30-35。蒴果卵球形，(2或)3或4瓣裂。花期3-5月，果期5-7月。生海拔800-3000米的山谷或河边。产华北、西北、四川、辽宁和云南。

Trees . Crown broadly ovoid. Branchlets yellowish green or grayish yellow. Leaves ovate or elliptic, apex mucronate or acuminate, margin glandular, crenate-serrate. Staminate catkins 5-6 cm long; pistillate catkins 4-5 cm long; stamens 30-35. Capsules ovoid, (2 or)3- or 4-valved. Fl. Mar-May. Fr. May-Jul. Valleys or riversides at 800-3000 m. Distributed in N and NW China, Sichuan, Liaoning and Yunnan.

冬瓜杨

Populus purdomii Rehd.

乔木。树皮幼时灰绿色，老时暗灰色，纵裂。叶卵形或宽卵形，先端渐尖，基部圆形或近心形，上面亮绿色，下面带白色，沿脉有毛，后渐脱落。蒴果球状卵形，(2-)3-4瓣裂。花期4-5月，果期5-6月。生海拔700-2600米的山地或河边。产河北、河南、陕西、甘肃、湖北和四川。

Trees. Bark grayish green when young, becoming dark gray, furrowed. Leaf blade ovate or broadly ovate, apex acuminate, base rounded or cordate, margin glandular serrulate or crenate-serrate, ciliate, adaxially bright green, abaxially shiny, pilose along veins, glabrescent or not. Capsule globose-ovoid, (2-)3-4-valved. Fl. Apr-May. Fr. May-Jun. Mountains or streamsides at 700-2600 m. Distributed in Hebei, Henan, Shaanxi, Gansu, Hubei and Sichuan.

三脉青杨

Populus trinervis C. Wang et Tung

乔木。树皮灰色，沟裂。短枝叶阔卵形或卵形，先端长尾尖或长渐尖，基部圆形，下面灰白色，光滑，上面绿色，具短柔毛，基部一对侧脉成弧形，为明显3出弧形脉。蒴果长卵形，长达5毫米，2瓣裂。花期3月，果期4月。生海拔2100-3000米的河边。产四川西部。

Trees. Bark gray, furrowed. Leaves of short branchlets broadly ovate or ovate, apex long caudate or acuminate, base rounded, abaxially grayish white and smooth, adaxially green, pilose along veins, veins 3, lateral veins curved at base. Capsule long ovoid, up to 5 mm, 2-valved. Fl. Mar. Fr. Apr. Along streams at

三脉青杨 *Populus trinervis*

2100-3000 m. Distributed in W Sichuan.

香杨
Populus koreana Rehd.

乔木。树皮幼时灰绿色，光滑，老时暗灰色，具深沟裂。芽大，具香气。短枝叶椭圆状长圆形，先端钝尖，上面暗绿色，有明显皱纹，下面白色；叶柄先端有短毛。蒴果卵圆形，无毛，2或4瓣裂。花期4-5月，果期6月。产河北、黑龙江、吉林、辽宁东部和内蒙古。朝鲜半岛亦有。

Trees. Bark grayish green and smooth when young, dull gray and deeply furrowed when old. Buds large, odoriferous. Leaves of short branchlets elliptic-oblong, apex obtuse, adaxially dull green, wrinkled, abaxially pale; petiole distally pubescent. Capsule ovoid-globose, 2- or 4-valved. Fl. Apr-May. Fr. Jun. Distributed in Hebei, Heilongjiang, Jilin, E Liaoning and Neimenggu. Also in Korean Peninsula.

辽杨 *Populus maximowiczii*

辽杨
Populus maximowiczii A. Henry

乔木。小枝圆柱形，粗壮。叶阔卵形、倒卵状椭圆形、椭圆形或阔卵形，下面具白粉，两面沿脉具柔毛，基部近心形或近圆形，边缘具腺状圆锯齿，具缘毛，先端短渐尖或锐尖，常扭曲。花序轴无毛；雄蕊30-40。蒴果卵球形。花期4-5月，果期5-6月。生海拔500-2000米的林地。产华北和东北。俄罗斯、朝鲜半岛和日本亦有。

Trees. Branchlets terete, stout. Leaves broadly elliptic, obovate-elliptic, elliptic, or broadly ovate, abaxially glaucous, both surfaces pubescent along veins, base subcordate or subrounded, margin glandular crenate-serrate, ciliate, apex shortly acuminate or acute, usually twisted. Catkin rachis glabrous; stamens 30-40. Capsules ovoid. Fl. Apr-May. Fr. May-Jun. Woods at 500-2000 m. Distributed in N and NE China. Also in Russia, Korean Peninsula and Japan.

香杨 *Populus koreana*

滇杨 *Populus yunnanensis*

滇杨
Populus yunnanensis Dode

乔木。树皮灰色，纵裂。叶纸质，卵形至广卵形，先端长渐尖，基部宽楔形或圆形，上面绿色，有光泽，沿中脉上稍有柔毛，下面灰白色，无毛。雄花序长12-20厘米，轴光滑；雌花序长10-15厘米。蒴果3-4瓣裂。花期4月，果期4-5月。生海拔1300-3700米的山地。产云南、贵州和四川。

Trees. Bark gray, furrowed. Leaf papery, blade ovate to broadly ovate, apex long acuminate, base broadly cuneate or rounded, adaxially green, shiny, pilose along midvein, abaxially grayish white and glabrous or pubescent along veins. Staminate catkin 12-20 cm; rachis glabrous; pistillate catkin 10-15 cm. Capsule 3-4-valved. Fl. Apr. Fr. Apr-May. Mountains, forests at 1300-3700 m. Distributed in Yunnan, Guizhou and Sichuan.

德钦杨
Populus haoana Cheng et C. Wang

乔木。树皮灰色，光滑。短枝叶卵形至卵状长椭圆形，先端短渐尖，常扭曲，基部心形，上面暗绿色，沿脉具柔毛，下面苍白色，被疏柔毛，沿脉密柔毛；叶柄圆，被密柔毛。果序长8-40厘米，轴被柔毛。蒴果卵圆形，3-4瓣裂。海拔2200-3600米的林中习见。产云南西北部。

Trees. Bark gray, smooth. Leaves of short branchlets ovate to long ovate-elliptic, apex shortly acuminate, often twisted, base cordate or deeply so, adaxially dull green, downy along veins, abaxially pale, pilose, densely downy along veins; petiole terete, densely downy. Fruiting catkin 8-40 cm, rachis pubescent. Capsule ovoid, 3-4-valved. Forests at 2200-3600 m. Distributed in NW Yunnan.

缘毛杨
Populus ciliata Wall.

乔木。树皮灰色。芽大，卵形。叶卵状心形，先端急尖，基部心形或圆形，边缘具腺状圆齿，有密缘毛，上面暗绿色，无毛，下面灰绿色，至少沿脉有柔毛。雄花序长6厘米；雌花序长达22厘米。蒴果4瓣裂。花期5月，果期6月。生海拔3300-3400米山地。产西藏

德钦杨 *Populus haoana*

缘毛杨 *Populus ciliata*

和云南等地。尼泊尔和印度亦有。

Trees. Bark gray. Buds ovoid, large. Leaf blade ovate-cordate, apex acute to acuminate, base cordate or rounded, margin glandular crenate, densely ciliate, adaxially dull green, glabrous, abaxially grayish green, downy at least along veins. Staminate catkin 6 cm long; pistillate catkin up to 22 cm long. Capsule 4-valved. Fl. May. Fr. Jun. Mountains at 3300-3400 m. Distributed in Xizang and Yunnan. Also in Nepal and India.

亚东杨

Populus yatungensis (C. Wang et P. Y. Fu) C. Wang et Tung

乔木。树皮灰绿色至淡灰色，纵裂。叶长卵形至广卵形，先端渐尖至长渐尖，基部浅心形至心形，边缘有毛至光滑，上面暗绿色，下面苍白色，两面沿脉被疏柔毛；叶柄圆柱形，被长柔毛。果序仅基部有毛。蒴果圆状卵形，无毛，4瓣裂。生海拔2400-3600米的山坡。产西藏、云南和四川西南部。

Trees. Bark grayish green to grayish, furrowed. Leaf blade long ovate to broadly ovate, apex acuminate to long acuminate, base cordate, margin ciliate or glabrous, adaxially dull green, abaxially glaucous, sometimes with coarse, long hairs, both surfaces pilose along veins; petiole terete, villous. Fruiting catkin peduncle pilose. Capsule globose-ovoid, glabrous, 4-valved. Mountain slopes at 2400-3600 m. Distributed in Xizang, Yunnan and SW Sichuan.

亚东杨 *Populus yatungensis*

胡杨

Populus euphratica Oliv.

乔木。小枝稀具毛。叶形多变化，卵圆形、圆形或三角状卵形，基部具2腺体，顶端具齿；幼苗和萌条叶条形、披针形、条状披针形或倒披针形。雄花序长2-3厘米；雄花花药紫红色；雌花柱头3，黄绿色。蒴果2瓣裂。花期5月，果期7-8月。生海拔200-2400米的沙漠、山谷或盆地。产内蒙古西部、甘肃、青海和新疆。中亚和西南亚亦有。

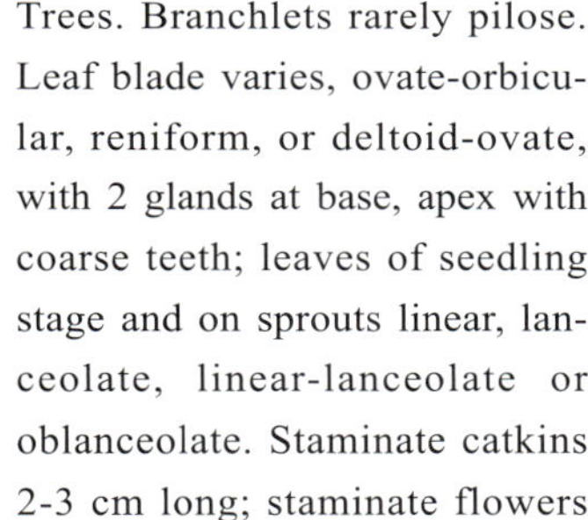

Trees. Branchlets rarely pilose. Leaf blade varies, ovate-orbicular, reniform, or deltoid-ovate, with 2 glands at base, apex with coarse teeth; leaves of seedling stage and on sprouts linear, lanceolate, linear-lanceolate or oblanceolate. Staminate catkins 2-3 cm long; staminate flowers with purplish red anthers; pistillate flowers with 3 yellowish green stigmas. Capsules 2-valved. Fl. May. Fr. Jul-Aug. Deserts, valleys or basins at 200-2400 m. Distributed in W Neimenggu, Gansu, Qinghai and Xinjiang. Also in C and SW Asia.

胡杨 *Populus euphratica*

钻天柳 *Salix arbutifolia*

钻天柳

Salix arbutifolia Pallas

乔木。树皮剥落状。叶披针形。花序先叶开放；雄花序下垂；苞片不脱落；雄花：雄蕊5；雌花序直立；苞片脱落；雌花：子房有短柄；花柱脱落性。花期5月。生海拔300-1000米的河流两岸。产内蒙古、黑龙江、吉林和辽宁。朝鲜半岛、日本和俄罗斯亦有。

Trees. Bark exfoliating. Leaf blade lanceolate. Catkins before leaves emerge; staminate catkin pendulous; bracts persistent; staminate flower: stamens 5; pistillate catkin erect; bracts caducous; pistillate flower: ovary shortly stipitate; styles caducous. Fl. May. Along streams at 300-1000 m. Distributed in Neimenggu, Heilongjiang, Jilin and Liaoning. Also in Korean Peninsula, Japan and Russia.

大白柳

Salix cardiophylla Trautv. et Mey.

乔木。叶卵状长圆形。花序与叶同时开放；雄花序直立；雄花：雄蕊5，腺体2；雌花序下垂；苞片脱落；雌花：腹腺2；子房有柄，无毛，花柱果期脱落。花期5-6月。生海拔300-800米的河边。产中国东北。朝鲜半岛和俄罗斯亦有。

Trees. Leaf blade ovate-oblong. Catkins as leaves emerge; staminate catkin erect; staminate flower: stamens 5, glands 2; pistillate catkin pendulous; bracts caducous; pistillate flower: adaxial glands 2; ovary stipitate; styles caducous in fruit. Fl. May-Jun. Along streams at 300-800 m. Distributed in NE China. Also in Korean Peninsula and Russia.

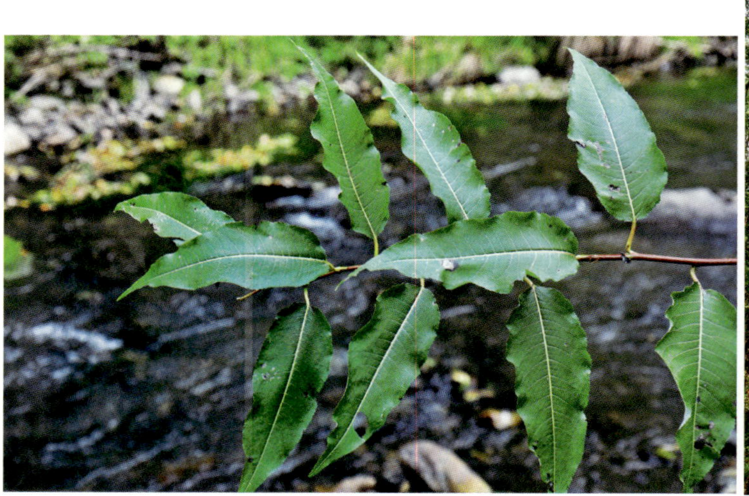

大白柳 *Salix cardiophylla*

滇大叶柳

Salix cavaleriei H. Lévl.

乔木。叶宽披针形，长5.5-8.7(-11)厘米，宽3.4-3.6(-4)厘米；叶柄先端有腺点。花与叶同时开放，花序直立；雄花：雄蕊6-8(-12)，腺体2；雌花：子房有长柄，腹腺宽，背腺常2-3裂。花期3-4月。生海拔1800-2500米的河边和林缘等湿润处。产广西、贵州、云南和四川。

Trees. Leaf blade broadly lanceolate, 5.5-8.7(-11) × 3.4-3.6(-4) cm; petiole with spherical glands distally. Catkins as leaves emerge, erect; staminate flower: stamens 6-8(-12), glands 2; pistillate flower: ovary with long stipe, adaxial gland broad, abaxial gland usually 2-3-lobed. Fl. Mar-Apr. Riversides and damp forest edges at 1800-2500 m. Distributed in Guangxi, Guizhou, Yunnan and Sichuan.

腺柳

Salix chaenomeloides Kimura

小乔木。叶片椭圆形或卵形至椭圆状披针形，两面无毛，基部楔形，边缘有腺锯齿或牙齿。花序花期疏松；子房窄卵球形，无毛，具柄；柱头头状或微凹。蒴果卵球状椭圆体形。花期4月，果期5月。生海

滇大叶柳 *Salix cavaleriei*

拔约1000米的山谷。产四川、河北、陕西、江苏和辽宁。朝鲜半岛和日本亦有。

Trees small. Leaf blade elliptic or ovate to elliptic-lanceolate, both surfaces glabrous, base cuneate, margin glandular serrate or dentate. Catkin laxly flowered at anthesis; ovary narrowly ovoid, glabrous, long stipitate; stigmas capitate or retuse. Capsules ovoid-ellipsoid. Fl. Apr. Fr. May. Valleys at ca. 1000 m. Distributed in Sichuan, Hebei, Shaanxi, Jiangsu and Liaoning. Also in Korean Peninsula and Japan.

紫柳

Salix wilsonii Seemen ex Diels

乔木。叶椭圆形或宽椭圆形至长圆形，边缘具圆锯齿或圆齿；叶柄具柔毛，常无腺。花序与叶同时开放；雄柔荑花序具梗，疏花；雄蕊3-5(-6)；雌柔荑花序，疏花；苞片长椭圆形。蒴果卵球状长圆形。花期3-4月，果期5月。生平原或河岸边。产中国东南、华中和华东。

Trees. Leaf blade elliptic or broadly elliptic to oblong, margin crenate or crenate-serrate; petioles pubescent, usually eglandular. Flowering coetaneous; staminate catkins pedunculate, laxly flowered; stamens 3-5 (-6); pistillate catkins laxlyflowered; bracts long elliptic. Capsules ovoid-oblong. Fl. Mar-Apr. Fr. May. Plains or riverbanks. Distributed in SE, C and E China.

南川柳

Salix rosthornii Seemen

乔木或灌木。叶披针形、椭圆状披针形或长圆形，基部楔形，边缘具规则腺锯齿。花序与叶同时开放；雄性柔荑花序疏花；花梗具3(-6)小叶；花柱短，2裂；雄蕊(3-)6。蒴果卵球形，长5-6毫米。花期3-4月，果期5月。生河边或溪边。产中国西南、东南、华中、华西和华东。

Trees or shrubs. Leaf blade lanceolate, elliptic-lanceolate or oblong, base cuneate, margin regularly glandular serrate. Flowering coetaneous; staminate catkins loosely flowered; peduncles with 3(-6) leaflets; styles short, 2-lobed; stamens (3-)6. Capsules ovoid, 5-6 mm long. Fl. Mar-Apr. Fr. May. Along rivers or streamsides. Distributed in SW, SE, C, W and E China.

南川柳 *Salix rosthornii*

腺柳 *Salix chaenomeloides*

紫柳 *Salix wilsonii*

长梗柳 Salix dunnii

五蕊柳 Salix pentandra

长梗柳

Salix dunnii C. K. Schneid.

灌木或小乔木。叶片椭圆形或椭圆状披针形，背面灰白色，贴生长柔毛，腹面疏被柔毛且幼时较密。花序长约5厘米，花疏松；总花梗长约1厘米，具3-5小叶；花柱短，柱头2裂。花期4月，果期5月。生溪边。产广东、福建、浙江和江西。

Shrubs or small trees. Leaf blade elliptic or elliptic-lanceolate, abaxially grayish white, densely appressed villous, adaxially pilose, more densely so when young. Catkin ca. 5 cm long, loosely flowered; peduncles ca. 1 cm long, with 3-5 leaflets; styles short, stigmas 2-lobed. Fl. Apr. Fr. May. Near streams. Distributed in Guangdong, Fujian, Zhejiang and Jiangxi.

五蕊柳

Salix pentandra L.

灌木或乔木，高达5米。叶宽披针形、卵状长圆形或椭圆状披针形，革质，幼时具黏液，边缘具腺齿，基部钝或楔形。雄蕊(5或)6-9(-12)；子房卵球状圆锥形，近无柄。蒴果长约9毫米，光亮，具短柄。花期6月，果期8-9月。生海拔500-1700米的水边林缘、林中沼泽地和草甸，或山谷。产华北、西北和东北。俄罗斯、蒙古和欧洲亦有。

Shrubs or trees, up to 5 m tall. Leaf blade broadly lanceolate, ovate-oblong or elliptic-lanceolate, leathery, viscid when young, margin glandular serrate, base obtuse or cuneate. Stamens (5 or)6-9(-12); ovary ovoid-conical, subsessile. Capsules ca. 9 mm long, shiny, shortly stipitate. Fl. Jun. Fr. Aug-Sep. Edges of woods near rivers, marshes and meadows within forests, or valleys at 500-1700 m. Distributed in N, NW and NE China. Also in Russia, Mongolia and Europe.

康定柳

Salix paraplesia C. K. Schneid.

乔木。叶倒卵状椭圆形或椭圆状披针形，长3.8-10.3厘米，宽1.6-3.6厘米，边缘有明显的细腺锯齿；叶柄先端有腺点。花叶同时开放，花序直立；雄花：腺体2；雄蕊5-7；雌花：雌花具腹腺；子房有短柄，无毛。花期4-5月。生海拔1500-3900米的山沟。产华西。

Trees. Leaf blade obovate-elliptic or elliptic-lanceolate, 3.8-10.3 × 1.6-3.6 cm, margin conspicuously glandular serrulate. Catkins as leaves emerge, erect; staminate flower: glands 2; stamens 5-7; pistillate flower with adaxial glands; ovary stipitate, glabrous. Fl. Apr-May. Valleys at 1500-3900 m. Distributed in W China.

旱柳

Salix matsudana Koidz.

乔木。叶披针形，下面具灰白色或稍白，幼时具长柔毛，边缘具腺锯齿，先端长渐尖。花序具短柄；苞片卵形，基部多毛；雄花：腺体2；雄蕊2；花药黄色；子房长椭圆体形，无毛，无柄。花期4月，果期5月。通常栽

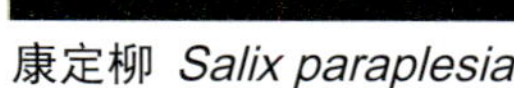
康定柳 Salix paraplesia

培平原或河岸。产中国东南、华北、华西、华中和华东。

Trees. Leaf blade lanceolate, abaxially glaucous or slightly white, villous when young, margin glandular serrulate, apex long acuminate. Catkins shortly pedunculate; bracts ovate, pilose at base; staminate flower with 2 glands; stamens 2; anthers yellow; ovary long ellipsoid, glabrous, subsessile. Fl. Apr. Fr. May. Commonly planted on plains or riverbanks. Distributed in SE, N, W, C and E China.

银叶柳
Salix chienii W. C. Cheng

灌木或乔木。叶长椭圆形、披针形或倒披针形，两面幼时具丝状短绒毛。雄花序圆柱状，长1.5-2厘米；总花梗长3-6毫米；雄花具2雄蕊，基部连合，被毛；子房卵球形，无毛，无柄；花柱明显；柱头2裂。蒴果卵球状长圆形，长约3毫米。花期4月，果期5月。生海拔500-600米的溪边或灌丛中。产福建、浙江、湖北、湖南、江苏和江西。

Shrubs or trees. Leaves long elliptic, lanceolate, or oblanceolate, both surfaces silky downy when young. Staminate catkin cylindric, 1.5-2 cm long; peduncles 3-6 mm long; staminate flower with 2 stamens, connate and pilose at base; ovary ovoid, glabrous, sessile; styles conspicuous; stigmas 2-lobed. Capsules ovoid-oblong, ca. 3 mm long. Fl. Apr. Fr. May. Along streams or thickets at 500-600 m. Distributed in Fujian, Zhejiang, Hubei, Hunan, Jiangsu and Jiangxi.

垂柳
Salix babylonica L.

乔木。小枝下垂。雄柔荑花序长1.5-2(-3)厘米，总花梗短；苞片披针形；雄花具2腺体；雄蕊2，约等长或长于苞片；苞片披针形；雌柔荑花序长2-3(-6)厘米，总花梗具3-4叶；子房无毛或基部稍具毛；花柱短，柱头2-4深裂。花期3-4月，果期4-5月。广布于中国。亚洲和欧洲亦有。

Trees. Branchlets pendulous. Staminate catkins 1.5-2(-3) cm long, peduncles short; staminate flower with 2 glands; stamens 2, ca. as long as to or longer than bracts; bracts lanceolate; pistillate catkins 2-3(-6) cm long, with 3-4 leaves on peduncles; ovary glabrous or slightly pilose at base; styles short; stigmas 2-4-parted. Fl. Mar-Apr. Fr. Apr-May. Widespread throughout China. Also in Asia and Europe.

银叶柳 *Salix chienii*

旱柳 *Salix matsudana*

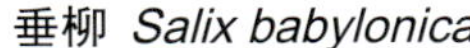

垂柳 *Salix babylonica*

巴郎柳 *Salix sphaeronymphe*

巴郎柳

Salix sphaeronymphe Goerz

灌木或小乔木。小枝细。叶披针形，长3.2-6厘米，宽0.9-1.4厘米。雄花：腺体2；雄蕊2；雌花：腺体1，子房密被白色柔毛，无柄，花柱短。花期4-5月。生海拔2600-3700米的水沟旁或山坡路旁。产甘肃、四川和西藏。

Shrubs or trees. Branchlets thin. Leaf blade lanceolate, 3.2-6 × 0.9-1.4 cm. Staminate flower: glands 2; stamens 2, distinct; pistillate flower: gland 1, ovary downy, sessile, styles short. Fl. Apr-May. Along rivers and roadsides of mountain slopes at 2600-3700 m. Distributed in Gansu, Sichuan and Xizang.

大叶柳

Salix magnifica Hemsl.

灌木或小乔木。小枝粗壮。椭圆形，卵形或倒卵状长圆形，长达20厘米，宽达11厘米，革质，常全缘。雄花：腺体2；雄蕊2，离生或部分合生；雌花：子房具柄长。果序长达23厘米。花期5-6月。生海拔2100-3000米的山地。产四川西部和北部。

Shrubs or small trees. Branchlets stout. Leaf blade elliptic, ovate, or obovate-oblong, up to 20 × 11 cm, leathery, margin usually entire. Staminate flower: glands 2; stamens 2, distinct or partly connate; pistillate flower: ovary with long stipe. Fruiting catkin up to 23 cm. Fl. May-Jun. Mountain slopes at 2100-3000 m. Distributed in W and N Sichuan.

大叶柳 *Salix magnifica*

宝兴柳

Salix moupinensis Franch.

乔木。叶长圆形，椭圆形，倒卵形或卵形，叶幼时下面特别是脉上密被白色丝状毛，边缘

宝兴柳 *Salix moupinensis*

长穗柳 *Salix radinostachya*

有腺锯齿。雄花：腺体2；雄蕊2，花丝离生；雌花：子房无毛，有短柄。花期4月。生海拔1500-3000米的山地。产四川西部和云南北部。

Trees. Leaf blade oblong, elliptic, obovate, or ovate, abaxially villous along veins when young, margin glandular serrate. Staminate flower: glands 2; stamens 2, distinct; pistillate flower: ovary glabrous, stipe short. Fl. Apr. Mountains at 1500-3000 m. Distributed in W Sichuan and N Yunnan.

长穗柳

Salix radinostachya C. K. Schneid.

高大灌木。叶披针形或长圆状披针形，长(10-)15-20厘米，宽3-4.5厘米，全缘或有不明显的腺锯齿。雄花：雄蕊2，花丝离生；腺体2；雌花：仅1腹腺；子房有短柄。果序长达20厘米。花期4-6月。生海拔2600-3200米的山坡。产四川西部、西藏东部和云南西北部。印度北部亦有。

Shrubs large. Leaf blade lanceolate or oblong-lanceolate, (10-)15-20 × 3-4.5 cm, margin entire or indistinctly glandular serrate. Staminate flower: stamens 2, distinct; glands 2. Pistillate flower: gland 1; ovary with short stipe. Fruiting catkin up to 20 cm long. Fl. Apr-Jun. Mountain slopes at 2600-3200 m. Distributed in W Sichuan, E Xizang and NW Yunnan. Also in N India.

异型柳

Salix dissa C. K. Schneid.

灌木。叶长圆形或长圆状椭圆形至长圆状卵形，长1-3(-4)厘米，宽0.5-1.8厘米，基部圆形(萌枝叶近心形)，全缘；叶柄短。苞片两面无毛；雄花：腺体2；雄蕊2，花丝中下部有柔毛；雌花：腺体1；子房无柄，无毛；花柱短。花期4-6月。生海拔900-3000米的溪流旁的开阔地和山坡。产甘肃南部、四川西部和云南西北部。

Shrubs. Leaf blade oblong or oblong-elliptic to oblong-ovate, 1-3(-4) × 0.5-1.8 cm, base rounded (subcordate on shoots), margin entire. Bracts glabrous; staminate flower: glands 2; stamens 2, free, filaments basally downy; pistillate flower: gland 1 ovary glabrous, sessile; styles short. Fl. Apr-Jun. Near streams, open places and mountain slopes at 900-3000 m. Distributed in S Gansu, W Sichuan and NW Yunnan.

异型柳 *Salix dissa*

迟花矮柳

Salix oreinoma C. K. Schneid.

矮灌木，高达20厘米。叶倒卵状椭圆形、椭圆形或长圆形，长1.5-3.5厘米，宽1-1.8厘米，边缘有不明显的疏腺锯齿。花序生当年生枝的顶端；雄花：腺体2；雄蕊2，花丝下部被毛；雌花：腺体1；子房无柄；花柱明显。花期6-7月。生海拔2400-3950米的山坡。产四川(康定县、泸定县)。

Dwarf shrubs, up to 20 cm tall. Leaf blade obovate-elliptic, elliptic, or oblong, 1.5-3.5 × 1-1.8 cm, margin indistinctly sparsely glandular serrate. Catkins terminal; staminate flower: glands 2; stamens 2, basally pilose; pistillate flower: gland 1; ovary sessile; styles conspicuous. Fl. Jun-Jul. Mountains at 2400-3950 m. Distributed in Sichuan (Kangding County, Luding County).

迟花矮柳 *Salix oreinoma*

小叶柳

Salix hypoleuca Seemen

灌木，高达3.6米。叶椭圆形或椭圆状长圆形，长2-4厘米，宽1.2-2.4厘米，无毛，全缘。苞片无毛；雄花：腺体1；雄蕊2，花丝中下部有长柔毛；雌花：仅1腹腺；子房近无柄，柱头短。花期5月。生海拔1400-2700米的山坡、路旁和山沟。产山西、陕西、甘肃、湖北和四川。

Shrubs, up to 3.6 m tall. Leaf blade elliptic or elliptic-oblong, 2-4 × 1.2-2.4 cm, glabrous, margin entire. Floral bracts glabrous; staminate flower: gland 1; stamens 2, filaments distinct, villous proximally; fistillate flower: gland 1; ovary subsessile; stigmas short. Fl. May. Mountain slopes, roadsides and valleys at 1400-2700 m. Distributed in Shanxi, Shaanxi, Gansu, Hubei and Sichuan.

小叶柳 *Salix hypoleuca*

中华柳

Salix cathayana Diels

灌木，高达1.5米。叶椭圆形或椭圆状披针形，基部钝或锐尖，全缘。雄性葇荑花序密花；花梗具3或更多小叶，总花梗长5-15毫米；雌性葇荑花序密花。子房椭圆体形，无毛，无柄；花柱短，2裂。蒴果近球形。花期5月，果期6-7月。生海拔1800-3000米的山坡灌丛或山谷。产中国西南、华中、华北和华西。

Shrubs, up to 1.5 m tall. Leaves elliptic or elliptic-lanceolate, base obtuse or acute, margin entire.

中华柳 *Salix cathayana*

Staminate catkins densely flowered; peduncles with 3 or more leaflets, peduncles 5-15 mm long; pistillate catkins densely flowered. Ovary ellipsoid, glabrous, sessile; styles short, 2-cleft. Capsules subglobose. Fl. May. Fr. Jun-Jul. Thickets on mountain slopes or valleys at 1800-3000 m. Distributed in SW, C, N and W China.

簇毛柳
Salix maerkangensis N. Chao

灌木，高达2.5米。小枝密被短柔毛。叶椭圆形或矩圆形，长2.2-3.3厘米，宽0.9-1.5厘米，背面基部被一簇丝状长柔毛，全缘。苞片被长柔毛；雄花：具1腺腺；雄蕊2，花丝分离，下部被毛；雌花：具1腹腺；子房无毛，近无柄。花期4月。生海拔2600-3000米的山坡。产四川(丹巴县、金川县和马尔康)。

Shrubs, up to 2.5 m tall. Branchlets pubescent. Leaf blade elliptic or oblong, 2.2-3.3 × 0.9-1.5 cm, abaxially greenish, tufted villous only near base, margin entire. Floral bract villous; staminate flower: gland 1; stamens 2, filaments distinct, pubescent basally; pistillate flower: gland 1; ovary glabrous, subsessile. Fl. Apr. Mountains at 2600-3000 m. Distributed in Sichuan (Danba County, Jinchuan County and Maerkang).

岷江柳
Salix minjiangensis N. Chao

灌木，高1-2米。小枝被长柔毛。叶宽椭圆形或宽倒卵形，稀圆形，长2.8-4.3厘米，宽1.5-3.5厘米，叶背被长柔毛，全缘或具腺点。苞片被长柔毛；雄花：雄蕊2，花丝分离；雌花：子房柄约0.3毫米，子房无毛；花柱0.7-1.1毫米。花期5-6月。生海拔1800-3200米的山坡林缘。产四川北部和甘肃南部。

Shrubs, 1-2 m tall. Branchlets villous or pilose. Leaf blade broadly elliptic or broadly obovate, sometimes orbicular, 2.8-4.3 × 1.5-3.5 cm, abaxially pilose, margin entire, or gland-dotted. Floral bract villous; staminate flower: stamens 2, filaments distinct; pistillate flower: stipe ca. 0.3 mm, ovary glabrous; styles 0.7-1.1 mm. Fl. May-Jun. Forest edges at 1800-3200 m. Distributed in N Sichuan and S Gansu.

簇毛柳 *Salix maerkangensis*

岷江柳 *Salix minjiangensis*

藏南柳 *Salix austrotibetica*

藏匐柳 *Salix faxonianoides*

藏南柳
Salix austrotibetica N. Chao

灌木，高0.2-0.5(-1.5)米。叶倒卵状椭圆形或倒披针形，长2-5.4厘米，宽1.1-2.5厘米，两面被疏长柔毛，后无毛，下面常被白粉，全缘或有不规则的腺齿。花序生枝端；雄花：雄蕊2，花丝分离；雌花：子房柄长0.2-0.6毫米，子房无毛，被白粉，部分或全部被毛；花柱长0.8-1.9毫米。花期6-8月。生海拔2400-4400米的山坡。产西藏东南部和云南西北部。

Shrubs, 0.2-0.5(-1.5) m tall. Leaf blade obovate-elliptic, or oblanceolate, 2-5.4 × 1.1-2.5 cm, abaxially pruinose, villous when young, glabrescent, adaxially villous, glabrescent, margin irregularly glandular dentate or entire. Catkins at apex of branchlets; staminate flower: stamens 2, filaments distinct; pistillate flower: stipe 0.2-0.6 mm, ovary glabrous, pruinose, partly downy, or densely downy; styles 0.8-1.9 mm. Fl. Jun-Aug. Mountain slopes at 2400-4400 m. Distributed in SE Xizang and NW Yunnan.

藏匐柳
Salix faxonianoides C. Wang et P. Y. Fu

矮灌木，高达0.5米。叶矩圆形、狭倒卵形、倒卵形或椭圆形，长1.3-3.2厘米，宽0.8-1.8厘米，无毛，边缘有圆齿或圆锯齿。花序生于枝端；苞片2.3-3.7(-7)毫米；雄花：雄蕊2，花丝离生；雌花：子房近无柄，无毛，稀部分被毛；花柱2.1-2.3(-4.5)毫米。花期6-8月。生海拔3150-4600米的灌丛中。产西藏南部。尼泊尔亦有。

Shrubs low, up to 0.5 m tall. Leaf blade oblong, narrowly obovate, obovate or elliptic, 1.3-3.2 × 0.8-1.8 cm, glabrous, margin crenate or crenate-serrate. Catkins at apex of branchlets; floral bract 2.3-3.7 (-7) mm; staminate flower: stamens 2, filaments distinct; pistillate flower: ovary subsessile, glabrous, rarely partly pilose; styles 2.1-2.3(-4.5) mm. Fl. Jun-Aug. Thickets at 3150-4600 m. Distributed in S Xizang. Also in Nepal.

宝兴矮柳
Salix microphyta Franch.

矮灌木，高达20(-30)厘米。枝直立或斜上，较细。叶倒卵

宝兴矮柳 *Salix microphyta*

形，长圆形或匙形，长1.4-2.4厘米，宽0.9-1.3厘米，无毛，边缘具的圆腺齿。花序生枝端；苞片1-2毫米；雄花：雄蕊2，花丝离生；雌花：子房近无柄，无毛；花柱0.4-0.6毫米。花期5-7月。生海拔2250-2700米的灌丛中。产四川西部和北部。

Dwarf Shrubs, up to 20(-30) cm tall. Branches erect or ascending, slender. Leaf blade obovate, oblong or spatulate, 1.4-2.4 × 0.9-1.3 cm, glabrous, margin crenate. Catkins at apex of branchlet; floral bract 1-2 mm; staminate flower: stamens 2, filaments distinct; pistillate flower: ovary glabrous, subsessile; styles 0.4-0.6 mm. Fl. May-Jul. Thickets at 2250-2700 m. Distributed in W and N Sichuan.

迟花柳 *Salix opsimantha*

藏截苞矮柳

Salix resectoides Hand.-Mazz.

灌木，高达0.5(-2)米。新枝被锈色短柔毛。叶狭倒卵形、倒卵形、宽倒卵形至圆状椭圆形，长2-6.8厘米，宽1.4-3.4厘米，上面沿脉被锈色短柔毛，全缘或具疏腺锯齿。花序生枝端；苞片1.1-1.7毫米；雄花：雄蕊2，花丝离生；雌花：子房柄长0.2-0.4毫米，子房无毛或基部有毛。花期5-7月。生海拔2800-4200米的山坡。产云南西北部和西藏东南部。

Shrubs, up to 0.5(-2) m tall. Branchlets pubescent, hairs ferruginous and white. Leaf blade narrowly obovate, obovate, broadly obovate, or orbicular-elliptic, 2-6.8 × 1.4-3.4 cm, adaxially pubescent along veins, hairs ferruginous and white, margin entire or sparsely glandular serrulate. Catkins at apex of branchlets; floral bract 1.1-1.7 mm; staminate flower: stamens 2, filaments distinct; pistillate flower: stipe 0.2-0.4 mm, ovary glabrous or basally pilose. Fl. May-Jul. Mountains at 2800-4200 m. Distributed in NW Yunnan and SE Xizang.

迟花柳

Salix opsimantha C. K. Schneid.

灌木，高0.2-0.5(-3)米。叶椭圆形或倒卵状椭圆形，长1.5-5.9厘米，宽1-3.8厘米，叶背幼时被疏长毛，腹面中脉被短柔毛，边缘有腺锯齿或全缘。花序生枝端；苞片1.7-3.2毫米；雄花：雄蕊2，花丝离生；雌花：子房柄0-0.4毫米，子房无毛或具疏柔毛。花期6-7月。生海拔3400-4700米的灌丛。产四川西部、云南西北部和西藏东南部。

Shrubs, 0.2-0.5(-3) m. Leaf blade elliptic or obovate-elliptic, 1.5-5.9 × 1-3.8 cm, abaxially pilose when young, adaxially pubescent along midvein, margin sparsely glandular serrate or entire. Catkins terminal on juvenile branchlets; floral bract 1.7-3.2 mm; staminate flower: stamens 2, filaments distinct; pistillate flower: stipe 0-0.4 mm, ovary glabrous, sometimes pilose. Fl. Jun-Jul. Thickets at 3400-4700 m. Distributed in W Sichuan, NW Yunnan and SE Xizang.

藏截苞矮柳 *Salix resectoides*

察隅矮柳 *Salix thomsoniana*

察隅矮柳

Salix thomsoniana Andersson

小灌木，高达0.5米。新枝密被长柔毛。叶椭圆形、倒卵状长圆形、长圆状椭圆形或椭圆状倒披针形，长1.2-5.7厘米，宽0.7-2厘米，叶背伏生长柔毛，全缘或有不明显的腺齿尖。花序生枝端；雄花：雄蕊2，花丝离生；雌花：子房密被柔毛或无毛，近无柄。花期5-7月。生海拔2400-3600米的山坡。产西藏东南部。印度北部亦有。

Shrubs low, up to 0.5 m tall. Branchlets densely villous. Leaf blade elliptic, obovate-oblong, oblong-elliptic or elliptic-oblanceolate, 1.2-5.7 × 0.7-2 cm, abaxially densely appressed villous, margin entire or gland-dotted. Catkins at apex of branchlets; staminate flower: stamens 2, filaments distinct; pistillate flower: ovary villous or glabrous, subsessile. Fl. May-Jul. Mountain slopes at 2400-3600 m. Distributed in SE Xizang. Also in N India.

怒江矮柳

Salix coggygria Hand.-Mazz.

灌木，高达2米。幼枝被灰色或褐色绒毛。叶长圆状倒卵形或倒卵圆形，长3-8.5厘米，宽1.5-3.5厘米，上面密被锈色柔毛，全缘，稀有疏锯齿，先端常有一皱褶。花序生枝端；苞片1.7-3.4毫米；雄花：雄蕊2，花丝离生；雌花：子房柄0-0.3毫

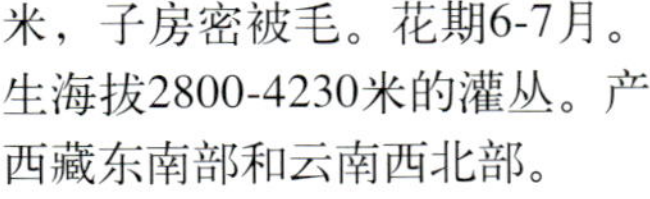

米，子房密被毛。花期6-7月。生海拔2800-4230米的灌丛。产西藏东南部和云南西北部。

Shrubs, up to 2 m tall. Branchlets densely gray pubescent. Leaf blade oblong-obovate or obovate-orbicular, 3-8.5 × 1.5-3.5 cm, adaxially pubescent, hairs ferruginous and white, margin entire, rarely sparsely gland-dotted, apex often with a pleated mucro. Catkins at apex of branchlets; floral bract 1.7-3.4 mm; staminate flower: stamens 2, filaments distinct; pistillate flower: stipe 0-0.3 mm, ovary densely grayish white pubescent. Fl. Jun-Jul. Thickets at 2800-4230 m. Distributed in SE Xizang and NW Yunnan.

环纹矮柳

Salix annulifera Marquand et Airy Shaw

矮灌木，高达0.5米。叶宽倒卵形、倒卵形、倒卵状椭圆形或倒披针形，长3.7-10.5厘米，宽1.8-4.8厘米，边缘有疏圆锯齿。花序生枝端；苞片先端近截形，有不规则的齿；雄花：雄蕊2，花丝离生；雌花：子房被密毛；花柱明显。花期7-8月。生海拔3000-4200米的山坡灌丛或草丛中。产西藏东南部和云南西北部。

Shrubs low, up to 0.5 m tall. Leaf blade broadly obovate, obovate, obovate-elliptic or oblanceolate, 3.7-10.5 × 1.8-4.8 cm, margin remotely crenate. Catkins at apex of branchlets; floral bract margin slightly irregularly toothed, apex

怒江矮柳 *Salix coggygria*

环纹矮柳 *Salix annulifera*

truncate; staminate flower: stamens 2, filaments distinct; pistillate flower: ovary pubescent; styles conspicuous. Fl. Jul-Aug. Shrubs or grasses on mountain slopes at 3000-4200 m. Distributed in SE Xizang and NW Yunnan.

小垫柳

Salix brachista C. K. Schneid.

垫状灌木。树干和侧枝匍匐；幼枝疏具长毛或无毛。叶椭圆形、倒卵状椭圆形或卵形，(5-)10(-20) × 4-8毫米，下面密具平伏长柔毛，叶缘在中部以上具疏腺锯齿，下部齿疏或全缘。花序卵圆形；苞片无毛；子房近无梗。花期6-7月，果期7-8月。生海拔2600-3900米的山坡。产云南、四川和西藏。

Shrubs cushion-shaped. Trunk and lateral branches creeping; juvenile branchlets with sparse, long hairs or glabrous. Leaves elliptic, obovate-elliptic, or ovate, (5-)10(-20) × 4-8 mm, abaxially densely appressed villous, margin sparsely glandular serrate distally, remotely dentate or entire proximally. Catkins ovate-orbicular; bracts glabrous; ovary subsessile. Fl. Jun-Jul. Fr. Jul-Aug. Mountain slopes at 2600-3900 m. Distributed in Yunnan, Sichuan and Xizang.

栅枝垫柳

Salix clathrata Hand.-Mazz.

垫状灌木，高达0.2米。主干及枝条匍匐生长。枝条极多而互相交错呈栅栏状。叶矩圆形、椭圆形或倒卵形，长1-1.7厘米，宽0.6-1厘米，革质，全缘，反卷，具疏缘毛。花先叶开放；花序椭圆形，长约6毫米，粗约3毫米；雌花：子房无毛。花期4-6月。生海拔3600-4600米的裸露岩石上。产四川西部和云南西北部。

栅枝垫柳 *Salix clathrata*

Shrubs procumbent, up to 0.2 m tall. Trunk and branches spreading. Branches many, appearing fence-like. Leaf blade oblong, elliptic, or obovate, 1-1.7 × 0.6-1 cm, leathery, margin entire, revolute, ciliate. Catkins just before leaves emerge; catkin ellipsoid, ca. 6 × ca. 3 mm; pistillate flower: ovary glabrous. Fl. Apr-Jun. Rocks at 3600-4600 m. Distributed in W Sichuan and NW Yunnan.

小垫柳 *Salix brachista*

毛枝垫柳 *Salix hirticaulis*

毛小叶垫柳 *Salix pilosomicrophylla*

毛枝垫柳

Salix hirticaulis Hand.-Mazz.

垫状灌木，高达4厘米。主干匍匐生根。新枝密被黄褐色长硬毛。叶椭圆形、倒卵状长圆形或圆形，长7-14毫米，宽3-8毫米，全缘，或中上部有腺锯齿。花序头状；花序轴密生黄褐色短硬毛。花期7-8月。生海拔3500-4200米的岩石缝上或草丛中。产云南西北部和西藏东南部。

Shrubs procumbent, up to 4 cm tall. Trunk creeping, rooting. Branchlets densely yellowish brown hirsute. Leaf blade elliptic, obovate-oblong, or orbicular, 7-14 × 3-8 mm, margin glandular serrate distally, entire proximally or throughout. Catkin capitates; rachis yellowish brown hispidulous. Fl. Jul-Aug. Rock crevices or grassy meadows at 3500-4200 m. Distributed in NW Yunnan and SE Xizang.

毛小叶垫柳

Salix pilosomicrophylla C. Wang et P. Y. Fu

垫状灌木，高达12厘米。新枝被长柔毛。叶卵状长圆形、矩圆形、倒卵状椭圆形或椭圆形，长3.6-14.6(-24)毫米，宽1.3-5.2(-9.6)毫米，下面被长柔毛，全缘，稀先端具腺齿。雄花序头状，具7-17花；雌花序具5-23花。花期7-8月。生海拔3430-5000米的湿润草甸或溪流旁。产西藏东南部和云南西北部。

Shrubs creeping, up to 12 cm tall. Branchlets villous. Leaf blade ovate-oblong, oblong, obovate-elliptic, or elliptic, 3.6-14.6(-24) × 1.3-5.2(-9.6) mm, abaxially villous, margin entire, rarely distantly serrulate distally. Staminate catkin subglobose 7-17 flowers; pistillate catkin with 5-23 flowers. Fl. Jul-Aug. Moist mountain meadows, or along streams at 3430-5000 m. Distributed in SE Xizang and NW Yunnan.

多花小垫柳

Salix fruticulosa Andersson

垫状灌木，高达10厘米。新枝散生短柔毛。叶矩圆形或倒卵状矩圆形或长椭圆形，长4-10毫米，宽1.5-4毫米，无毛，叶缘上部有疏腺锯齿，下部全缘。雌花序密花或下部较疏。花期6-7月。生海拔3500-4200米的河岸和林缘等处。产西藏南部。尼泊尔中部和印度北部亦有。

Shrubs creeping, up to 10 cm tall. Branchlets pubescent. Leaf blade oblong or obovate-oblong or long elliptic, 4-10 × 1.5-4 mm, glabrous, margin crenulate distally, entire proximally. Pistillate catkin densely flowered, proximal flowers sparser. Fl. Jun-Jul. Along streams and forest edges at 3500-4200 m. Distributed in S Xizang. Also in C Nepal and N India.

墨竹柳

Salix maizhokunggarensis N. Chao

灌木，高2米。叶椭圆形、长椭圆形或宽椭圆形，长2.2-6.6厘米，宽1.1-2.7(-4)厘米，边缘全缘、具腺点或具不明显的锯齿。花序梗长0.3-1.4厘米；雌花：子房无毛，稀局部被长柔毛，子房柄长0.4-1.2毫米，被毛或无；花柱长0.8-2.1毫米。花期5-7月。生海拔3800-4600米

多花小垫柳 *Salix fruticulosa*

山生柳 *Salix oritrepha*

的山坡。产四川西部、西藏南部和东部。

Shrubs up to 2 m tall. Leaf blade elliptic, long elliptic or broadly elliptic, 2.2-6.6 × 1.1-2.7(-4) cm, margin entire, gland-dotted or indistinctly serrulate. Peduncle 0.3-1.4 cm; pistillate flower: ovary glabrous, rarely partly villous, stipe 0.4-1.2 mm, pubescent or glabrous; styles 0.8-2.1 mm. Fl. May-Jul. Mountain slopes at 3800-4600 m. Distributed in W Sichuan, and S and E Xizang.

山生柳

Salix oritrepha C. K. Schneid.

灌木，高达2米。叶椭圆形或卵状椭圆形，长1.9-3.9厘米，宽0.7-2厘米，全缘。苞片宽倒卵形，两面具绵毛，毛白或杂有锈色。雌花和雄花具2腺体，腺体基部结合，腹腺常2裂；雌花：子房被绵毛。花期6-8月。生海拔3300-5000米的山坡灌丛。产甘肃东南部、青海东南部、四川和西藏东部。

Shrubs, up to 2 m tall. Leaf blade elliptic or ovate-elliptic, 1.9-3.9 × 0.7-2 cm, margin entire. Floral bract broadly obovate, woolly, hairs white, or white and ferruginous. Staminate flower and pistillate flower: glands 2, usually basally connate, adaxial gland usually 2-lobed; pistillate flower: ovary woolly. Fl. Jun-Aug. Thickets at 3300-5000 m. Distributed in SE Gansu, SE Qinghai, Sichuan and E Xizang.

墨竹柳 *Salix maizhokunggarensis*

奇花柳
Salix atopantha C. K. Schneid.

灌木，高1-4米。叶卵状矩圆形、椭圆状矩圆形或矩圆形，稀披针形，长1.8-3.9(-4.8)厘米，宽0.7-1.5(-1.8)厘米，边缘具腺锯齿或稀全缘，先端急尖或钝。花序梗长0.3-1.8厘米；苞片先端圆截形，全缘或有浅圆齿；雌花：子房被密绵毛。花期5-6月。生海拔2800-4600米的山坡或山谷。产甘肃南部、青海东南部、四川西部和北部、西藏东部和云南西北部。

木里柳 *Salix muliensis*

Shrubs, 1-4 m tall. Leaf blade ovate-oblong, elliptic-oblong or oblong, rarely lanceolate, 1.8-3.9 (-4.8) × 0.7-1.5(-1.8) cm, margin indistinctly glandular, serrate, rarely entire, apex acute or obtuse. Peduncle 0.3-1.8 cm; ploral bract entire or toothed, apex rounded-truncate; pistillate flower: ovary woolly. Fl. May-Jun. Mountain slopes and valleys at 2800-4600 m. Distributed in S Gansu, SE Qinghai, W and N Sichuan, E Xizang and NW Yunnan.

木里柳
Salix muliensis Goerz

大灌木或小乔木，高达5米。叶倒卵状长圆形、倒卵状椭圆形、椭圆形或倒披针形，长2.2-4.6厘米，宽0.9-1.8厘米，下面被密被伏锈毛，全缘或有不明显的腺齿；叶柄密被污色短柔毛。雌花：子房被白或锈色密毛。花期4-6月。生海拔3060-4600米的山坡灌丛。产四川西南部和云南西北部。

Large shrubs or small trees, up to 5 m tall. Leaf blade obovate-oblong, obovate-elliptic, elliptic, or oblanceolate, 2.2-4.6 × 0.9-1.8 cm, abaxially densely appressed dull red or rust-colored pubescent, margin entire or indistinctly glandular serrulate. Pistillate flower: ovary pubescent, hairs white or ferruginous. Fl. Apr-Jun. Thickets at 3060-4600 m. Distributed in SW Sichuan and NW Yunnan.

大苞柳
Salix pseudospissa Goerz

灌木，高1-2米。小枝较粗壮。叶倒卵形，长3.6-5.3厘米，宽1.5-2.3厘米，无毛，边缘有明显的锯齿。花序粗，花序梗0-1.1厘米；苞片3.2-5.6毫米；雌花：子房密被短毛。花期5-6月。生海拔3400-4860米的灌丛中。产甘肃东南部、青海、四川北部和西藏东部。

Shrubs, 1-2 m tall. Branchlets stout. Leaf blade obovate, 3.6-5.3 × 1.5-2.3 cm, glabrous, margin distinctly serrate. Catkins stout, peduncle 0-1.1 cm; floral bract 3.2-5.6 mm; pistillate flower: ovary pubescent. Fl. May-Jun. Thickets at 3400-4860 m. Distributed in SE Gansu, Qinghai, N Sichuan and E Xizang.

奇花柳 *Salix atopantha*

大苞柳 *Salix pseudospissa*

3000-4800米的山坡。产青海东南部、四川西部、西藏东部和云南西北部。

Shrubs, up to 3 m tall. Leaf blade obovate-elliptic, obovate, or elliptic, 3-5.1(-6.9) × 1.5-3 (-3.8) cm, abaxially glabrous, margin glandular serrate or subentire. Peduncle 0.3-2 cm; floral bract 1.3-2.3 mm; pistillate flower: ovary densely downy. Fl. Jul-Aug. Mountain slopes at 3000-4800 m. Distributed in SE Qinghai, W Sichuan, E Xizang and NW Yunnan.

川鄂柳

Salix fargesii Burk.

乔木或灌木。叶椭圆形或狭卵形，边缘具腺锯齿。葇荑花序长6-8厘米；花梗长1-3厘米，具小叶；花丝无毛；苞片狭倒卵形。果序长达12厘米。蒴果长圆状卵球形，具短梗。生海拔1400-1600米的山地。产四川、湖北、陕西和甘肃。

Trees or shrubs. Leaf blade elliptic or narrowly ovate, margin glandular serrulate. Catkins 6-8 cm long; peduncles 1-3 cm long, with small leaves; filaments glabrous; bracts narrowly obovate. Fruiting catkins up to 12 cm long. Capsules oblong-ovoid, shortly stipitate. Mountains at 1400-1600 m. Distributed in Sichuan, Hubei, Shaanxi and Gansu.

吉拉柳

Salix gilashanica C. Wang et P. Y. Fu

灌木，高达3米。叶倒卵状椭圆形，倒卵形或椭圆形，长3-5.1(-6.9)厘米，宽1.5-3(-3.8)厘米，叶背无毛，边缘具腺锯齿或近全缘。花序梗0.3-2厘米；苞片1.3-2.3毫米；雌花：子房密被柔毛。花期7-8月。生海拔

吉拉柳 *Salix gilashanica*

川鄂柳 *Salix fargesii*

银背柳

Salix ernestii C. K. Schneid.

大灌木，高1.5-4米。叶椭圆形，狭椭圆形或倒卵状披针形，长6.3-11.9厘米，宽2-4.6厘米，上面被疏绒毛，下面被丝状绒毛，全缘或上部具不明显的腺细齿。花序梗0.4-3.6厘米；苞片3.1-3.7毫米，背被长柔毛；雌花：子房被柔毛。花期5-6月。生海拔2000-4350米的山坡。产四川和云南。

Tall shrubs, 1.5-4 m tall. Leaf blade elliptic, narrowly elliptic or obovate-lanceolate, 6.3-11.9 × 2-4.6 cm, abaxially tomentose, adaxially sparsely tomentose, margin entire, or gland-dotted distally. Peduncle 0.4-3.6 cm; floral bract 3.1-3.7 mm, abaxially villous; pistillate flower: ovary downy. Fl. May-Jun. Mountain slopes at 2000-4350 m. Distributed in Sichuan and Yunnan.

银背柳 *Salix ernestii*

长叶柳

Salix phanera C. K. Schneid.

乔木或大灌木，高达7米。叶卵状披针形、椭圆状披针形或椭圆形，长8.7-14.4(-18.6)厘米，宽2-3.9(-4.6)厘米，下面初被白色厚绵毛，后毛渐脱落，边缘有细腺齿。苞片顶端通常有腺齿；雌花：子房有密绒毛。花期5-6月。生海拔2200-3250米山区溪流旁、杂木林中。产甘肃南部、四川西部和北部。

长叶柳 *Salix phanera*

Trees, or large shrubs, up to 7 m tall. Leaf blade ovate-lanceolate, elliptic-lanceolate, or elliptic, 8.7-14.4(-18.6) × 2-3.9(-4.6) cm, abaxially at first densely tomentose, glabrescent, margin glandular denticulate. Floral bract glandular dentate; pistillate flower: ovary tomentose. Fl. May-Jun. Streamsides, woods at 2200-3250 m. Distributed in S Gansu, W and N Sichuan.

纤柳

Salix phaidima C. K. Schneid.

乔木或灌木，高1-7米。叶线状披针形至卵状披针形，长6.3-14.5厘米，宽1.6-2.3(-2.7)厘米，下面初密被白色丝状绒毛，有光泽，全缘，基部圆至楔形，先端急尖至渐尖。雌花：子房被密毛。花期5-6月。生海拔1600-2800米山区。产四川。

纤柳 *Salix phaidima*

灰叶柳 *Salix floccosa*

Trees or shrubs, 1-7 m tall. Leaf blade linear-lanceolate to ovate-lanceolate, 6.3-14.5 × 1.6-2.3 (-2.7) cm, abaxially at first white tomentose, shiny, margin entire, base rounded to cuneate, apex acute to acuminate. Pistillate flower: ovary with hairs. Fl. May-Jun. Mountains at 1600-2800 m. Distributed in Sichuan.

灰叶柳
Salix floccosa Burkill

灌木，高1-3米。成叶革质，倒卵状椭圆形或倒披针形，稀倒卵形，长4-9(-11.5)厘米，宽1-2.9(-3.6)厘米，下面常有簇生的长柔毛，上面无毛或稍有蜘丝状柔毛，全缘。花序梗0.5-1.5厘米；雌花：子房无毛或被疏长柔毛。花期4-7月。生海拔2500-4600米的山坡，灌丛或林缘。产四川、西藏东南部和云南西北部。

Shrubs, 1-3 m tall. Leaf blade leathery, obovate-elliptic or oblanceolate, rarely obovate, 4-9(-11.5) × 1-2.9(-3.6) cm, abaxially woolly, adaxially glabrous or with sparse, filamentous hairs, margin entire. Peduncle 0.5-1.5 cm; pistillate flower: ovary glabrous or pilose. Fl. Apr-Jul. Mountain slopes, thickets, forest edges at 2500-4600 m. Distributed in Sichuan, SE Xizang and NW Yunnan.

银光柳
Salix argyrophegga C. K. Schneid.

乔木或灌木，高1.5-6米。新枝有密灰色长柔毛。叶椭圆形，卵状椭圆形或长圆形，长7.7-14(-17.2)厘米，宽4-7.2(-9.9)厘米，下面被密而厚的灰色或淡黄灰色绒毛，边缘有腺锯齿。花序具密花或疏花；雌花：子房密被毛。花期4-5月。生海拔1260-3350米的山坡和灌丛。产四川西部。

Trees or shrubs, 1.5-6 m tall. Branchlets gray villous. Leaf blade elliptic, ovate-elliptic, or oblong, 7.7-14(-17.2) × 4-7.2(-9.9) cm, abaxially gray or yellowish tomentose, margin glandular serrate. Catkins densely or moderately densely flowered; pistillate flower: ovary villous. Fl. Apr-May. Mountain slopes and thickets at 1260-3350 m. Distributed in W Sichuan.

银光柳 *Salix argyrophegga*

对叶柳 *Salix salwinensis*

对叶柳

Salix salwinensis Hand.-Mazz.

灌木，高1-4米。叶对生；披针形、倒披针形或倒卵状披针形，7.5-12.7厘米长，1.5-3(-3.6)厘米宽，两面都密被绒毛，全缘。雌花：子房密被柔毛。花期4-6月。生海拔1840-3900米的山坡、林中和溪流旁。产云南西北部和西藏南部。尼泊尔东部亦有。

Shrubs, 1-4 m tall. Leaves opposite; leaf blade lanceolate, oblanceolate or obovate-lanceolate, 7.5-12.7 × 1.5-3(-3.6) cm, abaxially tomentose, margin entire. Pistillate flower: ovary downy. Fl. Apr-Jun. Mountains, woods and riversides at 1840-3900 m. Distributed in NW Yunnan and S Xizang. Also in E Nepal.

大理柳

Salix daliensis C. F. Fang et S. D. Zhao

灌木。叶披针形、长圆披针形或狭椭圆形，背面密被白色绢毛。花序粗4-6毫米；苞片倒三角形或三角状倒卵形，密具白色绒毛；子房卵球形，密被白色柔毛。蒴果密被柔毛。花期4月，果期6月。生海拔1500-2700米的灌丛、山谷溪旁或阔叶林中。产湖北、云南和四川。

大理柳 *Salix daliensis*

Shrubs. Leaves lanceolate, oblong-lanceolate or narrowly elliptic, abaxially densely white-sericeous. Catkins 4-6 mm diam; bracts obdeltoid or deltoid-obovate, densely white downy; ovary ovoid, densely white-pubescent. Capsules densely pubescent. Fl. Apr. Fr. Jun. Thickets, by streams in valleys or broad-leaved forests at 1500-2700 m. Distributed in Hubei, Yunnan and Sichuan.

锡金柳

Salix sikkimensis Andersson

灌木或乔木，高1-5米。叶椭圆形、狭椭圆形或倒卵状披针形，长4.8-8.9厘米，宽1.8-3.3厘米，下面密被绢质伏生锈色柔毛，叶全缘具腺点。花序先叶开放，稀近同时开放；花序梗0-1.4厘米；苞片倒卵形，先端凹；雌花：子房被柔毛。花期6-7月。生海拔3700-4600米的河滩地和山坡灌丛。产西藏。印度北部亦有。

Large shrubs or trees, 1-5 m tall. Leaf blade elliptic, narrowly elliptic, or obovate-lanceolate, 4.8-8.9 × 1.8-3.3 cm, abaxially velvety, hairs ferruginous, margin entire and gland-dotted. Catkins before leaves emerge or just before leaves emerge; peduncle 0-1.4 cm; floral bract obovate, apex retuse; pistillate flower: ovary downy. Fl. Jun-Jul. Streamsides, mountains, thickets

锡金柳 *Salix sikkimensis*

双柱柳 *Salix bistyla*

at 3700-4600 m. Distributed in Xizang. Also in N India.

双柱柳
Salix bistyla Hand.-Mazz.

灌木，高0.3-1.5(-3)米。叶狭椭圆形或倒卵状狭椭圆形，长4.4-9.8(-11)厘米，宽1.6-2.5(-3.4)厘米，上面被疏绒毛，下面密绒毛，边缘有腺锯齿。苞片5.1-5.5毫米，先端截形，有不明显腺齿；雌花：子房被密绒毛；花柱长4-5毫米。花期6-8月。生海拔2650-4000米的山坡、溪流旁和草甸。产西藏东部和云南西部。缅甸北部亦有。

Shrubs, 0.3-1.5(-3) m tall. Leaf blade narrowly elliptic or narrowly obovate-elliptic, 4.4-9.8(-11) × 1.6-2.5(-3.4) cm, adaxially sparsely tomentose, abaxially densely tomentose, margin glandular serrate. Floral bract 5.1-5.5 mm, margin indistinctly dentate, apex truncate; pistillate flower: ovary tomentose; styles 4-5 mm long. Fl. Jun-Aug. Mountains, along rivers and alpine meadows at 2650-4000 m. Distributed in E Xizang and W Yunnan. Also in N Myanmar.

贡嘎山柳
Salix gonggashanica C. F. Fang et A. K. Skvortsov

灌木，高0.5-2.5米。叶椭圆状倒披针形或椭圆形，长4.1-7厘米，宽1.6-2.2厘米，纸质，叶背被短丝毛，叶全缘具腺点。花序纤细，具疏花；苞片长1.3-1.5毫米；雌花：子房被密毛。花期6-7月。生海拔2500-3850米的灌丛和溪流旁。产四川西部。

Shrubs, 0.5-2.5 m tall. Leaf blade elliptic-oblanceolate or elliptic, 4.1-7 × 1.6-2.2 cm, papery, abaxially short-sericeous, margin entire and gland-dotted. Catkins slender, moderately densely flowered; foral bract 1.3-1.5 mm long; pistillate flower: ovary pubescent. Fl. Jun-Jul. Thickets and along rivers at 2500-3850 m. Distributed in W Sichuan.

贡嘎山柳 *Salix gonggashanica*

秦岭柳 *Salix alfredii*

秦岭柳
Salix alfredii Goerz

灌木或乔木，高达8米。小枝细。叶椭圆形、卵状椭圆形或披针形，长3.4-6.4厘米，宽1.3-2厘米，下面有短柔毛，后无毛，全缘具腺点。花序梗0.2-0.9厘米；苞片约1毫米；雄花：雄蕊2，离生；雌花：子房被柔毛。花期5-6月。生海拔2400-3300米的山坡。产青海东部、甘肃南部、四川北部、湖北和陕西。

Shrubs or trees, up to 8 m tall. Branches thin. Leaf blade elliptic, ovate-elliptic, or lanceolate, 3.4-6.4 × 1.3-2 cm, abaxially pubescent, glabrescent, margin entire, or gland-dotted. Peduncle 0.2-0.9 cm; floral bract ca. 1 mm; staminate flower: stamens 2, filaments distinct; pistillate flower: ovary villous. Fl. May-Jun. Mountain slopes at 2400-3300 m. Distributed in E Qinghai, S Gansu, N Sichuan, Hubei and Shaanxi.

多腺柳
Salix nummularia Andersson

匍匐小灌木。叶片革质，近圆形、倒卵状椭圆形或椭圆形，长0.5-1.7厘米，基部圆形、心形或圆楔形，全缘，先端圆形、钝或微凹；无托叶。花序苞片黄褐色，色泽一致；花序与叶同时开放。花果期7月。生海拔2200-2600米的高山苔原。产吉林。俄罗斯、朝鲜半岛和北美洲亦有。

Creeping shrublets. Leaves leathery, suborbicular, obovate-elliptic, or elliptic, 0.5-1.7 cm long, base rounded, cordate, or rounded-cuneate, margin entire, apex rounded, obtuse, or retuse; stipules absent. Bracts of catkins yellowish brown, uniform in color; flowering coaetaneous. Fl. and fr. Jul. Alpine tundras at 2200-2600 m. Distributed in Jilin. Also in Russia, Korean Peninsula and North America.

蔓柳
Salix turczaninowii Laksch.

匍匐灌木。叶椭圆形、倒卵状椭圆形或卵状椭圆形、无毛，边缘具锯齿，长1.5-2厘米。花序苞片淡黄色，先端紫红色；花序与叶同期开放；葇荑花序顶生，疏花；子房具短梗；近无花柱。花期6-7月，果期7月。生海拔2600米以上的高山冻原。产新疆(阿尔泰山、天山)。哈萨克斯坦、俄罗斯(南西伯利亚)和蒙古亦有。

Shrubs creeping. Leaves elliptic, obovate-elliptic or ovate-elliptic, glabrous, margin serrulate, 1.5-2 cm long. Bracts of catkins pale yellow, apex purplish red; flowering coaetaneous; catkins terminal, laxlyflowered; ovary shortly stipitate; styles nearly absent. Fl. Jun-Jul. Fr. Jul. Alpine tundras above 2600 m. Distributed in Xinjiang (Altay Mountain, Tianshan Mountain). Also in Kazakhstan, Russia (S Siberia) and Mongolia.

多腺柳 *Salix nummularia*

褐色或褐色；雌花：子房无毛；柄长约0.5毫米；花柱明显。花期5-6月。生海拔1300-1700米的河谷和林缘。产黑龙江北部、内蒙古东部和新疆北部。蒙古、俄罗斯北部和欧洲亦有。

Large shrubs or trees. Leaf blade orbicular, ovate, or ovate-elliptic, 2-8 × 1.5-6 cm, abaxially whitish, glabrous, margin serrulate; stipules large, reniform. Peduncle short; floral bract russet or brown; pistillate flower: ovary glabrous; stipe ca. 0.5 mm; styles conspicuous. Fl. May-Jun. Valleys and edges of woods at 1300-1700 m. Distributed in N Heilongjiang, E Neimenggu and N Xinjiang. Also in Mongolia, N Russia and Europe.

蔓柳 *Salix turczaninowii*

越橘柳

Salix myrtilloides L.

灌木，高0.3-0.8米。叶椭圆形或长椭圆形，长1-3.5厘米，宽0.7-1.5厘米，两面无毛，全缘，稀有齿。雄花：雄蕊2，花丝离生；雌花：子房有长柄，无毛；花柱短。花期5月。生海拔300-500米的沼泽化草甸。产吉林、辽宁、黑龙江和内蒙古。朝鲜半岛、蒙古、俄罗斯和欧洲亦有。

Shrubs, 0.3-0.8 m tall. Leaf blade elliptic or long elliptic, 1-3.5 × 0.7-1.5 cm, glabrous, margin entire, rarely dentate. Staminate flower: stamens 2, filaments distinct; pistillate flower: ovary glabrous; stipe conspicuous. Fl. May. Marshes at 300-500 m. Distributed in Jilin, Liaoning, Heilongjiang and Neimenggu. Also in Korean Peninsula, Mongolia, Russia and Europe.

鹿蹄柳

Salix pyrolifolia Ledeb.

大灌木或乔木。叶圆形、卵形或卵状椭圆形，长2-8厘米，宽1.5-6厘米，下面带白色，两面无毛，边缘有细锯齿；托叶大，肾形。花序梗短；苞片棕

越橘柳 *Salix myrtilloides*

鹿蹄柳 *Salix pyrolifolia*

兴安柳 *Salix hsinganica*

兴安柳

Salix hsinganica Y. L. Chang et Skvortzov

灌木，高达1米。叶卵形，椭圆形或倒卵形，长1-4.5厘米，宽0.5-2.5厘米，质坚厚，稍发皱，两面有柔毛，全缘或有不整齐的疏浅齿。花先叶开放，花序椭圆形至短圆柱形；苞片长圆形，上端紫红色；雌花：子房被柔毛，子房柄长；花柱短或无。花期5-6月。生林间隙地或山坡。产黑龙江和内蒙古东部。

Shrubs, up to 1 m tall. Leaf blade ovate, elliptic, or obovate, 1-4.5 × 0.5-2.5 cm, thick, slightly wrinkled, both surfaces downy, margin entire or irregularly shallowly dentate. Catkin before leaves emerge, ellipsoid to shortly cylindric; floral bract oblong, apex purplish red; pistillate flower: ovary villous, long stipitate; styles short or absent. Fl. May-Jun. Open woodlands or mountain slopes. Distributed in Heilongjiang and E Neimenggu.

黄花柳

Salix caprea L.

灌木或小乔木。小枝黄绿色至黄红色。先叶开花；雄葇荑花序椭圆形或宽椭圆形；雌葇荑花序短圆柱形；苞片2色，下面色浅，上面黑色；花丝长约为苞片的4-5倍；子房狭圆形，被柔毛；柱头较短。蒴果长达9毫米。花期4月，果期5-6月。生山坡或林地。产内蒙古、黑龙江、吉林和辽宁。北亚和欧洲亦有。

Shrubs or small trees. Branchlets yellowish green to yellowish red. Flowering precocious; staminate catkins ellipsoid or broadly ellipsoid; pistillate catkins shortly cylindric; bracts 2-colored: pale proximally, black distally; filaments 4-5 times longer than bracts; ovary narrowly conical, downy; styles short. Capsules to 9 mm long. Fl. Apr. Fr. May-Jun. Mountain slopes or woods. Distributed in Neimenggu, Heilongjiang, Jilin and Liaoning. Also in N Asia and Europe.

中国黄花柳

Salix sinica (K. S. Hao ex C. F. Fang et A. K. Skvortsov) G. H. Zhu

灌木或小乔木。叶常椭圆形，椭圆状披针形、椭圆状菱形或倒卵状椭圆形，叶全缘或具齿。花序先叶开放；雌性柔荑花序基部具2个鳞片状小叶；子房狭圆锥形；柱头2裂。蒴果条状圆锥形。花期4月，果期5月。生山坡或林地。产华北、华西和西北。

Shrubs or small trees. Leaf blade usually elliptic, elliptic-lanceolate, elliptic-rhomboid or obovate-elliptic, margin entire or dentate. Flowering precocious; pistillate catkins with 2 scalelike small leaves at base; ovary narrowly conical; stigmas 2-cleft. Capsules linear-conical. Fl. Apr. Fr. May. Mountain slopes or woods. Distributed in N, W and NW China.

黄花柳 *Salix caprea*

中国黄花柳 *Salix sinica*

米；花柱长1-2毫米。花期5月。生林中和溪边。产黑龙江、吉林、辽宁、河北和内蒙古。朝鲜半岛、日本、蒙古和俄罗斯亦有。

Trees, up to 15 m tall. Two-year-old branchlets pruinose. Leaf blade lanceolate or oblanceolate, 8-12 × 1-2 cm, glabrous, margin glandular serrate. Catkin before leaves emerge, sessile; ploral bract black distally; pistillate flower: ovary glabrous; stipe 1-1.5 mm; styles 1-2 mm. Fl. May. Woods and streamsides. Distributed in Heilongjiang, Jilin, Liaoning, Hebei and Neimenggu. Also in Korean Peninsula, Japan, Mongolia and Russia.

皂柳
Salix wallichiana Andersson

灌木或乔木。小枝红褐色，黑褐色或绿褐色。叶背面有柔毛或无毛，叶基被丝状毛或疏柔毛。花序先叶或与叶同时开放；雄葇荑花序长圆形或倒卵形；雌葇荑花序圆柱形；花丝部分合生。花期4-5月，果期5月。生山坡、林缘或河边。产中国西南、华中、华北和华西。印度、尼泊尔和不丹亦有。

Shrubs or trees. Branchlets reddish brown, blackish brown, or greenish brown. Leaves abaxially silky pubescent or glabrous. Flowering precocious or coaetaneous; staminate catkins oblong or obovate; pistillate catkins cylindric; pilaments partly connate. Fl. Apr-May. Fr. May. Mountain slopes, forests edges or riversides. Distributed in SW, C, N and W China. Also in India, Nepal and Bhutan.

粉枝柳
Salix rorida Lacksch.

乔木，高达15米。二年生小枝常具白粉。叶披针形或倒披针形，长8-12厘米，宽1-2厘米，无毛，缘有锯齿。花序先叶开放，无梗；苞片先端黑色。雌花：子房无毛；柄长1-1.5毫

粉枝柳 *Salix rorida*

皂柳 *Salix wallichiana*

川滇柳 *Salix rehderiana*

川滇柳

Salix rehderiana C. K. Schneid

灌木或小乔木。叶披针形至倒披针形，长4.7-9.2(-11)厘米，宽1.2-2.5厘米，下面疏丝毛或无，全缘或有腺圆齿，叶缘翻卷。花序先叶开放；梗0.2-0.7厘米；苞片棕色或异色；雌花：子房疏毛或无毛；柄长0.3-0.6毫米；花柱长约1.3毫米。花期3-6月。生海拔920-4250米的山坡、灌丛中和溪流旁。产云南、四川、西藏东部、青海、甘肃、宁夏和陕西。

Shrubs or small trees. Leaf blade lanceolate to oblanceolate, 4.7-9.2(-11) × 1.2-2.5 cm, abaxially sparsely sericeous or glabrous, margin entire or glandular crenate, revolute. Catkins just before leaves emerge; peduncle 0.2-0.7 cm; floral bract brown or bicolor; pistillate flower: ovary downy or subglabrous; stipe 0.3-0.6 mm; styles ca. 1.3 mm. Fl. Mar-Jun. Mountain slopes, thickets and stream-sides at 920-4250 m. Distributed in Yunnan, Sichuan, E Xizang, Qinghai, Gansu, Ningxia and Shaanxi.

蒿柳

Salix schwerinii E. L. Wolf

灌木或小乔木。叶条状披针形，下面密具长丝状柔毛，银色光泽，全缘或稍波状。花序先叶开放或同时开放；雄葇荑花序长圆状卵球形，2-3 × 约1.5厘米；雌葇荑花序圆柱形。花期4-5月，果期5-6月。生海拔300-600米的水边或溪岸。产河北、内蒙古、黑龙江、吉林和辽宁。俄罗斯、朝鲜半岛、日本和欧洲亦有。

Shrubs or small trees. Leaf blade linear-lanceolate, abaxially densely long silky pubescent, silvery shiny, margin entire or slightly sinuolate. Flowering precocious or coaetaneous; staminate catkins oblong-ovoid, 2-3 × ca. 1.5 cm; pistillate catkins cylindric. Fl. Apr-May. Fr. May-Jun. Watersides or riverbank at 300-600 m. Distributed in Hebei, Neimenggu, Heilongjiang, Jilin and Liaoning. Also in Russia, Korean Peninsula, Japan and Europe.

蒿柳 *Salix schwerinii*

细叶沼柳

Salix rosmarinifolia L.

灌木，高达1米。小枝纤细。叶条状披针形或披针形，长2-6厘米，宽0.3-1厘米，下面苍白色、或有白柔毛或白绒毛，全缘。花序近无梗；苞片先端暗褐色；雌花：子房被柔毛；柄较长；花柱短。花期5月。生海拔300-600米的沼泽化草甸。产黑龙江、吉林、辽宁东部、内蒙古东部和新疆。朝鲜半岛、蒙古、俄罗斯和欧洲亦有。

Shrubs, up to 1 m tall. Branchlets slender. Leaf blade linear-lanceolate or lanceolate, 2-6 × 0.3-1 cm, abaxially pale, downy or white tomentose, margin entire. Catkin subsessile; floral bract brown distally; pistillate flower: ovary villous; long stipitate; styles short. Fl. May. Marshes at 300-600 m. Distributed in Heilongjiang, Jilin, E Liaoning, E Neimenggu and Xinjiang. Also in Korean Peninsula, Mongolia, Russia and Europe.

细叶沼柳 *Salix rosmarinifolia*

川柳

Salix hylonoma C. K. Schneid.

乔木，高达6米。叶椭圆形，长圆状披针形，卵状披针形或卵形，长(2.5-)4.7-7(-8.5)厘米，宽(1.5-)1.9-3.4厘米，下面幼时常杂有金色柔毛，边缘有细腺齿，稀全缘，先端渐尖。雄花：腺体1，狭圆柱形；雄蕊2，完全合生或仅花丝不同程度的合生；雌花：子房被柔毛。花期5-6月。生海拔1600-3800米的山坡林中。产云南西北部、四川、贵州和甘肃东南部。

Trees, up to 6 m tall. Leaf blade elliptic, oblong-lanceolate, ovate-lanceolate, or ovate, (2.5-)4.7-7(-8.5) × (1.5-)1.9-3.4 cm, abaxially golden-downy when young, margin indistinctly serrulate, rarely entire, apex acuminate. Staminate flower: gland 1, narrowly terete; stamens 2, filaments distinct, connate in part or throughout; pistillate flower: ovary pubescent. Fl. May-Jun. Woods on mountain slopes at 1600-3800 m. Distributed in NW Yunnan, Sichuan, Guizhou and SE Gansu.

川柳 *Salix hylonoma*

坡柳 *Salix myrtillacea*

坡柳

Salix myrtillacea C. K. Schneid.

灌木或小乔木，高达5米。叶倒卵状长圆形或倒披针形，稀为倒卵状椭圆形，长3-6厘米，宽1.1-2.2厘米，两面无毛，全缘，枝端叶有细齿。花序先叶开放，无花序梗；苞片黑色；雄花：雄蕊2，花丝合生，花药紫红色；雌花：子房被毛。花期3-5月。生海拔2700-4800米的溪流旁或灌丛。产云南西北部、四川、西藏、青海和甘肃东南部。印度北部亦有。

Shrubs or small trees, up to 5 m tall. Leaf blade obovate-oblong or oblanceolate, rarely obovate-elliptic, 3-6 × 1.1-2.2 cm, glabrous, margin entire, or distal leaves entire and gland-dotted. Catkins before leaves emerge, sessile; floral bract black; staminate flower: stamens 2, connate throughout, anthers purplish red; pistillate flower: ovary densely pubescent. Fl. Mar-May. Streamsides or thickets in valleys at 2700-4800 m. Distributed in NW Yunnan, Sichuan, Xizang, Qinghai and SE Gansu. Also in N India.

白毛柳

Salix lanifera C. F. Fang et S. D. Zhao

矮灌木，高达50厘米。叶倒卵形或椭圆形，长1.5-2.6厘米，宽0.9-1.6厘米，叶两面被白色绵毛，边缘具腺圆齿或全缘。花序先叶开放，无梗；苞片褐黑色；雄花：雄蕊2，花丝全部合生，花药红色；雌花：子房被毛。花期5月。生海拔3430-4500米的山坡上。产四川西部。

Shrubs low, up to 50 cm tall. Leaf blade obovate or elliptic, 1.5-2.6 × 0.9-1.6 cm, abaxially white woolly, margin entire, or gland-dotted. Catkins before leaves emerge, sessile; floral bract brownish black; staminate flower: stamens 2, filaments united throughout, anthers red; pistillate flower: ovary woolly. Fl. May. Mountain slopes at 3430-4500 m. Distributed in W Sichuan.

白毛柳 *Salix lanifera*

细柱柳 *Salix gracilistyla*

细柱柳
Salix gracilistyla Miq.

灌木。叶椭圆状长圆形或倒卵状长圆形，稀长圆形，长5(-12)厘米，宽1.5-2(-3.5)厘米，下面有绢质柔毛，边缘有锯齿。花序先叶放，无花序梗；苞片上部黑色；雄花：雄蕊2，花丝全部合生，花药红色或红黄色；雌花：子房被毛；花柱细长。花期4月。生山区溪流旁。产黑龙江、吉林和辽宁。朝鲜半岛、日本和俄罗斯亦有。

Shrubs. Leaf blade elliptic-oblong or obovate-oblong, rarely oblong, 5(-12) × 1.5-2(-3.5) cm, abaxially silky downy, margin serrate. Catkins before leaves emerge, sessile; floral bract black distally; staminate flower: stamens 2, filaments connate throughout, anthers red or russet; pistillate flower: ovary tomentose; styles slender. Fl. Apr. Along streams. Distributed in Heilongjiang, Jilin and Liaoning. Also in Korean Peninsula, Japan and Russia.

秋华柳
Salix variegata Franch.

灌木，通达3米。叶通常为长圆形、长圆状倒披针形或倒卵状长圆形，长0.9-3.8厘米，宽0.5-0.9厘米，下面有伏生绢毛，上面散生柔毛，全缘或有锯齿。花叶后开放；苞片先端常具一腺点；雄花：雄蕊2，花丝合生；雌花：子房被毛。花期4-10月。生海拔570-2550米的河边和溪流旁。产云南北部、贵州、四川、湖北西部、甘肃东南部、陕西南部和河南。

Shrubs, up to 3 m tall. Leaf blade usually oblong, oblong-oblanceolate or obovate-oblong, 0.9-3.8 × 0.5-0.9 cm, abaxially sericeous, adaxially pubescent, margin entire or serrulate. Catkins after leaves emerge; floral bract usually 1-gland-dotted distally; staminate flower: stamens 2, filaments connate; pistillate flower: ovary densely downy. Fl. Apr-Oct. Riverbanks and streamsides at 570-2550 m. Distributed in N Yunnan, Guizhou, Sichuan, W Hubei, SE Gansu, S Shaanxi and Henan.

秋华柳 *Salix variegata*

乌柳 *Salix cheilophila*

乌柳

Salix cheilophila C. K. Schneid.

灌木或小乔木，高5米。叶条形、条状倒披针形或倒披针形，基部渐狭，上部具腺锯齿。花序与叶同时开放；雄蕊2，完全合生；苞片倒卵状长圆形。蒴果长约3毫米。花期4-5月，果期5月。生海拔700-3000米的河岸、溪边或栽培。产中国西南、华北、华西和西北。

Shrubs or small trees, up to 5 m tall. Leaves linear, linear-oblanceolate or oblanceolate, base attenuate, margin glandular serrate distally. Flowering coaetaneous; stamens 2, connate throughout; bracts obovate-oblong. Capsules ca. 3 mm long. Fl. Apr-May. Fr. May. Riverbanks, streamsides or cultivated at 700-3000 m. Distributed in SW, N, W and NW China.

小红柳

Salix microstachya Turcz. ex Trautv. var. **bordensis** (Nakai) C. F. Fang

矮灌木，高达1米。叶条形，条状倒披针形或镰状倒披针形，基部渐狭，近全缘或具不明显锯齿。柔荑花序基部具1或2个鳞片状叶；雄蕊2，完全合生；苞片长圆形，先端截形；子房卵球状圆锥形，无柄；柱头红褐色，2裂。花期5月，果期6月。生河岸或沙丘间湿地。产河北、内蒙古东部、黑龙江、吉林和辽宁。

Dwarf shrubs, up to 1 m tall. Leaf blade linear, linear-oblanceolate or falcate-oblanceolate, base attenuate, margin subentire or indistinctly serrulate. Catkins with 1 or 2 small leaves at base; stamens 2, connate throughout; bracts oblong, apex truncate; ovary ovoid-conical, sessile; stigmas russet, 2-lobed. Fl. May. Fr. Jun. Riverbanks or damp places between dunes. Distributed in Hebei, E Neimenggu, Heilongjiang, Jilin and Liaoning.

小红柳 *Salix microstachya* var. *bordensis*

杨梅科 Myricaceae

毛杨梅

Myrica esculenta Buch.-Ham. ex. D. Don

乔木，高(2-)4-10米。小枝和叶柄被绒毛。叶狭椭圆状倒卵形或披针状倒卵形至楔状倒卵形，长4-18厘米，革质。雄花序多分枝，长4-9厘米。每个果序上核果多数，熟时红色，常椭圆体形。花期8月至翌年2月，果期11月至翌年5月。生海拔300-2500米的杂木林或干燥的山坡。产云南、四川、广西、贵州和广东。南亚亦有。

Trees, (2-)4-10 m tall. Branchlets and petioles tomentose. Leaves narrowly elliptic-obovate or lanceolate-obovate to cuneate-obovate, 4-18 cm long, leathery. Staminate inflorescences much branched, 4-9 cm long. Drupes many per infructescences, red at maturity, usually ellipsoid. Fl. Aug to next Feb. Fr. Nov to next May. Mixed forests or dry slopes at 300-2500 m. Distributed in Yunnan, Sichuan, Guangxi, Guizhou and Guangdong. Also in S Asia.

云南杨梅 *Myrica nana*

云南杨梅

Myrica nana Cheval.

灌木，高0.5-2米，雌雄异株。叶狭椭圆状倒卵形至楔状倒卵形，长2-8厘米，背面具金黄色腺点。雄穗状花序单个生于叶腋，长1-1.5厘米。核果红色，球形至椭圆体形，直径1-1.5厘米，具乳突。花期2-4月，果期5-7月。生海拔1900-3000米的山坡林缘或灌丛中。产云南和贵州。

Shrubs, 0.5-2 m tall, dioecious. Leaves narrowly elliptic-obovate to cuneate-obovate, 2-8 cm long, abaxially golden glandular-punctate. Staminate spikes simple, solitary in leaf axils, 1-1.5 cm long. Drupes red, globose to ellipsoid, 1-1.5 cm diam, papilliferous. Fl. Feb-Apr. Fr. May-Jul. Forest edges on slopes or thickets at 1900-3000 m. Distributed in Yunnan and Guizhou.

杨梅

Myrica rubra Sieb. et Zucc.

常绿乔木，高可达15米，雌雄异株。叶楔状倒卵形或狭椭圆状倒卵形，长5-14厘米，革质，无毛。小枝和叶柄无毛。穗状花序不分枝，单一或有时几个集生于叶腋。核果球形，乳状凸起长达3.5毫米。花期3-4月，果期5-7月。生海拔100-1500米的山坡林下或山谷。产中国西南、华南、东南和华东。菲律宾、朝鲜半岛和日本亦有。

Evergreen trees, up to 15 m tall, dioecious. Leaves cuneate-obovate or narrowly elliptic-obovate, 5-14 cm long, leathery, glabrous. Branchlets and petioles glabrous. Spikes simple, solitary or sometimes few together in leaf axils. Drupes globose, papillae to 3.5 mm long. Fl. Mar-Apr. Fr. May-Jul. Forests on slopes or valleys at 100-1500 m. Distributed in SW, S, SE and E China. Also in the Philippines, Korean Peninsula and Japan.

毛杨梅 *Myrica esculenta*

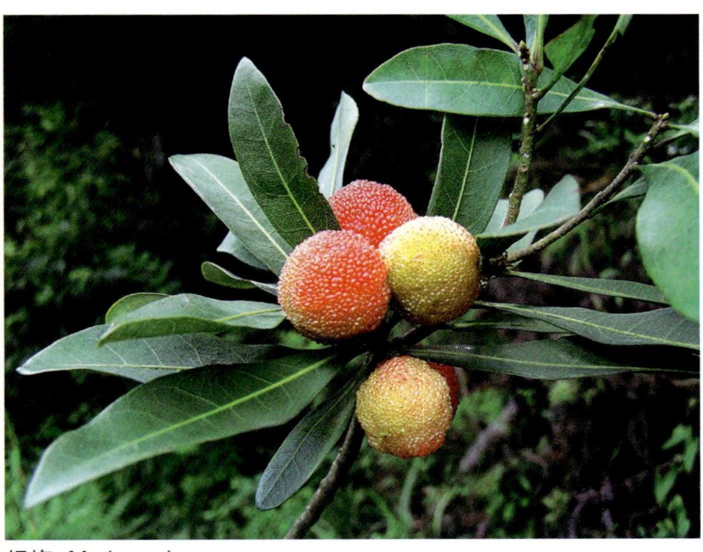

杨梅 *Myrica rubra*

胡桃科
Juglandaceae

圆果化香树 *Platycarya longipes*

化香树
Platycarya strobilacea Sieb. et Zucc.

乔木或灌木。叶长(6-)8-30厘米，叶轴无毛；小叶1-15(-23)。穗状花序雌雄兼有，基部中央为雌性穗状花序，顶端为雄性，或有时缺无。小坚果近球形至倒卵球形，3-6 × 3-6毫米。花期5-7月，果期7-10月。生海拔400-1400(-2200)米的山坡杂木林或有时生石灰岩上。产中国西南、华南、东南、华中、华北、华西和华东。越南、朝鲜半岛和日本亦有。

Trees or shrubs. Leaves (6-)8-30 cm long, rachises glabrous; leaflets 1-15(-23). Central spikes female in basal, male in apical, or sometimes absent. Nutlets subglobose to obovoid, 3-6 × 3-6 mm. Fl. May-Jul. Fr. Jul-Oct. Mixed forests on mountain slopes or sometimes on limestones at 400-1400(-2200) m. Distributed in SW, S, SE, C, N, W and E China. Also in Vietnam, Korean Peninsula and Japan.

圆果化香树
Platycarya longipes Wu

小乔木，高5-10米。叶长8-15厘米；小叶3-5(-7)；叶柄与叶轴近等长或较长。果序球形，直径1.2-2厘米。小坚果长宽均约3毫米，有狭翅。花期5月。生海拔450-800米的山地林中。产广东、广西和贵州。

Small trees, 5-10 m tall. Leaves 8-15 cm long; leaflets 3-5(-7); petioles nearly as long as leaf rachis or longer. Infructescences globose, 1.2-2 cm diam. Nutlets ca. 3 × 3 mm, with narrow wings. Fl. May. Montane forests at 450-800 m. Distributed in Guangdong, Guangxi and Guizhou.

化香树 *Platycarya strobilacea*

单叶化香树 *Platycarya simplicifolia*

单叶化香树

Platycarya simplicifolia G. R. Long

乔木，高达15米。小枝细，顶部被短柔毛。叶为单叶，具长柄，无毛；叶片卵形或椭圆状卵形，(3-)5-11.5 × (1.3-)2.7-6.8厘米，具10-16对侧脉。果序球形，直径1.9厘米；宿存苞片木质，长5-10毫米。小坚果宽倒卵形，约3 × 3毫米，两侧均有狭翅。花期5-6月。生海拔650米的石灰岩山林中。特产广西(融安县)。

Trees, up to 15 m tall. branchlets slender, at apex puberulous. Leaves simple, long petiolate, glabrous; blades ovate or elliptic-ovate, (3-)5-11.5 × (1.3-)2.7-6.8 cm, with 10-16 pairs of lateral nerves. Infructescences globose, 1.9 cm diam; persistent bracts woody, 5-10 mm long. Nutlets broadly obovate, ca. 3 × 3 mm, at each side narrowly winged. Fl. May-Jun. Forests in limestone hills at 650 m. Endemic to Guangxi (Rong'an County).

黄杞

Engelhardia roxburghiana Wall.

乔木，具橙色圆形腺体。偶数羽状复叶具2-10小叶；叶片无毛，背面具黄色盾状鳞片。雄花序穗状，下垂；雌花序直立；穗状果序不为圆锥状。果及苞片基部无刚毛；果球形，具果柄。花期2-8月，果期6-12月。生海拔200-1500米的林中。产中国西南、华南、华中和东南。东南亚和南亚亦有。

Trees, with orange round glands. Leaves even-pinnate, leaflets 2-10; blades glabrous but coated with yellowish, peltate scales abaxially. Male flowering spikes pendulous; female spike erect; fruiting spikes not conelike. Nuts and bracts glabrous at base; nuts globose, stiped. Fl. Feb-Aug. Fr. Jun-Dec. Forests at 200-1500 m. Distributed in SW, S, C and SE China. Also in SE and S Asia.

黄杞 *Engelhardia roxburghiana*

少叶黄杞

Engelhardia fenzelii Merr.

乔木，高3-10米。枝无毛。偶数羽状复叶无毛；小叶2-4。果序长7-12厘米，下垂；苞片长2-3.5厘米，3裂。坚果球形，直径3-4毫米，密被橙色腺体。花期7月，果期9-10月。生海拔400-1000米的山谷林中。产广西、广东、福建、浙江、江西和湖南。

Trees, 3-10 m tall. Branches glabrous. Leaves even-pinnate, glabrous; leaflets 2-4. Infructescences 7-12 cm long, pendulous, glabrous; bracts 2-3.5 cm long, 3-parted, glabrous. Nuts globose; 3-4 mm diam, with dense orange glands. Fl. Jul. Fr. Sep-Oct. Forests in valley at 400-1000 m. Distributed in Guangxi, Guangdong, Fujian, Zhejiang, Jiangxi and Hunan.

云南黄杞 *Engelhardia spicata*

少叶黄杞 *Engelhardia fenzelii*

云南黄杞

Engelhardia spicata Lesch. ex Blume

大乔木，高达15-20米。小枝变无毛。偶数羽状复叶长25-35厘米；小叶8-14，顶端短渐尖，边缘全缘，上面无毛，下面中脉疏被短柔毛；小叶柄长0.5-1厘米。果序长30-45厘米。坚果球形，直径约3.5毫米，上部有刚毛。花期11月，果期翌年1-2月。生海拔550-2100米的山坡杂木林中。产云南和广西。印度、泰国、越南、菲律宾和印度尼西亚亦有。

Large trees, up to 15-20 m tall. Branchlets glabrescent. Leaves even-pinnate, 25-35 cm long; leaflets 8-14, shortly acuminate at apex, entire at margin, adaxially glabrous, abaxially on midrib sparsely puberulous; with petiolules 0.5-1 cm long. Infructescences 30-45 cm long. Nuts globose, ca. 3.5 mm diam, above setose. Fl. Nov. Fr. next Jan-Feb. Mixed forests on slopes at 550-2100 m. Distributed in Yunnan and Guangxi. Also in India, Thailand, Vietnam, the Philippines and Indonesia.

毛叶黄杞

Engelhardia spicata Lesch. ex Blume var. **colebrookeana** (Lindl.) Koord. et Valeton

乔木。小枝、叶密被短柔毛。偶数羽状复叶具4-14小叶；小叶顶端钝或圆，下面具不明显腺状鳞片，全缘；叶柄长2-6厘米。穗状果序长(22-)30-45(-60)厘米，无毛。果球形，果翅基部具硬毛。花期1-4月，果期3-8(-10)月。生海拔1400(-2000)米以下的山腰或山谷疏林。产广东、广西、贵州、海南、西藏和云南。南亚和东南亚亦有。

Trees. Branchlets and leaves densely pubescent. Leaves even-pinnate, with 4-14 leaflets; leaflets obtuse or rotund at apex, with glandular scales inconspicuous abaxially, margin entire. acute; petioles 2-6 cm long. Fruiting

毛叶黄杞 *Engelhardia spicata* var. *colebrookeana*

齿叶黄杞 *Engelhardia serrata* var. *cambodica*

spikes (22-)30-45(-60) cm, glabrous. Nutlets globose, wings hispid at base. Fl. Jan-Apr. Fr. Mar-Aug(-Oct). Forests on mid-slopes or in valleys below 1400 (-2000) m. Distributed in Guangdong, Guangxi, Guizhou, Hainan, Xizang and Yunnan. Also in S and SE Asia.

爪哇黄杞

Engelhardia spicata Lesch. ex Blume var. **aceriflora** (Reinw.) Koord. et Valeton

乔木。小枝灰褐色，疏被短柔毛。叶轴常稍具柔毛；小叶4-10，腹面疏被短柔毛；小叶无柄至长5毫米的短柄；叶柄长4.5-9厘米，无毛或具柔毛。穗状果序长15-30(-40)厘米。果卵球形；果翅基部具硬毛。花期3月，果期4-5月。生海拔1500-1700米的山坡或林中。产云南。南亚和东南亚亦有。

Trees. Branchlets grayish brown, sparsely pubescent. Rachis usually slightly pubescent; leaflets 4-10, sparsely pubescent; petiolule absent or to 5 mm long; petioles 4.5-9 cm long, glabrous to puberulent. Fruiting spikes 15-30 (-40) cm long. Fruits ovoid; wings hispid at base. Fl. Mar. Fr. Apr-May. Mountain slopes or forests at 1500-1700 m. Distributed in Yunnan. Also in S and SE Asia.

齿叶黄杞

Engelhardia serrata Blume var. **cambodica** W. E. Manning

乔木。幼枝暗褐色，密被短柔毛和腺体。叶轴具绒毛；偶数羽状复叶具6-14小叶；小叶腹面被柔毛和腺体，边缘具不规则锯齿或全缘；叶柄长1-2厘米，具绒毛。果球形；果翅基部具硬毛。花期2月，果期4月。生海拔700-1000米的山坡林中。产云南西南部。南亚和东南亚亦有。

Trees. Young branches drak brown, densely clothed with pubescence and glands. Rachis tomentose; leaves even-pinnate, leaflets 6-14; adaxial side clothed with villus and glands, margin irregularly serrate or entire; petioles 1-2 cm long, tomentose . Fruits globose; wings hispid at base. Fl. Feb. Fr. Apr. Forests on mountain slopes at 700-1000 m. Distributed in SW Yunnan. Also in S and SE Asia.

爪哇黄杞 *Engelhardia spicata* var. *aceriflora*

青钱柳 *Cyclocarya paliurus*

青钱柳
Cyclocarya paliurus (Batal.) Iljinsk.

乔木。奇数羽状复叶，具有(5-)7-9(-11)小叶；小叶椭圆状卵形至宽披针形。花为风媒；雄花序3-5个簇生；果期穗状花序长25-30厘米。坚果扁球形，中部围有一盘状翅。花期5-6月，果期7-9月。生海拔400-2500米的山地湿润森林。产中国西南、东南、华中和华东。

Trees. Leaves odd-pinnate, with (5-)7-9(-11) leaflets; leaflets elliptic-ovate to broadly lanceolate. Flowers anemophilous; male spikes in clusters of 3-5; fruiting spikes 25-30 cm long. Nutlets depressed-globose, surrounded by a disc-like wing at middle. Fl. May-Jun. Fr. Jul-Sep. Moist forests on mountains at 400-2500 m. Distributed in SW, SE, C and E China.

湖北枫杨
Pterocarya hupehensis Skan

乔木。奇数羽状复叶，长(18-)20-25厘米；叶轴无翅，无毛；小叶5-11(-15)，长椭圆形至卵状椭圆形。结果穗状花序长30-45厘米。小坚果近球形，稍具棱，无毛；翅宽，椭圆状卵形。花期4-5月，果期8月。生海拔700-2000米的湿润林中河岸边。产四川、贵州、湖北和陕西。

Trees. Leaves odd-pinnate, (18-)20-25 cm long; rachises wingless, glabrous; leaflets 5-11(-15), long elliptic to ovate-elliptic. Fruiting spikes 30-45 cm long. Nutlets subglobose, slightly ribbed, glabrous; wings broad, elliptic-ovate. Fl. Apr-May. Fr. Aug. Streambanks in moist forests at 700-2000 m. Distributed in Sichuan, Guizhou, Hubei and Shaanxi.

枫杨
Pterocarya stenoptera C. DC.

乔木。芽具柄，密被锈褐色盾状着生的腺体。偶数羽状复叶具(6-)11-21(-25)无柄小叶；叶轴常具翅或有时仅具脊或沟槽；小叶椭圆形或椭圆状披针形，6-12 × 2-3厘米，先端钝至锐尖。果长圆形；果翅狭，条形或阔条形。花期4-5月，果期8-9月。生海拔1500米以下的河滩或山坡林中。产中国大部分地区。朝鲜半岛和日本亦有。

Trees. Buds stipitate, densely clothed with rubiginous brown peltate glands. Leaves even-pin-

湖北枫杨 *Pterocarya hupehensis*

枫杨 *Pterocarya stenoptera*

nate, leaflets (6-)11-21(-25), sessile; rachises often winged or sometimes only ridged or sulcate on some leaves; leaflets elliptic or elliptic-lanceolate, 6-12 × 2-3 cm, apex obtuse to acute. Fruits oblong; wings narrow, linear or wide linear. Fl. Apr-May. Fr. Aug-Sep. Flood lands or forests on mountain slopes below 1500 m. Distributed in most parts of China. Also in Korean Peninsula and Japan.

越南枫杨

Pterocarya tonkinensis (Franch.) Dode

乔木。偶数羽状复叶具6-14小叶，卵形或矩圆状卵形，9-17 × 3-7厘米，先端锐至渐尖；叶轴无翅，有时稍具脊或具沟槽。穗状果序长13-30厘米。坚果菱形；翅条形。花期3月，果期5-10月。生海拔200-1200米的沟边或湿地。产云南。老挝和越南亦有。

Trees. Leaves even-pinnate, leaflets 6-14, leaflets ovate or oblong-ovate, 9-17 × 3-7 cm apex acute to acuminate; rachis wingless, sometimes ridged or sulcate. Fruiting spikes 13-30 cm long. Nuts rhombic; wings linear. Fl. Mar. Fr. May-Oct. Streams or wet places at 200-1200 m. Distributed in Yunnan. Also in Laos and Vietnam.

华西枫杨

Pterocarya macroptera Rehd. et Wils. var. **insignis** Batal. (Rehd. et Wils.) W. E. Manning

乔木，高达25米。叶长(20-)30-45厘米，小叶(5-)7-13。结果穗状花序长达45厘米，轴无毛或近无毛。坚果长约8毫米，无毛；翅圆卵形，1.5(-2) × 2(-2.5)厘米。花期5-6月，果期8-9月。生海拔1100-2700米山地林中。产云南西北部、四川、浙江、湖北和陕西。

Trees, up to 25 m tall. Leaves (20-)30-45 cm long, leaflets (5-)7-13. Fruiting spikes to 45 cm long, axis glabrous or nearly so. Nutlets ca. 8 mm long, glabrous; wings orbicular-ovate, 1.5(-2) × 2(-2.5) cm. Fl. May-Jun. Fr. Aug-Sep. Forests on mountain slopes at 1100-2700 m. Distributed in NW Yunnan, Sichuan, Zhejiang, Hubei and Shaanxi.

越南枫杨 *Pterocarya tonkinensis*

华西枫杨 *Pterocarya macroptera* var. *insignis*

胡桃 *Juglans regia*

glabrous, irregularly dehiscent; shells wrinkled, without prominent ridges. Fl. Apr-May. Fr. Oct. Slopes or hills at 500-1800 (-4000) m. Distributed in SW, NW, C and E China. Also in SW Asia and Europe.

胡桃

Juglans regia L.

乔木。小枝被盾状着生腺体。小叶5-9，椭圆状卵形或狭长圆形，全缘，顶端钝或急尖，侧脉11-15对，背面侧脉脉腋内具毡毛，余近无毛。果序具1-3(-38)果。外皮无毛，不规则开裂；核壳皱纹状，无凸起的脊。花期4-5月，果期10月。生海拔500-1800(-4000)米山坡或丘陵地。产中国西南、西北、华中和华东。西南亚和欧洲亦有。

Trees. Branchlets with peltate-glands. Leaflets 5-9, oblong-ovate or narrowly oblong, entire, obtuse or acute at apex, with 11-15-paired nerves, almost glabescent except tufts of hairs on axils of nerves. Fruiting spikes usually with 1-3(-38) nuts. Husk

胡桃楸

Juglans mandshurica Maxim.

乔木或有时为灌木。树冠扁圆形。幼枝密被短茸毛。小叶下面具柔毛或稀渐无毛，边缘具锯齿或稀具锯齿。果序俯垂，通常具5-10(-13)果。外皮密被腺状柔毛，不开裂；核壳粗糙脊状且具深凹坑。花期4-5月，果期8-10月。生海拔500-2800米沟谷或山坡林中。产中国东北和华北。朝鲜半岛亦有。

Trees or sometimes shrubs. Crown oblate. Young branches densely clothed with short hairs. Leaflets abaxially pubescent or rarely glabrescent, margin serrate or rarely serrulate. Infructescences droop, usually with 5-10(-13) fruits. Husk densely

胡桃楸 *Juglans mandshurica*

野核桃 *Juglans cathayensis*

glandular pubescent, indehiscent; shells rough ridged and deeply pitted. Fl. Apr-May. Fr. Aug-Oct. Valleys or forests on mountain slopes at 500-2800 m. Distributed in NE and N China. Also in Korean Peninsula.

野核桃

Juglans cathayensis Dode

乔木，有时为灌木，高达12-25米。幼枝被腺毛。小叶9-17，卵状长圆形，渐尖，边缘有细锯齿，两面被星状毛，侧脉11-17对。果序有6-10(-13)果。果实卵球形，长3-4.5(-6)厘米，外皮密被腺毛。花期4-5月，果期8-10月。生海拔800-2000米的山地杂木林中。产云南、广西、湖南、贵州、湖北、四川、甘肃、陕西、河南和山西。

Trees, sometimes shrubs, up to 12-25 m tall. Young branches glandular-pubescent. Leaflets 9-17, ovate-oblong, acuminate, at margin serrulate, on both surfaces with stellate hairs, with 11-17 pairs of lateral nerves. Infructescences with 6-10(-13) fruits. Fruits ovoid, 3-4.5(-6) cm long, densely glandular-pubescent. Fl. Apr-May. Fr. Aug-Oct. Montane mixed forests at 800-2000 m. Distributed in Yunnan, Guangxi, Hunan, Guizhou, Hubei, Sichuan, Gansu, Shaanxi, Henan and Shanxi.

山核桃

Carya cathayensis Sarg.

乔木。顶芽裸露。小叶5或7，披针形或卵状披针形，具大量盾状鳞片。雄穗状花序长10-15厘米。坚果椭圆体形，外壳具翅至基部；果壳具4条不明显的纵棱。花期4-6月，果期8-9月。生海拔400-1500米的山坡林中、山谷或河岸。产贵州、浙江、安徽和江西。

Trees. Terminal buds naked. Leaflets 5 or 7, lanceolate or ovate-lanceolate, with abundant, peltate scales. Male spikes 10-15 cm long. Nuts ellipsoid, husk winged to base; shells with 4 faint, longitudinal ridges. Fl. Apr-Jun. Fr. Aug-Sep. Forests on mountain slopes, valleys or riverbanks at 400-1500 m. Distributed in Guizhou, Zhejiang, Anhui and Jiangxi.

湖南山核桃

Carya hunanensis W. C. Cheng et R. H. Chang ex Chang et Lu

乔木。奇数羽状复叶，长20-30厘米；小叶(5-)7或9，椭圆形至椭圆状披针形，具大量盾状鳞片。坚果倒卵球形；果自顶部至中部具翅；果壳具4条不明显的纵棱，基部分为4室。花期3-4月，果期9-11月。生海拔900-1000米的山谷林中或河岸边。产贵州、广西和湖南。

Trees. Leaves imparipinnately compound, 20-30 cm long; leaflets (5-)7 or 9, elliptic to elliptic-lanceolate, with abundant, peltate scales. Nuts obovoid; husks winged to middle; shells with 4 faint, longitudinal ridges, 4-chambered at base. Fl. Mar-Apr. Fr. Sep-Nov. Forests in valleys or riverbanks at 900-1000 m. Distributed in Guizhou, Guangxi and Hunan.

山核桃 *Carya cathayensis*

湖南山核桃 *Carya hunanensis*

桦木科 Betulaceae

华榛

Corylus chinensis Franch.

乔木，高达40米。叶卵形、卵状椭圆形或倒卵状椭圆形，基部斜心形，边缘具不规则重细齿。雄花序圆柱状，4-6个簇生；苞片具柔毛；雌花序2-6个簇生，果苞管状。小坚果被果苞包被，卵球形，无毛。生海拔1200-3500米的湿润山坡林中。产中国西南、华中和华西。

Trees, up to 40 m tall. Leaves ovate, ovate-elliptic or obovate-elliptic, base obliquely cordate, margin irregularly and doubly serrate. Male inflorescences 4-6 in a cluster, cylindric; bracts pubescent; female inflorescences 2-6 in a cluster, bracts forming a tubular sheath. Nutlets enclosed by bracts, ovoid, glabrous. Forests on moist mountain slopes at 1200-3500 m. Distributed in SW, C and W China.

刺榛 *Corylus ferox*

华榛 *Corylus chinensis*

刺榛

Corylus ferox Wall.

乔木或小乔木。芽鳞被白色柔毛。叶片卵状长圆形或倒卵状长圆形，边缘具重细密锐尖锯齿。苞片合成钟形，在坚果上部不收缩，苞片裂形成密的具分枝的翅。坚果扁球形。花期5-7月，果期7-9月。生海拔1700-3800米的山坡林中。产云南、四川和贵州。印度、尼泊尔、不丹和缅甸亦有。

Trees or small trees. Bud scales white villous. Leaves ovate-oblong or obovate-oblong, margin sharply, regularly minutely, and doubly serrate. Bracts campanulate, not constricted above nut, lobes of bracts forming dense, branching spines. Nuts oblate. Fl. May-Jul. Fr. Jul-Sep. Forests on mountain slopes at 1700-3800 m. Distributed in Yunnan, Sichuan and Guizhou. Also in India, Nepal, Bhutan and Myanmar.

藏刺榛 *Corylus ferox* var. *thibetica*

藏刺榛

Corylus ferox Wall. var. **thibetica** (Batal.) Franch.

与刺榛的区别在于本变种芽鳞片无毛。叶倒卵形或椭圆形。花期5-7月，果期7-9月。生海拔1500-3600米的杂木林中。产中国西南和华西。

This variety differs from the typical variety in its glabrous bud scales. Leaves obovate or elliptic. Fl. May-Jul. Fr. Jul-Sep. Mixed forests at 1500-3600 m. Distributed in SW and W China.

榛 *Corylus heterophylla*

榛

Corylus heterophylla Fisch. ex Trautv.

灌木或小乔木。小枝疏被长柔毛。叶长圆形或倒卵形，顶端具小尖头至尾尖，几截形，下面沿脉具毛，边缘上部有小裂片或锯齿；叶柄长1-3厘米。苞片的裂片全缘，稀具疏齿，长于果实。坚果卵球形。花期5-7月，果期7-8月。生海拔400-2500米的温带阔叶林中。产华北和东北。俄罗斯(东西伯利亚)、朝鲜半岛和日本亦有。

Shrubs or small trees. Branchlets sparsely villous. Leaf blades oblong or obovate, apex mucronate to caudate, nearly truncate, pilose along veins abaxially, margin lobulate or incised distally; petioles 1-3 cm long. Lobes of bracts entire, rarely sparsely dentate, longer than fruit. Nuts ovoid. Fl. May-Jul. Fr. Jul-Aug. Temperate broad-leaved forests at 400-2500 m. Distributed in N and NE China. Also in Russia (E Siberia), Korean Peninsula and Japan.

滇榛

Corylus yunnanensis (Franch.) A. Camus

灌木或小乔木。小枝具黄色绒毛。叶近圆形至阔卵形或倒卵形，下面具绒毛，边缘具不规则的重锯齿；叶柄长0.7-2.2厘米，具黄色绒毛。雄花序2-3枚排成总状，下垂。果苞钟状。坚果卵球形，密被绒毛。花期5-7月，果期7-9月。生海拔1600-3700米的山坡灌丛中。产四川、云南、贵州和湖北。

Shrubs or small trees. Branchlets yellow tomentose. Leaf blades suborbicular to broadly ovate or obovate, abaxially tomentose, margin irregularly and doubly serrate; petioles 0.7-2.2 cm long, yellow tomentose. Male inflorescences 2-3 in a cluster, pendulous. Bracts of fruits campanulate. Nuts ovoid, densely tomentose. Fl. May-Jul. Fr. Jul-Sep. Thickets on slopes at 1600-3700 m. Distributed in Sichuan, Yunnan, Guizhou and Hubei.

滇榛 *Corylus yunnanensis*

毛榛 *Corylus mandshurica*

千金榆 *Carpinus cordata*

毛榛

Corylus mandshurica Maxim.

灌木。小枝黄褐色，被长柔毛，下部的毛较密。叶阔卵形、长圆形或长圆状倒卵形，下面多毛，尤其脉上，边缘具不规则糙锯齿，先端刺状渐尖至尾尖。苞片形成一管状鞘，于坚果上部收缩。坚果近球形，被白色柔毛，由苞片包被。花期5-7月，果期7-9月。生海拔400-2600米的山坡灌丛中或林下。产华北和东北。俄罗斯、朝鲜半岛和日本亦有。

Shrubs. Branchlets yellowish brown, villose, denser at lower parts. Leaves broadly ovate, oblong, or oblong-obovate, abaxially pilose especially along veins, margin irregularly and coarsely serrate, apex mucronate-acuminate or caudate. Bracts forming a tubular sheath, constricted above nut. Nuts almost globose, white pubescent, enclosed by bracts. Fl. May-Jul. Fr. Jul-Sep. Shrubs of mountain slopes or forests at 400-2600 m. Distributed in N and NE China. Also in Russia, Korean Peninsula and Japan.

虎榛子

Ostryopsis davidiana Decne.

灌木。叶近革质，下面密被白色柔毛，侧脉7-9对。雄花序单生；苞片被柔毛，无小苞片；花药红色；雌花序顶生，总状头状；苞片成筒状鞘，密被柔毛，顶端浅裂。小坚果褐色。花期5-6月，果期7-8月。生海拔800-2800米疏林或灌丛中。产华北和华西。

Shrubs. Leaves subcoreaceous, abaxially densely white-pubescent, lateral veins 7-9 pairs. Male inflorescences solitary; bracts pubescent, without bracteoles; anthers red; female inflorescences terminal, racemose-capitulate; bracts forming a tubular sheath, densely pubescent, lobed at apex. Nutlets brown. Fl. May-Jun. Fr. Jul-Aug. Sparse forests or thickets at 800-2800 m. Distributed in N and W China.

千金榆

Carpinus cordata Blume

乔木。树皮灰色或黑灰色。小枝、叶柄、花序轴幼时无毛或稀被柔毛。叶基部偏心形，具(10-)15-20对侧脉。外侧苞片边缘上部具锯齿及反折，内侧边缘上部具锯齿。小坚果长圆形，具不明显的肋。花期5-6月，果期7-8月。生海拔200-2500米潮湿的山坡林中。产华西、华北和东北。俄罗斯、朝鲜半岛和日本亦有。

Trees. Bark gray or black-gray. Branchlets, petioles and rachises of inflorescences glabrous or sparsely villous when young. Leaf base unequally cordate, with (10-)15-20 lateral veins on each side of midvein. Bracts with outer margin remotely serrate and inflexed, and inner margin remotely serrate distally. Nutlets oblong, faintly ribbed. Fl. May-

虎榛子 *Ostryopsis davidiana*

Jun. Fr. Jul-Aug. Forests on moist mountain slopes at 200-2500 m. Distributed in W, N and NE China. Also in Russia, Korean Peninsula and Japan.

川黔千金榆

Carpinus fangiana Hu

乔木，高达20米。叶卵状披针形或椭圆形状披针形，侧脉24-34对，基部心形，近圆形或宽楔形，边缘具不规则刚毛状重锯齿。雌性花序45-50 × 3-4厘米；苞片纸质，内侧边缘上部具疏锯齿。小坚果长圆形。花期5-6月，果期7-9月。生海拔900-2000米的山谷林中或背阴山坡。产云南、四川、贵州和广西。

Trees, up to 20 m tall. Leaves ovate-lanceolate or elliptic-lanceolate, lateral veins 24-34 on each side of midvein, base cordate, subrounded or broadly cuneate, margin irregularly and doubly setiform serrate. Female inflorescences 45-50 × 3-4 cm; bracts papery, inner margin remotely minutely serrate proximally. Nutlets oblong. Fl. May-Jun. Fr. Jul-Sep. Forests in mountain valleys or on shaded slopes at 900-2000 m. Distributed in Yunnan, Sichuan, Guizhou and Guangxi.

海南鹅耳枥

Carpinus londoniana H. Winkl. var. **lanceolata** (Hand.-Mazz.) P. C. Li

乔木。叶披针形，6-8 × 1.7-2.6厘米；叶柄粗壮，长4-7毫米，密具柔毛。苞片稀疏重叠，无毛，3浅裂；中间裂片宽4-5毫米；外侧苞片全缘或具波状锯齿。小坚果宽卵球形，具有密的褐色树脂腺和亮黄色树脂，有凸起的棱。花期4-6月，果期7-9月。生海拔600-800米亚热带林中。产海南。

Trees. Leaf blades lanceolate, 6-8 × 1.7- 2.6 cm; petioles robust, 4-7 mm long, densely pubescent. Bracts loosely overlapping, glabrous, 3-lobed; middle lobe 4-5 mm broad; outer margin of bracts entire or undulate-serrate. Nutlets broadly ovoid, with dense, brown, resinous glands and light-yellow resin, prominently ribbed. Fl. Apr-Jun. Fr. Jul-Sep. Subtropical forests at 600-800 m. Distributed in Hainan.

海南鹅耳枥 *Carpinus londoniana* var. *lanceolata*

雷公鹅耳枥

Carpinus viminea Wall.

乔木。小枝密生白色皮孔，无毛。叶缘具规则或不规则重锯齿，先端急尖、锐尖或尾状渐尖；叶柄细弱，长(1-)1.5-3厘米，无毛。外侧苞片具牙齿。小坚果宽卵球形，无毛，有时上部疏生小树脂腺体和细柔毛。花期4-6月，果期7-9月。生海拔400-900米的亚热带阔叶林中。产中国西南、华南、东南和华中。南亚亦有。

Trees. Branchlets densely with white lenticels, glabrous. Leaf blades regularly or irregularly doubly mucronate serrate, apex acute, acuminate, or caudate; petioles slender, (1-)1.5-3 cm long, glabrous. Outer margin of bract dentate. Nutlets wide ovoid, glabrous, sometimes with small resin glands and pubescence. Fl. Apr-Jun. Fr. Jul-Sep. Subtropical broad-leaved forests at 400-900 m. Distributed in SW, S, SE and C China. Also in S Asia.

川黔千金榆 *Carpinus fangiana*

雷公鹅耳枥 *Carpinus viminea*

鹅耳枥

Carpinus turczaninowii Hance

乔木。叶卵形、阔卵形、卵状椭圆形、卵状菱形或卵状披针形，2-6 × 1.3-4厘米。苞片半卵形、半长圆形或宽半卵形，散生柔毛，外侧边缘不规则缺刻状齿，内面边缘全缘或具疏细齿。小坚果无毛，或有时顶端疏生长柔毛。花期5-7月，果期7-9月。生海拔500-2400米的温带林中。产华北和华东。朝鲜半岛和日本亦有。

Trees. Leaves ovate, broadly ovate, ovate-elliptic, ovate-rhombic, or ovate-lanceolate, 2-6 × 1.3-4 cm. Bracts semi-ovate, semi-oblong or broadly semi-ovate, sparsely pubescent, outer margin irregularly incised-dentate, inner margin entire or remotely minutely dentate. Nutlets glabrous or sometimes sparsely villose at apex. Fl. May-Jul. Fr. Jul-Sep. Temperate forests at 500-2400 m. Distributed in N and E China. Also in Korean Peninsula and Japan.

云南鹅耳枥

Carpinus monbeigiana Hand.-Mazz.

乔木，高8-15米。叶厚纸质，长圆状披针形或长椭圆形，5-11 × 2.8-4厘米，边缘具重锯齿。果苞半卵形，长16-20毫米，内侧边缘全缘，基部无裂片，外侧边缘有少数小齿。小坚果宽卵球形，长3-4毫米，密生橙色或褐色腺体，上部被柔毛。生海拔1700-2800米的山地林中。产云南西北部和中部。

云南鹅耳枥 *Carpinus monbeigiana*

Trees, 8-15 m tall. Leaves thickly papery, oblong-lanceolate or long elliptic, 5-11 × 2.8-4 cm, margin double-serrate. Bracts semi-ovate, 16-20 mm long, inner margin entire, at base without lobe, outer margin with a few small teeth. Nutlets broadly ovoid, 3-4 mm long, with dense orange or brown glands, above pubescent. Montane forests at 1700-2800 m. Distributed in NW and C Yunnan.

云贵鹅耳枥

Carpinus pubescens Burk.

乔木，高达17米。叶纸质，长圆形、长圆状披针形、卵状披针形或卵状椭圆形，基部近圆形至楔形、近圆形或近心形，边缘具规则的重细锯齿，侧脉在上面不凸起。苞片半卵形，内侧全缘。小坚果宽卵球形。

鹅耳枥 *Carpinus turczaninowii*

云贵鹅耳枥 *Carpinus pubescens*

花期5-6月，果期7-9月。生海拔450-2000米的山谷林中、山顶灌丛或石灰岩上。产云南东部、四川、贵州和陕西。越南北部亦有。

Trees, up to 17 m tall. Leaves papery, oblong, oblong-lanceolate, ovate-lanceolate or ovate-elliptic, base subrounded-cuneate, subrounded or subcordate, margin regularly and doubly minutely serrate, lateral veins not impressed adaxially. Bracts semi-ovate, inner margin entire. Nutlets broadly ovoid. Fl. May-Jun. Fr. Jul-Sep. Forests in valleys, thickets on mountain summits, or on limestones at 450-2000 m. Distributed in E Yunnan, Sichuan, Guizhou and Shaanxi. Also in N Vietnam.

湖北鹅耳枥
Carpinus hupeana Hu

乔木，高达18米。叶卵状披针形、卵状椭圆形或狭椭圆形，叶背生疣状凸起，边缘具重锯齿；叶柄具灰色长柔毛。雌性花序6-11 × 2-3厘米；苞片半卵形，内侧边缘全缘。小坚果宽卵球形。花期5-6月，果期7-8月。生海拔700-1800米的亚热带林中。产中国东南、华中、华北、华西和华东。

Trees, up to 18 m tall. Leaf blades ovate-lanceolate, ovate-elliptic or narrowly elliptic, abaxially glandular punctate, margin doubly dentate-serrate; petioles gray villous. Female inflorescences 6-11 × 2-3 cm; bracts semi-ovate, inner margin entire. Nutlets broadly ovoid. Fl. May-Jun. Fr. Jul-Aug. Subtropical forests at 700-1800 m. Distributed in SE, C, N, W and E China.

香港鹅耳枥
Carpinus insularis N. H. Xia, K. S. Pang et Y. H. Tong

落叶乔木，高3米。小枝疏被柔毛。叶披针形，4-6厘米长，边缘有小锯齿，在两面中脉上被柔毛，侧脉13-16对。果序长3-4厘米；苞片半卵形，内缘全缘。小坚果顶端密被柔毛。果期8-10月。生海拔190米的山坡丛林中。特产香港。

Deciduous shrubs, 3 m tall. Branchlets sparsely pubescent. Leaves lanceolate, 4-6 cm long, at margin serrulate, on midribs of both surfaces pubescent, lateral nerves 13-16 pairs. Infructescences 3-4 cm long; bracts semi-ovate, inner margin entire. Nutlets at apex densely pubescent. Fr. Aug-Oct. Thickets on hill slopes at 190 m. Endemic to Hong Kong.

香港鹅耳枥 *Carpinus insularis*

湖北鹅耳枥 *Carpinus hupeana*

铁木 *Ostrya japonica*

日本桤木（赤杨） *Alnus japonica*

铁木
Ostrya japonica Sarg.

乔木。叶互生；叶片长圆状卵形或长圆状披针形，长3.5-12厘米，渐尖，边缘具重锯齿，下面脉上被短柔毛，侧脉10-15对；叶柄长1-1.5厘米。果期总状花序具4或更多果实；总苞膜质，倒卵球状长圆形，长1-2厘米，被贴伏柔毛。小坚果长圆状卵球形，长6毫米，淡褐色，无毛。果期8-9月。生山坡林中。产四川、甘肃、陕西、河南和河北。朝鲜半岛和日本亦有。

Trees. Leaves alternate; blades oblong-ovate or oblong-lanceolate, 3.5-12 cm long, acuminate, margin double-serrulate, abaxial surface on nerves puberulous, lateral nerves 10-15 pairs; petioles 1-1.5cm long. Fruiting raceme with 4 or more fruits; involucres membranous, obovoid-oblong, 1-2 cm long, appressed-pubescent. Nutlets oblong-ovoid, 6 mm long, pale brown, glabrous. Fr. Aug-Sep. Forests on slopes. Distributed in Sichuan, Gansu, Shaanxi, Henan and Hebei. Also in Korean Peninsula and Japan.

尼泊尔桤木
Alnus nepalensis D. Don

乔木。叶倒卵状披针形、倒卵状长圆形、卵形或椭圆形，全缘或上部具密锯齿。雌花序圆锥状，多数；雌花序梗强壮，2-8毫米，无毛。小坚果长圆形，长约2毫米，具有膜质翅。花期5-6月，果期7-9月。生海拔200-2800米的湿润坡地、村边或山谷台地林中。产中国西南。南亚亦有。

Trees. Leaves obovate-lanceolate, obovate-oblong, ovate, or elliptic, margin entire or remotely minutely serrate. Female inflorescences numerous, in a panicle; peduncle of female inflorescence robust, 2-8 mm long, glabrous. Nutlets oblong, ca. 2 mm long, with membranous wings. Fl. May-Jun. Fr. Jul-Sep. Moist slopes, by villages or forests on terraces of valleys at 200-2800 m. Distributed in SW China. Also in S Asia.

日本桤木（赤杨）
Alnus japonica (Thunb.) Steud.

乔木。树皮灰褐色，光滑。叶倒卵形、倒卵状椭圆形或倒卵状披针形，基部楔形，边缘上部具细锯齿。果序长圆形；雌花序梗长10-20毫米。小坚果卵球形或倒卵球形。花期5-7月，果期7-8月。生海拔800-1500米的温带森林、河边或路旁。产台湾、河南、河北、安徽、江苏、山东、吉林和辽宁。俄罗斯、朝鲜半岛和日本亦有。

Trees. Bark gray-brown, smooth. Leaves obovate, obovate-elliptic, or obovate-lanceolate, base cuneate, margin remotely minutely serrate. Infrutescences oblong; peduncles of female inflorescence 10-20 mm long. Nutlets ovoid or obovoid. Fl. May-Jul. Fr. Jul-Aug. Temperate forests, riversides or roadsides at 800-1500 m. Distributed in Taiwan, Henan, Hebei, Anhui, Jiangsu, Shandong, Jilin and Liaoning. Also in Russia, Korean Peninsula and Japan.

尼泊尔桤木 *Alnus nepalensis*

辽东桤木 *Alnus hirsuta*

辽东桤木

Alnus hirsuta Turcz. ex Rupr.

乔木。叶近圆形，下面具浅褐色柔毛，边缘具波状齿。雌花序总状2-8，近球形或长圆形，长1-2厘米；苞片木质，顶端圆，5浅裂。小坚果宽卵球形，长约3毫米，具有厚纸质翅，约1/4宽于小坚果。花期5-6月，果期7-8月。生海拔700-1500米的温带森林或沟边。产内蒙古、山东、黑龙江、吉林和辽宁。俄罗斯(西伯利亚)、朝鲜半岛和日本亦有。

Trees. Leaves suborbicular, pale brown pubescent abaxially, margin undulate-incised. Female inflorescences 2-8, in a raceme, subglobose or oblong, 1-2 cm long; bracts woody, apex rounded, 5-lobed. Nutlets broadly ovoid, ca. 3 mm long, with thick-papery wings, wings ca. 1/4 as wide as nutlets. Fl. May-Jun. Fr. Jul-Aug. Temperate forests or along stream banks at 700-1500 m. Distributed in Neimenggu, Shandong, Heilongjiang, Jilin and Liaoning. Also in Russia (Siberia), Korean Peninsula and Japan.

东北桤木 *Alnus mandshurica*

东北桤木

Alnus mandshurica (Callier ex C. K. Schneid.) Hand.-Mazz.

灌木或小乔木。芽无柄，具3-6芽鳞。叶阔卵形、卵形、椭圆形或阔椭圆形，边缘具细密的重或单锯齿。雌花序3-6枚呈总状排列，长方形或球形；雌花梗下垂。小坚果卵球形，膜质翅与果近等宽。花期5-7月，果期7-8月。生海拔200-1900米的林边、河岸或山坡林中。产中国东北。俄罗斯和朝鲜半岛亦有。

Shrubs or small trees. Buds sessile, with 3-6 scales. Leaves broadly ovate, ovate, elliptic, or broadly elliptic, margin densely minutely doubly or simply serrate. Female inflorescences 3-6 in a raceme, oblong or globose; peduncles of female inflorescences pendulous. Nutlets ovoid, widths of membranous wings subequal to those of fruits. Fl. May-Jul. Fr. Jul-Aug. Forest edges, river banks or forests of mountain slopes at 200-1900 m. Distributed in NE China. Also in Russia and Korean Peninsula.

江南桤木

Alnus trabeculosa Hand.-Mazz.

乔木。芽具柄，具2个无毛的具棱的芽鳞。叶倒卵状长圆形、倒披针状长圆形或长圆形，基部近圆形、心形或阔楔形。雌花序2-4，成总状，长圆形。小坚果宽卵球形，有纸质的翅。花期5-6月，果期7-8月。生海拔200-1000米的山谷林下或河边。产中国西南、华南、东南、华中和华东。日本亦有。

Trees. Buds stipitate, with 2 glabrous, ribbed scales. Leaves obovate-oblong, oblanceolate-oblong, or oblong, base subrounded, subcordate, or broadly cuneate. Female inflorescences 2-4 in a raceme, oblong. Nutlets broadly ovoid, with paper wing. Fl. May-Jun. Fr. Jul-Aug. Forests on mountain valleys or riverbanks at 200-1000 m. Distributed in SW, S, SE, C and E China. Also in Japan.

江南桤木 *Alnus trabeculosa*

桤木
Alnus cremastogyne Burk.

乔木。小枝、叶柄、总花梗和叶幼时下面疏具白色柔毛。叶倒卵形、倒卵状长圆形、长圆形或倒披针形，侧脉8-11对。雌花序单生叶腋；雌花序梗长4-8厘米，下垂。小坚果具宽而膜质的翅。花期5-7月，果期8-9月。生海拔500-3000米的湿润坡地或河边台地。产四川、贵州、浙江、甘肃、陕西和江苏。

Trees. Branchlets, petioles, peduncles and leaves sparsely white pubescent abaxially when young. Leaves obovate, obovate-oblong, oblong, or oblanceolate, lateral nerves 8-11 pairs. Female inflorescences solitary, axillary; peduncles of female inflorescences 4-8 cm long, pendulous. Nutlets with broad and membranous wings. Fl. May-Jul. Fr. Aug-Sep. Moist mountain slopes or terraces of river banks at 500-3000 m. Distributed in Sichuan, Guizhou, Zhejiang, Gansu, Shaanxi and Jiangsu.

西桦
Betula alnoides Buch.-Ham. ex D. Don

高大乔木。树皮片状脱落。小枝密被白色长柔毛和树脂腺体。叶下面密被腺点，沿脉散生长柔毛。果序3-5枚组成总状；苞片下面密具柔毛。小坚果倒卵球形，膜质翅宽约为小坚果宽的2倍。花期10月至翌年1月，果期翌年3-5月。生海拔700-2100米的山坡杂木林中。产云南、四川、广西、海南、福建和湖北。南亚亦有。

Trees. Bark exfoliating. Branchlets densely white villous and resinous glandular. Leaves abaxially densely glandular-punctate, sparsely villose along nerves. Infructescences 3-5, in racemes; bracts densely pubescent abaxially. Nutlets obovoid, with membranous wings ca. 2 × as wide as nutlet. Fl. Oct to next Jan. Fr. next Mar-May. Forests on slopes at 700-2100 m. Distributed in Yunnan, Sichuan, Guangxi, Hainan, Fujian and Hubei. Also in S Asia.

西桦 *Betula alnoides*

亮叶桦
Betula luminifera H. Winkl.

乔木。树皮深褐色，光滑。叶长圆形、阔长圆形或长圆状披针形，下面密生树脂腺点，边缘具不规则和重刚毛状锯齿。雌花序1(或2)；果序长圆柱形。小坚果倒卵球形，膜质翅宽为小坚果的1-2倍。花期5-6月，果期6-8月。生海拔200-2900米的林中。产中国西南、华南、东南、华西和华中。

Trees. Bark dark brown, smooth. Leaves oblong, broadly oblong, or oblong-lanceolate, abaxially densely resinous punctate, margin irregularly and doubly setiform serrate. Female inflorescences 1(or 2); infrutescences long cylindric. Nutlets obovoid, with membranous wings 1-2 × as wide as nutlet. Fl. May-Jun. Fr. Jun-Aug. Forests at 200-2900 m. Distributed in SW, S, SE, W and C China.

桤木 *Alnus cremastogyne*

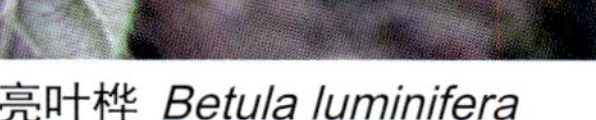
亮叶桦 *Betula luminifera*

白桦
Betula platyphylla Suk.

乔木。树皮灰白色，成层剥裂。小枝不下垂。叶片三角形卵状、三角形菱形或三角形，边缘具重锯齿或单锯齿，或锐锯齿。小坚果狭长圆形、长圆形或卵球形，果翅与果约等宽。花期6-7月，果期7-9月。生海拔700-4200米的针阔混交林或自成纯林。产中国西南、华北、西北和东北。俄罗斯、蒙古、朝鲜半岛和日本亦有。

Trees. Bark grayish white, exfoliating in sheets. Branches not pendulous. Leaves deltoid-ovate, deltoid-rhombic or deltoid, margin doubly or simply serrate, or incised-serrate. Nutlets narrowly oblong, oblong, or ovoid, wings of nutlets about as wide as nutlets. Fl. Jun-Jul. Fr. Jul-Sep. Coniferous and broad-leaved mixed forests, or forming pure forests at 700-4200 m. Distributed in SW, N, NW and NE China. Also in Russia, Mongolia, Korean Peninsula and Japan.

垂枝桦 *Betula pendula*

垂枝桦
Betula pendula Roth

乔木。树皮灰白色。成熟树上小枝常下垂。叶三角状卵形或菱状卵形，下面密具树脂腺点，边缘具糙或锐的重锯齿。果序长圆形至长圆状圆柱形。小坚果倒卵球状椭圆体形，具膜质翅；小坚果翅宽约为其2倍。花期6-7月，果期7-8月。生海拔500-2300米的温带阔叶林中。产新疆北部。哈萨克斯坦、俄罗斯、蒙古和欧洲亦有。

Trees. Bark grayish white. Branches usually pendulous in mature trees. Leaves triangular-ovate or rhombic-ovate, abaxially densely resinous punctate, margin coarsely or incised doubly serrate. Infrutescences oblong to oblong-cylindric. Nutlets obovoid-ellipsoid, with membranous wings; wings of nutlet ca. 2 × as wide as nutlets. Fl. Jun-Jul. Fr. Jul-Aug. Temperate broad-leaved forests at 500-2300 m. Distributed in N Xinjiang. Also in Kazakhstan, Russia, Mongolia and Europe.

白桦 *Betula platyphylla*

小叶桦 *Betula microphylla*

小叶桦
Betula microphylla Bunge

小乔木，高5-6米。树皮灰白色。叶菱形或菱状倒卵形，基部楔形，边缘具重或单锯齿，变无毛。果序直立，长圆状圆柱形，长1-2.5厘米。果苞长5-6毫米，上部具3裂片。小坚果约2.5 × 1.5毫米，密被短柔毛，翅与果近等宽。花期6-7月。生海拔1200-1600米的河岸或山谷杂木林中。产新疆北部。哈萨克斯坦和蒙古亦有。

Small trees, 5-6 m tall. Bark grayish white. Leaves rhombic or rhombic-obovate, at base cuneate, at margin doubly or simply serrate, glabrescent. Infructescenes erect, oblong-terete, 1-2.5 cm long. Bracts 5-6 mm long, above 3-lobed. Nutlets ca. 2.5 × 1.5 mm, densely puberulous, with wings nearly as broach as the fruits. Fl. Jun-Jul. Mixed forests on river banks or in valley at 1200-1600 m. Distributed in N Xinjiang. Also in Kazakhstan and Mongolia.

黑桦
Betula dahurica Pall.

乔木，可达20米高。树皮灰褐色，龟裂。小枝密生树脂腺体。叶边缘具不规则和锐重锯齿。雌花序直立或下垂；苞片有缘毛，3浅裂。坚果宽椭圆体形，无毛，具膜质翅，翅宽约为小坚果的1/2。花期6-7月，果期7-8月。生海拔400-1300米杂木林或针叶林下、阳坡或山顶岩石上。产中国东北和华北。俄罗斯(远东地区)、蒙古东部、朝鲜半岛和日本亦有。

Trees, up to 20 m tall. Bark black-brown, fissured. Branchlets with dense resinous glands. Leaves margin irregularly and acutely doubly serrate. Female inflorescences erect or pendulous; bracts ciliate, 3-lobed. Netlets broadly ellipsoid, glabrous, with membranous wings, wings of nutlet ca. 1/2 as wide as nutlet. Fl. Jun-Jul. Fr. Jul-Aug. Mixed or coniferous forests, sunny slopes or rocks on mountain summits at 400-1300 m. Distributed in NE and N China. Also in Russia (Far East), E Mongolia, Korean Peninsula and Japan.

糙皮桦
Betula utilis D. Don

乔木。树皮红褐色或暗红褐色，薄片状剥落。叶通常卵形或长卵形，边缘具不规则尖锐重锯齿，脉腋间具密髯毛。雌花序1，或2-4组成总状；苞片疏具柔毛。小坚果卵球形，膜质翅与小坚果近等宽。花期6-7月，果期7-8月。生海拔2500-3800米的温带阔叶林中。产华西和华北。印度、尼泊尔和阿富汗亦有。

Trees. Bark red-brown or dark red-brown, exfoliating in thin flakes. Leaves ovate or narrowly ovate, margin irregularly acutely double-serrate, blades densely bearded in axils of lateral veins abaxially. Female inflorescence 1, or 2-4 in a raceme; bracts sparsely pubescent. Nutlets obovoid, with membranous wings ca. as wide as nutlet. Fl. Jun-Jul. Fr. Jul-Aug. Temperate broad-leaved forests at 2500-3800 m. Distributed in W and N China. Also in India, Nepal and Afghanistan.

黑桦 *Betula dahurica*

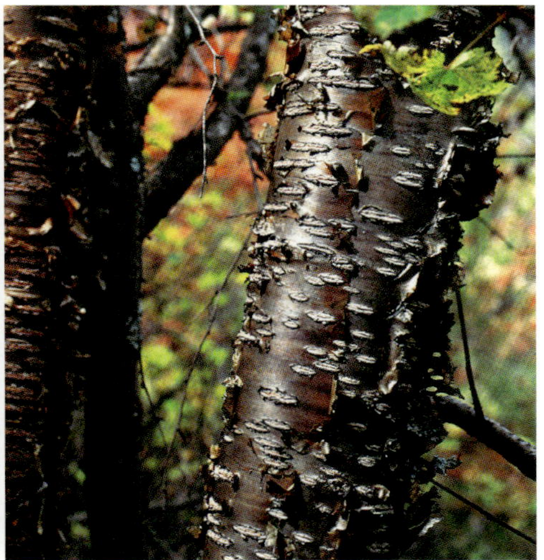

糙皮桦 *Betula utilis*

半岛)、朝鲜半岛和日本亦有。

Trees. Bark grayish white, exfoliating in thin sheets. Bud scales densely silky villous. Leaves triangular-ovate, broadly ovate, or ovate, acute or acuminate at apex. Lateral lobes of bracts erect, apex. acute, without a lobule at base. Female inflorescences oblong, peduncles 3-6 mm long. Nutlets obovoid or ovoid-ellipsoid, with membranous wings ca. 1/2 as wide as nutlet. Fl. Jun-Jul. Fr. Jul-Aug. Pure forest stands or in mixed coniferous and broad-leaved trees at 1000-1700 m. Distributed in NE China and Neimenggu. Also in Russia (Kamchatka Peninsula.), Korean Peninsula and Japan.

红桦

Betula albosinensis Burk.

高大乔木。树皮橙红色或橙色，有光泽，纸片状剥落。叶卵形、卵状椭圆形或卵状长圆形，边缘具不规则的重刺尖状锯齿。果序圆柱状。小坚果卵球形，膜质翅与小坚果近等宽。花期5-6，果期7-8月。生海拔1000-3400米温带阔叶林中、溪岸或路边。产华中、华北和华西。俄罗斯、朝鲜半岛和日本亦有。

Large trees. Bark orange-red or orange, shiny, exfoliating in papery flakes. Leaves ovate, ovate-elliptic, or ovate-oblong, margin irregularly and doubly mucronate serrate, usually not bearded in axils of lateral veins abaxially. Infrutescences cylindric. Nutlets ovate, with membranous wings ca. as wide as nutlet. Fl. May-Jun. Fr. Jul-Aug. Temperate forests, streambanks or roadsides at 1000-3400 m. Distributed in C, N and W China. Also in Russia, Korean Peninsula and Japan.

岳桦

Betula ermanii Cham.

乔木。树皮灰白色，薄片状脱落。芽鳞密具丝状长柔毛。叶片三角状卵形、阔卵形或卵形，先端急尖或锐尖。外侧苞片直立，急尖，基部无小叶。雌花序长圆形，雌花序梗3-6毫米。小坚果倒卵球形或倒卵球状椭圆体形，膜质翅宽约为小坚果的1/2。花期6-7月，果期7-8月。生海拔1000-1700米纯林或针阔叶混交林下。产中国东北和内蒙古。俄罗斯(堪察加

红桦 *Betula albosinensis*

岳桦 *Betula ermanii*

硕桦 *Betula costata*

硕桦

Betula costata Trautv.

乔木。树皮灰褐色，纸片状脱落。叶片卵形或卵圆形，叶缘有不规则重锯齿，侧脉9-16对。雌花序1；苞片具缘毛；果序长圆形。小坚果倒卵球形，膜质翅宽约为小坚果的1/2。花期5-6月，果期7-9月。生海拔600-2500米的针阔混交林中。产河北、内蒙古、黑龙江、吉林和辽宁。俄罗斯和朝鲜半岛亦有。

Trees. Bark grayish brown, exfoliating in papery flakes. Leaves ovate or ovate-elliptic, margin irregularly and doubly minutely serrate, lateral veins 9-16 on each side of midvein. Female inflorescence 1; bracts ciliate; infrutescences oblong. Nutlets obovoid, with membranous wings ca. 1/2 as wide as nutlet. Fl. May-Jun. Fr. Jul-Sep. Mixed forests of coniferous and broad-leaved trees at 600-2500 m. Distributed in Hebei, Neimenggu, Heilongjiang, Jilin and Liaoning. Also in Russia and Korean Peninsula.

甸生桦

Betula humilis Schrank

灌木，高1-2米。小枝被短柔毛和黄色腺体。叶卵形，1-2.5(-4)×0.6-2厘米，顶端急尖，边缘具重锯齿，近无毛。果序长圆形，长1-1.5厘米；果苞长3-4毫米，被缘毛，上部3裂。小坚果长椭圆形，约2 × 1.5毫米，顶端被短柔毛，翅膜质，比果狭。花期6-7月。生海拔1400-1800米的草甸上。产新疆东北部。蒙古、哈萨克斯坦、俄罗斯(西伯利亚)和欧洲亦有。

Shrubs, 1-2 m tall. Branchlets puberulous and with yellow glandular glands. Leaves ovate, 1-2.5(-4) × 0.6-2 cm, at apex acute, at margin double-serrate, subglabrous. Infructescences oblong, 1-1.5 cm long; bracts 3-4 mm long, ciliate, above 3-lobed. Nutlets long elliptic, ca. 2 × 1.5 mm, at apex puberulous, with very narrow membranous wings. Fl. Jun-Jul. Meadows at 1400-1800 m. Distributed in NE Xinjiang. Also in Mongolia, Kazakhstan, Russia (Siberia) and Europe.

甸生桦 *Betula humilis*

圆叶桦
Betula rotundifolia Spach

灌木，高达2米。枝条密被黄色树脂腺。叶近圆形，基部圆形或宽楔形，边缘具钝锯齿。雌花序直立；总花序梗长3-4毫米；苞片具缘毛，3裂。小坚果长圆形，翅与小坚果近等宽或稍狭。花期6-7月，果期7-8月。生海拔约2300米的山坡灌丛。产新疆(阿尔泰山)。哈萨克斯坦、俄罗斯(西伯利亚)和蒙古北部亦有。

Shrubs, up to 2 m tall. Branches densely yellow resinous glandular. Leaves suborbicular, base rounded or broadly cuneate, margin obtusely serrate. Female inflorescences erect; peduncles 3-4 mm long; bracts ciliate, 3-lobed. Nutlets oblong, wings ca. as wide as or slightly narrower than nutlet. Fl. Jun-Jul. Fr. Jul-Aug. Thickets on mountain slopes at ca. 2300 m. Distributed in Xinjiang (Altay Mountain). Also in Kazakhstan, Russia (Siberia) and N Mongolia.

高山桦 *Betula delavay*

坚桦 *Betula chinensis*

高山桦
Betula delavayi Franch.

乔木，高3-15米。小枝褐色，密被黄色柔毛。叶厚纸质，椭圆形或卵形，3-7 × 2-4厘米，基部圆形，边缘具重锯齿，上面变无毛，下面脉上被柔毛。果序长圆状圆柱形，长1.5-2.5厘米；果苞长6-10毫米，被缘毛，上部3裂。小坚果椭圆形，约3 × 2毫米，上部有短柔毛，翅膜质，极狭。生海拔2400-4000米的山坡或山谷丛林中或石上。产云南西北部、西藏东南部和四川西部。

Trees, 3-15 m tall. Branchlets brown, densely yellow-pubescent. Leaves thickly papery, elliptic or ovate, 3-7 × 2-4 cm, at base rounded, at margin double-serrate, adaxially glabrescent, abaxially on nerves pubescent. Infructescences oblong-terete, 1.5-2.5 cm long; bracts 6-10 mm long, ciliate, above 3-lobed. Nutlets elliptic, ca. 3 × 2 mm, above puberulous, with very narrow membranous wings. Thickets on slopes or valley or stones at 2400-4000 m. Distributed in NW Yunnan, SE Xizang and W Sichuan.

圆叶桦 *Betula rotundifolia*

坚桦
Betula chinensis Maxim.

灌木或小乔木。树皮黑灰色。叶卵形、阔卵形或卵状椭圆形，每侧具8或9对侧脉。雌花序近球形，稀长圆形；苞片裂片顶端常反折；果序通常球形。小坚果倒卵球形或卵球形，具狭翅，有时顶端成角状。花期5-6月，果期7-8月。生海拔700-3000米的山谷、石山坡阔叶林中。产华北和东北。朝鲜半岛亦有。

Shrubs or small trees. Bark black-gray. Leaves ovate, broadly ovate, or ovate-elliptic, with 8 or 9 lateral veins on each side of midvein. Female inflorescences subglobose, rarely oblong; lobes of bracts with apex usually reflexed; infrutescences usually globose. Nutlets obovoid or ovoid, narrowly winged, sometimes horn-shaped at apex. Fl. May-Jun. Fr. Jul-Aug. Broad-leaved forests in mountain valleys, shaded, rocky mountain slopes at 700-3000 m. Distributed in N and NE China. Also in Korean Peninsula.

壳斗科
Fagaceae

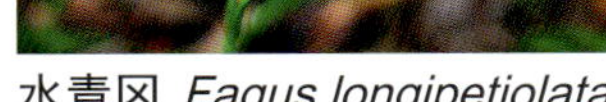
水青冈 *Fagus longipetiolata*

水青冈
Fagus longipetiolata Seem.

乔木。冬芽长2厘米。叶片卵形至长卵形，下面具细密柔毛及带苍白色。壳斗长2-2.5厘米，被线形反折的苞片。坚果比壳斗稍短或等长，近顶部有狭而略伸延的薄翅。花期4-5月，果期8-10月。生海拔300-2400米的山坡常绿阔叶林或混交林下。产中国西南、华南、东南和华中。越南亦有。

Trees. Winter buds to 2 cm long. Leaves ovate to ovate-oblong, abaxially finely densely pubescent and glaucescent. Cupules 2-2.5 cm long, covered with filiform recurved bracts. Nuts as long as or slightly shorter than cupules, with narrowed wings near apex. Fl. Apr-May. Fr. Aug-Oct. Broad-leaved evergreen forests or mixed forests at 300-2400 m. Distributed in SW, S, SE and C China. Also in Vietnam.

光叶水青冈
Fagus lucida Rehd. et Wils.

乔木。冬芽大约1.5厘米。叶片卵形至椭圆状卵形，亮绿色，除叶下面中脉及侧脉处被丝状柔毛外无毛。壳斗长1-1.5厘米；苞片具瘤，紧密贴生，具三角状钻尖。坚果顶端具小翅或翅不明显。花期4-5月，果期9-10月。生海拔800-2000米的山坡混交林中。产中国西南、华南、东南和华中。

Trees. Winter buds ca. 1.5 cm. Leaves ovate to elliptic-ovate, lustrous green and glabrous except for silky pubescence abaxially on midveins. Cupules 1-1.5 cm long; bracts tuberculate, closely appressed, triangular mucronate. Nuts with minute or inconspicuous wings near apex. Fl. Apr-May. Fr. Sep-Oct. Mixed mesophytic forests on slopes at 800-2000 m. Distributed in SW, S, SE and C China.

锥栗
Castanea henryi (Skan) Rehd. et Wils.

乔木。叶长圆形或披针形，下面被黄褐色鳞片状腺体及幼时沿脉疏具毛，渐无毛，顶端长渐尖至尾状长尖。壳斗生于短穗状花序上。每壳斗有坚果1，卵球形，长多于宽。花期5-7月，果期9-10月。生海拔100-1800米的丘陵或山地。产中国除东北和西北以外大部分地区。

Trees. Leaves oblong-ovate or lanceolate, abaxially covered with yellowish brown scalelike glands and sparsely pilose along veins when young, glabrescent, apex long acuminate to caudate. Cupules on a short spike. Nut 1 per cupule, ovoid, longer than wide. Fl. May-Jul. Fr. Sep-Oct. Hills or mountains at 100-1800 m. Distributed in most parts of China, except NE and NW China.

光叶水青冈 *Fagus lucida*

锥栗 *Castanea henryi*

茅栗
Castanea seguinii Dode

灌木或小乔木。叶长圆状卵形、长圆状披针形或披针形，背面被黄褐色或浅灰色鳞状腺，幼时沿脉散生柔毛。壳斗短穗状，覆以稍具柔毛的针状苞片。每个壳斗具2或3小坚果，卵球形，长大于宽。花期5-7月，果期9-11月。生海拔400-2000米的山坡或平坝灌丛中。产中国西南、华南、华中和华北。

Shrubs or small trees. Leaves oblong-ovate, oblong-lanceolate, or lanceolate, abaxially covered with yellowish brown or grayish scalelike glands, sparsely hairy along nerves when young. Cupules on a short spike, covered with slightly pubescent spinelike bracts. Nuts 2 or 3 per cupule, ovoid, longer than wide. Fl. May-Jul. Fr. Sep-Nov. Thickets on slopes or flat lands at 400-2000 m. Distributed in SW, S, C and N China.

板栗
Castanea mollissima Blume

乔木。叶椭圆状长圆形至长圆状披针形，腹面具鳞腺，有时消失，叶背有软贴绒毛，顶端短尖至渐尖。壳斗密覆柔毛及针状苞片。每壳斗有坚果2或3，直径2-3厘米。花期4-6月，果期8-10月。生海拔2800米以下的山地，野生或栽培。产中国大部分地区。

Trees. Leaves elliptic-oblong to oblong-lanceolate, adaxially scalelike glands sometimes absent, abaxially tomentose to softly pubescent, apex acute to acuminate. Cupules densely covered with pubescent spinelike bracts. Nuts usually 2 or 3 per cupule, 2-3 cm diam. Fl. Apr-Jun. Fr. Aug-Oct. Mountains below 2800 m, wild or cultivated. Distributed in most parts of China.

茅栗 *Castanea seguinii*

板栗 *Castanea mollissima*

枹丝锥 *Castanopsis calathiformis*

枹丝锥

Castanopsis calathiformis (Skan) Rehd. et Wils.

乔木，高5-10米。枝无毛。叶螺旋状排列，无毛；叶片长椭圆形或倒卵状椭圆形，背面有棕红色紧密的腊鳞层。壳斗杯状，高6-10毫米；苞片小，密集。坚果1，卵球形，长10-15毫米。子叶脑皱状。花期3-5月。生海拔700-3200米的山地林中。产云南南部和西藏东南部。越南北部、老挝、泰国和缅甸亦有。

Trees, 5-10 m tall. Branches glabrous. Leaves spirally arranged, glabrous; blades long elliptic or obovate-elliptic, abaxially densely brown-red-scurfy. Cupules cupular, 6-10 mm tall; with dense small triangular bracts. Nut 1, ovoid, 10-15 mm long. Cotyledons cerebriform rugose. Fl. Mar-May. Montane forests at 700-3200 m. Distributed in S Yunnan and SE Xizang. Also in N Vietnam, Laos, Thailand and Myanmar.

黧蒴锥

Castanopsis fissa (Champ. ex Benth.) Rehd. et Wils.

乔木。叶螺旋状排列，长圆形或倒卵状椭圆形，边缘具钝锯齿或波状齿，背面被灰黄色的鳞秕。壳斗椭圆体形至卵球形，完全或几乎完全包覆坚果。坚果球形至椭圆体形，直径1.1-1.6厘米。花期4-6月，果期10-12月。生海拔1600米以下的常绿阔叶林或山坡。产中国西南、华南和东南。泰国北部和越南北部亦有。

Trees. Leaves spirally arranged, oblong or obovate- elliptic, margin obtusely serrate or undulate-serrate, abaxially gray-yellow scurfy. Cupules ellipsoid to ovoid, completely or almost completely enclosing nuts. Nuts globose to ellipsoid, 1.1-1.6 cm diam. Fl. Apr-Jun. Fr. Oct-Dec. Broad-leaved evergreen forests or slopes below 1600 m. Distributed in SW, S and SE China. Also in N Thailand and N Vietnam.

龙州锥

Castanopsis longzhouica C. C. Huang et Y. T. Chang

常绿乔木。叶8-10 × 3-4厘米，2列；叶缘下半部有裂齿，中脉在叶面明显凸起，侧脉每边9-12条，两面同色；叶背有紧实的腊鳞层；叶柄长1-1.5厘米。雌花序1-2厘米；雌花单生；花柱3。果序与雌花序等长。壳斗浅碗状，包着坚果不到1/4；小苞片覆瓦状排列。坚果1。花期2-3月，果期8-9月。生海拔450-600米常绿阔叶林中。产广西(龙州县)。

Evergreen trees. Leaf blade 8-10 × 3-4 cm, in two rows; margin from middle to apex dentate to crenate, midvein abaxially prominent, secondary veins 9-12 on each side of midvein; surfaces abaxially with closely adherent grayish waxy scalelike trichomes; petiole 1-1.5 cm. Female inflorescences 1-2 cm; flowers solitary; styles 3. Infructescences almost as long as pistillate inflorescences. Cupule shallowly cupular, covering no more than basal 1/4 of nut; bracts imbricate. Nut 1 of a cupule. Fl. Feb-Mar.

黧蒴锥 *Castanopsis fissa*

龙州锥 *Castanopsis longzhouica*

Fr. Aug-Sep. Broad-leaved evergreen forests at 450-600 m. Distributed in Guangxi (Longzhou County).

苦槠

Castanopsis sclerophylla (Lindl.) Schott.

常绿乔木。叶片无毛，革质。果序长8-15厘米；壳斗球形至近球形，全部或几全部包被坚果，不规则瓣裂，外面微被带黄色柔毛；苞片鳞状。每壳斗内含1(-3)枚坚果，近球形，被绒毛。花期4-5月，果期10-11月。生海拔200-1000米常绿阔叶林中。产中国西南、东南、华中和华东。

Evergreen trees. Leaves glabrous, leathery. Infructescences 8-15 cm long; cupules globose to subglobose, completely or almost completely enclosing nuts, irregularly valved, outside yellowish-brown puberulent; bracts scalelike. Nuts 1(-3) per cupule, subglobose, tomentose. Fl. Apr-May. Fr. Oct-Nov. Broad-leaved evergreen forests at 200-1000 m. Distributed in SW, SE, C and E China.

红锥 (刺栲)

Castanopsis hystrix Hook. f. et Thoms. ex DC.

乔木。叶纸质或薄革质，宽披针形或窄卵形，叶背面被红褐色鳞秕和短柔毛。雌花序单生叶腋间。果序长约15厘米；壳斗球形，裂成4瓣；苞片针状。每壳斗有1坚果，阔圆锥形，无毛。花期4-6月，果期翌年8-11月。生海拔1600米以下的常绿阔叶林。产中国西南和华南。印度、缅甸、老挝、越南和柬埔寨亦有。

Trees. Leaves papery or thinly leathery, broadly lanceolate or narrowly ovate, abaxially rufous scurfy and pubescent. Female inflorescences solitary in leaf axil. Infructescences ca. 15 cm long; cupule globose, splitting into 4 segments; bracts spinelike. Nut 1 per cupule, broadly conical, glabrous. Fl. Apr-Jun. Fr. next Aug-Nov. Evergreen broad-leaved forests below 1600 m. Distributed in SW and S China. Also in India, Myanmar, Laos, Vietnam and Cambodia.

红锥 (刺栲) *Castanopsis hystrix*

苦槠 *Castanopsis sclerophylla*

湄公锥（湄公栲） *Castanopsis mekongensis*

毛锥（南岭栲） *Castanopsis fordii*

湄公锥（湄公栲）

Castanopsis mekongensis A. Camus

乔木。叶长椭圆形或卵状披针形，全缘，背面被黄色绒毛。果序长10厘米或稍长；壳斗球形；苞片针状，完全包被壳斗，疏具柔毛，基部合生成刺束。壳斗有1坚果，扁圆形。花期3-4月，果期翌年8-10月。生海拔600-2000米的林缘或山坡。产云南南部至西南部。老挝亦有。

Trees. Leaves narrowly elliptic or ovate-lanceolate, entire, abaxially yellow tomentose. Infructescences 10 cm or rarely longer; cupules globose; bracts spinelike, entirely covering cupule, sparsely pubescent, base connate into bundles. Nut 1 per cupule, oblate. Fl. Mar-Apr. Fr. next Aug-Oct. Forest edges or slopes at 600-2000 m. Distributed in S to SW Yunnan. Also in Laos.

圆芽锥

Castanopsis globigemmata Chun et Huang

乔木，高达25米。叶近革质，卵状椭圆形或披针形，10-15 × 3.5-5厘米，多全缘，背面沿中脉疏被短毛；中脉在背面隆起。果序长5-7厘米；壳斗球形，连刺直径6-7厘米；苞片刺状，完全包围壳斗，长1-1.5厘米。坚果每壳斗1，宽圆锥形，直径2-3.5厘米，密被短伏毛。花期8-9月。生海拔1400米的山地密林中。特产云南(屏边县、元阳县)。

Trees, up to 25 m tall. Leaves subcoriaceous, ovate-elliptic or lanceolate, 10-15 × 3.5-5 cm, mostly entire, abaxially along midrib sparsely puberulous; midribs abaxially prominent. Infructescences 5-7 cm long; cupules globose, including bracts 6-7 cm diam; bracts spine-like, entirely covering cupule, 1-1.5 cm long. Nut 1 per cupule, broadly conical, 2-3.5 cm diam, densely appressed-puberulous. Fl. Aug-Sep. Dense montane forests at 1400 m. Endemic to Yunnan (Pingbian County, Yuanyang County).

毛锥（南岭栲）

Castanopsis fordii Hance

乔木。叶长圆形、披针形或倒披针状长圆形，基部心形或浅耳形，叶全缘。壳斗4(-5)规则瓣裂；苞片刺状，完全包被壳斗，基部贴生成刺束。壳斗具1坚果，近圆锥形；果脐约占坚果面积的1/3。花期3-4月，果期翌年9-10月。生海拔1200米以下的常绿阔叶林。产广西、广东、福建、浙江、湖南和江西。

Trees. Leaves oblong, lanceolate or oblanceolate-oblong, base cordate or shallowly auriculate, margin entire. Cupule splitting into 4 (-5) regular segments; bracts spinelike, entirely covering cupules, base connate into many bundles. Nut 1 per cupule, subconical; scars covering ca. 1/3 of nuts. Fl. Mar-Apr. Fr. next Sep-Oct. Broad-leaved evergreen forests below 1200 m. Distributed in Guangxi, Guangdong, Fujian, Zhejiang, Hunan and Jiangxi.

圆芽锥 *Castanopsis globigemmata*

钩锥（钩栲） *Castanopsis tibetana*

印度锥 *Castanopsis indica*

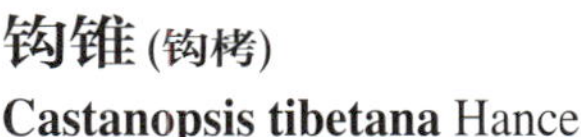

钩锥(钩栲)

Castanopsis tibetana Hance

乔木。叶卵状椭圆形、卵形、长圆形或倒卵状椭圆形，叶缘除基部全缘外具锯齿。壳斗球形，规则开裂成4(或5)瓣；苞片针状，完全包被坚果，基部合生成刺束。每壳斗有1坚果，密被毛；果脐约占坚果面积1/4。花期4-5月，果翌年8-10月成熟。生海拔1500米以下的常绿阔叶林中。产中国西南、华南、东南和华东。

Trees. Leaves ovate-elliptic, ovate, oblong or obovate-elliptic, margin serrate except basally entire. Cupule globose, splitting into 4(or 5) regular segments; bracts spinelike, entirely covering cupule, base usually connate into bundles. Nut 1 per cupule, hairy; scars covering ca. 1/4 of nuts. Fl. Apr-May. Fr. next Aug-Oct. Broad-leaved evergreen forests below 1500 m. Distributed in SW, S, SE and E China.

印度锥

Castanopsis indica (Roxb. ex Lindl.) A. DC.

常绿乔木。当年生枝、叶柄、叶背及花序轴均被黄棕色短柔毛。叶9-20 × 4-10厘米，叶缘下半部有锯齿状锐齿，侧脉每边15-25条；叶柄长5-10毫米。雄花序多为圆锥花序；花柱3。壳斗连刺径3.5-4厘米，4瓣开裂。坚果1，密被毛。花期3-5月，果期翌年9-11月。生海拔1500米以下常绿阔叶林中。产广东南部、海南南部、广西南部、云南南部和西藏东南部。亚洲热带地区广布。

Evergreen trees. Young shoots, petioles, leaf blades abaxially, and rachis of inflorescences yellowish brown puberulent. Leaf blade 9-20 × 4-10 cm, margin serrate except basally entire, secondary veins 15-25 on each side of midvein; petiole 5-10 mm. Male inflorescence paniculate; styles 3. Cupule 3.5-4 cm diam, splitting into 4 segments. Nut 1 per cupule, densely hairy. Fl. Mar-May. Fr. next Sep-Nov. Broad-leaved evergreen forests below 1500 m. Distributed in S Guangdong, S Hainan, S Guangxi, S Yunnan and SE Xizang. Also in tropical regions of Asia.

棱刺锥

Castanopsis clarkei King ex Hook. f.

乔木，高10-20米。幼枝密被柔毛。叶椭圆形或长圆形，10-20 × 5-9厘米，边缘有尖齿，侧脉14-20对；叶柄长1.5-3厘米。雌花序长达20厘米。壳斗近球形，连刺直径3.5-4厘米；苞片刺状，极密，长1-1.5厘米，横切面具3或4棱。坚果1，宽圆锥形，直径1.4-1.6厘米。花期3-5月。生海拔500-800米的山地常绿阔叶林中。产云南南部和西藏东南部。不丹、印度北部和缅甸东北部亦有。

Trees, 10-20 m tall. Young branches densely hairy. Leaves elliptic or oblong, 10-20 × 5-9 cm, at margin acutely denticulate, lateral nerves 14-20 pairs; petioles 1.5-3 cm long. Female inflorescences up to 20 cm long. Cupules subglobose, including bracts 3.5-4 cm diam; bracts spine-like, very dense, 1-1.5 cm long, with trigonous or tetragonous cross sections. Nut 1 per cupule, broadly conical, 1.4-1.6 cm diam. Fl. Mar-May. Montane evergreen broad-leaved forests at 500-800 m. Distributed in S Yunnan and SE Xizang. Also in Bhutan, N India and NE Myanmar.

棱刺锥 *Castanopsis clarkei*

银叶锥 *Castanopsis argyrophylla*

银叶锥

Castanopsis argyrophylla King ex Hook. f.

乔木。枝叶均无毛。叶片厚纸质，背面银灰色。果序轴被浅灰褐色、毡状短毛，渐无毛。壳斗球形；苞片针刺状，排列成不连续的环或螺旋状。每个壳斗有坚果1-3，近球形。花期5-6月，果期10-12月。生海拔1000-1500米的常绿阔叶林中。产云南南部。印度、缅甸、老挝、泰国和越南亦有。

Trees. Branches and leaves glabrous. Leaves thickly papery, abaxially silvery-gray. Rachis of infructescences covered with pale grayish brown, feltlike, short hairs, glabrescent. Cupules globose; bracts spinelike, arranged in discontinuous rings or spirals. Nut 1-3 per cupule, subglobose. Fl. May-Jun. Fr. Oct-Dec. Broad-leaved evergreen forests at 1000-1500 m. Distributed in S Yunnan. Also in India, Myanmar, Laos, Thailand and Vietnam.

薄叶锥(薄叶栲)

Castanopsis tcheponensis Hick. et A. Camus

乔木。叶椭圆形至卵状椭圆形，纸质。果序轴无毛或具极短的稀疏的粉状柔毛。壳斗幼时基部有甚短的柄，成熟时近圆球形；苞片针刺状，几乎全部包被壳斗。每壳斗有1坚果，阔卵球形或近圆球形。花期3-4月，果期10-11月。生海拔900-1400米的山谷常绿阔叶林中。产云南南部。缅甸、老挝和越南亦有。

Trees. Leaves elliptic to ovate-elliptic, papery. Rachis of infructescences glabrous or very shortly and sparsely mealy puberulent. Cupules shortly stalked when young, subglobose when mature; bracts spinelike, almost entirely covering cupule. Nut 1 per cupule, broadly ovoid or almost globose. Fl. Mar-Apr. Fr. Oct-Nov. Valleys of broad-leaved evergreen forests at 900-1400 m. Distributed in S Yunnan. Also in Myanmar, Laos and Vietnam.

大叶锥

Castanopsis megaphylla Hu

乔木。芽鳞、新生枝及花序轴被灰棕色微柔毛及细片状蜡鳞。叶26-45 × 8-18厘米，薄革质，椭圆形，有时兼有倒卵状椭圆形，基部常一侧少偏斜，全缘，中脉在上面凹陷。每壳斗有1坚果，近球形。花期5-7月，果翌年成熟。生海拔1100-1500米常绿阔叶林中。产云南(屏边县)。

Trees. Bud scales, young shoots and rachis of inflorescences grayish brown puberulent with small, lamellate, waxy scalelike trichomes. Leaves 26-45 × 8-18 cm, thinly leathery, elliptic, sometimes obovate-elliptic, base often oblique, margin entire, midvein adaxially impressed. Nut subglobose, 1 per cupule. Fl. May-Jul. Fr. next year. Broad-leaved evergreen forests at 1100-1500 m. Distributed in Yunnan (Pingbian County).

疏齿锥

Castanopsis remotidenticulata Hu

常绿乔木。叶硬纸质，6-12

薄叶锥(薄叶栲) *Castanopsis tcheponensis*

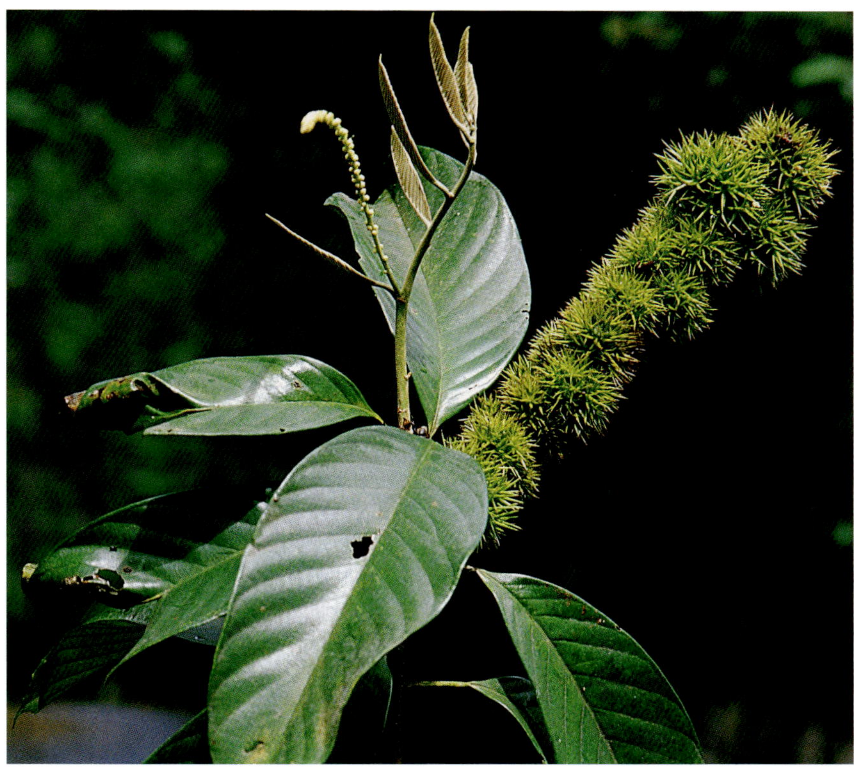
大叶锥 *Castanopsis megaphylla*

密刺锥

Castanopsis densispinosa Y. C. Hsu et H. W. Jen

常绿乔木。叶13-18 × 3.5-6厘米，两面同色或叶面深绿，叶背浅绿，具裂齿；中脉在叶面明显凹陷；侧脉每边10-13条；叶柄长1-2厘米。雄穗状花序或圆锥花序，雌花序生于当年生枝的顶部。壳斗连刺径2-3厘米，3瓣开裂。坚果1，卵形，顶部不明显的3-4棱。果期12月。生海拔约1700米的常绿阔叶林中。产云南(金平县)。

Evergreen trees. Leaf blade 13-18 × 3.5-6 cm, with same color on both surfaces, or adaxially deep green, abaxially pale green, margin dentate; midvein abaxially raised and adaxially impressed; secondary veins 10-13 on each side of midvein; petiole 1-2 cm. Male inflorescences paniculate or spicate, female inflorescences on terminal of annual shoot. Cupule 2-3 cm diam, splitting into 3 segments. Nut 1 per cupule, ovoid, apex obscurely 3-4 ridged. Fr. Dec. Broad-leaved evergreen forests at ca. 1700 m. Distributed in Yunnan (Jinping County).

× 3-4厘米，叶缘有裂齿；中脉及侧脉在叶面呈细肋状凸起；侧脉每边10-13条；叶背有带灰白色紧实的蜡鳞层；叶柄长不超过1厘米。果序轴横切面径4-5毫米。壳斗连刺径3-3.5厘米。坚果1，扁圆锥形，被伏毛。花期4-5月，果期9-11月。生在海拔1000-2200米的常绿阔叶林中。产云南中部至东南部。

Evergreen trees. Leaf blade 6-12 × 3-4 cm, firmly papery, margin serrate; midvein and secondary veins adaxially prominent; secondary veins 10-13 on each side of midvein; abaxially covered with tight grayish waxy scalelike trichomes; petiole to 1 cm. Infructescences rachis 4-5 mm thick. Cupule 3-3.5 cm diam. Nut 1 per cupule, flat-conical, puberulent. Fl. Apr-May. Fr. Sep-Nov. Broad-leaved evergreen forests at 1000-2200 m. Distributed in C to SE Yunnan.

疏齿锥 *Castanopsis remotidenticulata*

密刺锥 *Castanopsis densispinosa*

矩叶锥 *Castanopsis oblonga*

甜槠（甜槠栲） *Castanopsis eyrei*

矩叶锥

Castanopsis oblonga Y. C. Hsu et H. W. Jen

乔木，高8-10米。叶卵形，长椭圆形或披针形，6-9 × 2-3.5厘米，边缘有疏齿或全缘；中脉基部在叶面稍隆起；侧脉10-14对；叶柄长6-10毫米。果序长5-10厘米。壳斗宽倒卵球形，连苞片直径约3厘米；苞片刺状，长4-7毫米，下部被短柔毛。坚果宽圆锥形，直径1-1.8厘米，疏被伏毛。果期10-11月，生海拔2000米的山谷密林中。特产云南(元江县、绿春县)。

Trees, 8-10 m tall. Leaves ovate, long elliptic or lanceolate, 6-9 × 2-3.5 cm, margin sparsely denticulate or entire; midrib adaxially near base slightly preminent; lateral nerves 10-14 pairs; petioles 6-10 mm long. Infructescences 5-10 cm long. Cupules broadly obovoid, including bracts ca. 3 cm diam; bracts spinelike, 4-7 mm long, below puberulous. Nut broadly conical, 1-1.8 cm diam, sparsely appressed-puberulous. Fr. Oct-Nov. Dense forests in valley at 2000 m. Endemic to Yunnan (Yuanjiang County, Lüchun County).

甜槠 (甜槠栲)

Castanopsis eyrei (Champ. ex Benth.) Tutch.

乔木。叶革质。壳斗阔卵球形至近球形，2-4瓣开裂，外壁和苞片被浅灰色或灰黄色微柔毛，先端急尖至钝尖；苞片常不成束状。壳斗有1坚果，阔圆锥状，无毛；果脐位于坚果底部。花期4-6月，果期翌年9-11月。生海拔300-1700米的常绿阔叶混交林中。产中国西南、华南、东南、华中和华西。

Trees. Leaves leathery. Cupules broadly ovoid to subglobose, splitting into 2-4 segments, outside and bracts grayish to yellowish gray puberulent, apically acute to obtuse; bracts usually not in bundles. Nut 1 per cupule, broadly conical, glabrous; scar basal. Fl. Apr-Jun. Fr. next Sep-Nov. Mixed and broad-leaved evergreen forests at 300-1700 m. Distributed in SW, S, SE, C and W China.

栲 (丝栗栲)

Castanopsis fargesii Franch.

常绿乔木。叶边缘全缘或近顶部有凸起的齿。雌花序单生，无毛，长达30厘米。壳斗散生在轴上，壳斗球形至宽卵球形，直径2.5-3厘米，不规则开裂；苞片刺状。每壳斗内1坚果，圆锥形至近球形。花期4-6月和8-10月，果期翌年4-10月。生海拔200-2100米的常绿阔叶林中。产中国西南、华南、东南和华中。

Evergreen trees. Leaf margin entire or sometimes with few

栲（丝栗栲） *Castanopsis fargesii*

秀丽锥（东南栲） *Castanopsis jucunda*

shallow teeth from middle to apex. Female inflorescences solitary, glabrous, up to 30 cm long. Cupules scattered on rachises, globose to broadly ovoid, 2.5-3 cm diam, splitting irregularly; bracts spinelike. Nut 1 per cupule, conical to subglobose. Fl. Apr-Jun and Aug-Oct. Fr. next Apr-Oct. Broad-leaved evergreen forests at 200-2100 m. Distributed in SW, S, SE and C China.

秀丽锥 (东南栲)

Castanopsis jucunda Hance

乔木。一年生小枝和叶上面干后黑褐色。叶卵形、卵状椭圆形或长圆形，宽4-8厘米，基部圆形至宽楔形，叶中部至顶部边缘具锯齿。雌花序单个腋生；壳斗近球形，3-5瓣裂。果宽圆锥形至球形；果脐生坚果基部。花期4-5月，果期翌年8-10月。生海拔1500米以下的常绿阔叶林中。产中国西南、华南、东南和华中。越南亦有。

Trees. 1st-year branchlets and leaves adaxially brownish black when dry. Leaves ovate, ovate-elliptic or oblong, 4-8 cm wide, base rounded to broadly cuneate, margin at least from middle to apex serrate. Female inflorescences axillary, solitary; cupule subglobose, splitting into 3-5 segments. Nuts broadly conical to ovoid; scars basal. Fl. Apr-May. Fr. next Aug-Oct. Broad-leaved evergreen forests below 1500 m. Distributed in SW, S, SE and C China. Also in Vietnam.

红壳锥

Castanopsis rufotomentosa Hu

乔木。当年生枝、叶背、叶柄和壳斗外壁及刺均被暗红褐色易抹落的细片状蜡鳞。叶狭披针形至长圆形，革质。壳斗圆球形，有1坚果；刺粗壮，离生或数条在基部合生成刺束；坚果扁圆形或近圆球形，顶端尖。果期11-12月。生海拔约1300米的常绿阔叶林中。产云南东南部(西畴县)。

Trees. 1st-year branchlets, leaves abaxially, petioles and cupules outside covered with caducous, dark reddish brown, small, lamellate, waxy scalelike trichomes. Leaves narrowly lanceolate to oblong, leathery. Cupule globose, with 1 nut; bracts spine-like, free or several with base connate into bundles; nuts oblate-globose to subglobose, apex pointed. Fr. Nov-Dec. Broad-leaved evergreen forests ca. 1300 m. Distributed in SE Yunnan (Xichou County).

红壳锥 *Castanopsis rufotomentosa*

高山锥（高山栲） *Castanopsis delavayi*

高山锥（高山栲）

Castanopsis delavayi Franch.

乔木。叶近革质，顶端锐尖至圆形，中脉在上面稍凸起，最宽部位为中部至顶端。成熟壳斗阔卵球形或近圆球形，开裂成2或3瓣，外面具黄褐色、蜡质鳞片状毛及贴生的柔毛；坚果阔卵球形。花期4-5月，果期翌年9-11月。生海拔1500-2800米的混交林或常绿阔叶林中。产云南、四川、贵州和广东。

Trees. Leaves subleathery, apex acute to rounded, midvein adaxially slightly raised, widest usually from middle to apex. Mature cupule broadly ovoid or almost globose, splitting into 2 or 3 segments, outside with yellowish brown, waxy scalelike trichomes and appressed pubescence; nuts broadly ovoid. Fl. Apr-May. Fr. next Sep-Nov. Mixed and broad-leaved evergreen forests at 1500-2800 m. Distributed in Yunnan, Sichuan, Guizhou and Guangdong.

短刺锥

Castanopsis echinocarpa Miq.

常绿乔木。叶厚纸质，一侧略短，具裂齿；中脉在叶面微凹；侧脉每边9-13条；嫩叶叶背有红或黄棕色的蜡鳞层。雄花序穗状或圆锥花序；雌花的花柱3。壳斗连刺径1.5-2厘米，少开裂，刺粗短，横切面具棱角，近木质，全被灰黄色微柔毛。坚果1。花期4-5月，果期翌年9-10月。生海拔500-2300米山地杂木林中。产云南南部和西藏东南部。

Evergreen trees. Leaf blades thickly papery, one side a litter shorter than the other one, margin serrate; midveiv adaxially slightly impressed to rarely nearly flat; secondary veins 9-13 on each side of midvein; abaxially covered with a very tight layer of waxy scalelike trichomes and reddish brown to yellowish brown when young. Male inflorescences paniculate or spicate; styles 3. Cupule 1.5-2 cm diam, rarely splitting, angular in cross section, nearly woody, pale yellow pubescent; bracts spinelike, thick and short. Nut 1 per cupule. Fl. Apr-May. Fr. next Sep-Oct. Broad-leaved evergreen forests at 500-2300 m. Distributed in S Yunnan and SE Xizang.

米槠

Castanopsis carlesii (Hemsl.) Hayata

乔木。新生枝及花序轴有稀少的红棕色片状蜡鳞。叶披针形至卵形，革质；叶柄基部枕状。壳斗外壁有疣状体，或有时顶部被长1-2毫米的刺状苞片所包被，或部分连生但不形成簇，成熟后自顶端开裂。花期3-6月，果期翌年9-11月。生海拔1700米以下的常绿阔叶混交

短刺锥 *Castanopsis echinocarpa*

米槠 *Castanopsis carlesii*

林或混交中生林中。产中国西南、华南和华中。

Trees. Young shoots and rachis of inflorescences sparsely covered with reddish brown, lamellate, waxy scalelike trichomes. Leaf blades lanceolate to ovate, leathery; petioles base pillow-shaped. Cupules with tubercles or sometimes apically with spine-like bracts 1-2 mm long, sometimes connate but not forming fascicles, splitting at apex after maturation. Fl. Mar-Jun. Fr. next Sep-Nov. Mixed or broad-leaved evergreen forests or mixed mesophytic forests below 1700 m. Distributed in SW, S and C China.

瓦山锥(瓦山栲)

Castanopsis ceratacantha Rehd. et Wils.

乔木。叶披针形、长圆形或有时倒披针形，略呈硬纸质，10-18 × 2-5厘米，侧脉13-17对。成熟壳斗近圆球形；苞片针刺状，数个自中部或顶端贴生成刺束，有时鸡冠状。每个壳斗有坚果1或2，稀3，阔圆锥形。花期4-5月，果期秋季至翌年初冬。生海拔1500-2500米的山地林中。产云南、四川、贵州和湖北。老挝、泰国和越南亦有分布。

Trees. Leaves lanceolate, oblong, or sometimes oblanceolate, slightly hard papery, 10-18 × 2-5 cm, secondary veins 13-17 on each side of midvein. Mature cupules subglobose; bracts spine-like, several connate into bundles from middle or apex, sometimes cristate. Nuts 1 or 2(-3) per cupule, broadly conical. Fl. Apr-May. Fr. autumn to early winter next year. Montane forests at 1500-2500 m. Distributed in Yunnan, Sichuan, Guizhou and Hubei. Also in Laos, Thailand and Vietnam.

罗浮锥(罗浮栲)

Castanopsis fabri Hance

乔木。叶革质。每壳斗内2-3雌花。壳斗球形、阔椭圆体形或卵球形，不规则开裂；苞片针状，长5-10毫米，基部合生，顶部如鹿角状分枝。每壳斗内(1-)2(-3)坚果，圆锥状，无毛或仅幼时在柱座周围具稀疏短毛。花期4-5月，果期翌年9-11月。生海拔100-2000米的常绿阔叶林中。产华南、西南、中南和东南。老挝和越南亦有。

Trees. Leaves leathery. Pistillate flowers 2-3 per cupule. Cupules globose, broadly ellipsoid, or ovoid, splitting irregularly; bracts spine-like, 5-10 mm long, base connate into bundles, apex branched. Nuts (1-)2(-3) per cupule, conical, glabrous or with sparse short hairs only around scar when young. Fl. Apr-May. Fr. next Sep-Nov. Broad-leaved evergreen forests at 100-2000 m. Distributed in S, SW, SC and SE China. Also in Laos and Vietnam.

瓦山锥(瓦山栲) *Castanopsis ceratacantha*

罗浮锥(罗浮栲) *Castanopsis fabri*

鹿角锥(红勾栲) *Castanopsis lamontii*

鹿角锥(红勾栲)
Castanopsis lamontii Hance

乔木。枝、叶和花序轴均无毛。叶厚纸质或近革质，先端锐尖至渐尖，全缘或有时上部具少许浅锯齿；叶柄长1.5-3厘米。雌花序腋生，每个壳斗具3(-7)花。壳斗有坚果2-3个。花期3-5月，果期翌年9-11月。生海拔500-2500米的山地常绿阔叶林中。产中国西南、华南、东南和华中。越南北部亦有。

Trees. Branches, leaf blades and rachis of inflorescences glabrous. Leaf blades thickly papery to subleathery, apex acute to acuminate, margin entire or sometimes apically with few shallow teeth; petioles 1.5-3 cm. Female inflorescences axillary, flowers 3(-7) per cupule. Nuts 2-3 per cupule. Fl. Mar-May. Fr. next Sep-Nov. Montane evergreen broad-leaved forests at 500-2500 m. Distributed in SW, S, SE and C China. Also in N Vietnam.

元江锥(毛果栲)
Castanopsis orthacantha Franch.

乔木。叶片卵形或卵状披针形，较小，长6-10厘米，干后绿色；第一年生叶无毛。每壳斗有雌花2至3朵；壳斗近球形，幼时阔卵球形。每壳斗有坚果1到3个，圆锥形，密被柔毛。花期4-5月，果期翌年9-11月。生海拔1500-3200米的常绿阔叶混交林中。产云南、四川和贵州。

Trees. Leaf blades ovate or ovate-lanceolate, small, 6-10 cm long, green when dry; first-year leaves glabrous. Pistillate flowers 2-3 per cupule; cupules subglobose, broadly ovoid when young. Nuts 1-3 per cupule, conical, densely pubescent. Fl. Apr-May. Fr. next Sep-Nov. Mixed and broad-leaved evergreen forests at 1500-3200 m. Distributed in Yunnan, Sichuan and Guizhou.

元江锥(毛果栲) *Castanopsis orthacantha*

扁刺锥(扁刺栲)
Castanopsis platyacantha Rehd. et Wils.

乔木。树枝光滑。叶革质，侧脉9-13对，幼时具早落的、红褐色、小的、片层状的蜡质鳞片状的柔毛。成熟壳斗近圆球形或阔椭圆形。每壳斗有坚果1-3个，坚果阔圆锥形。花期5-6月，果期翌年9-11月。生海拔1500-2500米的常绿阔叶林中。产云南、四川和贵州。

Trees. Branches glabrous. Leaves leathery, secondary veins 9-13 on each side of midvein, covered with caducous, reddish brown, small, lamellate, waxy scalelike trichomes when young. Mature cupule almost globose or broadly ellipsoid. Nuts 1-3 per cupule, broadly conical. Fl.

扁刺锥(扁刺栲) *Castanopsis platyacantha*

猴面柯 *Lithocarpus balansae*

May-Jun. Fr. next Sep-Nov. Broad-leaved evergreen forests at 1500-2500 m. Distributed in Yunnan, Sichuan and Guizhou.

猴面柯

Lithocarpus balansae (Drake) A. Camus

乔木，高达30米；枝无毛。叶长椭圆形，长10-38厘米，全缘，无毛，侧脉9-12对；叶柄长1.5-2.5厘米。果序长达15厘米。壳斗宽倒卵球形，直径达8厘米，全包坚果，壁厚0.5-1.5厘米；苞片退化成螺旋状或环状线纹。坚果近球形，直径2-3厘米；果脐占坚果面积一半以上，凸起。花期4-5月。生海拔400-1900米的山地常绿阔叶林中。产云南(金平县、屏边县)。老挝和越南亦有。

Trees, up to 30 m tall; branches glabrous. Leaves long elliptic, 10-38 cm long, entire, glabrous, lateral nerves 9-12 pairs; petioles 1.5-2.5 cm long. Infructescences up to 15 cm long. Cupules broadly obovoid, up to 8 cm diam, completely enclosing nut, wall 0.5-1.5 cm thick; bracts all reduced to spiral or concentric lines. Nut subglobose, 2-3 cm diam; scar covering more than 1/2 of nut, convex. Fl. Apr-May. Montane evergreen broad-leaved forests at 400-1900 m. Distributed in Yunnan (Jinping County, Pingbian County). Also in Laos and Vietnam.

瘤果柯

Lithocarpus handelianus A. Camus

乔木，高达28米；当年生枝被灰棕色短毛。叶厚革质，长椭圆形，长15-20厘米，全缘，变无毛，有褐色紧密的鳞秕层，侧脉12-19对；叶柄长2-3厘米。壳斗近球形，直径2-3.2厘米，被灰黄色短伏毛。花期5月和8-10月。生海拔400-1000米的山地常绿阔叶林中。特产海南。

Trees, up to 28 m tall; hornotinous branches grayish-brown-pubescent. Leaves thickly coriaceous, long elliptic, 15-20 cm long, entire, glabrescent, with brown scalelike trichomes, lateral nerves 12-19 pairs; petioles 2-3 cm long. Cupules subglobose, 2-3.2 cm diam, with tawny appressed minute hairs. Fl. May and Aug-Oct. Montane evergreen broad-leaved forests at 400-1000 m. Endemic to Hainan.

瘤果柯 *Lithocarpus handelianus*

白穗柯 *Lithocarpus craibianus*

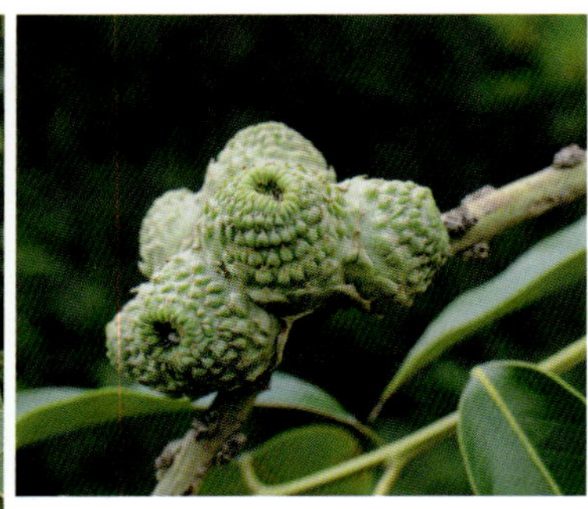

白穗柯

Lithocarpus craibianus Barn.

乔木。幼枝、叶背及雌花序轴均有棕黄色或灰白色蜡鳞层。叶侧脉8-12对。雄花序穗状腋生。壳斗圆球形或略扁，顶端常呈乳头状短凸起，全包坚果；小苞片覆瓦状排列，三角形，钻尖状，贴伏于壳壁。花期8-9月，果期翌年8-9月。生海拔1500-2700米山地杂木林中，多见于较干燥坡地。产云南南部和西南部、四川西南部。老挝和泰国北部亦有。

Trees. Young branchlets, leaves abaxially, and rachises of female inflorescences covered with tawny to grayish, waxy scalelike trichomes. Leaves secondary veins 8-12 on each side of midvein. Male inflorescences axillary. Cupules globose to slightly depressed, apically with a short projection, completely enclosing nuts; bracts imbricate, triangular, subulate, appressed, with tawny, lamellate waxy scalelike trichomes. Fl. Aug-Sep. Fr. next Aug-Sep. Mixed mesophytic forests, usually on dry slopes at 1500-2700 m. Distributed in S and SW Yunnan, SW Sichuan. Also in Laos and N Thailand.

截果柯

Lithocarpus truncatus (King ex Hook. f.) Rehd. et Wils.

乔木。叶薄革质，无毛或下面被紧贴的蜡质鳞片状柔毛。壳斗陀螺状或圆锥形，长2.5-3.5厘米；苞片覆瓦状，三角形，有鳞片。坚果近球形，顶端稍平或稍凹陷；果脐占坚果的2/3-4/5。花期6-8月，果期翌年8-10月。生海拔700-2200米的山地常绿阔叶林或河谷两岸。产西藏和云南。印度东北部、缅甸东北部、泰国北部和越南北部亦有。

Trees. Leaves thin leathery, glabrous or abaxially covered with tightly adherent waxy scalelike trichomes. Cupules turbinate or conical, 2.5-3.5 cm long; bracts imbricate, triangular, squamose. Nuts subglobose, apex ± flat or slightly convex; scar covering 2/3-4/5 of nut. Fl. Jun-Aug. Fr. next Aug-Oct. Montane evergreen forests or by river banks in valleys at 700-2200 m. Distributed in Xizang and Yunnan. Also in NE India, NE Myanmar, N Thailand and N Vietnam.

包果柯 (包槲柯)

Lithocarpus cleistocarpus (Seem.) Rehd. et Wils.

乔木。第二年或第三年生枝具灰色片层状蜡质鳞片状毛。新叶通常干后有油润的树脂，叶革质，侧脉7-10对。壳斗近圆球形，包着坚果绝大部分，壳壁被黄褐色鳞毛。坚果被疏毛或无毛；果脐占坚果面积的2/3以上。花期6-10月，果期翌年秋冬季。生海拔1000-1900米的山地林中。产华南、东南和华中。

Trees. Branches of 2nd- or 3rd-year growth with grayish lamelliform waxy scalelike trichomes. Young leaves usually with oily resin when dry, leaves leathery, secondary veins 7-10 on each side of midvein. Cupules almost globose, enclosing most parts of nuts, wall with tawny scalelike trichomes. Nuts sparsely puberulent or glabrous; scars covering more than 2/3 of nuts. Fl. Jun-Oct. Fr. next autumn-winter. Montane forests at 1000-1900 m. Distributed in S, SE and C China.

截果柯 *Lithocarpus truncatus*

包果柯(包槲柯) *Lithocarpus cleistocarpus*

薄叶柯 *Lithocarpus tenuilimbus*

Oct. Fr. next autumn and winter. Montane forests at 1000-1900 m. Distributed in S, SE and C China.

薄叶柯

Lithocarpus tenuilimbus H. T. Chang

乔木，高达25米。幼枝疏被长柔毛。叶长椭圆形或椭圆状披针形，长12-20厘米，全缘；背面中脉疏被长柔毛，变无毛；侧脉12-16对。壳斗陀螺形，直径2-2.8厘米；顶部苞片三角形，覆瓦状排列，紧贴壳壁，其他苞片与壳壁愈合。坚果近球形，近顶端被短伏毛；果脐占坚果面积的3/4-5/6。花期5-6月。生海拔700-1200米的山地常绿阔叶林中。产云南东南部、广西南部和广东西部。越南北部亦有。

Trees, up to 25 m tall. Young branches sparsely villous. Leaves long elliptic or elliptic-lanceolate, 12-20 cm long, entire; abaxially on midrib sparsely villous, glabrescent; lateral nerves 12-16 pairs. Cupules turbinate, 2-2.8 cm diam; apical bracts triangular, imbricate, appressed to cupule wall, other bracts fused with cupule wall. Nuts subglobose, near apex appressed-puberulous; scars covering 3/4-5/6 of nut. Fl. May-Jun. Montane evergreen broad-leaved forests at 700-1200 m. Distributed in SE Yunnan, S Guangxi and W Guangdong. Also in N Vietnam.

厚叶柯

Lithocarpus pachyphyllus (Kurz) Rehd.

乔木。叶薄革质，侧脉明显，急弯，于近边缘汇合。壳斗直径1.5-4.6厘米，嫩时包坚果绝大部分，成熟时包坚果1/3-1/2；苞片与壳壁愈合，呈三角形或不规则多边形。果脐仅位于坚果底部。花期5-6月，果期翌年8-9月。生海拔800-2000(-3200)米的常绿阔叶混交林。产云南西南部和西藏东南部。印度东北部、尼泊尔、不丹和缅甸东北部亦有。

Trees. Leaves thinly leathery, secondary veins conspicuous, abruptly curving, and anastomosing near margin. Cupules 1.5-4.6 cm diam, enclosing most parts of nuts when young, but only 1/3-1/2 when mature; bracts fused with cupules and reduced to scales, triangular or irregularly multilateral. Scar only at base of nuts. Fl. May-Jun. Fr. next Aug-Sep. Broadleaved evergreen and mixed forests at 800-2000(-3200) m. Distributed in SW Yunnan and SE Xizang. Also in NE India, Nepal, Bhutan and NE Myanmar.

厚叶柯 *Lithocarpus pachyphyllus*

麻子壳柯（多变柯）*Lithocarpus variolosus*

密脉柯 *Lithocarpus fordianus*

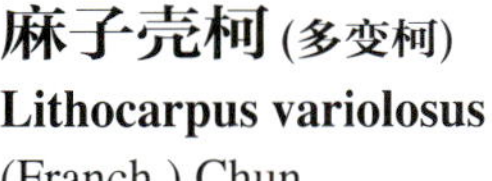

麻子壳柯(多变柯)

Lithocarpus variolosus
(Franch.) Chun

乔木。叶芽、幼枝和雌花蕾有树脂。叶革质至厚纸质，侧脉6-10对，干后下面灰色。雄花序单生叶腋或成圆锥花序；雌花序常生枝顶。壳斗杯形，半包至近全包坚果。坚果扁球形，无毛。花期5-7月，果期翌年7-9月。生海拔2500-3000米的杂木林中，常和云冷、冷杉、亚高山栎类一起生长。产云南西北部和四川西南部。

Trees. Leaf buds, young branchlets, and female flower buds with resin. Leaves leathery to thick papery, lateral nerves 6-10 pairs, abaxially glaucous when dry. Male inflorescences solitary in axils of leaves or paniculate; female inflorescences usually terminal clusters. Cupules cupular, enclosing 1/2 to most parts of nuts. Nuts depressed globose, glabrous. Fl. May-Jul. Fr. next Jul-Sep. Mixed mesophytic forests, usually in association with *Picea*, *Abies*, and subalpine *Quercus* at 2500-3000 m. Distributed in NW Yunnan and SW Sichuan.

金毛柯

Lithocarpus chrysocomus
Chun et Tsiang

乔木。叶卵形或长圆形，硬革质，全缘，下面密具疏松黄褐色至红褐色屑鳞状毛。雄花序单穗腋生或多穗排成圆锥状；雌花序约3个聚为一簇。壳斗近球形，包被坚果大部分。坚果近球形，上宽下窄；果脐约占坚果的1/3，凸出。花期6-8月，果期翌年8-10月。生海拔600-1400米的常绿阔叶林或中生杂木林中。产广西东北部、广东北部和湖南南部。

Trees. Leaves ovate or oblong, rigidly leathery, margin entire, abaxially densely covered with lax yellowish brown to reddish brown, scurfy scalelike trichomes. Male inflorescences solitary in axils of leaves or in paniculate clusters; female inflorescences in clusters of ca. 3. Cupules subglobose, enclosing most parts of nuts. Nuts subglobose but broadest apically; scars covering ca. 1/3 of nuts, convex. Fl. Jun-Aug. Fr. next Aug-Oct. Broad-leaved evergreen or mixed mesophytic forests at 600-1400 m. Distributed in NE Guangxi, N Guangdong and S Hunan.

密脉柯

Lithocarpus fordianus
(Hemsl.) Chun

乔木，雌雄同株。叶厚纸质，背面密被星状毛，侧脉16-25对。壳斗杯状，包被坚果的2/3-3/4。坚果陀螺状，被毛，先端圆或平；果脐占据坚果多于1/2的面积，凸起。花期5-9月，果期翌年8-10月。生海拔700-1500米的常绿阔叶林中或常湿润地。产云南南部和贵州西南部。越南亦有。

Trees, androgynous. Leaves thick papery, abaxially with dense stellate hairs, lateral nerves 16-25 pairs. Cupules cupular, enclosing 2/3-3/4 of nut. Nuts turbinate, hairy, apex rounded or flat; scar covering more than 1/2 of nut, convex. Fl.

金毛柯 *Lithocarpus chrysocomus*

May-Sep. Fr. next Aug-Oct. Broad-leaved evergreen forests or frequent in moist sites at 700-1500 m. Distributed in S Yunnan and SW Guizhou. Also in Vietnam.

烟斗柯

Lithocarpus corneus (Lour.) Rehd.

常绿乔木。叶4-20 × 1.5-7厘米，叶缘有裂齿，两面同色，侧脉每边9-26条；叶柄长0.5-4.5厘米。雌花常生于雄花序轴的下段；3朵一簇或单朵散生；花柱斜展。壳斗22-45 × 25-55毫米。果壁比壳壁厚。子叶4-8浅裂。花期几乎全年，果翌年同期成熟。生海拔1000米以下的山地常绿阔叶林中。产台湾南部、福建南部、湖南南部、贵州南部、广西、广东和云南东南部。

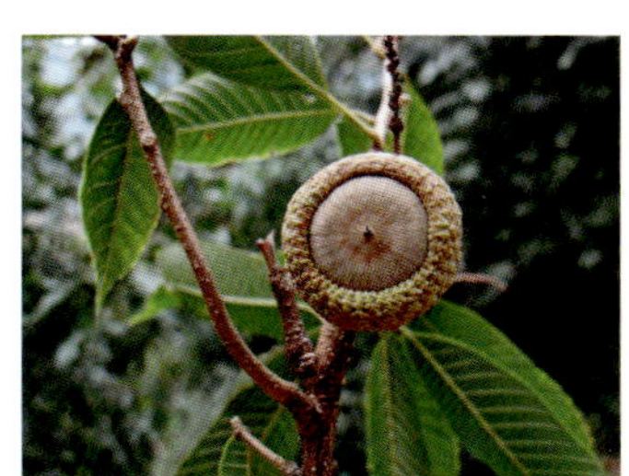

Evergreen trees. Leaf blade 4-20 × 1.5-7 cm, concolorous, margin dentate, secondary veins 9-26 on each side of midvein; petiole 0.5-4.5 cm. Male inflorescences often with female flowers borne at base of rachis, in clusters of 3 or solitary; styles inclined. Cupule 22-45 × 25-55 mm. Nut wall usually thicker than wall of cupule wall. Cotyledons 4-8-lobed. Fl. almost all year around. Fr. maturing on 1st-year-old branchlets. Broad-leaved evergreen forests below 1000 m. Distributed in S Taiwan, S Fujian, S Hunan, S Guizhou, Guangxi, Guangdong and SE Yunnan.

烟斗柯 *Lithocarpus corneus*

厚鳞柯

Lithocarpus pachylepis A. Camus

乔木，高10-20米。当年生枝。叶柄，叶背脉上和花序轴均被星状毛；叶倒披针形或长圆形，长20-35厘米，边缘具小牙齿，侧脉25-30对。壳斗盘状，直径4-6.5厘米；苞片覆瓦状排列，卵状三角形。坚果扁圆锥形，密被黄棕色短柔毛。花期4-6月。生海拔900-1800米的山地常绿阔叶林中。产云南东南部和广西西部。越南北部亦有。

厚鳞柯 *Lithocarpus pachylepis*

Trees, 10-20 m tall. Hornotinous branches. petioles, leaf blades abaxially on nerves and inflorescence rachis all with stellate hairs; leaves oblanceolate or oblong, 20-35 cm long, denticulate, lateral nerves 25-30 pairs. Cupules disciform, 4-6.5 cm diam; bracts imbricate, ovate-triangular. Nuts depressed-conical, densely tawny-puberulous. Fl. Apr-Jun. Montane evergreen broad-leaved forests at 900-1800 m. Distributed in SE Yunnan and W Guangxi. Also in N Vietnam.

滑皮柯 *Lithocarpus skanianus*

榄叶柯 *Lithocarpus oleifolius*

滑皮柯

Lithocarpus skanianus (Dunn) Rehd.

乔木。芽鳞、小枝、叶柄及花序轴密被黄棕色柔毛。叶厚纸质；侧脉9-14对。雄圆锥花序顶生或稀穗状花序单生于叶腋。壳斗扁球形至近球形，几全包坚果。坚果扁球形至广圆锥形，光滑，直径1.4-2.2厘米。花期9-10月，果期翌年9-10月。生海拔500-1000米的林中。产中国西南、华南和华中。

Trees. Bud scales, branchlets, petioles, and rachis of inflorescences tawny tomentose. Leaves thick papery; secondary veins 9-14 on each side of midvein. Male panicles terminal or rarely solitary spikes in axils of leaves. Cupules depressed globose to subglobose, almost completely enclosing nuts. Nuts depressed globose to broadly conical, glabrous, 1.4-2.2 cm diam. Fl. Sep-Oct. Fr. next Sep-Oct. Forests at 500-1000 m. Distributed in SW, S and C China.

榄叶柯

Lithocarpus oleifolius A. Camus

乔木。小枝、叶柄、叶背和花序分枝均被锈色柔毛。叶宽2-4厘米，坚纸质，下面具平伏毛和蜡质鳞片状毛。雄花序圆锥状；雌花序三朵一簇。壳斗球形，完全包被坚果或有时包被3/4。坚果扁球形至近球形。花期8-9月，果期翌年10-11月。生海拔500-1200米的杂木林中。产贵州、广西、广东、福建、湖南和江西。越南亦有。

Trees. Branchlets, petioles, leaves abaxially, and rachis of inflorescences rusty pilose. Leaves 2-4 cm wide, rigidly papery, abaxially with appressed hairs and waxy scalelike trichomes. Male inflorescences a panicle; female inflorescences with cupules in clusters of 3. Cupules globose, completely enclosing nut or sometimes 3/4 of nut. Nuts depressed globose to subglobose. Fl. Aug-Sep. Fr. next Oct-Nov. Mixed forests at 500-1200 m. Distributed in Guizhou, Guangxi, Guangdong, Fujian, Hunan and Jiangxi. Also in Vietnam.

光叶柯

Lithocarpus mairei (Schott.) Rehd.

乔木。叶披针形至椭圆形，革质至纸质，侧脉7-10对。花序轴具鳞片状毛。壳斗碗状，包被坚果的一半。坚果宽圆锥形至略扁圆形，光滑，顶端略扁；果脐略凹。花期8-9月，果期翌年8-9月。生海拔1500-2500米的干燥山地杂木林中。产云南中部至北部。

光叶柯 *Lithocarpus mairei*

水仙柯 *Lithocarpus naiadarum*

Trees. Leaves lanceolate to elliptic, leathery to papery, lateral nerves 7-10 pairs. Rachis of inflorescences covered with scalelike trichomes. Cupules bowl-shaped, enclosing half of nuts. Nuts broadly conical to slightly depressed globose, glabrous, apex ± flat; scars slightly concave. Fl. Aug-Sep. Fr. next Aug-Sep. Mixed forests on dry mountains at 1500-2500 m. Distributed in C to N Yunnan.

水仙柯

Lithocarpus naiadarum (Hance) Chun.

乔木，高4-10米；枝和叶均无毛。叶硬纸质，狭长椭圆形或长披针形，长达25厘米，宽1-3厘米，在顶端急尖，在基部渐狭，全缘，侧脉在叶每侧11-15条；叶柄短。雄穗状花序多条排成圆锥花序；雌花序长达20厘米；雌花3朵簇生；壳斗浅碟状，宽12-18毫米；苞片小，密集，三角形，排列成环状。坚果宽圆锥形，高10-20毫米，宽15-25毫米，无毛。花期7-8月。生低海拔河岸湿润处或溪边。产海南。

Trees, 4-10 m tall; branches with leaves glabrous. Leaves hard-chartaceous, long narrow-elliptic or long lanceolate up to 25 cm long, 1-3 cm broad, at apex acute, at base attenuate, entire, lateral nerves 11-15 on each leaf side; peticles short. Staminate spikes many in panicles; pistillate spikes up to 20 cm long; pistillate flowers 3 fascicled; cupules shallowly cotyliform, 12-18 mm diam; bracts small, very dense, triangular, arranged in several rings. Nuts broadly conical, 10-20 mm tall, 15-25 mm diam, glabrous. Fl. Jul-Aug. Moist places on river banks or by streams of low elevation. Distributed in Hainan.

硬壳柯

Lithocarpus hancei (Benth.) Rehd.

乔木，高不超过15米。叶长5-10厘米，两面绿色，侧脉不显著，短而密。雄花序常排成圆锥花序，有时下生雌花，上生雄花；雌花序2至多穗聚生枝顶；果序轴粗0.2-0.3厘米。壳斗3-5个一簇。花期7-8月，果期翌年8-11月。生海拔2600米以下的各种生境中。产中国西南、华南、东南和华中。

Trees, usually less than 15 m tall. Leaves 5-10 cm long, green on both surfaces, lateral nerves obscure, short and dense. Male inflorescences usually in a panicle, sometimes with pistillate flowers from base to middle; female inflorescences 2-many congested at branch apices; rachis of infructescences 0.2-0.3 cm thick. Cupules usually in clusters of 3-5. Fl. Jul-Aug. Fr. next Aug-Nov. Various habitats below 2600 m. Distributed in SW, S, SE and C China.

硬壳柯 *Lithocarpus hancei*

柯 (石栎)

Lithocarpus glaber (Thunb.) Nakai

乔木。叶最宽部位位于中至上部。雄花序圆锥状或单生叶腋，长达15厘米；雌花序常具有少数雄花。壳斗3(-5)簇生，盘至杯状，包被坚果1/5-2/5；苞片覆瓦状或连合成同心环，三角形，贴生，微被柔毛。坚果椭圆体形。花期7-11月，果期翌年7-11月。生海拔1500米以下坡地杂木林中，阳坡较常见。产中国西南、华南和华中。日本亦有。

Trees. Leaves usually broadest from middle to apex. Male inflorescences in a panicle or solitary in leaf axils, up to 15 cm long; female inflorescences often with a few staminate flowers. Cupules in clusters of 3(-5), plate-shaped to cupular, enclosing 1/5-2/5 of nuts; bracts imbricate or connate in concentric rings, triangular, appressed, densely puberulent. Nuts ellipsoid. Fl. Jul-Nov. Fr. next Jul-Nov. Mixed mesophytic forests, frequent on sunny slopes below 1500 m. Distributed in SW, S and C China. Also in Japan.

耳叶柯 *Lithocarpus grandifolius*

柯 (石栎) *Lithocarpus glaber*

耳叶柯

Lithocarpus grandifolius (D. Don) Biswas

乔木，高10-15米；枝无毛。叶革质，倒披针形或长椭圆形，长15-40厘米，全缘，无毛，基部浅耳形，钝或楔形。壳斗浅碗状，直径2.2-2.8厘米；苞片覆瓦状排列，宽卵形或宽菱形，被短柔毛。坚果扁球形，直径2-2.6厘米；果脐直径1.6-2厘米。花期4-5月。生海拔600-1900米的山地常绿阔叶林中。产云南南部和西南部。印度、缅甸东北部和老挝北部亦有。

Trees, 10-15 m tall; branches glabrous. Leaves coriaceous, oblanceolate or long elliptic, 15-40 cm long, entire, glabrous, base shallowly auriculate, obtuse or cuneate. Cupules shallowly bowl-shaped. 2.2-2.8 cm diam; bracts imbricate, broadly ovate or broadly rhombic, puberulous. Nuts depressed-globose, 2-2.6 cm diam; scars 1.6-2 cm diam. Fl. Apr-May. Montane evergreen broad-leaved forests at 600-1900 m. Distributed in S and SW Yunnan. Also in India, NE Myanmar and N Laos.

短尾柯

Lithocarpus brevicaudatus (Skan) Hayata

大乔木。叶革质，卵形，椭圆形或长圆形，长6-15厘米，全缘，侧脉9-13对。壳斗浅碗状，直径1.4-2厘米；苞片覆瓦状排列，三角形或近菱形，被短柔毛。坚果宽圆锥形，直径1.4-2.2厘米。花期5-7月。生海拔300-1900米的山地杂木林中。产广西、海南、广东、台湾、福建、浙江、江西、湖南、贵州、四川、湖北和安徽。

Large trees. Leaves coriaceous, ovate, elliptic or oblong, 6-15 cm long, entire, lateral nerves 9-13 pairs. Cupules shallowly bowl-shaped, 1.4-2 cm diam; bracts imbricate, triangular or subrhombic, puberulous. Nuts broadly conical, 1.4-2.2 cm diam. Fl. May-Jul. Montane mixed forests at 300-1900 m. Distributed in Guangxi, Hainan, Guangdong, Taiwan, Fujian, Zhejiang, Jiangxi, Hunan, Guizhou, Sichuan, Hubei and Anhui.

大叶柯

Lithocarpus megalophyllus Rehd. et Wils.

乔木，高15-25米；枝无毛。叶革质，长圆形或倒卵形，长14-30厘米，全缘，无毛，侧脉14-18对。壳斗浅碗状，高4-10毫米，直径2-3厘米；苞片覆瓦状排列，长1-1.4毫米。坚果圆锥形，2.4-2.8 × 2-2.5厘米，或扁球形，1.6-1.8 × 2.8-3.2厘米。花期5-6月。生海拔900-2200米的山地杂木林中。产云南东部、广西西部、贵州、四川西部和湖北西部。越南东北部亦有。

短尾柯 *Lithocarpus brevicaudatus*

Trees, 15-25 m tall; branches glabrous. Leaves coriaceous, oblong or obovate, 14-30 cm long, entire, glabrous, lateral nerves 14-18 pairs. Cupules shallowly bowl-shaped, 4-10 mm × 2-3 cm; bracts imbricate, 1-1.4 mm long. Nuts conical, 2.4-2.8 × 2-2.5 cm or depressed-globose, 1.6-1.8 × 2.8-3.2 cm. Fl. May-Jun. Montane mixed forests at 900-2200 m. Distributed in E Yunnan, W Guangxi, Guizhou, W Sichuan and W Hubei. Also in NE Vietnam.

木姜叶柯 *Lithocarpus litseifolius*

木姜叶柯

Lithocarpus litseifolius

(Hance) Chun

乔木。小枝和花序轴无毛。叶椭圆形、倒卵状椭圆形或卵形，纸质至近革质，全缘。壳斗盘状，直径0.8-1.4厘米；小苞片覆瓦状排列，在基部形成同心环带，三角形，贴伏于壳斗上。花期5-9月，果期翌年6-10月。生海拔2200米以下的常绿阔叶林中。产中国西南、华南、东南和华中。老挝、缅甸东北部和越南北部亦有。

Trees. Branchlets and rachis of inflorescences glabrous. Leaves elliptic, obovate-elliptic or ovate, papery to subleathery, margin entire. Cupules plate-shaped, 0.8-1.4 cm diam; bracts imbricate but basal ones connate into concentric rings, triangular, appressed. Fr. May-Sep. Fl. next Jun-Oct. Broad-leaved evergreen forests below 2200 m. Distributed in SW, S, SE and C China. Also in Laos, NE Myanmar and N Vietnam.

大叶柯 *Lithocarpus megalophyllus*

菱果柯 *Lithocarpus taitoensis*

窄叶柯 *Lithocarpus confinis*

菱果柯

Lithocarpus taitoensis (Hayata) Hayata

乔木。小枝具灰黄色的小片层状蜡质鳞片状毛。叶片厚革质，侧脉7-10对，三级脉在下面不明显。壳斗浅碟状，包被坚果基部，壳壁基部坚硬，木质。坚果圆锥状，常有白粉，先端短尖。花期5-9月，果期翌年8-12月。生海拔约1500米的混交林中。产中国西南、华南、东南和华中。

Trees. Branchlets with tawny minute lamellate waxy scalelike trichomes. Leaves thick leathery, lateral nerves 7-10 pairs, tertiary veins abaxially inconspicuous. Cupules plate-shaped, covering base of nuts, wall woody and basally thickened. Nuts conical, often white farinose, apex shortly pointed. Fl. May-Sep. Fr. next Aug-Dec. Mixed mesophytic forests at ca. 1500 m. Distributed in SW, S, SE and C China.

窄叶柯

Lithocarpus confinis C. C. Huang ex Y. C. Hsu et H. W. Jen

乔木，除花序外全株无毛。叶长圆状至披针形，厚纸质，全缘，宽1.5-3.5厘米，侧脉12-16对，短、密、不显著，先端短渐尖至钝。雄花序单生或成圆锥花序；雌花2-6朵成簇聚生枝顶。坚果扁圆形。花期6-8月，果期翌年8-10月。生海拔1500-2400米的干燥山坡次生林中。产云南和贵州。

Trees, glabrous except for inflorescences. Leaves oblong to lanceolate, thickly papery, entire, 1.5-3.5 cm wide, lateral nerves 12-16 pairs, obscure, short and dense, apex shortly acuminate to obtuse. Male inflorescences solitary or in a panicle; female inflorescences in clusters of 2-6 at apex of branches. Nuts depressed globose. Fl. Jun-Aug. Fr. next Aug-Oct. Secondary forests on dry slopes at 1500-2400 m. Distributed in Yunnan and Guizhou.

轮叶三棱栎

Trigonobalanus verticillata Forman

常绿乔木。3叶轮生。花序腋生，穗状。壳斗，3-5裂，近轴一裂片常退化，外壁被横向排列的鳞片。每1壳斗内有坚果1-3(-7)，坚果三棱形，顶端有宿存的花被和花柱，果实内壁被绒毛。生山坡林中。产海南。马来西亚和印度尼西亚亦有。

轮叶三棱栎 *Trigonobalanus verticillata*

麻栎 *Quercus acutissima*

Trees evergreen. Leaves in whorls of 3. Inflorescences axillary and spike-like. Cupules splitting into 3-5 valves, bracts scale-like, transversely arranged. Nuts 1-3(-7) per cupule, prismatic, apex with persistent perianth and styles, endocarp tomentose. Forests on mountains. Distributed in Hainan. Also in Malaysia and Indonesia.

麻栎
Quercus acutissima Carruth.

落叶乔木。幼枝具灰黄色绒毛，渐无毛。老叶无毛或仅背面脉腋有毛。壳斗1-2个生二年生枝，杯形，包着坚果1/4-1/2；苞片钻形至扁条形。坚果卵球形至椭圆体形，顶端圆形。花期3-4月，果期翌年9-10月。生海拔100-2200米的阔叶林中。产中国除西北以外大部分地区。南亚和东亚亦有。

Deciduous trees. Young branchlets yellowish gray tomentose, glabrescent. Leaves tomentose, glabrous or only veins abaxially tomentose with age. Cupules on previous year's branchlets, 1-2, cupular to discoid, including bracts, enclosing 1/4-1/2 of nuts; bracts subulate to ligulate. Nuts ovoid to ellipsoid, apically rounded. Fl. Mar-Apr. Fr. next Sep-Oct. Deciduous forests at 100-2200 m. Distributed in most parts of China, except NW China. Also in S and E Asia.

匙叶栎
Quercus dolicholepis A. Camus

常绿乔木。叶椭圆形、倒卵状匙形或倒卵状椭圆形，革质，全缘或仅顶端具锯齿；侧脉7或8对。壳斗杯状，包被坚果的2/3-3/4；小苞片条状披针形。坚果卵球形至近球形；果脐凸出；柱座直径约1毫米，易碎。花期3-5月，果期翌年10月。生海拔500-2800米的山地林中。产中国西南、华中、华北和华西。

Evergreen trees. Leaves elliptic, obovate-spatulate or obovate-elliptic, leathery, margin entire or apically serrate; secondary veins 7 or 8 on each side of midvein. Cupules cupular, enclosing 2/3-3/4 of nuts; bracts linear-lanceolate. Nuts ovoid to subglobose; scars raised; stylopodia ca. 1 mm diam, easily broken. Fl. Mar-May. Fr. next Oct. Forests in mountains at 500-2800 m. Distributed in SW, C, N and W China.

匙叶栎 *Quercus dolicholepis*

栓皮栎
Quercus variabilis Blume

落叶乔木。树皮木栓质发达。小枝无毛。叶卵状披针形至狭椭圆形，老叶背面密被灰白色星状毛，基部圆形或宽楔形，叶缘具刺芒状锯齿。壳斗杯形，包着坚果2/3。坚果近球形至阔卵球形，先端圆。花期3-4月，果期翌年9-10月。生海拔3000米以下的常绿或落叶林中。产中国除西北以外大部分地区。朝鲜半岛和日本亦有。

Deciduous trees. Cork of bark developed. Branchlets glabrous. Leaves ovate-lanceolate to narrowly elliptic, abaxially densely grayish white stellate-tomentose, base rounded to broadly cuneate, margin with spiniform teeth. Cupules cupular, enclosing 2/3 of nuts. Nuts subglobose to broadly ovoid, apex rounded. Fl. Mar-Apr. Fr. next Sep-Oct. Evergreen or deciduous forests below 3000 m. Distributed in most parts of China, except NW China. Also in Korean Peninsula and Japan.

栓皮栎 *Quercus variabilis*

槲树
Quercus dentata Thunb.

落叶乔木。小枝被灰黄色星状绒毛。叶倒卵形至狭卵形，叶基圆形，叶缘两边有少许波状牙齿。壳斗杯状，包被坚果的1/2-2/3；苞片红褐色，狭披针形，长约1厘米，弯曲或直立。小坚果卵球形或阔卵球形。花期4-5月，果期9-10月。生海拔100-2700米的山坡杂木林或松林中。产中国除西北和华南以外大部分地区。朝鲜半岛和日本亦有。

槲树 *Quercus dentata*

Deciduous trees. Branchlets yellowish gray stellate tomentose. Leaves obovate to narrowly obovate, base rounded, margin with a few undulate to rough teeth on each side. Cupules cupular, enclosing 1/2-2/3 of nut; bracts reddish brown, narrowly lanceolate, ca. 1 cm long, inflexed or erect. Nuts ovoid to broadly so. Fl. Apr-May. Fr. Sep-Oct. Mixed forests or *Pinus* forests on slopes at 100-2700 m. Distributed in most parts of China, except NW and S China. Also in Korean Peninsula and Japan.

白栎
Quercus fabri Hance

落叶乔木或偶为大灌木，高达20米。小枝密生灰白色或灰褐色绒毛。叶倒卵形至椭圆状倒卵形，边缘波状至具锯齿。壳斗杯形，包着坚果约1/3；小苞片卵状披针形，排列紧密。坚果窄椭圆体形至卵球形。花期4

白栎 *Quercus fabri*

月，果期10月。生海拔100-1900米的杂木林中。产中国除西北和东北以外大部分地区。

Deciduous trees or occasionally large shrubs, up to 20 m tall. Branchlets densely grayish white or fawn-tomentose. Leaves obovate to elliptic-obovate, margin undulate to denticulate. Cupules cupular, enclosing ca. 1/3 of nut; bracts ovate-lanceolate, crowded. Nuts narrowly ellipsoid to ovoid. Fl. Apr. Fr. Oct. Mixed mesophytic forests at 100-1900 m. Distributed in most parts of China, except NW and NE China.

槲栎

Quercus aliena Blume

落叶乔木。小枝灰褐色，渐无毛。叶狭椭圆状倒卵形至倒卵形，边缘具波状钝齿，叶背面灰棕色。壳斗杯形，包被坚果的一半；苞片卵状披针形，密集。坚果椭圆体形至卵球形。花期4-5月，果期9-10月。生海拔100-2000米的向阳山坡。产中国除西北以外大部分地区。朝鲜半岛和日本亦有。

Deciduous trees. Branchlets grayish brown, glabrescent. Leaves narrowly elliptic-obovate to obovate, margin undulate-dentate, abaxially grayish brown. Cupules cupulate, enclosing ca. 1/2 of nut; bracts ovate-lanceolate, crowded. Nut ellipsoid to ovoid. Fl. Apr-May. Fr. Sep-Oct. Sunny mountain slopes at 100-2000 m. Distributed in most parts of China, except NW China. Also in Korean Peninsula and Japan.

锐齿槲栎

Quercus aliena Blume var. **acutiserrata** Maxim. ex Wenz.

本变种与槲栎的区别在于本变种的叶具锯齿，顶端具锐锯齿。花期3-5月，果期9-11月。生海拔100-2700米的山地林中。产中国除西北以外大部分地区。

This variety differs from the typical variety in its serrate leaves, acute apices of serrations. Fl. Mar-May. Fr. Sep-Nov. Montane forests at 100-2700 m. Distributed in most parts of China, except NW China.

槲栎 *Quercus aliena*

锐齿槲栎 *Quercus aliena* var. *acutiserrata*

大叶栎 *Quercus griffithii*

大叶栎

Quercus griffithii Hook. f. et Thoms. ex Miq.

落叶乔木。小枝具黄灰色疏毛或柔毛，渐无毛。叶10-20(-30) × 4-10厘米，下面密被灰色星状毛，侧脉12-18对。壳斗杯状，包坚果1/3-1/2；苞片狭卵状三角形。坚果椭圆体形至卵球状椭圆体形。果期9-10月。生海拔700-2800米的森林中。产云南、四川、西藏和贵州。印度东北部、不丹、斯里兰卡、缅甸和泰国北部亦有。

Deciduous trees. Branchlets yellowish gray pilose or pubescent, glabrescent. Leaves 10-20(-30) × 4-10 cm , with dense gray stellate hairs abaxially, lateral nerves 12-18 pairs. Cupules cupular, enclosing 1/3-1/2 of nut; bracts narrowly ovate-triangular. Nuts ellipsoid to ovoid-ellipsoid. Fr. Sep-Oct. Forests at 700-2800 m. Distributed in Yunnan, Sichuan, Xizang and Guizhou. Also in NE India, Bhutan, Sri Lanka, Myanmar and N Thailand.

枹栎

Quercus serrata Murray

落叶乔木，高达25米。叶有或无柄，卵披针形或倒卵形，薄革质，下面无毛或极稀具星状绒毛，边缘具腺状牙齿。壳斗杯状，包坚果1/4-1/3；苞片三角形，贴生，边缘具有柔毛。坚果卵球形至卵球形。花期3-4月，果期9-10月。生海拔100-2000米的山地或沟谷林中。产中国西南、华中、华北、华西和华东。朝鲜半岛和日本亦有。

Trees, up to 25 m tall, deciduous. Leaves subsessile to petiolate, ovate-lanceolate, or obovate, thinly leathery, abaxially glabrous or occasionally stellate tomentose, margin glandular dentate. Cupules cupular, enclosing 1/4-1/3 of nut; bracts triangular, appressed to cupule wall, margin pilose. Nuts ovoid to ovoid-globose. Fl. Mar-Apr. Fr. Sep-Oct. Forests in valleys or on mountains at 100-2000 m. Distributed in SW, C, N, W and E China. Also in Korean Peninsula and Japan.

蒙古栎

Quercus mongolica Fisch. ex Ledeb.

落叶乔木，高达30米。小枝紫褐色，具棱。叶倒卵形至狭倒卵形，边缘具(5-)7-10个波状至粗牙齿，侧脉(5-)10-18对；叶柄长2-8毫米。壳斗小苞片呈半球形瘤状凸起。坚果卵球形至长卵球形。花期5-6月，果期9-10月。生海拔200-2500米的中生杂木林中。产中国东北、华西、华东和华北。俄罗斯、朝鲜半岛和日本亦有。

Deciduous trees, up to 30 m tall. Branchlets purple-brown, angular. Leaf blades obovate to narrowly so, margin with (5-)7-10 undulate to rough teeth on each side, se-

枹栎 *Quercus serrata*

蒙古栎 *Quercus mongolica*

蒙古栎 *Quercus mongolica*

帽斗栎 *Quercus guajavifolia*

condary veins (5-)10-18 on each side of midvein; petioles 2-8 mm long. Bracts of cupules with semi-globose tuberculate processes. Nuts ovoid to narrowly ovoid. Fl. May-Jun. Fr. Sep-Oct. Mixed mesophytic forests at 200-2500 m. Distributed in NE, W, E and N China. Also in Russia, Korean Peninsula and Japan.

高山栎
Quercus semecarpifolia Sm.

常绿乔木。叶5-12 × 3-6.5厘米，叶背被棕色星状毛和糠秕状粉末，侧脉每边8-14条；叶柄长2-6毫米。雄花序生于新枝基部，花序轴被灰褐色长毛。壳斗浅碗形或碟形，通常近于平展，包着坚果基部；小苞片披针形。坚果近球形，直径2-3厘米，紫褐色。坚果1-2。花期5-6月，果期翌年8-10月。生海拔2600-4000米的山地林中。产西藏南部。

Evergreen trees. Leaf blade 5-12 × 3-6.5 cm, abaxially with brown stellate hairs and scurfy powder, secondary veins 8-14 on each side of midvein; petiole 2-6 mm. Male inflorescence at young shoots base, rachis pale brown puberulent. Cupule shallowly bowl-shaped to discoid, nearly flattened; bracts lanceolate. Nut subglobose, 2-3 cm diam, purple-brown. Nut 1-2. Fl. May-Jun.Fr. next Aug-Oct. Montane forests at 2600-4000 m. Distributed in S Xizang.

帽斗栎
Quercus guajavifolia Lévl.

常绿灌木或乔木，高达15米。叶下面具宿存的毛。壳斗较厚，帽斗状，直径2-3厘米；苞片顶端红褐色。坚果卵球形至近球形，直径1.5-1.8厘米，无毛，顶端较钝。花期5-7月，果期9月至翌年11月。生海拔2500-4000米的山地或云杉、冷杉林下。产云南、四川和贵州。

Evergreen shrubs or trees, up to 15 m tall, Leaves abaxially persistently hairy. Cupules cuculliform, thick, 2-3 cm diam; bracts reddish brown at apex. Nuts ovoid to subglobose, 1.5-1.8 cm diam, glabrous, apex obtuse. Fl. May-Jul. Fr. Sep to next Nov. Mountains or *Picea* or *Abies* forests at 2500-4000 m. Distributed in Yunnan, Sichuan and Guizhou.

高山栎 *Quercus semecarpifolia*

黄背栎 *Quercus pannosa*

川滇高山栎 *Quercus aquifolioides*

黄背栎

Quercus pannosa Hand.-Mazz.

常绿灌木或小乔木，高达15米；小枝被褐色绒毛。叶卵形，倒卵形或椭圆形，长2-6厘米，全缘或有刺状齿，背面密被褐色腺毛，星状毛和单毛。壳斗碗形，直径1-2厘米；苞片覆瓦状排列，狭卵形，被绒毛。坚果卵球形，长1.5-2厘米。花期5-6月。生海拔2500-3900米的山坡栎林或松栎林中。产云南、四川和贵州。

Evergreen shrubs or small trees, up to 15 m tall; branchlets brown-tomentose. Leaves ovate, obovate or elliptic, 2-6 cm long, margin entire or spinate-serrate, abaxially with dense tawny glandular hairs, stellate hairs and simple hairs. Cupules bowl-shaped, 1-2 cm diam; bracts imbricate, ovate, tomentose. Nuts ovoid, 1.5-2 cm long. Fl. May-Jun. *Quercus* or *Pinus-Quercus* thickets on slopes at 2500-3900 m. Distributed in Yunnan, Sichuan and Guizhou.

川滇高山栎

Quercus aquifolioides Rehd. et Wils.

常绿乔木，有时灌木状，高达20米；幼枝被星状绒毛。叶椭圆形或倒卵形，长2.5-7厘米，边缘具刺齿，被黄棕色腺质星状毛。壳斗浅碗状，直径0.9-1.2厘米，高5-6毫米，被短柔毛；苞片覆瓦状排列，狭卵形。坚果卵球形，直径1-1.5厘米，高1.2-2厘米。花期5-6月。生海拔2000-4500米的山坡上或高山松林中。产西藏东南部、云南、四川和贵州。不丹和缅甸亦有。

Evergreen trees, sometimes shrublike, up to 20 m tall; young branches stellate-tomentose. Leaves elliptic or obovate, 2.5-7 cm long, margin spinate-serrate, with tawny glandular stellate hairs. Cupules shallowly bowl-shaped, 0.9-1.2 cm diam, 5-6 mm tall, puberulous; bracts imbricate, narrowly ovate. Nuts ovoid, 1-1.5 cm diam, 1.2-2 cm tall. Fl. May-Jun. Slopes or *Pinus densata* forests at 2000-4500 m. Distributed in SE Xizang, Yunnan, Sichuan and Guizhou. Also in Bhutan and Myanmar.

长穗高山栎

Quercus longispica (Hand.-Mazz.) A. Camus

常绿乔木。叶4-11 × 2-5厘米，叶背有腺毛，中脉呈之字形曲折，侧脉每边4-8条；叶柄长3-5毫米。雄花序长8-11厘米，花序轴及花被均被星状绒毛；雌花序长3.6-16厘米。果序长6-16厘米。壳斗杯形；小苞片线状披针形，长约1.5毫米。坚果直径1-1.2厘米。花期5-6月，果期10-11月。生海拔2000-3800米的常绿阔叶林中。产四川和云南。

长穗高山栎 *Quercus longispica*

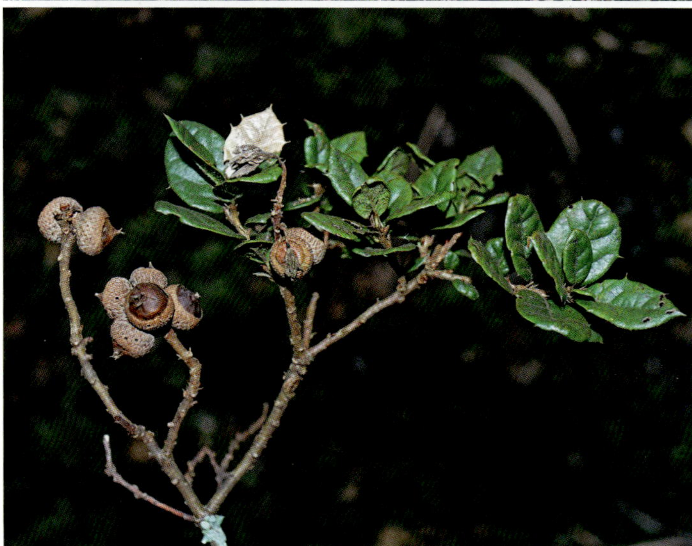

矮高山栎 *Quercus monimotricha*

Evergreen trees. Leaf blade 4-11 × 2-5 cm, abaxially with brown stellate hairs, secondary veins 4-8 on each side of midvein; petiole 3-5 mm. Male inflorescence 8-11 cm, rachis and tepal puberulent; female inflorescence 3.6-16 cm. Infructescences 6-16 cm. Cupule cup-shaped; bracts linear-lanceolate, ca. 1.5 mm. Nut 1-1.2 cm diam. Fl. May-Jun. Fr. Oct-Nov. Evergreen broad-leaved forests at 2000-3800 m. Distributed in Sichuan and Yunnan.

矮高山栎

Quercus monimotricha

Hand.-Mazz.

常绿灌木，高0.5-2米；小枝被褐色绒毛。叶椭圆形或倒卵形，长2-3.5厘米，边缘有刺齿，背面有褐色星状毛。壳斗碗状，直径约1厘米，高3-4毫米；苞片覆瓦状排列，狭卵形，被短柔毛。坚果卵球形，直径0.8-1厘米，高1-1.3厘米。花期5-6月。生海拔2000-3500米的山坡上。产云南西北部和四川西部。

Evergreen shrubs, 0.5-2 m tall; branchlets brown-tomentose. Leaves elliptic or obovate, 2-3.5 cm long, margin spinate-serrate, abaxidlly with brown stellate hairs. Cupules bowl-shaped, ca. 1 cm diam, 3-4 mm tall; bracts imbricate, narrowly ovate, puberulous. Nuts ovoid, 0.8-1 cm diam, 1-1.3 cm tall. Fl. May-Jun. Slopes at 2000-3500 m. Distributed in NW Yunnan and W Sichuan.

毛脉高山栎

Quercus rehderiana

Hand.-Mazz.

常绿乔木。叶倒卵状椭圆形或长椭圆形，长3-8厘米，全缘或有几个刺状齿，侧脉6-8(-12)对，下面渐无毛，但中脉基部有星状毛。雌花序长3.5-16厘米。壳斗浅杯状，包被坚果小于1/2；苞片条状披针形至三角状卵形。坚果卵球形。花期5-6月，果期翌年10-11月。生海拔1500-4000米的山地森林中。产云南、四川、西藏和贵州。泰国亦有。

Evergreen trees. Leaves obovate-elliptic or narrowly elliptic, 3-8 cm long, entire or with several spiny serratures, lateral nerves 6-8(-12) pairs, abaxially glabrescent but usually retaining some stellate hairs along the midvein. Female inflorescences 3.5-16 cm long. Cupules shallowly cupular, enclosing less than 1/2 of nut; bracts linear-lanceolate to triangular-ovate. Nuts ovoid. Fl. May-Jun. Fr. next Oct-Nov. Montane forests at 1500-4000 m. Distributed in Yunnan, Sichuan, Xizang and Guizhou. Also in Thailand.

毛脉高山栎 *Quercus rehderiana*

光叶高山栎
Quercus pseudosemecarpifolia A. Camus

常绿小乔木或灌木，高达12米。叶椭圆形或长倒卵形，长3-7(-13)厘米，全缘或有数刺齿，变无毛。壳斗碗状，直径0.6-1.2厘米，高4-6毫米；苞片覆瓦状排列，三角状卵形，被短柔毛。坚果卵球形，直径0.7-1.2厘米，高约1.2厘米。花期5-6月。生海拔1500-4000米的山坡灌丛中。产西藏东部、云南西北部和北部、四川西部。

Evergreen trees or shrubs, up to 12 m tall. Leaves elliptic or long obovate, 3-7(-13) cm long, entire or several spinate-serrate, glabrescent. Cupules bowl-shaped, 0.6-1.2 cm diam, 4-6 mm tall; bracts imbricate, triangular-ovate, puberulous. Nuts ovoid, 0.7-1.2 cm diam, ca. 1.2 cm tall. Fl. May-Jun. Thickets on slopes at 1500-4000 m. Distributed in E Xizang, NW and N Yunnan, and W Sichuan.

岩栎
Quercus acrodonta Seem.

常绿乔木或灌木。叶椭圆形、椭圆状披针形或狭倒卵形，中部以上具散生针刺，侧脉7-11对；成熟者下面密具灰黄色星状绒毛。雌花序生于枝顶叶腋。壳斗杯状，包被坚果约1/2。坚果狭椭圆体形；果脐稍凸出；柱座约1毫米。花期5月，果期9-10月。生海拔300-2300米的山谷或山地。产中国西南、华中和华西。

Evergreen trees or sometimes shrubs. Leaves elliptic, elliptic-lanceolate or narrowly obovate, margin with spiniform scattered teeth from middle to apex, secondary veins 7-11 on each side of midvein; mature ones abaxially densely yellowish gray stellate tomentose. Female inflorescences axillary on branches toward apex of trees. Cupules cupular, enclosing ca. 1/2 of nut. Nuts narrowly ellipsoid; scars slightly raised; stylopodia ca. 1 mm diam. Fl. May. Fr. Sep-Oct. Valleys or mountains at 300-2300 m. Distributed in SW, C and W China.

岩栎 *Quercus acrodonta*

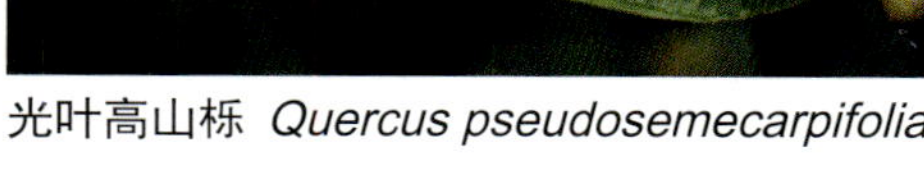
光叶高山栎 *Quercus pseudosemecarpifolia*

铁橡栎
Quercus cocciferoides Hand.-Mazz.

半常绿乔木。叶纸质，侧脉6-8对，基圆形或楔形，常偏斜，叶缘中部以上有小牙齿。壳斗杯形或壶形，包坚果2/3-3/4；小苞片三角形，常不贴生于壳斗，具贴伏浅灰色柔毛。花期4-6月，果期9-11月。生海拔1000-2600米的阳坡或干燥河谷地。产云南、四川和陕西。

Semievergreen trees. Leaves papery, secondary veins 6-8 on each side of midvein, base rounded to cuneate and often oblique, margin denticulate from middle to apex. Cupules cupular to kettle-shaped, enclosing 2/3-3/4 of nut; bracts triangular, usually not appressed to cupules, with appressed grayish hairs. Fl. Apr-Jun. Fr. Sep-Nov. Sunny mountains slopes or dry river valleys at 1000-2600 m. Distributed in Yunnan, Sichuan and Shaanxi.

乌冈栎
Quercus phillyreoides A. Gray

常绿灌木或小乔木，高达10米。叶革质，侧脉8-13对，基部圆形或近心型，边缘1/4以上有小牙齿。壳斗杯状，包着坚果1/3-1/2；小苞片三角形，紧密排列，除顶端外被灰白色柔毛。坚果椭圆体形；果脐平坦或微凸起。花期3-4月，果期9-10月。生海拔300-1200米的阳坡、山顶和山谷密林中。产

铁橡栎 *Quercus cocciferoides*

坝王栎 *Quercus bawanglingensis*

中国除西北和东北以外大部分地区。朝鲜半岛和日本亦有。

Evergreen shrubs or trees, up to 10 m tall . Leaves leathery, secondary veins 8-13 on each side of midvein, base rounded or subcordate, margin denticulate above 1/4 from base. Cupules cupular, enclosing 1/3-1/2 of nut; bracts triangular, crowded, grayish pubescent except for apices. Nuts ellipsoid; scars flat or slightly raised. Fl. Mar-Apr. Fr. Sep-Oct. Sunny slopes, mountain tops or dense forests in valleys at 300-1200 m. Distributed in most parts of China, except NW and NE China. Also in Korean Peninsula and Japan.

坝王栎

Quercus bawanglingensis Huang, Li et Xing

常绿乔木，高6-8米。叶纸质，卵形或椭圆形，长4-6厘米，宽1.5-2.5厘米，顶端急尖或短渐尖，基部宽楔形至圆形，稍不对称，边缘有小锯齿，中脉在叶面平或稍隆起，侧脉每侧6-9条；叶柄长5-8毫米。雄花序下垂。果序长3-6毫米，通常有成熟壳斗1个；壳斗浅碗形，高3-5毫米，直径9-12毫米；苞片被灰白色微柔毛。坚果宽椭圆体形，高10-12毫米，无毛。生海拔900米石灰岩山地。特产海南(昌江县坝王岭)。

Evergreen trees, 6-8 m tall. Leaves papery, ovate to elliptic, 4-6 cm long, 1.5-2.5 cm wide, at apex acute or shortly acuminate, at base broadly cuneate or rounded, slightly asymmetrical, lateral nerves 6-9 on each leaf side; petioles 5-8 mm long. Staminate inflorescences pendulous. Infructescences 3-6 mm long, usually with 1 mature cupule; cupule shallowly bowl-shaped, 3-5 mm tall, 9-12 mm diam; bracts grey-white -puberulous. Nut broadly ellipsoid, 10-12 mm tall, glabrous. Limestone hills at 900 m. Endemic to Hainan (Bawang Mountain, Changjiang County).

乌冈栎 *Quercus phillyreoides*

锥连栎
Quercus franchetii Skan

常绿乔木。叶倒卵形至椭圆形，薄革质，上面光滑，下面具灰黄色腺毛，中部以上具腺状小牙齿；侧脉8-12对。果序着生5或6果；壳斗杯状，或有时碟形，包被坚果的一半；苞片三角形。坚果近球形；果脐凸出；柱座直径约2毫米。花期2-3月，果期9月。生海拔800-2600米的中生杂木林中。产云南和四川。泰国北部亦有。

Evergreen trees. Leaves obovate to elliptic, thinly leathery, adaxially glabrous , abaxially with yellowish gray glandular hairs, margin glandular-tipped denticulate from middle to apex; secondary veins 8-12 on each side of midvein. Infructescence with 5 or 6 fruits; cupules cupular to sometimes discoid, enclosing to 1/2 of nut; bracts triangular. Nuts subglobose; scars raised; stylopodia ca. 2 mm diam. Fl. Feb-Mar. Fr. Sep. Mixed mesophytic forests at 800-2600 m. Distributed in Yunnan and Sichuan. Also in N Thailand.

麻栗坡栎 *Quercus marlipoensis*

麻栗坡栎
Quercus marlipoensis Hu et Cheng

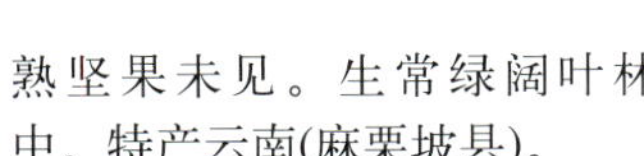

常绿乔木，高达18米；小枝被黄色绒毛。叶革质，长椭圆形或倒卵状长椭圆形，长15-22厘米，边缘有疏锯齿或全缘，叶背脉上有星状绒毛。壳斗碗状，直径1.4厘米，高8毫米，内壁被绒毛；苞片覆瓦状排列，卵形，被紫红色绒毛。成熟坚果未见。生常绿阔叶林中。特产云南(麻栗坡县)。

Evergreen trees, up to 18 m tall; branchlets yellow-tomentose. Leaves coriaceous, long elliptic or long obovate-elliptic, 15-22 cm long, margin sparsely serrate or entire, abaxially on nerves stellate-tomentose. Cupules bowl-shaped, 1.4 cm diam, 8 mm tall, on inner wall tomentose; bracts imbricate, ovate, purple-red-tomentose. Mature nuts not seen. Evergreen broad-leaved forests. Endemic to Yunnan (Malipo County).

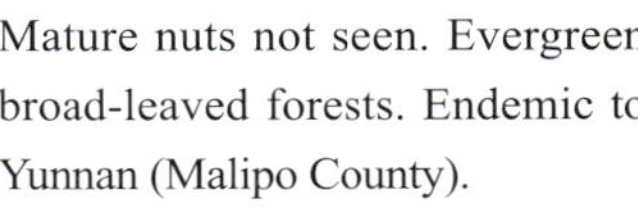

饭甑青冈
Cyclobalanopsis fleuryi (Hick. et A. Camus) Chun ex Q. F. Zheng

乔木。小枝密被棕色长绒毛，后渐无毛。叶片革质，基部楔形，全缘或顶端有波状锯齿，幼时密被黄棕色绒毛，老时无毛，中脉在叶面微凸起。壳斗钟形或近圆筒形，包着坚果约2/3；小苞片合生成10-13条同心环带。花期3-4月，果期10-12月。生海拔500-1500米的山地密林中。产中国西南、华南、东南和华中。老挝和越南亦有。

Trees. Branchlets densely brown tomentose, glabrescent. Leaves leathery, base cuneate, margin entire or apically undulate and

锥连栎 *Quercus franchetii*

饭甑青冈 *Cyclobalanopsis fleuryi*

岭南青冈 *Cyclobalanopsis championii*

serrulate, densely orangish brown tomentose when young but glabrescent, midveins adaxially slightly raised. Cupules campanulate to cylindric, enclosing ca. 2/3 of nut; bracts in 10-13 rings. Fl. Mar-Apr. Fr. Oct-Dec. Dense forests in mountains at 500-1500 m. Distributed in SW, S, SE and C China. Also in Laos and Vietnam.

法斗青冈

Cyclobalanopsis camusiae

(Trel. ex Hick. et A. Camus) Y. C. Hsu et H. W. Jen

常绿乔木，高达15米；小枝被绒毛。叶片革质，披针形或狭长圆形，长9-17厘米，全缘，无毛。壳斗浅碗形，直径2-2.5厘米，高8毫米；苞片合生成5-7条环带。坚果近球形，直径和高度均约1.7厘米。生海拔1400-2000米的山地密林中。特产云南(西畴县法斗乡)。

Evergreen trees, up to 15 m tall; branchlets tomentose. Leaves coriaceous, lanceolate or narrowly oblong, 9-17 cm long, entire, glabrous. Cupules shallowly bowl-shaped, 2-2.5 cm diam, 8 mm tall; bracts connate in 5-7 concentric rings. Nuts subglobose, ca. 1.7 cm diam and tall. Montane dense forests at 1400-2000 m. Endemic to Yunnan (Fadou Village, Xichou County).

岭南青冈

Cyclobalanopsis championii

(Benth.) Oerst.

乔木。树皮暗黑色，薄片状开裂。叶片厚革质，下面具浅黄棕色星状毛，上面深绿色，无毛；叶柄长0.8-1.5厘米。雌花序具花3-10朵；壳斗碗形，包被坚果的1/3-1/2；苞片合生成4-7条同心环带。坚果宽卵球形至扁圆形。花期12月至翌年3月，果期翌年11-12月。生海拔100-1700米的山地常绿阔叶林中。产云南、广西、广东、海南、福建和台湾。

Trees. Bark dark ashy and slice splitting. Leaf blades thickly leathery, abaxially pale orangish brown stellate-tomentose, adaxially dark green and glabrous; petiole 0.8-1.5 cm. Female inflorescences with cupules 3-10; cupules bowl-shaped, enclosing 1/3-1/2 of nut; bracts in 4-7 rings. Nuts broadly ovoid to oblate. Fl. Dec to next Mar. Fr. next Nov-Dec. Evergreen broad-leaved forests in mountains at 100-1700 m. Distributed in Yunnan, Guangxi, Guangdong, Hainan, Fujian and Taiwan.

法斗青冈 *Cyclobalanopsis camusiae*

福建青冈 *Cyclobalanopsis chungii*

福建青冈
Cyclobalanopsis chungii (Metc.) Y. C. Hsu et H. W. Jen ex Q. F. Zheng

乔木。叶片背面密被灰褐色星状绒毛，毛8-10分叉；侧脉每边10-15；叶柄长(0.5-)1-2厘米雌花序长1.5-2厘米，轴和苞片密被褐色绒毛；壳斗2-6，浅碟状，包被小坚果基部，外面和内部被灰褐色绒毛。小坚果扁圆形。果期10-11月。生海拔200-800米的山坡、山谷和常绿阔叶林中。产广西、广东、福建、湖南和江西。

Trees. Leaf blades abaxially densely grayish-brown stellate-tomentose, with 8-10-forked hairs; lateral veins 10-15 on each side of midrein; petiole (0.5-)1-2 cm. Female inflorescences 1.5-2 cm long, rachis and bracts densely brown tomentose; cupules 2-6, saucer-shaped, covering base of nut, outside and inside grayish brown tomentose. Nuts oblate. Fr. Oct-Nov. Evergreen broad-leaved forests on mountain slopes and in valleys at 200-800 m. Distributed in Guangxi, Guangdong, Fujian, Hunan and Jiangxi.

雷公青冈
Cyclobalanopsis hui (Chun) Chun ex Y. C. Hsu et H. W. Jen

乔木。小枝密被黄褐色卷曲绒毛，后渐无毛。叶长圆状椭圆形，倒披针形或椭圆状披针形，近革质，边缘全缘、反卷或仅先端具不明显锯齿；侧脉6-10对。壳斗浅碗形至深碟形，包被坚果底部；小苞片合生成4-6个环带。坚果扁球形；果脐压扁。花期4-5月，果期10-12月。生海拔300-1200米的山地杂木林或密湿常绿阔叶林中。产广西、广东和湖南。

Trees. Branchlets densely curly orangish brown tomentose, glabrescent. Leaves oblong-elliptic, oblanceolate or elliptic-lanceolate, subleathery, margin recurved and entire or indistinctly serrulate toward apex; secondary veins 6-10 on each side of midvein. Cupules shallowly bowl-shaped to deeply discoid, covering base of nut; bracts in 4-6 rings. Nuts oblate; scars impressed. Fl. Apr-May. Fr. Oct-Dec. Mixed or dense wet evergreen broad-leaved forests in mountains at 300-1200 m. Distributed in Guangxi, Guangdong and Hunan.

上思青冈
Cyclobalanopsis delicatula (Chun et Tsiang) Y. C. Hsu et H. W. Jen

乔木。叶纸质，卵形、长圆状椭圆形，或有时倒卵状椭圆形，全缘或仅顶部具浅圆锯齿；侧脉7-8对。壳斗杯形，直径1.6-1.8厘米，包着坚果1/3；小苞片合生成7或8条环带。坚果椭圆体形，两端均圆形；果脐平。花期4-5月，果期10-11月。生海拔300-700米的山地中生杂木林中。产广西、广东和湖南。

Trees. Leaves papery, ovate, oblong-elliptic or sometimes obovate-elliptic, margin entire or shallowly crenate toward apex; secondary veins 7-8 on each side of midvein. Cupules cupular, 1.6-1.8 cm diam, enclosing ca. 1/3 of nut; bracts in 7 or 8 rings. Nuts ellipsoid, base and apex rounded; scars flat. Fl. Apr-May. Fr. Oct-Nov. Mixed mesophytic forests in mountains at 300-700 m. Distributed in Guangxi, Guangdong and Hunan.

薄片青冈
Cyclobalanopsis lamellosa (Sm) Oerst.

常绿大乔木，高达40米；幼枝有绒毛。叶革质，长椭圆形，16-30(-39)厘米，边缘有锯齿，叶面无毛，叶背有星状毛，侧脉18-25(-33)对。壳斗扁球形，直径3-5厘米，高2-3厘米，被绒毛；苞片合生成7-10条同心环带。坚果扁球形，直径3-4厘米，高2-3厘米，被绒毛。花期4-5月。生海拔1200-2500米的山地杂木林中。产西藏东南部、云南西部和东南部、广西西部。缅甸北部、印度和尼泊尔亦有。

Large evergreen trees, up to 40 m tall; young branches tomentose. Leaves coriaceous, long elliptic, 16-30(-39) cm long, serrate,

雷公青冈 *Cyclobalanopsis hui*

上思青冈 *Cyclobalanopsis delicatula*

薄片青冈 *Cyclobalanopsis lamellosa*

adaxially glabrous, abaxially with stellate hairs, lateral nerves 18-25(-33) pairs. Cupules depressed-globose, 3-5 cm diam, 2-3 cm tall, tomentose; bracts connate in 7-10 concentric linear rings. Nuts depressed-globose, 3-4 cm diam, 2-3 cm tall, tomentose. Fl. Apr-May. Montane mixed forests at 1200-2500 m. Distributed in SE Xizang, W and SE Yunnan, and W Guangxi. Also in N Myanmar, India and Nepal.

毛枝青冈

Cyclobalanopsis helferiana (A. DC.) Oerst.

常绿乔木，高达20米；幼枝被黄色绒毛。叶长椭圆形或椭圆状披针形，长12-15(-22)厘米，边缘具钝齿，叶面中脉基部被绒毛，背面密被绒毛。壳斗盘形，直径1.8-2.5厘米，高5-10毫米；苞片合生成8-12条同心环带，被绒毛。坚果扁球形，直径1.5-2.2厘米，高1-1.6厘米，被短柔毛。花期3-4月。生海拔900-2000米的山地林中。产云南南部和西南部、广西、贵州和广东。越南、老挝、泰国、缅甸和印度亦有。

Evergreen trees, up to 20 m tall; young branches yellow-tomentose. Leaves long elliptic or elliptic-lanceolate, 12-15(-22) cm long, margin obtusely serrate, adaxially on midrib base tomentose, abaxially densely tomentose. Cupules disciform, 1.8-2.5 cm diam, 5-10 mm tall; bracts connate in 8-12 concentric rings, tomentose. Nuts depressed-globose, 1.5-2.2 cm diam, 1-1.6 cm tall, puberulous. Fl. Mar-Apr. Montane forests at 900-2000 m. Distributed in S and SW Yunnan, Guangxi, Guizhou and Guangdong. Also in Vietnam, Laos, Thailand, Myanmar and India.

毛枝青冈 *Cyclobalanopsis helferiana*

毛曼青冈 *Cyclobalanopsis gambleana*

毛曼青冈

Cyclobalanopsis gambleana (A. Camus) Y. C. Hsu et H. W. Jen

常绿乔木，高达20米；幼枝被绒毛。叶长椭圆形或椭圆状披针形，长12-20厘米，边缘具小牙齿，背面被星状绒毛。壳斗碗状，直径1.5-1.8厘米，高约1厘米；苞片合生成5-7条同心环带，被绒毛。坚果卵球形，直径约1厘米，高约2厘米，变无毛。花期4-5月。生海拔1100-3000米的山地杂木林中。产西藏南部和东南部、云南、贵州、四川和湖北。印度东北部亦有。

Evergreen trees, up to 20 cm tall; young branches tomentose. Leaves long elliptic or ellptic-lanceolate, 12-20 cm long, margin denticulate, abaxially stellate-tomentose. Cupules bowl-shaped, 1.5-1.8 cm diam, ca. 1 cm tall; bracts connate in 5-7 concentric rings, tomentose. Nuts ovoid, ca. 1 cm diam, ca. 2 cm tall, glabrescent. Fl. Apr-May. Montane mixed forests at 1100-3000 m. Distributed in S and SE Xizang, Yunnan, Guizhou, Sichuan and Hubei. Also in NE India.

曼青冈 *Cyclobalanopsis oxyodon*

滇南青冈 *Cyclobalanopsis austroglauca*

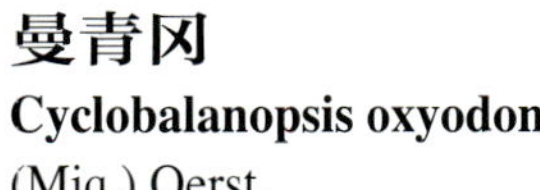

曼青冈
Cyclobalanopsis oxyodon (Miq.) Oerst.

乔木。叶椭圆形至长圆状披针形，下面具灰白粉至白色粉末和贴生的毛，迅速脱落，边缘具锯齿；侧脉16-24对。壳斗杯状，包被坚果1/2；小苞片合生成6-8条环带。坚果卵球形至近球形，无毛；果脐稍凸起。花期5-6月，果期9-10月。生海拔700-2800米的山坡或山谷中生杂木林中。产中国西南、华南、东南、华中、华西和华东。印度东北部、尼泊尔、不丹和缅甸亦有。

Trees. Leaves elliptic to oblong-lanceolate, abaxially pruinose to whitish farinose and with appressed hairs but soon glabrescent, margin serrate; secondary veins 16-24 on each side of midvein. Cupules cupular, enclosing ca. 1/2 of nut; bracts in 6-8 rings. Nuts ovoid to subglobose, glabrous; scars slightly convex. Fl. May-Jun. Fr. Sep-Oct. Mixed mesophytic forests on mountain slopes or valleys at 700-2800 m. Distributed in SW, S, SE, C, W and E China. Also in NE India, Nepal, Bhutan and Myanmar.

滇南青冈
Cyclobalanopsis austroglauca Y. T. Chang ex Y. C. Hsu et H. W. Jen

常绿乔木。小枝无毛，被皮孔。叶缘上部具芒状内弯锯齿，先端渐尖。果序长4-5厘米，着生2-3果；壳斗碗形，包着坚果1/2；小苞片合生成7条同心环带，除上部1-2环全缘外均有裂齿。坚果宽卵球形，柱座宿存，有4-5环纹。生海拔800-1500米的山地常绿阔叶林中。产云南东南部。

Evergreen trees. Branchlets glabrous, lenticellate. Leaf margin with awn-like and incurved serrations apically, apex acuminate. Infructescences 4-5 cm, with 2-3 fruits; cupules bowl-shaped, enclosing ca. 1/2 of nut; bracts in 7 rings, margin of apical 1-2 entire, others dentate. Nuts broadly ovoid, stylopodium persistent, with 4-5 rings. Evergreen broad-leaved forests in mountains at 800-1500 m. Distributed in SE Yunnan.

多脉青冈
Cyclobalanopsis multinervis W. C. Cheng et T. Hong

乔木。叶长圆状椭圆形至椭圆形披针形，叶上部2/3具锐锯齿；侧脉10-15对。壳斗杯状，包被少于坚果的1/2；小苞片合生成6或7个环带。坚果长圆状卵球形；果脐平坦或凸起；柱座宿存。果期在花期翌年的10-11月。生海拔1000-2000米的山地，常组成纯林。产中国西南、东南、华中和华东。

Trees. Leaves oblong-elliptic to elliptic-lanceolate, margin apical

多脉青冈 *Cyclobalanopsis multinervis*

青冈 *Cyclobalanopsis glauca*

2/3 sharply serrate; secondary veins 10-15 on each side of midvein. Cupules cupular, enclosing less than 1/2 of nut; bracts in 6 or 7 rings. Nuts oblong-ovoid; scars flat or convex; stylopodia persistent. Fr. next Oct-Nov from Fl. Often in pure stands in mountains at 1000-2000 m. Distributed in SW, SE, C and E China.

细叶青冈

Cyclobalanopsis gracilis (Rehd. et Wils.) W. C. Cheng et T. Hong

常绿乔木。小枝幼时被柔毛，后渐脱落。叶长圆状卵形至卵状披针形。雌花序长1-1.5厘米，花序轴和苞片被毛；顶生壳斗2-3，壳斗碗形，直径1-1.3厘米。坚果椭圆体形，直径约1厘米。花期4-6月，果期10-11月。生海拔500-2600米的山地杂木林中。产中国西南、华南、华中和华西。

Evergreen trees. Branchlets tomentose when young, glabrescent. Leaves oblong-ovate to ovate-lanceolate. Female inflorescences 1-1.5 cm long, rachis and bracts tomentose; cupules 2-3, borne apically, bowl-shaped, 1-1.3 cm diam. Nuts ellipsoid, ca. 1 cm diam. Fl. Apr-Jun. Fr. Oct-Nov. Mixed mesophytic forests in mountains at 500-2600 m. Distributed in SW, S, C and W China.

细叶青冈 *Cyclobalanopsis gracilis*

青冈

Cyclobalanopsis glauca (Thunb.) Oerst.

乔木。叶下具单毛或鳞片状毛，上部边缘具锯齿。果序长1.5-3厘米，具2-3果；壳斗碗状，包被坚果的1/3-1/2，外面微被白色柔毛或无毛，内面被白色绢毛；苞片成5-6环，密集，边缘全缘或有细齿。花期4-5月，果期10月。生海拔100-2600米的山坡或山谷常绿阔叶林或常绿阔叶与落叶阔叶混交林中。产中国西南、华南、华中至东南。南亚和东亚亦有。

Trees. Leaves abaxially with simple or lepidote hairs, margin apically remotely serrate. Infructescences 1.5-3 cm, with 2-3 fruits; cupules bowl-shaped, enclosing 1/3-1/2 of nut, outside white-puberulent or glabrous, inside white sericeous; bracts in 5-6 rings, crowded, margin entire or denticulate. Fl. Apr-May. Fr. Oct. Evergreen broad-leaved forests and mixed mesophytic forests on mountain slopes or in valleys at 100-2600 m. Distributed in SW, S, C to SE China. Also in S and E Asia.

小叶青冈 *Cyclobalanopsis myrsinifolia*

小叶青冈
Cyclobalanopsis myrsinifolia (Blume) Oerst.

乔木。叶卵形至椭圆状披针形，叶中部以上具锯齿；侧脉9-14对。壳斗杯状，包被坚果1/3-1/2；小苞片合生成6-9个环带。坚果卵球形至椭圆体形；果脐平；柱座明显，具5或6条环纹。花期6月，果期10月。生海拔200-2500米的山谷中生杂木林中。产中国西南、华南、东南、华中、华北、华西和华东。老挝、泰国北部、越南、朝鲜半岛和日本亦有。

Trees. Leaves ovate to elliptic-lanceolate, margin apical 1/2 serrulate; secondary veins 9-14 on each side of midvein. Cupules cupular, enclosing 1/3-1/2 of nut; bracts in 6-9 rings. Nuts ovoid to ellipsoid; scars flat; stylopodiums conspicuous, 5- or 6-ringed. Fl. Jun. Fr. Oct. Mixed mesophytic forests in mountain valleys at 200-2500 m. Distributed in SW, S, SE, C, N, W and E China. Also in Laos, N Thailand, Vietnam, Korean Peninsula and Japan.

赤皮青冈
Cyclobalanopsis gilva (Blume) Oerst.

乔木。叶倒披针形至倒卵状椭圆形，下面具星状绒毛，叶缘中部以上具短芒状锯齿，侧脉11-18对。壳斗碗形，包被坚果约1/4。坚果倒卵球状椭圆体形；果脐稍凸起；柱座宿存。花期5月，果期10月。生海拔300-1500米的山地常绿阔叶林中。产中国西南、华南、东南和中南。日本亦有。

Trees. Leaves oblanceolate to obovate-elliptic, abaxially stellate velutinous, margin apical 1/2 with short awnlike serrations, secondary veins 11-18 on each side of midvein. Cupules bowl-shaped, enclosing ca. 1/4 of nut. Nuts obovoid-ellipsoid; scars slightly raised; stylopodiums persistent. Fl. May. Fr. Oct. Evergreen broad-leaved forests in mountains at 300-1500 m. Distributed in SW, S, SE and SC China. Also in Japan.

毛果青冈
Cyclobalanopsis pachyloma (Seem.) Schott.

乔木。叶革质，幼时具橙色卷毛，渐无毛。雌花序长1.5-3厘米，密被褐色绒毛，具有2-5壳斗；壳斗半球形或钟状，包被坚果的1/3-2/3，外面被茶色绒毛；苞片成7-8环，边全缘或具齿。坚果幼时密被茶色绒毛，之后脱落。花期3月，果期9-10月。生海拔200-1000米山坡或山谷潮湿林中。产中国西南、华南至中南。

Trees. Leaves leathery, orangish woolly hairy when young, glabrescent. Female inflorescences 1.5-3 cm, densely brown tomentose, with 2-5 cupules; cupules semiglobose to campanulate, enclosing ca. 1/3-2/3 of nut, outside usually densely tawny tomentose; bracts in 7-8 rings, margin entire or dentate. Nuts densely tawny tomentose when young, but then glabrescent. Fl. Mar. Fr. Sep-Oct. Wet forests on mountain slopes or valleys at 200-1000 m. Distributed in SW, S to SC China.

赤皮青冈 *Cyclobalanopsis gilva*

毛果青冈 *Cyclobalanopsis pachyloma*

黄毛青冈 *Cyclobalanopsis delavayi*

黄毛青冈
Cyclobalanopsis delavayi (Franch.) Schott.

常绿乔木。小枝密被黄褐色柔毛。叶长圆形至卵状椭圆形，近革质，腹面光滑，侧脉10-14对。雌花序单生叶腋，壳斗2-3；壳斗浅碗状，包被坚果的1/2，外面被淡橙褐色绒毛。坚果椭圆体形或卵球形，被绒毛，后脱落。花期4-5月，果期翌年9-10月。生海拔1000-2800米的山地樟树林或混交栎林和松林中。产云南、四川、贵州、广西和湖北。

Evergreen trees. Branchlets densely pale orangish brown stellate tomentose . Leaves oblong to ovate-elliptic, subleathery, adaxially glabrescent, secondary veins 10-14 on each side of midvein. Female inflorescences axillary, solitary, with cupules 2-3; cupules shallowly bowl-shaped, enclosing ca. 1/2 of nut, outside pale orangish brown tomentose. Nuts ellipsoid or ovoid, tomentose, later glabrescent. Fl. Apr-May. Fr. next Sep-Oct. *Lauraceous* or mixed *Quercus* and *Pinus* forests in mountains at 1000-2800 m. Distributed in Yunnan, Sichuan, Guizhou, Guangxi and Hubei.

滇青冈
Cyclobalanopsis glaucoides Schott.

乔木。叶革质，长椭圆形或倒卵状倒披针形，1/3以上有锯齿，侧脉8-12对。壳斗碗形，外壁被淡棕色绒毛，直径0.8-1.2厘米；苞片具6-8环纹。坚果椭圆体形至卵球形，初被柔毛，后渐脱落；果脐直径5-6毫米，微凸起。花期5月，果期10月。生海拔1500-2500米的常绿阔叶林下。产云南、四川和贵州。

Trees. Leaves leathery, narrowly elliptic to obovate-oblanceolate, margin beyond basal 1/3 serrate, secondary veins 8-12 on each side of midvein. Cupules bowl-shaped, outside pale brown tomentose, 0.8-1.2 cm diam; bracts in 6-8 rings. Nuts ellipsoid to ovoid, pubescent when young, glabrescent; scars 5-6 mm diam, slightly convex. Fl. May. Fr. Oct. Evergreen broad-leaved forests at 1500-2500 m. Distributed in Yunnan, Sichuan and Guizhou.

滇青冈 *Cyclobalanopsis glaucoides*

榆科
Ulmaceae

大果榆 *Ulmus macrocarpa*

长序榆
Ulmus elongata L. K. Fu et C. S. Ding

落叶乔木。树皮棕灰色，片状脱落。次级叶脉16-30对。总状聚伞花序；总轴伸长，下垂；花被6；花梗长为花被的2.4倍。核果梭形，基部具一长雌蕊柄，边缘密被白色缘毛。种子位于翅果中部或稍偏上部。花期2月，果期3月。散生海拔200-900米的常绿阔叶林中。产福建、浙江、安徽和江西。

Deciduous trees. Bark brownish gray, exfoliating. Secondary veins of leaves 16-30 on each side of midvein. Inflorescences racemose cymes; rachises elongated, pendulous; tepals 6; pedicels 2.4 × as long as perianth. Samaras shuttle-shaped, basally with a long gynophore, margin densely white ciliate. Seeds at center or slightly toward apex of samara. Fl. Feb. Fr. Mar. Scattered in evergreen broad-leaved forests at 200-900 m. Distributed in Fujian, Zhejiang, Anhui and Jiangxi.

长序榆 *Ulmus elongata*

大果榆
Ulmus macrocarpa Hance

乔木。树皮灰色至灰黑色，纵裂。小枝无毛或疏具柔毛。叶先端渐尖或短尾尖。花序为簇生聚伞状；花被钟状，5裂，边有缘毛。翅果茶色至亮褐色，1.5-4.7 × 1.3-9厘米，被柔毛，顶端微凹或圆形；翅厚；柱头被毛。花期3月，果期4月。生海拔700-1800米的山坡或山谷。产华东、华中、华北和东北。朝鲜半岛、蒙古和俄罗斯(远东地区、东西伯利亚)亦有。

Trees. Bark gray to blackish gray, longitudinally fissured. Branchlets glabrous or sparsely pubescent. Leaves apex acuminate or shortly caudate. Inflorescences fascicled cymes; perianth campanulate, 5-lobed, margin ciliate. Samaras tan to light brown, 1.5-4.7 × 1.3-9 cm, pubescent, apically concave or rounded; wings thick; stigmas pubescent. Fl. Mar. Fr. Apr. Slopes or valleys at 700-1800 m. Distributed in E, C, N and NE China. Also in Korean Peninsula, Mongolia and Russia (Far East, E Siberia).

脱皮榆
Ulmus lamellosa Wang et S. L. Chang ex L. K. Fu

落叶小乔木。树皮灰色或灰白色，裂成不规则薄片脱落。幼枝密生腺状毛或柔毛；小枝无木栓翅。冬芽芽鳞背面多少被毛。叶先端尾尖或骤凸，叶面

脱皮榆 *Ulmus lamellosa*

粗糙，密生硬毛或有毛迹，边缘锯齿较圆。花常自混合芽抽出。翅果散生于新枝的近基部，有密毛。果核位于翅果的中部；宿存花被钟状。花期3-4月，果期4-5月。生海拔1200米的山谷。特产华北。

Deciduous trees. Bark gray to grayish white, exfoliating in irregular flakes. Sprout densely pubescent; branchlets unwinged. Winter buds scales ± pubescent. Leaf apex caudate to cuspidate, adaxially scabrous, densely pubescent, margin blunt serrate. Flowers from mixed buds. Samara scattered near base of branchlets; pubescent; perianth persistent. Fl. Mar-Apr. Fr. Apr-May. Mountain ravines at 1200 m. Endemic to N China.

兴山榆

Ulmus bergmanniana C. K. Schneid.

乔木，高达26米。叶下面无毛或具柔毛，基部明显倾斜，边缘具重锯齿，侧脉(15-)17-26对。花序在次年生枝上簇成聚伞花序；宿存花被钟形，4-6裂。翅果宽倒卵形、倒卵状圆形或长圆形。果核位于翅果的中部或稍基部。花果期2-4月。生海拔1500-2900米的林中。产中国西南、东南、华中、华北和华西。

Trees, up to 26 m tall. Leaves abaxially glabrous or pubescent, base distinctly oblique, margin doubly serrate, secondary veins (15-)17-26 on each side of mid-vein. Inflorescences fascicled cymes on second year's branchlets; persistent perianth campanulate, 4-6-lobed. Samaras broadly obovate, obovate-orbicular or long orbicular. Seeds at center or slightly toward base of samaras. Fl. and fr. Feb-Apr. Forests at 1500-2900 m. Distributed in SW, SE, C, N and W China.

裂叶榆

Ulmus laciniata (Trautv.) Mayr

乔木。树皮深灰褐色至灰色。叶下具柔毛，边缘具缘毛和深重锯齿，先端通常截形，渐尖或尾尖，通常每边1-3个尾状叶。聚伞花序簇生；花梗无毛。翅果椭圆体形或长圆状椭圆体形。花果期4-5月。生海拔700-2200米的山坡、山谷、河边或林中。产华北、华东和东北。俄罗斯(远东地区、东西伯利亚)、朝鲜半岛和日本亦有。

Trees. Bark dark grayish brown to gray. Leaves abaxially pubescent, margin ciliate and deeply doubly serrate, apex ± truncate, abruptly caudate-acuminate, and often with 1-3 caudate lobes on each side. Cymes clustered; pedicels glabrous. Samaras ellipsoid or oblong-ellipsoid. Fl. and fr. Apr-May. Slopes, valleys, riversides or forests at 700-2200 m. Distributed in N, E and NE China. Also in Russia (Far East, E Siberia), Korean Peninsula and Japan.

裂叶榆 *Ulmus laciniata*

榆树

Ulmus pumila L.

乔木。树皮灰色，不规则纵裂。叶椭圆状卵形或椭圆状披针形，光滑无毛或脉腋具簇毛，边缘具单锯齿或疏具重锯齿，基部对称至稍倾斜；叶柄无毛至具柔毛。聚伞花序簇生，先叶开放。翅果近圆形，稀倒卵圆形。花果期3-6月。生海拔1000-2500米的山坡、山谷、川地、丘陵或沙岗。产中国西南、华中、华北、华西、西北和东北。中亚和东北亚亦有。

Trees. Bark dark gray, irregularly longitudinally fissured. Leaves elliptic-ovate to elliptic-lanceolate, smooth and glabrous or with tufts of hair in vein axils, margin simply serrate or sparsely doubly serrate, base symmetric to ± oblique; petioles glabrous or pubescent. Inflorescences fascicled cymes, appearing before leaves. Samaras almost orbicular, rare obovate. Fl. and fr. Mar-Jun. Mountain slopes, valleys, plains, hills or sand hills at 1000-2500 m. Distributed in SW, C, N, W, NW and NE China. Also in C and NE Asia.

兴山榆 *Ulmus bergmanniana*

榆树 *Ulmus pumila*

旱榆 *Ulmus glaucescens*

春榆 *Ulmus davidiana* var. *japonica*

旱榆
Ulmus glaucescens Franch.

乔木或灌木。树皮浅纵裂。叶宽1.3厘米，两面具柔毛或近无毛，常具簇毛。聚伞花序簇生；花被钟状，约4裂，无毛或裂片有缘毛。翅果茶色，除顶端缺口柱头面有毛，余处无毛；翅厚；花被宿存。花果期3-5月。生海拔2000-2600米的河岸或山坡。产华北、东北和西北。

Trees or shrubs. Bark shallowly longitudinally fissured. Leaves 1.3 cm wide, both surfaces pubescent or subglabrous, often with tufts of hairs. Inflorescences fascicled cymes; perianth campanulate, ca. 4-lobed, glabrous or lobe margin ciliate. Samaras brown, glabrous except stigmatic surface pubescent in notch; wings thick; perianth persistent. Fl. and fr. Mar-May. Along rivers or mountain slopes at 2000-2600 m. Distributed in N, NE and NW China.

黑榆
Ulmus davidiana Planch.

乔木或灌木。树皮淡灰色或灰色。小枝有时具木栓质层。叶倒卵形至倒卵状披针形，下面幼时密具柔毛，边缘具重锯齿，先端尾尖状渐尖至渐尖，侧脉12-22对。聚伞花序簇生。翅果倒卵状或近倒卵状，果核部分常被密毛。花果期3-5月。生海拔2300米以下的石灰岩山地或山谷。产河南、河北、山西、陕西和辽宁。

Trees or shrubs. Bark light gray or gray. Branchlets sometimes with a corky layer. Leaves obovate to obovate-elliptic, abaxially densely pubescent when young, margin doubly serrate, apex caudate-acuminate to acuminate, secondary veins 12-22 on each side of midvein. Cymes clustered. Samaras obovate to subobovate, densely pubescent over seeds. Fl. and fr. Mar-May. Calcacious mountains or valleys below 2300 m. Distributed in Henan, Hebei, Shanxi, Shaanxi and Liaoning.

春榆
Ulmus davidiana Planch. var. **japonica** (Rehd.) Nakai

本变种与黑榆的区别在于本变种的树皮黑色。瘦果无毛。花果期2-5月。生海拔2300米以下的山坡、湿地、河边或山谷。产中国东南、华中、华北、西北和东北。俄罗斯(远东地区、东西伯利亚)、蒙古、朝鲜半岛和日本亦有。

黑榆 *Ulmus davidiana*

春榆 *Ulmus davidiana* var. *japonica*

This variety differs from the typical variety in its blackish bark. Samaras glabrous. Fl. and fr. Feb-May. Slopes, wetlands, by streams, or valleys below 2300 m. Distributed in SE, C, N, NW and NE China. Also in Russia (Far East, E Siberia), Mongolia, Korean Peninsula and Japan.

多脉榆

Ulmus castaneifolia Hemsl.

乔木，高达20米。树皮厚，浅灰色至黑褐色，木栓层发达，纵裂成条状脱落。小枝较粗，无木栓翅及膨大的木栓层。叶基部明显偏斜，下面密具柔毛。簇状聚伞花序。翅果黄褐色，长圆状倒卵形、倒三角状倒卵形或倒卵形。花果期2-4月。生海拔500-1600米的阔叶林中。产中国西南、华南、东南、华中和华东。

Trees, up to 20 m tall. Bark thick, pale gray to blackish brown, with a corky layer, longitudinally fissured. Branchlets thick, unwinged and usually without a corky layer. Leaf base distinctly oblique, abaxially densely pubescent. Inflorescences fascicled cymes. Samaras tan, oblong-obovate, obtriangular-obovate, or obovate. Fl. and fr. Feb-Apr. Broad-leaved forests at 500-1600 m. Distributed in SW, S, SE, C and E China.

榔榆

Ulmus parvifolia Jacq.

乔木。树冠广圆形。小枝无翅。叶光亮，披针状卵形或窄椭圆形，边缘有钝而不规则的单锯齿。簇生聚伞花序具3-6花。花和果于晚夏至初秋生出。翅果椭圆体形或卵球状椭圆体形。花果期8-10月。生海拔800米以下的平原、丘陵、山坡或山谷。产华南、东南、华中、华北和华东。印度、越南、朝鲜半岛和日本亦有。

多脉榆 *Ulmus castaneifolia*

Trees. Crown broadly globose. Branchlets never winged. Leaves lustrous, lanceolate-ovate to narrowly elliptic, margin obtusely and irregularly simply serrate. Inflorescences fascicled cymes, 3-6-flowered. Flowers and fruit appearing late summer to early autumn. Samaras ellipsoid or ovoid-ellipsoid. Fl. and fr. Aug-Oct. Plains, hills, slopes or valleys below 800 m. Distributed in S, SE, C, N and E China. Also in India, Vietnam, Korean Peninsula and Japan.

榔榆 *Ulmus parvifolia*

刺榆 *Hemiptelea davidii*

柔毛。翅果状坚果近圆形或近四方形，至少一侧具阔翅。花期3-5月，果期8-10月。生海拔100-1500米的山谷溪边或石灰岩疏林中。产华南、华中、华北、华东、西北和东北。

Trees. Leaves papery, broadly ovate to oblong, base oblique, margin irregularly serrate, 3-veined from base, secondary veins curving inward, not reaching margin. Flowers unisexual; anthers apically pubescent. Samaroid nuts almost orbicular or square, broadly winged at least on one side. Fl. Mar-May. Fr. Aug-Oct. Valley rivers or limestone sparse forests at 100-1500 m. Distributed in S, C, N, E, NW and NE China.

刺榆

Hemiptelea davidii (Hance) Planch.

灌木或乔木。小枝被棘刺。叶椭圆形或椭圆状长圆形，基部近心形至圆形，边缘具钝齿，先端锐尖至钝；侧脉8-12对。果不对称，带黄绿色，卵球形，仅一边有翅。种柄细长，长2-4毫米。花期4-5月，果期9-10月。生海拔2000米以下的小山坡、小径旁或栽植在房屋周围。产华中、华北、西北和东北。朝鲜半岛亦有。

Shrubs or trees. Branchlets with spines. Leaves elliptic or elliptic-oblong, base ± cordate to rounded, margin with teeth obtuse, apex acute to obtuse; secondary veins 8-12 on each side of midvein. Fruits asymmetric, yellowish-green, ovoid, winged only one side. Seed stalks slender, 2-4 mm long. Fl. Apr-May. Fr. Sep-Oct. Hill slopes, trail sides or planted around houses below 2000 m. Distributed in C, N, NW and NE China. Also in Korean Peninsula.

青檀

Pteroceltis tatarinowii Maxim.

乔木。叶纸质，阔卵形至长方形，叶基不对称，边缘有不整齐锯齿，基出3脉，侧脉内弯不达边缘。花单性；花药顶端具

榉树

Zelkova serrata (Thunb.) Makino

乔木。树皮片状剥落。幼枝紫褐色至褐色。叶纸质至厚纸质，基部稍倾斜，边缘具细齿至圆锯齿；侧脉9-15条。雄花具短梗；花被(5-)7(-8)裂至中部；雌花近无柄；花被4或5(-6)裂。核果表面具不规则网状浅肋。花期4月，果期9-11月。生海拔500-2000米的山谷或溪边。产华南、华中、华西、华北和华东。千岛群岛、朝鲜半岛和日本亦有。

青檀 *Pteroceltis tatarinowii*

榉树 *Zelkova serrata*

大叶榉树 *Zelkova schneideriana*

Trees. Bark exfoliating. Young branchlets brownish purple to brown. Leaves papery to thickly papery, base slightly oblique, margin serrate to crenate; secondary veins 9-15 on each side of midvein. Staminate flowers shortly pedicellate; perianth (5-)7(-8)-parted to middle; pistillate flowers subsessile; perianth 4- or 5(-6)-parted. Drupes surface covered by an irregular network of low ridges. Fl. Apr. Fr. Sep-Nov. Valleys or by streams at 500-2000 m. Distributed in S, C, W, N and E China. Also in Kuril Islands, Korean Peninsula and Japan.

大叶榉树

Zelkova schneideriana Hand.-Mazz.

乔木，高达35米。树皮灰褐色至深灰色，呈不规则的片状剥落。小枝灰色至灰褐色，密具灰白色柔毛。叶基部稍偏斜。雄花单生或2-3朵簇生，花序轴短；雌花或两性花常单生于小枝上部叶腋。核果表面被不规则网状矮脊状凸起。花期4月，果期9-11月。生海拔200-1100米(西藏和云南1800-2800米)的溪边。产中国西南、华南、华中和华东。

Trees, up to 35 m tall. Bark grayish brown to dark gray, exfoliating. Young branchlets gray to grayish brown, densely covered with grayish white pubescence. Leaf base slightly oblique. Staminate flowers solitary or 2-3 clustered, shortly pedicellate; pistillate and bisexual flowers usually solitary in distal leaf axil of young branchlets. Drupes surface covered by an irregular network of low ridges. Fl. Apr. Fr. Sep-Nov. Beside streams at 200-1100 m (1800-2800 m in Xizang and Yunnan). Distributed in SW, S, C and E China.

白颜树

Gironniera subaequalis Planch.

落叶乔木。叶椭圆形至椭圆长圆形，革质；托叶对生，披针形，基部常合生，包被芽。雄花序多分枝；花序轴疏具硬毛。果序有1-5个核果，果不偏斜；宿存柱头2，有梗。花期2-4月，果期全年。生海拔100-800米的山谷或溪边。产云南、广西、广东和海南。缅甸、老挝、泰国、越南、柬埔寨和马来西亚亦有。

Deciduous trees. Leaf blades elliptic to elliptic-oblong, leathery; stipules opposite, lanceolate, usually basally connate, enclosing bud. Staminate inflorescences highly branched; rachis sparsely strigose. Infructescences with 1-5 drupes, drupes not oblique; with 2 persistent stigmata, stipitate. Fl. Feb-Apr. Fr. most of the year. Valleys or by streams at 100-800 m. Distributed in Yunnan, Guangxi, Guangdong and Hainan. Also in Myanmar, Laos, Thailand, Vietnam, Cambodia and Malaysia.

白颜树 *Gironniera subaequalis*

滇糙叶树 *Aphananthe cuspidata*

滇糙叶树

Aphananthe cuspidata (Blume) Planch.

乔木。叶革质，狭卵形至长圆状或卵状披针形，光滑无毛，边缘全缘或有疏锯齿，具羽状脉，在边缘网结。雄花对生或形成3-7厘米的聚伞状；雌花单生。核果包括喙长1.3-2厘米，无毛。花期3-4月和9-11月，果期7-9月和11-12月。生海拔100-900(-1800)米的山坡混交林中。产云南、广东和海南。南亚和东南亚亦有。

Trees. Leaves leathery, narrowly ovate to oblong or ovate-lanceolate, glabrous, margin entire or inconspicuously serrate, pinnately veined, secondary veins anastomosing before reaching margin. Staminate flowers in pairs or in 3-7 cm cymes; pistillate flowers solitary. Drupes 1.3-2 cm including beaks, glabrous. Fl. Mar-Apr and Sep-Nov. Fr. Jul-Sep and Nov-Dec. Mixed forests on slopes at 100-900 (-1800) m. Distributed in Yunnan, Guangdong and Hainan. Also in S and SE Asia.

糙叶树

Aphananthe aspera (Thunb.) Planch.

落叶乔木。幼枝和叶柄被细伏毛。叶纸质，叶背被疏生细伏毛，叶面被刚伏毛，边缘具锐锯齿。雄花生幼枝近轴叶腋；雌花单生。核果近球形、椭圆体形或卵球形，8-13 × 6-9毫米，具柔毛。花期3-5月，果期8-10月。生海拔100-600米的山谷、溪边或山坡。产中国西南、华南、东南、华中、华西和华东。越南、朝鲜半岛和日本亦有。

Deciduous trees. Young branchlets and petioles sparsely pubescent. Leaves papery, abaxially sparsely pubescent, adaxially scabrous with bristles, margin sharply serrate. Staminate flowers in proximal leaf axil of young branchlets; pistillate flowers solitary. Drupes almost globose, ellipsoid or ovoid, 8-13 × 6-9 mm, pubescent. Fl. Mar-May. Fr. Aug-Oct. Valleys, by streams or slopes at 100-600 m. Distributed in SW, S, SE, C, W and E China. Also in Vietnam, Korean Peninsula and Japan.

山黄麻

Trema tomentosa (Roxb.) H. Hara

小乔木或灌木。树皮灰褐色。叶两面同色，下面具灰褐色柔毛，上面极粗糙，具直立刚毛。雄花序2-4。核果成熟时褐紫色至黑紫色，球形，直径2-3毫米。种子具棱。花期3-6月，果期9-11月。生海拔100-2000米的林中、山谷或山坡。产中国西南和华南。南亚、东南亚、日本南部、东非、澳大利亚和太平洋岛屿亦有。

Small trees or shrubs. Bark grayish brown. Leaves ± concolor, abaxially with rayish brown pubescence, adaxially very scabrous with erect bristles. Staminate inflorescences 2-4. Drupes brownish purple to blackish purple when mature, globose, 2-

糙叶树 *Aphananthe aspera*

山黄麻 *Trema tomentosa*

3 mm diam. Seeds ribbed. Fl. Mar-Jun. Fr. Sep-Nov. Forests, valleys or slopes at 100-2000 m. Distributed in SW and S China. Also in S and SE Asia, S Japan, E Africa, Australia and Pacific Islands.

银毛叶山黄麻
Trema nitida C. J. Chen

乔木。叶披针形至狭披针形，薄纸质，下面完全被银灰色至灰黄色平伏的光亮柔毛，边缘具小齿。聚伞花序短于叶柄；雌花具短柄；花被片5。核果成熟时黑紫色，近球形至宽椭圆体形，多少压扁，无毛；花被宿存。花期4-7月，果期8-11月。生海拔600-1800米的石灰岩山坡上的潮湿林中。产云南、四川、贵州和广西。

Trees. Leaves lanceolate to narrowly lanceolate, thinly papery, abaxially completely covered with silver gray to grayish yellow appressed shiny pubescence, margin denticulate. Cymes shorter than petioles; pistillate flowers shortly pedicellate; tepals 5. Drupes blackish purple when mature, subglobose to broadly ellipsoid, subcompressed, glabrous; perianth persistent. Fl. Apr-Jul. Fr. Aug-Nov. Moist forests on limestones slopes at 600-1800 m. Distributed in Yunnan, Sichuan, Guizhou and Guangxi.

银毛叶山黄麻 *Trema nitida*

异色山黄麻
Trema orientalis (L.) Blume

乔木或灌木。叶背面干后灰白色或淡绿灰色，密被绒毛，革质易碎。花序与叶柄等长或稍长于叶柄。果卵球形或近球形，熟时黑色，直径3-5毫米；花被宿存。花期3-5(-6)月，果期6-11月。生海拔400-1900米的林中或灌丛中。产中国西南和华南。南亚、东南亚、日本、澳大利亚，太平洋岛屿和热带非洲亦有。

Trees or shrubs. Leaves abaxially greyish white or greenish gray when dry, densely tomentose, leathery and fragile. Inflorescences as long as or longer than petioles. Drupes ovoid or subglobose, black when mature, 3-5 mm diam; perianth persistent. Fl. Mar-May(-Jun). Fr. Jun-Nov. Forests or thickets at 400-1900 m. Distributed in SW and S China. Also in S and SE Asia, Japan, Australia, Pacific Islands and tropical Africa.

异色山黄麻 *Trema orientalis*

光叶山黄麻

Trema cannabina Lour.

灌木或小乔木。小枝黄绿色。叶近膜质，干后浅黄绿色或黄棕色；叶柄被贴生的短柔毛。聚伞花序一般长不过叶柄；花被无毛或疏生柔毛。核果成熟后橙色，稍压扁，直径2-3毫米；花被宿存。花期3-6月，果期9-10月。生海拔100-1100米的向阳湿润林、向阳坡灌丛、河边或旷野。产中国西南、华南、东南、华中和华东。南亚、东南亚、日本、澳大利亚和太平洋岛屿亦有。

Shrubs or small trees. Branchlets greenish yellow. Leaf blades ± membranous, drying light yellow-green or yellow-brown; petioles with appressed pubescence, glabrescent. Cymes as long as or shorter than petioles; tepals glabrous or sparsely tomentose. Drupes reddish orange when mature, ± compressed, 2-3 mm diam; perianth persistent. Fl. Mar-Jun. Fr. Sep-Oct. Sunny moist forests, scrubs on sunny slopes, riversides or open places at 100-1100 m. Distributed in SW, S, SE, C and E China. Also in S and SE Asia, Japan, Australia and Pacific Islands.

紫弹树（紫弹朴） *Celtis biondii*

紫弹树 (紫弹朴)

Celtis biondii Pamp.

乔木。芽鳞密具糙毛。叶薄革质，两面常散生不明显的平伏毛。花序1-3簇生叶腋，细弱，被毛。核果黄色至红橙色，近球形；核微扁，直径约4毫米，有4肋，网状蜂窝状。花期4-5月，果期9-10月。生海拔2000米以下的林下、山地灌丛或石灰岩山地。产中国西南、华南、华中和东南。朝鲜半岛和日本亦有。

Trees. Bud scales densely strigose. Leaves thin leathery, usually with inconspicuous scattered appressed hairs on both surfaces. Inflorescences 1-3 per leaf axil, slender, pubescent. Drupes yellow to reddish orange, subglobose; stones slightly compressed, ca. 4 mm diam, 4-ribbed, reticulately foveolate. Fl. Apr-May. Fr. Sep-Oct. Forests, shrubs on mountains or limestone areas below 2000 m. Distributed in SW, S, C and SE China. Also in Korean Peninsula and Japan.

珊瑚朴

Celtis julianae C. K. Schneid.

落叶乔木，可达30米高。顶芽败育。叶顶端1/3-2/3边缘具细牙齿，齿短于1毫米；托叶早落；叶柄和叶下具金色柔毛。核果单生叶腋，金黄色至橘黄色，椭圆体形至近球形；核乳白色。花期3-4月，果期9-10月。生海拔300-1300米的山坡、山谷林中或林缘。产中国西南、华南、东南和华中。

Deciduous trees, up to 30 m tall. Terminal buds abortive. Leaves with margin finely toothed in apical 1/3-2/3, teeth less than 1 mm long; stipules caducous; petioles and leaf blades abaxially golden pubescent. Drupes solitary, axillary, golden to orange-yellow, ellipsoid to ± globose; stones milky white. Fl. Mar-Apr. Fr. Sep-Oct. Mountain slopes, forests of valleys or forest edges at 300-1300 m. Distributed in SW, S, SE and C China.

西川朴

Celtis vandervoetiana C. K. Schneid.

乔木。叶厚纸质，上部1/3-2/3具细牙齿，叶下面无毛或仅脉腋具簇毛，侧脉(2-)3或4对；叶柄无毛。果序单生，较粗壮，不分枝。每个果序具1个核果，黄色至橘黄色，椭圆体形至球形。果核具网孔状凹陷及4条纵肋。花期4月，果期9-10月。生海拔600-1400米的林中或山谷，常生于背阴区。

光叶山黄麻 *Trema cannabina*

珊瑚朴 *Celtis julianae*

西川朴 *Celtis vandervoetiana*

大叶朴 *Celtis koraiensis*

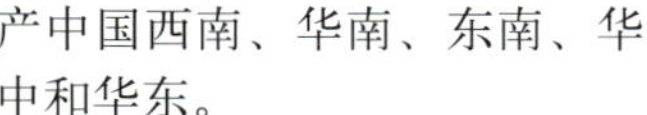
产中国西南、华南、东南、华中和华东。

Trees. Leaves thickly papery, margin finely toothed on apical 1/3-2/3, abaxially with tufts of hairs in vein axils or glabrous, secondary veins (2-)3 or 4 on each side of midvein; petioles glabrous. Infructescences solitary, unbranched, robust. Drupes 1 per infructescence, yellow to orange, ellipsoid to globose. Stones reticulately foveolate, 4-ribbed. Fl. Apr. Fr. Sep-Oct. Forests or valleys, usually in shaded places at 600-1400 m. Distributed in SW, S, SE, C and E China.

大叶朴
Celtis koraiensis Nakai

落叶乔木。冬芽内部鳞片具棕色柔毛。叶椭圆形至倒卵状椭圆形，先端具尾状长尖，长尖由平截状先端伸出，具粗锯齿。花单生或簇生。果序单生叶腋，果梗长1.5-2.5厘米；果近球形至球状椭圆形；核球状椭圆形，4条纵肋，表面具网孔状凹陷。花期4-5月，果期9-10月。生海拔100-1500米的山坡、沟谷林中。产辽宁、河北、山东、安徽北部、山西南部、河南西部、陕西南部和甘肃东部。

Deciduous tree. Winter buds brown, inner scales brown pubescent. Leaf blade elliptic, obovate-elliptic, apex truncate with a caudate tip, margin deeply laciniate-toothed. Flowers solitary or fascicled. Infructescence solitary, pedicel 1.5-2.5 cm long; drupe globose to ellipsoid; stone ovoid-elliptic, 4-ribbed, reticulately foveolate. Fl. Apr-May. Fr. Sep-Oct. Forests, valleys, slopes at 100-1500 m. Distributed in Liaoning, Hebei, Shandong, N Anhui, S Shanxi, W Henan, S Shaanxi and E Gansu.

四蕊朴
Celtis tetrandra Roxb.

落叶或半常绿乔木。叶厚纸质或纸质，卵状椭圆形、卵状披针形或多少呈菱形，基部明显偏斜，先端渐尖至短尾尖状渐尖。果较大，直径7-8毫米，成熟时黄色至橙黄色，果梗长0.4-1.5厘米。花期3-4月，果期9-10月。生海拔700-1500米的林中或灌丛中。产中国西南和华南。南亚和东南亚亦有。

Trees, deciduous or semievergreen. Leaves, thick papery or papery, ovate-elliptic, ovate-lanceolate, or ± rhombic, base distinctly oblique, apex acuminate to shortly caudate-acuminate. Drupes large, 7-8 mm diam, yellow to orange-yellow when mature, fruiting pedicels 0.4-1.5 cm long. Fl. Mar-Apr. Fr. Sep-Oct. Forests or thickets at 700-1500 m. Distributed in SW and S China. Also in S and SE Asia.

四蕊朴 *Celtis tetrandra*

朴树
Celtis sinensis Pers.

落叶乔木，可达20米高。叶厚纸质，卵形至卵状椭圆形，基部稍倾斜或否，先端短渐尖。花簇生在叶腋或茎基部。核果近球形，直径5-8毫米；核白色，多少呈球形。花期3-4月，果期9-10月。生海拔100-1500米的路边或山坡。产华南、华中和华东。日本亦有。

Deciduous trees, up to 20 m tall, deciduous. Leaves thick papery, ovate to ovate-elliptic, base slightly oblique or not, apex shortly acuminate. Flowers fascicled in leaf axils or stem bases. Drupes subglobose, 5-8 mm diam; stones white, ± globose. Fl. Mar-Apr. Fr. Sep-Oct. Roadsides or slopes at 100-1500 m. Distributed in S, C and E China. Also in Japan.

黑弹树 *Celtis bungeana*

朴树 *Celtis sinensis*

黑弹树
Celtis bungeana Blume

乔木。叶厚纸质，无毛，叶缘中部以上疏具不规则浅齿，有时全缘。果序单生，细弱，无毛；果梗长约为叶柄的1.7-4倍。每个果序具1或2个核果，近球形，成熟时蓝黑色。花期4-5月，果期10-11月。生海拔100-2300米的路边、山坡或灌丛中。产华中、华北、华西、华东和东北。朝鲜半岛亦有。

Trees. Leaves thick papery, glabrous, margin irregularly shallowly serrate on apical half, sometimes entire. Infructescence solitary, slender, glabrous; fruiting pedicel 1.7-4 × as long as subtending petiole. Drupes 1 or 2 per infructescence, subglobose, blue-black when mature. Fl. Apr-May. Fr. Oct-Nov. Roadsides, slopes or thickets at 100-2300 m. Distributed in C, N, W, E and NE China. Also in Korean Peninsula.

马尾树科 Rhoipteleaceae

马尾树
Rhoiptelea chiliantha Diels et Hand.-Mazz.

乔木。幼树、托叶、叶轴、叶柄和花序均密被黄白色腺体和柔毛。小叶常9或11。花序分枝15-30(-38)厘米。小坚果倒梨形，略扁，幼时绿色，成熟时紫红色，满布稀疏的灰褐色腺体。花期10月至翌年1月，果期翌年7-8月。生海拔700-2500米的山坡、河谷或溪边林中。产云南、贵州和广西。越南北部亦有。

Trees. Young branches, stipules, rachises, petioles, and inflorescences densely covered with yellowish white glands and pubescence. Leaflets usually 9 or 11. Inflorescence branches 15-30 (-38) cm long. Nutlets obpyriform, slightly flattened, green to purple-red, with sparse gray-brown glands. Fl. Oct to next Jan. Fr. next Jul-Aug. Slopes, valleys or forests along streams at 700-2500 m. Distributed in Yunnan, Guizhou and Guangxi. Also in N Vietnam.

马尾树 *Rhoiptelea chiliantha*

桑科 Moraceae

水蛇麻

Fatoua villosa (Thunb.) Nakai

草本。茎直立。叶卵形至宽卵形，膜质，具平伏毛，基部心形至截形，下延至叶柄，边缘具钝牙齿，先端渐尖，侧脉3或4对。聚伞花序两性；雄蕊反折。瘦果卵球形，红褐色，具3棱。花期5-8月。生灌丛、草地、铁路边或石堆中。产中国西南、华南、东南、华中和华北。东南亚、朝鲜半岛、日本和澳大利亚亦有。

Herbs. Stems erect. Leaves ovate to broadly ovate, membranous, appressed hirsute, base cordate to truncate and decurrent on petioles, margin crenate-toothed, apex acute, secondary veins 3 or 4 on each side of midvein. Inflorescences bisexual, cymose; stamens exserted. Achenes ovoid, reddish brown, 3-angled. Fl. May-Aug. Scrubs, grassy areas, trail sides or rocks. Distributed in SW, S, SE, C and N China. Also in SE Asia, Korean Peninsula, Japan and Australia.

桑 *Morus alba*

水蛇麻 *Fatoua villosa*

桑

Morus alba L.

灌木或乔木。叶卵形至宽卵形，不规则浅裂，长5-15厘米，边缘有粗锯齿，先端急尖或钝尖，中脉疏具柔毛或中脉腋处及初级侧脉具簇毛。雌花为葇荑花序。聚花果卵球状椭圆体形，熟时暗紫色。花期4-5月，果期5-8月。栽培于海拔200-2800米。原产华中和华北，现栽培于中国各地。世界各地亦有栽培。

Shrubs or trees. Leaves ovate to broadly ovate, irregularly lobed, 5-15 cm long, margin coarsely serrate, apex acuate or obtuse, sparsely pubescent along midvein or in tufts in axil of midvein and primary lateral veins. Pistillate flowers in catkins. Syncarps blackish purple when mature, ovoid to ellipsoid. Fl. Apr-May. Fr. May-Aug. Planted at 200-2800 m. Native to C and N China, now cultivated throughout China. Also cultivated worldwide.

华桑
Morus cathayana Hemsl.

小乔木或灌木。树皮灰白色，光滑。叶厚纸质，基部心形或截形，略偏斜，背面密被白色柔毛，边缘具疏浅锯齿或钝锯齿。柱头具短柔毛。聚花果圆柱状，白色、红色，长2-3厘米，成熟时深紫色。花期4-5月，果期5-6月。生海拔900-1300米的阳坡或山谷。产华南、华中、华北和华东。朝鲜半岛和日本亦有。

Small trees or shrubs. Bark grayish white, smooth. Leaves thick papery, base cordate to truncate and ± oblique, abaxially white pubescent, margin shallowly to coarsely serrate. Stigma shortly pubescent. Syncarps cylindric, 2-3 cm long, white, red, or dark purple when mature. Fl. Apr-May. Fr. May-Jun. Sunny slopes or valleys at 900-1300 m. Distributed in S, C, N and E China. Also in Korean Peninsula and Japan.

奶桑 *Morus macroura*

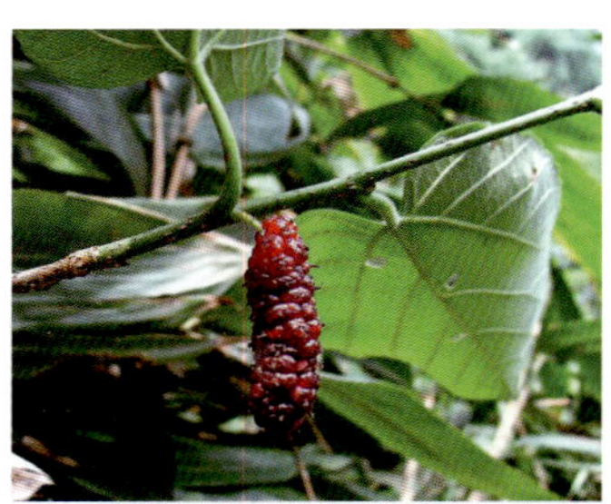

奶桑
Morus macroura Miq.

乔木，雌雄异株。叶卵形至阔卵形，膜质，先端渐尖至尾尖，边缘具细密锯齿，下面幼时灰绿色，沿中脉和侧脉具短柔毛。果序长6-12厘米，肉质。聚花果成熟时黄白色。花期3-4月，果期4-5月。生海拔300-1300(-2200)米的沟谷或平坝的季雨林中。产云南南部和西藏东部。南亚和东南亚亦有。

Trees, dioecious. Leaves ovate to broadly ovate, membranous, apex acuminate to caudate, margin serrulate, abaxially pale green and with short soft hairs along midvein and lateral veins when young. Infructescences 6-12 cm long, fleshy. Syncarps yellowish white when mature. Fl. Mar-Apr. Fr. Apr-May. Monsoon forests in valleys or flat lands at 300-1300(-2200) m. Distributed in S Yunnan and E Xizang. Also in S and SE Asia.

鸡桑
Morus australis Poir.

灌木或小乔木。冬芽大圆锥形至卵球形。叶卵圆形至斜卵圆形，边缘具锯齿，不分裂或3-5裂。雄花萼裂片绿色，卵形。聚花果成熟后红色至紫黑色，短圆柱状，直径约1厘米。花期3-4月，果期4-5月。生海拔500-2000米的山坡、林缘、灌丛或岩崖。产中国除西北以外大部分地区。印度、老挝、越南、柬埔寨、朝鲜半岛和日本亦有。

华桑 *Morus cathayana*

鸡桑 *Morus australis*

构树 *Broussonetia papyrifera*

Small trees or shrubs. Winter buds conic to ovoid, large. Leaves ovate to oblique-ovate, margin serrate, unlobed or 3-5-lobed. Calyx lobes of staminate flowers green, ovate. Syncarps red to dark purple when mature, shortly cylindric, ca. 1 cm diam. Fl. Mar-Apr. Fr. Apr-May. Slopes, forest edges, thickets or rocky cliffs at 500-2000 m. Distributed in most parts of China, except NW China. Also in India, Laos, Vietnam, Cambodia, Korean Peninsula and Japan.

构树

Broussonetia papyrifera (L.) L'Hér. ex Vent.

高大乔木，雌雄异株。叶卵圆形至椭圆状卵形，不裂或3-5裂，背面被细毛；叶柄长2.3-8厘米。雄花序粗壮。聚花果成熟后橙红色，直径1.5-3厘米，多具柔毛及散生粗壮髯毛，肉质。花期4-5月，果期6-7月。生海拔200-2800米山谷灌丛中、溪边或山坡路旁。产中国除西北和东北以外大部分地区。南亚、东南亚、东北亚和太平洋岛屿亦有。

Large trees, dioecious. Leaf blades ovate to elliptic-ovate, undivided or 3-5-lobed, abaxially pubescent; petioles 2.3-8 cm long Staminate inflorescences robust. Syncarps orange-red when mature, 1.5-3 cm diam, mostly pubescent with scattered stout and ± barbed hairs, fleshy. Fl. Apr-May. Fr. Jun-Jul. Thickets in valleys, along streams or roadsides of slopes at 200-2800 m. Distributed in most parts of China, except NW and NE China. Also in S, SE and NE Asia, and Pacific Islands.

楮 (小构树)

Broussonetia kazinoki Siebold et Zucc.

灌木，雌雄同株。单叶，互生，叶片卵形或斜卵形，不开裂或3裂，下面近无毛。雄花序球形，直径0.8-1厘米。聚花果直径0.8-1厘米。瘦果扁球形；外果皮壳质，表面具瘤体。花期4-5月，果期5-6月。生中海拔以下山坡林缘或沟边。产中国西南、华南、东南、华中和华东。朝鲜半岛和日本亦有。

Shrubs, monoecious. Leaves simple, alternate, ovate or oblique-ovate, undivided or 3-lobed, abaxially subglabrous. Staminate inflorescences globose, 0.8-1 cm diam. Syncarps 0.8-1 cm diam. Drupelets verrucate, compressed; exocarp crustaceous. Fl. Apr-May. Fr. May-Jun. Forest edges on slopes or along streams below middle altitudes. Distributed in SW, S, SE, C and E China. Also in Korean Peninsula and Japan.

楮 (小构树) *Broussonetia kazinoki*

藤构

Broussonetia kaempferi Siebold var. **australis** T. Suzuki

攀援灌木，雌雄异株。叶互生，螺旋状排列，椭圆形至卵状椭圆形，不分裂或偶2或3裂，粗糙无毛。雄花序穗状，长1.5-2.5厘米；花柱条形，延长。聚花果直径约1厘米。花期4-6月，果期5-7月。生海拔300-1000米的山谷灌丛中，或沟边和山坡路旁。产中国西南、华南、华中和东南。越南北部、朝鲜半岛和日本亦有。

Climbing shrubs, dioecious. Leaves alternate, spirally arranged, elliptic to ovate-elliptic, undivided or occasionally 2- or 3-lobed, scabrous and glabrous. Staminate inflorescences spicate, 1.5-2.5 cm long; styles linear, exserted. Syncarps ca. 1 cm diam. Fl. Apr-Jun. Fr. May-Jul. Thickets in valleys or along streams and roadsides of slopes at 300-1000 m. Distributed in SW, S, C and SE China. Also in N Vietnam, Korean Peninsula and Japan.

牛筋藤

Malaisia scandens (Lour.) Planch.

攀援灌木。叶纸质，下面稍粗糙，上面光滑，窄椭圆形至椭圆状倒卵形。雄花无梗；花被裂片三角形，被柔毛；雄蕊与裂片同数；花药近球形；退化雌蕊小。核果卵球形，红色。花期春夏季。生海拔100-300米的丘陵灌木丛中。产中国西南和华南。南亚、东南亚和澳大利亚亦有。

藤构 *Broussonetia kaempferi* var. *australis*

牛筋藤 *Malaisia scandens*

Climbing shrubs. Leaves papery, abaxially ± scabrous, adaxially smooth, narrowly elliptic to elliptic-obovate. Staminate flowers sessile; calyx lobes triangular, pubescent; stamens the same number as calyx lobes; anthers ± globose; pistillode small. Drupes ovoid, red. Fl. spring to summer. Among bushes of hills at 100-300 m. Distributed in SW and S China. Also in S and SE Asia, and Australia.

鹊肾树

Streblus asper Lour.

乔木或灌木，雌雄同株或异株。叶粗糙，革质，椭圆状倒卵形或椭圆形，边缘具不规则钝锯齿或全缘。雄花序头状，单生或腋生，具1-8花。宿存花被包核果。花期2-4月，果期5-6月。生海拔200-1000米的村旁或林中。产云南南部、广西、广东和海南。南亚和东南亚亦有。

Trees or shrubs, monoecious or dioecious. Leaves rugged, leathery, elliptic-obovate to elliptic, margin irregularly obtuse-serrate or entire. Staminate inflorescences solitary or paired, capitates, 1-8-flowered. Drupes enclosed by enlarged calyx lobes. Fl. Feb-Apr. Fr. May-Jun. By villages or forests at 200-1000 m. Distributed in S Yunnan, Guangxi, Guangdong and Hainan. Also in S and SE Asia.

鹊肾树 *Streblus asper*

假鹊肾树 *Streblus indicus*

米扬噎

Streblus tonkinensis (Dub et Eberh) Corner

常绿乔木。叶纸质，光滑，中部以上有锯齿。雌雄同株或同序。雄花序腋生；雄花花被片4-5；花丝长；花药外向；退化雌蕊四方形；雌花单生叶腋或生于雄花序中部，花被片4，边缘连合成鞘状；花柱分枝。核果长7-10毫米，基部一边非肉质，成熟时开裂。花期春夏。生海拔500米的石灰岩山地。产广西西南部、云南东南部和海南。

Evergreen trees, monoecious. Leaf papery, glabrous, with teeth apically from middle. Staminate inflorescences axillary; staminate flowers: calyx lobes 4-5; filaments long; anthers extrorse; pistillode cubic; pistillate inflorescences 1-flowered, axillary or at middle of staminate inflorescences; pistillate flower; calyx lobes 4, basally adnate with ovary; style apically branched. Drupes 7-10 mm diam, without a fleshy base, dehiscent. Fl. spring to summer. Limestone areas in shade at 500 m. Distributed in SW Guangxi, SE Yunnan and Hainan.

假鹊肾树

Streblus indicus (Bureau) Corner

乔木，雌雄同株或异株。树皮棕色，平滑。叶革质，2列，全缘，无毛，椭圆状披针形。雄花蝎尾形，腋生，单生或成对，雄花5数；雌花序具单花。核果不开裂。花期10-11月。生海拔650-1400米的林中或阴湿地区。产云南南部、广西、广东和海南。印度东北部和泰国北部亦有。

Trees, monoecious or dioecious. Bark brown, smooth. Leaves leathery, distichous, margin entire, glabrous, elliptic-lanceolate. Staminate inflorescences scorpioid, axillary, solitary or paired, staminate flowers 5-merous; pistillate inflorescences 1-flowered. Drupes indehiscent. Fl. Oct-Nov. Forests or moist shady places at 650-1400 m. Distributed in S Yunnan, Guangxi, Guangdong and Hainan. Also in NE India and N Thailand.

米扬噎 *Streblus tonkinensis*

波罗蜜

Artocarpus heterophyllus Lam.

常绿乔木。小枝无毛。托叶抱茎卵状，痕迹明显；成熟树上叶全缘。花生老茎或短枝上；雄花序长2-7厘米。聚花果圆柱状或近球形，长30-100厘米，直径25-50厘米，熟时黄褐色。花期2-3月，果期4-8月。广泛栽培于海拔120-1300米。产云南南部、广西、广东和海南。原生印度，热带地区广栽培。

Evergreen trees. Branchlets glabrous. Stipules amplexicaul, ovate, scars annular and conspicuous; leaf blades on mature trees entire. Inflorescences on old stems or brachyblasts; staminate inflorescences 2-7 cm long. Aggregate fruits terete or subglobose, 30-100 cm long, 25-50 cm diam, yellowish brown when mature. Fl. Feb-Mar. Fr. Apr-Aug. Commonly planted at 120-1300 m. Distributed in S Yunnan, Guangxi, Guangdong and Hainan. Native to India, cultivated throughout the tropics.

波罗蜜 *Artocarpus heterophyllus*

二色波罗蜜

Artocarpus styracifolius Pierre

乔木。树皮暗灰色，粗糙。叶革质或纸质，干后赭色，4-8 × 2.5-3厘米。花单生叶腋；雄花序总梗长6-12毫米。聚花果干后为黄色或红棕色，球形，被凸起的弯曲圆柱状乳突。花期夏秋季，果期秋冬季。生海拔200-1500米的林中。产云南东南部、广西、广东、海南和湖南西南部。越南和老挝亦有。

Trees. Bark dark gray, rough. Leaves leathery to papery, umber when dry, 4-8 × 2.5-3 cm. Inflorescences axillary, solitary; staminate inflorescence peduncle 6-12 mm long. Syncarps yellow or reddish brown when dry, globose, covered with prominent curved cylindric papillae. Fl. summer to autumnr. Fr. autumn to winter. Forests at 200-1500 m. Distributed in SE Yunnan, Guangxi, Guangdong, Hainan and SW Hunan. Also in Vietnam and Laos.

二色波罗蜜 *Artocarpus styracifolius*

白桂木

Artocarpus hypargyreus Hance

乔木。树皮暗紫色，片状剥落。叶互生，革质，侧脉6-7对。花序单生于叶腋；雄花具4萼片，匙形，密被柔毛。聚花果灰白色至金黄色，近球形，直径3-4厘米，具褐色柔毛及乳突。花果期春夏。生海拔100-1700米的常绿阔叶林下。产华南。

Trees. Bark dark purple, exfoliating. Leaves distichous, leathery, with 6-7 lateral veins on each side of midvein. Inflorescences axillary, solitary; staminate flowers with 4 calyx lobes, spatulate, densely pubescent. Fruiting syncarp pale to golden yellow, subglobose, 3-4 cm diam, brown pubescent, papillate. Fl. and fr. spring-summer. Evergreen broad-leaved forests at 100-1700 m. Distributed in S China.

野波罗蜜

Artocarpus lakoocha Roxb.

乔木。叶椭圆形、长椭圆形或卵形，25-30 × 15-20厘米，叶缘全缘或具小齿，顶端钝，背面密被锈色或灰色柔毛，侧脉9-18对。雄花序外面苞片顶端具盾状纤毛。聚花果干后红褐色，近球形，直径约7厘米，具弯刚毛。花期夏秋季，果期冬季。生海拔100-1300(-1800)米的石灰岩山地林中。产云南南部。南亚和东南亚亦有。

Trees. Leaves elliptic, narrowly elliptic or ovate, 25-30 × 15-20 cm, margin entire or with small teeth, apex obtuse, abaxially densely ferruginous or grey pubescent, lateral nerves 9-18 pairs. Bract apex of staminate inflorescence peltate-ciliate.

白桂木 *Artocarpus hypargyreus*

构棘 *Maclura cochinchinensis*

Syncarps reddish brown when dry, ± globose, ca. 7 cm diam, with bent bristles. Fl. summer to autumn. Fr. winter. Forests on limestone mountains at 100-1300 (-1800) m. Distributed in S Yunnan. Also in S and SE Asia.

胭脂

Artocarpus tonkinensis A. Chev. ex Gagnep.

乔木。小枝浅红褐色。托叶锥形，托落后有明显痕迹；叶全缘，有时顶端有浅锯齿，先端凸尖，每边有侧脉6-9条；叶柄长0.4-1厘米。花序单个腋生。聚花果熟时黄色，干后红棕色，近球形。花期夏秋季，果期冬季。生海拔800米以下的山坡阳处。产中国西南和华南。越南和柬埔寨亦有。

Trees. Branchlets pale reddish brown. Stipules pyramidal, often caducous and leaving a scar; leaf blades with margin entire or sometimes apically with a few shallow teeth, apex mucronate, with 6-9 lateral nerves on each side; petioles 0.4-1 cm long. Inflorescences axillary, solitary. Syncarps yellow when mature, reddish brown when dry, subglobose. Fl. summer to autumn. Fr. winter. Sunny slopes below 800 m. Distributed in SW and S China. Also in Vietnam and Cambodia.

构棘

Maclura cochinchinensis (Lour.) Corner

直立或攀援状灌木。茎无毛或近无毛。叶革质，椭圆状披针形或长圆形，宽2-2.5厘米，基部楔形，侧脉7-10对。雄花序头状，具梗。聚花果熟时橙红色，肉质，直径2-5厘米，具柔毛。花期4-5月，果期6-7月。生海拔600-1600米的荒地或林边。产中国西南、华南、东南和华中。南亚、东南亚、日本和澳大利亚亦有。

Shrubs, erect or scandent. Stems glabrous or nearly so. Leaves leathery, elliptic-lanceolate to oblong, 2-2.5 cm wide, base cuneate, secondary veins 7-10 on each side of midvein. Staminate inflorescences a capitulum, pedunculate. Syncarp reddish orange when mature, succulent, 2-5 cm diam, pubescent. Fl. Apr-May. Fr. Jun-Jul. Wastelands or forest edges at 600-1600 m. Distributed in SW, S, SE and C China. Also in S and SE Asia, Japan and Australia.

野波罗蜜 *Artocarpus lakoocha*

胭脂 *Artocarpus tonkinensis*

柘藤 *Maclura fruticosa*

柘藤

Maclura fruticosa (Roxb.) Corner

木质藤本，刺弯曲。叶宽椭圆形，膜质，基部2条侧脉伸至叶长1/3处，侧脉4-5对。花序成对腋生。聚花果稍球形，直径约2厘米；熟核果梨形，包于宿存的花萼中。花期4-5月，果期7-8月。生海拔1000-1700米的季雨林。产云南南部。印度东北部、孟加拉国、缅甸、泰国和越南亦有。

Woody vines, spines curved. Leaves broadly elliptic, membranous, basal lateral veins 2 and extending to 1/3 of leaf length, secondary veins 4-5 on each side of midvein. Inflorescences axillary, in pairs. Syncarp ± globose, ca. 2 cm diam; mature drupes pear-like, enclosed by persistent calyx lobes. Fl. Apr-May. Fr. Jul-Aug. Monsoon forests at 1000-1700 m. Distributed in S Yunnan. Also in NE India, Bangladesh, Myanmar, Thailand and Vietnam.

毛柘藤

Maclura pubescens (Trécul) Z. K. Zhou et M. G. Gilbert

木质灌木，雌雄异株。小枝圆柱形，具腋生刺。叶椭圆形，长4-12厘米，宽2.5-5.5厘米，全缘，背面密被黄褐色柔毛，中脉隆起；叶柄长1.5厘米，密被柔毛。雄花序成对腋生，密被柔毛；雄花花被4；雄蕊4；花丝短；退化雌蕊圆锥形。聚花果近球形肉质。小核果卵圆形。生海拔500-1100米的林缘。产广东北部、广西、贵州南部和云南南部。

Woody vines, dioecious. Branchlets cylindric, spines axillary. Leaves oval, 4-12 × 2.5-5.5 cm, margin entire, abaxially densely yellowish brown pubescent, midvein adaxially conspicuous; petiole ca. 1.5 cm, densely pubescent. Staminate inflorescences axillary, in pair, densely pubescent; Staminate flower: perianth 4-lobed; stamenr 4; filaments short; pistillode conic. Syncarp ± globose. Drupes ovoid. Forest margins at 500-1100 m. Distributed in N Guangdong, Guangxi, S Guizhou and S Yunnan.

柘

Maclura tricuspidata Carr.

落叶灌木或小乔木，雌雄异株。叶卵形至菱状卵形，基部圆形至楔形，全缘，先端渐尖，侧脉4-6对。花序腋生，单

毛柘藤 *Maclura pubescens*

柘 *Maclura tricuspidata*

生或成对；雄花序头状，直径约5毫米。聚花果成熟时橙红色，近球形，直径约2.5厘米。花期5-6月，果期6-7月。生海拔500-2200米的山地或阳坡林缘。产中国西南、华南、华中、华北、华西和华东。朝鲜半岛和日本亦有。

Deciduous Shrubs or small trees, dioecious. Leaves ovate to rhombic-ovate, base rounded to cuneate, margin entire, apex acuminate, secondary veins 4-6 on each side of midvein. Inflorescences axillary, single or in pairs; staminate inflorescences capitulate, ca. 5 mm diam. Syncarps orange red when mature, ± globose, ca. 2.5 cm diam. Fl. May-Jun. Fr. Jun-Jul. Mountains or sunny forest edges at 500-2200 m. Distributed in SW, S, C, N, W and E China. Also in Korean Peninsula and Japan.

见血封喉 *Antiaris toxicaria*

见血封喉

Antiaris toxicaria Lesch.

乔木，大树偶见有板根。托叶披针形，早落；叶椭圆形至倒卵形，密具厚长毛，边缘具锯齿，先端渐尖，侧脉10-13对。雄花雄蕊3-4；雌花单生藏于梨形花托内。核果梨形，鲜红至紫红色。花期3-4月，果期5-6月。生海拔1500米以下的热带雨林。产云南南部、广西、广东和海南。南亚和东南亚亦有。

Trees, occasionally with buttresses when large. Stipules lanceolate, caducous; leaf blades elliptic to obovate, densely covered with long thick hairs, margin serrate, apex acuminate, secondary veins 10-13 on each side of midvein. Staminate flowers with stamens 3-4; pistillate flowers solitary, included in pyriform receptacles. Drupes bright red to purple red, pear-shaped. Fl. Mar-Apr. Fr. May-Jun. Tropical rain forests below 1500 m. Distributed in S Yunnan, Guangxi, Guangdong and Hainan. Also in S and SE Asia.

笔管榕

Ficus subpisocarpa Gagnep.

落叶乔木。枝条有少许气生根。叶互生或簇生，椭圆形至长圆形，光滑；叶柄长3-7厘米，近无毛；托叶外面被微毛。榕果熟时紫黑色；雄蕊1；雄花、瘿花和雌花生于同一榕果内。花期4-6月。生海拔100-1400米的平坝或村旁。产中国西南、华南和东南。南亚、东南亚和日本亦有。

Deciduous trees. Branches with few aerial roots. Leaves alternate or fasciculate, elliptic to oblong, glabrous; petioles 3-7 cm long, subglabrous; stipules abaxially puberulous. Figs purplish black when mature; stamen 1; staminate, gall and pistillate flowers within the same fig. Fl. Apr-Jun. Flat lands or near villages at 100-1400 m. Distributed in SW, S and SE China. Also in S and SE Asia, and Japan.

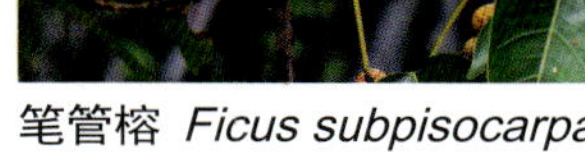
笔管榕 *Ficus subpisocarpa*

黄葛树
Ficus virens Aiton

落叶或半落叶乔木，幼时附生，有板根或支柱根。叶卵状披针形至椭圆状卵形，先端渐尖至短渐尖；叶柄长达5厘米。榕果腋生于多叶小枝，成对或单生或成簇生，成熟后紫红色；雄花、瘿花、雌花生于同一榕果内。瘦果有皱纹。花期4-8月。生海拔300-2700米的丛林中。产中国西南、华南、东南和华中。南亚、东南亚、日本和澳大利亚亦有。

Trees, deciduous or semideciduous, epiphytic when young, with buttress or prop roots. Leaves ovate-lanceolate to elliptic-ovate, apex acuminate to shortly acuminate; petioles up to 5 cm long. Figs axillary on leafy branchlets, paired or solitary or in clusters, purple red when mature; staminate, gall and pistillate flowers within the same fig. Achenes wrinkled on surface. Fl. Apr-Aug. Forests at 300-2700 m. Distributed in SW, S, SE and C China. Also in S and SE Asia, Japan and Australia.

龙州榕 *Ficus cardiophylla*

龙州榕
Ficus cardiophylla Merr.

乔木或灌木。叶卵形至心形，近革质，基部圆形至近心形；叶柄长2-4厘米。榕果单生或成对生于枝腋，球形，光滑，直径5-7毫米，顶孔不凹陷；雄花、瘿花和雌花均生于同一榕果内；雄花生近顶孔处；雄蕊1；瘿花与雌花多数。花期5-7月。产广西(龙州县)。越南北部亦有。

Trees or shrubs. Leaves ovate to cordate, subleathery, base rounded to ± cordate; petioles 2-4 cm long. Figs axillary on leafy branchlets, solitary or paired, globose, smooth, 5-7 mm diam, apical pore not concave; staminate, gall and pistillate flowers within the same fig; staminate flowers near apical pore; stamen 1; gall and pistillate flowers many. Fl. May-Jul. Distributed in Guangxi (Longzhou County). Also in N Vietnam.

菩提树
Ficus religiosa L.

乔木，幼时附生。叶革质，三角状卵形，先端骤尖，顶部延伸为尾状，基部宽截形至浅心形，全缘或为波状，基出叶脉三出；叶柄长于9厘米。榕果成对或单生于多叶小枝叶腋处，球形至扁球形，光滑；总苞片卵形。花期3-4月，果期5-6月。云南南部、广西和广东有栽培。原产印度北部、尼泊尔和巴基斯坦，热带地区广泛栽培。

Trees, epiphytic when young. Leaves triangular-ovate, leathery, apex acute to caudate, base broadly cuneate to cordate, margin entire or undulate, basal late-

黄葛树 *Ficus virens*

菩提树 *Ficus religiosa*

ral veins 3; petioles more than 9 cm long. Figs axillary on leafy branchlets, paired or solitary, globose to depressed globose, smooth; involucral bracts ovate. Fl. Mar-Apr. Fr. May-Jun. Cultivated in S Yunnan, Guangxi and Guangdong. Native to N India, Nepal and Pakistan, cultivated throughout the tropics.

大青树

Ficus hookeriana Corner

高大乔木，树干通直。叶长椭圆形或广卵状椭圆形，多少呈革质，侧脉6-9对，干后两面明显；托叶红色，披针形。总苞片合生成杯状。榕果成对生叶腋小枝，倒卵球形椭圆形至圆筒形。花期4-10月。生海拔500-2000米的石灰岩山地或栽于寺庙内。产广西、贵州和云南。印度东北部、尼泊尔和不丹亦有。

Tall trees, trunk straight. Leaves narrowly elliptic or broadly ovate-elliptic, ± leathery, secondary veins 6-9 on each side of midvein, conspicuous on both surfaces when dry; stipules red, lanceolate. Involucral bracts connate into a cup. Figs axillary on leafy branchlets, paired, obovoid-ellipsoid to cylindric. Fl. Apr-Oct. Limestone regions at 500-2000 m, also cultivated around temples. Distributed in Guangxi, Guizhou and Yunnan. Also in NE India, Nepal and Bhutan.

大青树 *Ficus hookeriana*

直脉榕

Ficus orthoneura Lévl. et Vaniot

乔木。小枝圆柱形，干后略具纵槽。叶革质，全缘，侧脉7-15对，平行直出。榕果成对或单个腋生，球形至卵球形，直径1-2厘米；花柱侧生，条形，宿存；柱头浅2裂。瘦果球形，光滑。花期4-9月。生海拔200-1700米的石灰岩山地。产云南、贵州和广西。缅甸、泰国西北部和越南北部亦有。

Trees. Branchlets terete, longitudinally fissured when dry. Leaves leathery, entire, lateral nerves 7-15 pairs, parallel and straight. Figs axillary, paired or solitary, globose to ovoid, 1-2 cm diam; styles lateral, linear, persistent; stigma slightly 2-lobed. Achenes globose, smooth. Fl. Apr-Sep. Limestone hills at 200-1700 m. Distributed in Yunnan, Guizhou and Guangxi. Also in Myanmar, NW Thailand and N Vietnam.

直脉榕 *Ficus orthoneura*

高山榕
Ficus altissima Blume

乔木。叶厚革质，广卵形至卵状椭圆形，两面无毛，侧脉5-7对；托叶厚革质，具银灰色毛。榕果成对腋生，熟时红色或带黄色，椭圆体卵球形；花柱近顶生。瘦果具瘤。花期3-4月，果期5-7月。生海拔100-2000米的山地或平原。产云南、广西、广东和海南。南亚和东南亚亦有。

Trees. Leaf blades thickly leathery, broadly ovate or ovate-elliptic, glabrous on both surfaces, secondary veins 5-7 on each side of midvein; stipules thickly leathery, with gray silky hairs. Figs axillary on leafy branchlets, paired, red or yellow when mature, ellipsoid-ovoid; styles subapical. Achenes tuberculate. Fl. Mar-Apr. Fr. May-Jul. Mountains or plains at 100-2000 m. Distributed in Yunnan, Guangxi, Guangdong and Hainan. Also in S and SE Asia.

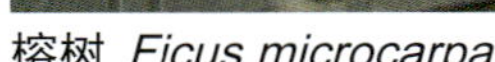
榕树 *Ficus microcarpa*

高山榕 *Ficus altissima*

榕树
Ficus microcarpa L. f.

乔木。老树常有锈褐色气根。叶薄革质，狭椭圆形至倒卵状椭圆形，全缘，上面深绿色且光亮，但是干后深褐色，基部脉长，不凸起。榕果扁球形，直径6-8毫米，熟时黄或微红色；基生苞片广卵形，宿存。花期5-7月。生海拔1900米以下的林中或村边。产中国西南、华南和东南。南亚、东南亚、东亚和大洋洲亦有。

Trees. Branches producing rust-colored aerial roots when old. Leaves thinly leathery, narrowly elliptic to obovate-elliptic, entire, adaxially dark green and shiny, but dark brown when dry, basal lateral veins long, not raised. Figs depressed globose, 6-8 mm diam, yellow to slightly

钝叶榕 *Ficus curtipes*

red when mature; involucral bracts broadly ovate, persistent. Fl. May-Jul. Forests or near villeges below 1900 m. Distributed in SW, S and SE China. Also in S, SE and E Asia, and Oceania.

钝叶榕
Ficus curtipes Corner

乔木。幼时多附生，茎下部多分枝。叶长12-18厘米，厚革质，全缘，先端圆形。榕果球形至扁球形，无毛，成熟时深红至紫红，内壁无刚毛；顶生苞片小而关闭；总苞片绿色，广卵形。花果期9-11月。生海拔500-1400米的石灰岩山地或村边。产云南和贵州。南亚和东南亚亦有。

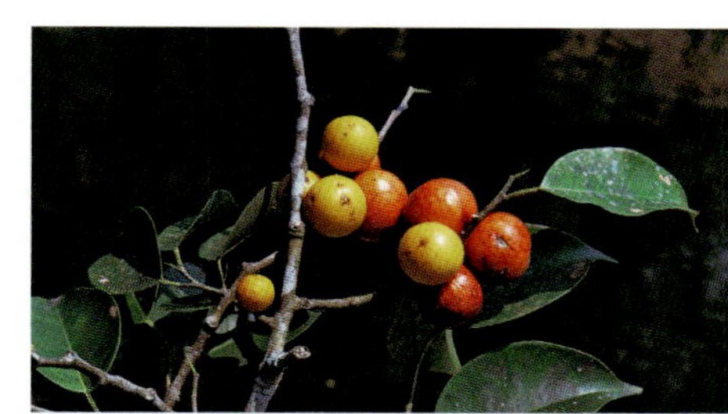

Trees. Stems often basally many branched, epiphytic when young. Leaves 12-18 cm long, thickly leathery, margin entire, apex rounded. Figs globose to depressed-globose, glabrous, dark red or purple-red when mature, inside without bristles, apical pore flat; bracts small, closed, sessile; involucral bracts green, broadly ovate. Fl. and fr. Sep-Nov. Limestone hills or near villages at 500-1400 m. Distributed in Yunnan and Guizhou. Also in S and SE Asia.

垂叶榕
Ficus benjamina L.

乔木。小枝下垂。叶卵形至广椭圆形，呈革质，无毛，基部圆形或楔形，全缘，先端短渐尖。总苞不明显；榕果直径0.8-1.5厘米，球形，表面无毛，成熟时红色至黄色。花果期8-11月。生海拔500-800米的湿润杂木林或村旁。产中国西南和华南。南亚、东南亚、澳大利亚和太平洋岛屿亦有。

Trees. Branches pendulous. Leaves ovate to broadly elliptic, coriaceous, glabrous, base rounded to cuneate, margin entire, apex shortly acuminate. Involucral bracts inconspicuous; figs 0.8-1.5 cm diam, globose, glabrous, red to yellow when mature. Fl. and fr. Aug-Nov. Moist mixed forests or near villages at 500-800 m. Distributed in SW and S China. Also in S and SE Asia, Australia and Pacific Islands.

白肉榕
Ficus vasculosa Wall. ex Miq.

乔木。树皮灰色，光滑。叶椭圆形至长披针形，长4-11厘米，革质，全缘或不规则分裂。榕果成熟后黄色至黄红色，球形，干后形成纵迹，基部缢缩为短柄。瘦果光滑，通常在顶一侧有龙骨。花果期5-7月。生海拔800米以下的季雨林中。产中国西南和华南。缅甸、泰国、越南和马来西亚亦有。

Trees. Bark gray, smooth. Leaves elliptic to oblong-lanceolate, 4-11 cm long, leathery, entire or irregularly lobed. Figs yellow to yellowish red when mature, globose, obscurely longitudinally ridged when dry, base attenuate into a short stalk. Achenes smooth, usually with one row of small apical tubercles. Fl. and fr. May-Jul. Monsoon forests below 800 m. Distributed in SW and S China. Also in Myanmar, Thailand, Vietnam and Malaysia.

垂叶榕 *Ficus benjamina*

白肉榕 *Ficus vasculosa*

硬皮榕

Ficus callosa Willd.

乔木，树干通直。托叶卵状披针形，被毛；叶广椭圆形或椭圆状倒卵形，长15-30厘米，革质，全缘。榕果单生或成对生叶腋，熟时黄色，椭圆体形，基部渐狭成长1厘米的柄。瘦果倒卵球形。花期9-10月。生海拔600-800米的季雨林或栽培。产广东和云南。南亚和东南亚亦有。

Trees, trunk straight. Stipules ovate-lanceolate, pubescent; leaf blades broadly elliptic or elliptic-obovate, 15-30 cm long, leathery, entire. Figs axillary on normal leafy stems, paired or solitary, yellow when mature, ellipsoid, base attenuate into a 1 cm long stalk. Achenes obovoid. Fl. Sep-Oct. Monsoon forests or planted at 600-800 m. Distributed in Guangdong and Yunnan. Also in S and SE Asia.

硬皮榕 *Ficus callosa*

聚果榕

Ficus racemosa L.

乔木，雌雄同株。小枝、幼叶和榕果被贴伏柔毛或近无毛。叶近革质，全缘，侧脉4-8对；叶柄长2-3厘米。榕果呈瘤状聚生于老茎短枝上，稀成对生于落叶枝叶腋，熟时橙红色。花期5-7月，果期秋季。生海拔100-1700米的溪边或河畔。产云南、贵州和广西。南亚、东南亚和澳大利亚亦有。

Trees, monoecious. Branchlets, young leaves and figs adpressed-pubescent or subglabrous. Leaf blades ± leathery, margin entire, secondary veins 4-8 on each side of midvein; petioles 2-3 cm long. Figs in a tumor-like aggregate on short branchlets of old stems, occasionally axillary on leafy shoots or on older leafless branchlets, paired, reddish orange when mature. Fl. May-Jul. Fr. Autumn. By streams or river banks at 100-1700 m. Distributed in Yunnan, Guizhou and Guangxi. Also in S and SE Asia, and Australia.

歪叶榕

Ficus cyrtophylla (Wall. ex Miq.) Miq.

灌木或小乔木。小枝、叶柄和榕果密被短硬毛。叶两边极不

聚果榕 *Ficus racemosa*

歪叶榕 *Ficus cyrtophylla*

尖叶榕 *Ficus henryi*

对称，纸质，排成两列，腹面极粗糙，具有乳突状钟乳体，背面密被褐色短硬毛。榕果成对或簇生于叶腋，成熟后橘黄色，卵球形。花期5-6月。生海拔500-1300米的山地疏林中。产云南、西藏、广西和贵州。印度、不丹、缅甸、泰国和越南北部亦有。

Shrubs or trees. Branchlets, petioles and figs covered with short stiff hairs. Leaves very unsymmetrical on both sides, papery, distichous, adaxially very rough, with papillate cystoliths, densely brown-hirsute abaxially. Figs axillary on normal leafy shoots, paired or clustered, yellowish orange when mature, ovoid. Fl. May-Jun. Montane open forests at 500-1300 m. Distributed in Yunnan, Xizang, Guangxi and Guizhou. Also in India, Bhutan, Myanmar, Thailand and N Vietnam.

无花果
Ficus carica L.

落叶灌木，多分枝。茎与叶具不明显的灰色短柔毛。叶互生，厚纸质，广卵圆形，通常3-5裂。榕果单生于叶腋，大而梨形，顶部下陷，熟时紫红色或黄色，具极不明显的毛。瘦果透镜状。花果期5-7月。现中国各地均有栽培，新疆栽培尤多。原产地中海地区，东至阿富汗。

Deciduous shrubs, many branched. Stems and leaves with inconspicuous short gray pubescence. Leaves alternate, thickly papery, broadly ovate, usually with 3-5 ovate lobes. Figs axillary on normal leafy shoots, solitary, purplish red to yellow when mature, with very inconspicuous hairs. Achenes lenslike. Fl. and fr. May-Jul. Throughout China, particularly cultivated in Xinjiang. Native to Mediterranean regions, eastward to Afghanistan.

尖叶榕
Ficus henryi Warb. ex Diels

乔木。叶倒卵状长圆形至狭披针形，先端渐尖或尾尖。榕果单生于叶腋，球形至椭圆体形，成熟时橘红色，直径1-2厘米，光滑。雄花生近顶孔处或散生；雄蕊3-5；瘿花在榕果中生雌花基部。瘦果具一排小结节。花期5-6月，果期7-9月。生海拔600-1300(-1600)米的山坡或溪边。产中国西南、华中和华西。越南北部亦有。

Trees. Leaves obovate-oblong to narrowly lanceolate, apex acuminate to caudate. Figs axillary on normal leafy shoots, solitary, globose to ellipsoid, reddish orange when mature, 1-2 cm diam, smooth. Staminate flowers near apical pore or scattered; stamens 3-5; gall flowers basal to pistillate flowers in figs. Achenes with 1 row of small tubercles. Fl. May-Jun. Fr. Jul-Sep. Mountain slopes or by streams at 600-1300 (-1600) m. Distributed in SW, C and W China. Also in N Vietnam.

无花果 *Ficus carica*

棒果榕

Ficus subincisa Buch.-Ham. ex Sm.

灌木。小枝有薄翅。叶纸质，先端骤尖为尾状，两面有钟乳体；叶柄长4-6毫米。榕果单生叶腋，基生苞片3；雌雄异株。雄花和瘿花生于一榕果内壁；雄花生近口部；花被片4；雄蕊1；瘿花子房光滑；花柱侧生；雌花，花柱侧生；柱头2裂。瘦果透镜状，光滑。花期5-7月，果期9-10月。生海拔400-2400米的山谷、沟边或疏林中。产广西西部和云南。

Shrubs. Branchlets narrowly winged. Leaf papery, apex long-caudate, with cystoliths on each side papery; petiole 4-6 mm. Figs axillary, basal bracts 3; staminate flowers and gall flowers born in the same fig wall; staminate flower near apical pore; calyx lobes 4; stamen 1; gall flowers ovary smooth; style lateral; pistillate flower: style lateral; stigma 2-branched. Achenes lenslike, smooth. Fl. May-Jul. Fr. Sep-Oct. Valleys, along streams or dense and sparse forests at 400-2400 m. Distributed in W Guangxi and Yunnan.

青藤公

Ficus langkokensis Drake

乔木。树皮红褐色或灰黄色。叶椭圆状披针形，互生，纸质，长7-15厘米，宽3-4厘米，基出侧脉会长至叶长的1/4-1/2。榕果成对或单生于叶腋，球形，被锈色糠屑状毛。花期5-9月。生海拔150-2000米的亚热带季雨林中。产中国西南、华南和华中。印度、老挝和越南亦有。

Trees. Bark reddish brown to grayish yellow. Leaves elliptic-lanceolate, alternate, papery, 7-15 cm long, 3-4 cm wide, with basal lateral veins extending to 1/4-1/2 of leaf length. Figs axillary on normal leafy shoots, paired or solitary, globose, with rust-colored scurfy hairs. Fl. May-Sep. Subtropical monsoon forests at 150-2000 m. Distributed in SW, S and C China. Also in India, Laos and Vietnam.

天仙果 *Ficus erecta*

天仙果

Ficus erecta Thunb.

落叶小乔木或灌木。小枝无毛或密具褐色绒毛。托叶早落，红棕色，阔卵形至卵状三角形，膜质；叶全缘或上部偶有疏齿；叶柄长1-4厘米。榕果单生叶腋，球形或梨形，熟时黄红至紫黑色或红色。花果期5-6月。生山坡林下或溪边。产中国西南、华南、东南和华中。越南、朝鲜半岛和日本亦有。

Deciduous small trees or shrubs. Branchlets glabrous or densely brown tomentose. Stipules caducous, reddish brown, broadly ovate or triangular-lanceolate, membranous; leaf margin entire or sparsely crenate at upper part; petioles 1-4 cm long. Figs solitary, axillary, globose or pyriform; reddish yellow to blackish purple or red when mature. Fl. and fr. May-Jun. Montane forests or by streams. Distributed in SW, S, SE and C China. Also in Vietnam, Korean Peninsula and Japan.

棒果榕 *Ficus subincisa*

青藤公 *Ficus langkokensis*

变叶榕 *Ficus variolosa*

菱叶冠毛榕 *Ficus gasparriniana* var. *laceratifolia*

变叶榕

Ficus variolosa Lindl. ex Benth.

灌木或小乔木。树皮灰棕色，光滑。叶薄革质，窄椭圆形至椭圆状披针形，叶缘全缘，顶端钝。榕果单生或成对生于叶腋，球形，直径1-1.2厘米，具瘤。瘦果表面有瘤。花期12月至翌年6月。生林中或潮湿处。产中国西南、华南和东南。老挝和越南亦有。

Shrubs or small trees. Bark grayish brown, smooth. Leaves thin leathery, narrowly elliptic to elliptic-lanceolate, margin entire, apex obtuse. Figs axillary on normal leafy shoots, paired or solitary, globose, 1-1.2 cm diam, tuberculate. Achenes tuberculate. Fl. Dec to next Jun. Forests or wet areas. Distributed in SW, S and SE China. Also in Laos and Vietnam.

楔叶榕

Ficus trivia Corner

灌木或乔木。叶卵状椭圆形、狭倒卵形或菱状倒卵形，纸质。榕果单生或成对或有时密集簇生于叶腋，成熟后红色至紫色，近球形，直径1-2厘米。雄花生近顶孔处或散生，雄蕊2；瘿花与雌花花被片4。瘦果卵球形，光滑。花期9月至翌年4月，果期5-8月。生溪边。产云南、贵州、广西和广东。越南北部亦有。

Shrubs or trees. Leaves ovate-elliptic, narrowly obovate or rhombic-obovate, papery. Figs axillary on normal leafy shoots, solitary or paired, or sometimes densely clustered, red to purple when mature, subglobose, 1-2 cm diam. Staminate flowers near apical pore or scattered, stamens 2; gall flowers and pistillate flowers with 4 calyx lobes. Achenes ovoid, smooth. Fl. Sep to next Apr. Fr. May-Aug. Along streams. Distributed in Yunnan, Guizhou, Guangxi and Guangdong. Also in N Vietnam.

菱叶冠毛榕

Ficus gasparriniana Miq. var. **laceratifolia** (Lévl. et Vant.) Corner

灌木或乔木。小枝具糙毛，渐无毛。叶倒卵形，厚纸质至亚革质，叶背白绿色，微被柔毛或近无毛，叶面稍粗糙具密糙伏毛；叶上半部具1-4个不规则齿裂；二级叶脉5-7对。总苞片阔卵形。瘦果直径2.5-3.5毫米。花期5-7月。生海拔600-1300米林中或路边。产云南、四川、贵州和广西。不丹和缅甸亦有。

Shrubs or trees. Branchlets coarsely hairy, glabrescent. Leaves obovate, thickly papery to leathery, abaxially greenish white and glabrous or sparsely pubescent, adaxially scabrid and minutely strigose; margin with 1-4 irregular teeth near apex; secondary veins 5-7 on each side of midvein. Involucral bracts broadly ovate. Achenes 2.5-3.5 mm diam. Fl. May-Jul. Forests or along trails at 600-1300 m. Distributed in Yunnan, Sichuan, Guizhou and Guangxi. Also in Bhutan and Myanmar.

楔叶榕 *Ficus trivia*

异叶榕 *Ficus heteromorpha*

异叶榕

Ficus heteromorpha Hemsl.

落叶灌木或小乔木。叶提琴形、椭圆形或椭圆状披针形，全缘或微波状，常纸质，上面粗糙。榕果成对生短枝叶腋，稀单生，成熟后紫黑色，球形或圆锥状球形，直径6-10毫米，光滑。花期4-5月，果期5-7月。生山谷、坡地或林中。产中国西南、华南、华中、华北和华东。缅甸亦有。

Deciduous shrubs or small trees. Leaves lyrate, elliptic, or elliptic-lanceolate, entire or slightly sinuate, usually papery, adaxially scabrous. Figs axillary on short branchlets, paired, occasionally solitary, purplish black when mature, globose to conic-globose, 6-10 mm diam, smooth. Fl. Apr-May. Fr. May-Jul. Valleys, slopes or forests. Distributed in SW, S, C, N and E China. Also in Myanmar.

壶托榕

Ficus ischnopoda Miq.

灌木状小乔木。叶椭圆状披针形至倒卵状披针形，纸质。榕果单生叶腋，稀成对腋生，或生于落叶小枝上，圆锥形或纺锤形，具纵脊，表面具槽纹。瘦果肾形，具瘤。花果期5-8月。生海拔100-2200米的河滩或灌丛中。产云南和贵州。南亚和东南亚亦有。

Shrublike trees. Leaves elliptic-lanceolate to obovate-lanceolate, papery. Figs axillary on leafy or older leafless branches, solitary or occasionally paired, conic to spindle-shaped, with longitudinal ridges, surface sulcate. Achenes reniform, tuberculate. Fl. and fr. May-Aug. River banks or thickets at 100-2200 m. Distributed in Yunnan and Guizhou. Also in S and SE Asia.

竹叶榕

Ficus stenophylla Hemsl.

灌木。小枝具散生灰白色刚毛。托叶红色；叶纸质，线状披针形、披针形或倒披针形，下面具小瘤。榕果单生，成熟时红色。雄花生近顶孔处；花被片3或4，红色；瘿花具短柄；雌花花被片4，条形。瘦果透镜状。花期5-7月。生溪边。产中国西南、华南、华中和华东。老挝、泰国和越南亦有。

Shrubs. Branchlets with scattered grayish white bristles. Stipules red; leaf blades papery, linear-lanceolate, lanceolate, or oblanceolate, abaxially with small tubercles. Figs solitary, dark red when mature. Staminate flowers near apical pore; calyx lobes 3 or 4, red; gall flowers pedicellate; pistillate flowers: calyx lobes 4, li-

壶托榕 *Ficus ischnopoda*

竹叶榕 *Ficus stenophylla*

石榕树 *Ficus abelii*

near. Achenes lenslike. Fl. May-Jul. Along streams. Distributed in SW, S, C and E China. Also in Laos, Thailand and Vietnam.

琴叶榕

Ficus pandurata Hance

灌木。叶倒卵形，有时为中部缢缩的提琴形、披针形或条状披针形。榕果成对生于叶腋，鲜红色，椭圆体形至球形，直径4-10毫米，顶部脐状凸起。雄花生近顶孔处，雄蕊(2-)3；雌花花柱长，侧生；柱头漏斗状。花期6-8月。生山地林中或灌丛。产中国西南、华南、东南和华中。泰国和越南亦有。

Shrubs. Leaves obovate, sometimes violin-shaped with middle constricted, lanceolate or linear-lanceolate. Figs axillary on normal leafy shoots, paired, fresh red, ellipsoid to globose, 4-10 mm diam, apical pore navel-like. Staminate flowers near apical pore, stamens (2-)3; styles of pistillate flowers lateral, long; stigmas funnelform. Fl. Jun-Aug. Montane forests or scrubs. Distributed in SW, S, SE and C China. Also in Thailand and Vietnam.

琴叶榕 *Ficus pandurata*

石榕树

Ficus abelii Miq.

藤状灌木。叶狭椭圆形至倒披针形，纸质。榕果单生于叶腋，成熟后黑色至棕红色，近梨形，长1.5-2厘米，顶部脐状凸起。雄花散生于榕果壁；瘿花同生一榕果内，花被合生；雌花无花被。瘦果肾形，外被一层黏膜。花期5-7月。产中国西南、华南和华中。南亚和东南亚亦有。

Scandent shrubs. Leaves narrowly elliptic to oblanceolate, papery. Figs axillary on normal leafy shoots, solitary, purplish black to brown red when mature, ± pear-shaped, 1.5-2 cm long, apical pore navel-like. Staminate flowers scattered; gall flowers together with staminate flowers, calyx lobes connate; pistillate flowers without calyx. Achenes reniform, covered by a sticky membrane. Fl. May-Jul. Distributed in SW, S and C China. Also in S and SE Asia.

地果

Ficus tikoua Bureau

匍匐木质藤本，节膨大。叶坚纸质，倒卵状椭圆形，宽1.5-4厘米，边缘具密齿。榕果簇生于无叶的短枝上，常生地下，无柄，成熟后深红色，球形或卵球形，直径1-2厘米，表面具圆瘤。花期5-6月，果期7月。生海拔800-1400米的荒地、草坡或岩石缝中。产中国西南、华南、华中和华西。印度东北部、老挝和越南亦有。

Prostrating woody vines, nodes enlarged. Leaves thickly papery, obovate-elliptic, 1.5-4 cm wide, margin finely toothed. Figs clustered on short stems without leaves, usually underground, sessile, dark red when mature, globose to ovoid, 1-2 cm diam, surface with rounded tubercles. Fl. May-Jun. Fr. Jul. Wastelands, grassy slopes or rock crevices at 800-1400 m. Distributed in SW, S, C and W China. Also in NE India, Laos and Vietnam.

黄毛榕 *Ficus esquiroliana*

地果 *Ficus tikoua*

黄毛榕

Ficus esquiroliana Lévl.

小乔木或灌木。树皮灰褐色至灰绿色，具纵棱。叶广卵形，幼叶常3-5裂，厚纸质，叶背面密被黄褐色长柔毛和硬毛。榕果单个腋生，卵球形，疏被浅褐长毛。瘦果斜卵球形，具瘤。花期5-6月，果期6-7月。生海拔500-1900(-2100)米的密林中。产中国西南和华南。缅甸、老挝、泰国、越南和印度尼西亚亦有。

Small trees or shrubs. Bark grayish brown to grayish green, with longitudinal ridges. Leaves broadly ovate, juvenile leaves often 3-5-lobed, thickly papery, densely fulvous-villose and hispid abaxially. Figs axillary on normal leafy shoots, solitary, ovoid, sparsely pale brown hirsute. Achenes obliquely ovoid, tuberculate. Fl. May-Jun. Fr. Jun-Jul. Dense forests at 500-1900(-2100) m. Distributed in SW and S China. Also in Myanmar, Laos, Thailand, Vietnam and Indonesia.

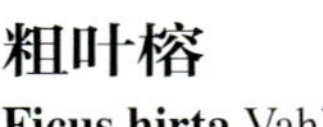

粗叶榕

Ficus hirta Vahl

灌木或小乔木。小枝中间无叶。叶纸质，形状变化极大，不裂或掌状分裂，无毛或具金

粗叶榕 *Ficus hirta*

色粗毛。榕果被金黄色至棕色硬毛；幼时顶部苞片形成脐状凸起，无梗；总苞片被贴伏柔毛，先端急尖。生村寨附近、山坡林边或附生于其他树干。产中国西南、华南和东南。南亚和东南亚亦有。

Shrubs or small trees. Branchlets leafless in middle. Leaves papery, variable in shape, undivided or palmately lobed, glabrous or golden yellow hirsute. Figs with long stiff spreading golden yellow or brown hairs; apical pore navel-like when young, sessile; involucral bracts ovate-lanceolate, with bent hairs, apex acute. Near villages, forest edges on slopes or on big trees, epiphytic. Distributed in SW, S and SE China. Also in S and SE Asia.

纸叶榕

Ficus chartacea Wall. ex King

灌木。叶狭卵形至倒卵形，被微柔毛，侧脉3-4对。榕果单生或成对生叶腋，球形，直径5-7毫米，具总梗。雌花：萼片3或4，卵状披针形；子房卵球形；花柱侧生，宿存。花果期5-7月。生海拔1400-1800(-2100)米的山坡或河边。产云南东南部。缅甸、泰国、越南、马来西亚和印度尼西亚亦有。

纸叶榕 *Ficus chartacea*

Shrubs. Leaves narrowly ovate to obovate, puberulent, with 3-4 lateral veins on each side of midvein. Figs axillary on normal leafy stem, paired or solitary, globose, 5-7 mm diam, pedunculate. Pistillate flower: calyx lobes 3 or 4, ovate-lanceolate; ovary ovoid; styles, lateral, persistent. Fl. and fr. May-Jul. Slopes or along streams at 1400-1800(-2100) m. Distributed in SE Yunnan. Also in Myanmar, Thailand, Vietnam, Malaysia and Indonesia.

大果榕

Ficus auriculata Lour.

乔木。叶互生，厚纸质，卵状心形，背面被短柔毛，边缘具整齐细锯齿，先端钝且凸尖。榕果簇生于树干基部或老茎短枝上，大而梨形或扁球形至陀螺形，具8-12个明显的纵棱。瘦果有黏液。花期8月至翌年3月，果期5-8月。生海拔100-1700(-2100)米的沟谷林中。产中国西南和华南。南亚亦有。

Trees. Leaves alternate, thick papery, ovate-cordate, abaxially pubescent, margin regularly serrulate, apex obtuse and mucronate. Figs on specialized leafless branchlets at base of trunk and main branches, pear-shaped, depressed globose, or turbinate, with 8-12 conspicuous longitudinal ridges. Achenes with adherent liquid. Fl. Aug to next Mar. Fr. May-Aug. Forests in valleys at 100-1700(-2100) m. Distributed in SW and S China. Also in S Asia.

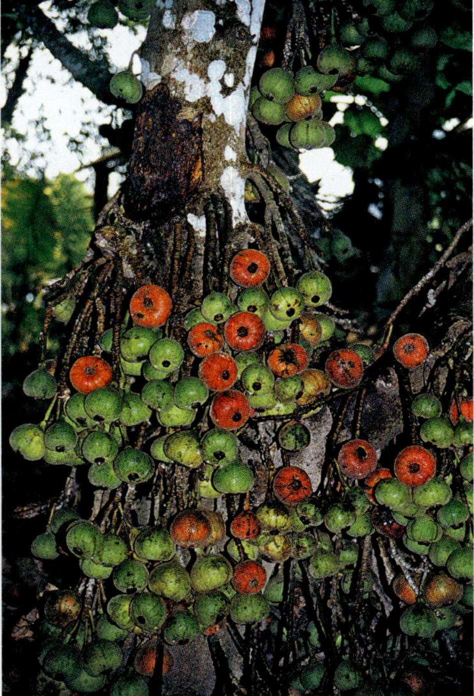

大果榕 *Ficus auriculata*

苹果榕

Ficus oligodon Miq.

小乔木。叶纸质，下面密生小瘤，边缘在1/3以上具不规则粗锯齿数对。榕果簇生于老茎发出的短枝上，成熟时深红色，梨形或近球形，具4-6个纵脊和小瘤，直径2-3.5厘米，具柔毛。花期9月至翌年4月，果期翌年5-6月。生海拔200-2100米的山谷或溪边湿润土壤区。产中国西南和华南。南亚和东南亚亦有。

Small trees. Leaves papery, abaxially densely small tuberculate, margin with several pairs of irregular coarse crenation above 1/3. Figs clustered on short branchlets of old stems, dark red when mature, pear-shaped to ± globose, with 4-6 longitudinal ridges and small tubercles, 2-3.5 cm diam, puberulent. Fl. Sep to next Apr. Fr. next May-Jun. Valleys or damp soils along streams at 200-2100 m. Distributed in SW and S China. Also in S and SE Asia.

杂色榕

Ficus variegata Blume

乔木，雌雄异株。叶厚纸质，幼时被柔毛。雄花具3或4萼片，雄蕊2；雌花花柱宿存，等长于瘦果；基生苞片早落，残存环状疤痕。榕果簇生于老茎生出的具瘤小枝，成熟后红色，有绿色条纹和斑点。花期冬季。生低至中海拔山谷。产中国西南和华南。南亚、东南亚、日本和大洋洲亦有。

Trees, dioecious. Leaves thickly papery, pubescent when young. Staminate flower with 3 or 4 calyx lobes, and 2 stamens; styles of pistillate flowers persistent, as long as achenes; involucral bracts caducous, scars ring-like. Figs clustered on shortly tuberculate branchlets from old stem, red, with green stripes and spots when mature. Fl. winter. Valleys at low to medium elevation. Distributed in SW and S China. Also in S and SE Asia, Japan and Oceania.

岩木瓜 *Ficus tsiangii*

岩木瓜

Ficus tsiangii Merr. ex Corner

灌木或小乔木，分枝稀疏。叶对称，螺旋状排列；叶柄细长。雄蕊2，稀为1，花丝基部有毛。榕果簇生于老茎基部或落叶瘤状短枝上，熟时红色；花序梗长2-4厘米。花果期5-8月。生海拔200-2400米的山谷或沟边。产云南、四川、广西、贵州、湖北和湖南。

Shrubs or trees, few-branched. Leaves symmetrical, spirally arranged; petioles thin. Stamens 2, rarely 1, filament base with hairs. Figs clustered at base of old stems or on leafless shortly tuberculate branchlets, red when mature; peduncles 2-4 cm long. Fl. and fr. May-Aug. Valleys or by streams at 200-2400 m. Distributed in Yunnan, Sichuan, Guangxi, Guizhou, Hubei and Hunan.

苹果榕 *Ficus oligodon*

杂色榕 *Ficus variegata*

鸡嗉子榕

Ficus semicordata Buch.-Ham. ex Sm.

乔木。树冠平展，伞状。托叶红色，长2-3.5厘米；叶边缘具细锯齿或全缘，基部偏心形，一侧耳状。雄花雄蕊1-2。榕果单生于老茎发出的无叶小枝上，果枝下垂至根部或穿入土中，熟时紫红色。花期5-10月。生海拔600-1900(-2800)米的路旁或林缘。产云南、西藏、贵州和广西。南亚和东南亚亦有。

Trees. Crown flat, spreading and umbrella-like. Stipules red, 2-3.5 cm long; leaf blades with margin serrulate or entire, base oblique-cordate, auriculate on one side. Staminate flower with 1-2 stamens. Figs on pendulous, eventually prostrate, leafless branchlets ± underground at maturity, solitary, reddish purple when mature. Fl. May-Oct. Roadsides or forest edges at 600-1900(-2800) m. Distributed in Yunnan, Xizang, Guizhou and Guangxi. Also in S and SE Asia.

鸡嗉子榕 *Ficus semicordata*

山榕

Ficus heterophylla L. f.

匍匐状灌木。幼叶常羽状裂，叶边缘有粗锯齿，两面被短硬毛。榕果单生叶腋或落叶枝上，熟时黄橙色，光滑，球形至梨形，直径1-2厘米，被粗毛和小瘤体。花果期7-11月。生海拔400-800米的潮湿山谷或溪边。产云南、广东和海南。南亚和东南亚亦有。

Shrubs, procumbent. Juvenile leaves usually pinnately lobed, leaf margin scattered-serrate, hirsute on both surfaces. Figs axillary on leafy or older leafless branches, solitary, yellowish orange and smooth when mature, globose to pear-shaped, 1-2 cm diam, tuberculate, coarsely hairy. Fl. and fr. Jul-Nov. Moist valleys or along streams at 400-800 m. Distributed in Yunnan, Guangdong and Hainan. Also in S and SE Asia.

山榕 *Ficus heterophylla*

染料榕

Ficus tinctoria G. Forst.

小乔木，幼时附生。叶椭圆形至卵状椭圆形，一侧稍宽，革质，背面略粗糙，无毛，基部宽楔形，全缘，先端钝至锐尖，侧脉5-8对。榕果单生或成对生于叶枝叶腋，球形或梨形，稍粗糙，直径约1厘米，具石细胞。花果期冬季至翌年6月。生山谷潮湿岩石上。产海南和台湾。印度尼西亚、菲律宾、澳大利亚和新几内亚岛亦有。

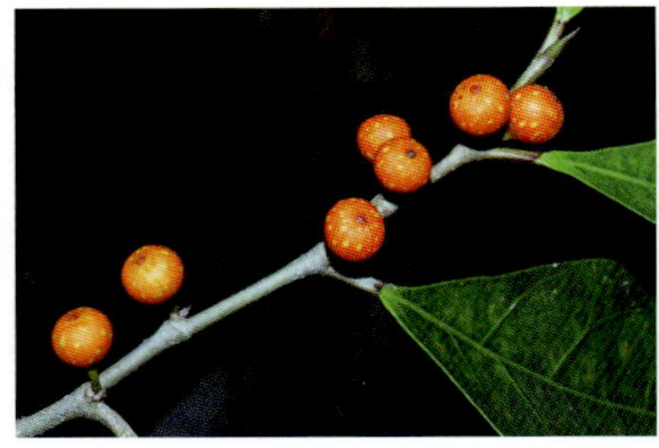

Small trees, epiphytic when young. Leaves elliptic to ovate-elliptic, one side wider, leathery, abaxially slightly coarse, glabrous, base broadly cuneate, margin entire, apex obtuse to acute, secondary veins 5-8 on each side of midvein. Figs axillary on normal leafy shoots, solitary or paired, globose to globose-pear-shaped, slightly rough, ca. 1 cm diam, with stone cells. Fl. and fr. winter to next Jun. Moist valleys or rocks. Distributed in Hainan and Taiwan. Also in Indonesia, the Philippines, Australia and New Guinea.

斜叶榕 *Ficus tinctoria* subsp. *gibbosa*

染料榕 *Ficus tinctoria*

斜叶榕

Ficus tinctoria G. Forst. subsp. **gibbosa** (Blume) Corner

本亚种与染料榕的区别在于本亚种为乔木、小乔木、灌木或附生植物。榕果直径达8毫米。花果期6-7月。潮湿山谷或岩石上。产中国西南和华南。南亚和东南亚亦有。

This subspecies differs from the typical subspecies in its arborescent or shrubby habits and obliquely cuneate bases of leaves. Figs to 8 mm diam. Fl. and fr. Jun-Jul. Moist valleys or rocks. Distributed in SW and S China. Also in S and SE Asia.

对叶榕

Ficus hispida L. f.

灌木或小乔木，被糙毛。叶对生，密被糙毛；托叶常4枚交互

对叶榕 *Ficus hispida*

对生于无叶的果枝上，卵状披针形。榕果单生或成对腋生于叶腋，有时生无叶小枝或主轴分枝上，成熟后黄色或红色，陀螺形，直径1.2-3厘米。花无花被。花期6-7月。生海拔120-1600米的沟谷中。产中国西南和华南。南亚、东南亚和澳大利亚亦有。

Shrubs or small trees, coarsely hairy. Leaves opposite, densely hispid; stipules usually 4 and decussate on leafless fruiting branchlets, ovate-lanceolate. Figs axillary on normal leafy shoots, sometimes on leafless branchlets or branchlets from main branches, solitary or paired, yellow or red when mature, turbinate, 1.2-3 cm diam. Flowers without perianth. Fl. Jun-Jul. Valleys at 120-1600 m. Distributed in SW and S China. Also in S and SE Asia, and Australia.

水同木

Ficus fistulosa Reinw. ex Blume

常绿小乔木，雌雄异株。叶纸质，下面疏具柔毛或黄色瘤，先端凸尖。雄花少，近顶端有孔，有短柄；萼片3或4；雌花花柱宿存，长，棍棒状。榕果簇生于老干发出的瘤状枝上，熟时橘红色，近球形，近无毛。花期5-7月。生海拔200-600米林下、河边或石堆。产中国西南和华南。南亚和东南亚亦有。

Evergreen small trees, dioecious. Leaves papery, abaxially sparsely pubescent or yellow tuberculate, apex mucronate. Staminate flowers few, near apical pore, shortly pedicellate; calyx lobes 3 or 4; styles of pistillate flowers persistent, long, clavate. Figs on short conic branchlets on main branches, reddish orange when mature, subglobose, subglabrous. Fl. May-Jul. Forests, along streams or on rocks at 200-600 m. Distributed in SW and S China. Also in S and SE Asia.

藤榕

Ficus hederacea Roxb.

藤状灌木。茎、枝节上生气生根。叶椭圆形至卵状椭圆形，两面均有乳突状钟乳体，先端钝至偶为圆形。榕果单生或成对腋生，熟时黄绿色至红色，球形。总苞片3，下半部合生。花果期5-7月。生海拔500-700(-1500)米的山地林中。产中国西南和华南。南亚亦有。

Scandent shrubs. Stems and branchlets with aerial roots at nodes. Leaves elliptic to ovate-elliptic, with papillose cystoliths on both surfaces, apex obtuse to occasionally rounded. Figs axillary on leafy or on leafless branchlets, solitary or paired, yellowish green to red when mature, globose. Involucral bracts 3, connate for basal half. Fl. and fr. May-Jul. Montane forests at 500-700 (-1500) m. Distributed in SW and S China. Also in S Asia.

水同木 *Ficus fistulosa*

藤榕 *Ficus hederacea*

光叶榕 *Ficus laevis*

光叶榕

Ficus laevis Blume

攀援藤状灌木。常附生，通常光滑无毛。叶螺旋状排列，薄纸质；叶柄长3.5-7厘米。榕果单生或成对生于叶腋，熟时紫色，球形。瘦果椭圆体形，有龙骨。花果期4-6月。生海拔800-1900米的雨林中或沟谷中。产贵州和广西。南亚和东南亚亦有。

Scandent shrubs, often epiphytic, usually glabrous. Leaf blades spirally arranged, thinly papery; petioles 3.5-7 cm long. Figs axillary on normal leafy branches, solitary or paired, purple when mature, globose. Achenes ellipsoid, keeled. Fl. and fr. Apr-Jun. Rain forests or valleys at 800-1900 m. Distributed in Guizhou and Guangxi. Also in S and SE Asia.

薜荔

Ficus pumila L.

攀援或匍匐灌木。叶两型；在结果枝上无叶柄，卵心形；结果枝上叶柄长5-10毫米；叶全缘。果单生叶腋；瘿花果梨形；雌花果近球形，直径3-5厘米；总梗短粗；顶孔截形或舟状。花果期5-8月。产中国西南、华南、华中、华北和华东。越南和日本亦有。

Climbing or trailing shrubs. Leaves of two types; leaves on sterile branchlets subsessile, ovate-cordate; leaves on fertile branchlets with 5-10 mm long petioles; leaves integer. Figs solitary, axillary; figs of gall flowers pyriform; figs of pistillate flowers almost globose, 3-5 cm diam; peduncles short, thick; apical pore truncate or navel-like. Fl. and fr. May-Aug. Distributed in SW, S, C, N and E China. Also in Vietnam and Japan.

爱玉子

Ficus pumila L. var. **awkeotsang** (Makino) Corner

本变种与的区别在于本变种的叶长圆状卵形。榕果圆柱形，顶孔渐尖。花果期5-8月。产台

薜荔 *Ficus pumila*

爱玉子 *Ficus pumila* var. *awkeotsang*

尾尖爬藤榕 *Ficus sarmentosa* var. *lacrymans*

湾、福建和浙江东南部。

This variety differs from the typical variety in its oblong-ovate leaves. Figs cylindric, apical pore acuminate. Fl. and fr. May-Aug. Distributed in Taiwan, Fujian and SE Zhejiang.

爬藤榕

Ficus sarmentosa Buch.-Ham. ex Sm. var. **impressa** (Champ. ex Benth.) Corner

藤状或匍匐灌木。叶背白色至灰白色，披针形，革质，基部圆形，脉明显；侧脉6-8对。榕果成对生于叶腋或无叶枝条上，成熟后紫黑色，球形，直径7-10毫米。花期5-7月。攀援于树上或岩石上。产中国西南、华南、华中、华北、华西和华东。

Procumbent or scandent shrubs. Leaves abaxially white to pale grayish brown, lanceolate, leathery, base rounded, veins conspicuous; secondary veins 6-8 on each side of midvein. Figs axillary on leafy or on leafless branchlets, paired, blackish purple when mature, globose, 7-10 mm diam. Fl. May-Jul. Trees or rocks. Distributed in SW, S, C, N, W and E China.

尾尖爬藤榕

Ficus sarmentosa Buch.-Ham. ex Sm var. **lacrymans** (Lévl.) Corner

藤状或匍匐灌木。叶披针状卵形，近革质，先端渐尖或尾尖，干后绿色、浅绿色至黄绿色，基部楔形，脉平坦；侧脉5-6对。榕果成对生于多叶或无叶的枝条上，球形，直径5-9毫米。花期5-7月。产中国西南、华南和华中。越南北部亦有。

Procumbent or scandent shrubs. Leaves lanceolate-ovate, subleathery, apex acuminate to caudate, green, pale green to yellow green when dry, base cuneate, veins flat; secondary veins 5-6 on each side of midvein. Figs axillary on leafy or on leafless branchlets, paired, globose, 5-9 mm diam. Fl. May-Jul. Distributed in SW, S and C China. Also in N Vietnam.

爬藤榕 *Ficus sarmentosa* var. *impressa*

珍珠莲

Ficus sarmentosa Buch.-Ham. ex Sm. var. **henryi** (King ex Oliv.) Corner

藤状或匍匐灌木。叶卵状椭圆形，革质。榕果成对生于叶腋，圆锥形，直径1-1.5厘米。顶部苞片直立，明显；总苞长3-6毫米；雄花生近顶孔处；雄蕊2；瘿花花柱短；雌花花柱近顶生。瘦果卵球状椭圆体形，具黏液。花期5-7月。生常绿阔叶林中或灌丛中。产中国西南、华南、东南、华中和华西。

Procumbent or scandent shrubs. Leaves ovate-elliptic, leathery. Figs axillary on normal leafy branches, paired, conic, 1-1.5 cm diam. Apical bracts erect, conspicuous; involucral bracts 3-6 mm long; staminate flowers near apical pore; stamens 2; gall flowers with short styles; pistillate flowers with subapical styles. Achenes ovoid-ellipsoid, with adherent liquid. Fl. May-Jul. Evergreen broad-leaved forests or scrubs. Distributed in SW, S, SE, C and W China.

广西榕

Ficus guangxiensis S S Chang

灌木状藤本。叶革质，两列；托叶宿存；叶先端钝尖，全缘，叶脉明显隆起；叶柄长4-6毫米，密被褐色短柔毛。榕果腋生，长6-7毫米，宽4-5毫米，密被褐色长柔毛。基生苞片3，卵状三角形；总梗长3毫米。雄花被片4，倒卵状披针形，红色；雄蕊2，花药具短尖，长约1.5毫米，花丝短。生海拔400-500米的石灰岩山地上。特产广西。

Scandent vines. Leaves leathery, two rows; stipules persistent; blade with apex obtuse, margin entire, veins abaxially raised and adaxially impressed; petiole 4-6 mm, densely covered with short brown pubescence. Figs axillary, 6-7 × 4-5 mm, densely covered with brown long pubescence. Basal bracts 3, ovate-triangular; peduncle 3 mm. Staminate flowers calyx lobes 4, obovate-lanceolate, red; stamens 2; anthers ca. 1.5 mm, mucronate, filaments short. Limestone areas at 400-500 m. Endemic to Guangxi.

广西榕 *Ficus guangxiensis*

珍珠莲 *Ficus sarmentosa* var. *henryi*

葎草

Humulus scandens (Lour.) Merr.

缠绕草本。茎、枝、叶柄均具倒钩刺。叶掌状(3-)5-9裂，下面具柔毛却不密集，边缘有锯齿。至少于花序中部每个苞片中具2花；苞片长7-10毫米，具小刺。瘦果成熟时露出苞片外。花期春夏，果期秋季。生沟边、荒地或林缘。产中国大部分地区。越南、朝鲜半岛、日本、欧洲和北美洲东部亦有。

Twining herbs. Stems, branches, petioles with barbs. Leaves palmately (3-)5-9-lobed, adaxially pubescent but not densely so, mar-

葎草 *Humulus scandens*

gin serrate. Flowers 2 per bract at least in middle of inflorescence; bracts 7-10 mm long, spinulose. Achenes exerted from bracts when mature. Fl. spring to summer. Fr. autumn. By valleys, wastelands or forest edges. Distributed in most parts of China. Also in Vietnam, Korean Peninsula, Japan, Europe and E North America.

啤酒花

Humulus lupulus L.

多年生草本。叶3-5(-7)裂，有时单生，下面无毛或具散生的柔毛，边缘具糙锯齿。果序球形，直径3-4厘米；苞片卵圆形，长1.5-2厘米，干燥，膜质，顶端尖。瘦果平，被苞片包住。花期秋季。生沟边、荒地或林缘。产四川北部、甘肃和新疆。北非、北亚、东北亚、欧洲和北美洲东部亦有。在中国广栽培，特别是山东(青岛)和新疆。

Perennial herbs. Leaves 3-5(-7)-lobed, sometimes simple, abaxially glabrous or with scattered soft pubescence, margin coarsely serrate. Infructescences globose, 3-4 cm diam; bracts ovoid, 1.5-2 cm long, dry, membranous, apex acute. Achenes flat, included in bracts. Fl. Autumn. By valleys, wastelands or forest edges. Distributed N Sichuan, Gansu and Xinjiang. Also in N Africa, N and NE Asia, Europe and E North America. Cultivated throughout China, especially in Shandong (Qingdao) and Xinjiang.

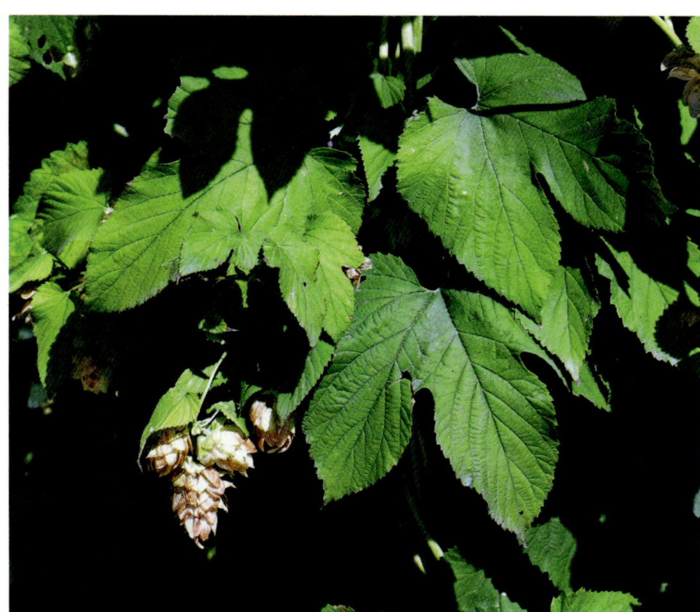

啤酒花 *Humulus lupulus*

大麻

Cannabis sativa L.

一年生草本。小枝密生灰白色贴伏毛。叶互生，掌状全裂，下面白绿色，具硬毛及散生的褐色脂粒，全裂片常披针形至条形。雌花序丛生于叶腋处；宿存苞片黄色。瘦果扁卵球形。花期5-6月，果期7月。中国新疆有野生。原产印度、不丹和中亚。中国广泛栽培。

Annual herbs. Branchlets densely white pubescent. Leaves alternate, palmatisect, abaxially whitish green, strigose, and with scattered brownish resinous dots, segments usually lanceolate to linear. Pistillate inflorescences crowded in apical leaf axils among leaflike bracts and bracteoles; persistent bracts yellow. Achenes flattened ovoid. Fl. May-Jun. Fr. Jul. Naturalized in Xinjiang. Native to India, Bhutan and C Asia. Cultivated throughout China.

大麻 *Cannabis sativa*

荨麻科 Urticaceae

三角叶荨麻
Urtica triangularis Hand.-Mazz.

多年生草本。根状茎木质。叶狭三角形或三角状披针形，叶缘具粗牙齿，有时下部为重锯齿，下面疏具刺毛和柔毛。雄花序圆锥状；雌花序穗状。瘦果淡褐色，卵球形，具疣状凸起。花期7-8月，果期8-9月。生海拔2500-4100米的灌丛、溪边、沟谷、近村庄处或路边。产云南、四川、西藏、甘肃和青海。

Perennial herbs. Rhizomes woody. Leaf narrowly triangular or triangular-lanceolate, sometimes ovate, margin coarsely dentate or sharply serrate, sometimes double-serrate in the lower part, abaxial surface with sparse stinging and pubescent hairs. Staminate inflorescences paniculate; pistillate inflorescences spicate. Achenes brownish, ovoid, verrucose. Fl. Jul-Aug. Fr. Aug-Sep. Thickets, along streams, valleys, near villages or roadsides at 2500-4100 m. Distributed in Yunnan, Sichuan, Xizang, Gansu and Qinghai.

麻叶荨麻
Urtica cannabina L.

多年生草本。根状茎木质。叶轮廓5角形，下面具柔毛和沿脉疏具刺毛，掌状3全裂或深裂；钟乳体点状。花序单性；雄花序圆锥状；雌花序穗状。瘦果灰褐色，卵球形，具疣。花期7-8月，果期8-10月。生海拔800-2800米的灌丛、草地、沙地、河岸或路边。产华北、华西、西北和东北。西南亚、中亚、东北亚和欧洲亦有。

Perennial herbs. Rhizomes woody. Leaves pentagonal in outline, abaxial surface puberulent and with sparse stinging hairs on veins, 3-palmatisect or paimatipartite; cystoliths punctiform. Inflorescences unisexual; staminate inflorescences paniculate; pistillate inflorescences spicate. Achenes grayish-brown, ovoid, verrucose. Fl. Jul-Aug. Fr. Aug-Oct. Thickets, grasslands, sandy places, river banks or roadsides at 800-2800 m. Distributed in N, W, NW and NE China. Also in SW, C and NE Asia, and Europe.

宽叶荨麻 *Urtica laetevirens*

宽叶荨麻
Urtica laetevirens Maxim.

多年生草本，雌雄同株。根状茎匍匐。叶卵形或披针形，常膜质，两面疏具刺毛和柔毛，边缘具牙齿或细锯齿；钟乳体常杆状，有时点状；托叶每节4。花序单性；雄花序生远轴处，穗状。瘦果灰褐色，卵球形或狭卵球形。花期6-8月，果期8-9月。生海拔100-3500米的

三角叶荨麻 *Urtica triangularis*

麻叶荨麻 *Urtica cannabina*

狭叶荨麻 *Urtica angustifolia*

山谷溪边、灌丛或山坡林下阴湿处。产中国西南、华北、华西、华东和东北。俄罗斯(远东地区)、朝鲜半岛和日本亦有。

Perennial herbs, monoecious. Rhizomes creeping. Leaves ovate or lanceolate, often membranous, both surfaces with sparse stinging and hirtellous hairs, margin dentate or serrate; cystoliths often bacilliform, sometimes punctiform; stipules 4 at each node. Inflorescences unisexual; staminate inflorescences in distal axils, spicate. Achenes gray-brown, ovoid or narrowly ovoid. Fl. Jun-Aug. Fr. Aug-Sep. By valley rivers, thickets or shady and damp sites of montane forests at 100-3500 m. Distributed in SW, N, W, E and NE China. Also in Russia (Far East), Korean Peninsula and Japan.

狭叶荨麻

Urtica angustifolia Fisch. ex Hornem.

多年生草本。有木质化根状茎。茎和叶的两面疏生刺毛和稀疏的细糙毛。叶披针形至披针状条形，常草质，边缘具9-15(-19)粗牙齿或锯齿；钟乳体常点状。花序圆锥状，有时具少量具短分枝的穗状花序。瘦果褐灰色，卵球形或宽卵球形。花期6-8月，果期8-10月。生海拔800-2200米的林下潮湿地、灌丛或河边。产华北和东北。东北亚亦有。

Perennial herbs. Rhizomes woody, stoloniferous. Stems and both surfaces of leaves sparsely hirtellous and armed with stinging hairs. Leaves lanceolate to lanceolate-linear, often herbaceous, margin coarsely 9-15(-19)-dentated or serrated; cystoliths often punctiform. Inflorescences paniculate, sometimes with few, short branchlike spikes. Achenes brownish gray, ovoid or broadly ovoid. Fl. Jun-Aug. Fr. Aug-Oct. Moist places in forests, thickets or stream banks at 800-2200 m. Distributed in N and NE China. Also in NE Asia.

荨麻

Urtica fissa E. Pritzel

多年生草本，雌雄同株。叶近膜质或草质，轮廓五角形或近圆形，边缘具5-7浅裂或掌状2裂(不规则2-4再裂)；钟乳体杆状或近点状。花序单性，雄花序位于近轴处，稀疏圆锥状或有时近穗状。瘦果宽卵球形或近球形，明显具疣。花期7-10月，果期9-11月。生海拔100-2000米的林中半阴湿处、灌丛、溪边或路边。产中国西南、东南、华中和华西。越南亦有。

Perennial herbs, monoecious. Leaves submembranous or herbaceous, pentagonal or suborbicular in outline, margin shallowly 5-7-lobed or palmately bilobed (irregularly 2-4-lobed again); cystoliths bacilliform or subpunctiform. Inflorescences unisexual, staminate inflorescences usually in proximal axils, paniculate with a few branches or sometimes subspicate. Achenes broadly ovoid or subglobose, conspicuously verrucose. Fl. Jul-Oct. Fr. Sep-Nov. Partly shady, moist places in forests, thickets, along streams or roadsides at 100-2000 m. Distributed in SW, SE, C and W China. Also in Vietnam.

荨麻 *Urtica fissa*

滇藏荨麻

Urtica mairei Lévl.

多年生草本，雌雄同株。叶对生，宽卵形，边有多数有规则小裂片和缺刻状重牙齿，被刺毛。花序圆锥状，开展，长过叶柄；雄花序生于下部叶腋；雌花序生于上部叶腋。瘦果浅褐色，长圆状球形或近球形。花期5-8月，果期7-12月。生海拔1500-3400米山地林下阴湿处、灌丛中、溪边或路旁。产云南、四川西南部和西藏东南部。印度北部、不丹和缅甸亦有。

Perennial herbs, monoecious. Leaves opposite, broadly ovate, margin regularly many-lobulate and incisidely double-denticulate with stinging hairs. Inflorescences paniculate with many long branches, longer than petioles; staminate flowers in proximal axils; pistillate flowers in distal axils. Achenes brownish, oblong-globose or subglobose. Fl. May-Aug. Fr. Jul-Dec. Partly shady, moist places in forests, thickets, along streams or roadsides at 1500-3400 m. Distributed in Yunnan, SW Sichuan and SE Xizang. Also in N India, Bhutan and Myanmar.

毛花点草 *Nanocnide lobata*

滇藏荨麻 *Urtica mairei*

毛花点草

Nanocnide lobata Wedd.

多年生草本。茎稍肉质，具反曲粗毛。叶呈3-5出脉，边缘具不等大的4-5(-7)钝齿或裂片状粗齿。雄花亮绿色，花被裂片(4-)5，卵形；雌花绿色，花被裂片4。瘦果卵球形，扁，长约1毫米，有疣点。花期4-6月，果期6-8月。生海拔1400米以下的林下阴湿地、草地、石缝或沿河岸。产中国西南、华南、东南和华中。越南亦有。

Perennial herbs. Stems somewhat succulent, retrorsely hirtellous. Leaves 3-5-veined, margin unequally 4-5(-7)-crenated or incised-dentated. Staminate flowers light green, perianth lobes (4-)5, ovate; pistillate flowers greenish, with 4 perianth lobes. Achenes ovoid, compressed, ca. 1 mm long, verrucose. Fl. Apr-Jun. Fr. Jun-Aug. Shady, moist places in forests, grasslands, rock crevices or along streams below 1400 m. Distributed in SW, S, SE and C China. Also in Vietnam.

珠芽艾麻

Laportea bulbifera (Sieb. et Zucc.) Wedd.

草本，雌雄同株。茎上部常“之”字形弯曲，叶腋具1-3个木质化珠芽。叶卵形至披针形，纸质；钟乳体细点状。雄花序生近轴处，圆锥状；雌花序顶生或近顶生，花沿一侧生长。瘦果稍具紫斑，宽倒卵球形或半球形。花期6-8月，果期8-12月。生海拔700-3500米的半遮阴、潮湿的林缘、灌丛或路边。产中国除西北以外大部分地区。南亚、东南亚和东北亚亦有。

珠芽艾麻 *Laportea bulbifera*

拉格艾麻 *Laportea tageensis*

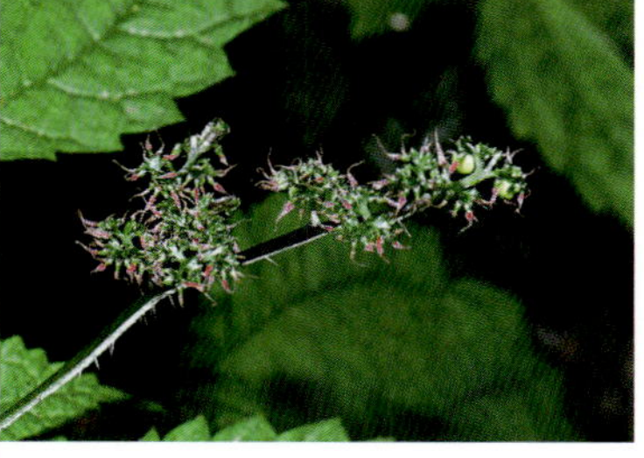

Herbs, monoecious. Upper part of stem often zigzag, axils often with 1-3 woody bulbils. Leaves ovate to lanceolate, papery; cystoliths minutely punctiform. Staminate inflorescences in proximal axils, paniculate; pistillate inflorescences terminal or in subterminal axils, often with flowers along one side. Achenes minutely purplish spotted, broadly obovoid or semiorbicular. Fl. Jun-Aug. Fr. Aug-Dec. Forest edges, thickets or roadsides at often partly shady and moist places at 700-3500 m. Distributed in most parts of China, except NW of China. Also in S, SE and NE Asia.

拉格艾麻

Laportea tageensis W. T. Wang

多年生草本。茎高约25厘米。叶互生，卵形，两面疏被刺毛，具三出脉，基部宽楔形。雌聚伞圆锥花序长约6厘米；花梗在果期无翅；雌花的4花被片极不等大，长0.25-0.45毫米。瘦果光滑。花期9-10月。生海拔3000米的山谷林下。特产西藏(墨脱县拉格)。

Perennial herbs. Stems ca. 25 cm tall. Leaves alternate, ovate, on both surfaces with sparse sting hairs, trinerved, base broadly cuneate. Pistillate thyrses ca. 6 cm long; pedicels at fruiting time not winged; pistillate flower with 4 tepals strongly unequal in size, 0.25-0.45 mm long. Achenes smooth. Fl. Sep-Oct. Under forests in valleys at 3000 m. Endemic to Xizang (Lage, Mêdog County).

艾麻

Laportea cuspidata (Wedd.) Friis

多年生草本。根纺锤状。茎常丛生，基部多少木质化。叶膜质或纸质，卵形、椭圆形或近圆形，基出脉3条，稀离基三出脉；钟乳体小点状，在上面明显。雄花序圆锥状；雌花序长穗状。瘦果卵球形。花期6-7月，果期8-9月。生海拔800-2700米的林缘、灌丛中或路边。产中国西南、华中、华东和华北。缅甸和日本亦有。

Perennial herbs. Roots fusiform. Stems often caespitose, slightly woody at base. Leaves membranous or papery, ovate, elliptic or suborbicular, 3-veined, rarely tripliveined; cystoliths minutely punctiform. Staminate inflorescences paniculate; pistillate inflorescences long spicate. Achenes ovoid. Fl. Jun-Jul. Fr. Aug-Sep. Forest edges, thickets or roadsides at 800-2700 m. Distributed in SW, C, E and N China. Also in Myanmar and Japan.

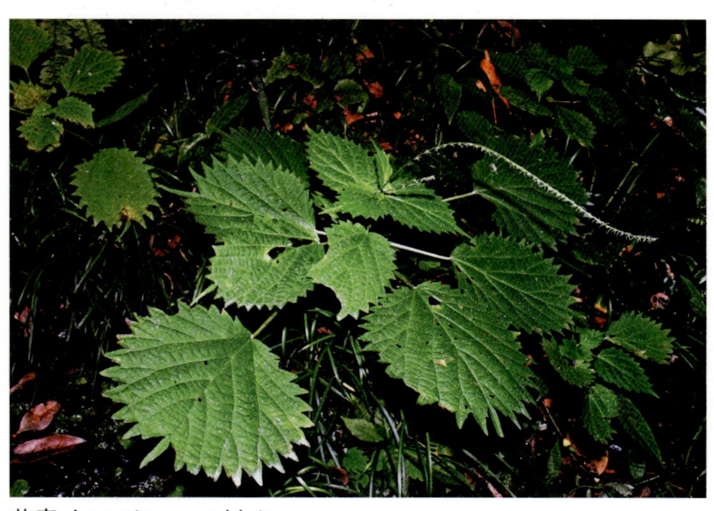

艾麻 *Laportea cuspidata*

火麻树

Dendrocnide urentissima (Gagnep.) Chew

乔木。托叶革质，三角状卵形，长约1厘米；叶片心形，有刺毛，叶背被短茸毛和极小的红色腺点，脉上疏生刺毛。花序顶生于小枝腋处，长圆锥状，短于叶；雌花花被片4，不等大，1个较大。瘦果深红色，近球形。花期9-10月或翌年1-2月，果期10-12月或翌年4-5月。生海拔800-1300米石灰岩山混交林或灌丛中。产云南南部和广西西南部。越南亦有。

Trees. Stipules leathery, triangular-ovate, ca.1 cm long; leaf blades cordate, with stinging hairs, abaxial surface puberulous, with red glandular dots, armed with stinging hairs on veins. Inflorescences in subterminal axils of branchlets, long paniculate, shorter than leaves; pistillate flower perianth lobes 4, unequal, 1 large. Achenes dark reddish, subglobose. Fl. Sep-Oct or next Jan-Feb. Fr. Oct-Dec or next Apr-May. Mixed forests or bushes on limestone hills at 800-1300 m. Distributed in S Yunnan and SW Guangxi. Also in Vietnam.

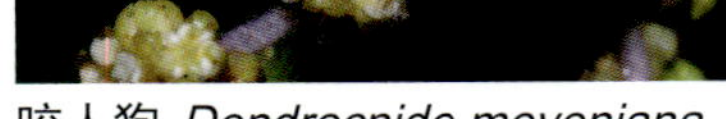

咬人狗 *Dendrocnide meyeniana*

咬人狗

Dendrocnide meyeniana (Walp.) Chew

常绿乔木，高5-7米，雌雄异株；小枝、叶柄、叶片和花序被刺毛。叶互生；叶片卵形或椭圆形，长达55厘米，宽达27厘米，具羽状脉；叶柄长5-18厘米。圆锥花序，分枝密被刺毛；雌花序苞片条形；雄花具4(-5)花被片和4(-5)雄蕊；雌花无梗，其4花被片下部合生成杯状。花期4-7月。生低山林中或灌丛中。产台湾。菲律宾亦有。

Evergreen trees, 5-7 m tall, dioecious; branchlets, petioles, leaf blades and panicles all with sting hairs. Leaves alternate; blades ovate or elliptic, up to 55 cm long and 27 cm broad, penninerved; petioles 5-18 cm long. Panicles, branches with dense sting hairs; bracts of pistillate panicle linear; staminate flower with 4(-5) tepals and 4(-5) stamens; pistillate flower sessile, its 4 tepals below connate into a short cup. Fl. Apr-Jul. Forests or bushes in hills at low altitude. Distributed in Taiwan. Also in the Philippines.

大蝎子草

Girardinia diversifolia (Link) Friis (=*Girardinia formosana* Hayata, *Girardinia chingiana* Chien)

多年生高大草本，雌雄同株或异株，多分枝。叶通常掌状(3-)5-7深裂，有毒刺毛；托叶先端2裂。花序单性，二叉状分枝；雄花序总状或近圆锥状；雌花序生茎末梢叶腋处。瘦果近心形至阔卵球形，明显具疣。花期9-10月，果期10-11月。生海拔1500-2800米的林下、溪边、灌丛和林缘湿处。产西藏、云南、贵州、四川和湖北。南亚、东南亚和非洲亦有。

Perennial tall herbs, dioecious or monoecious, many branched. Leaves usually palmately (3-)5-7-lobed, with poisonous stinging hairs; stipules 2-lobed at apex. Inflorescences unisexual, dichotomous; staminate inflorescences cymose-racemose or subpaniculate; pistillate ones in distal axils of stems. Achenes subcordate to broadly ovoid, conspicuously verrucose. Fl. Sep-Oct. Fr. Oct-Nov. Forests, along streams, thickets and wet places

火麻树 *Dendrocnide urentissima*

大蝎子草 *Girardinia diversifolia*

of forest edges at 1500-2800 m. Distributed in Xizang, Yunnan, Guizhou, Sichuan and Hubei. Also in S and SE Asia, and Africa.

蝎子草

Girardinia diversifolia subsp. **suborbiculata** (C. J. Chen) C. J. Chen et Friis (=*Girardinia suborbiculata* C. J. Chen)

多年生高大草本。叶近圆形，稀3裂，基部圆形或截形，叶缘8-13粗牙齿或重牙齿，有毒刺毛；托叶大，披针形。雄花序穗状；雌花序腋部具刚毛。瘦果基部不凸起。花期7-9月，果期9-11月。生海拔(100-)400-800米的山谷、溪边、山地林缘或林下。产华北和东北。朝鲜半岛亦有。

Perennial tall herbs. Leaves often suborbicular, rarely 3-lobed, base rounded or truncate, margin coarsely 8-13-dentate or doubly dentate, with poisonous stinging hairs; stipules big, lanceolate. Staminate inflorescences spicate; pistillate inflorescences setulose in axil. Achenes not projected at base. Fl. Jul-Sep. Fr. Sep-Nov. Valleys, by rivers, montane forest edges or under forests at (100-) 400-800 m. Distributed in N and NE China. Also in Korean Peninsula.

山冷水花

Pilea japonica (Maxim.) Hand.-Mazz.

草本。茎肉质。叶对生或在茎顶部近轮生；叶片菱状卵形或卵形，稍不对称，大小不等，边缘具圆齿状锯齿或圆齿。花序有时为杂性；雄花序常为头状；雄花有5花被片；雌花序具长梗。瘦果卵球形，有疣状突起。花期7-9月，果期8-11月。生海拔500-1900米的潮湿岩石上或溪边。产中国西南、华南、华中、华北、华西、华东和东北。俄罗斯、朝鲜半岛和日本亦有。

Herbs. Stems succulent. Leaves opposite or subwhorled at top of stem; blades rhombic-ovate or ovate, slightly asymmetric, unequal in size, margin crenate-serrate or dentate. Inflorescences sometimes of mixed sexes; staminate inflorescences often a capitulum; staminate flower with 5 tepals; pistillate inflorescences long pedunculate. Achenes ovoid, verrucose. Fl. Jul-Sep. Fr. Aug-Nov. Wet rocks or streamsides at 500-1900 m. Distributed in SW, S, C, N, W, E and NE China. Also in Russia, Korean Peninsula and Japan.

蝎子草 *Girardinia diversifolia* subsp. *suborbiculata*

山冷水花 *Pilea japonica*

五萼冷水花 *Pilea boniana*

五萼冷水花
Pilea boniana Gagnep.

多年生草本，雌雄异株或雌雄同株。根状茎匍匐。茎无毛，肉质，下部木质化。叶膜质或薄纸质，三出脉，边缘具细锯齿或上部有不明显波状圆齿。花序聚伞状，总状或圆锥状，开展，长6-16厘米；雄花具5花被片。瘦果菱状卵球形，具疣状凸起。花期7月至翌年3月，果期9月至翌年7月。生海拔300-2200米的石灰岩林下岩隙。产云南、贵州和广西。越南北部亦有。

Perennial herbs, dioecious or monoecious. Rhizome creeping. Stems glabrous, succulent, woody at base. Leaves membranous or thin papery, 3-veined, margin crenate-serrulate or indistinctly undulate-crenate distally. Inflorescences a cyme, racemelike or paniculate, spreading, 6-16 cm long; staminate flower with 5 tepals. Achenes rhomboid-ovoid, verrucose. Fl. Jul to next Mar. Fr. Sep to next Jul. Rock crevices of limestone forests at 300-2200 m. Distributed in Yunnan, Guizhou and Guangxi. Also in N Vietnam.

基心叶冷水花
Pilea basicordata W. T. Wang ex C. J. Chen

矮小灌木或亚灌木，无毛，雌雄同株。茎灰绿色，密布短杆状钟乳体。叶肉质，两面钟乳体明显，基部心形或深心形，边缘自中部以上浅波状或近全缘。花序单生，疏松的聚伞圆锥状，长8-13厘米；雄花具4花被片；雌花也具4花被片。瘦果长圆状卵球形，表面有皱纹。花期3-4月，果期4-5月。生海拔约900米的石灰岩山坡林阴处石上。产广西。

Short shrublets or subshrubs, glabrous, monoecious. Stems gray-green, densely covered with bacilliform cystoliths. Leaves succulent, cystoliths conspicuous on both surfaces, base cordate or deeply cordate, margin repand or subentire from middle. Inflorescences solitary, a lax cymose-panicle, 8-13 cm long; staminate flower with 4 tepals; pistillate flower also with 4 tepals. Achenes oblong-ovoid, corrugate at surface. Fl. Mar-Apr. Fr. Apr-May. Shady rocks of limestone montane forests at ca. 900 m. Distributed in Guangxi.

花叶冷水花
Pilea cadierei Gagnep. et Guill.

多年生草本或亚灌木，雌雄异株。茎、托叶、叶柄和叶片密被钟乳体。叶不等大，纸质，叶表面具2条间断的白斑。花序成对着生；雄花序头状；宿存花被片为瘦果长的一半。瘦果卵球形，长约1.5毫米，压扁。花期9-11月，果期11-12月。生海拔500-1500米的林中阴湿处。在云南和贵州栽培。原产越南。

Perennial herbs or subshrubs, dioecious. Stems, stipules, petioles and leaves densely covered with cystoliths. Leaves subequal in size, papery, adaxial

基心叶冷水花 *Pilea basicordata*

花叶冷水花 *Pilea cadierei*

surface with 2 interrupted white blotches. Inflorescences in pairs; staminate inflorescences a capitulum; persistent perianth lobes 1/2 as long as achenes. Achenes ovoid, ca. 1.5 mm long, compressed. Fl. Sep-Nov. Fr. Nov-Dec. Shaded wet places in forests at 500-1500 m. Cultivated in Yunnan and Guizhou. Native to Vietnam.

疣果冷水花

Pilea gracilis Hand.-Mazz.

多年生草本，近无毛。叶基部圆形或宽楔形；托叶三角形，长约1毫米。雄花序长2-5厘米，有少数分枝；雌花序长0.7-2厘米；雄花有4花被片；雌花有3花被片。瘦果有疣状凸起。花期4-5月，果期5-7月。生海拔400-1600米山谷阴湿处。产云南、贵州、四川、重庆南部、湖北西南部、湖南、广西、广东东北部和江西西部。

Perennial herbs, subglarous. Leaf base rounded or broadly cuneate; stipules triangular, ca. 1 mm long. Staminate cymes few-branched, 2-5 cm long; pistillate cymes 0.7-2 cm long; staminat flower with 4 tepals; pistilate flower with 3 tepals. Achenes verrucose. Fl. Apr-May. Fr. May-Jul. Shady places in valleys at 400-1600 m. Distributed in Yunnan, Guizhou, Sichuan, S Chongqing, SW Hubei, Hunan, Guangxi, NE Guangdong and W Jiangxi.

湿生冷水花

Pilea aquarum Dunn

多年生草本，雌雄异株或有时雌雄同株。茎常淡红色。叶宽椭圆形或卵状椭圆形，叶缘圆齿状，先端锐尖，膜质。花序单生；雄花序为聚伞状圆锥花序；雌花序为二歧聚伞花序，紧缩成簇状。瘦果棕绿色，斜卵球形，稍压扁，具疣状凸起。花期3-5月，果期4-6月。生海拔350-1500米的沟谷湿地或溪边。产福建、江西、广东北部、湖南和重庆。

Perennial herbs, dioecious or sometimes monoecious. Stems often reddish. Leaves broadly elliptic or ovate-elliptic, margin crenate, apex acute, membranous. Inflorescences solitary; staminate inflorescence a cymose panicle; pistillate cymes dichotomously branched, compacted into clusters. Achenes greenish brown, obliquely ovoid, slightly compressed, verrucose. Fl. Mar-May. Fr. Apr-Jun. Wet places along ditches or streams at 350-1500 m. Distributed in Fujian, Jiangxi, N Guangdong, Hunan and Chongqing.

湿生冷水花 *Pilea aquarum*

翅茎冷水花

Pilea subcoriacea (Hand.-Mazz.) C. J. Chen

多年生草本，雌雄同株或雌雄异株。茎常带紫色，单生，具数个纵向波状翅，肉质。叶近等大。花序单生；雄花序为聚伞状圆锥花序；雄花具4花被片；雌花序为多回二歧聚伞状；雌花具3花被片。瘦果宽卵形，稍压扁，具疣。花期4月，果期5-6月。生海拔800-1800米的沟边阴湿处、溪边或溪流附近。产云南北部、四川、贵州、广西和湖南。

Perennial herbs, dioecious or monoecious. Stems often purplish, simple, with several longitudinal wavy wings, succulent. Leaves subequal in size. Inflorescences solitary; staminate inflorescence a cymose panicle; staminate flower with 4 tepals; pistillate cymes dichotomously branched many times; pistillate flower with 3 tepals. Achenes broadly ovoid, slightly compressed, verrucose. Fl. Apr. Fr. May-Jun. Shaded wet places along ditches, streams or near streams at 800-1800 m. Distributed in N Yunnan, Sichuan, Guizhou, Guangxi and Hunan.

疣果冷水花 *Pilea gracilis*

翅茎冷水花 *Pilea subcoriacea*

大叶冷水花

Pilea martinii (Lévl.) Hand.-Mazz.

多年生草本，常雌雄异株。叶膜质，三出脉，侧脉多数，边缘具粗锯齿状牙齿，狭卵形或披针形，大小不等，长7-20厘米。花序单生；雌花具3花被片。瘦果带褐色，窄卵球形、扁、斜、平滑。花期5-9月，果期8-10月。生海拔1100-3500米的沟边阴湿处或河边。产中国西南、华中和华西。印度北部、尼泊尔和不丹亦有。

Perennial herbs, often dioecious. Leaves membranous, 3-veined, lateral veins many, margin coarsely serrate-dentate, narrowly ovate or lanceolate, unequal in size, 7-20 cm long. Inflorescences solitary; pistillate flower with 3 tepals. Achenes brownish, narrowly ovoid, compressed, oblique, smooth. Fl. May-Sep. Fr. Aug-Oct. Shaded wet places along ditches or streams at 1100-3500 m. Distributed in SW, C and W China. Also in N India, Nepal and Bhutan.

镰叶冷水花

Pilea semisessilis Hand.-Mazz.

多年生草本。茎无毛。叶卵状披针形，同对的不等大；托叶卵状三角形，长2-5毫米。雄聚伞圆锥花序与叶近等长；雌聚伞圆锥花序比叶短；雄花有4花被片；雌花有3花被片。瘦果光滑。花期7-9月，果期9-10月。生海拔1000-2800(-3400)米山谷林下或山坡路边草丛中。产江西西部、湖南、广西北部、云南北部、四川和西藏东南部。

Perennial herbs. Stems glabrous. Leaves ovate-lanceolate, those of the same pair unequal in size; stipules ovate-triangular, 2-5 mm long. Staminate thyrses nearly as long as leaves; pistillate thyrses shorter than leaves; staminate flower with 4 tepals; pistillate flower with 3 tepals. Achenes smooth. Fl. Jul-Sep. Fr. Sep-Oct. Forests in Valleys, among grasses by roads on slopes at 1000-2800(-3400) m. Distributed in W Jiangxi, Hunan, N Guangxi, N Yunnan, Sichuan and SE Xizang.

大叶冷水花 *Pilea martinii*

镰叶冷水花 *Pilea semisessilis*

石筋草

Pilea plataniflora C. H. Wright

多年生草本。茎肉质，不分枝或分枝。叶卵形、披针形或倒卵状长圆形，不对称，钟乳体纺锤形，叶缘全缘或有时波状。花序单生；雄花序聚伞圆锥花序或头状花序；雌聚伞花序有花梗。瘦果褐色，卵球形。花期(4-)6-9月，果期7-10月。生海拔200-2400米的林中、湿地、石灰岩地区或溪边。产中国西南、华南和华西。泰国和越南亦有。

Perennial herbs. Stems succulent, simple or branched. Leaves ovate, lanceolate, or obovate-oblong, asymmetric, cystoliths fusiform, margin entire or sometimes undulate. Inflorescences solitary; staminate inflorescences a cymose panicle or raceme-like; pistillate cymes pedunculate. Achenes brownish, ovoid. Fl. (Apr-)Jun-Sep. Fr. Jul-Oct. Forests, wet places, limestone areas or by streams at 200-2400 m. Distributed in SW, S and W China. Also in Thailand and Vietnam.

圆瓣冷水花

Pilea angulata (Blume) Blume

草本，无毛。叶椭圆形，基部圆形；托叶长圆形，长1-2.5厘米。雄聚伞圆锥花序长1-2厘米；雄花具4花被片；花被片背

石筋草 *Pilea plataniflora*

圆瓣冷水花 *Pilea angulata*

长柄冷水花 *Pilea angulata*

面顶端之下有1长角状凸起。瘦果具小刺；宿存花被3浅裂。花期6-9月，果期9-11月。生海拔800-2300米的山坡阴湿处。产广东、广西、贵州、云南、西藏东南部、四川和陕西南部。越南、印度尼西亚、印度和斯里兰卡亦有。

Herbs, glabrous. Leaves elliptic, base rounded; stipules oblong, 1-2.5 cm long. Staminate thyrses 1-2 cm long; staminate flower with 4 tepals; tepals abaxily below apex long corniculate. Achenes spinulose; persistent perianth 3-lobed. Fl. Jun-Sep. Fr. Sep-Nov. Shady places on slopes at 800-2300 m. Distributed in Guangdong, Guangxi, Guizhou, Yunnan, SE Xizang, Sichuan and S Shaanxi. Also in Vietnam, Indonesia, India and Sri Lanka.

长柄冷水花

Pilea angulata (Blume) Blume subsp. **petiolaris** (Sieb. et Zucc.) C. J. Chen

本亚种与模式亚种的区别在于本亚种雌雄同株，雄花序念珠状，具稀疏花簇，雄花片长圆形，具短角状凸起。花期7-10月。生海拔750-2700米的林下。产云南、四川、贵州、湖南、广西、广东、江西、浙江、福建和台湾。日本亦有。

This subspecies differs from subsp. *angulata* in its monoecious plants, moniliform staminate inflorescences with sparse flower fascicles, and oblong shortly corniculate staminate tepals. Fl. Jul-Oct. Forests at 750-2700 m. Distributed in Yunnan, Sichuan, Guizhou, Hunan, Guangxi, Guangdong, Jiangxi, Zhejiang, Fujian and Taiwan. Also in Japan.

冷水花

Pilea notata C. H. Wright

多年生草本，具根状茎，雌雄异株或同株。叶卵形或卵状披针形，近等大，膜质，三出脉，侧脉8-13对。花序单生；雄花黄绿色，4花被片1/2合生；雌花具3花被片。瘦果长圆状卵球形，压扁，斜，有刺状疣点。花期6-9月，果期9-11月。生海拔300-1500米的阔叶林下阴湿地。产中国西南、华南、东南、华中和华东。日本亦有。

Perennial herbs, stoloniferous, dioecious or monoecious. Leaves ovate or ovate-lanceolate, subequal in size, membranous, 3-veined, lateral veins 8-13 each side. Inflorescences solitary; staminate flower yellow-green, 4 tepals connate 1/2 of length; pistillate flower with 3 tepals. Achenes oblong-ovoid, compressed, oblique, spinulose-verrucose. Fl. Jun-Sep. Fr. Sep-Nov. Shaded moist places in broad-leaved forests at 300-1500 m. Distributed in SW, S, SE, C and E China. Also in Japan.

冷水花 *Pilea notata*

粗齿冷水花 *Pilea sinofasciata*

镜面草 *Pilea peperomioides*

粗齿冷水花
Pilea sinofasciata C. J. Chen

多年生草本，雌雄异株或同株。茎肉质。叶近等大，草质，上面沿中脉具2条白色条纹，钟乳体蠕虫状，叶缘具粗齿。圆锥状聚伞花序；雌花近无柄，退化雄蕊3，小，鳞片状。瘦果宽卵球形，具疣。花期6-7月，果期8-10月。生海拔700-2500米的林中阴湿处。产中国西南、华南、华中、华西和华东。

Perennial herbs, dioecious or monoecious. Stems succulent. Leaves with 2 whitish striae along midrib adaxially, subequal in size, herbaceous, cystoliths worm-shaped, margin coarsely dentate. Inflorescence a paniculate cyme; pistillate flowers subsessile, staminodes 3, small, scalelike. Achenes broadly ovoid, verrucose. Fl. Jun-Jul. Fr. Aug-Oct. Shaded moist places in forests at 700-2500 m. Distributed in SW, S, C, W and E China.

念珠冷水花
Pilea monilifera Hand.-Mazz.

多年生草本，雌雄同株或雌雄异株。茎节间膨大。叶不等大，近膜质，基稍倾斜，钟乳体纺锤形。花序单生，雄花序为3-8疏生于单一轴上的团伞花序构成的念珠状复穗状花序；雄花具4花被片；雌花序长1-3.5厘米；雌花的3花被片不等大。瘦果棕色，宽卵球形，光滑。花期6-8月，果期7-9月。生海拔(900-)1400-2400(-3500)米的阴湿处或岩石上。产中国西南和华中。

Perennial herbs, monoecious or dioecious. Stem with swollen nodes. Leaves often unequal in size, submembranous, slightly oblique at base, cystoliths fusiform. Inflorescences solitary, staminate one a moniliform spike of 3-8 loose glomerules on simple axis; staminate flower with 4 tepals; pistillate inflorescences 1-3.5 cm long; pistillate flower with 3 tepals unequal in size. Achenes brownish, broadly ovoid, smooth. Fl. Jun-Aug. Fr. Jul-Sep. Shaded moist places or rocks at (900-)1400-2400 (-3500) m. Distributed in SW and C China.

镜面草
Pilea peperomioides Diels

多年生肉质草本，具根状茎，无毛。叶对生在上部节上，叶近圆形，盾状，大小不等，肉质，钟乳体杆状，常于上面不明显，全缘。花序单生，生上部节间；雄花具梗，带紫红色。瘦果带紫色，宽卵球形，微扁，被宿存3花被片包被。花期4-7月，果期7-9月。生海拔1500-3000米的林下阴湿地石上。产云南和四川西南部。

Perennial herbs, rhizomatous, glabrous. Leaves opposite, crowded on upper nodes, suborbicular, peltate, unequal in size, succulent, cystoliths bacilliform, often conspicuous adaxially, margin entire. Inflorescences solitary in upper nodes; staminate flowers purplish, pedicellate. Achenes purplish, broadly ovoid, slightly compressed, enclosed by 3 persistent tepals. Fl. Apr-Jul. Fr. Jul-Sep. Shaded moist rocks in forests at 1500-3000 m. Distributed in Yunnan and SW Sichuan.

念珠冷水花 *Pilea monilifera*

攀枝花冷水花 *Pilea panzhihuaensis*

攀枝花冷水花

Pilea panzhihuaensis C. J. Chen, A. K. Monro et L. Chen

多年生草本。茎肉质，无毛。叶交互对生，具柄，无毛，圆卵形或近圆形，基部盾状；托叶长1-2厘米。雄聚伞圆锥花序长8-20厘米；雄花四基数；雌花具3花被片。花期4-6月。生海拔1500-1900米的林中石上或陡崖上。产四川西南部和贵州西部。

Perennial herbs. Stems fleshy, glabrous. Leaves decussate, petiolate, glabrous, orbicular-ovate or suborbicular, base peltate; stipules 1-2 cm long. Staminate thyrses 8-20 cm long; staminate flower tetramerous; pistillate flower with 3 tepals. Fl. Apr-Jun. Rocks or cliffs in forests at 1500-1900 m. Distributed in SW Sichuan and W Guizhou.

卵形盾叶冷水花

Piea peltata Hance var. **ovatifolia** C. J. Chen

肉质草本，无毛。茎高5-27厘米。叶卵形，基部盾形；托叶三角形，长约1毫米。团伞花序数个生于一条花序轴上，呈串珠状；雄花具4花被片，花被片背面顶端之下有正三角形凸起；雌花具3花被片。瘦果光滑。花期4-7月，果期10-11月。生海拔300-400米山谷灌丛。产广东北部和江西西南部。

Succulent herbs, glabrous. Stems 5-27 cm tall. Leaves ovate, base peltate; stipules triangular, ca. 1 mm long. Glomerules several along the rachis moniliformly arranged; staminate flower with 4 tepals, tepals abaxially below apex each with a deltoid projection; pistillate flower with 3 tepals. Achenes smooth. Fl. Apr-Jul. Fr. Oct-Nov. Bushes of hill valley at 300-400 m. Distributed in N Guangdong and SW Jiangxi.

波缘冷水花(石油菜)

Pilea cavaleriei Lévl.

多年生草本，雌雄异株。茎密被钟乳体。叶菱状卵形，或近圆形，大小近相等，肉质，钟乳体线形。花序在上部节单生；雄花序头状；团伞花序花少，有时仅一朵发育；雌花序具少量花成一簇；雌花具3不等大花被片。瘦果卵球形，长约0.7毫米，压扁，光滑。花期5-8月，果期8-10月。生海拔200-1500米的阴湿处或林下岩石上。产中国西南、华南、东南和华中。不丹亦有。

Perennial herbs, dioecious. Stem with dense cystoliths. Leaves rhombic-ovate, or suborbicular, subequal in size, succulent, cystoliths linear. Inflorescences solitary in upper nodes; staminate inflorescences a capitulum; glomerules few-flowered, sometimes only 1 developed; pistillate inflorescences a cluster of a few flowers; pistillate flower with 3 tepals unequal in size. Achenes ovoid, ca. 0.7 mm long, compressed, smooth. Fl. May-Aug. Fr. Aug-Oct. Shaded moist places or rocks in forests at 200-1500 m. Distributed in SW, S, SE and C China. Also in Bhutan.

卵形盾叶冷水花 *Piea peltata* var. *ovatifolia*

波缘冷水花(石油菜) *Pilea cavaleriei*

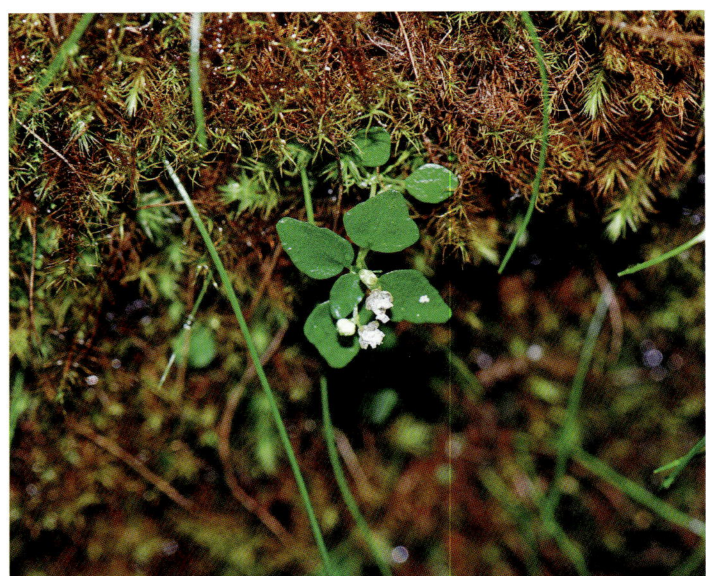
小冷水花 *Pilea minima*

小冷水花
Pilea minima W. T. Wang

一年生小草本，无毛。茎高2-6厘米，上部有2-3对叶。叶宽卵形，长3-10毫米，具三出脉。雄聚伞花序有2-4花；雄花的花被片4，长1毫米；雄蕊4，白色；雌聚伞花序有约3花；雌花无梗，花被片3，不等大，长0.2-0.4毫米；雌蕊长0.7毫米。花期7月。生海拔3400米的山谷草地阴处。特产云南(漾濞县)。

Small annual herbs, glabrous. Stems 2-6 cm tall, above with 2-3 pairs of leaves. Leaves broadly ovate, 3-10 mm long, trinerved. Staminate cyme 2-4-flowered; staminate flower with tepals 4, 1 mm long; stamens 4, white; pistillate cyme ca. 3-flowered; pistillate flower sessile, with tepals 3, unequal in size, 0.2-0.4 mm long; pistil 0.7 mm long. Fl. Jul. Shaded grassy places in valley at 3400 m. Endemic to Yunnan (Yangbi County).

厚叶冷水花
Pilea sinocrassifolia C. J. Chen

平卧草本，无毛，雌雄同株，密被钟乳体。叶不等大，肉质，钟乳体纺锤状，叶缘全缘且反卷。雄聚伞花序由少数几朵花集成头状，在节上部单生；雄花有4花被片；雌花序未见；雄花黄绿色，具梗。花期11月至翌年3月。生海拔200-1000米的阴湿处或溪边岩石上。产云南东南部、贵州、广东、福建和湖南。

Prostrate herbs, glabrous, monoecious, with dense cystoliths. Leaves subequal in size, succulent, cystoliths fusiform, margin entire and revolute. Staminate inflorescences solitary in upper nodes, consisting of a few dense flowers; staminate flower with 4 tepals; pistillate inflorescence not seen; staminate flowers yellow-green, pedicellate. Fl. Nov to next Mar. Shaded moist places or rocks along streams at 200-1000 m. Distributed in SE Yunnan, Guizhou, Guangdong, Fujian and Hunan.

亚高山冷水花
Pilea racemosa (Royle) Tuyama

草本，无毛。块茎球形，直径3-20毫米。叶长0.5-2厘米；托叶三角形，长约1毫米。聚伞花序密集；雄花的花被片4，背面顶端之下有1短角状凸起；雌花具3花被片。瘦果光滑。花期5-6月，果期7-9月。生海拔2200-5400米的林下石上或房屋墙上。产西藏南部、云南西北部和四川西部。尼泊尔、不丹和印度北部亦有。

Herbs, glabrous. Tubers globose, 3-20 mm diam. Leaves 0.5-2 cm long; stipules triangular, ca. 1 mm long. Cymes dense; staminate flower with 4 tepals, tepals abaxially below apex shortly corniculate; pistillate flower with 3 tepals. Achenes smooth. Fl. May-Jun. Fr. Jul-Sep. On rocks in forests, on house walls at 2200-5400 m. Distributed in S Xizang, NW Yunnan and W Sichuan. Also in Nepal, Bhutan and N India.

透茎冷水花
Pilea pumila (L.) A. Gray

一年生草本，近无毛，雌雄同株。茎肉质。叶菱状卵形或宽卵形，叶缘有锯齿，先端渐尖或急尖。雌花花被片近等大，条形，果期比果实短或近等长；雄花序为蝎尾状的聚伞花序；雄花有2(-3-4)花被片；雌花有3花被片。瘦果三角状卵球形。花期6-8月，果期8-11月。生海拔300-2200(-2900)米的林中、山谷或田中。产中国

厚叶冷水花 *Pilea sinocrassifolia*

亚高山冷水花 *Pilea racemosa*

透茎冷水花 *Pilea pumila*

大部分地区。东北亚和北美洲亦有。

Annual herbs, subglabrous, monoecious. Stems succulent. Leaves rhombic-ovate or ovate, margin serrate, apex acuminate or acute. Pistillate perianth lobes subequal, linear, shorter than or subequal to achenes; staminate inflorescences scorpioid cymose; staminate flower with 2(-3-4) tepals; pistillate flower with 3 tepals. Achenes triangular-ovoid. Fl. Jun-Aug. Fr. Aug-Nov. Forests, valleys or fields at 300-2200 (-2900) m. Distributed in most parts of China. Also in NE Asia and North America.

盾基冷水花

Pilea insolens Wedd.

草本，无毛。茎高20-50厘米。叶卵形，基部盾形，同一对叶极不等大；托叶长1.5-2毫米。聚伞圆锥花序长5-10厘米；雄花花被片4；雌花花被片3，不等大。瘦果有疣点。花期6-8月，果期9-10月。生海拔1600-2700米的山谷林下或灌丛下。产西藏东南部。不丹和印度东北部亦有。

Herbs, glabrous. Stems 20-50 cm tall. Leaves ovate, those of the same pair very unequal in size, base peltate; stipules 1.5-2 mm long. Thyrses 5-10 cm long; staminate flower with 4 tepals; pistillate flower with 3 tepals very unequal in size. Achenes verrucose. Fl. Jun-Aug. Fr. Sep-Oct. Forests or bushes at 1600-2700 m. Distributed in SE Xizang. Also in Bhutan and NE India.

盾基冷水花 *Pilea insolens*

perianths. Fl. Jun-Aug. Fr. Sep-Oct. Rocks or planted at 600-1500 m. Naturalized and cultivated in SW, S and E China. Native to tropical South America, introduced and naturalized in tropical Asia and tropical Africa.

异叶冷水花 *Pilea anisophylla*

异叶冷水花

Pilea anisophylla Wedd.

多年生草本，具匍匐地下茎。同一对叶异形，膜质，基部深或浅心形，全缘或先端具1-3个浅细锯齿。雄花序穗状，有时内向拳卷，很少分枝，花或团伞花序零散地分布在花序轴的一侧。瘦果棕色，卵球形，稍倾斜。花期6-9月，果期9-12月。生海拔900-2400米的常绿阔叶混交林较阴湿处或溪边。产云南南部和西部、西藏东南部。印度北部、尼泊尔、不丹和缅甸亦有。

Perennial herbs, stoloniferous. Leaves of the same pair heteromorphic, membranous, base deeply or shallowly cordate, margin entire or shallowly 1-3-serrate distally. Staminate inflorescences spike-like, somewhat coiled distally, few branched, flowers or glomerules loosely arranged along one side of axis. Achenes brownish, ovoid, slightly oblique. Fl. Jun-Sep. Fr. Sep-Dec. Shaded moist places in evergreen and broad-leaved mixed forests or near streams at 900-2400 m. Distributed in S and W Yunnan, and SE Xizang. Also in N India, Nepal, Bhutan and Myanmar.

小叶冷水花

Pilea microphylla (L.) Liebm.

纤细小草本，无毛，雌雄同株。茎直立或斜升，肉质，密被钟乳体。叶椭圆形、倒卵形或匙形，全缘，大小不等，长2-7毫米，肉质。花序常雌雄同株，紧密聚伞状头状；团伞花序具少花；雄花具梗。瘦果卵球形，压扁，光滑，由宿存花被包被。花期6-8月，果期9-10月。生海拔600-1500米的石上或栽培。归化和栽培于中国西南、华南和华东。原产热带南美洲，后引进归化于热带亚洲和热带非洲。

Small herbs, glabrous, monoecious. Stems erect or ascending, succulent, cystoliths dense. Leaves elliptic, obovate or spathulate, entire, unequal in size, 2-7 mm long, succulent. Inflorescences often androgynous, compactly cymose-capitate; glomerules few flowered; staminate flowers pedicellate. Achenes ovoid, compressed, smooth, enclosed by persistent perianths. Fl. Jun-Aug. Fr. Sep-Oct. Rocks or planted at 600-1500 m. Naturalized and cultivated in SW, S and E China. Native to tropical South America, introduced and naturalized in tropical Asia and tropical Africa.

矮冷水花

Pilea peploides (Gaudich.) Hook. et Arn.

一年生小草本，无毛。茎高3-20厘米，不分枝或有少数分枝。叶对生，宽卵形或菱状圆形，长3.5-18毫米，宽3-16毫米，全缘或波关。雄和雌花序均有明显的梗；雄花有4花被片；雌花有2花被片。瘦果长0.5毫米，平滑。花期4-7月。生海拔200-950米的山坡阴湿处石上。产湖南、江西、安徽、河南、河北、内蒙古东部和辽宁。俄罗斯(东西伯利亚)、朝鲜半岛、印度尼西亚(爪哇)和美国(夏威夷)亦有。

Small annual herbs, glabrous. Stems 3-20 cm tall, simple or few-branched. Leaves opposite, broadly ovate or rhombic-orbicular, 3.5-18 mm long, 3-16 mm broad, margin entire or repand. Staminate and pistillate inflorescences conspicuously pedunculate; staminate flower with 4 tepals; pistillate flower with 2 tepals. Achenes 0.5 mm long,

小叶冷水花 *Pilea microphylla*

矮冷水花 *Pilea peploides*

三角形冷水花 *Pilea swinglei*

smooth. Fl. Apr-Jul. On rocks in shaded places on slopes at 200-950 m. Distributed in Hunan, Jiangxi, Anhui, Henan, Hebei, E Neimenggu and Liaoning. Also in Russia (E Siberia), Korean Peninsula, Indonesia (Java) and USA (Hawaii).

齿叶矮冷水花

Pilea peploides (Gaudich.) Hook. et Arn. var. **major** Wedd.

本变种与模式变种的区别在于本变种的茎分枝较多，叶边缘上部有小钝齿，花序近无梗或有短梗，瘦果有稀疏小疣点。茎高5-30厘米。叶常菱状扁圆形，长(7-)10-21毫米，宽(7-)11-23毫米。花期4-5月。生海拔150-1300米的山坡路边或林中。产广西、广东、贵州、湖北、湖南、江西、浙江、福建和台湾。印度北部、缅甸、越南、印度尼西亚、日本和美国(夏威夷)亦有。

This variety differs from var. *peploides* in its more branched stems, crenate leaves, subsessile or shortly pedunculate inflorescences and verruculose achenes. Stems 5-30 cm tall. Leaves usually depressed-rhombic-orbicular, (7-)10-21 mm long, (7-)11-23 mm broad. Fl. Apr-May. By roads or in forests at 150-1300 m. Distributed in Guangxi, Guangdong, Guizhou, Hubei, Hunan, Jiangxi, Zhejiang, Fujian and Taiwan. Also in N India, Myanmar, Vietnam, Indonesia, Japan and USA (Hawaii).

三角形冷水花

Pilea swinglei Merr.

草本，无毛。茎高7-30厘米。团伞花序小，2-4个生于花序轴上；雄花花被片4，背面顶端之下有2小凸起；雌花花被片2(-3)，极不等大。瘦果光滑。花期6-8月，果期8-11月。生海拔400-1500米的山谷溪边或石上。产广西、广东、福建、浙江、江西、湖南、贵州东部、湖北西部和安徽南部。

Herbs, glabrous. Stems 7-30 cm tall. Glomerules small, 2-4 along rachis arranged; staminate flower with 4 tepals, each abaxially below apex with 2 small projections; pistilate flower with 2(-3) tepals very unequal in size. Achenes smooth. Fl. Jun-Aug. Fr. Aug-Nov. By streams or on rocks in valley at 400-1500 m. Distributed in Guangxi, Guangdong, Fujian, Zhejiang, Jiangxi, Hunan, E Guizhou, W Hubei and S Anhui.

齿叶矮冷水花 *Pilea peploides* var. *major*

假楼梯草

Lecanthus peduncularis (Wall. ex Royle) Wedd.

多年生草本，雌雄同株或异株。茎肉质，通常分枝。对生叶不等大，卵形，稀披针形，三出脉；钟乳体线形，两面均明显。花序具盘状花序托。瘦果椭圆状卵球形，棕灰色，上部背腹侧有一条略隆起的脊，具疣。花期7-10月，果期9-11月。生海拔1300-2700米的山谷林下或溪边阴处。产中国西南和华南。南亚、东南亚和非洲亦有。

Perennial herbs, monoecious or dioecious. Stems succulent, usually branched. Leaves unequal in a same pair, ovate to lanceolate, 3-veined; cystoliths linear, conspicuous on both surfaces. Inflorescences with discoid receptacles. Achenes brownish gray, ellipsoid-ovoid, above with a slightly elevated longitudinal ridge along dorsi-ventral edge, verrucose. Fl. Jul-Oct. Fr. Sep-Nov. Under forests in mountain valleys or shady sites along streams at 1300-2700 m. Distributed in SW and S China. Also in S and SE Asia, and Africa.

吐烟花

Pellionia repens (Lour.) Merr.

多年生草本。茎平卧，长达60厘米。叶斜椭圆形，边缘浅波状，具半离基三出脉；托叶三角形；退化叶长1毫米。雄聚伞花序有长梗。瘦果具瘤状凸起。花期5-10月。生海拔800-1100米的山谷林中或石上阴湿处。产云南南部和海南。越南、老挝和柬埔寨亦有。

Perennial herbs. Stems prostrate, up to 60 cm long. Leaves obliquely elliptic, semitriplinerved, margin repand; stipules triangular; reduced leaves 1 mm long. Staminate cymes long pedunculate. Achenes tuberculate. Fl. May-Oct. Forests or on rocks in valley at 800-1100 m. Distributed in S Yunnan and Hainan. Also in Vietnam, Laos and Cambodia.

吐烟花 *Pellionia repens*

曲毛赤车 *Pellionia retrohispida*

假楼梯草 *Lecanthus peduncularis*

曲毛赤车

Pellionia retrohispida W. T. Wang

多年生草本。茎渐升，上部密被反曲糙伏毛。叶椭圆形，具半离基三出脉，顶端短渐尖；托叶三角形，长3.2-6.5毫米。雄聚伞花序具长梗。瘦果具瘤状凸起。花期4-6月。生海拔350-1550米的山谷林中。产四川、甘肃南部、重庆、湖北西南部、贵州东南部、湖南、江西北部、福建西北部和浙江西部。

Perennial herbs. Stems ascending, above densely and retrorsely strigose. Leaves elliptic, semitriplinerved, apex shortly acuminate; stipules triangular, 3.2-6.5 mm long. Staminate cymes long pedunculate. Achenes tuberculate. Fl. Apr-Jun. Forests in valley at 350-1550 m. Distri-

蔓赤车 *Pellionia scabra*

短叶赤车 *Pellionia brevifolia*

buted in Sichuan, S Gansu, Chongqing, SW Hubei, SE Guizhou, Hunan, N Jiangxi, NW Fujian and W Zhejiang.

蔓赤车

Pellionia scabra Benth.

亚灌木。茎上部被开展糙硬毛。叶斜长圆形或倒披针形，具半离基三出脉，顶端长渐尖；托叶钻形。雄聚伞花序有细梗。瘦果有瘤状凸起。花期3-7月。生海拔500-1200米的山谷溪边或林下。产云南东南部、广西、香港、广东、贵州、重庆、湖南、江西、安徽南部、浙江、福建和台湾。越南和日本亦有。

Subshrubs. Stems above spreading hispid. Leaves obliquely oblong or oblanceolate, semitriplinerved, apex long acuminate; stipules subulate. Staminate cymes slenderly pedunculate. Achenes tuberculate. Fl. Mar-Jul. By streams or under forests in valley at 500-1200 m. Distributed in SE Yunnan, Guangxi, Hong Kong, Guangdong, Guizhou, Chongqing, Hunan, Jiangxi, S Anhui, Zhejiang, Fujian and Taiwan. Also in Vietnam and Japan.

赤车

Pellionia radicans (Sieb. et Zucc.) Wedd.

多年生草本。茎渐升，无毛或有极短毛。叶斜狭卵形，具半离基三出脉，顶端渐尖或长渐尖；托叶钻形。雄聚伞花序具细梗。瘦果具瘤状凸起。花期5-10月。生海拔200-1500米的山谷林下、灌丛中或溪边。产云南东南部、广西、广东、福建、台湾、江西、安徽南部、湖北、湖南、贵州、四川和甘肃南部。越南北部、朝鲜半岛和日本亦有。

Perennial herbs. Stems ascending, glabrous or with very short hairs. Leaves obliquely narrow-ovate, semitriplinerved, apex acuminate or long acuminate; stipules subulate. Staminate cymes slenderly pedunculate. Achenes tuberculate. Fl. May-Oct. Forests, bushes or by streams in valley at 200-1500 m. Distributed in SE Yunnan, Guangxi, Guangdong, Fujian, Taiwan, Jiangxi, S Anhui, Hubei, Hunan, Guizhou, Sichuan and S Gansu. Also in N Vietnam, Korean Peninsula and Japan.

赤车 *Pellionia radicans*

短叶赤车

Pellionia brevifolia Benth.

小草本。茎平卧，有短硬毛。叶斜椭圆形或倒卵形，长0.5-3.2厘米，具半离基三出脉，顶端钝；托叶钻形。雄聚伞花序具长梗。瘦果有瘤状凸起。花期5-7月。生海拔350-1500米的山谷林中或溪边。产广西、广东、香港、福建、江西、湖南和湖北西南部。

Small herbs. Stems prostrate, hispidulous. Leaves obliquely elliptic or obovate, 0.5-3.2 cm long, semitriplinerved, apex obtuse; stipules subulate. Staminate symes long pedunculate. Achenes tuberculate. Fl. May-Jul. Forests or by streams in valley at 350-1500 m. Distributed in Guangxi, Guangdong, Hong Kong, Fujian, Jiangxi, Hunan and SW Hubei.

绿赤车 *Pellionia viridis*

绿赤车

Pellionia viridis C. H. Wright

多年生草本或半灌木。茎高达70厘米，无毛。叶狭长圆形或披针形，长约10厘米，无毛，具不等离基三出脉，边缘上部有波状浅齿。雄聚伞花序宽约2厘米，有密集的花，花序梗长达1.8厘米；雌聚伞花序宽约4毫米，具短梗；雄花和雌花均五基数。瘦果长1毫米，具瘤状凸起。花期6-8月。生海拔650-1200米的山谷林中或沟边。产云南东北部、四川、重庆和湖北西部。

Perennial herbs or subshrubs. Stems up to 70 cm tall, glabrous. Leaves narrow-oblong or lanceolate, ca. 10 cm long, glabrous, unequally triplinerved, base peltate margin above repand-crenate. Staminate cyme ca. 2 cm broad, dense-flowered, with peduncle up to 1.8 cm long; pistillate cyme ca. 4 mm broad, shortly pedunculate; staminate and pistillate flowers all pentamerous. Achenes 1 mm long, tuberculate. Fl. Jun-Aug. Forest or by streams in valley at 650-1200 m. Distributed in NE Yunnan, Sichuan, Chongqing and W Hubei.

滇南赤车

Pellionia paucidentata (H. Schroter) Chien

多年生草本。茎高达50厘米。叶斜长椭圆形，具半离基三出脉，顶端渐尖，边缘具浅齿；托叶钻形。雄聚伞花序具长梗。瘦果有瘤状突起。花期9-10月。生海拔230-1000米的山谷溪边或林中。产云南南部和广西西南部。

Perennial herbs. Stems up to 50 cm tall. Leaves obliquely long elliptic, semitriplinerved, apex acuminate, margin crenate; stipules subulate. Staminate cymes long pedunculate. Achenes tuberculate. Fl. Sep-Oct. Forests or by streams in valley at 230-1000 m. Distributed in S Yunnan and SW Guangxi.

华南赤车

Pellionia grijsii Hance

多年生草本。茎高达70厘米，被开展糙毛。叶斜长椭圆形或倒披针形，具近羽状脉，顶端长渐尖；托叶钻形。雄聚伞花序有长梗。瘦果有瘤状凸起。花期冬季至翌年春季。生海拔250-1400米的山谷林下、石上或沟边。产云南东南部、广西、海南、广东、香港、福建南部和江西西部。

华南赤车 *Pellionia grijsii*

滇南赤车 *Pellionia paucidentata*

大叶赤车 *Pellionia macrophylla*

Perennial herbs. Stems up to 70 cm tall, above hispid. Leaves obliquely long elliptic or oblanceolate, subpenninerved, apex long acuminate; stipules subulate. Staminate cymes long pedunculate. Achenes tuberculate. Fl. winter to next spring. Under forests, on rocks or by streams in valley at 250-1400 m. Distributed in SE Yunnan, Guangxi, Hainan, Guangdong, Hong Kong, S Fujian and W Jiangxi.

大叶赤车

Pellionia macrophylla W. T. Wang

多年生草本。茎高约80厘米。叶倒卵状长圆形，长14-25厘米，具羽状脉，边缘具浅钝齿。雄聚伞花序具长梗；雄花和雌花均为五基数。瘦果具纵列短线纹。花期9-10月。生海拔1700-2000米的山谷林下或溪边。特产云南东南部。

Perennial herbs. Stems ca. 80 cm tall. Leaves obovate-oblong,

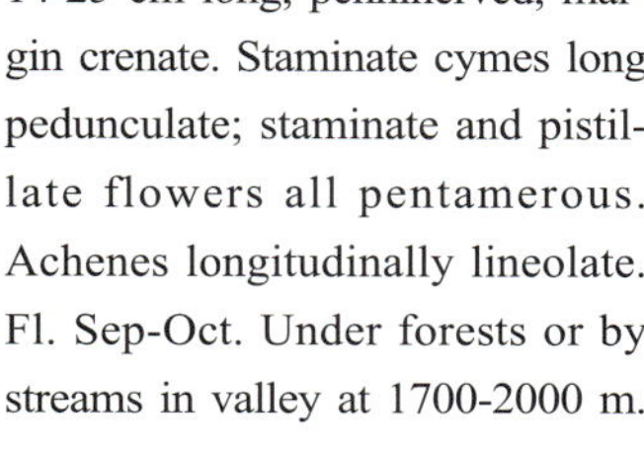

14-25 cm long, penninerved, margin crenate. Staminate cymes long pedunculate; staminate and pistillate flowers all pentamerous. Achenes longitudinally lineolate. Fl. Sep-Oct. Under forests or by streams in valley at 1700-2000 m. Endemic to SE Yunnan.

光果赤车

Pellionia leiocarpa W. T. Wang

多年生草本。茎高达1.5米，无毛。叶交互对生，无毛，同一对叶极不等大；较大叶斜狭长圆形，长达8厘米，具羽状脉，顶端长渐尖，较小叶菱形，长0.6-1.5厘米。雄聚伞花序具长梗。瘦果光滑。花期4-5月。生海拔1100-1600米的石灰岩山林中石上。产云南东南部和广西西南部。

Perennial herbs. Stems up to 1.5 m tall, glabrous. Leaves decussate, glabrous; larger ones obliquely narrow-oblong up to 8 cm long, penninerved, apex long acuminate, smaller ones rhombic, 0.6-1.5 cm long. Staminate cymes long pedunculate. Achenes smooth. Fl. Apr-May. On rocks under forests in limestone hills at 1100-1600 m. Distributed in SE Yunnan and SW Guangxi.

光果赤车 *Pellionia leiocarpa*

长圆楼梯草 *Elatostema oblongifolium*

长圆楼梯草

Elatostema oblongifolium Fu ex W. T. Wang

多年生草本。茎无毛。叶斜长圆形或椭圆形，长达20厘米，无毛，具羽状脉，钟乳体不明显。雄聚伞花序具短梗；雌头状花序近方形，长达8毫米。瘦果具6条白色纵肋。花期4-5月。生海拔300-1000米的山谷林中。产四川、重庆、湖北西部、湖南、贵州、云南东南部、广西和福建西北部。

Perennial herbs. Stems glabrous. Leaves obliquely or elliptic, up to 20 cm long, glabrous, penninerved, cystoliths inconspicuous. Staminate cymes shortly pedunculate; pistillate capitula rectangular, up to 8 mm long. Achenes longitudinally 6-white-ribbed. Fl. Apr-May. Forests in valley at 300-1000 m. Distributed in Sichuan, Chongqing, W Hubei, Hunan, Guizhou, SE Yunnan, Guangxi and NW Fujian.

花葶楼梯草

Elatostema scaposum Q. Lin et L. D. Duan

多年生草本。雄茎的叶强烈退化，叶片消失，托叶尚存在。雄聚伞花序7-11枚成对着生于雄茎上部的2-5个结上；雄花五基数。雌茎的叶正常发育，长圆形，长7.5-19.5厘米，具羽状脉；雌头状花序无梗或近无梗，成对腋生，具小花序托。瘦果具8条纵肋。花期10-11月。生海拔750-800米的石灰岩山洞中。特产贵州(荔波县)。

Perennial herbs. Leaves of staminate stem strongly reduced, with blades disappeared and only stipules still present. Staminate cymes 7-11 in pairs borne on upper 2-5 nodes of staminate stem; staminate flowers pentamerous. Leaves of pistillate stem normally developed, oblong, 7.5-19.5 cm long, penninerved; pistillate capitula sessile or subsessile, in pairs axillary, with small receptacles. Achenes longitudinally 8-ribbed. Fl. Oct-Nov. In caves of limestone hills at 750-800 m. Endemic to Guizhou (Libo County).

细尾楼梯草

Elatostema tenuicaudatum W. T. Wang

亚灌木。茎高达100厘米，多分枝，无毛。叶斜长圆形，具三出脉或离基三出脉，无毛，顶端尾状骤尖。雄、雌头状花序均小，无梗，具不明显花序托；苞片无任何凸起。瘦果具6条纵肋和瘤状凸起。花期3-4月。生海拔300-2000米的山谷林下或溪边。产云南东南部和中部、广西西北部和贵州南部。越南北部亦有。

Subshrubs. Stems up to 100 cm tall, ramose, glabrous. Leaves obliquely oblong, trinerved or triplinerved, glabrous, apex caudate-cuspidate. Staminate and pistillate capitula small, sessile, with obscure receptacles; involucral bracts without any projections. Achenes longitudinally 6-ribbed and tuberculate. Fl. Mar-Apr. Under forests or by streams in valley at 300-2000 m. Distributed in SE and C Yunnan, NW Guangxi and S Guizhou. Also in N Vietnam.

狭叶楼梯草

Elatostema lineolatum Wight var. **majus** Wedd.

亚灌木。茎多分枝，枝密被短柔毛。叶斜狭倒披针形，具半离基三出脉，顶端骤尖。雄头状花序无梗，直径5-10毫米，花序托不明显。瘦果有7条纵肋和小瘤状凸起。花期1-5月。生海拔200-1800米的山地林中或灌丛中或沟边。产西藏东南部、云南南部、广西、广东、福建和台湾。尼泊尔、不丹、印度、缅甸和泰国亦有。

Subshrubs. Stems ramose, branches densely puberulous. Leaves obliquely narrow-oblanceolate, semitriplinerved, apex cuspidate. Staminate capitula sessile, 5-10 mm diam, receptacles inconspicuous. Achenes longitu-

花葶楼梯草 *Elatostema scaposum*

细尾楼梯草 *Elatostema tenuicaudatum*

狭叶楼梯草 *Elatostema lineolatum* var. *majus*

多枝楼梯草 *Elatostema ramosum*

dinally 7-ribbed and minutely tuberulate. Fl. Jan-May. Forests, bushes or by streams at 200-1800 m. Distributed in SE Xizang, S Yunnan, Guangxi, Guangdong, Fujian and Taiwan. Also in Nepal, Bhutan, India, Myanmar and Thailand.

多枝楼梯草

Elatostema ramosum W. T. Wang

多年生草本。茎多分枝，上部有疏毛。叶狭长椭圆形，长达3.3厘米，具半离基三出脉，顶端骤尖，无毛或下面疏被短柔毛。雌头状花序无梗，直径约3毫米。花期7-9月。生海拔300-1500米的山地林下。产云南东南部和贵州西南部。

Perennial herbs. Stems ramose, above sparsely pubescent. Leaves narrowly long elliptic, up to 3.3 cm long, semitriplinerved, apex cuspidate, glabrous or abaxially sparsely puberulous. Pistillate capitula sessile, ca. 3 mm diam. Fl. Jul-Sep. Under forests in valley at 300-1500 m. Distributed in SE Yunnan and SW Guizhou.

异叶楼梯草

Elatostema monandrum (D. Don) Hara

小草本。茎高5-20厘米，下部有稀疏白色柔毛。上部茎生叶较大，下部茎生叶较小，长0.8-4(-6.5)厘米，具三出脉；退化叶长2-4毫米。雄、雌头状花序小，无梗或近无梗。瘦果狭长圆形，有6条纵肋。花期6-8月。生海拔1900-2800米的山地林中、沟边石上或附生树上。产西藏南部和东南部、四川西南部、云南、贵州西部和陕西南部。尼泊尔、不丹、印度和缅甸北部亦有。

Small herbs. Stems 5-20 cm, below with sparse white hairs. Leaves 0.8-4(-6.5) cm long, tri-nerved, upper cauline ones larger, lower cauline ones smaller; reduced leaves 2-4 mm long. Staminate and pistillate capitula small, sessile or subsessile. Achenes narrow-oblong, longitudinally 6-ribbed. Fl. Jun-Aug. Forest, on rocks by streams or epiphytic to tree trunks in valley at 1900-2800 m. Distributed in S and SE Xizang, SW Sichuan, Yunnan, W Guizhou and S Shaanxi. Also in Nepal, Bhutan, India and N Myanmar.

异叶楼梯草 *Elatostema monandrum*

羽裂楼梯草 *Elatostema monandrum* var. *pinnatifidum*

羽裂楼梯草

Elatostema monandrum var. **pinnatifidum** (Hook. f.) Murti

本变种与异叶楼梯草的区别：本变种的茎被锈色短柔毛，叶羽状分裂。生海拔2000-2800米的山谷林中。产云南西部和东南部。印度北部亦有。

From var. *monandrum* this variety differs in its rusty-puberulous stems and in its pinnatifid leaves. Forests in valley at 2000-2800 m. Distributed in W and SE Yunnan. Also in N India.

瘤茎楼梯草

Elatostema myrtillus (Lévl.) Hand.-Mazz.

多年生草本。茎高约35厘米，多分枝，无毛。叶狭卵形，长1.3-2.8厘米，无毛，具三出脉，无侧脉。雄、雌头状花序小，无梗；花序托不明显；苞片无任何凸起。瘦果有5-7条纵肋。花期5-10月。生海拔300-1000米的石灰岩山林下或沟边石上。产云南东南部、广西西部、贵州南部、湖南西北部、湖北西南部和重庆南部。

Perennial herbs. Stems ca. 35 cm tall, ramose, glabrous. Leaves narrow-ovate, 1.3-2.8 cm long, glabrous, trinerved, without lateral nerves. Staminate and pistillate capitula small, sessile; receptacles obscure; involucral bracts without any projections. Achenes longitudinally 5-7-ribbed. Fl. May-Oct. Under forests or on rocks by streams in limestone hills at 300-1000 m. Distributed in SE Yunnan, W Guangxi, S Guizhou, NW Hunan, SW Hubei and S Chongqing.

瘤茎楼梯草 *Elatostema myrtillus*

七花楼梯草

Elatostema septemflorum W. T. Wang

多年生小草本。茎高11-22厘米，无毛。叶斜椭圆状长圆形，无毛，具半离基三出脉，基部宽侧耳形；托叶长0.8-2毫米。雄头状花序具短梗和7花；花序托不明显；苞片2，无毛；雄花五基数。花期2月。生海拔1450米的山谷阴湿处。特产云南(富宁县)。

Small perennial herbs. Stems 11-22 cm tall, glabrous. Leaves obliquely elliptic-oblong, glabrous, semitriplinerved, base at leaf broad side auriculate; stipules 0.8-2 mm long. Staminate capitula shortly pedunculate, 7-flowered; receptacles obscure; involucral bracts 2, glabrous; staminate flowers pentamerous. Fl. Feb. Shady, damp places of valley at 1450 m. Endemic to Yunnan (Funing County).

粗齿楼梯草

Elatostema grandidentatum W. T. Wang

多年生草本。茎高约45厘米。叶斜椭圆形，具半离基三出

七花楼梯草 *Elatostema septemflorum*

粗齿楼梯草 *Elatostema grandidentatum*

脉；托叶长约3.5毫米。雄头状花序具短梗；花序托不明显；苞片2，顶端具角状凸起；雄花四基数；雌头状花序有盘状花序托。瘦果有13条纵肋。花期7-8月。生海拔2000-3100米的山地林中。产西藏东南部(波密县)。不丹亦有。

Perennial herbs. Stems ca. 45 cm tall. Leaves obliquely elliptic, semitriplinerved; stipules ca. 3.5 mm long. Staminate capitula shortly pedunculate; receptacles obscure; involucral bracts 2, at apex corniculate; staminate flower tetramerous; pistillate capitula with discoid receptacles. Achenes longitudinally 13-ribbed. Fl. Jul-Aug. Forests in valley at 2000-3100 m. Distributed in SE Xizang (Bomê County). Also in Bhutan.

小叶楼梯草

Elatostema parvum (Blume) Miq.

多年生草本。茎高8-30厘米，密被反曲糙伏毛。叶斜狭倒卵形，长1.5-8厘米，具三出脉或半离基三出脉；退化叶长3-9毫米。雄、雌头状花序均小，无梗，花序托不明显。瘦果有4条纵肋。花期7-8月。生海拔1000-2800米的山谷林下或沟边。产云南南部、重庆南部、贵州西南部、广西西北部、广东北部和台湾。尼泊尔、印度北部和印度尼西亚(爪哇)亦有。

Perennial herbs. Stems 8-30 cm tall, densely retrorsely strigose. Leaves obliquely narrow-obovate, 1.5-8 cm long, trinerved or semi-triplinerved; reduced leaves 3-9 mm long. Staminate or pistillate capitula small, sessile and their receptacles inconspicuous. Achenes longitudinally 4-ribbed. Fl. Jul-Aug. Under forests or by streams in valley at 1000-2800 m. Distributed in S Yunnan, S Chongqing, SW Guizhou, NW Guangxi, N Guangdong and Taiwan. Also in Nepal, N India and Indonesia (Java).

小叶楼梯草 *Elatostema parvum*

对叶楼梯草 *Elatostema sinense*

对叶楼梯草

Elatostema sinense H. Schröter

多年生草本。茎高达40厘米，上部被反曲短柔毛。叶斜长圆形，长3.5-9.5厘米，具半离基三出脉，顶端长渐尖；退化叶长3-5毫米。雄、雌头状花序小，具短梗。瘦果具5条纵肋和瘤状凸起。花期6-9月。生海拔500-2000米的山地沟边或林中。产云南、广西、贵州、四川、重庆、陕西南部、湖北、湖南、江西、安徽南部和福建北部。

Perennial herbs. Stems up to 40 cm tall, above retrorsely puberulous. Leaves narrow-oblong, 3.5-9.5 cm long, semitriplinerved, apex long acuminate; reduced leaves 3-5 mm long. Staminate and pistillate capitula small, shortly pedunculate. Achenes longitudinally 5-ribbed and tuberculate. Fl. Jun-Sep. By streams or under forests in valley at 500-2000 m. Distributed in Yunnan, Guangxi, Guizhou, Sichuan, Chongqing, S Shaanxi, Hubei, Hunan, Jiangxi, S Anhui and N Fujian.

密齿楼梯草

Elatostema pycnodontum W. T. Wang

多年生草本。茎高10-30厘米。叶互生，斜长圆形，长2.4-5.8厘米，顶端通常长渐尖，边缘有密锯齿。雄头状花序1.2-4毫米，有小的不明显花序托；苞片2或6。雌头状花序直径约5毫米；花序托长约2毫米；苞片约20，三角形。瘦果长约0.8毫米，有5条纵肋和小瘤状凸起。花期8-9月。生海拔500-1400米的林中或溪边。产云南东部、广西西部、贵州、四川东南部、重庆南部、湖南西北部和湖北西南部。

Perennial herbs. Stems 10-30 cm tall. Leaves alternate, obliquely oblong, 2.4-5.8 cm long, apex usually long acuminate, margin densely serrate. Staminate capitula 1.2-4 mm diam, with small inconspicuous receptacles; bracts 2 or 6. Pistillate capitula ca. 5 mm diam; receptacles ca. 2 mm diam; bracts ca. 20, triangular. Achenes ca. 0.8 mm long, longitudinally 5-ribbed and minutely tuberculate. Fl. Aug-Sep. Bushes or by streams at 500-1400 m. Distributed in E Yunnan, W Guangxi, Guizhou, SE Sichuan, S Chongqing, NW Hunan and SW Hubei.

疏晶楼梯草

Elatostema hookerianum Wedd.

多年生草本。茎高15-35厘米，无毛。叶斜倒卵状长圆形，稍镰状弯曲，具三出脉，无毛，边缘上部有少数牙齿。雄、雌头状花序小，近无梗。瘦果具10条纵肋和瘤状凸起。花期6月。

密齿楼梯草 *Elatostema pycnodontum*

疏晶楼梯草 *Elatostema hookerianum*

厚叶楼梯草 *Elatostema crassiusculum*

生海拔800-2400米的山地林中。产西藏东南部、云南西部和广西西南部。印度北部和东北部亦有。

Perennial herbs. Stems 15-35 cm tall, glabrous. Leaves obliquely obovate-oblong, slightly falcate, trinerved, glabrous, margin above few-dentate. Staminate and pistillate capitula small, subsessile. Achenes longitudinally 10-ribbed and tuberculate. Fl. Jun. Forests in valley at 800-2400 m. Distributed in SE Xizang, W Yunnan and SW Guangxi. Also in N and NE India.

厚萼楼梯草

Elatostema apicicrassum W. T. Wang

多年生草本。茎高15-34厘米，无毛。叶斜狭倒卵形，长达10厘米，具半离基三出脉，侧脉2-3对。雄、雌头状花序小，具短梗；雌总苞苞片条形或宽条形，下部白色，上部绿色，变厚。瘦果具5条纵肋和瘤状凸起。花期5-8月。生海拔2300米的山谷常绿阔叶林中。特产云南(贡山县)。

Perennial herbs. Stems 15-34 cm tall, glabrous. Leaves obliquely narrow-obovate, up to 10 cm long, semitrplinerved, with 2-3 pairs of lateral nerves. Staminate and pistillate capitula small, shortly pedunculate; pistillate involucral bracts linear or broadly linear, below white, above green and thickened. Achenes longitudinally 5-ribbed and tuberculate. Fl. May-Aug. Evergreen broad-leaved forests in valley at 2300 m. Endemic to Yunnan (Gongshan County).

厚叶楼梯草

Elatostema crassiusculum W. T. Wang

多年生小草本。茎高5-24厘米，疏被短柔毛。叶斜倒卵形或椭圆形，具三出脉或半离基三出脉，边缘有浅钝齿。雄、雌头状花序小，无梗。瘦果具5条纵肋和瘤状凸起。花期5月。生海拔450-700米的山地林下石上。产云南东南部。

Small perennial herbs. Stems 5-24 cm tall, sparsely puberulous. Leaves obliquely obovate or elliptic, trinerved or semitriplinerved, margin crenate. Staminate or pistillate capitula small, sessile. Achenes longitudinally 5-ribbed and tuberculate. Fl. May. On rocks under forests in valley at 450-700 m. Distributed in SE Yunnan.

厚萼楼梯草 *Elatostema apicicrassum*

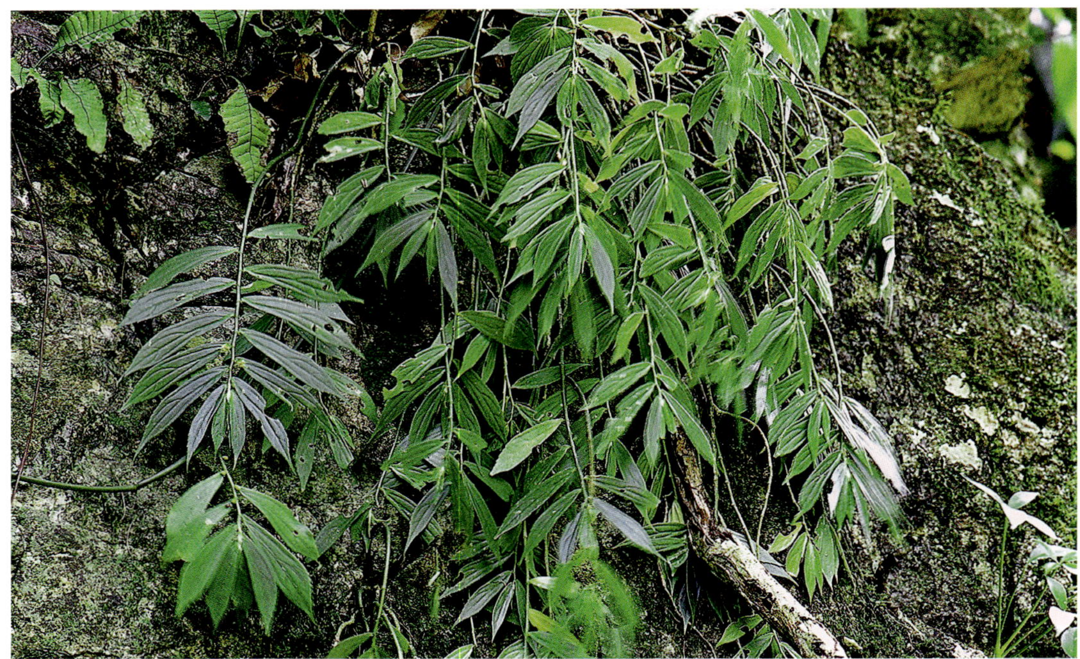
拟渐尖楼梯草 *Elatostema attenuatoides*

拟渐尖楼梯草
Elatostema attenuatoides W. T. Wang

多年生草本。茎无毛或近顶端疏被短柔毛。叶斜倒披针形，具半离基三出脉，全缘或近顶有1-4小齿。雄头状花序具短梗和不明显花序托；雌头状花序近无梗，具小花序托。瘦果有5条纵肋和极小瘤状凸起。花期2月。生海拔200-250米的山谷雨林中。特产云南(河口县)。

Perennial herbs. Stems glabrous or near apex sparsely puberulous. Leaves obliquely oblanceolate, semitriplinerved, margin entire or near apex 1-4-denticulate. Staminate capitula shortly pedunculate, with inconspicuous receptacles; pistillate capitula subsessile, with small receptacles. Achenes longitudinally 5-ribbed and minutely tuberculate. Fl. Feb. Rain forests in valley at 200-250 m. Endemic to Yunnan (Hekou County).

定结楼梯草
Elatostema dingjieense W. T. Wang

多年生草本。茎高约16厘米，无毛。叶无柄，斜长圆形，长达7厘米，渐尖，边缘具牙齿，具三出脉。雄头状花序腋生，无梗，长2-4毫米，有3-5花；花序托不明显；总苞苞片2层，外层2，背面有1角状凸起或1鸡冠状凸起，内层2-4。花期7月。生海拔2400米的山谷林下。特产西藏(定结县)。

Perennial herbs. Stem ca. 16 cm tall, glabrous. Leaves sessile, obliquely oblong, up to 7 cm long, acuminate, at margin dentate, trinerved. Staminate capitula sessile, 2-4 mm long, 3-5-flowered; receptacle inconspicuous; involucral bracts 2-seriate, outer bracts 2, opposite, abaxially 1-corniculate or 1-crested, inner bracts 2-4. Fl. Jul. Forests in valley at 2400 m. Endemic to Xizang (Dinggyê County).

宜昌楼梯草
Elatostema ichangense H. Schröter

多年生草本。茎高约30厘米，无毛。叶斜倒卵状长圆形，无毛，具半离基三出脉。雄、雌头状花序小，无梗或有短梗；雄总苞苞片6，排成2层，外层2苞片较大，顶端具长3.5-7毫米的条形凸起。瘦果有6-8条纵肋和瘤状凸起。花期8-9月。生海拔300-900米的山谷林中或石上。产贵州、湖南西部、湖北西部和重庆。

Perennial herbs. Stems ca. 30 cm tall, glabrous. Leaves obliquely obovate-oblong, glabrous, semitriplinerved. Staminate and pistillate capitula small, sessile or shortly pedunculate; staminate involucral bracts 6, 2-seriate, 2 outer ones larger, apex with linear projections 3.5-7 mm long. Achenes longitudinally 6-8-ribbed and tuberculate. Fl. Aug-Sep. Forests or on rocks in valley at 300-900 m. Distributed in Guizhou, W Hunan, W Hubei and Chongqing.

定结楼梯草 *Elatostema dingjieense*

宜昌楼梯草 *Elatostema ichangense*

滇黔楼梯草 *Elatostema backer*

竹桃楼梯草 *Elatostema neriifolium*

滇黔楼梯草

Elatostema backer H. Schröter

多年生草本。茎高20-50厘米，上部疏被反曲短毛。叶斜椭圆形，具半离基三出脉，顶端长渐尖。雄、雌头状花序小，具短梗或无梗。瘦果具瘤状凸起。花期6-7月。生海拔500-2000(-2700)米的山谷林中或林边。产云南、广西西部、贵州、四川和重庆南部。印度尼西亚(爪哇)亦有。

Perennial herbs. Stems 20-50 cm tall, above sparsely retrorsely puberulous. Leaves obliquely elliptic, semitriplinerved, apex long acuminate. Staminate and pistillate capitula small, shortly pedunculate or sessile. Achenes tuberculate. Fl. Jun-Jul. Forests or at forest margins in valleys at 500-2000(-2700) m. Distributed in Yunnan, W Guangxi, Guizhou, Sichuan and S Chongqing. Also in Indonesia (Java).

渐狭楼梯草

Elatostema attenuatum W. T. Wang

多年生草本。茎高约25厘米，无毛。叶倒披针形，全缘，无毛，具半离基三出脉，基部渐狭。雄、雌头状花序小，具短梗。瘦果具5条纵肋和瘤状凸起。花期6月。生海拔300-800米的山谷林中或沟边。产云南东南部(马关县、河口县)。

Perennial herbs. Stems ca. 25 cm tall, glabrous. Leaves oblanceolate, entire, glabrous, semitriplinerved, base long attenuate. Staminate and pistillate capitula small, shortly pedunculate. Achenes longitudinally 5-ribbed and tuberculate. Fl. Jun. Forests or by streams at 300-800 m. Distributed in SE Yunnan (Maguan County, Hekou County).

竹桃楼梯草

Elatostema neriifolium W. T. Wang et Zeng Y. Wu

多年生草本。茎高约35厘米，上部被短柔毛。叶狭长圆形或狭倒披针形，具三出脉或半离基三出脉，全缘。雄头状花序具短梗；花序托不明显；苞片6，无任何凸起。瘦果具多数纵列短线纹。花期10-11月。生海拔120-170米的石灰岩山热带雨林中。产云南东南部(麻栗坡县、马关县和河口县)。越南北部亦有。

Perennial herbs. Stems ca. 35 cm tall, above puberulous. Leaves narrow-oblong or narrow-oblanceolate, trinerved or semitriplinerved, margin entire. Staminate capitula shortly pedunculate; receptacles obscure; involucral bracts 6, without any projections. Achenes longitudinally multilineolate. Fl. Oct-Nov. Rain forests in limestone hills at 120-170 m. Distributed in SE Yunnan (Malipo County, Maguan County and Hekou County). Also in N Vietnam.

渐狭楼梯草 *Elatostema attenuatum*

迭叶楼梯草

Elatostema salvinioides W. T. Wang

小草本。茎长12-17厘米，下部密生三角形低出叶。叶二列，长圆形，长1-1.9厘米，宽4-6毫米，无毛，具1条脉；托叶心形或三角形；退化叶长4-6毫米。雄、雌头状花序很小，无梗。瘦果长0.3毫米，有多数纵列短线纹。花期4-5月。生海拔700-1600米的山谷雨林下石上或树干上。产云南南部和西南部。泰国北部和缅甸北部亦有。

Small herbs. Stems 12-17 cm long, below with dense triangular cataphylls. Leaves distichous, oblong, 1-1.9 cm long, 4-6 mm broad, glabrous, 1-nerved; stipules cordate or triangular; reduced leaves 4-6 mm long. Staminate and pistillate capitula very small, sessile. Achenes 0.3 mm long, longitudinally multi-lineolate. Fl. Apr-May. Epiphytic on rocks or tree trunks under rain forests at 700-1600 m. Distributed in S and SW Yunnan. Also in N Thailand and N Myanmar.

粗壮迭叶楼梯草

Elatostema salvinioides var. **robustum** W. T. Wang

本变种与模式变种迭叶楼梯草的区别在于本变种的茎粗壮，叶较大，长2.6厘米，宽1厘米；托叶披针形。瘦果较大，长0.5毫米。生海拔300-700米的热带雨林中。特产云南(马关县)。

From var. *salvinioides* this variety differs in its robust stems, larger leaves, 2.6 cm long, 1 cm broad; lanceolate stipules. Larger achenes, 0.5 mm long. Rain forests in valley at 300-700 m. Endemic to Yunnan (Maguan County).

丝梗楼梯草

Elatostema filipes W. T. Wang

多年生草本。茎高约25厘米，无毛。叶斜长椭圆形，具三出脉，无毛，顶端渐尖。雄头状花序直径3-5毫米；花序梗丝形，长0.9-3厘米；花序托不明显；苞片约4，顶端有角状凸起。瘦果具10条纵肋。花期6月。生海拔860-1200米的山谷石上阴处。特产广西(龙胜县)。

Perennial herbs. Stems ca. 25 cm tall, glabrous. Leaves obliquely long elliptic, trinerved, glabrous, apex acuminate. Staminate capitula 3-5 mm diam; peduncles filiform, 0.9-3 cm long; receptacles obscure; involucral bract ca. 4, apex corniculate. Achenes longitudinally 10-ribbed. Fl. Jun. Shady rocks in valley at 860-1200 m. Endemic to Guangxi (Longsheng County).

丝梗楼梯草 *Elatostema filipes*

托叶楼梯草

Elatostema nasutum Hook. f.

多年生草本，干燥时变黑色。茎高达40厘米，无毛。叶斜椭圆形，具三出脉，边缘具牙齿；托叶长9-18毫米。雄头状花序有细梗；花序托不明显；苞片6，顶端有角状凸起。瘦果有10-12条纵肋。花期7-10月。生海拔600-2400米的山谷林下。产西藏东南部、云南、广西、贵州、四川、重庆南部、湖北西南部、湖南和江西西部。不丹、尼泊尔和印度北部亦有。

Perennial herbs, turning black when drying. Stems up to 40 cm tall, glabrous. Leaves obliquely elliptic, trinerved, margin dentate; stipules 9-18 mm long. Staminate capitula slenderly pedunculate; receptacles inconspicuous; involucral bracts 6, apex corniculate. Achenes longitudinally 10-12-ribbed. Fl. Jul-Oct. Under forests in valley at 600-2400 m. Distributed in SE Xizang, Yunnan, Guangxi, Guizhou, Sichuan, S Chongqing, SW Hubei, Hunan and W Jiangxi. Also in Bhutan, Nepal and N India.

迭叶楼梯草 *Elatostema salvinioides*

粗壮迭叶楼梯草
Elatostema salvinioides var. *robustum*

托叶楼梯草 *Elatostema nasutum*

海南楼梯草 *Elatostema nasutum* var. *hainanense*

海南楼梯草

Elatostema nasutum Hook. f. var. **hainanense** (W.T. Wang) W.T. Wang

本变种与模式变种的区别在于其雄头状花序苞片在背面顶端之下具角状凸起。叶长8-12厘米。雄头状花序直径约10毫米；花序梗长1-4.8厘米。花期3-4月。生密林中石上。特产海南。

This variety differs from var. *nasutum* in its staminate involucral bracts, which are abaxially below apex corniculate. Leaves 8-12 cm long. Staminate capitula ca. 10 mm diam; peduncles 1-4.8 cm long. Fl. Mar-Apr. Rocks under dense forests. Endemic to Hainan.

瀑布楼梯草

Elatostema cataractum L. D. Duan et Q. Lin

多年生小草本。茎高8-18厘米，无毛。叶斜椭圆形，无毛，具半离基三出脉；托叶长1-1.5毫米。雄头状花序直径约5毫米；花序梗长5-13毫米；苞片6，排成2层，2外层苞片背面具1条纵翅；雌头状花序无梗，具小花序托。瘦果具4-5条纵肋。花期5-6月。生海拔750米的石灰岩山瀑布侧石上。特产贵州(荔波县)。

Small perennial herbs. Stems 8-18 cm tall, glabrous. Leaves obliquely elliptic, glabrous, semi-triplinerved; stipules 1-1.5 mm long. Staminate capitula ca. 5 mm diam; peduncles 5-13 mm long; involucral bracts 6, 2-seriate, 2 outer ones abaxially longitudinally 1-winged; pistillate capitula sessile, with small receptacles. Achenes longitudinally 4-5-ribbed. Fl. May-Jun. Rocks near waterfall in limestone hill at 750 m. Endemic to Guizhou (Libo County).

瀑布楼梯草 *Elatostema cataractum*

褐脉楼梯草 *Elatostema brunneinerve*

钝叶楼梯草 *Elatostema obtusum*

褐脉楼梯草

Elatostema brunneinerve W. T. Wang

多年生草本。茎渐升，长约60厘米，近顶部被柔毛。叶斜倒卵形，具半离基三出脉，背面脉上被短柔毛；托叶披针形，长1.5-1.9厘米。雄头状花序直径约5毫米；花序梗长5-10毫米；花序托不明显；苞片6，2层，2外层苞片背面有5条纵肋，顶端有1-4枚角状凸起。花期6月。生海拔1300-1400米的石灰岩山林下。产广西西南部和云南东南部。

Perennial herbs. Stems ascending, ca. 60 cm long, near apex pubescent. Leaves obliquely obovate, semitriplinerved, abaxially on nerves puberulous; stipules lanceolate, 1.5-1.9 cm long. Staminate capitula ca. 5 mm diam; peduncles 5-10 mm long; involucral bracts 6, 2-seriate, 2 outer ones abaxially 5-ribbed, apex 1-4-corniculate. Fl. Jun. Under forests in limestone hills at 1300-1400 m. Distributed in SW Guangxi and SE Yunnan.

钝叶楼梯草

Elatostema obtusum Wedd.

小草本。茎平卧或渐升，长10-14厘米，被反曲短毛。叶斜倒卵形，具三出脉。雄头状花序小，具细梗；苞片2；雌头状花序亦小，无梗。瘦果狭卵球形，长约2毫米，光滑。花期6-9月。生海拔1500-3000米的山地林下、沟边或石上。产西藏南部和东南部、云南、贵州、四川、重庆、湖南西部、湖北西部、甘肃南部和陕西南部。不丹、尼泊尔和印度北部亦有。

Small herbs. Stems prostrate or ascending, 10-14 cm long, retrorse-hirtellous. Leaves obliquely obovate, trinerved. Staminate capitula small, slenderly pedun-culate; involucral bracts 2; pistillate capitula also small, sessile. Achenes narrow-ovoid, ca. 2 mm long, smooth. Fl. Jun-Sep. Under forests, by streams or on rocks at 1500-3000 m. Distributed in S and SE Xizang, Yunnan, Guizhou, Sichuan, Chongqing, W Hunan, W Hubei, S Gansu and S Shaanxi. Also in Bhutan, Nepal and N India.

光茎钝叶楼梯草

Elatostema obtusum var. **trilobulatum** (Hayata) W. T. Wang

本变种与模式变种钝叶楼梯草的区别在于本变种的茎疏被反曲短毛或无毛。生海拔680-2000米的山谷溪边或林中。产

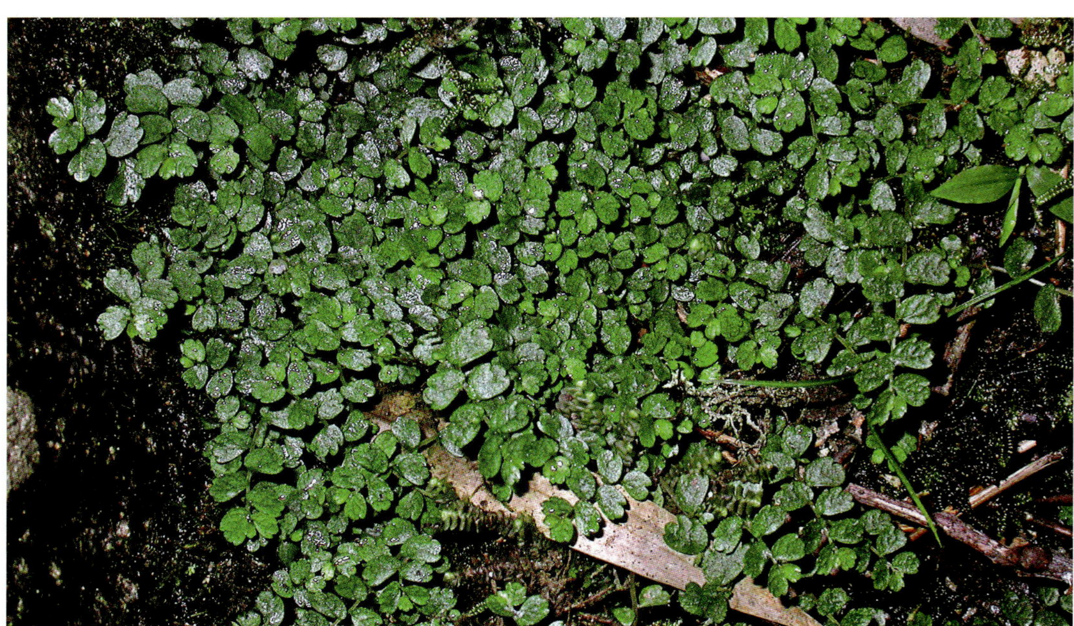
光茎钝叶楼梯草（变种） *Elatostema obtusum* var. *trilobulatum*

疣果楼梯草 *Elatostema trichocarpum*

柔毛楼梯草 *Elatostema villosum*

四川、贵州、重庆、湖北西南部、湖南、广西北部、广东北部、江西、浙江、福建和台湾。

From var. *obtusum* this variety differs in its sparsely retrorse-hirtellous or glabrous stems. By streams or under forests in valley at 680-2000 m. Distributed in Sichuan, Guizhou, Chongqing, SW Hubei, Hunan, N Guangxi, N Guangdong, Jiangxi, Zhejiang, Fujian and Taiwan.

疣果楼梯草

Elatostema trichocarpum Hand.-Mazz.

多年生草本。茎高12-25厘米。叶斜椭圆状卵形，长1.5-5.5厘米，具三出脉或半离基三出脉。雄、雌头状花序小，无梗或具细长梗。瘦果有4条纵肋和瘤状凸起。花期5-8月。生海拔450-1700米的山谷林下或沟边。产云南东部、贵州、四川、重庆、湖北西部和湖南。

Perennial herbs. Stems 12-25 cm tall. Leaves obliquely elliptic-ovate, 1.5-5.5 cm long, trinerved or semitriplinerved. Staminate and pistillate capitula small, sessile or slenderly long pedunculate. Achenes longitudinally 4-ribbed and tuberculate. Fl. May-Aug. Under forests or by streams in valley at 450-1700 m. Distributed in E Yunnan, Guizhou, Sichuan, Chongqing, W Hubei and Hunan.

柔毛楼梯草

Elatostema villosum Shih et Yang

多年生小草本。茎高7-15(-45)厘米，密被柔毛。叶斜狭倒卵形，长2-5.5厘米，具锯齿和羽状脉，下面密被柔毛。雌头状花序单生叶腋，近无梗；花序托长约2毫米；总苞片约40，条形，长2毫米，密被短柔毛。瘦果长0.6毫米，有12条纵肋。生海拔400-1100米的山谷中，有时生林下。特产台湾南部。

Small perennial herbs. Stems 7-15(-45) cm tall, densely puberulous. Leaves obliquely narrow-obovate, 2-5.5 cm long, serrate, abaxially densely puberulous, penninerved. Pistillate capitula singly axillary, subsessile; receptacle ca. 2 mm long; involucral bracts ca. 40, linear, 2 mm long, densely puberulous. Achenes 0.6 mm long, longitudinally 12-ribbed. Valley, sometimes under forests at 400-1100 m. Endemic to S Taiwan.

台湾楼梯草

Elatostema acuteserratum Shih et Yang

多年生草本。茎高达80厘米。叶斜狭长圆形，长9-18厘米，具小牙齿和羽状脉。雄头状花序，长约6毫米，近无梗；花序托不明显；外层苞片2枚对生，内层苞片约14；雄花五基数；雌头状花序似雄头状花序，也具2层苞片；雌花具4枚小花被片。瘦果长0.75毫米，具12条纵肋。生山谷林中。特产台湾(兰屿)。

Perennial herbs. Stems up to 80 cm tall. Leaves obliquely narrow-oblong, 9-18 cm long, denticulate, penninerved. Staminate capitula ca. 6 mm long, subsessile; receptacle inconspicuous; outer bracts 2, opposite, inner bracts ca. 14; staminate flowers pentamerous; pistillate capitula similar to staminate ones, also with 2-seriate bracts; pistillate flower with 4 small tepals. Achenes 0.75 mm long, longitudinally 12-ribbed. Forests in valley. Endemic to Taiwan (Lan Yu).

台湾楼梯草 *Elatostema acuteserratum*

齿尾楼梯草 *Elatostema dentatocaudatum*

ones larger, corniculate. Achenes longitudinally multi-lineolate and tuberculate. Fl. Jul-Sep. Under forests or by streams at 580-1400 m. Distributed in Chongqing, S Gansu, S Shaanxi, W Hubei, Hunan, Jiangxi, N Fujian, Zhejiang and S Anhui.

齿尾楼梯草

Elatostema dentatocaudatum W. T. Wang et Zeng Y. Wu

多年生草本，雌雄异株。茎高约30厘米。叶斜倒披针形，长约10厘米，顶端长尾状，边缘具牙齿，叶脉羽状。雌头状花雄单生叶腋；花序托小；苞片约16。瘦果长约0.7毫米，中部之上具4条纵肋。花期9-10月。生海拔1680米的山谷溪边阴处。特产云南(贡山县独龙江乡)。

Perennial herbs, dioecious. Stems ca. 30 cm tall. Leaves obliquely oblanceolate, ca 10 cm long, penninerved, apex long caudate, margin dentate. Pistillate capitula singly axillary; receptacle small; bracts ca. 16. Achenes ca. 0.7 mm long, above the middle longitudinally 4-ribbed. Fl. Sep-Oct. Shady places by streams in valley at 1680 m. Endemic to Yunnan (Dulongjiang Village, Gongshan County).

庐山楼梯草

Elatostema stewardii Merr.

多年生草本。茎高达40厘米，常具珠芽。叶斜椭圆状倒卵形或长圆形，具羽状脉。雄头状花序具短梗；花序托不明显；苞片6，排列成2层，外层2枚较大，具角状凸起。瘦果具多数纵列短线纹或瘤状凸起。花期7-9月。生海拔580-1400米的山谷林下或沟边。产重庆、甘肃南部、陕西南部、湖北西部、湖南、江西、福建北部、浙江和安徽南部。

Perennial herbs. Stems up to 40 cm tall, often with bulbils. Leaves obliquely elliptic-obovate or oblong, penninerved. Staminate capitula shortly pedunculate; receptacle inconspicuous; involucral bracts 6, 2-seriate, 2 outer

条叶楼梯草

Elatostema sublineare W. T. Wang

多年生草本。茎高约25厘米，上部被开展柔毛。叶斜倒披针形或条状倒披针形，具羽状脉。雄头状花序有细梗；花序梗长6-10毫米；花序托不明显。瘦果有8条纵肋。花期3-5月。生海拔300-850米的山谷林下或沟边石上。产广西西部、贵州东部、湖南、湖北西南部和重庆南部。越南北部亦有。

Perennial herbs. Stems ca. 25 cm tall, above spreading-pubescent. Leaves obliquely oblanceolate or linear-oblanceolate, penninerved. Staminate capitula slenderly pedunculate; peduncles 6-10 mm long; receptacles inconspicuous. Achenes longitudinally 8-ribbed. Fl. Mar-May. Under forests or on rocks by streams in valley at 300-850 m. Distributed in W Guangxi, E Guizhou, Hunan, SW Hubei and S Chongqing. Also in N Vietnam.

庐山楼梯草 *Elatostema stewardii*

条叶楼梯草 *Elatostema sublineare*

楼梯草 *Elatostema involucratum*

细角楼梯草 *Elatostema tenuicornutum*

楼梯草

Elatostema involucratum Franch. et Sav.

多年生草本。茎高达60厘米，通常无毛。叶斜长圆形或倒卵状长圆形，具羽状脉。雄头状花序有细长梗；花序梗长0.7-3厘米；花序托不明显；雌头状花序近无梗。瘦果具6条纵肋和小瘤状凸起。花期5-10月。生海拔200-2000米的山谷林中、灌丛中或沟边石上。产云南、广西、广东北部、福建、浙江、江苏南部、安徽南部、江西、湖南、贵州、四川、重庆、湖北西部、甘肃南部、陕西南部和河南南部。日本亦有。

Perennial herbs. Stems up to 60 cm tall, usually glabrous. Leaves obliquely oblong or obovate-oblong, penninerved. Staminate capitula slenderly pedunculate; peduncles 0.7-3 cm long; receptacles inconspicuous; pistillate capitula subsessile. Achenes longitudinally 6-ribbed and tuberculate. Fl. May-Oct. Forests, bushes or on rocks by streams in valley at 200-2000 m. Distributed in Yunnan, Guangxi, N Guangdong, Fujian, Zhejiang, S Jiangsu, S Anhui, Jiangxi, Hunan, Guizhou, Sichuan, Chongqing, W Hubei, S Gansu, S Shaanxi and S Henan. Also in Japan.

细角楼梯草

Elatostema tenuicornutum W. T. Wang

多年生草本。茎高约30厘米，上部密被短伏毛。叶斜长圆形或倒卵状长圆形，具羽状脉。雄头状花序直径6-8毫米；花序梗细，长6-12毫米；花序托不明显；苞片约5，2层，2外层苞片背面顶端之下有角状凸起。瘦果有6条纵肋和小瘤状凸起。花期4-6月。生海拔1130-1800米的山谷林中或石上。特产四川西部。

Perennial herbs. Stems ca. 30 cm tall, above appressed-puberulous. Leaves obliquely oblong or obovate-oblong, penninerved. Staminate capitula 6-8 mm diam; peduncles 6-12 mm long; receptacles obscure; involucral bracts ca.5, 2-seriate, 2 outer ones larger, abaxially below apex corniculate. Achenes longitudinally 6-ribbed and tuberculate. Fl. Apr-Jun. Forests or rocks in valley at 1130-1800 m. Endemic to W Sichuan.

曲毛楼梯草

Elatostema retrohirtum Dunn

多年生草本。茎渐升，长约30厘米，被反曲短糙毛。叶斜椭圆形，具三出脉，顶端急尖或短渐尖；托叶长4-6毫米。雌头状花序具短梗；花序托直径3-5.5毫米；苞片被短柔毛，无任何凸起。瘦果具6条纵肋。花期6-8月。生海拔460-1200米的山谷林中或林边。产云南东部、四川南部、贵州东南部、广西和广东北部。

Perennial herbs. Stems ascending, ca. 30 cm long, retrorsely strigillose. Leaves obliquely elliptic, trinerved, apex acute or shortly acuminate; stipules 4-6 mm long. Pistillate capitula shortly pedunculate; receptacles 3-5.5 mm diam; involucral bracts puberulous, without any projections. Achenes longitudinally 6-ribbed. Fl. Jun-Aug. Forests or forest edges in valley at 460-1200 m. Distributed in E Yunnan, S Sichuan, SE Guizhou, Guangxi and N Guangdong.

曲毛楼梯草 *Elatostema retrohirtum*

多序楼梯草 *Elatostema macintyrei*

食用楼梯草 *Elatostema edule*

多序楼梯草

Elatostema macintyrei Dunn

多年生草本。茎高达100厘米，无毛或近无毛。叶斜椭圆形或狭倒卵形，具半离基三出脉，边缘具浅齿。雄头状花序直径约2毫米，数个腋生，有短梗；雌头状花序直径3-5毫米，5-9个腋生。瘦果有10条纵肋。花期3-4月。生海拔160-2000米的山谷林中或沟边阴湿处。产西藏东南部、云南西部和南部、贵州、重庆南部、湖南、广西和广东西部。泰国和越南亦有。

Perennial herbs. Stems up to 100 cm tall, glabrous or subglabrous. Leaves obliquely elliptic or narrow-obovate, semitriplinerved, margin crenate. Staminate capitula ca. 2 mm diam, several axillary, shortly pedunculate; pistillate capitula 3-5 mm diam, 5-9 axillary. Achenes longitudinally 10-ribbed. Fl. Mar-Apr. Forests or by streams in valley at 160-2000 m. Distributed in SE Xizang, W and S Yunnan, Guizhou, S Chongqing, Hunan, Guangxi and W Guangdong. Also in Thailand and Vietnam.

大新楼梯草

Elatostema daxinense W. T. Wang et Zeng Y. Wu

多年生草本。茎高达70厘米，被短柔毛。叶斜长圆形，长达17厘米，具三出脉，顶端尾状尖头边缘有小齿。雄头状花序有短梗；花序托长方形；总苞苞片6，有密集钟乳体。花期11-12月。生海拔350米的山谷溪边。特产广西(大新县)。

Perennial herbs. Stems up to 70 cm tall, puberulous. Leaves obliquely oblong, up to 17 cm long, trinerved, with tail-like apexes at margin densely denticulate. Staminate capitula shortly pedunculate; receptacles rectangular; involucral bracts 6, with dense cystoliths. Fl. Nov-Dec. By streams in valley at 350 m. Endemic to Guangxi (Daxin County).

食用楼梯草

Elatostema edule C. B. Robinson

多年生草本。茎高达2米，无毛。叶斜长椭圆形，长10-25厘米，无毛，具半离基三出脉，基部在宽侧耳形；托叶长1.5厘米。雄头状花序具短梗；花序托盘状；苞片6，背面有1-2条纵肋。花期2-3月。生海拔10-110米的林中或溪边。产台湾(兰屿、绿岛乡)。菲律宾(巴丹岛)亦有。

Perennial herbs. Stems up to 2 m tall, glabrous. Leaves obliquely long elliptic, 10-25 cm long, glabrous, semitriplinerved, base at leaf broad side auriculate; stipules 1.5 cm long. Staminate capitula shortly pedunculate; receptacle discoid; involucral bracts 6, abaxially longitudinally 1-2-ribbed. Fl. Feb-Mar. Forests or by streams at 10-110 m. Distributed in Taiwan (Lan Yu, Lüdao Village). Also in the Philippines (Batan Islands).

大新楼梯草 *Elatostema daxinense*

似宽叶楼梯草 *Elatostema platyphylloides*

似宽叶楼梯草
Elatostema platyphylloides Shih et Yang

多年生草本。茎高达1.5米，被硬毛，变无毛。叶斜长圆形，长达30厘米，疏被硬毛，具半离基三出脉；托叶长达3厘米，被硬毛。雄头状花序具盘状花序托；苞片合生。生海拔250-600米的山地林中或溪边。产台湾。日本南部亦有。

Perennial herbs. Stems up to 1.5 m tall, hirsute, glabrescent. Leaves obliquely oblong, up to 30 cm long, sparsely hirsute, semitriplinerved; stipules up to 3 cm long, hirsute. Staminate capitula with discoid receptacles; involucral bracts connate. Forests or by streams at 250-600 m. Distributed in Taiwan. Also in S Japan.

锐齿楼梯草
Elatostema cyrtandrifolium (Zoll. et Mor.) Miq.

多年生草本。茎高达40厘米。叶斜椭圆形，具三出脉或离基三出脉，顶端骤尖，基部斜楔形；托叶长约4毫米。雄、雌头状花序具短梗和盘状花序托。瘦果具6条纵肋。花期4-9月。生海拔450-1400米的山谷林中或溪边石上。产云南东南部、广西、广东北部、台湾、福建、江西、湖南、贵州、湖北西部、重庆、四川和甘肃南部。尼泊尔、印度北部、泰国和印度尼西亚亦有。

Perennial herbs. Stems up to 40 cm tall. Leaves obliquely elliptic, trinerved or semitriplinerved, apex cuspidate, base obliquely cuneate; stipules ca. 4 mm long. Staminate and pistillate capitula shortly pedunculate and with discoid receptacles. Achenes longitudinally 6-ribbed. Fl. Apr-Sep. Forests or on rocks by streams in valley at 450-1400 m. Distributed in SE Yunnan, Guangxi, N Guangdong, Taiwan, Fujian, Jiangxi, Hunan, Guizhou, W Hubei, Chongqing, Sichuan and S Gansu. Also in Nepal, N India, Thailand and Indonesia.

锐齿楼梯草 *Elatostema cyrtandrifolium*

华南楼梯草
Elatostema balansae Gagnep.

多年生草本。茎高20-40(-80)厘米。叶斜椭圆形，长6-17厘米，顶端渐尖头边缘有小齿，边缘有牙齿，具半离基三出脉或三出脉。雌头状花序近无梗；花序托长方形，长3-9毫米；苞片有角状突起。瘦果椭圆体形，长约0.6毫米，有8条纵肋。花期4-6月。生海拔300-2100米的林中或沟边。产广东北部、广西、湖南、贵州南部、四川西南部、云南和西藏东南部。泰国南部和越南北部亦有。

Perennial herbs. Stems 20-40 (-80) cm tall. Leaves obliquely elliptic, 6-17 cm long, at apex acuminate (acumens at margin denticulate), dentate, semitriplinerved or trinerved. Pistillate capitula subsessile; receptacles rectangular, 3-9 mm long; bracts corniculate. Achenes ellipsoidal, ca. 0.6 mm long, longitudinally 8-ribbed. Fl. Apr-Jun. Forests or by streams at 300-2100 m. Distributed in N Guangdong, Guangxi, Hunan, S Guizhou, SW Sichuan, Yunnan and SE Xizang. Also in S Thailand and N Vietnam.

华南楼梯草 *Elatostema balansae*

尖牙楼梯草 *Elatostema oxyodontum*

骤尖楼梯草 *Elatostema cuspidatum*

尖牙楼梯草

Elatostema oxyodontum W. T. Wang

多年生草本。茎高约40厘米，近顶端被糙硬毛。叶斜椭圆形，具半离基三出脉，基部楔形；托叶三角形长8-16毫米。雄头状花序具短梗和盘状花序托；苞片5-6，其中3-4枚较大，顶端具角状凸起。花期1-2月。生海拔1500米的山谷灌丛中。特产云南(贡山县)。

Perennial herbs. Stems ca. 40 cm tall, near apex hispid. Leaves obliquely elliptic, semitriplinerved, base cuneate; stipules triangular, 8-16 mm long. Staminate capitula shortly pedunculate, with discoid receptacles; involucral bracts 5-6, of them 3-4 larger and at apex corniculate. Fl. Jan-Feb. Bushes in valley at 1500 m. Endemic to Yunnan (Gongshan County).

保山楼梯草

Elatostema baoshanense W. T. Wang et Zeng Y. Wu

多年生草本，雌雄异株。茎高约35厘米。叶斜长椭圆形，长9-18厘米，具牙齿和半离基三出脉，下面脉上被短柔毛。雄头状花序单生叶腋；花序托近长圆形，长10-16毫米；苞片约3，小；雄花四基数。花期6-7月。生海拔2230米的常绿阔叶林中。特产云南(保山)。

Perennial herbs, dioecious. Stems ca. 35 cm tall. Leaves obliquely long elliptic, 9-18 cm long, dentate, abaxially on nerves puberulous, semitriplinerved. Staminate capitula singly axillary; receptacle suboblong, 10-16 mm long; bracts ca. 3, small; staminate flowers tetramerous. Fl. Jun-Jul. Evergreen broad-leaved forests at 2230 m. Endemic to Yunnan (Baoshan).

骤尖楼梯草

Elatostema cuspidatum Wight

多年生草本。茎高达90厘米，无毛。叶斜长圆形，具半离基三出脉，顶端骤尖，基部宽楔形；托叶白色，条形，长约1厘米，有1条绿色脉。雄、雌头状花序具短梗和盘状花序托。花期5-8月。生海拔450-2800米的山谷林下或溪边石上。产西藏南部、云南、四川、重庆南部和东部、湖北西南部、贵州、广西西部、湖南和江西西南部。尼泊尔和印度东北部亦有。

Perennial herbs. Stems up to 90 cm tall, glabrous. Leaves obliquely oblong, semitriplinerved, apex cuspidate, base broadly cuneate; stipules white, linear, ca. 1 cm long, 1-green-nerved. Staminate and pistillate capitula shortly pedunculate and with discoid receptacles. Fl. May-Aug. Under forests or on rocks by streams in valley at 450-2800 m. Distributed in S Xizang, Yunnan, Sichuan, S and E Chongqing, SW Hubei,

保山楼梯草 *Elatostema baoshanense*

粗角楼梯草 *Elatostema pachyceras*

光序楼梯草 *Elatostema leiocephalum*

Guizhou, W Guangxi, Hunan and SW Jiangxi. Also in Nepal and NE India.

粗角楼梯草
Elatostema pachyceras W. T. Wang

多年生草本。茎高约50厘米，无毛。叶斜长圆形，具三出脉或半离基三出脉；托叶披针形，长1.3-1.7厘米。雄头状花序具短梗和盘状花序托。瘦果具9条纵肋。花期4-6月。生海拔1100-2400米的山谷林中或林边。特产云南东南部至西部。

Perennial herbs. Stems ca. 50 cm tall, glabrous. Leaves obliquely oblong, trinerved or semi-triplinerved; stipules yellowish-greenish, lanceolate, 1.3-1.7 cm long. Staminate capitula shortly pedunculate and with discoid receptacles. Achenes longitudinally 9-ribbed. Fl. Apr-Jun. Under forests or at forest edges at 1100-2400 m. Endemic to SE to W Yunnan.

光序楼梯草
Elatostema leiocephalum W. T. Wang

多年生草本。茎高约35厘米，无毛或近无毛。叶斜长椭圆形，具半离基三出脉，基部楔形；托叶长5毫米。雄头状花序具短梗；花序托长方形；苞片6，2层，2外层苞片背面顶端之下有短凸起；雌头状花序无梗；总苞片约50，顶端有长角状凸起。花期7月。生海拔1300-1500米的山谷溪边。产贵州西南部和广西北部。

Perennial herbs. Stems ca. 35 cm tall, glabrous or subglabrous. Leaves obliquely long elliptic, semitriplinerved, base cuneate; stipules 5 mm long. Staminate capitula shortly pedunculate; receptacles rectangular; involucral bracts 6, 2-seriate, 2 outer ones abaxially below apex shortly corniculate; pistillate capitula sessile; involucral bracts ca. 50, at apex long corniculate. Fl. Jul. By streams in valley at 1300-1500 m. Distributed in SW Guizhou and N Guangxi.

显脉楼梯草
Elatostema longistipulum Hand.-Mazz.

多年生草本。茎高约50厘米，被糙伏毛。叶披针形，具三出脉，背面脉上密被糙伏毛；托叶长7-11毫米。雄、雌头状花序具短梗和长方形花序托。瘦果具多数纵列短线纹。花期12月。生海拔120-1300米的山谷林中或溪边。产云南东南部、广西西部和贵州南部。越南北部亦有。

Perennial herbs. Stems ca. 50 cm tall, strigose. Leaves lanceolate, trinerved, abaxially on nerves densely strigose; stipules 7-11 mm long. Staminate and pistillate capitula shortly pedunculate and with rectangular receptacles. Achenes longitudinally multi-lineolate. Fl. Dec. Forests or by streams in valley at 120-1300 m. Distributed in SE Yunnan, W Guangxi and S Guizhou. Also in N Vietnam.

显脉楼梯草 *Elatostema longistipulum*

耳状楼梯草 *Elatostema auriculatum*

峨眉楼梯草 *Elatostema omeiense*

耳状楼梯草

Elatostema auriculatum W. T. Wang

多年生草本。茎高约80厘米，无毛。叶斜长圆状倒卵形，具三出脉，基部宽侧耳形。雄头状花序具短梗和盘状花序托；雌头状花序具细长梗。瘦果具8条纵肋和瘤状凸起。花期5-6月。生海拔850-2200米的山谷林中。产西藏东南部(林芝)和云南西北部(贡山县、福贡县)。

Perennial herbs. Stems ca. 80 cm tall, glabrous. Leaves obliquely oblong-ovate, trinerved, base at leaf broad side auriculate. Staminate capitula shortly pedunculate and with discoid receptacles; pistillate capitula slenderly long pedunculate. Achenes longitudinally 8-ribbed and tuberculate. Fl. May-Jun. Forests in valley at 850-2200 m. Distributed in SE Xizang (Linzhi) and NW Yunnan (Gongshan County, Fugong County).

盘托楼梯草

Elatostema dissectum Wedd.

多年生草本。茎高约40厘米，无毛。叶斜长圆形，无毛，具半离基三出脉，基部斜楔形；托叶长3-5毫米。雄头状花序具细长梗和盘状花序托。瘦果具8条纵肋。花期5-6月。生海拔500-2200米的山谷林中或溪边。产云南西部、南部和东南部、广西和广东西部。不丹和印度北部亦有。

Perennial herbs. Stems ca. 40 cm tall, glabrous. Leaves obliquely oblong, glabrous, semitriplinerved, base cuneate. Staminate capitula slenderly long pedunculate and with discoid receptacles. Achenes longitudinally 8-ribbed. Fl. May-Jun. Forests or by streams in valley at 500-2200 m. Distributed in W, S and SE Yunnan, Guangxi and W Guangdong. Also in Bhutan and N India.

峨眉楼梯草

Elatostema omeiense W. T. Wang

多年生草本。茎高约30厘米，无毛。叶斜椭圆形，长4-8厘米，具近三出脉。雄头状花序具长梗；花序梗长达7厘米，花序托长方形，长约1.2厘米；总苞片10余枚；雄花四基数；雌头状花序具短梗；花序托2.5毫米；总苞片约30；雌花无花被片。花期3-4月。生海拔1100米的山谷林下。特产四川(峨眉山)。

Perennial herbs. Stems ca. 30 cm

盘托楼梯草 *Elatostema dissectum*

球柱楼梯草 *Elatostema globosostigmatum*

tall, glabrous. Leaves obliquely elliptic, 4-8 cm long, nearly trinerved. Staminate capitula long pedunculated; peduncle up to 7 cm long; receptacle rectangular, ca. 1.2 cm long; involucral bracts more than 10; staminate flowers tetramerous; pistillate capitula shortly pedunculate; receptacle 2.5 mm diam; involucral bracts ca. 30; pistillate flower without tepals. Fl. Mar-Apr. Under forests in valley at 1100 m. Endemic to Sichuan (Emei Mountain).

球柱楼梯草

Elatostema globosostigmatum W. T. Wang et Zeng Y. Wu

多年生草本，雌雄同株。茎高约30厘米，约有5分枝。叶斜椭圆形，长约8厘米，具牙齿和半离基三出脉。雄头状花序单生叶腋，具长梗；花序梗长达2.5厘米；花序托长方形，长5-20毫米。雄花四基数；雌头状花序也单生叶腋，近无梗；花序托长2-4毫米。瘦果长0.8毫米，具8条纵肋。花期8-10月。生海拔1685米的山谷溪边阴处。特产云南(贡山县独龙江乡)。

Perennial herbs, monoecious. Stems ca. 30 cm tall, from below to apex ca. 5-branched. Leaves obliquely elliptic, ca. 8 cm long, dentate, semi-triplinerved. Staminate capitula singly axillary, long pedunculate; peduncle up to 2.5 cm long; receptacle rectangular, 5-20 mm long; staminate flowers tetramerous; pistillate capitula also singly axillary, subsessile; receptacle 2-4 mm long. Achenes 0.8 mm long, longitudinally 8-ribbed. Fl. Aug-Oct. Shady places by streams in valley at 1685 m. Endemic to Yunnan (Dulongjiang Village, Gangshan County).

独龙楼梯草

Elatostema dulongense W. T. Wang

多年生草本。茎高约30厘米，近顶端被疏短柔毛。叶斜长圆状倒卵形，具半离基三出脉，顶端尾状，基部斜楔形；托叶长4毫米。雄头状花序具细长梗；花序托盘状；总苞不存在。花期8月。生海拔1350米的山谷林中。特产云南(贡山县独龙江乡)。

Perennial herbs. Stems ca. 30 cm tall, near apex puberulous. Leaves obliquely oblong-obovate, semitriplinerved, apex caudate, base obliquely cuneate; stipules 4 mm long. Staminate capitula slenderly long pedunculate; receptacles discoid; involucres wanting. Fl. Aug. Forests in valley at 1350 m. Endemic to Yunnan (Dulongjiang Village, Gongshan County).

腺托楼梯草

Elatostema adenophorum W. T. Wang

多年生草本。茎高35-100厘米，无毛。叶长圆形，长9-21厘米，边缘有小牙齿，具半离基三出脉，无毛。雄头状花序具长梗；花序托有密集小腺点；苞片扁卵形或横条形；雌头状花序具短梗，长约1厘米。瘦果长0.8毫米，有瘤状凸起。花期4-5月。生海拔约570米的山谷雨林中。特产云南(马关县)。

Perennial herbs. Stems 35-100 cm tall, glabrous. Leaves elliptic, 9-21 cm long, at margin denticulate, semitriplinerved, glabrous. Staminate capitula long pedunculate; receptacles densely glandular-puncticulate; bracts depressed-ovate or transversely linear; pistillate capitula shortly pedunculate, ca. 1 cm long. Achenes 0.8 mm long, tuberculate. Fl. Apr-May. Rain forests in valleys, at ca. 570 m. Endemic to Yunnan (Maguan County).

独龙楼梯草 *Elatostema dulongense*

腺托楼梯草 *Elatostema adenophorum*

多脉楼梯草 *Elatostema pseudoficoides*

异晶楼梯草 *Elatostema heterogrammicum*

多脉楼梯草

Elatostema pseudoficoides W. T. Wang

多年生草本。茎高达100厘米，无毛。叶斜长圆形，具羽状脉，侧脉9-12对，顶端骤尖，基部宽侧近耳形。雄、雌头状花序无梗或有短梗，具盘状花序托。花期8-9月。生海拔1200-2200米的山谷林中或溪边。产云南东北部、四川、重庆、湖北西部和湖南西部。

Perennial herbs. Stems up to 100 cm tall, glabrous. Leaves obliquely oblong, penninerved with 9-12 pairs of lateral nerves, apex cuspidate, base at leaf broad side subauriculate. Staminate and pistillate capitula sessile or shortly pedunculate, and with discoid receptacles. Fl. Aug-Sep. Forests or by streams at 1200-2200 m. Distributed in NE Yunnan, Sichuan, Chongqing, W Hubei and W Hunan.

二脉楼梯草

Elatostema binerve W. T. Wang

多年生草本。茎高达55厘米，上部被短糙伏毛。叶倒披针形，具羽状脉，背面脉上被短糙伏毛；托叶狭披针形，长10-12毫米，有2条纵脉。雄头状花序有短梗和盘状花序托；苞片约7，顶端有角状凸起；雄花五基数。花期7月。生海拔500-900米的石灰岩山林中。特产云南(马关县)。

Perennial herbs. Stems up to 55 cm tall, above strigillose. Leaves oblanceolate, penninerved, abaxially on nerves strigillose; stipules narrow-lanceolate, 10-12 mm long, longitudinally 2-nerved. Staminate capitula shortly pedunculate, with discoid receptacles; involucral bracts ca. 7, at apex corniculate; staminate flowers pentamerous. Fl. Jul. Forests in limestone forests at 500-900 m. Endemic to Yunnan (Maguan County).

异晶楼梯草

Elatostema heterogrammicum W. T. Wang

多年生草本。茎高约35厘米，无毛。叶长达17厘米，具羽状脉，有些叶具明显密集钟乳体，有些叶的钟乳体则不明显。雄头状花序具短梗；花序托直径4-7毫米；总苞片约8，有时具角状凸起。花期7月。生海拔2000米的山谷林下。特产云南(漾濞县)。

Perennial herbs. Stems ca. 35 cm tall, glabrous. Leaves up to 17 cm long, penninerved, some leaves with dense conspicuous cystoliths, and some leaves with inconspicuous cystoliths. Staminate capitula shortly pedunculate; receptacles 4-7 mm diam; involucral bracts ca. 8, sometimes corniculate. Fl. Jul. Under forests in valley at 2000 m. Endemic to Yunnan (Yangbi County).

黑叶楼梯草

Elatostema melanophyllum W. T. Wang

多年生草本，干燥时变黑色。茎高达50厘米，被短硬毛。叶斜长圆形或倒披针形，具羽状脉，基部楔形。雄、雌头状花序具短梗或无梗，具盘状花序托；雄总苞苞片多具角状凸起。花期1-4月。生海拔200-1850米的山谷林中。特产云南东南部(马关县、河口县)。

Perennial herbs, turning black when drying. Stems up to 50 cm tall, hirtellous. Leaves obliquely

二脉楼梯草 *Elatostema binerve*

oblong or oblanceolate, penninerved, base cuneate. Staminate and pistillate capitula shortly pedunculate or sessile, and with discoid receptacles; staminate involucral bracts mostly corniculate. Fl. Jan-Apr. Forests in valley at 200-1850 m. Endemic to SE Yunnan (Maguan County, Hekou County).

黑叶楼梯草 *Elatostema melanophyllum*

南川楼梯草

Elatostema nanchuanense W. T. Wang

多年生草本，干燥时多少变黑色。茎高约40厘米，被短毛。叶斜长圆形或倒披针形，具羽状脉，侧脉9-12对。雄、雌头状花序均具短梗、盘状花序托和具角状突起的凸片。瘦果有8条纵肋。花期5-6月。生海拔600-1200米的山谷林下或石壁上。产贵州东部、湖南西北部、湖北西南部和重庆南部。

Perennial herbs, more or less turning black when drying. Stems ca. 40 cm tall, puberulous. Leaves obliquely oblong or oblanceolate, penninerved with 9-12 pairs of lateral nerves. Staminate and pistillate capitula all shortly pedunculate, and with discoid receptacles and corniculate involucral bracts. Achenes longitudinally 8-ribbed. Fl. May-Jun. Under forests or on damp rocky walls in valley at 600-1200 m. Distributed in E Guizhou, NW Hunan, SW Hubei and S Chongqing.

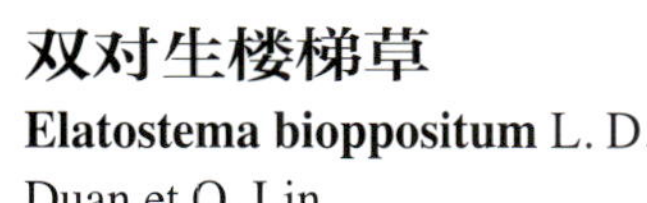

双对生楼梯草

Elatostema bioppositum L. D. Duan et Q. Lin

多年生草本，雌雄异株。茎高达80厘米，无毛，具软鳞片。叶无毛，长12-25厘米，具羽状脉；托叶披针状条形，长1.2-2.5厘米。雄、雌头状花序均成对，一个生叶腋，另一个腋外生；雄花序托蝴蝶形。花期4-5月。生海拔410-550米的石灰岩山常绿阔叶林下。特产广西(龙州县)。

Perennial herbs, dioecious. Stems up to 80 cm tall, furfuraceous. Leaves glabrous, 12-25 cm long, penninerved; stipules lanceolate-linear, 1.2-2.5 cm long. Staminate and pistillate capitula in pairs, in each pair one arising from leaf axil and the other extra-axillary; staminate receptacles butterfly-like. Fl. Apr-May. Evergreen broad-leaved forests in limestone hills at 410-550 m. Endemic to Guangxi (Longzhou County).

南川楼梯草 *Elatostema nanchuanense*

双对生楼梯草 *Elatostema biopposítum*

拟长圆楼梯草 *Elatostema pseudooblongifolium*

紫花楼梯草 *Elatostema sinopurpureum*

拟长圆楼梯草

Elatostema pseudooblongifolium W. T. Wang

多年生草本。茎高约40厘米，无毛。叶斜长圆形或狭倒卵形，无毛，具羽状脉，侧脉5-7对，钟乳体明显。雄头状花序具短梗和盘状花序托；雄总苞只有1枚苞片。瘦果有6条纵肋和瘤状凸起。花期9-10月。生海拔180米的石灰岩山林中或洞中。特产广西西南部(宁明县、大新县、靖西和龙州县)。

Perennial herbs. Stems ca. 40 cm tall, glabrous. Leaves obliquely oblong or narrow-obovate, glabrous, peninnerved with 5-7 pairs of lateral nerves; cystoliths conspicuous. Staminate capitula with short peduncles and discoid receptacle; involucres only 1-bracteate. Achenes longitudinally 6-ribbed and tuberculate. Fl. Sep-Oct. Forests or in caves in limestone hills at 180 m. Endemic to SW Guangxi (Ningming County, Daxin County, Jingxi and Longzhou County).

滇南楼梯草

Elatostema austroyunnanense W. T. Wang

多年生草本。茎高约30厘米，无毛。叶斜椭圆形或狭倒卵形，具羽状脉，基部宽楔形。雌头状花序近无梗；花序托盘状；苞片多数，无任何凸起。瘦果具6条纵肋和瘤状凸起。花期1月。生海拔580米的灌丛阴湿处。特产云南(勐腊县、勐仑镇)。

Perennial herbs. Stems ca. 30 cm tall, glabrous. Leaves obliquely elliptic or narrow-obovate, penninerved, base broadly cuneate. Pistillate capitula subsessile; receptacles discoid; involucral bracts very numerous, without any projections. Achenes longitudinally 6-ribbed and tuberculate. Fl. Jan. Bushes at 580 m. Endemic to Yunnan (Mengla County, Menglun Town).

紫花楼梯草

Elatostema sinopurpureum W. T. Wang

多年生草本。茎高约30厘米，无毛。叶斜长圆形，具羽状脉，背面脉上被短柔毛；托叶长5-6毫米。雄头状花序具长梗和盘状花序托；苞片6，2层；花被片5，紫红色；雌头状花序无梗或近无梗。瘦果具5条纵肋和小瘤状凸起。花期6-7月。生海拔700-800米的石灰岩山岩石间或山坑溪边阴湿处。特产贵州(荔波县)。

Perennial herbs. Stems ca. 30 cm tall, glabrous. Leaves obliquely oblong, penninerved, abaxially on nerves puberulous; stipules 5-6 mm long. Staminate capitula with long peduncles and discoid receptacles; involucral bracts 6, 2-seriate; tepals 5, purple-red; Pistillate capitula sessile or subsessile. Achenes longitudinally 5-ribbed and minutely tuberculate. Fl. Jun-Jul. Among rocks or in shady, damp places by streams of large pit in limestone hills at 700-800 m. Endemic to Guizhou (Libo County).

拟疏毛楼梯草

Elatostema albopilosoides Q. Lin et L. D. Duan

多年生草本，干燥后变黑色。茎高达120厘米，被短柔毛。叶斜倒卵状长圆形，具羽状脉；托叶披针形，长13-16毫米。雄、雌头状花序均具长梗和盘状花序托；雄花五基数。瘦果具5-7条纵肋和小瘤状凸起。花期6-7月。生海拔750-800米的石灰岩山岩石间或巨坑溪边阴湿处。特产贵州(荔波县)。

Perennial herbs, turning black when drying. Stems up to 120 cm tall, puberulous. Leaves obliquely obovate-oblong, penninerved; stipules lanceolate, 13-16 mm long. Staminat and pistillate capitula all with long peduncles and discoid receptacles; staminate flowers pentamerous. Achenes longitudinally 5-7-ribbed and minutely tuberculate.

滇南楼梯草 *Elatostema austroyunnanense*

拟疏毛楼梯草 *Elatostema albopilosoides*

梨序楼梯草 *Elatostema ficoides*

Fl. Jun-Jul. Among rocks or in shady, damp places by streams of large pit in limestone hills at 750-800 m. Endemic to Guizhou (Libo County).

梨序楼梯草

Elatostema ficoides Wedd.

多年生草本。茎高达1米，无毛。叶斜长圆形，具羽状脉；托叶条形，长7-10毫米。雄隐头花序具长梗；花序托初梨形，花期开裂呈蝴蝶形；雌头状花序无梗；花序托直径约1.5毫米；苞片约16，无任何凸起。瘦果具6条纵肋。花期8-9月。生海拔560-2100米的山谷林中或沟边。产西藏南部、云南、四川、重庆南部、贵州、湖南南部、广西和海南。尼泊尔、不丹、印度北部和越南北部亦有。

Perennial herbs. Stems up to 1 m tall, glabrous. Leaves obliquely oblong, penninerved; stipules linear, 7-10 mm long. Staminate hypanthodia long pedunculate; receptacles at beginning pyriform, at anthesis divided and papilionaceous in shape; pistillate capitula sessile; receptacles ca. 1.5 mm diam; involucral bracts ca. 16, without any projections. Achenes longitudinally 6-ribbed. Fl. Aug-Sep. Forests or by streams in valley at 560-2100 m. Distributed in S Xizang, Yunnan, Sichuan, S Chongqing, Guizhou, S Hunan, Guangxi and Hainan. Also in Nepal, Bhutan, N India and N Vietnam.

藤麻

Procris crenata C. B. Robinson

多年生草本。茎高达80厘米，无毛。叶二列，狭长圆形，无毛。具羽状毛，边缘有浅钝齿；退化叶狭长圆形，长5-17毫米。雄聚伞花序直径约5毫米，位于雌头状花序之下；雌头状花序直径1.5-3毫米；花序托近球形。瘦果常具小点。花期7-8月。生海拔300-1750米的山地林下石上或附生树上。产西藏东南部、云南西部和南部、四川南部、贵州西南部、广西、广东、福建南部和台湾。菲律宾、加里曼丹岛、越南、柬埔寨、泰国、不丹、尼泊尔、印度、斯里兰卡和非洲亦有。

Perennial herbs. Stems up to 80 cm tall, glabrous. Leaves distichous, narrow-oblong, glabrous, penninerved, margin crenate; reduced leaves narrow-oblong, 5-17 mm long. Staminate cymes borne below pistillate capitula; pistillate capitula 1.5-3 mm diam; receptacles subglobose. Achenes often minutely punctate. Fl. Jul-Aug. Rocks or epiphytic to tree trunks under forests in valley at 300-1750 m. Distributed in SE Xizang, W and S Yunnan, S Sichuan, SW Guizhou, Guangxi, Guangdong, S Fujian and Taiwan. Also in the Philippines, Kalimantan Island, Vietnam, Cambodia, Thailand, Bhutan, Nepal, India, Sri Lanka and Africa.

藤麻 *Procris crenata*

舌柱麻 *Archiboehmeria atrata*

苎麻 *Boehmeria nivea*

舌柱麻

Archiboehmeria atrata (Gagnep.) C. J. Chen

灌木或亚灌木。枝条具贴生柔毛，迅速脱落。叶互生。雄聚伞花序四至六回二歧分枝；雌花无柄，花被管与子房离生，下面具柔毛；柱头舌状。瘦果卵球形，淡绿色，有疣状凸起。花期5-8月，果期8-10月。生海拔300-1500米的半阴湿润林中或岩缝。产广西、广东、海南和湖南南部。越南北部亦有。

Shrubs or subshrubs. Branches appressed pubescent, soon glabrescent. Leaves alternate. Staminate cymes 4-6 times dichotomously branched; pistillate flowers sessile, perianth tubes free from ovary, puberulent abaxially; stigmas ligulate. Achenes ovoid, greenish, verrucose. Fl. May-Aug. Fr. Aug-Oct. Partly shady, moist places in forests or rock crevices at 300-1500 m. Distributed in Guangxi, Guangdong, Hainan and S Hunan. Also in N Vietnam.

腋球苎麻

Boehmeria glomerulifera Miq.

灌木。茎高3-5米；枝被短柔毛。叶互生，椭圆状卵形，长9-19厘米，两面均有毛，边缘有小牙齿。团伞花序单性，单生叶腋，直径2-7毫米；雄花四基数；雌花被狭椭圆形，长1毫米，顶端具2小齿；柱头长1.4毫米。花期冬季。生海拔850-1000米的林中或林边。产云南南部和西藏东南部。印度、缅甸、泰国、越南和印度尼西亚亦有。

Shrubs. Stems 3-5 m tall; branches puberulous. Leaves alternate, elliptic-ovate, 9-19 cm long, hairy, margin denticulate. Glomerules unisexual, singly axillary, 2-7 mm diam; staminate flowers tetramerous; pistillate perianth narrowly elliptic, 1 mm long, apex 2-denticulate; stigma 1.4 mm long. Fl winter. Forests or at forest margins at 850-1000 m. Distributed in S Yunnan and SE Xizang. Also in India, Myanmar, Thailand, Vietnam and Indonesia.

苎麻

Boehmeria nivea (L.) Gaudich.

亚灌木或灌木，雌雄同株。茎密被开展的长硬毛。叶互生；托叶分生；叶片圆形或阔卵形，下面密被白色或灰色毡毛。圆锥花序生当年或最近落叶的结果枝腋处。瘦果近卵球形，基部具柄。花期7-8月，果期9-10月。生海拔200-1700米的次生林、灌丛或路边。产中国西南、华南、东南、华中和华东。南亚、东南亚和日本亦有。

Subshrubs or shrubs, monoecious. Stems densely patent-hirsute. Leaves alternate; stipules free; blades often orbicular or broadly ovate, abaxial surface densely white or gray tomentose. Panicles on flowering branches

腋球苎麻 *Boehmeria glomerulifera*

in the axils of current or recently fallen leaves. Achenes subovoid, base stipitate. Fl. Jul-Aug. Fr. Sep-Oct. Edges of secondary forests, thickets or roadsides at 200-1700 m. Distributed in SW, S, SE, C and E China. Also in S and SE Asia, and Japan.

青叶苎麻

Boehmeria nivea var. **tenacissima** (Gaudich.) Miq.

灌木。茎和叶柄密被或疏被短伏毛。叶片多为卵形或椭圆状卵形，顶端骤尖或渐尖，基部多为圆形、宽楔形或突狭楔形；托叶基部合生。圆锥花序生当年或近期落叶的结果枝腋处。瘦果近卵球形，基部具柄。花期5-8月，果期9-11月。生海拔200-1200米的林缘、灌丛或溪边潮湿处，偶有栽培。产华南和华东。东南亚亦有。

Shrubs. Stems and petioles appressed strigose. Leaf blades usually ovate or elliptic-ovate, apex cuspidate or acuminate, base rounded, broadly cuneate or abruptly narrowly cuneate; Stipules connate at base Panicles on flowering branches in the axils of current or recently fallen leaves. Achenes subovoid, base stipitate. Fl. May-Aug. Fr. Sep-Nov. Forest edges, thickets, moist places along streams, or occasionally cultivated at 200-1200 m. Distributed in S and E China. Also in SE Asia.

序叶苎麻

Boehmeria clidemioides Miq. var. **diffusa** (Wedd.) Hand.-Mazz.

多年生草本或亚灌木，多分枝。叶互生或有时在茎基部对生，叶型与大小多变，边缘具牙齿。穗状花序单性，顶端常具一簇2-4片小叶；雄花4数，无柄；花被片4，基部合生；果期雌花被基部较钝。花期6-8月，果期8-9月。生海拔200-2400米的林缘、路边或受干扰区的较干燥处。产中国西南、华南、东南、华中、华西和华东。南亚和东南亚亦有。

Perennial herbs or subshrubs, well branched. Leaves alternate or sometimes opposite in lower stems, shape and size variable, margin dentate. Spikes unisexual, always with a tuft of 2-4 small foliage leaves at the apex; staminate flowers 4-merous, sessile; tepals 4, connate at base; fruiting pistillate perianth base often obtuse. Fl. Jun-Aug. Fr. Aug-Sep. Forest edges, roadsides or often somewhat dry places in disturbed areas at 200-2400 m. Distributed in SW, S, SE, C, W and E China. Also in S and SE Asia.

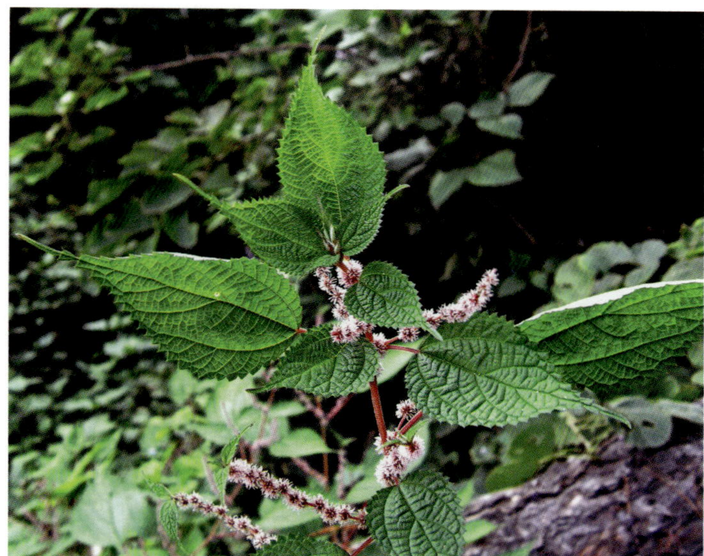

序叶苎麻 *Boehmeria clidemioides* var. *diffusa*

水苎麻

Boehmeria macrophylla Hornem.

半灌木或多年生草本。茎高1-2(-3.5)米。叶对生，卵形，长6.5-14厘米，边缘有牙齿。穗状花序单性，长7-15厘米，有短分枝，团伞花序直径1-2.5毫米；雄花四基数；雌花被纺锤形，长1毫米，顶端有2小齿；柱头长1-1.6毫米。花期7-9月。生海拔1800-3000米的山谷林下或沟边。产西藏东南部、云南、广西和广东北部。越南、缅甸、印度和尼泊尔亦有。

Subshrubs or perennial herbs. Stems 1-2(-3.5) m tall. Leaves opposite, ovate, 6.5-14 cm long, margin dentate. Spikes unisexual, 7-15 cm long, with short branches; glomerules 1-2.5 mm diam; staminate flower tetramerous; pistillate perianth fusiform, 1 mm long, at apex 2-denticulate; stigma 1-1.6 mm long. Fl. Jul-Sep. Under forests or by streams in valley at 1800-3000 m. Distributed in SE Xizang, Yunnan, Guangxi and N Guangdong. Also in Vietnam, Myanmar, India and Nepal.

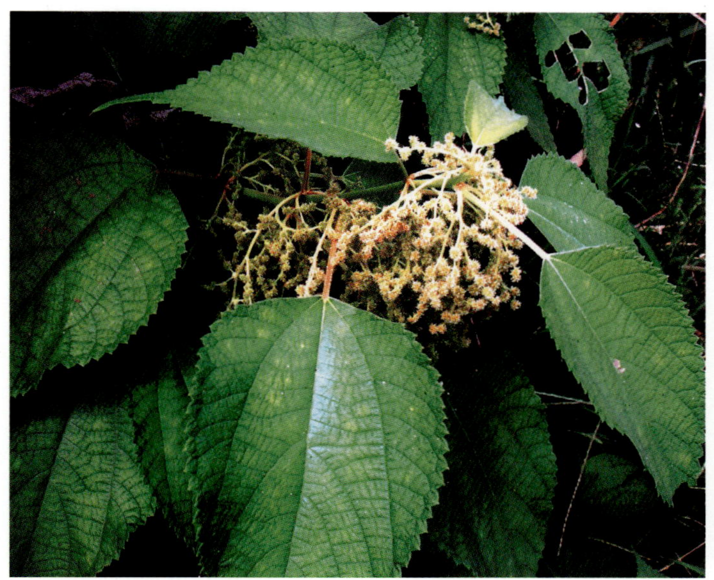

青叶苎麻 *Boehmeria nivea* var. *tenacissima*

水苎麻 *Boehmeria macrophylla*

海岛苎麻 *Boehmeria formosana*

野线麻 *Boehmeria japonica*

海岛苎麻
Boehmeria formosana Hayata

亚灌木或多年生草本。茎通常不分枝，具四棱。叶对生，卵形至披针形，基部钝或圆形。穗状花序腋生于上部，有时基部分枝；果期雌花被管阔菱形或倒卵球形，稍压扁，先端2(-4)齿裂。花期6-8月，果期8-10月。生海拔100-1400米的林中、灌丛中、溪边或路旁。产华南、东南和华东。日本亦有。

Subshrubs or perennial herbs. Stems often simple, 4-angled. Leaves opposite, ovate to lanceolate, base obtuse or rounded. Spikes sometimes branched at base; fruiting pistillate perianth tube broadly rhomboid or obovoid, slightly compressed, apex 2(-4)-toothed. Fl. Jun-Aug. Fr. Aug-Oct. Forests, thickets, along streams or roadsides at 100-1400 m. Distributed in S, SE and E China. Also in Japan.

野线麻
Boehmeria japonica (L. f.) Miq.

亚灌木或多年生草本。叶对生，近圆形、圆卵形或卵形，边缘有牙齿。穗状花序腋生，不分枝或有时具少量分枝；果期雌花被菱状倒卵球形，压扁，具2齿。花期6-8月，果期9-11月。生海拔300-1300米的林缘、灌丛、低地或丘陵溪边。产中国西南、华南、东南、华中和华东。日本亦有。

Subshrubs or perennial herbs. Leaves opposite, suborbicular, orbicular-ovate or ovate, dentate at margin. Spikes unbranched, or sometimes few-branched; fruiting pistillate perianth rhomboid-obovoid, compressed, 2-toothed. Fl. Jun-Aug. Fr. Sep-Nov. Forest edges, thickets, low mountains or by streams of hills at 300-1300 m. Distributed in SW, S, SE, C and E China. Also in Japan.

八角麻
Boehmeria tricuspis (Hance) Makino

亚灌木或多年生草本。叶对生稀互生，先端3骤尖或3浅裂，叶缘具粗牙齿。穗状花序常具分枝；果期雌花被呈倒卵状菱形，扁长，光滑，顶端有2个浅锯齿。花期7-8月，果期9-10月。生海拔500-1400米的林缘、灌丛或丘陵及山地溪边。产中国西南、华南、东南、华中、华北、华西和华东。朝鲜半岛和日本亦有。

Subshrubs or perennial herbs. Leaves opposite or rarely alternate, apex 3-cuspidate or 3-lobed, margin coarsely dentate. Spikes often branched; fruiting pistillate perianth rhomboidobovoid, compressed, smooth, apex 2-toothed. Fl. Jul-Aug. Fr.

八角麻 *Boehmeria tricuspis*

密球苎麻 *Boehmeria densiglomerata*

Sep-Oct. Forest edges, thickets or mountains and along streams in hills at 500-1400 m. Distributed in SW, S, SE, C, N, W and E China. Also in Korean Peninsula and Japan.

密球苎麻

Boehmeria densiglomerata W. T. Wang

多年生草本。叶对生，近圆形或圆卵形，草质，叶背浅紫色。雄性穗状花序常于基部分枝，雌性穗状花序不分枝，均具密集的团伞花序；雄花4数，花被片基部合生；果期雌花被菱状倒卵形，顶端具2齿。花期6-8月，果期9-10月。生海拔200-700(-1200)米的林中、灌丛或溪边。产中国西南、华南和华中。

Perennial herbs. Leaves opposite, suborbicular or orbicular-ovate, herbaceous, abaxial surface purplish. Staminate spikes often branched at base, pistillate spikes unbranched, all with very dense glomerules; staminate flowers 4-merous, tepals connate at base; fruiting pistillate perianth rhomboid-obovoid, apex 2-toothed. Fl. Jun-Aug. Fr. Sep-Oct. Forests, thickets or along streams at 200-700(-1200) m. Distributed in SW, S and C China.

细野麻

Boehmeria gracilis C. H. Wright

多年生草本或亚灌木。茎高达1.2米。叶对生，草质，圆卵形或卵形，顶端骤尖，基部圆形或宽楔形，边缘每侧有8-13正三角形牙齿。花期6-8月。生海拔100-1600米的山坡草地、石上或沟边。产广东北部、广西、贵州、湖南、江西、福建、浙江、安徽、湖北、重庆、陕西南部、河南、山西东南部、山东东部、河北西部和北部、辽宁南部和吉林东南部。朝鲜半岛和日本亦有。

Perennial herbs or subshrubs. Stems up to 1.2 m tall. Leaves opposite, herbaceous, orbicular-ovate or ovate, apex cuspidate, base rounded or broadly cuneate, margin at easide with 8-13 deltoid teeth. Fl. Jun-Aug. Grassy slopes, on rocks or by streams at 100-1600 m. Distributed in N Guangdong, Guangxi, Guizhou, Hunan, Jiangxi, Fujian, Zhejiang, Anhui, Hubei, Chongqing, S Shaanxi, Henan, SE Shanxi, E Shandong, W and N Hebei, S Liaoning and SE Jilin. Also in Korean Peninsula and Japan.

赤麻

Boehmeria silvestrii (Pamp.) W. T. Wang

多年生草本或亚灌木。茎高达1米。叶对生，草质，近五角形或圆卵形，顶端有3或5骤尖头，基部近截形或宽楔形。花期6-8月。生海拔700-2600米的低山草坡或山谷沟边或丘陵。产四川、重庆、湖北、甘肃南部、陕西南部、河南西南部、山东东部、河北西部和北部、辽宁南部和吉林东南部。朝鲜半岛和日本亦有。

Perennial herbs or subshrubs. Stems up to 1 m tall. Leaves opposite, herbaceous, subpentagonal or orbicular-ovate, apex 3- or 5-cuspidate, base subtruncate or broadly cuneate. Fl. Jun-Aug. Grassy slopes or by streams in hills or low mountains at 700-2600 m. Distributed in Sichuan, Chongqing, Hubei, S Gansu, S Shaanxi, SW Henan, E Shandong, W and N Hebei, S Liaoning and SE Jilin. Also in Korean Peninsula and Japan.

细野麻 *Boehmeria gracilis*

赤麻 *Boehmeria silvestrii*

小赤麻 *Boehmeria spicata*

八棱麻 *Boehmeria siamensis*

小赤麻

Boehmeria spicata (Thunb.) Thunb.

多年生草本或亚灌木。茎高达1米。叶对生，草质，卵状菱形，顶端骤尖，基部宽楔形，边缘每侧有3-9(-16)牙齿，上部牙齿狭三角形或三角形。花期6-8月。生低山草坡或丘陵。产浙江、江苏、江西、湖北、湖南西部和山东东部。朝鲜半岛和日本亦有。

Perennial herbs or subshrubs. Stems up to 1 m tall. Leaves opposite, herbaceous, ovate-rhombic, apex cuspidate, base broadly cuneate, margin at each side with 3-9(-16) teeth, upper teeth narrow-triangular or triangular. Fl. Jun-Aug. Grassy slopes of hills or low mountains. Distributed in Zhejiang, Jiangsu, Jiangxi, Hubei, W Hunan and E Shandong. Also in Korean Peninsula and Japan.

八棱麻

Boehmeria siamensis Craib

灌木。茎不分枝或分枝。叶对生，宽长圆形至宽椭圆形，边缘在基部以上有小牙齿。穗状花序在当年生枝顶部单生，在其下，2-4条生落叶叶腋处，具密集的团伞花序；果期雌花被纺锤形或狭菱状倒卵球形，先端具3小齿。花期3-4月，果期9-10月。生海拔400-1800米的山坡疏林下、灌丛中或路旁。产中国西南。缅甸、老挝、泰国和越南亦有。

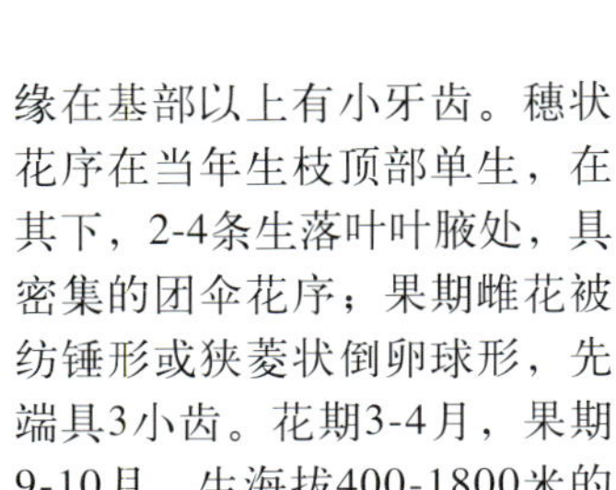

Shrubs. Stems simple or branched distally. Leaves opposite, broadly oblong to broadly elliptic, margin denticulate above base. Spikes solitary in distal part of twigs or in groups of 2-4 in axils of fallen leaves, with densely congested glomerules; fruiting pistillate perianth fusiform or narrowly rhomboid-obovoid, apex 3-denticulate. Fl. Mar-Apr. Fr. Sep-Oct. Sparse forests, bushes or by roads at 400-1800 m. Distributed in SW China. Also in Myanmar, Laos, Thailand and Vietnam.

长叶苎麻

Boehmeria penduliflora Wedd. ex Long

灌木。茎高1.5-4.5米，小枝被短糙伏毛。叶对生，披针形，长达25厘米。雄、雌穗状花序长达32厘米，具密集的团伞花序；雌花被顶端具2齿。花期7-10月。生海拔500-2000米的山谷林中、灌丛，林边或溪边。产西藏东南部、四川西南部、云南、贵州西南部、广西西部和南部。越南、泰国北部、缅甸北部、印度北部、不丹和尼泊尔亦有。

Shrubs. Stems 1.5-4.5 m tall, branchlets strigillose. Leaves opposite, lanceolate, up to 25 cm long. Staminate and pistillate spikes up to 32 cm long, with dense glomerules; pistillate perianth at apex 2-dentate. Fl. Jul-Oct. Forests, bushes, forest edges or by streams at 500-2000 m. Distributed in SE Xizang, SW Sichuan, Yunnan, SW Guizhou, W and S Guangxi. Also in Vietnam, N Thailand, N Myanmar, N India, Bhutan and Nepal.

密花苎麻

Boehmeria densiflora Hook. et Arn.

灌木或小乔木。茎高达4米。枝

长叶苎麻 *Boehmeria penduliflora*

条密被短糙伏毛。叶对生，披针形，长5-19(-24)厘米，宽1.2-5(-6.4)厘米，两面被糙伏毛。雌穗状花序长4-12(-18)厘米；雌花被纺锤形或倒披针形，长1-1.5毫米，顶端渐狭；柱头长0.7-1毫米。花期12月至翌年9月。生路边、河岸或多石山坡。产广东和台湾。菲律宾和日本南部亦有。

Shrubs or small trees. Stems up to 4 m tall. branches densely strigose. Leaves opposite, lanceolate, 5-19(-24) cm long, 1.2-5(-6.4) cm broad, on both surfaces strigose. Pistillate spikes 4-12(-18) cm long; pistillate perianth fusiform or oblanceolate, 1-1.5 mm long, apex attenuate; stigma 0.7-1 mm long. Fl. Dec to next Sep. By roads, on river banks or rocky slopes. Distributed in Guangdong and Taiwan. Also in the Philippines and S Japan.

微柱麻

Chamabainia cuspidata Wight

直立或斜升草本，具细长、丝状越冬根状茎；茎常紫色。叶对生，同节上的叶近等大，卵形或菱状卵形，草质。团伞花序直径3-10毫米；雄花花被片合生至中部。果期雌花被倒卵球形，顶端常截形。花期6-8月，果期8-11月。生海拔1000-2900米的林缘、灌丛、山谷、溪边或石缝中。产中国西南、华南、东南和华中。南亚和东南亚亦有。

Herbs erect or ascending, producing slender, filiform overwintering stolons; Stems often purplish. Leaves opposite, subequal in size at the same node, ovate or rhombic-ovate, herbaceous. Glomerules 3-10 mm diam; tepals of male flowers connate to middle; fruiting pistillate perianth obovoid, often truncate at top. Fl. Jun-Aug. Fr. Aug-Nov. Forest edges, thickets, valleys, along streams or in rock crevices at 1000-2900 m. Distributed in SW, S, SE and C China. Also in S and SE Asia.

红雾水葛 *Pouzolzia sanguinea*

红雾水葛

Pouzolzia sanguinea (Blume) Merr.

灌木。茎红色。叶卵形，2.6-11厘米长，先端渐尖，纸质，叶背密或疏被贴伏毛，基部圆形或宽楔形。团伞花序腋生，有时也生于无叶的侧枝上。瘦果卵球形，长约1.6毫米。花期5-6月，果期7-8月。生海拔300-2300米的温暖常绿阔叶林、山坡灌丛、林缘、干谷或路边。产中国西南和华南。南亚和东南亚亦有。

Shrubs. Stems reddish. Leaves ovate, 2.6-11 cm long, apex acuminate, papery, abaxial surface sparsely or densely pubescent, base rounded or broadly cuneate. Glomerules axillary, sometimes also on leafless lateral shoots. Achenes ovoid, ca. 1.6 mm long. Fl. May-Jun. Fr. Jul-Aug. Warm evergreen broad-leaved forests, thickets on slopes, edges of woods, dry valleys or roadsides at 300-2300 m. Distributed in SW and S China. Also in S and SE Asia.

密花苎麻 *Boehmeria densiflora*

微柱麻 *Chamabainia cuspidata*

菱叶雾水葛 *Pouzolzia elegans* var. *delavayi*

菱叶雾水葛

Pouzolzia elegans Wedd. var. **delavayi** (Gagnep.) W. T. Wang

灌木。茎高约2米，小枝被开展短硬毛。叶菱状卵形，长1-5厘米，被短硬毛，边缘具小齿。团伞花序单性或两性，无梗；雄花被片4，顶端具角状凸起；雌花被顶端具4齿。花期5-7月。生海拔1500-2200米的干燥河谷的灌丛中或林边。产西藏东南部、云南西北部和四川西部。

Shrubs. Stems ca. 2 m tall, branchlets spreading-hirtellous. Leaves rhombic-ovate, 1-5 cm long, hirtellous, margin denticulate. Glomerules unisexual or bisexual, sessile; staminate tepals 4, apex corniculate; pistillate perianth at apex 4-denticulate. Fl. May-Jul. Bushes or forest edges in dry valley at 1500-2200 m. Distributed in SE Xizang, NW Yunnan and W Sichuan.

雾水葛

Pouzolzia zeylanica (L.) Benn.

多年生草本。茎直立或斜升。叶对生或仅茎顶部对生，卵形，全缘，下面具硬毛或沿脉具糙毛，侧脉一对。团伞花序常两性，两性者生下面节上；雌花序生上部腋处；苞片三角形。瘦果白色、黄色或浅褐色，卵球形。花期7-8月，果期8-10月。生海拔100-800(-1300)米的草地、田边、灌丛或林缘。产中国西南、华南、东南和华中。热带和亚热带亚洲、澳大利亚和太平洋岛屿亦有。

Perennial herbs. Stems erect or ascending. Leaves opposite or alternate on lower part of stems, ovate, entire, abaxial surface strigillose or strigose along veins, lateral nerves 1 pair. Glomerules often bisexual, bisexual ones in proximal nodes; pistillate ones in distal axils; bracts triangular. Achenes white, yellow or light brown, ovoid. Fl. Jul-Aug. Fr. Aug-Oct. Grasslands, fields, thickets or woods at 100-800 (-1300) m. Distributed in SW, S, SE and C China. Also in tropical and subtropical Asia, Australia, and Pacific Islands.

多枝雾水葛

Pouzolzia zeylanica var. **microphylla** (Wedd.) W. T. Wang

多年生草本或半灌木。茎长40-100(-200)厘米，有时铺地，多分枝；末回小枝多数，互生，长2-10厘米，具多数长约5毫米的叶子。下部茎生叶对生，上部茎生叶互生，卵形，狭卵形或披针形，长0.5-5厘米。花期秋季。生海拔500米的草地、田边或草坡上。产云南东南部、广西、海南、广东、江西南部、福建和台湾。在亚洲热带地区广布。

Perennial herbs or subshrubs. Stems 40-100(-200) cm long, sometimes prostrate, much branched; ultimate branchlets numerous, alternate, 2-10 cm long, with many small leaves ca. 5 mm long. Lower cauline leaves opposite, upper ones alternate, ovate, narrowly ovate or lanceolate, 0.5-5 cm long. Fl. Autumn. Grasslands, grassy slopes, or by fields at 500 m. Distributed in SE Yunnan, Guangxi, Hainan, Guangdong, S Jiangxi, Fujian and Taiwan. Widespread in tropical Asia.

糯米团

Gonostegia hirta (Blume) Miq.

多年生草本。叶对生，叶片草质或纸质，狭卵形，具3(或5)脉。团伞花序常两性或单性；雄花花被片5，倒披针形；雌花无梗，花被管卵球形，具10纵翅，先端具2齿。瘦果卵球形，有光泽。花期5-7月，果期8-9月。生海拔100-1000(-2700)米的丘陵、沟边灌丛或草地。产中国西南、华南、华中和华东。热带和亚热带亚洲及大洋洲亦有。

雾水葛 *Pouzolzia zeylanica*

多枝雾水葛 *Pouzolzia zeylanica* var. *microphylla*

糯米团 *Gonostegia hirta*

瘤冠麻 *Cypholophus moluccanus*

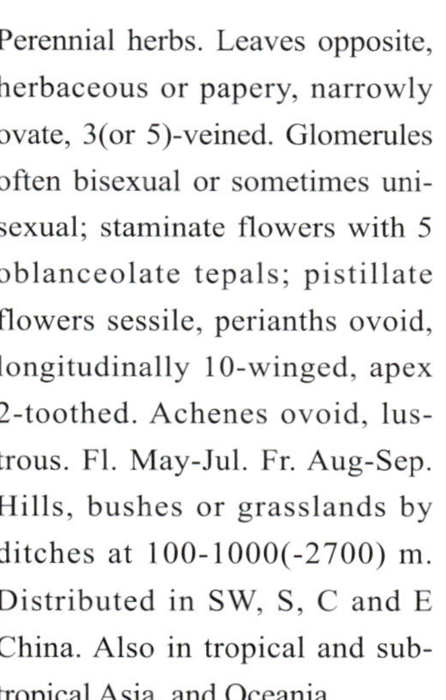

Perennial herbs. Leaves opposite, herbaceous or papery, narrowly ovate, 3(or 5)-veined. Glomerules often bisexual or sometimes unisexual; staminate flowers with 5 oblanceolate tepals; pistillate flowers sessile, perianths ovoid, longitudinally 10-winged, apex 2-toothed. Achenes ovoid, lustrous. Fl. May-Jul. Fr. Aug-Sep. Hills, bushes or grasslands by ditches at 100-1000(-2700) m. Distributed in SW, S, C and E China. Also in tropical and subtropical Asia, and Oceania.

瘤冠麻

Cypholophus moluccanus
(Blume) Miq.

灌木。茎直立。小枝被毡毛。叶对生，宽卵形，长10-25厘米，被糙伏毛，具三出脉。团伞花序无梗；雄花4基数；雌花被顶端有4小齿；果期肉质并包围瘦果；子房直立；柱头丝形，下弯，有长毛。花期4-7月。生低山地区阴湿处。产台湾东部。产印度尼西亚(苏门答腊)、菲律宾、密克罗尼西亚和美国(夏威夷)亦有。

Shrubs. Stems erect. branchlets pannose. Leaves opposite, broadly ovate, 10-25 cm long, strigose, trinerved. Glomerules sessile; staminate flowers tetramerous; pistillate perianth at apex 4-denticulate; at fruiting time carnose and surrounding achene; ovary erect; style filiform, recurved, with long hairs. Fl. Apr-Jul. Shady damp places in low mountains. Distributed in E Taiwan. Also in Indonesia (Pulau Sumatera), the Philippines, Micronesia and USA (Hawaii).

毛叶锥头麻

Poikilospermum lanceolatum
(Trèc.) Merr.

攀援灌木。叶披针形或椭圆形，下密被柔毛或近无毛。雄聚伞花序三至六回二歧状分枝；雌花序二至三回二歧状分枝；雄花无柄，倒圆锥状；花被片4，深红色；雄蕊4；雌花具短柄，柱头短舌状。瘦果长圆状椭圆体形。花期2-5月，果期5-7月。生海拔700-1800米的季雨林或溪边湿地。产西藏(墨脱县)和云南西南部。印度东北部和缅甸亦有。

Climbing shrubs. Leaves lanceolate or elliptic, densely pubescent or subglabrous abaxially. Staminate cymes 3-6 times dichotomously branched; pistillate cymes 2-3 times dichotomously branched; staminate flowers sessile, obpyramidal; tepals 4, dark red; stamens 4; pistillate flowers shortly pedicellate, stigma shortly ligulate. Achenes oblong-ellipsoid. Fl. Feb-May. Fr. May-Jul. Monsoon forests or wet places near streams at 700-1800 m. Distributed in Xizang (Mêdog County) and SW Yunnan. Also in NE India and Myanmar.

毛叶锥头麻 *Poikilospermum lanceolatum*

锥头麻 *Poikilospermum suaveolens*

锥头麻
Poikilospermum suaveolens (Blume) Merr.

攀援灌木。枝、叶无毛。叶片菱状近圆形或卵圆形。雄聚伞花序二至三回二歧分枝；苞片船形；雌花序一或二回二歧分枝；雄花无柄，倒卵球形，花被裂片4；雌花柱头短舌形。瘦果长圆状倒卵球形，具疣。花期4-5月，果期5-6月。生海拔500-600米的雨林、季雨林潮湿处或溪边。产云南南部。南亚和东南亚亦有。

Climbing shrubs. Branches and leaves glabrous. Leaves rhomboid-suborbicular or ovate. Staminate cymes 2-3 times dichotomously branched; bracts boat-shaped; pistillate cymes 1 or 2 times dichotomously branched; staminate flowers sessile, obovoid, perianth lobes 4; pistillate flowers with stigma shortly ligulate. Achenes oblong-obovoid, verrucose. Fl. Apr-May. Fr. May-Jun. Moist places, rain forests, monsoon forests, near streams at 500-600 m. Distributed in S Yunnan. Also in S and SE Asia.

落尾木
Pipturus arborescens (Link) C. B. Robinson

常绿灌木或小乔木；小枝和叶柄被白色短绒毛。叶互生；叶片卵形，长6-20厘米，具三出脉，下面密被白色短绒毛，边缘有细锯齿；叶柄长2-9.5厘米；托叶卵形，长约1厘米。团伞花序雌雄异株，腋生，球形；雄花4或5基数；雌花被筒状，顶端有4-5齿；果期稍肉质；子房贴生于花被；柱头丝形，脱落。花期4-6月。生低山林中。产台湾。菲律宾和日本南部亦有。

Evergreen shrubs or small trees; branchlets with petioles white-velutinous. Leaves alternate; blades ovate, 6-20 cm long, tri-nerved, abaxially densely white-velutinous, margin serrulate; petioles 2-9.5 cm long; stipules ovate, ca. 1 cm long. Glomerules dioecious, axillary, globose; staminate flower tetra- or pentamerous; perianth of pistillate flower tubular, apex 4-5-dentate; at fruiting time slightly fleshy; ovary adnate to perianth; stigma filiform, deciduous. Fl. Apr-Jun. Forests or in hills at low altitude. Distributed in Taiwan. Also in the Philippines and S Japan.

长梗紫麻
Oreocnide pedunculata (Shirai) Masam.

常绿灌木或小乔木。小枝被绵毛。叶卵状长圆形或卵状披针形，长达15厘米，具三出脉；托

落尾木 *Pipturus arborescens*

长梗紫麻 *Oreocnide pedunculata*

叶披针形。雄花簇生，3基数；雌花序长6-10毫米，二回二歧分枝。瘦果卵形，下部被壳斗状花托包围。花期3-5月。生低山林下。产台湾。日本南部亦有。

Evergreen shrubs or small trees. Branchlets lanate. Leaves ovate-oblong or ovate-lancolate, up to 15 cm long, trinerved; stipules lanceolate. Staminate flowers fascicled, trimerous; pistillate cyme 6-10 mm long, 2 times dichotomously branched. Achene ovate, below surrounded by a cupule-like receptacle. Fl. Mar-May. Under forests in hills. Distributed in Taiwan. Also in S Japan.

紫麻

Oreocnide frutescens (Thunb.) Miq.

灌木。小枝和叶柄紫褐色；花先叶开放或与叶同时。叶片卵形或窄倒卵形，先端渐尖或尾尖，草质或有时纸质，叶背常被灰白色毛。雌花序无梗；雄花花被片3，基部合生。瘦果卵球形，常具疣。花期2-4月，果期6-10月。生海拔300-2500米的混生林缘、沟谷、灌丛或路边。产中国西南、华南、东南、华中、华西和华东。南亚、东南亚和日本亦有。

Shrubs. Branchlets and petioles purplish brown; flowering before or with new leaf flush. Leaves ovate to narrowly obovate, apex acuminate or caudate, herbaceous, sometimes papery, often gray-white tomentose abaxially. Pistillate glomerules sessile; staminate flower with tepals 3, connate at base. Achenes ovoid, often verrucose. Fl. Feb-Apr. Fr. Jun-Oct. Mixed forest edges, valleys, thickets or roadsides at 300-2500 m. Distributed in SW, S, SE, C, W and E China. Also in S and SE Asia, and Japan.

鳞片水麻

Debregeasia squamata King ex Hook. f.

落叶矮生灌木。小枝粗壮，叶痕明显；皮刺肉质，弯生。叶卵圆形至心形，边缘具小牙齿。聚伞花序两性，生当年生枝和老枝上，二至三回二歧分枝。瘦果由膜质花被包被但不贴生。花期8-10月，果期10-12月(至翌年1月)。生海拔100-1500米的灌丛、山谷或阴湿地。产中国西南和华南。泰国、越南、马来西亚和加里曼丹岛北部亦有。

Deciduous dwarf shrubs. Branches stout, leaf scars conspicuous; protuberances fleshy, recurved. Leaves ovate to cordate, margin denticulate. Cymes androgynous, borne on current and previous years' branches, 2-3-dichotomously branched. Achenes enclosed by membranous perianth but not adnate to it. Fl. Aug-Oct. Fr. Oct-Dec (to next Jan). Thickets, valleys or shady and wet places at 100-1500 m. Distributed in SW and S China. Also in Thailand, Vietnam, Malaysia and N Kalimantan Island.

紫麻 *Oreocnide frutescens*

鳞片水麻 *Debregeasia squamata*

长叶水麻 *Debregeasia longifolia*

长叶水麻
Debregeasia longifolia (Burm. f.) Wedd.

灌木或小乔木，雌雄异株或同株。小枝纤细，淡红色或褐紫色，被开展微硬毛。叶长圆形至长圆状披针形或披针形，边缘有细齿。花序生当年和去年生枝上，二至四回二歧分枝。瘦果带红色或橙色，宿存花被与果实贴生。花期8-12月，果期9月至翌年2月。生海拔500-3200米河谷、溪边或林缘潮湿地上。产中国西南和华南。南亚和东南亚亦有。

Shrubs or small trees, dioecious or monoecious. Branchlets slender, reddish or purplish brown, spreading hirtellous. Leaves oblong to oblong-lanceolate or lanceolate, margin denticulate. Inflorescences borne on current and previous years' branches, 2-4-dichotomously branched. Achenes reddish or orange, enclosed by persistent perianth and adnate to it. Fl. Aug-Dec. Fr. Sep to next Feb. Valleys, by streams or wet places of forest edges at 500-3200 m. Distributed in SW and S China. Also in S and SE Asia.

水麻
Debregeasia orientalis C. J. Chen

灌木。小枝和叶柄密被贴生或近贴生的短毛。叶长圆状披针形或条状披针形，边缘有细齿。花序常生于去年枝上，开花早于叶芽，一至二回二歧分枝。瘦果与宿存肉质花被呈浆果状，橙色。花期2-4月，果期5-9月。生海拔300-2800米的溪谷阴湿处或溪边。产中国西南、东南和华中。印度、尼泊尔、不丹和日本亦有。

Shrubs. Young branches and petioles densely adpressed- or subadpressed-puberulous. Leaves oblong-lanceolate or linear-lanceolate, margin denticulate. Inflorescences borne always on previous years' branches, often flowering before leaf flush, 1-2-dichotomously branched. Achenes with persistent fleshy perianths, baccate, orange. Fl. Feb-Apr. Fr. May-Sep. Shady and wet places in valleys or along streams at 300-2800 m. Distributed in SW, SE and C China. Also in India, Nepal, Bhutan and Japan.

水丝麻
Maoutia puya (Hook.) Wedd.

灌木。茎高1-2米。小枝被硬毛。叶互生，椭圆形或卵形，具三出脉，背面被白色毡毛。聚伞花序分枝，成对腋生；雄花五基数；雌花具2合生的小花被片，子房直立，柱头画笔头状。瘦果具3棱。花期6-8月。生海拔400-2000米的山谷疏林或灌丛中。产西藏东南部、云南、广西、贵州和四川西南部。不丹、尼泊尔、印度北部和越南亦有。

水麻 *Debregeasia orientalis*

水丝麻 *Maoutia puya*

Shrubs. Stems 1-2 m tall. Branchlets hispid. Leaves alternate, elliptic or ovate, trinerved, abaxially white-pannose. Cymes ramose, in pairs axillary; staminate flower pentamerous; pistillate flower with 2 small connate tepals, ovary erect, stigma penicillate. Achenes triquetrous. Fl. Jun-Aug. Sparse forests or bushes at 400-2000 m. Distributed in SE Xizang, Yunnan, Guangxi, Guizhou and SW Sichuan. Also in Bhutan, Nepal, N India and Vietnam.

墙草

Parietaria micrantha Ledeb.

一年生铺散草本。茎平卧或渐升，长10-40厘米。叶卵形，长0.5-3厘米，无托叶。聚伞花序有少数花；花杂性；两性花4基数；雌花4花被片下部合生；柱头画笔头状。瘦果长约1.2毫米，光滑。花期6-7月。生海拔700-3500米的山坡阴湿草地、墙上或石下。自中国西南至西北和东北广布。在西亚、北非、大洋洲和南美洲广布。

Annual diffuse herbs. Stems prostrate or ascending, 10-40 cm long. Leaves ovate, 0.5-3 cm long, without stipules. Cymes few-flowered; flowers polygamous; bisexual flowers tetramerous; pistillate flower with 4 tepals connate below; stigma penicillate. Achenes smooth. Fl. Jun-Jul. Shady places on grassy slopes, on walls or below rocks at 700-3500 m. Widespread SW to NW and NE China. Also in W Asia, N Africa, Oceania and South America.

单蕊麻

Droguetia iners (Forssk.) Schweinf. subsp. **urticoides** (Wight) Friis et Wilmot-Dear

多年生草本。茎渐升，长20-40厘米。叶对生，卵形，长4.5-6厘米，具三出脉，边缘具小齿。团伞花序直径4毫米，无梗，成对腋生；雄花具1花被片和1雄蕊；雌花无花被，子房直立，柱头丝形。瘦果扁压，卵形。花期9-10月。生山地林中。产西藏东南部、云南西南部和台湾。印度东北部和印度尼西亚(爪哇)亦有。

Perennial herbs. Stems ascending, 20-40 cm long. Leaves opposite, ovate, 4.5-6 cm long, trinerved, margin denticulate. Glomerules 4 mm diam, sessile, in pairs axillary; staminate flower with 1 tepal and 1 stamen; pistillate flower without perianth, ovary erect, stigma filiform. Achenes compressed, ovate. Fl. Sep-Oct. Forests. Distributed in SE Xizang, SW Yunnan and Taiwan. Also in NE India and Indonesia (Java).

墙草 *Parietaria micrantha*

单蕊麻 *Droguetia iners* subsp. *urticoides*

川苔草科 Podostemonaceae

川藻
Terniopsis sessilis H. C. Chao

多年生草本。根暗红褐色至黄绿色，扁平，宽1-1.5毫米，具羽状分枝。茎侧生，具5-10叶。叶三列，卵形，长0.5-1毫米。花腋生；花被片3，基部合生；雄蕊(2)3；柱头3，垫状。蒴果椭圆体形，3瓣裂。花期11月至翌年1月，果期12月至翌年2月。生海拔380-410米急流中季节性淹没的岩石和木桩上。特产福建。

Perennial herbs. Roots dark red-brown to yellowish green, flattened, 1-1.5 mm broad, pinnately branched. Stems arising laterally from root, with 5-10 leaves. Leaves tristichous, ovate, 0.5-1 mm long. Flowers axillary; tepals 3, basally connate; stamens (2)3; stigmas 3, cushion-shaped. Capsules ellipsoid, 3-valved. Fl. Nov to next Jan. Fr. Dec to next Feb. On seasonally submerged rocks and woody piles in rapids at 380-410 m. Endemic to Fujian.

川藻 *Terniopsis sessilis*

川苔草 *Cladopus doianus*

川苔草
Cladopus doianus (Koidz.) Kôriba

多年生草本。根暗绿至黄绿色，带状，宽(0.8-)2.5(-2.9)毫米，单轴式分枝。叶狭条形，长3-3.5毫米。苞片近半圆形至近圆形，6-9掌状半裂；花单朵顶生；花被片2，条形；雄蕊1或2，等长于或短于子房；柱头2，倒卵状匙形。蒴果近球形，不等2瓣裂。花期9月至翌年2月，果期1-3月。生海拔200-430米的急流中岩石上。产福建。日本南部亦有。

Perennial herbs. Roots dark green to yellow green, ribbon-like, (0.8-)2.5(-2.9) mm broad, branching monopodially. Leaves narrowly linear, 3-3.5 mm long. Bracts semicircular to suborbicular, 6-9 palmately divided to middle; flowers solitary, terminal; tapals 2, linear; stamens 1 or 2, as long as or shorter than ovary; stigmas 2, obovate-spatulate. Capsule subglobose, unequally 2-valved. Fl. Sep to next Feb. Fr. Jan-Mar. On submerged rocks in rapids at 200-430 m. Distributed in Fujian. Also in S Japan.

水石衣
Hydrobryum griffithii (Wall. ex Griff.) Tulasne

多年生草本。根绿色，近圆形，干后革质。叶狭条形，长15-20毫米。苞片6，船形；花单朵顶生；花被片2，条形；雄蕊2；柱头2，鸡冠状，先端截形或微凹。蒴果椭圆体形；具纵肋，2瓣裂。花期8-11月，果期10月至翌年1月。生海拔220-1450米溪中岩石上。产云南南部。印度、尼泊尔、不丹、缅甸和越南亦有。

水石衣 *Hydrobryum griffithii*

Perennial herbs. Roots green, suborbicular, leathery when dry. Leaves narrowly linear, 15-20 mm long. Bracts 6, boat-shaped; flower solitary, terminal; tepals 2, linear; stamens 2; stigmas 2, subcristate, truncate or emarginate at apex. Capsules ellipsoid; longitudinally ribbed, 2-valved. Fl. Aug-Nov. Fr. Oct to next Jan. On rocks in streams at 220-1450 m. Distributed in S Yunnan. Also in India, Nepal, Bhutan, Myanmar and Vietnam.

日本水石衣

Hydrobryum japonicum Imamura

多年生草本。根绿色，近圆形，直径5-10厘米或超过。叶狭条形，长6-10毫米。苞片4(-6)，船形；花单朵顶生；花被片2，条形；雄蕊2；柱头2，狭条形，先端尖。蒴果椭圆体形，具纵肋，2瓣裂。花期10-11月，果期11-12月。生海拔1000-1100米的林缘溪中岩石上。产云南南部。缅甸、泰国和日本南部亦有。

Perennial herbs. Roots green, suborbicular, 5-10 cm (or more) diam. Leaves narrowly linear, 6-10 mm long. Bracts 4(-6), boat-shaped; flower solitary, terminal; tepals 2, linear; stamens 2; stigmas 2, linear, pointed at apex. Capsule ellipsoid, longitudinally ribbed, 2-valved. Fl. Oct-Nov. Fr. Nov-Dec. On rocks in streams at forest edges at 1000-1100 m. Distributed in S Yunnan. Also in Myanmar, Thailand and S Japan.

日本水石衣 *Hydrobryum japonicum*

山龙眼科 Proteaceae

网脉山龙眼 *Helicia reticulata*

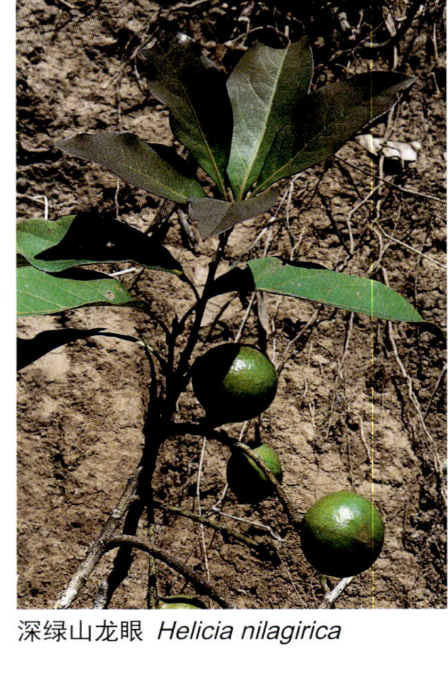
深绿山龙眼 *Helicia nilagirica*

大山龙眼
Helicia grandis Hemsl.

乔木，3-10米高。小枝、幼叶和花序具锈色绒毛。叶倒披针形，背面被绒毛。花序腋生或生小枝已落叶腋部；花被片红色。果暗棕色，椭圆体形至近球形；果皮厚约0.5毫米，革质。花期3-8月，果期10-12月。生海拔1100-2400米的林下潮湿处。产云南南部至东南部。越南北部亦有。

Trees, 3-10 m tall. Branchlets, young leaves, and inflorescences rust-colored tomentose. Leaves oblanceolate, abaxially tomentose. Inflorescences axillary or ramiflorous; perianth red. Fruits dull brown, ellipsoid to ± globose; pericarps ca. 0.5 mm thick, leathery. Fl. Mar-Aug. Fr. Oct-Dec. Damp places in forests at 1100-2400 m. Distributed in S to SE Yunnan. Also in N Vietnam.

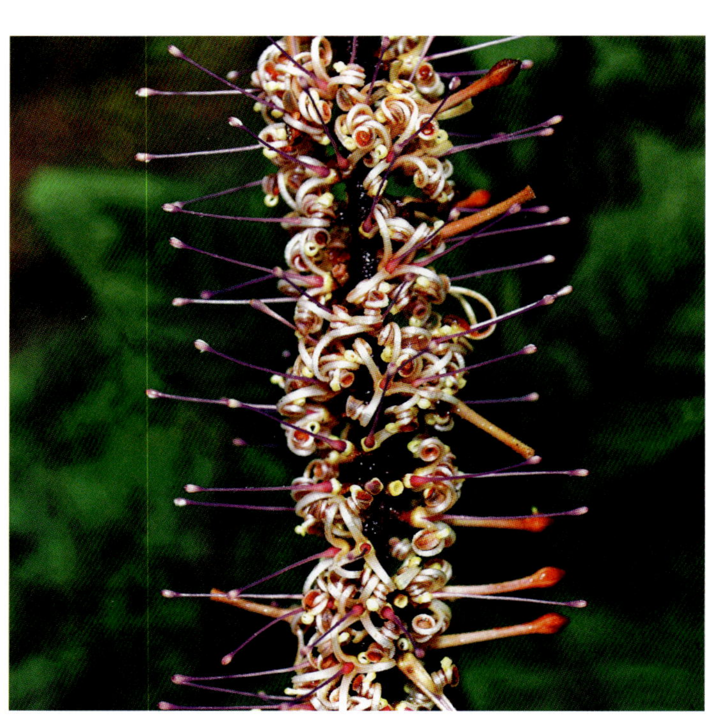
大山龙眼 *Helicia grandis*

深绿山龙眼
Helicia nilagirica Bedd.

乔木。小枝顶端密具锈色柔毛，渐无毛。叶纸质或多少革质，倒卵状长圆形或椭圆形；叶柄长0.5-3.5厘米。花序腋生或生小枝已落叶腋部；花被白色或浅黄色。果实深绿色，卵球形至球形；果皮厚1.5-3毫米，质硬，革质。花期5-8月，果期11月至翌年7月。生海拔1000-2000米的疏林、沟谷或阳坡。产云南。南亚亦有。

Trees. Branchlets apically densely rust-colored pilosulose, glabrescent. Leaves papyraceous or subleathery, obovate-oblong or elliptic; petioles 0.5-3.5 cm long. Inflorescences axillary or ramiflorous; perianth white or yellowish. Fruits dark green, ovoid to globose; pericarps 1.5-3 mm thick, hard, leathery. Fl. May-Aug. Fr. Nov to next Jul. Open forests, valleys or sunny slopes at 1000-2000 m. Distributed in Yunnan. Also in S Asia.

网脉山龙眼
Helicia reticulata W. T. Wang

灌木或小乔木。小枝和成熟叶渐无毛。叶革质或近革质，侧脉6-12对，网脉两面明显。花序腋生或生小枝已落叶腋部，无毛；花被白色或带黄色，长1.3-1.6厘米。果黑色，椭圆体形；果皮约1毫米厚，革质。花期5-7月，果期10-12月。生海拔300-2100米林下、次生灌丛、山坡或山脊。产中国西南、华南和东南。

Shrubs or small trees. Branchlets and mature leaves glabrescent. Leaves leathery to ± leathery, lateral veins 6-12 on each side; reticulate veins conspicuous on both surfaces. Inflorescences axillary or ramiflorous, glabrous; perianth white or yellowish, 1.3-1.6 cm long. Fruits black, ellipsoid; pericarps ca. 1 mm thick, leathery. Fl. May-Jul. Fr. Oct-Dec. Forests, secondary thickets, hilly slopes or mountain ridges at 300-2100 m. Distributed in SW, S and SE China.

瑞丽山龙眼
Helicia shweliensis W. W. Sm.

灌木或小乔木。小枝顶端具锈色柔毛。叶倒卵状长圆形，长圆形或倒披针形。花序生小枝已落叶腋部；苞片大，叶状，披针形或线状披针形，长5-20毫米；花被淡黄色。果暗绿色，球形，直径约1.5厘米；果皮厚约1毫米，革质。花期8-9月，果期9-12月。生海拔300-2800米的林中。产云南。

Shrubs or small trees. Branchlets apically rust-colored pilose. Leaves obovate-oblong, oblong, or oblanceolate. Inflorescences ramiflorous; bracts large, leaf-like, lanceolate to linear-lanceolate, 5-20 mm long; perianth yellowish. Fruits dark green, globose, ca. 1.5 cm diam; pericarps

瑞丽山龙眼 *Helicia shweliensis*

小果山龙眼 *Helicia cochinchinensis*

ca. 1 mm thick, leathery. Fl. Aug-Sep. Fr. Sep-Dec. Forests at 300-2800 m. Distributed in Yunnan.

小果山龙眼

Helicia cochinchinensis Lour.

灌木或乔木。小枝与叶无毛。叶纸质，长圆形；叶柄长6-15毫米。花序腋生；花被白色或淡黄色。果蓝黑色至黑色，椭圆体形，1-1.5 × 0.8-1厘米；果皮厚约0.5毫米，革质。花期6-10月，果期11月至翌年3月。生海拔1300米以下的林中湿润处、平原或山坡。产云南、广西、广东和海南。泰国、越南北部、柬埔寨和日本亦有。

Shrubs or trees. Branchlets and leaves glabrous. Leaves papyraceous, oblong; petioles 6-15 mm long. Inflorescences axillary; perianth whitish or yellowish. Fruits bluish black to black, ellipsoid, 1-1.5 × 0.8-1 cm; pericarps ca. 0.5 mm thick, leathery. Fl. Jun-Oct. Fr. Nov to next Mar. Damp places in forests, plains or mountain slopes below 1300 m. Distributed in Yunnan, Guangxi, Guangdong and Hainan. Also in Thailand, N Vietnam, Cambodia and Japan.

镰叶山龙眼

Helicia falcata C. Y. Wu

乔木。小枝顶端具褐色微柔毛，渐无毛。叶镰状披针形或长圆状披针形，全缘或极稀疏生浅锯齿。花序腋生，无毛；花被绿白色至浅红色。果绿黑色，椭圆体形，先端细尖；果皮厚约1毫米，革质。花期5-7月，果期10-11月。生海拔1200-1900米的林下潮湿处。产云南。越南北部亦有。

Trees. Branchlets apically brown pilosulose, glabrescent. Leaves falcate-lanceolate or oblong-lanceolate, entire, or rarely remotely serrulate. Inflorescences axillary, glabrous; perianth greenish white to reddish. Fruits greenish black, ellipsoid, apex apiculate; pericarps ca. 1 mm thick, leathery. Fl. May-Jul. Fr. Oct-Nov. Damp places in forests at 1200-1900 m. Distributed in Yunnan. Also in N Vietnam.

镰叶山龙眼 *Helicia falcata*

铁青树科
Olacaceae

疏花铁青树 *Olax austrosinensis*

疏花铁青树
Olax austrosinensis Y. R. Ling

灌木，有时攀援。叶片长卵形或椭圆形，革质，无毛，基部楔形或圆形。花序具5-6花；花萼小，具不明显锯齿；花瓣白色，2裂。核果红色，椭圆体形至倒卵球形，基部一半被增大的花萼包被。花期3-5月，果期4-9月。生海拔100-1600米的林中或山谷。产广西和海南。

Shrubs, sometimes climbers. Leaves oblong-ovate to elliptic, coriaceous, glabrous, base cuneate to rounded. Inflorescences 5-6 flowered; calyx small, obscurely dentate; petals white, 2-lobed. Drupes red, ellipsoid to obovoid, basal 1/2 covered by accrescent calyx. Fl. Mar-May. Fr. Apr-Sep. Forests or mountain valleys at 100-1600 m. Distributed in Guangxi and Hainan.

青皮木
Schoepfia jasminodora Sieb. et Zucc.

灌木或小乔木。叶卵形、狭卵形、椭圆形或卵状椭圆形。聚伞花序穗状，具2-9朵花；总花梗基部不具芽鳞；无花梗；花冠白色或浅黄色，管状。核果椭圆体形，红色至淡紫色。花期3-5月，果期4-6月。生海拔500-2600米的林中、山坡或山谷。产中国西南、华南、华中、华西和华东。泰国、越南和日本亦有。

Shrubs or small trees. Leaves ovate, narrowly ovate, elliptic, or ovate-elliptic. Cymes spicate, 2-9-flowered; peduncles base without bud scales; pedicels absent; corolla white or pale yellow, tubular. Drupes ellipsoid, red to purplish. Fl. Mar-May. Fr. Apr-Jun. Forests, slopes or valleys at 500-2600 m. Distributed in SW, S, C, W and E China. Also in Thailand, Vietnam and Japan.

青皮木 *Schoepfia jasminodora*

华南青皮木 *Schoepfia chinensis*

华南青皮木

Schoepfia chinensis Gardn. et Champ.

灌木或乔木。树皮棕紫色，具白色皮孔。叶长圆状披针形至椭圆形，基部不对称。花序具(1-) 2-4花。花冠粉色，坛状，喉部缢缩。核果红色至紫色，成熟后变黑，椭圆体形。花期2-4月，果期4-7月。生海拔2000米以下的林中、山坡或山谷。产中国西南、华南、东南和华东。

Shrubs or trees. Bark purplish brown with white lenticels. Leaves oblong-lanceolate to elliptic, base unequal. Inflorescences (1-)2-4-flowered. Corolla pink, urceolate, throat narrower. Drupes red to purple, blackish in age, ellipsoid. Fl. Feb-Apr. Fr. Apr-Jul. Forests, mountain slopes or valleys below 2000 m. Distributed in SW, S, SE and E China.

赤苍藤

Erythropalum scandens Blume

无毛藤本。小枝基部具宿存的芽鳞。叶腋有卷须；叶卵形、长圆状卵形或三角状卵形，叶基长盾形，先端锐尖，基出脉3-5。聚伞花序6-18厘米，多花；花萼杯状，具5个锯齿。核果椭圆体形至倒卵球形。花果期3-9月。生海拔100-1500米的河流淤积所形成的林中。产中国西南和华南。南亚和东南亚亦有。

Lianas, glabrous. Branchlets with persistent acute bud scales at base. Leaf axils with axillary tendrils; Leaves ovate, oblong-ovate, or triangular-ovate, base usually peltate, apex acuminate, basal veins 3-5. Cymes 6-18 cm long, many-flowered; calyx cupular, 5-dentate. Drupes ellipsoid to obovoid. Fl. and fr. Mar-Sep. Alluvial and riverine forests at 100-1500 m. Distributed in SW and S China. Also in S and SE Asia.

赤苍藤 *Erythropalum scandens*

山柚子科
Opiliaceae

山柑藤 *Cansjera rheedei*

山柑藤
Cansjera rheedei J. F. Gmelin

攀援藤本或直立灌木。小枝被黄色绒毛。叶近革质；侧脉4-6对。花序束状，长1-3厘米，具柔毛；花被片黄色，坛状；子房圆柱形；柱头头状，4裂。核果橘红色，椭圆体形，无毛。花期10-12月，果期翌年1-4月。生海拔1400米以下的林中或灌丛中。产云南、广西、广东和海南。南亚、东南亚、澳大利亚和太平洋岛屿亦有。

Lianas climbing or erect shrubs. Branchlets yellowish tomentose. Leaves subcoriaceous, veins 4-6 on each side of midvein. Inflorescences fascicled, 1-3 cm long, tomentulose; perianth yellowish, urceolate; ovary cylindric; stigmas capitate, 4-lobed. Drupes orange-red, ellipsoid, glabrous. Fl. Oct-Dec. Fr. next Jan-Apr. Forests or thickets below 1400 m. Distributed in Yunnan, Guangxi, Guangdong and Hainan. Also in S and SE Asia, Australia and Pacific Islands.

甜菜树
Yunnanopilia longistaminea (W. Z. Li) H. S. Kiu

小乔木。小枝绿色，有皮孔。叶椭圆形，长10-12厘米，宽3-3.5厘米，基部锲形，全缘，无毛，密被小凸起。花序在主干和老枝上簇生，在小枝上单生于叶腋，被毛；花瓣4，淡绿色；雄蕊与花瓣同数且对生；花药卵形，黄色；花盘微4裂，裂瓣与雄蕊互生。核果浅黄色，椭圆形。花期3-4月，果期4-7月。生海拔300-1300米的森林、灌木丛等地。产广西和云南。

Small trees. Branchlets green, with lenticels. Leaves elliptic, 10-12 × 3-3.5 cm, base wedge shaped, entire, glabrous, with dense small bulges. Inflorescence in trunks and older branches fascicled, axillary on branchlets, pubescent; petals 4, pale green; stamen and petal with same number and opposite; anthers ovate, yellow; disk 4-cleft, with lobes alternate with stamens. Drupe yellow, ovate. Fl. Mar-Apr. Fr. Apr-Jul. Forests, thickets at 300-1300 m. Distributed in Guangxi and Yunnan.

甜菜树 *Yunnanopilia longistaminea*

檀香科
Santalaceae

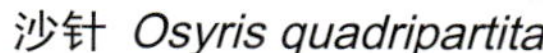
沙针 *Osyris quadripartita*

檀梨
Pyrularia edulis (Wall.) A. DC.

落叶灌木或小乔木。树皮灰色，皮孔长方形。小枝黄褐色，有或无刺毛。叶纸质或带肉质。圆锥状聚伞花序顶生(或腋生)；花杂性，下部花组成伞形或聚伞花序，顶端花单生。核果梨形。花期12月至翌年4月，果期翌年8-11月。生海拔1200-2700米常绿阔叶林中。产中国西南、华南和华中。印度、尼泊尔、不丹和缅甸亦有。

Shrubs or small trees, deciduous. Bark gray, lenticels oblong. Branches yellowish brown, with or without spines. Leaves papery or slightly fleshy. Inflorescences terminal (or axillary), thyrses; flowers polygamous, lower ones in umbels or cymes apical ones solitary. Drupes pear-shaped. Fl. Dec to next Apr. Fr. next Aug-Nov. Evergreen broad-leaved forests at 1200-2700 m. Distributed in SW, S and C China. Also in India, Nepal, Bhutan and Myanmar.

沙针
Osyris quadripartita Salzm. ex Decne.

直立灌木。枝常呈三棱形。叶浅灰绿色，革质，有时两面发皱，颇密集。雄花序具2-4(-13)花；雌花序具1-3花。核果浆果状，球形，成熟时橙色至红色，干后浅黑色。花期4-6月，果期10月。生海拔600-2700米的灌丛。产云南、四川、西藏和广西。南亚亦有。

Erect shrubs. Branches 3-angled. Leaves grayish green, coriaceous, sometimes rugose on both surfaces, with dense glands. Staminate inflorescences 2-4(-13)-flowered; pistillate inflorescences 1-3-flowered. Drupes baccate, globose, orange to red when ripe, after drying pale blackish. Fl. Apr-Jun. Fr. Oct. Thickets at 600-2700 m. Distributed in Yunnan, Sichuan, Xizang and Guangxi. Also in S Asia.

重寄生
Phacellaria fargesii Lecomte

亚灌木或草本，常重寄生于桑寄生科和寄生藤属的茎上。茎密集，不分枝。叶几无。花单生，两性，小且无柄，常极密生；苞片大，长1-1.3毫米，顶端反折。核果卵状椭圆形，具5-6条纵沟。花果期7-8月。生海拔1000-1400米的林中。产四川、贵州、广西和湖北。

Subshrubs or herbs, usually epiparasitic on the stems of species of Loranthaceae and Dendrotrophe. Stems in dense clusters, unbranched. Leaves almost absent. Flowers solitary, bisexual, minute, sessile, often very densely clustered; bracts big, 1-1.3 mm long, apex reflexed. Drupes ovoid-ellipsoid, 5-6-furrowed. Fl. and fr. Jul-Aug. Forests at 1000-1400 m. Distributed in Sichuan, Guizhou, Guangxi and Hubei.

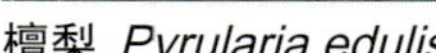
檀梨 *Pyrularia edulis*

重寄生 *Phacellaria fargesii*

聚果重寄生

Phacellaria glomerata D. D. Tao

亚灌木或草本，重寄生于柳树寄生。茎簇生，明显不分枝，具柔毛。花单性，密集成伸长的簇状；苞片覆瓦状排列，密被灰褐色毛；雌花花被片倒卵形，裂片5；柱头盾状。核果球形，无毛。种子黑色，卵形。生海拔约2400米的杂木林中。产云南(景东县)。

Subshrubs or herbs, epiparasitic on *Taxillus delavayi*. Stems in clusters, apparently unbranched, puberulous. Flowers unisexual, in dense, elongated clusters; bracts imbricate, densely brownish gray pilose; female flowers with obovoid perianth, lobes 5; stigmas peltate. Drupes globose, glabrous. Seeds black, ovoid. Mixed forests at ca. 2400 m. Distributed in Yunnan (Jingdong County).

聚果重寄生 *Phacellaria glomerata*

寄生藤

Dendrotrophe varians (Blume) Miq.

木质藤本，雌雄异株，常呈灌木状。枝深灰黑色，嫩时黄绿色，三棱形，扭曲。叶厚，具掌状3-9(-11)脉。花通常单性。核果浅红色，成熟时黄褐色至红褐色，卵球形。花期1-3月，果期6-8月。生海拔100-300米山坡灌丛或疏林中，常攀援于树上。产广西、广东、海南和福建。东南亚亦有。

Woody vines, usually shrubby. Stems darkish black, yellowish green when young, 3-ribbed, twisted. Leaves thick, palmately 3-9(-11)-veined. Flowers usually unisexual; dioecious. Drupes reddish, brownish yellow to reddish brown when mature, ovoid. Fl. Jan-Mar. Fr. Jun-Aug. Thickets on slopes or open forests at 100-300 m, often climbing on trees. Distributed in Guangxi, Guangdong, Hainan and Fujian. Also in SE Asia.

寄生藤 *Dendrotrophe varians*

无刺硬核

Scleropyrum wallichianum (Wight et Arn.) Arn. var. **mekongense** (Gagnep.) Lecomte

乔木。茎或枝无刺。叶被疏毛，基部楔形。花序单生、成对或数个成簇，具黄色绒毛；花被片黄色至黄红色，下面具长柔毛。核果成熟后橙色或橙红色，顶端乳头状，宿存花被片不明显膨大。花期4-5月，果期8-9月。生海拔600-1700米的密林中。产云南南部。老挝、越南和柬埔寨亦有。

Trees. Stems or branches spineless. Leaves pilose, base cuneate. Inflorescences solitary, paired or a few in fascicles, yellow tomentose; perianth pale yellow to reddish yellow, abaxially villous. Drupes orange or orange-red when mature, apex nipple-like, persistent perianth not conspicuously enlarged. Fl. Apr-May. Fr.

无刺硬核 *Scleropyrum wallichianum* var. *mekongense*

Aug-Sep. Dense forests at 600-1700 m. Distributed in S Yunnan. Also in Laos, Vietnam and Cambodia.

长梗百蕊草

Thesium chinense Turcz. var. **longipedunculatum** Y. C. Chu

多年生草本。茎簇生。叶线形。花序总状，多花；苞片线状披针形；小苞片2，线形；花被5数，绿白色；雄蕊不伸出。小坚果淡绿色，椭圆体形或近球形；宿存花被近球形；果梗长达8毫米。花期4-5月，果期6-7月。生山谷、山坡、草地或田中。产中国大部分地区。蒙古、朝鲜半岛和日本亦有。

Perennial herbs. Stems clustered. Leaves linear. Inflorescences racemelike, many flowered; bracts linear-lanceolate; bracteoles 2, linear; perianth 5-merous, greenish white; stamens not exserted. Nutlets light green, ellipsoid or subglobose; persistent perianth subglobose; Fruiting pedicel up to 8 mm. Fl. Apr-May. Fr. Jun-Jul. Valleys, slopes, grasslands or fields. Distributed in most parts of China. Also in Mongolia, Korean Peninsula and Japan.

急折百蕊草 *Thesium refractum*

长梗百蕊草 *Thesium chinense* var. *longipedunculatum*

急折百蕊草

Thesium refractum C. A. Mey.

多年生草本。叶无柄，披针形。花序总状，常分枝，花轴之字形，每腋具1花；苞片叶状；花被片5数，白色。小坚果椭圆体形或卵球形；果梗在果实成熟后反折。花期7月，果期9月。生草甸或沙质山坡。产华中、华西、西北、华北和东北。中亚和东亚亦有。

Perennial herbs. Leaves sessile, lanceolate. Inflorescences racemelike, often branched, rachis zigzag, with 1 flower per axil; bracts leaflike; perianth 5-merous, white. Nutlets ellipsoid or ovoid; fruit stalk reflexed in mature fruit. Fl. Jul. Fr. Sep. Meadows or sandy slopes. Distributed in C, W, NW, N and NE China. Also in C and E Asia.

桑寄生科 Loranthaceae

鞘花
Macrosolen cochinchinensis (Lour.) Tiegh.

灌木。叶革质，5-10 × 2.5-6厘米。花序单生或2-3个簇生叶腋，有时(2-)4-8生于小枝已落叶腋部；总花梗长15-20毫米；花冠橙黄；小苞片2，三角形，基部彼此合生。浆果橙色，近球形。花期2-6月，果期5-8月。生海拔1600米以下常绿阔叶林、沟谷或山坡，寄生于板栗、樟、杉木、黄葛树、枫香、木荷或木油桐。产中国西南、华南和东南。南亚、东南亚和新几内亚岛亦有。

Shrub. Leaves 5-10 × 2.5-6 cm, leathery. Inflorescences solitary or 2-3 fascicled, axillary, sometimes at older leafless nodes, (2-)4-8-flowered racemes; peduncles 15-20 mm long; corolla orange-yellow; bracteoles 2, triangular, connate at base. Berries orange, subglobose. Fl. Feb-Jun. Fr. May-Aug. Evergreen broad-leaved forests, valleys or mountain slopes below 1600 m, parasitic on *Castanea mollissima, Cinnamomum camphora, Cunninghamia lanceolata, Ficus virens, Liquidambar formosana, Schima superba* or *Vernicia montana.* Distributed in SW, S and SE China. Also in S and SE Asia, and New Guinea.

鞘花 *Macrosolen cochinchinensis*

双花鞘花
Macrosolen bibracteolatus (Hance) Danser

灌木，全株无毛。叶革质。伞形花序，1-4个腋生或小枝已落叶腋部；伞形花序具2花；雄花芽长3.2-3.5厘米；花冠红色，冠筒顶端带绿色条纹。浆果红色，椭圆体形。花期11-12月，果期12月至翌年4月。生海拔300-1800米的常绿阔叶林中，常寄生于五月茶、樟、木荷和山矾属植物上。产贵州、广东、广西、海南和云南。缅甸、越南和马来西亚亦有。

双花鞘花 *Macrosolen bibracteolatus*

Shrub, whole plant glabrous. Leaves coriaceous. Umbels 1-4, axillary or at deciduous leaf axil of branchlet, 2-flowered; mature buds 3.2-3.5 cm long; corolla red with green band at top of tube. Berries red, ellipsoid. Fl. Nov-Dec. Fr. Dec to next Apr. Evergreen broad-leaved forests at 300-1800 m, often parasitic on *Antidesma bunius*, *Cinnamomum camphora*, *Schima superba*, and *Symplocos* spp. Distributed in Guizhou, Guangdong, Guangxi, Hainan and Yunnan. Also in Myanmar, Vietnam and Malaysia.

离瓣寄生
Helixanthera parasitica Lour.

灌木。枝和叶均无毛。叶对生，纸质或薄革质。总状花序，具花40-60朵，花梗长1-2毫米被乳头状毛；副萼环状；花瓣上半部披针形，反折；花药4室；花柱柱状，具5棱；柱头头状。果椭圆状，红色，被乳头状毛。花期1-7月，果期5-8月。生海拔20-1800米的平原或山地常绿阔叶林，寄生于壳斗科、樟科等。产福建、广东、广西、贵州、海南、西藏(墨脱县)和云南。

Shrubs. Branch and leaves glabrous. Leaves opposite, papery to thinly leathery. Racemes 40-60-flowered; pedicel 1-2 mm, papillose; accesory calyx annu-

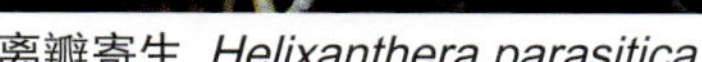
离瓣寄生 *Helixanthera parasitica*

lar; corolla, reflexed from above basal keels; anthers 4-loculed; style cylindric, 5-angled; stigma capitate. Berry red, ellipsoid, papillose. Fl. Jan-Jul. Fr. May-Aug. Forests in plains or mountain slopes at 20-1800 m, parasitic on Lauraceae, Fagaceae spp. Distributed in Fujian, Guangdong, Guangxi, Guizhou, Hainan, Xizang (Mêdog County) and Yunnan.

北桑寄生

Loranthus tanakae Franch. et Sav.

落叶灌木。枝常二歧分枝，黑色，具散生皮孔。叶倒卵形或椭圆形，纸质。穗状花序顶生，具10-20朵花；花两性，近对生；花冠绿色，花瓣披针形；花柱常具6棱；柱头稍头状。浆果黄色，球形。花期5-6月，果期9-11月。生海拔900-2000 (-2600) 米的阔叶林或人工林中。产华北、华西和华东。朝鲜半岛和日本亦有。

Deciduous shrubs. Branches usually dichotomous, black, scattered lenticellate. Leaves obovate or elliptic, papery. Spikes terminal, 10-20-flowered; flowers bisexual, subopposite; corolla greenish, petals lanceolate; styles usually 6-angled; stigmas slightly capitate. Berries orange, globose. Fl. May-Jun. Fr. Sep-Nov. Broad-leaved forests or plantations at 900-2000(-2600) m. Distributed in N, W and E China. Also in Korean Peninsula and Japan.

椆树桑寄生

Loranthus delavayi Tiegh.

灌木。小枝不分枝，淡黑色，具散生皮孔。叶卵形至椭圆形，纸质或革质。穗状花序腋生，8-16朵，单生或2-3簇；花冠黄绿色，花瓣条状匙形；花药4室。浆果椭圆形或卵形，果皮平滑。花期1-3月，果期9-10月。生海拔(200-)500-3000米的林中、山谷或山坡，常寄生于赤杨、沙梨和山毛榉科植物上，稀寄生云南油杉。产中国西南、华南、华中和东南。缅甸和越南亦有。

Shrubs. Branches not dichotomous, blackish, scattered lenticellate. Leaves ovate to elliptic, papery or leathery. Racemes axillary, 8-16-flowered, solitary or 2-3-fascicled; corolla yellowish green, petals linear-spatulate; anthers 4-loculed. Berries ellipsoid or ovoid, smooth. Fl. Jan-Mar. Fr. Sep-Oct. Forests, valleys or slopes at (200-)500-3000 m, parasitic on *Alnus japonica*, *Pyrus pyrifolia* and *Fagaceae* spp., rarely on *Keteleeria evelyniana*. Distributed in SW, S, C and SE China. Also in Myanmar and Vietnam.

北桑寄生 *Loranthus tanakae*

椆树桑寄生 *Loranthus delavayi*

五蕊寄生 *Dendrophthoe pentandra*

五蕊寄生

Dendrophthoe pentandra (L.) Miq.

灌木。叶披针形至椭圆形或近圆形，厚革质，互生或在短枝上近对生。总状花序，单生或2-3个簇生于小枝，具花3-10朵；花基数5，辐射对称；花冠橙色，基部一半稍膨大。浆果红色。花果期12月至翌年6月。生海拔100-1600米的亚热带山地常绿阔叶林，常寄生于乌榄、白榄、木油桐、芒果、黄皮和榕属多种植物上。产广东、广西和云南。南亚和东南亚亦有。

Shrubs. Leaves lanceolate to elliptic or suborbicular, thickly coriaceous, alternate or subopposite. Racemes solitary or 2-3 together, 3-10-flowered; flowers 5-merous, actinomorphic; corolla orange, basal 1/2 slightly inflated. Berries red. Fl. And fr. Dec to next Jun. Montane subtropical evergreen broad-leaved forests at 100-1600 m, parasitic on branches of *Canarium pimela, C. album, Vernicia montana, Mangifera indica, Clausena lansium* and *Ficus* spp. Distributed in Guangdong, Guangxi and Yunnan. Also in S and SE Asia.

红花寄生

Scurrula parasitica L.

灌木。叶卵形或长卵形，幼时密被锈色星状毛。总状花序或2或3个簇生叶腋，具3-5(-6)花；花冠红色或黄绿色，稍反折，长2-2.5厘米。浆果梨形，平滑。花果期10月至翌年1月。生海拔100-2100(-2800)米的平原或山坡，寄生于柚、柠檬、柑橘、黄皮、紫薇、桑、桃、梨、无患子、竹叶花椒，夹竹桃科、大戟科、山茶科、榆科或无患子科的植物上，稀寄生于干香柏和云南油杉。产中国西南、华南和东南。南亚和东南亚亦有。

Shrubs. Leaves ovate to ovate-oblong, ferruginous pubescent when young. Racemes solitary or 2 or 3 fascicled, axillary, 3-5(-6)-flowered; corolla red or greenish yellow, slightly curved, 2-2.5 cm long. Berries pyriform, smooth. Fl. and fr. Oct to next Jan. Evergreen broad-leaved forests at 100-2100 (-2800) m, parasitic on *Citrus grandis, C. limon, C. reticulata, Clausena lansium, Lagerstroemia indica, Morus alba, Prunus persica, Pyrus pyrifolia, Sapindus mukorossi, Zanthoxylum armatum*, or species of Apocynaceae, Euphorbiaceae, Theaceae, Ulmaceae and Sapindaceae, rarely parasitic on *Cupressus duclouxiana* and *Keteleeria evelyniana*. Distributed in SW, S and SE China. Also in S and SE Asia.

红花寄生 *Scurrula parasitica*

柳叶钝果寄生 *Taxillus delavayi*

小红花寄生

Scurrula parasitica L. var. **graciliflora** (Roxb. ex J. H. Schultes) H. S. Kiu

本变种与红花寄生的区别在于本变种的花冠黄绿色，裂片长约3毫米；雄花蕾长约1-1.2厘米。花果期2-12月。生海拔100-2100米的常绿阔叶林中，常寄生于云南羊蹄甲、普洱茶、锥栗、柚、杏、桃、沙梨、石榴和槐，稀寄生在松属植物上。产中国西南和华南。南亚亦有。

This variety differs from the typical variety in its corolla greenish yellow, lobes ca. 3 mm long; mature bud 1-1.2 cm long. Fl. and fr. Feb-Dec. Evergreen broad-leaved forests at 100-2100 m, parasitic on *Bauhinia yunnanensis, Camellia sinensis* var. *assamica, Castanea henryi, Citrus grandis, Prunus armeniaca, P. persica, Pyrus pyrifolia, Punica granatum* and *Sophora japonica*, rarely parasitic on species of *Pinus*. Distributed in SW and S China. Also in S Asia.

柳叶钝果寄生

Taxillus delavayi (Tiegh.) Danser

灌木，植株无毛。叶卵圆形，或椭圆形至披针形，3-5 × 1.5-2厘米。伞形花序单生或2个合生，具2-4花；花冠红色，稍反折，无毛。浆果黄色或橙色，椭圆体形。花期2-7月，果期5-9月。生海拔1500-3500米的阔叶林中，寄生于马桑、核桃、桃、微毛樱桃、沙梨、花楸或槭属、桦木属、杨属、栎属、杜鹃属和柳属，稀寄生云南油杉。产中国西南。缅甸和越南亦有。

Shrubs, glabrous. Leaves ovate, or elliptic to lanceolate, 3-5 × 1.5-2 cm. Umbels solitary or 2 together, 2-4-flowered; corolla red, slightly curved, glabrous. Berries yellow or orange, ellipsoid. Fl. Feb-Jul. Fr. May-Sep. Broad-leaved forests at 1500-3500 m, parasitic on *Coriaria sinica, Juglans cathayensis, Prunus persica, Prunus pilosiuscula, Pyrus pyrifolia, Sorbus wilsoniana* and species of *Acer, Betula, Populus, Quercus, Rhododendron* and *Salix* spp., rarely parasitic on *Keteleeria evelyniana*. Distributed in SW China. Also in Myanmar and Vietnam.

小红花寄生 *Scurrula parasitica* var. *graciliflora*

松柏钝果寄生 *Taxillus caloreas*

松柏钝果寄生
Taxillus caloreas (Diels) Danser

灌木，雌雄异株。小枝密具褐色星状毛，后无毛。叶对生或稀3枚轮生，近匙形至条形，厚革质或革质。伞形花序单生或双生，具2或3花；花冠红色，无毛。浆果紫色，近球形，具宿存花柱。花期7-8月，果期翌年4-5月。生海拔900-2800(-3100)米的山区林中，常寄生于铁坚油杉、台湾云杉、马尾松、台湾松、云南铁杉和雪松属的植物上。产中国西南、华南和华中。不丹亦有。

Shrubs, dioecious. Branchlets with dense brown stellate hairs, becoming glabrous. Leaves opposite or rarely 3 verticillate, subspatulate to linear, thick coriaceous or coriaceous. Umbels solitary or 2 together, 2- or 3-flowered; corolla red, glabrous. Berries purple, subglobose, with persistent styles. Fl. Jul-Aug. Fr. next Apr-May. Montane forests at 900-2800 (-3100) m, parasitic on *Keteleeria davidiana*, *Picea morrisonicola*, *Pinus massoniana*, *P. taiwanensis*, *Tsuga dumosa* and species of *Cedrus*. Distributed in SW, S and C China. Also in Bhutan.

木兰寄生
Taxillus limprichtii (Grüning) H. S. Kiu

灌木，具深棕色或棕黄色星状毛。小枝具散生皮孔。叶对生或近对生，革质，基部常下延到叶柄。伞状花序1-3，常生于无叶茎节处，具2-6朵花；花冠红色或橙色，基部膨大。浆果黄色至红黄色，椭圆形，具疣。花期10月至翌年3月，果期6-7月。生海拔200-2200米的林中、山坡或山谷，寄生于木兰科、壳斗科、金缕梅科、樟科和梧桐科。产中国西南、华南和华中。泰国和越南亦有。

Shrubs, with deep brown or yellowish brown stellate hairs. Branches scattered lenticellate. Leaves opposite or subopposite, coriaceous, base frequently decurrent into petiole. Umbels 1-3, sometimes at leafless node, 2-6-flowered; corolla red or orange, basal parts inflated. Berries yellowish or reddish yellow, ellipsoid, verrucose. Fl. Oct to next Mar. Fr. Jun-Jul. Forests, slopes or valleys at 200-2200 m, parasitic on Magnoliaceae, Fagaceae, Hamamelidaceae, Lauraceae and Sterculiaceae. Distributed in SW, S and C China. Also in Thailand and Vietnam.

木兰寄生 *Taxillus limprichtii*

龙陵钝果寄生
Taxillus sericus Danser

灌木。幼茎被绒毛。叶互生或近对生，长圆状卵形或阔长圆形，稍呈革质或革质。伞形花序1-3，有时生无叶节处，具2-4花；花冠红黄色，具平伏星状毛。浆果淡黄色，椭圆体形。花果期8月至翌年2月。生海拔1500-2700米的林中或山坡，常寄生于尼泊尔桤木、桦木属或壳斗科植物。产云南和西藏。印度北部亦有。

Shrubs. Young stems tomentose. Leaves alternate or subopposite, oblong-ovate or broadly oblong, slightly leathery to leathery. Umbels 1-3, sometimes at leafless nodes, 2-4-flowered; corolla reddish yellow, with adpressed stellate hairs. Berries yellowish, ellipsoid. Fl. and fr. Aug to next Feb. Forests or mountain slopes at 1500-2700 m, parasitic on *Alnus nepalensis*, *Betula* or Fagaceae spp. Distributed in Yunnan and Xizang. Also in N India.

桑寄生
Taxillus sutchuenensis (Lecomte) Danser

灌木。小枝黑色，具皮孔。叶近对生或互生。近伞形总状花序单生或成2或3束，具2-5朵

龙陵钝果寄生 *Taxillus sericus*

花；花冠红色，稍反折，基部膨大；花药多室；花柱红色；柱头圆锥形。浆果黄绿色，椭圆体形。花期6-9月，果期8-10月。生海拔500-1900米的林中、山坡或山谷。产中国西南、华南、东南、华中、华北和华西。

Shrubs. Branches black, lenticellate. Leaves subopposite or alternate. Subumbellate racemes solitary or 2 or 3 fascicled, 2-5-flowered; corolla red, slightly curved, basal part inflated; anthers multilocellate; styles red; stigmas cone-shaped. Berries greenish yellow, ellipsoid. Fl. Jun-Sep. Fr. Aug-Oct. Forests, slopes or valleys at 500-1900 m. Distributed in SW, S, SE, C, N and W China.

广寄生 *Taxillus chinensis*

广寄生

Taxillus chinensis (DC.) Danser

灌木。嫩枝、叶密被锈色星状毛。叶对生，厚纸质。伞形花序，常2朵，花序和花被星状毛；花梗长6-7毫米；苞片鳞片状。花褐色；副萼环状；花冠长2.5-2.7厘米，下半部膨胀，顶部卵球形，裂片4，匙形，反折；花柱红色，柱头头状。果皮密生小瘤体。花果期2月至12月。生海拔20-400米森林、平原、山坡或果园中。产海南、广西、广东和福建南部。

Shrubs. Young stems and leaves tomentose, reddish brown. Leaves opposite, thick papery. Umbels 2-flowered, pubescent; peduncle stellate hairy; pedicel 6-7 mm; bract scale. Flower brownish; accesory calyx annular; corolla 2.5-2.7 cm, basal part inflated, lobes 4, spatulate, reflexed; style red, stigma capitate. Berry verrucose. Fl. and fr. Feb-Dec. Forests, plains, mountain slopes or orchards at 20-400 m. Distributed in Hainan, Guangxi, Guangdong and S Fujian.

桑寄生 *Taxillus sutchuenensis*

大苞寄生 *Tolypanthus maclurei*

黔桂大苞寄生 *Tolypanthus esquirolii*

大苞寄生

Tolypanthus maclurei (Merr.) Danser

灌木。叶互生至近对生，或3-4片簇生于短枝，近革质。簇生花序单生或2-3个腋生，具3-5朵花；苞片粉色，卵形，基部圆形或心形；花冠淡红色或橘色，顶部膨大，具5棱。浆果黄色，椭圆体形。花期4-7月，果期8-10月。生海拔100-900(-1200)米的林中、山坡、山谷或溪边。产贵州、广西、广东、福建、湖南和江西。

Shrubs. Leaves alternate to subopposite or 3-4 leaves fascicled on short shoots, subcoriaceous. Fascicles solitary or 2-3 together, axillary, 3-5-flowered; bracts pink, ovate, base rounded or cordate; corolla reddish or orange, apical portion inflated, 5-angled. Berries yellow, ellipsoid. Fl. Apr-Jul. Fr. Aug-Oct. Forests, mountain slopes, valleys or along rivers at 100-900(-1200) m. Distributed in Guizhou, Guangxi, Guangdong, Fujian, Hunan and Jiangxi.

黔桂大苞寄生

Tolypanthus esquirolii (H. Lévl.) Lauener

灌木。叶互生或近对生，或2-3枚簇生于短枝，叶纸质。3-4朵花成束状腋生；苞片粉色，具1-3脉；萼檐杯状，具5齿；花冠红色，顶部膨大，具5棱。浆果黄色，椭圆体形，具星状毛。花期4-6月，果期5-8月。生海拔1100-1200米的林中、山坡或沟谷中，寄生于枇杷、油桐或山茶属植物。产广西和贵州。

Shrubs. Leaves alternate to subopposite, or 2-3 fascicled on short shoots, papery. Fascicles axillary, 3-4-flowered; bracts pink, 1-3-veined; calyx limbs cupular, 5-denticulate; corolla reddish, apical portion inflated, 5-angled. Berries yellow, ellipsoid, with stellate hairs. Fl. Apr-Jun. Fr. May-Aug. Forests, slopes or valleys at 1100-1200 m, parasitic on *Eriobotrya japonica*, *Vernicia fordii* or species of *Camellia*. Distributed in Guangxi and Guizhou.

栗寄生

Korthalsella japonica (Thunb.) Engl.

绿色亚灌木。小枝扁平，通常对生。叶退化成鳞片状，成2列，合生成环状。花序侧生于节上；花基部具毛围绕；花被片小，三角形；花药合生成聚药雄蕊。浆果淡黄色。花果期几乎全年。生海拔100-700(-2500)米的常绿阔叶林、山坡或沟谷，常寄生于栎属、柯属、杜鹃属、山茶科、樟科、桃金娘科和山矾科等植物上。

栗寄生 *Korthalsella japonica*

产中国西南、华南、东南、华中和华西。南亚和东南亚、日本、东非、澳大利亚和印度洋岛屿亦有。

Shrublets, green. Branchlets flat, usually opposite. Leaves degraded to scalelike, in 2 ranks, fused into a ring. Inflorescences lateral at node; flowers subtended by hairs; perianth lobes triangular, minute; anthers fused into synandrium. Berries yellowish. Fl. and fr. almost all around the year. Evergreen forests, mountain slopes or valleys at 100-700 (-2500) m, often parasitic on *Quercus, Pasania, Rhododendron,* Theaceae, Lauraceae, Myrtaceae and Symplocaceae spp. Distributed in SW, S, SE, C and W China. Also in S and SE Asia, Japan, E Africa, Australia and Indian Ocean Islands.

高山松寄生

Arceuthobium pini Hawksw. et Wiens

植株黄绿色或微绿色，高5-20厘米。侧枝交叉对生，稀3-4条轮生，节间5-15毫米。叶呈鳞状。单朵花生于短侧枝的腋部或顶部，黄色。浆果椭圆形，基部黄绿色。花期4-7月，果期翌年9-10月。生海拔2600-3500(-4000)米的松林中，常寄生于高山松或云南松枝条上。产云南西北部、四川西南部和西藏。

Plants greenish yellow or greenish, 5-20 cm tall. Branches opposite, rarely 3-4 verticillate, internodes 5-15 mm long. Leaves scalelike. Flowers solitary or paired, terminal on branchlets, yellow. Berries ellipsoid, basal portion greenish yellow. Fl. Apr-Jul. Fr. next Sep-Oct. On branches of pines (*Pinus densata, P. yunnanensis*), parasitic, at 2600-3500 (-4000) m. Distributed in NW Yunnan, SW Sichuan and Xizang.

高山松寄生 *Arceuthobium pini*

槲寄生

Viscum coloratum (Kom.) Nakai

灌木，雌雄异株。叶对生或稀3枚轮生，椭圆形或长圆状披针形，3-7 × 0.7- 1.5(-2)厘米，厚革质或革质。花序顶生；雄花序总状，常具3花。果球形，具宿存花柱。花期4-5月，果期9-11月。生海拔500-2200米的林中，常寄生于台湾桤木、山荆子、杏、枫杨、秋子梨、蒙古栎、榆和杨属、柳树和椴属等植物。产中国西南、华南、华中、华西和华东。俄罗斯东部、朝鲜半岛和日本亦有。

Shrubs, dioecious. Leaves opposite or rarely 3 verticillate, elliptic or oblong-lanceolate, 3-7 × 0.7-1.5 (-2) cm, thick coriaceous or coriaceous. Inflorescences terminal; male inflorescences cymose, usually 3-flowered. Fruits orbicular, with persistent styles. Fl. Apr-May. Fr. Sep-Nov. Forests at 500-2200 m, parasitic on *Alnus formosana, Malus baccata, Prunus armeniaca, Pterocarya stenoptera, Pyrus ussuriensis, Quercus mongolica, Ulmus pumila* and species of *Populus, Salix* and *Tilia*. Distributed in SW, S, C, W and E China. Also in E Russia, Korean Peninsula and Japan.

槲寄生 *Viscum coloratum*

扁枝槲寄生 *Viscum articulatum*

扁枝槲寄生
Viscum articulatum Burm. f.

亚灌木。雌雄同株，常下垂，绿色。枝交叉对生或二歧分枝，扁平，具纵肋3条。叶退化呈鳞片状。花序腋生，聚伞花序1-3；总花梗几无；具3花；总苞舟形。浆果白色或绿白色，球形，直径3-4毫米。花果期几乎全年。生海拔100-1200(-1700)米的林中、平原或山坡，常寄生于桑科、壳斗科、大戟科和樟科植物。产广东、广西、海南和云南。南亚、东南亚和澳大利亚亦有。

Subshrubs, monoecious, usually becoming pendulous, green. Branches opposite and decussate or dichotomous, flattened, longitudinally 3-ribbed. Leaves reduced to scales. Inflorescences axillary, cymes 1-3 together; peduncles ± absent; 3-flowered; involucres navicular. Berries whitish or greenish white, globose, 3-4 mm diam. Fl. and fr. almost all year. Forests, plains or slopes at 100-1200(-1700) m, often parasitic on Moraceae, Fagaceae, Euphorbiaceae and Lauraceae. Distributed in Guangdong, Guangxi, Hainan and Yunnan. Also in S and SE Asia, and Australia.

瘤果槲寄生
Viscum ovalifolium Wall. ex DC.

灌木，雌雄同株。叶对生，革质，3-5条脉。聚伞花序单生叶腋或簇生；具3花。浆果近球形，基部骤狭呈长1毫米的状，果皮具小瘤体，熟时淡黄色，果皮变平滑。花果期几全年。生海拔1100米以下平坝或山地常绿阔叶林，常寄生于海莲、板栗、柚、黄皮、柿子、无患子和柞木植物。产云南、广西、广东和海南。印度东北部和东南亚亦有。

Shrubs, monoecious. Leaves opposite, coriaceous, 3-5-nerved. Inflorescences axillary, cymes solitary or some fascicled;

瘤果槲寄生 *Viscum ovalifolium*

单叶细辛 *Asarum himalaicum*

细辛 *Asarum sieboldii*

long puberulent. Perianth purplish; adaxially dark red puberulent, lobes reflexed, triangular; ovary inferior; styles connate, apex 6-cleft. Fl. Apr-Jun. Mixed forests, along streams or moist places at 1300-3100 m. Distributed in Sichuan, Xizang, Guizhou, Hubei, Shaanxi and Gansu. Also in India, Nepal and Bhutan.

细辛

Asarum sieboldii Miq.

多年生草本。叶常2，叶单色，心形或卵状心形，下面仅沿脉具柔毛或密具柔毛。花被紫色；花被片基部子房贴生，裂片直立或开展，三角状卵形；花柱6，较短，顶端2裂。花期4-5月。生海拔1200-2100米林下阴湿处。产辽宁。朝鲜半岛和日本亦有。

Perennial herbs. Leaves usually 2, uniformly colored, cordate or ovate-cordate, abaxial surface pubescent only along veins or densely pubescent. Perianth dark purple, lobes erect or spreading, triangular-ovate; ovary superior; styles 6, short, apex 2-lobed. Fl. Apr-May. Wet forests at 1200-2100 m. Distributed in Liaoning. Also in Korean Peninsula and Japan.

香港细辛

Asarum hongkongense S. M. Hwang et T. P. Wong Siu

多年生草本。根状茎横生。叶单生，卵状心形至卵形，长6-11厘米，先端急尖，基部深心形，两面无毛；叶柄长12-30厘米。花序梗上升，长1.5-4厘米；花萼紫绿色，钟状瓮形，萼筒喉部稍缢缩，萼片宽卵形，合生超过贴生的子房；子房半上位；花柱合生，先端具6枚直立2半裂的臂。花期2-5月。生海拔500-700米的灌丛、山坡或阴湿地区。产香港。

Perennial herbs. Rhizomes horizontal. Leaves solitary, ovate-cordate to ovate, 6-11 cm long, apex acute, base deeply cordate, both surfaces glabrous; petiole 12-30 cm long. Peduncle ascending, 1.5-4 cm long; perianth purple-green, campanulate-urceolate, tube slightly constricted at throat, lobes broadly ovate; ovary subsuperior; styles connate, apex with 6, erect, slightly 2-cleft arms. Fl. Feb-May. Thickets, mountain slopes or moist shady areas at 500-700 m. Distributed in Hong Kong.

香港细辛 *Asarum hongkongense*

川滇细辛 *Asarum delavayi*

川滇细辛
Asarum delavayi Franch.

粗壮草本。根状茎横走。叶片长卵形、宽卵形或近戟形，下面沿脉疏具柔毛。花被紫绿色；基部与子房贴生，筒喉部强烈收缩，裂片宽卵形；子房近上位或半下位；花柱离生，先端2叉状。花期4-6月。生海拔800-1600米的林中、灌丛中、阴湿处或石坡上。产云南和四川。

Herbs robust. Rhizomes horizontal. Leaves narrowly ovate, broad-ovate or nearly hastate, adaxial surface sparsely pubescent along tube 2 cm long veins. Perianth purple-green, strongly constricted at throat, lobes broadly ovate; ovary nearly superior or half-inferior; styles free, apex 2-forked. Fl. Apr-Jun. Forests, thickets, moist shady areas or stony slopes at 800-1600 m. Distributed in Yunnan and Sichuan.

红金耳环
Asarum petelotii O. C. Schmidt

草本。根状茎长可达20厘米。叶单生，单色，长卵形、三角状卵形或窄卵形，下面沿脉具柔毛，后无毛。花被紫色或绿紫色；下部与子房贴生，筒喉部稍收缩；子房半下位；花柱离生，先端2裂。花期2-5月。生海拔1100-1700米的林中、山坡或潮湿处石上。产云南南部。越南亦有。

Herbs. Rhizomes, up to 20 cm long. Leaves solitary, uniformly colored, long-ovate, deltoid-ovate or narrow-ovate, abaxial surface pubescent along veins then glabrescent. Perianth purple or greenish purple; below adnate to ovary, slightly constricted at tube throat; ovary half-inferior; styles free, apex 2-fid. Fl. Feb-May. Forests, slopes or damp rocks at 1100-1700 m. Distributed in S Yunnan. Also in Vietnam.

南粤马兜铃
Aristolochia howii Merr. et Chun

攀援灌木。叶革质，宽长圆状倒披针形、线形或长圆形，长7-20厘米，先端急尖，基部狭耳形，湾缺深3-5毫米，边缘全缘至2-3浅裂。花1-2朵；花被管外面被棕色长柔毛，檐部呈盘状圆形，边缘3浅裂，裂片平展，宽卵形；花药长圆形；合蕊柱3浅裂。蒴果圆柱形，自基部向上开裂。花期5-9月；果期10-12月。生海拔200-600米的混交林中。产海南。

Climbing shrubs. Leaves leathery, broadly oblong-oblanceolate, linear, or oblong, 7-20 cm long, apex acute, base narrowly auriculate, sinus 3-5 mm deep, margin entire to shallowly 2-3-lobed. Flowers 1-2; perianth tube abaxially brown villous, limb discoid-orbicular, margin 3-lobed, lobes flat, broadly ovate; anthers oblong; gynostemium 3-lobed. Capsule cylindric, dehiscing basipetally. Fl. May-Sep. Fr. Oct-Dec. Mixed forests at 200-600 m. Distributed in Hainan.

异叶马兜铃
Aristolochia kaempferi Willd.

攀援状灌木。叶卵形至卵状披针形、倒卵状长圆形或线形，下面疏具白色长柔毛，侧脉每边3-6条。花单生，稀2朵聚生叶腋；花萼黄绿色，具紫脉，喉部黄色；花梗下垂；合蕊柱顶端3裂。蒴果长圆形或椭圆体形，有6条脉。花期3-5月，果期6-8月。生林中、灌丛或山坡。产华西、华中和东南。日本亦有。

红金耳环 *Asarum petelotii*

南粤马兜铃 *Aristolochia howii*

异叶马兜铃 *Aristolochia kaempferi*

关木通 *Aristolochia manshuriensis*

Climbing shrubs. Leaves ovate to ovate-lanceolate, obovate-oblong or linear, abaxially sparsely white villous, lateral veins 3-6 per side. Flowers solitary, rarely 2 clustered at leaf axils; perianth yellow-green with purple veins, throat yellow; peduncles pendulous; gynostemium 3-lobed at apex.Capsules oblong or ellipsoidal, 6-ribbed. Fl. Mar-May. Fr. Jun-Aug. Forests, thickets or mountain slopes. Distributed in W, C and SE China. Also in Japan.

广西马兜铃

Aristolochia kwangsiensis W. Y. Chun et F. C. How ex C. F. Liang

攀援灌木。叶心形至圆形，两面均被硬毛，脉掌状，自基部伸出2对，基部心形或耳状。总状花序生多叶枝条腋处，具2-3花；檐部深紫色，喉部黄色，管马蹄形。蒴果圆柱状。花期3-5月，果期8-9月。生海拔600-1600米的山坡林中。产中国西南和东南。

Climbing shrubs. Leaves cordate to orbicular, on both surfaces hirsute, veins palmate, 2 pairs from base, base cordate or auriculate. Racemes in axils of leafy shoots, 2-3 flowered; perianth limb dark purple, throat yellow, tube horseshoe-shaped. Capsules cylindric. Fl. Mar-May. Fr. Aug-Sep. Forests on mountain slopes at 600-1600 m. Distributed in SW and SE China.

关木通

Aristolochia manshuriensis Kom.

攀援状灌木。叶心形至圆形，革质，两面生白色长柔毛。花单生或2朵聚生叶腋，檐部紫色，长4.5-5.5厘米，筒马蹄状；外被白色硬毛；合蕊柱3裂。蒴果窄筒状。花期6-7月，果期8-9月。生海拔100-2200米混交林中或潮湿地区。产华西和东北。

Climbing shrubs. Leaves cordate to orbicular, leathery, both surfaces white villous. Flowers solitary, rarely 2 clustered at leaf axils; perianth limb purple, 4.5-5.5 cm long, tube horseshoe-shaped, outside white-hirsute; gynostemium 3-lobed. Capsules narrowly cylindric. Fl. Jun-Jul. Fr. Aug-Sep. Mixed forests or moist and shady areas at 100-2200 m. Distributed in W and NE China.

广西马兜铃 *Aristolochia kwangsiensis*

西藏马兜铃 *Aristolochia griffithii*

香港马兜铃 *Aristolochia westlandii*

西藏马兜铃

Aristolochia griffithii Hook. f. et Thoms ex Duch.

攀援状灌木。叶卵状心形或卵形，纸质，下面密具红褐色或白色长柔毛。花单生；花序梗下垂，长达10厘米；花被深紫色，管及檐部具黄色斑点，喉部血红色，冠筒马蹄形，檐部3浅裂。蒴果圆柱状。花期4-7月，果期8-10月。生海拔2100-2800米的林中。产西藏和云南。印度、尼泊尔、不丹和缅甸亦有。

Climbing shrubs. Leaves ovate-cordate or ovate, papery, abaxially densely red-brown or white villous. Flowers solitary; peduncles pendulous, up to 10 cm long; perianth dark purple with yellow spots on tube and limb, throat blood red, tube horse-shoe-shaped, limb shallowly 3-lobed. Capsules cylindric. Fl. Apr-Jul. Fr. Aug-Oct. Forests at 2100-2800 m. Distributed in Xizang and Yunnan. Also in India, Nepal, Bhutan and Myanmar.

淮通

Aristolochia moupinensis Franch.

攀援状灌木。叶卵形至卵状心形，下面密被黄棕色长柔毛，掌状脉。花腋生，单生或成对；花序梗稍下垂；花被浅黄色，具紫色脉纹；冠筒膝曲。蒴果圆柱形。花期5-6月，果期8-10月。生海拔2000-3200米的林中、灌丛或溪边。产中国西南至东南。

Climbing shrubs. Leaves ovate to ovate-cordate, yellow-brown-villose abaxially, veins palmate. Flowers axillary, solitary or paired; peduncles slightly pendulous; perianth yellowish with purple veins, tube geniculately curved. Capsules cylindric. Fl. May-Jun. Fr. Aug-Oct. Forests, thickets, along streams at 2000-3200 m. Distributed in SW to SE China.

淮通 *Aristolochia moupinensis*

香港马兜铃

Aristolochia westlandii Hemsl.

攀援状灌木。叶披针状长圆形至狭长圆形，革质或纸质，侧脉每边8-12条。总状花序生于小枝下部叶腋或老枝近基部；花被黄色，具紫色脉和斑点；子房圆柱形；合蕊柱肉质，顶端深3裂。花期3-4月。生海拔300-800米山谷林中。产广东和香港。

Climbing shrubs. Leaves lanceolate-oblong to narrowly oblong, leathery or papery, veins pinnate, 8-12 pairs. Racemes on branchlets below or on old shoots base; perianth yellow with purple veins and blotches; ovary terete; gynostemium fleshy, 3-lobed at apex. Fl. Mar- Apr. Forests in valleys at 300-800 m. Distributed in Guangdong and Hong Kong.

海南马兜铃

Aristolochia hainanensis Merr.

攀援状灌木。叶卵形或卵状披针形至狭卵形，革质，幼叶两面密被长柔毛，掌状脉。总状花序腋生或生于老茎近基部，有花3-6朵；小苞片钻形；花被浅黄色，具深紫色裂片，管膝

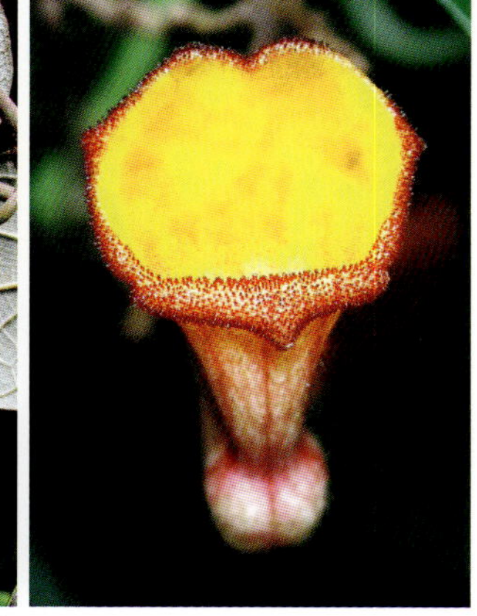

海南马兜铃 *Aristolochia hainanensis*

曲。蒴果窄圆筒状。花期10月至翌年2月，果期6-7月。生海拔800-1200米山谷林中。产广西和海南。

Climbing shrubs. Leaves ovate or ovate-lanceolate to narrowly ovate, leathery, both surfaces densely villose when young, veins palmate. Racemes in axils of leafy shoots or on old woody stems, 3-6-flowered; bracteoles subulate; perianth yellowish with dark purple lobes, tube geniculately curved. Capsules narrowly cylindric. Fl. Oct to next Feb. Fr. Jun-Jul. Forests in valleys at 800-1200 m. Distributed in Guangxi and Hainan.

苞叶马兜铃

Aristolochia chlamydophylla C. Y. Wu ex S. M. Hwang

缠绕草本。叶卵形至卵状三角形，革质或纸质，具芳香腺点，下面具柔毛。总状花序生具多叶小枝腋处，具8-10花；小苞片卵形；花萼黄绿色，檐部深紫色；冠筒直伸或稍弯曲；合蕊柱6裂。花期3-4月。生海拔1000-1300米的山坡林中。产云南和广西。

Twining herbs. Leaves ovate to ovate-deltate, leathery or papery, aromatic-punctate, abaxially puberulous. Racemes in axils of leafy shoots, 8-10-flowered; bractlets ovate; perianth yellowish green, limb dark purple; tube rectilinear or slightly curved; gynostemium 6-lobed. Fl. Mar-Apr. Forests on mountain slopes at 1000-1300 m. Distributed in Yunnan and Guangxi.

苞叶马兜铃 *Aristolochia chlamydophylla*

北马兜铃

Aristolochia contorta Bunge

缠绕草本。茎圆柱形。叶正三角形至三角状心形，纸质，两面无毛。总状花序生枝腋，具2-8花，近簇生；花萼黄绿色，具紫色脉，管直伸或稍弯曲；合蕊柱6裂。蒴果宽倒卵形至椭圆状倒卵形。花期5-7月，果期8-10月。生海拔500-1200米的山坡、灌丛中或山谷。产华北、华西和东北。俄罗斯(西伯利亚)、朝鲜半岛和日本亦有。

Twining herbs. Stems terete. Leaves deltate to deltate-cordate, papery, both surfaces glabrous. Racemes in axils of leafy shoots, 2-8-flowered, nearly fasciculate; perianth yellow-green with purple veins, tube rectilinear or slightly curved; gynostemium 6-lobed. Capsules broadly obovoid to ellipsoid-obovoid. Fl. May-Jul. Fr. Aug-Oct. Slopes, bushes or valleys at 500-1200 m. Distributed in N, W and NE China. Also in Russia (Siberia), Korean Peninsula and Japan.

北马兜铃 *Aristolochia contorta*

马兜铃 *Aristolochia debilis*

管花马兜铃 *Aristolochia tubiflora*

马兜铃
Aristolochia debilis Sieb. et Zucc.

缠绕草本。叶卵形或长圆状卵形至箭形，纸质，两面无毛。花单生或2朵聚生叶腋；花被黄绿色，长3-5.5厘米，喉部有紫斑，管直伸，檐部单侧，舌状。蒴果近球形。花期7-8月，果期9-10月。生海拔200-1500米的灌丛、山坡或潮湿山谷。产中国西南、东南、华中和华东。日本亦有。

Twining herbs. Leaves ovate or oblong-ovate to sagittate, papery, both surfaces glabrous. Flowers solitary or 2-clustered at leaf axils; perianth yellow-green, 3-5.5 cm long, throat purple-punctate, tubes rectilinear, limb unilateral, ligulate. Capsules subglobose. F1. Jul-Aug. Fr. Sep-Oct, Thickets, mountain slopes or moist valleys at 200-1500 m. Distributed in SW, SE, C and E China. Also in Japan.

管花马兜铃
Aristolochia tubiflora Dunn

缠绕草本。叶纸质或近膜质。花单生或2朵聚生叶腋；花萼紫黑色，筒直伸，长3-4厘米，基部膨大，檐部单侧生，舌状；合蕊柱顶端6裂。果黄褐色，6瓣开裂。花期4-8月，果期10-12月。生海拔100-1700米的林下阴处或湿山坡。产中国西南、东南、华中和华西。

Twining herbs. Leaves papery or submembranous. Flowers solitary or 2-clustered at leaf axils; perianth dark purple, tubes rectilinear, 3-4 cm long, dilated at base, limb unilateral, ligulate; gynostemium 6-lobed at apex. Fruits yellow-brown, 6-ribbed. Fl. Apr-Aug. Fr. Oct-Dec. Shady forests or moist mountain slopes at 100-1700 m. Distributed in SW, SE, C and W China.

背蛇生
Aristolochia tuberosa C. F. Liang et S. M. Hwang

缠绕草本。叶心形，纸质或近膜质，两面无毛。花1-3朵聚生于叶腋；花被黄绿色；喉暗红色，管直，檐部单侧，舌状；合蕊柱6裂。蒴果倒卵球形，顶端开裂。花期11月至翌年4月，果期翌年6-10月。生海拔100-1600米的灌丛、山谷或石灰岩山坡。产中国西南和华南。

Twining herbs. Leaves cordate, papery or submembranous, both surfaces glabrous. Flowers 1-3 per axil; perianth yellow-green, throat dark-purple, tube rectilinear, limb unilateral, ligulate; gynostemium 6-lobed. Capsules obovoid, dehiscing acropetally. Fl. Nov to next Apr. Fr. next Jun-Oct. Thickets, valleys or limestone mountain slopes at 100-1600 m.

背蛇生 *Aristolochia tuberosa*

多型马兜铃 *Aristolochia polymorpha*

dehiscing acropetally. Fl. Oct-Dec. Fr. Jun-Aug. Forests below 100-200 m. Distributed in Hainan.

Distributed in SW and S China.

多型马兜铃

Aristolochia polymorpha S. M. Hwang

缠绕草本。叶纸质，卵形或卵状三角形至戟形，长2.5-5.5厘米，边缘通常3深裂，两面无毛或被面具长柔毛；基出脉1-2对；叶柄长1-2厘米。总状花序腋生，有花2-4朵；花被基部收狭呈柄状，檐部仅具单侧，成舌状；花药椭圆形；合蕊柱6浅裂。蒴果近球形，由基部向上瓣裂。花期10-12月；果期6-8月。生海拔100-200米以下的林中。产海南。

Twining herbs. Leaves papery, ovate or ovate-deltate to sagittate, 2.5-5.5 cm long, margin usually 3-parted,both surfaces glabrous or abaxially villous; veins 1-2 pairs from base; petiole 1-2 cm long. Racemes axillary, 2-4 flowered; perianth base constricted into stalk, limb unilateral, ligulate; anthers elliptic; gynostemium 6-lobed. Capsule subglobose, dehiscing acropetally. Fl. Oct-Dec. Fr. Jun-Aug. Forests below 100-200 m. Distributed in Hainan.

凹脉马兜铃

Aristolochia impressinervis C. F. Liang

草质藤本。叶纸质或革质，卵状披针形至披针形。总状花序腋生，有花3-7朵；小苞片披针形；花被浅灰黄色或浅绿色，喉部紫色，管直伸或稍弯曲，檐部舌状，椭圆形；合蕊柱6裂。蒴果倒卵形或近球形。花期5-6月，果期8-10月。生海拔约400米林中、灌丛或石灰岩山坡。产广西。

Twining herbs. Leaves papery or leathery, ovate-lanceolate to lanceolate. Racemes in axils of leafy shoots, 3-7-flowered; bracteoles lanceolate; perianth pale yellow or pale green, throat purple, tube rectilinear or slightly curved, limb ligulate, elliptic; gynostemium 6-lobed. Capsules obovoid to subglobose, dehiscing acropetally. Fl. May-Jun. Fr. Aug-Oct. Forests, thickets or limestone mountain slopes at ca. 400 m. Distributed in Guangxi.

凹脉马兜铃 *Aristolochia impressinervis*

耳叶马兜铃 *Aristolochia tagala*

耳叶马兜铃
Aristolochia tagala Champ.

攀援草本。叶卵状心形或长卵形，两面无毛，基部深心形。总状花序生多叶枝腋处，具2或3花；花萼浅黄色或浅绿色，喉部深紫色，管直伸或稍弯曲，檐部舌状，长圆形；合蕊柱6裂。蒴果倒卵球状球形至卵球状圆柱形，顶端开裂。花期5-8月，果期10-12月。生海拔100-2000米的林中或山坡。产云南、贵州、广西、广东、台湾和福建。南亚、东南亚和日本亦有。

Twining herbs. Leaves ovate-cordate or long-ovate, both surfaces glabrous, base deeply cordate. Racemes in axils of leafy shoots, 2- or 3-flowered; perianth pale yellow or pale green, throat dark purple, tube rectilinear or slightly curved, limb ligulate, oblong; gynostemium 6-lobed. Capsules obovoid-globose to ovoid-cylindric, dehiscing acropetally. Fl. May-Aug. Fr. Oct-Dec. Forests or mountain slopes 100-2000 m. Distributed in Yunnan, Guizhou, Guangxi, Guangdong, Taiwan and Fujian. Also in S and SE Asia, and Japan.

竹叶马兜铃
Aristolochia bambusifolia C. F. Liang ex H. Q. Wen

木质藤本。叶线状披针形，革质，下面密被褐色长柔毛，上面无毛；羽状脉12-16对。花单生；花梗下垂；花萼下面黄绿色，具紫色脉纹，上面白色，具紫色斑，檐部深紫色；筒膝曲；合蕊柱3裂。生石灰岩山坡林中或石缝。产广西(隆林县)。

Woody lianas. Leaves linear-lanceolate, leathery, abaxially densely brown-villous, adaxially glabrous; veins pinnate, 12-16 pairs. Flowers solitary; peduncles pendulous; perianth abaxially yellow-green with purple veins, adaxially whitish with purple speckles, limb deep purple; tube geniculately curved; gynostemium 3-lobed at apex. Forests on limestones slopes or rock fissures. Distributed in Guangxi (Longlin County).

翅茎马兜铃
Aristolochia caulialata C. Y. Wu ex J. S. Ma et C. Y. Cheng

攀援灌木。茎直立，老茎纵裂，树皮木栓质，近无毛。叶片圆形，两面无毛；脉掌状2-3对。总状花序长13-16厘米，具2花；檐部紫色，喉黄色具紫点，瓣片近盾形；合蕊柱3裂。花期5月。生山谷密林中。产云南。

Climbing shrubs. Stems terete, old stems longitudinally fissured, bark corky, subglabrous. Leaves orbicular, glabrous on both surfaces; veins palmate, 2-3 pairs from base. Racemes 2-flowered, 13-16 cm long; perianth limb purple, throat yellow with purple spots, limb subpeltate; gynostemium 3-lobed Fl. May. Dense forests in valleys. Distributed in Yunnan.

竹叶马兜铃 *Aristolochia bambusifolia*

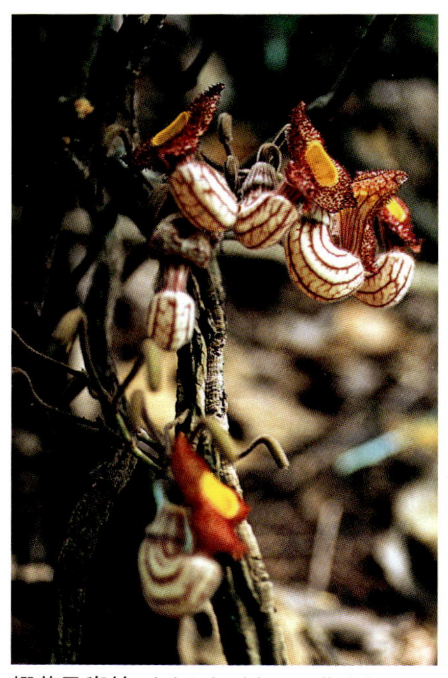
翅茎马兜铃 *Aristolochia caulialata*

大花草科 Rafflesiaceae

帽蕊草
Mitrastemon yamamotoi Makino

寄生肉质草本，株高3-8厘米；根状茎杯状，外被瘤状凸起，口部4-5齿裂。鳞片叶交互对生，共排成4列，顶端1对叶最大。花单朵顶生；花被杯状，白色，高5-6毫米，宿存；雄蕊合生成筒，花药极多数，顶端孔裂。花期2-4月，果期10月。寄生海拔约1200米的锥栗属、栎属植物的根部。产云南、广西、广东、福建和台湾。柬埔寨、日本、马来西亚、泰国和印度尼西亚亦有。

Fleshy parasitic herbs, 3-8 cm tall. Rhizome densely tuberculate, mouth shallowly 4- 5-lobed. Scalelike leaves decussate in 4 ranks, apicalmost largest. Inflorescence 1-flowered; perianth cupular, 5-6 mm, persistent, white; stamens very numerous, completely connate into a closed tube. Fl. Feb-Apr. Fr. Oct. Parasitic on roots of *Castanopsis* and *Quercus* at ca. 1200 m. Distributed in Yunnan, Guangxi, Guangdong, Fujian and Taiwan. Also in Cambodia, Japan, Malaysia, Thailand and Indonesia.

帽蕊草 *Mitrastemon yamamotoi*

蛇菰科 Balanophoraceae

盾片蛇菰
Rhopalocnemis phalloides Jungh.

草本，雌雄异序。鞘5裂，裂片不整齐，不规则三角形。鳞苞片螺旋状排列，部分散生，稍反折，有疣瘤。花茎无叶或叶鳞片状且不明显；花序由盾状鳞片包被；雌花无花被，花柱2。生海拔1000-2700米的密林或灌丛中。产广西西部和云南东南部。印度、尼泊尔、泰国、越南和印度尼西亚亦有。

Herbs, dioecious. Sheath lobes 5, irregularly deltoid. Leaves spirally arranged, ±scattered, warty, apically slightly reflexed. Flowering shoot leafless or leaves scalelike and indistinct; inflorescences covered by peltate scales; female flowers without perianth, with 2 styles. Dense forests or thickets at 1000-2700 m. Distributed in W Guangxi and SE Yunnan. Also in India, Nepal, Thailand, Vietnam and Indonesia.

印度蛇菰
Balanophora indica (Arn.) Griff.

草本。根状茎橙黄色至棕色，偶有明显的星芒状皮孔。鳞苞片10-20，橙黄色，散生或多少旋生，覆瓦状排列。雄花序淡红色，辐射对称，每个花序都被单一粗壮而短的苞片包被；雄花4-6数。花期10-12月。生海拔900-1500米的常绿阔叶林中。产云南、广西和海南。南亚、东南亚和太平洋岛屿(关岛)亦有。

Herbs. Rhizomes yellowish orange to brown, occasionally with conspicuous stellate lenticelles. Leaves 10-20, yellowish orange, spirally arranged, imbricate. Male inflorescences reddish, actinomorphic, each subtended by a single stout and truncate bract; perianth lobes 4-6 merous. Fl. Oct-Dec. Broad-leaved evergreen forests at 900-1500 m. Distributed in Yunnan, Guangxi and Hainan. Also in S and SE Asia, and Pacific Islands (Guam).

盾片蛇菰 *Rhopalocnemis phalloides*

印度蛇菰 *Balanophora indica*

杯茎蛇菰 *Balanophora subcupularis*

多蕊蛇菰 *Balanophora polyandra*

杯茎蛇菰

Balanophora subcupularis Tam

草本，雌雄同株。根状茎黄棕色，近球状，表面密被颗粒状小疣瘤和黄色分散的星芒状小皮孔。鳞苞片5-8，螺旋状排列或互生，阔卵形。花序椭圆体形；雄花仅着生于花序基部，由不同开头的苞片所包被；雌花淡黄色；花被常4。花期9-11月。生海拔800-1500米的林中。产云南、贵州、广西、广东、湖南和江西。

Herbs, monoecious. Rhizomes yellowish brown, subglobose, surface with granular warts and scattered yellow stellate lenticels. Leaves 5-8, spirally arranged or rather opposite, broadly ovate. Inflorescences ellipsoid; male flowers basally on androgynous inflorescences, nearly actinomorphic, subtended by variously shaped bract(s); female flowers yellowish; perianth lobes usually 4. Fl. Sep-Nov. Forests at 800-1500 m. Distributed in Yunnan, Guizhou, Guangxi, Guangdong, Hunan and Jiangxi.

多蕊蛇菰

Balanophora polyandra Griff.

草本，雌雄异株。根状茎块状，表面有星芒状小皮孔。鳞苞片4-12，下部的旋生，上部的互生。花茎深红色，雄花序狭椭圆体形；每个雄花由单个坚实的截形苞片所包被；花药短裂为20-60个小药室。花期8-10月。生海拔1000-2500米密林下。产中国西南和华中。印度、尼泊尔、不丹和缅甸亦有。

Herbs, dioecious. Rhizomes tuber-like, on surface with stellate lenticelles. Leaves 4-12, lower ones spirally arranged, upper ones alternate. Flowering stems deep-red, male inflorescences narrowly ellipsoid; male flowers each subtended by a single stout and truncate bract; anthers broken up into 20-60 locelli. Fl. Aug-Oct. Dense forests at 1000-2500 m. Distributed in SW and C China. Also in India, Nepal, Bhutan and Myanmar.

疏花蛇菰

Balanophora laxiflora Hemsl.

草本，雌雄同株或雌雄异株，全株红色至深红色，有时紫红色。根状茎分枝近球形，表面密被鳞片状斑点和黄色星状皮孔。雄花序圆柱形；花药由短缝开裂成多室；雄花之字形；雌花序卵球形至长圆状椭圆形；花被通常4-6裂。花期9-11月。生海拔(200-)600-1700米的密林中。产中国西南、华南、东南和华中。老挝、泰国和越南亦有。

Herbs, monoecious or dioecious, red to dark red, sometimes purplish. Rhizomes branched, subspherical, surface densely covered with scabrous speckles and yellowish stellate warts. Male inflorescences cylindric; anthers broken up into many locelli, dehiscent by short slits; male flowers zygomorphic; female inflorescences ovoid-spheroid to oblong-ellipsoid; perianth lobes 4-6. Fl. Sep-Nov. Dense forests at (200-)600-1700 m. Distributed in SW, S, SE and C China. Also in Laos, Thailand and Vietnam.

疏花蛇菰 *Balanophora laxiflora*

红菌(筒鞘蛇菰)

Balanophora involucrata Hook. f.

草本，雌雄异株。根状茎肥厚，表面密被星芒状皮孔。鳞苞叶2-5，轮生，基部连合成筒鞘状，顶端离生呈撕裂状。雄花序卵球形；苞片截形，具舌状膨大的边缘；雄花具梗，常3数；花药横向开裂。花期7-8月。生海拔2300-3600米的云杉、铁杉或栎木林中，常寄生于杜鹃属或其他植物的根上。产中国西南和华西。印度、尼泊尔和不丹亦有。

Herbs, dioecious. Rhizomes thick, on surface with dense stellate lenticelles. Leaves 2-5, verticillate, connate into a tube-sheath at base but free and laciniate at top. Male inflorescences ovoid-spheroid; bracts truncate with expanded liguliform margin; male flowers pedicellate, usually 3-merous; anthers transversely dehiscent. Fl. Jul-Aug. *Picea*, *Tsuga* or *Quercus* forests at 2300-3600 m, usually parasitic on roots of *Rhododenron* or other plants. Distributed in SW and W China. Also in India, Nepal and Bhutan.

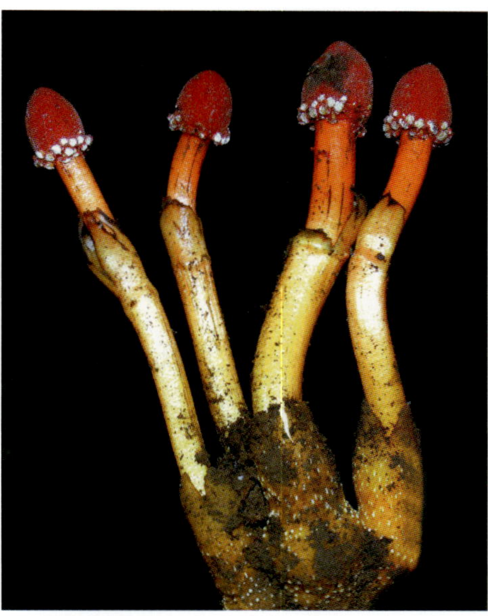

红菌(筒鞘蛇菰) *Balanophora involucrata*

川藏蛇菰 *Balanophora fargesii*

川藏蛇菰

Balanophora fargesii (Tiegh.) Harms

草本，雌雄同株。根状茎常呈球状卵圆形，黄褐色，分枝或不分枝。鳞苞片轮生，基部连生呈筒鞘状。花序头状，卵球形至近球形，直径1-3厘米；雄花着生花序基部，3数；花药横裂。花期7-8月。生海拔2700-3100米的松林中。产云南、四川和西藏。不丹亦有。

Herbs, monoecious. Rhizomes usually globose-ovoid, yellowish brown, branched or unbranched. Leaves verticillate and connate into a sheath-like whorl at base. Capitulums, ovoid to subglobose, 1-3 cm diam; male flowers clustered basally on inflorescence, perianth lobes 3; anthers transversely dehiscence. Fl. Jul-Aug. *Pinus* forests at 2700-3100 m. Distributed in Yunnan, Sichuan and Xizang. Also in Bhutan.

葛菌(红冬蛇菰)

Balanophora harlandii Hook. f.

草本，雌雄异序。鳞苞片6-12，黄色至淡红色，常簇生于花葶基部，交互对生、近对生或螺旋状，长圆状卵形。雄花序近球形至卵球状椭圆体形；苞片截形，具膨大的舌状边缘，每边合生，花被片3裂。花期9-11月。生海拔600-2100米的阔叶林或竹林下。产中国西南、华南、东南和华中。印度和泰国亦有。

Dioecious herbs. Leaves 6-12, yellow to reddish, usually clustered on base of scape, decussate, subopposite, or spiraled, oblong-ovate. Male inflorescences subspheroid to ovoid-ellipsoid; bracts truncate with expanded liguliform margin, fused side by side, perianthes 3-lobed. Fl. Sep-Nov. Broad-leaved forests or bamboo-forests at 600-2100 m. Distributed in SW, S, SE and C China. Also in India and Thailand.

葛菌(红冬蛇菰) *Balanophora harlandii*

蓼科 Polygonaceae

珊瑚藤 *Antigonon leptopus*

珊瑚藤

Antigonon leptopus Hook. et Arn.

多年生草质或基部有时木质的藤本。叶卵状心形，边缘全缘或波浪状，两面无毛，托叶鞘小。总状花序顶生或生上部叶腋，中轴延伸为卷须；花多数，丛生，花被片粉红色，有时白色。果实稍增大。花期春末至秋季。中国西南和华南栽培或归化。原产墨西哥和中美洲。

Perennial herbaceous or base some-times woody vine. Leaves ovate-cordate, glabrous on both surfaces, ocrea small. Racemes terminal or on the upper leaf axils, axis elongate into tendrils; flowers numerous; tepals pinkish, sometimes white. Fruits slightly enlarged. Fl. late spring to autumn. Cultivated or naturalized in SW and S China. Native to Mexico and Central America.

萹蓄

Polygonum aviculare L.

一年生草本。叶披针形或狭椭圆形，两面无毛。花1-5，簇生叶腋；花被绿色，边缘白色或粉红色，5裂至2/3-3/4；雄蕊8。瘦果包于或稍超出宿存花被，黑褐色，三棱状卵球形。花期5-9月，果期6-11月。生海拔4200米以下的田边、路边或沟边湿地。产中国各地。北温带广布。

Annual herbs. Leaves lanceolate or narrowly elliptic, both surfaces glabrous. Flowers 1-5, in axillary fascicles; perianth green, margin white or pinkish, 5-cleft to 2/3-3/4; stamens 8. Achenes included or slightly exceeding persistent perianth, black-brown, ovoid, trigonous. Fl. May-Sep. Fr. Jun-Nov. Field edges, roadsides or moist places along streams below 4200 m. Widespread in China. Also widely distributed in the north temperate zone.

习见蓼

Polygonum plebeium R. Br.

一年生草本。茎匍匐。叶狭椭圆形或倒披针形；托叶鞘白色，先端撕裂状。花3-6，簇生叶腋；花梗中部具关节；雄蕊8。瘦果平滑，被包在宿存花被中，黑褐色，宽卵球形。花期5-8月，果期6-9月。生海拔2200米以下的田边、路边或潮湿处。产中国大部分地区。南亚、东南亚、东北亚、大洋

萹蓄 *Polygonum aviculare*

习见蓼 *Polygonum plebeium*

丽蓼 *Polygonum pulchrum*

洲、非洲和欧洲亦有。

Annual herbs. Stems prostrate. Leaves narrowly elliptic or oblanceolate; ocrea white, apex lacerate. Flowers 3-6, in axillary fascicles; pedicels articulate at middle; stamens 8. Achenes smooth, included in persistent perianth, black-brown, broadly ovoid. Fl. May-Aug. Fr. Jun-Sep. Fieldsides, roadsides or wet areas below 2200 m. Distributed in most parts of China. Also in S, SE and NE Asia, Oceania, Africa and Europe.

丽蓼

Polygonum pulchrum Bl.

多年生草本，具根状茎。叶阔披针形，两面密被绢毛；托叶鞘筒状，膜质，顶端截形，具长缘毛。穗状花序排成圆锥状，顶生，直立；花被白色，5深裂；花柱2，离生。瘦果包于宿存花被内，黑色，光亮，圆形，双凸形。花期9-10月，果期10-11月。生海拔100-300米的浅水沼泽或潮湿地。产广西、广东和台湾。南亚、东南亚和澳大利亚亦有。

Perennial herbs, rhizomatous. Leaves broadly lanceolate, both surfaces densely sericeous; ocrea tubular, membranous, apex truncate, longi-ciliate. Inflorescences terminal, of panicled spikes, erect; perianth white, 5-parted; styles 2, free. Achenes included in persistent perianth, black, shiny, orbicular, biconvex. Fl. Sep-Oct. Fr. Oct-Nov. Shallow water of swamps or marshy areas at 100-300 m. Distributed in Guangxi, Guangdong and Taiwan. Also in S and SE Asia, and Australia.

香蓼(黏毛蓼)

Polygonum viscosum Buch.-Ham. ex D. Don

一年生草本，有香味，密被开展的长糙硬毛和腺毛。叶卵状披针形或椭圆状披针形。花序顶生或腋生，穗状，常数枚穗状花序聚集为圆锥状；花被粉红色，5深裂；雄蕊8，内藏。瘦果包于宿存花被内，黑褐色，光亮，三棱状卵球形。花期7-9月，果期8-10月。生海拔1900米以下的路边湿地或沟边草地。产中国西南、华南、华中、华东和东北。印度、尼泊尔、俄罗斯(远东地区)、朝鲜半岛和日本亦有。

Annual herbs, odoriferous, densely hirsute and glandular hairy. Leaves ovate-lanceolate or elliptic-lanceolate. Inflorescences terminal or axillary, spicate, usually several spikes aggregated to a panicle; perianth pinkish, 5-parted; stamens 8, included. Achenes included in persistent perianth, black-brown, shiny, trigonous ovoid. Fl. Jul-Sep. Fr. Aug-Oct. Moist places by roads or grassy places by streams below 1900 m. Distributed in SW, S, C, E and NE China. Also in India, Nepal, Russia (Far East), Korean Peninsula and Japan.

光蓼

Polygonum glabrum Willd.

一年生草本，全株完全无毛。茎直立，少分枝。叶披针形或长圆状披针形；托叶鞘筒状，具数条纵脉，顶端截形，无缘毛。穗状花序顶生，常数个聚集为圆锥状；花被白色或粉红色，5深裂；雄蕊6-8。瘦果藏于宿存花被内，黑褐色，光亮，卵球形，双凸形。花期6-8月，果期7-9月。生海拔100米以下的河岸、溪边或沼泽地。产华南至华中。南亚、大洋洲、非洲、美洲和太平洋岛屿亦有。

Annual herbs, glabrous throughout. Stems erect, rarely branched. Leaves lanceolate or oblong-lanceolate; ocrea tubular, veins numerous, apex truncate, not ciliate. Spikes terminal, usually aggregated and panicle-like; perianth white or pinkish, 5-parted; stamens 6-8. Achenes included in persistent perianth, black-brown, shiny, ovoid, biconvex. Fl. Jun-Aug. Fr. Jul-Sep. Riverbanks, streamsides or marshy areas below 100 m. Distributed in S to C China. Also in S Asia, Oceania, Africa, America and Pacific Islands.

香蓼(黏毛蓼) *Polygonum viscosum*

光蓼 *Polygonum glabrum*

毛蓼 *Polygonum barbatum*

酸模叶蓼 *Polygonum lapathifolium*

毛蓼
Polygonum barbatum L.

多年生草本。茎粗壮，直立，稀分枝。叶披针形或椭圆状披针形，两面具短柔毛；托叶鞘筒状，膜质，顶端截形，具长缘毛。花序顶生，穗状，直立，几个集生成圆锥状，稀单生；花被白色或淡绿色，5深裂；雄蕊5-8，花柱3，柱头头状。瘦果包于宿存花被内，黑色，光亮，三棱状卵球形。花期8-9月，果期9-10月。生海拔1300米以下河边或潮湿地。产中国西南、华南、东南和华中。南亚和东南亚亦有。

Perennial herbs. Stems robust, erect, rarely branched. Leaves lanceolate or elliptic-lanceolate, both surfaces pubescent; ocrea tubular, membranous, apex truncate, long-ciliate. Inflorescences terminal, spicate, erect, several spikes aggregated and panicle-like, rarely solitary; perianth white or greenish, 5-parted; stamens 5-8; styles 3, stigmas capitate. Achenes included in persistent perianth, black, shiny, ovoid. Fl. Aug-Sep. Fr. Sep-Oct. Streamsides or wet areas below 1300 m. Distributed in SW, S, SE and C China. Also in S and SE Asia.

酸模叶蓼
Polygonum lapathifolium L.

一年生草本。叶披针形或阔披针形，上面具心形黑斑，下面近无毛或密具绵毛；托叶鞘浅褐色，筒状，先端截形，无缘毛。花序顶生或腋生，穗状，直立或下垂，数个穗状花序聚集为圆锥状；花被粉色或白色，4深裂。瘦果宽卵球形，双凹。花期5-7月，果期6-10月。生海拔3900米以下的田地、路边、水边或荒地。产中国大部分地区。南亚、东南亚、中亚、东北亚、澳大利亚、北非、欧洲和北美洲亦有。

Annual herbs. Leaves lanceolate or broadly lanceolate, with a cordate blackish spot adaxially, abaxially nearly glabrous or densely lanose; ocrea brownish, tubular, glabrous, apex truncate, eciliate. Inflorescences terminal or axillary, spicate, erect or nodding, several spikes aggregated to a panicle; perianth pink or white, 4-parted. Achenes broadly ovoid, biconcave. Fl. May-Jul. Fr. Jun-Oct. Fields, roadsides, by waters or wastelands below 3900 m. Distributed in most parts of China. Also in S, SE, C and NE Asia, Australia, N Africa, Europe and North America.

红蓼
Polygonum orientale L.

一年生草本。茎直立，强壮。叶阔卵形，阔椭圆形或卵状披针形，两面密被柔毛，沿脉密生长柔毛；托叶鞘筒状，具长柔毛，先端截形，具长缘毛，常具环形绿色叶状翅。花序顶生或腋生，穗状，稍下垂；花粉红色或白色。瘦果近扁球形。花期6-9月，果期8-10月。生海拔3000米以下的路边或荒地。产中国大部分地区。亚洲、大洋洲和欧洲亦有。

Annual herbs. Stems erect, robust. Leaves broadly ovate, broadly elliptic, or ovate-lanceolate, both surfaces densely pubescent, along veins densely villous; ocrea tubular, villous, apex truncate, long ciliate, usually with annular green leaflike wing. Inflorescences terminal or axillary, spicate, slightly pendulous; flowers purple-red or white. Achenes nearly orbicular, biconcave. Fl. Jun-Sep. Fr. Aug-Oct. Roadsides or wastelands below 3000 m. Distributed in most parts of China. Also in Asia, Oceania and Europe.

蓼蓝
Polygonum tinctorium Ait.

一年生草本。茎直立，常分枝。叶卵形或宽椭圆形，绿色，干后蓝绿色；托叶鞘筒状，具贴伏柔毛，顶端截形，具长缘毛。花序顶生或腋生，穗状，密集；花被淡红色，5深裂；雄蕊6-8；花柱3，下部合生。瘦果包于宿存花被内，褐

红蓼 *Polygonum orientale*

蓼蓝 *Polygonum tinctorium*

丛枝蓼 *Polygonum posumbu*

色，三棱状宽卵球形。花期6-9月，果期8-10月。生海拔200-1000米湿山谷或河岸，亦广泛栽培。中国广布。中南半岛亦有；栽培或偶逸生于其他地方。

Annual herbs. Stems erect, usually branched. Leaves ovate or broadly elliptic, green, blue-green when dry; ocrea tubular, appressed pubescent, apex truncate, long-ciliate. Inflorescences terminal or axillary, spicate, dense; perianth reddish, 5-parted; stamens 6-8; styles 3, connate at lower part. Achenes included in persistent perianth, brown, trigonous, broadly ovoid. Fl. Jun-Sep. Fr. Aug-Oct. Moist valleys or stream banks, at 200-1000 m. also widely cultivated. Widespread in China. Also in Indo-China Peninsula; cultivated or occasionally naturalized elsewhere.

水蓼 (辣蓼)

Polygonum hydropiper L.

一年生草本，有辣味。叶披针形或椭圆披针形，全缘，两面无毛，密具褐色腺点；托叶鞘筒状，疏具平伏硬毛，先端截形，具缘毛。花序顶生或腋生，穗状，下垂，下部间断，常疏松，细弱；苞片绿色，漏斗状；花被浅绿色、白色或上部粉色。花期5-9月，果期6-10月。生海拔3500米以下河滩、水沟边或山谷湿地。产中国各地。亚洲、大洋洲、欧洲和北美洲亦有。

Annual herbs, spicy. Leaves lanceolate or elliptic-lanceolate, entire, both surfaces glabrous, densely brown punctate; ocrea tubular, sparsely appressed hispidulous, apex truncate, ciliate. Inflorescences terminal or axillary, spicate, pendulous, interrupted below, usually lax, slender; bracts green, funnel-shaped; perianth greenish, white or pink above. Fl. May-Sep. Fr. Jun-Oct. River banks, along streams or moist places in valleys below 3500 m. Widespread in China. Also in Asia, Oceania, Europe and North America.

丛枝蓼

Polygonum posumbu
Buch.-Ham. ex D. Don

一年生草本。茎细弱，无毛，基部多分枝，丛生状。叶卵形或卵状披针形，两面疏被硬伏毛或渐无毛，先端尾状渐尖；托叶鞘筒状，具硬伏毛，先端截形，具长缘毛。花序顶生或腋生，穗状，疏松，下部间断。花紫红色。花期6-9月，果期7-10月。生海拔150-3000米的山坡林下或山谷水边。产中国西南、华南、东南、华中、华西、华东和东北。南亚、东南亚和东北亚亦有。

Annual herbs. Stems slender, glabrous, branched at base, clustered. Leaves ovate or ovate-lanceolate, both surfaces sparsely appressed hispid or glabrescent, apex caudate-acuminate; ocrea tubular, appressed hispid, apex truncate, longi-ciliate. Inflorescences terminal or axillary, spicate, lax, interrupted below. Flowers purple-red. Fl. Jun-Sep. Fr. Jul-Oct. Forests on slopes or along streams in valleys at 150-3000 m. Distributed in SW, S, SE, C, W, E and NE China. Also in S, SE and NE Asia.

水蓼 (辣蓼) *Polygonum hydropiper*

长鬃蓼 *Polygonum longisetum*

长鬃蓼

Polygonum longisetum Bruijn

一年生草本。茎直立，基部斜升或匍匐。叶披针形或阔披针形，下面沿脉具硬伏毛；托叶鞘筒状，疏具柔毛，先端截形，具长缘毛。穗状花序顶生或腋生，下部间断；苞片斜漏斗状，先端具缘毛；花紫红色或白色。瘦果宽卵球形。花期5-8月，果期6-10月。生海拔3100米以下的潮湿山谷、溪岸或沟边阴处。产中国大部分地区。南亚、东南亚和东北亚亦有。

Annual herbs. Stems erect, ascending or prostrate at base. Leaves lanceolate or broadly lanceolate, abaxially appressed hispidulous along veins; ocrea tubular, sparsely pubescent, apex truncate, long-ciliate. Inflorescences terminal or axillary, spicate, interrupted below; bracts oblique funnel-shaped, apex cilicate; flowers purple-red or white. Achenes broadly ovoid. Fl. May-Aug. Fr. Jun-Oct. Moist valleys, stream banks or shaded places along ditches below 3100 m. Distributed in most parts of China. Also in S, SE and NE Asia.

珠芽蓼

Polygonum viviparum L.

多年生草本。根状茎粗壮，弯曲。叶线形、卵状披针形或长圆形，革质；托叶鞘筒状，下部绿色，上部褐色，膜质，先端偏斜，无缘毛。花序顶生，穗状，具珠芽；花被白色或粉红色。花期5-7月，果期7-9月。生海拔1200-5100米的林缘、草坡或高山草甸。产中国西南、华北、华西、西北和东北。西南亚、东亚、欧洲和北美洲亦有。

Perennial herbs. Rhizomes robust, contorted. Leaves linear, ovate-lanceolate, or oblong, leathery; ocrea tubular, green at lower part, brown at upper part, membranous, apex oblique, eciliate. Inflorescences terminal, spicate, with bulbils; perianth white or pinkish. Fl. May-Jul. Fr. Jul-Sep. Forest edges, grassy slopes or alpine meadow at 1200-5100 m. Distributed in SW, N, W, NW and NE China. Also in SW and E Asia, Europe and North America.

倒根蓼(倒根拳参)

Polygonum ochotense V. Petrov ex Kom.

多年生草本。根状茎弯曲；茎直立。叶绿色，卵状披针形或长圆状披针形，近革质，下面密被灰白色柔毛；托叶鞘筒状，下部绿色，上部褐色，具柔毛，先端无缘毛，裂至中部。穗状花序短；花被粉红色。瘦果卵球形。花期7-8月，果期8-9月。生海拔1500-2500米的山坡。产吉林。俄罗斯(远东地区)和朝鲜半岛亦有。

Perennial herbs. Rhizomes curved; stems erect. Leaves green, ovate-lanceolate or

珠芽蓼 *Polygonum viviparum*

倒根蓼(倒根拳参) *Polygonum ochotense*

大海蓼 *Polygonum milletii*

curved, stems erect. Leaves lanceolate or narrowly ovate, papery, both surfaces glabrous or abaxially pubescent; ocrea tubular, lower part green, upper part brown, apex oblique, cleft to middle. Inflorescences terminal, spicate; perianth pinkish or white, 5-parted. Achenes slightly exceeding persistent perianth, ellipsoid. Fl. Jun-Jul. Fr. Aug-Sep. Hilly grasslands or meadows at 800-3000 m. Distributed in C, N, W, E and NE China. Also in Kazakhstan, Russia, Mongolia, Japan and Europe.

oblong-lanceolate, subleathery, abaxially densely gray-white pubescent; ocrea tubular, lower part green, upper part brown, pubescent, apex not ciliate, cleft to middle. Inflorescences spicate, short; perianth pinkish. Achenes ovoid. Fl. Jul-Aug. Fr. Aug-Sep. Slopes at 1500-2500 m. Distributed in Jilin. Also in Russia (Far East) and Korean Peninsula.

大海蓼

Polygonum milletii (Lévl.) Lévl.

多年生草本。根状茎弯曲，茎直立。叶披针形或狭披针形，近革质，下面无毛或具柔毛；茎上部的叶不抱茎；托叶鞘下部绿色，上部褐色，筒状，先端偏斜，不具缘毛，裂至中部。花序顶生，穗状，密集；花被紫红色，5深裂；雄蕊伸出，花药紫黑色；花柱3，中下部合生。瘦果包在宿存花被内，三棱状卵球形，褐色，有光泽。花期7-8月，果期9-10月。生海拔1700-3900米山坡、草甸或湿山谷。产云南、四川、陕西南部和青海。印度北部、尼泊尔和不丹亦有。

Perennial herbs. Rhizomes curved, stems erect. Leaves lanceolate or narrowly lanceolate, subleathery, abaxially glabrous or pubescent; upper cauline leaves not clasping; ocrea tubular, lower part green, upper part brown, apex oblique, not ciliate, cleft to middle. Inflorescences terminal, spicate, dense; perianth purple-red, 5-parted; stamens exserted, anthers black-purple; styles 3, connate below middle. Achenes included in persistent perianth, trigonous ovoid, brown, shiny. Fl. Jul-Aug. Fr. Sep-Oct. Mountain slopes, meadows or wet valleys at 1700-3900 m. Distributed in Yunnan, Sichuan, S Shaanxi and Qinghai. Also in N India, Nepal and Bhutan.

拳参

Polygonum bistorta L.

多年生草本。根状茎弯曲，茎直立。叶披针形至狭卵形，纸质，两面无毛或下面具柔毛；托叶鞘筒状，下部绿色，上部褐色，先端偏斜，裂至中部。穗状花序顶生；花被粉红色或白色，5深裂。瘦果稍伸出宿存花被，椭球形。花期6-7月，果期8-9月。生海拔800-3000米的山坡草地或草甸。产华中、华北、华西、华东和东北。哈萨克斯坦、俄罗斯、蒙古、日本和欧洲亦有。

Perennial herbs. Rhizomes

拳参 *Polygonum bistorta*

中华抱茎蓼

Polygonum amplexicaule D. Don var. **sinense** Forbes et Hemsl. ex Steward

多年生草本。根状茎横生，不呈念珠状。基生叶卵形或长圆状卵形，下面有时沿脉具柔毛，基部心形，不下延；托叶鞘褐色，筒状，基部开裂，不具缘毛。花序顶生，穗状，疏散；花被红色，5深裂，裂片狭椭圆形。生海拔1000-3300米的草坡、山坡杂木林或林缘。产中国西南、华中和华西。印度、尼泊尔、不丹和巴基斯坦亦有。

Perennial herbs. Rhizomes horizontal, not torulose. Basal leaves ovate or oblong-ovate, abaxially sometimes pubescent along veins, base cordate, not decurrent; ocrea brown, tubular, dehiscent at base, not ciliate. Inflorescences terminal, spicate, lax; perianth red, 5-parted, lobes narrowly elliptic. Grassy slopes, mixed forests on mountain slopes or forest edges at 1000-3300 m. Distributed in SW, C and W China. Also in India, Nepal, Bhutan and Pakistan.

长梗蓼(长梗拳参)

Polygonum griffithii Hook. f.

多年生草本。根茎横生，长达20厘米。基生叶具长柄；叶具黄褐色柔毛；茎生叶具短柄，叶卵状椭圆形；托叶鞘筒状，先端偏斜，无缘毛。花序顶生或腋生，穗状，下垂，疏松，长3-5厘米；花梗长1-1.2厘米，中部具关节；花紫红色。瘦果黄褐色，光亮。花期7-8月，果期9-10月。生海拔3000-5000米的高山草甸或岩石缝中。产云南和西藏。不丹和缅甸北部亦有。

长梗蓼(长梗拳参) *Polygonum griffithii*

Perennial herbs. Rhizomes horizontal, up to 20 cm long. Basal leaves long petiolate; yellow-brown pubescent; cauline leaves shortly petiolate, ovate-elliptic; ocrea tubular, apex oblique, not ciliate. Inflorescences terminal or axillary, spicate, nutant, lax, 3-5 cm long; pedicels 1-1.2 cm long, articulate at middle; flowers purple-red. Achenes yellow-brown, shiny. Fl. Jul-Aug. Fr. Sep-Oct. Alpine meadows or rocky fissures at 3000-5000 m. Distributed in Yunnan and Xizang. Also in Bhutan and N Myanmar.

中华抱茎蓼 *Polygonum amplexicaule* var. *sinense*

草血竭

Polygonum paleaceum Wall. ex Hook. f.

多年生草本。根状茎弯曲，较大，直径2-3厘米；茎直立，不分枝。基生叶椭圆形或披针形，革质，两面无毛，基部不

草血竭 *Polygonum paleaceum*

下延；茎生叶向上渐狭，最上部叶线形；托叶鞘筒状，下部绿色，上部褐色，顶端无缘毛，开裂。花序穗状，密集；花被粉红色或白色，5深裂。花期7-8月，果期9-10月。生海拔1500-4000米的草地或疏林中。产四川、贵州、云南和广西。印度东北部和泰国北部亦有。

Perennial herbs. Rhizomes curved, large, 2-3 cm diam; stems erect, not branched. Basal leaves oblong or lanceolate, leathery, both surfaces glabrous; cauline leaves becoming narrow upward, upmost ones linear; orchreae tubular, lower part green, upper part brown, apex not ciliate, dehiscent. Inflorescences spicate, dense; perianth pinkish or white, 5-parted; anthers red-brown. Fl. Jul-Aug. Fr. Sep-Oct. Grasslands or sparse forests at 1500-4000 m. Distributed in Sichuan, Guizhou, Yunnan and Guangxi. Also in NE India and N Thailand.

革叶蓼

Polygonum coriaceum Sam.

多年生草本。根状茎弯曲，较大；茎直立，不分枝。叶革质，卵状椭圆形或卵状披针形，下面有时疏具柔毛；托叶鞘筒状，先端偏斜，开裂。花序顶生，穗状，紧密；花被紫红色，5深裂；雄蕊8，花药蓝黑色；花柱3，离生。花期7-8月，果期9-10月。生海拔2800-5000米的山坡草地、灌丛或林缘。产云南、四川、西藏和贵州。

Perennial herbs. Rhizomes curved, large; stems erect, not branched. Leaves leathery, ovate-elliptic or ovate-lanceolate, abaxially sometimes sparsely pubescent; ocrea tubular, apex oblique, dehiscent. Inflorescences terminal, spicate, dense; perianth purple-red, 5-parted; stamens 8, anthers blue-black; styles 3, free. Fl. Jul-Aug. Fr. Sep-Oct. Grassy slopes, bushes or forest edges at 2800-5000 m. Distributed in Yunnan, Sichuan, Xizang and Guizhou.

乌饭树叶蓼

Polygonum vacciniifolium Wall. ex Meisn.

亚灌木簇生。树皮黑褐色，纵向剥裂；老枝匍匐，小枝直立。叶椭圆形，薄革质；托叶鞘筒状，褐色，先端偏斜，不具缘毛，常撕裂状。穗状花序顶生，疏散；花紫红色。瘦果被包在宿存花被内，狭椭球形。花期8-9月，果期9-10月。生海拔3000-4200米的山坡灌丛或岩石缝中。产西藏。印度、尼泊尔、不丹和巴基斯坦亦有。

Subshrubs. Bark black-brown, longitudinally exfoliating; old branches nearly prostrate, small branches erect. Leaves elliptic, thinly leathery; ocrea tubular, brown, apex oblique, not ciliate, usually lacerate. Inflorescences terminal, spicate, lax; flowers purple-red. Achenes included in persistent perianth, narrowly ellipsoid. Fl. Aug-Sep. Fr. Sep-Oct. Thickets on mountain slopes or rock crevices at 3000-4200 m. Distributed in Xizang. Also in India, Nepal, Bhutan and Pakistan.

革叶蓼 *Polygonum coriaceum*

乌饭树叶蓼 *Polygonum vacciniifolium*

火炭母 *Polygonum chinense*

Perennial herbs. Stems glabrous, much branched. Leaves broadly ovate or elliptic, 6-10 cm broad, both surfaces glabrous, sometimes abaxially along veins sparsely pubescent; ocrea membranous, tubular, glabrous, much veined, apex oblique, not ciliate. Inflorescences terminal or axillary, capitate; perianth white or pinkish, 5-parted, lobes accrescent in fruit, becoming blue-black, fleshy. Fl. Jul-Sep. Fr. Jul-Nov. Forests on slopes at 1200-3000 m. Distributed in Yunnan, Xizang and Guizhou. Also in Himalaya and India.

火炭母

Polygonum chinense L.

多年生草本。茎光滑无毛，多分枝。叶卵形或长卵形，宽2-4厘米，两面无毛，有时下面沿脉疏生柔毛；托叶鞘膜质，筒状，光滑，具多脉，先端偏斜，无缘毛。花序顶生或腋生，头状；花被白色或粉红色，5深裂；花被片果期增大，后为蓝黑色，肉质。花期7-11月，果期7-12月。生海拔3000米以下的潮湿地、杂木林、灌丛或草地。产中国大部分地区。南亚、东南亚和日本亦有。

Perennial herbs. Stems glabrous, much branched. Leaves ovate, or narrowly ovate, 2-4 cm broad, both surfaces glabrous, sometimes abaxially along veins sparsely pubescent; ocrea membranous, tubular, glabrous, much veined, apex oblique, not ciliate. Inflorescences terminal or axillary, capitate; perianth white or pinkish, 5-parted; lobes accrescent in fruit, becoming blue-black, fleshy. Fl. Jul-Nov. Fr. Jul-Dec. Wet places, mixed forests, thickets or grasslands below 3000 m. Distributed in most parts of China. Also in S and SE Asia, and Japan.

宽叶火炭母

Polygonum chinensis L. var. **ovalifolium** Meisn.

多年生草本。茎无毛，多分枝。叶宽卵形或椭圆形，宽6-10厘米，两面无毛，有时下面沿脉疏生柔毛；托叶鞘膜质，筒状，光滑，具多脉，先端偏斜，无缘毛。花序顶生或腋生，头状；花被白色或粉红色，5深裂；花被片果期增大，后为蓝黑色，肉质。花期7-9月，果期7-11月。生海拔1200-3000米的山坡林下。产云南、西藏和贵州。喜马拉雅和印度亦有。

窄叶火炭母

Polygonum chinensis L. var. **paradoxum** (Lévl.) A. J. Li

多年生草本。茎无毛，多分枝。叶阔披针形，两面无毛，有时下面沿脉疏生柔毛；托叶鞘膜质，筒状，光滑，具多脉，先端偏斜，无缘毛。花序顶生或腋生，头状；花被白色或粉红色，5深裂；花被片果期增大，后为蓝黑色，肉质。花期7-11月，果期7-12月。生海拔900-2600米的草坡或山谷灌丛中。产贵州、四川和云南。

Perennial herbs. Stems glabrous, much branched. Leaves broadly lanceolate, both surfaces gla-

宽叶火炭母 *Polygonum chinensis* var. *ovalifolium*

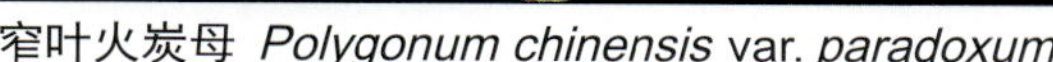
窄叶火炭母 *Polygonum chinensis* var. *paradoxum*

羽叶蓼 *Polygonum runcinatum*

brous, sometimes abaxially along veins sparsely pubescent; ocrea membranous, tubular, glabrous, much veined, apex oblique, not ciliate. Inflorescences terminal or axillary, capitate; perianth white or pinkish, 5-parted, lobes accrescent in fruit, becoming blue-black, fleshy. Fl. Jul-Nov. Fr. Jul-Dec. Grassy slopes or thickets in valleys at 900-2600 m. Distributed in Guizhou, Sichuan and Yunnan.

头花蓼

Polygonum capitatum

Buch.-Ham. ex D. Don

多年生草本，茎匍匐，丛生，基部木质。叶卵形或椭圆形，两面具腺毛，有时上面具黑斑；托叶鞘筒状，疏具腺毛，先端截形，具缘毛。花序顶生，头状，单生或成双；花被粉红色，5深裂。花期6-9月，果期8-10月。生海拔600-3500米的山坡、潮湿地或河堤上。产中国西南、华南至华中。南亚和东南亚亦有。

Perennial herbs. Stems creeping, tufted, ligneous at base. Leaves ovate or elliptic, both surfaces glandular hairy, sometimes with a blackish spot adaxially; ocrea tubular, sparsely glandular hairy, apex truncate, ciliate. Inflorescences terminal, capitate, solitary or geminate; perianth pinkish, 5-parted. Fl. Jun-Sep. Fr. Aug-Oct. Slopes, wet places or banks at 600-3500 m. Distributed in SW, S to C China. Also in S and SE Asia.

羽叶蓼

Polygonum runcinatum

Buch.-Ham. ex D. Don

多年生草本，茎近直立或斜升。叶羽裂，叶片两面疏被糙伏毛，侧生羽片1-3对。头状花序直径1-1.5厘米，通常成对着生；花被粉红色或白色，5深裂；雄蕊常8，内藏；花药紫色。花期4-8月，果期6-10月。生海拔800-3900米的草坡、沟谷灌丛或水边。产中国西南、东南、华中和华西。南亚和东南亚亦有。

Perennial herbs. Stems suberect or ascending. Leaves pinnatifid, sparsely strigose on both surfaces, lateral lobes 1-3 pairs. Capitulum 1-1.5 cm diam, usually geminate; perianth pinkish or white, 5-parted; stamens usually 8, included; anthers purple. Fl. Apr-Aug. Fr. Jun-Oct. Grassy slopes, thickets in valleys or by waters at 800-3900 m. Distributed in SW, SE, C and W China. Also in S and SE Asia.

头花蓼 *Polygonum capitatum*

尼泊尔蓼 *Polygonum nepalense*

尼泊尔蓼

Polygonum nepalense Meisn.

一年生草本。茎匍匐或斜升，自基部多分枝。叶卵形或三角状卵形，疏生黄色透明腺点，基部宽楔形，沿叶柄下延成翅；托叶鞘筒状，先端斜截形，不具缘毛，基部具反折刚毛。花序头状，被叶状总苞所包；苞片无毛；花被紫红色或白色，常4裂。花期5-8月，果期6-10月。生海拔200-4000米的山坡草地、潮湿沟谷或田间。产中国大部分地区。南亚、东南亚、东北亚和非洲亦有。

Annual herbs. Stems decumbent or ascending, much branched at base. Leaves ovate or triangular-ovate, sparsely yellow pellucid glandular punctate, base broadly cuneate, decurrent along petiole forming wing; ocrea tubular, brownish, apex obliquely truncate, not ciliate, with recurved seta at base. Inflorescences capitate, included by an involucral leaf; bracts glabrous; perianth purplish red or white, usually 4-parted. Fl. May-Aug. Fr. Jun-Oct. Grassy slopes, moist valleys or fields at 200-4000 m. Distributed in most parts of China. Also in S, SE and NE Asia, and Africa.

冰川蓼

Polygonum glaciale (Meisn.) Hook. f.

一年生草本。茎匍匐或斜升，自基部多分枝。叶卵形或宽卵形，基部近截形或宽楔形，两面无毛，全缘或基部2裂；托叶鞘筒状，疏松，无毛，先端截形或2裂。花序顶生或腋生，头状，小；花被白色或粉红色，5深裂；雄蕊常5。花期6-7月，果期7-8月。生海拔1300-4300米的山顶、山坡草地或山谷湿地。产中国西南、华北和华西。印度、尼泊尔和阿富汗亦有。

Annual herbs. Stems decumbent or ascending, much branched at base. Leaves ovate or broadly ovate, subtruncate or broadly cuneate at base, both surfaces glabrous, margin entire or 2-lobed at base; ocrea tubular, lax, glabrous, apex truncate or 2-cleft. Inflorescences terminal or axillary, capitate, small; perianth white or pinkish, 5-parted; stamens usually 5. Fl. Jun-Jul. Fr. Jul-Aug. Mountain tops, grassy slopes or moist places in valleys at 1300-4300 m. Distributed in SW, N and W China. Also in India, Nepal and Afghanistan.

冰川蓼 *Polygonum glaciale*

杠板归

Polygonum perfoliatum L.

一年生草本。茎攀援，多分枝，沿棱具倒刺。叶三角状盾形，下面常沿脉疏具倒刺；托叶鞘筒状，顶端有圆形绿色草质翅。花序顶生或腋生，穗状；花被白色或淡红色，5深裂；雄蕊8，2轮。瘦果包于肉质蓝色的宿存花被内，光亮，近球形。花期6-8月，果期7-10月。生海拔100-2300米的田边、路边或潮湿的山谷。产中国大部分地区。南亚、东南亚、西南亚和东北亚亦有；并引入北美洲。

杠板归 *Polygonum perfoliatum*

Annual herbs. Stems trailing, much branched, with retrose prickles along angles. Leaves triangular-peltate, abaxially usually sparsely retrorsely prickly along veins; ocrea tubular, with annular green herbaceous wings at apex. Inflorescences terminal or axillary, spicate; perianth white or pinkish, 5-parted; stamens 8, in 2 whorls. Achenes included in succulent, blue, persisitent perianth, shiny, subglobose. Fl. Jun-Aug. Fr. Jul-Oct. By fields, roadsides or wet valleys at 100-2300 m. Distributed in most parts of China. Also in S, SE, SW and NE Asia; introduced in North America.

长戟叶蓼

Polygonum maackianum Regel

一年生草本。茎直立或斜升，被倒刺或星状毛。叶长戟形，两面密被星状毛和疏具刺，基部心形；托叶鞘筒状，先端具绿色草质翅；翅环形，边缘具牙齿。花序顶生或腋生，头状。瘦果被包在宿存花被内，卵球形。花期6-9月，果期7-10月。生海拔100-1600米的山谷阴湿草地。产中国大部分地区。俄罗斯(远东地区)、朝鲜半岛和日本亦有。

Annual herbs. Stems erect or ascending, retrorsely prickly and densely stellate hairy. Leaves narrowly hastate, both surfaces densely stellate hairy and sparsely prickly, base cordate; ocrea tubular, apex with green, herbaceous wing; wings annular, margin dentate. Inflorescences terminal or axillary, capitate, Achenes included in persistent perianth, ovoid. Fl. Jun-Sep. Fr. Jul-Oct. Shaded grassy places in valleys at 100-1600 m. Distributed in most parts of China. Also in Russia (Far East), Korean Peninsula and Japan.

戟叶蓼 *Polygonum thunbergii*

戟叶蓼

Polygonum thunbergii Sieb. et Zucc.

多年生草本。茎具棱，直立或斜升，沿棱有倒刺。叶戟形，两面疏具刚毛；托叶鞘筒状，短，先端常具绿色草质翅，边缘全缘或具牙齿，具短缘毛。花序头状；花被粉红色或白色，5深裂；雄蕊8，2轮，内藏。瘦果黄褐色，宽卵球形。花期7-9月，果期8-10月。生海拔100-2400米的沟谷或草坡。产中国大部分地区。俄罗斯、朝鲜半岛和日本亦有。

Annual herbs. Stems angulate, erect or ascending, retrorsely prickly along angles. Leaves hastate, both surfaces sparsely bristly; ocrea tubular, short, usually apex with green herbaceous wing, margin entire or crenate, shortly ciliate. Inflorescences capitate; perianth pinkish or white, 5-parted; stamens 8, in 2 whorls, included. Achenes yellow-brown, broadly ovoid. Fl. Jul-Sep. Fr. Aug-Oct. Valleys or grassy slopes at 100-2400 m. Distributed in most parts of China. Also in Russia, Korean Peninsula and Japan.

长戟叶蓼 *Polygonum maackianum*

稀花蓼
Polygonum dissitiflorum Hemsl.

一年生草本。茎直立或基部匍匐，分枝，疏具倒刺，常具少量星状毛。叶卵状椭圆形，基部心形或戟形，边缘具短缘毛；托叶鞘筒状，顶端偏斜，具短缘毛。圆锥花序顶生或腋生；花序梗纤细，具紫红色腺毛；花被粉红色，5深裂；雄蕊7或8，2轮，内藏。花期6-8月，果期7-9月。生海拔100-1500米的山谷、丘陵草地或溪岸。产中国除华南和西北的大部分地区。俄罗斯(远东地区)和朝鲜半岛亦有。

Annual herbs. Stems erect or prostrate at base, branched, sparsely retrorsely prickly, usually with few stellate hairs. Leaves ovate-elliptic, base cordate or hastate, margin shortly ciliate; ocrea tubular, shortly ciliate, oblique. Inflorescences terminal or axillary, paniculate; peduncles slender, reddish purple glandular hairy; perianth pink, 5-parted; stamens 7 or 8, in 2 whorls, included. Fl. Jun-Aug. Fr. Jul-Sep. Valleys, hilly grasslands or stream banks at 100-1500 m. Distributed in most parts of China, except S and NW China. Also in Russia (Far East) and Korean Peninsula.

箭叶蓼
Polygonum sagittatum L.

一年生草本。茎攀援，分枝，沿棱具倒刺。叶宽披针形至长圆形，下面常于中脉近基部具倒刺，基部箭形。头状花序通常成对，顶生或腋生；花被白色至绿白色，常浅红色，5深裂；花被片不为肉质；雄蕊8，2轮。瘦果宽卵球形，黑色。花期6-9月，果期8-10月。生海拔100-2200米的山谷或水边。产中国西南、华中、华北、华东和东北。俄罗斯(远东地区)、朝鲜半岛和日本亦有。

Annual herbs. Stems scandent, branched, retrorsely prickly along angles. Leaves broadly lanceolate to oblong, abaxially usually retrorsely prickly near base of midvein, sagittate at base. Capitulums paired, terminal or axillary; perianth white to greenish white, often reddish, 5-parted; lobes not becoming fleshy; stamens 8, in 2 whorls. Achenes broadly ovoid, black. Fl. Jun-Sep. Fr. Aug-Oct. Valleys or by waters at 100-2200 m. Distributed in SW, C, N, E and NE China. Also in Russia (Far East), Korean Peninsula and Japan.

柳叶刺蓼
Polygonum bungeanum Turcz.

一年生草本。茎直立、斜升或匍匐，疏具倒刺。叶披针形或狭椭圆形，下面具硬毛；托叶鞘筒状，顶端截形，具长缘毛。花序顶生或腋生，穗状，常分枝，基部间断；花白色或粉红色；雄蕊8，2轮，内藏。瘦果黑色，扁球形，双凸。花期7-8月，果期8-9月。生海拔1700米以下山谷草地、田边或路边湿地。产华北、华西、华

稀花蓼 *Polygonum dissitiflorum*

箭叶蓼 *Polygonum sagittatum*

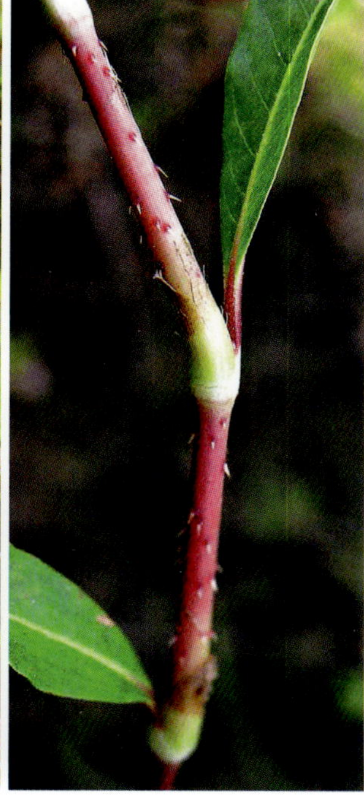

柳叶刺蓼 *Polygonum bungeanum*

东和东北。俄罗斯(远东地区)、朝鲜半岛和日本亦有。

Annual herbs. Stems erect, ascending or creeping. Leaves lanceolate or narrowly elliptic, abaxially hispidulous; ocrea tubular, apex truncate, long ciliate. Inflorescences terminal or axillary, spicate, usually branched, interrupted at base; flowers white or pinkish; stamens 8, in 2 whorls, included. Achenes black, orbicular, biconvex. Fl. Jul-Aug. Fr. Aug-Sep. Grassy valleys, by fields or moist places by roads below 1700 m. Distributed in N, W, E and NE China. Also in Russia (Far East), Korean Peninsula and Japan.

绢毛蓼(绢毛神血宁)

Polygonum molle D. Don

半灌木。茎直立，多分枝，具长硬毛。叶椭圆形或椭圆状披针形，下面疏生或密被绢毛；托叶鞘膜质，筒状，具柔毛，顶端偏斜。花序圆锥状，大型，开展；花被白色，5深裂；花被片椭圆形，果期增大，后变黑色，肉质；雄蕊8，内藏，花柱3，自基部分离。花期8-9月，果期9-11月。生海拔1200-3500米的山坡、林下或草地。产云南、西藏、贵州和广西。印度、尼泊尔、不丹、缅甸、泰国和印度尼西亚亦有。

Subshrubs. Stems erect, much branched, long-hispid. Leaves elliptic or elliptic-lanceolate, abaxially sparsely or densely sericeous; ocrea membranous, tubular, pubescent, apex oblique. Inflorescences paniculate, large, spreading; perianth white, 5-parted, lobes elliptic, accrescent in fruit, becoming black, fleshy; stamens 8, included; styles 3, free from base. Fl. Aug-Sep. Fr. Sep-Nov. Slopes, woods or grasslands at 1200-3500 m. Distributed in Yunnan, Xizang, Guizhou and Guangxi. Also in India, Nepal, Bhutan, Myanmar, Thailand and Indonesia.

大铜钱叶蓼(大铜钱叶神血宁)

Polygonum forrestii Diels

多年生草本。茎匍匐，丛生。叶近圆形或肾形，两面具长柔毛或近无毛，基部心形；托叶鞘膜质，疏松，筒状，具柔毛，顶端偏斜。伞房状聚伞花序顶生；花被白色或浅黄色；雄蕊6-8；花药紫色。瘦果狭椭球形，具3棱。花期7-8月，果期8-9月。生海拔3500-4800米的山坡或高山草甸。产云南、四川、西藏和贵州。印度北部、尼泊尔、不丹和缅甸亦有。

Perennial herbs. Stems creeping, tufted. Leaves nearly orbicular or reniform, both surfaces long pilose or nearly glabrous, base cordate; ocrea membranous, laxly, tubular, pilose, apex oblique. Inflorescences terminal, corymbose-cymose; perianth white or yellowish, 5(4)-parted; stamens 6-8; anthers purple. Achenes narrowly ellipsoid. Fl. Jul-Aug. Fr. Aug-Sep. Slopes or alpine meadows at 3500-4800 m. Distributed in Yunnan, Sichuan, Xizang and Guizhou. Also in N India, Nepal, Bhutan, and Myanmar.

大铜钱叶蓼(大铜钱叶神血宁)
Polygonum forrestii

绢毛蓼(绢毛神血宁) *Polygonum molle*

西伯利亚蓼(西伯利亚神血宁)

Polygonum sibiricum Laxm.

多年生草本。根状茎纤细；茎外倾或近直立，自基部分枝。叶狭椭圆形或披针形，基部戟形或楔形；托叶鞘筒状，膜质，先端偏斜，不具缘毛。圆锥花序顶生，疏松，间断；花被黄绿色，5深裂；雄蕊7或8，内藏。瘦果包于宿存花被或稍伸出，卵球形。花果期6-9月。生海拔5000米以下的路边、沙地、河岸或盐碱地。产中国西南、华中、华北、华西、华东、西北和东北。印度北部、尼泊尔、巴基斯坦、阿富汗、哈萨克斯坦、吉尔吉斯斯坦、塔吉克斯斯坦、俄罗斯和蒙古亦有。

钟花蓼(钟花神血宁) *Polygonum campanulatum*

Perennial herbs. Rhizomes slender; stems decumbent or suberect, branched from base. Leaves narrowly elliptic or lanceolate, base hastate or cuneate; ocrea tubular, membranous, apex oblique, not ciliate. Inflorescences terminal, paniculate, lax, interrupted; perianth yellow-green, 5-parted; stamens 7 or 8, included. Achenes included in persistent perianth or slightly exceeding them, ovoid. Fl. and fr. Jun-Sep. Roadsides, sandy areas, riverbanks or saline areas below 5000 m. Distributed in SW, C, N, W, E, NW and NE China. Also in N India, Nepal, Pakistan, Afghanistan, Kazakhstan, Kyrgyzstan, Tajikistan, Russia and Mongolia.

钟花蓼(钟花神血宁)

Polygonum campanulatum Hook. f.

多年生草本。茎近直立，基部近平卧。叶狭卵形或阔卵形，顶端渐尖或呈尾状，两面疏生柔毛；托叶鞘筒状，膜质，疏松，具柔毛，先端偏斜。花序圆锥状，小型；花被淡红色或白色，5深裂；雄蕊8，内藏；花药紫色。花期7-8月，果期9-10月。生海拔1400-4100米的山坡或潮湿地。产云南、四川、西藏、贵州和湖北。印度北部、尼泊尔、不丹和缅甸亦有。

Perennial herbs. Stems suberect, prostrate at base. Leaves narrowly ovate or broadly so, apex acuminate or caudate, both surfaces sparsely pilose; ocrea tubular, membranous, lax, pilose, apex oblique. Inflorescences paniculate, small; perianth pinkish or white, 5-parted; stamens 8, included; anthers purple. Fl. Jul-Aug. Fr. Sep-Oct. Slopes or wet places at 1400-4100 m. Distributed in Yunnan, Sichuan, Xizang, Guizhou and Hubei. Also in N India, Nepal, Bhutan and Myanmar.

西伯利亚蓼(西伯利亚神血宁)
Polygonum sibiricum

绒毛钟花蓼 *Polygonum campanulatum* var. *fulvidum*

绒毛钟花蓼

Polygonum campanulatum Hook. f. var. **fulvidum** Hook. f.

多年生草本。茎近直立，基部

近平卧。叶狭卵形或阔卵形，顶端渐尖或呈尾状，下面密具褐色绒毛；托叶鞘筒状，具柔毛，先端偏斜。花序圆锥状；花被淡粉色或白色；雄蕊8，内藏；花药紫色。生海拔1400-4000米的山坡或山谷。产云南、四川、西藏、贵州和湖北。不丹、缅甸、尼泊尔和印度北部亦有。

Perennial herbs. Stems suberect, prostrate at base. Leaves narrowly ovate or broadly so, apex acuminate or caudate, abaxially densely brown tomentose; ocrea tubular, pilose, apex oblique. Inflorescences paniculate; perianth pinkish or white; stamens 8, included; anthers purple. Slopes or valleys at 1400-4000 m. Distributed in Yunnan, Sichuan, Xizang, Guizhou and Hubei. Also in Bhutan, Myanmar, Nepal and N India.

齿翅蓼（齿翅首乌）*Fallopia dentatoalata*

齿翅蓼（齿翅首乌）

Fallopia dentatoalata (F. Schmidt) Holub

一年生草本。茎纤细，缠绕。叶卵状心形，两面无毛，沿脉稍具乳突；托叶鞘筒状，淡褐色，先端偏斜，不具缘毛。总状花序腋生或顶生，疏散；花被白色或浅绿色，5深裂，不等大，外轮3枚花被片果期增大，背面中脉上具翅；翅边缘具牙齿。瘦果包在宿存花被内，椭球形。花期6-9月，果期7-10月。生海拔200-2800米的林中或山坡。产中国西南、华北、华西、华东和东北。俄罗斯(远东地区)、朝鲜半岛和日本亦有。

Annual herbs. Stems slender, twining. Leaves ovate-cordate, both surfaces glabrous, along veins minutely papillate; ocrea tubular, brownish, apex oblique, not ciliate. Racemes axillary or terminal, lax; perianth white or greenish, 5-parted, unequal in size, outer 3 lobes accrescent, winged along midvein abaxially in fruit; wings dentate at margin. Achenes included in persistent perianth, ellipsoid. Fl. Jun-Sep. Fr. Jul-Oct. Forests or slopes at 200-2800 m. Distributed in SW, N, W, E and NE China. Also in Russia (Far East), Korean Peninsula and Japan.

何首乌

Fallopia multiflora (Thunb.) Harald.

多年生草本。块根黑褐色，狭椭球形，较大，木质。茎纤细，缠绕，分枝。叶卵状心形或长卵状心形；边缘全缘。花序顶生或腋生，圆锥状，开展；花白色或浅绿色，花被5深裂，花被片不等大，果期外轮3枚增大，沿背面中脉具翅；翅沿花梗下延。花期6-10月，果期7-11月。生海拔200-3000米的山谷灌丛中、山坡林下或沟边石隙。产中国大部分地区。日本亦有。

Perennial herbs. Root tubers black-brown, narrowly ellipsoid, large, ligneous. Stems slender, twining. Leaves ovate-cordate or long ovate-cordate; margin entire. Inflorescences terminal or axillary, paniculate, spreading; perianth white or greenish, 5-parted, unequal in size, 3 outer lobes larger, winged along midvein abaxially; wings decurrent along pedicels. Fl. Jun-Oct. Fr. Jul-Nov. Bushes of valleys, forests on slopes or rock crevices along streams at 200-3000 m. Distributed in most parts of China. Also in Japan.

何首乌 *Fallopia multiflora*

木藤蓼(木藤首乌) *Fallopia aubertii*

木藤蓼(木藤首乌)

Fallopia aubertii (L. Henry) Holub

半灌木。茎攀援。叶簇生，稀单生，叶狭卵形或卵形，两面近革质，无毛，基部近心形；托叶鞘筒状，褐色，膜质，先端偏斜，开裂。圆锥花序腋生或顶生，少分枝；花被淡绿色或白色，5深裂，不等大；外轮3枚花被片果期增大，沿背面中脉具翅；翅沿花梗下延。瘦果卵球形。花期7-8月，果期8-9月。生海拔900-3200米的山坡或山谷灌丛中。产中国西南、华中、华北和华西。

Subshrubs. Stems twining. Leaves clustered, rarely solitary, narrowly ovate or ovate, both surfaces subleathery, glabrous, base subcordate; ocrea tubular, brown, membranous, apex oblique, dehiscent. Inflorescences axillary or terminal, paniculate, few branched; perianth greenish or white, 5-parted, unequal in size, 3 outer lobes accrescent and winged along midvein abaxially in fruit; wings decurrent along pedicels. Achenes ovoid. Fl. Jul-Aug. Fr. Aug-Sep. Slopes or thickets in valleys at 900-3200 m. Distributed in SW, C, N and W China.

虎杖

Reynoutria japonica Houtt.

多年生草本，雌雄异株。茎多数，直立，丛生，表面常具红色或紫色斑点。叶宽卵形或卵状椭圆形，基部宽楔形、截形或圆形，近革质，两面无毛，沿脉具乳突。花序腋生，圆锥状，少分枝；花被白色或淡绿色，5深裂，不等大；雌花柱头流苏状。花期6-9月，果期7-10月。生海拔100-2000米的灌丛、山坡或田地。产中国大部分地区。俄罗斯(远东地区)、朝鲜半岛和日本亦有。

Perennial herbs dioecious. Stems numerous, erect, tufted, often with red or purple spots on surfaces. Leaves broadly ovate or ovate-elliptic, base broadly cuneate, truncate or rounded, subleathery, both surfaces glabrous, papillate along veins. Inflorescences axillary, paniculate, few branched; perianth white or greenish, 5-parted, lobes unequal in size; pistillate flowers with stigmas fimbriate. Fl. Jun-Sep. Fr. Jul-Oct. Shrubbery, slopes or fields at 100-2000 m. Distributed in most parts of China. Also in Russia (Far East), Korean Peninsula and Japan.

金线草

Antenoron filiforme (Thunb.) Rober et Vaut.

多年生草本。根状茎粗壮。茎直立，节具糙伏毛，膨大。叶椭圆形或狭椭圆形至卵形，两面具糙伏毛；托叶鞘筒状，膜质，顶端具短缘毛。花序穗状，顶生或腋生，纤细，花排列稀疏；花被玫瑰色，4深裂；花柱2，宿存，成熟时硬化，外折，顶端钩状，伸出花被外。花期8-10月，果期9-11月。生海拔100-2500米山坡林下、灌丛或山谷。产中国西南、华南、东南、华中、华西和华东。缅甸、俄罗斯(远东地区)、朝鲜半岛和日本亦有。

Perennial herbs. Rhizomes robust. Stems erect, appressed hispid, swollen at nodes. Leaves elliptic or narrowly elliptic to ovate, both surfaces appressed hispid; ocrea tubular, membranous, apex shortly ciliate. Inflorescences spicate, terminal or axillary, slender, laxly flowered; perianth rose, 4-parted; styles 2, persistent, and indurate at maturity, deflexed, hooked at apex, exserted from perianth. Fl. Aug-Oct. Fr. Sep-Nov. Forests on mountain slopes, thickets or val-

虎杖 *Reynoutria japonica*

金线草 *Antenoron filiforme*

细柄野荞麦 *Fagopyrum gracilipes*

leys at 100-2500 m. Distributed in SW, S, SE, C, W and E China. Also in Myanmar, Russia (Far East), Korean Peninsula and Japan.

金荞麦

Fagopyrum dibotrys (D. Don) Hara

多年生草本。根状茎黑褐色，粗壮，木质。茎直立，多分枝。叶三角形，两面具乳突，基部近戟形；托叶鞘筒状，膜质，褐色，先端偏斜，不具缘毛。花序顶生或腋生，伞房状；花被白色，5深裂。瘦果宽卵球形，明显伸出宿存花被。花期4-10月，果期5-11月。生海拔300-3200米的山谷湿地或山坡灌丛中。产中国西南、东南、华中和华东。印度、尼泊尔、不丹、缅甸、克什米尔地区、泰国和越南亦有。

Perennial herbs. Rhizomes black-brown, stout, ligneous. Stems erect, much branched. Leaves triangular, both surfaces papillate, base nearly hastate; ocrea tubular, brown, apex oblique, not ciliate. Inflorescences terminal or axillary, corymbose; perianth white, 5-parted. Achenes broadly ovoid, obviously exceeding persistent perianth. Fl. Apr-Oct. Fr. May-Nov. Moist places in valleys or bushes on slopes at 300-3200 m. Distributed in SW, SE, C and E China. Also in India, Nepal, Bhutan, Myanmar, Kashmir, Thailand and Vietnam.

细柄野荞麦

Fagopyrum gracilipes

(Hemsl.) Damm. ex Diels

一年生草本。茎直立，自基部分枝。叶卵状三角形，基部心形，两面疏具短硬毛，基部心形或戟形；托叶鞘膜质，偏斜，先端锐尖，具短糙伏毛。花序总状，腋生或顶生，极疏松，间断，纤细，俯垂；花被淡红色，5深裂。瘦果宽卵形，长3毫米，具3锐棱，有时沿棱具翅，稍伸出宿存花被。花期6-10月，果期7-11月。生海拔300-3400米的山坡草地、山谷湿地或田埂路边。产华中、华西和华北。

Annual herbs. Stems erect, branched from base. Leaves ovate-triangular, cordate at base, both surfaces sparsely shortly strigose, base cordate or hastate; ocrea membranous, tubular, oblique, apex acute, shortly strigose. Inflorescences racemose, axillary or terminal, very lax, interrupted, slender, pendulous; perianth pinkish, 5-parted. Achenes broadly ovoid, 3 mm long, sharply trigonous, sometimes winged along angles, slightly exceeding persistent perianth. Fl. Jun-Oct. Fr. Jul-Nov. Grassy slopes, moist places in valleys or field edges at 300-3400 m. Distributed in C, W and N China.

金荞麦 *Fagopyrum dibotrys*

荞麦 *Fagopyrum esculentum*

荞麦

Fagopyrum esculentum Moench

一年生草本。茎直立，自基部分枝。叶三角形，两面沿脉具乳突，基部心形或近截形；托叶鞘膜质，短筒状，顶端偏斜，不具缘毛，易脱落。花序总状或伞房状，腋生或顶生；花被粉色或白色，5深裂。瘦果卵球形，顶端渐尖，伸出宿存花被。花期5-9月，果期6-10月。中国各地有栽培或逸生。可能原生中国。亚洲、澳大利亚、欧洲和北美洲亦有栽培。

Annual herbs. Stems erect, branched from base. Leaves triangular, both surfaces papillate along veins, base cordate or nearly truncate; ocrea membranous, shortly tubular, apex oblique, not ciliate, caducous. Inflorescences racemose or corymbose, axillary or terminal; perianth pink or white, 5-parted. Achenes ovoid, apex acuminate, exceeding persistent perianth. Fl. May-Sep. Fr. Jun-Oct. Widely cultivated in China or escaped. Probably native to China. Also cultivated in Asia, Australia, Europe and North America.

泡果沙拐枣

Calligonum calliphysa Bunge

灌木，自基部多分枝，老枝扭曲。叶线形，对生；托叶鞘膜质，淡黄色，不与叶联合。花常2-4朵生叶腋，密集；花被片鲜时白色，背面中部绿色。果实膜质泡状果，球形或宽椭球形，幼时淡黄色、淡红色或红色，成熟后淡黄色、黄褐色或红褐色。瘦果椭球形，不扭转；肋较宽，每肋具3行刺，刺密，柔软，膜质。花期4-6月，果期5-7月。生海拔300-800米的砾石荒漠或草原。产内蒙古和新疆。西南亚和中亚亦有。

Shrubs, much branched from base, old branches tortuose. Leaves linear; ocrea light yellow, not united with leaf. Flowers often 2-4 at leaf axils, slightly dense; tepals green with a broad white margin abaxially. Fruits membranous-saccate, globose, or broadly ellipsoid, light red or red when young, light yellow, yellow-brown, or red-brown when mature. Achenes ellipsoid, not coiled; ribs broad, bristles in 3 rows per rib, dense, soft, membranous. Fl. Apr-Jun. Fr. May-Jul. Stony deserts or steppes at 300-800 m. Distributed in Neimenggu and Xinjiang. Also in SW and C Asia.

淡枝沙拐枣

Calligonum leucocladum (Schrenk) Bunge

灌木，高50-120厘米。老枝黄灰色或灰色，拐曲，斜展；新枝灰绿色，纤细。叶线形，易早落；托叶鞘膜质，浅黄色。花密集，2-4朵生叶腋；花被片绿色，具阔白色边缘。果实狭椭球形，不扭转或微扭转，4条肋各具2翅；翅浅黄色或黄褐色，近膜质。花期4-5月，果期5-6月。生海拔500-1200米的沙丘或荒漠。产新疆。西南亚和中亚亦有。

泡果沙拐枣 *Calligonum calliphysa*

淡枝沙拐枣 *Calligonum leucocladum*

Shrubs, 50-120 cm tall. Old branches yellow-gray, tortuose, ascending; young branchlets gray-green, slender. Leaves linear, easily deciduous; ocrea membranous, light yellow. Flowers dense, 2-4 at leaf axils; tepals green with broad white margin. Achenes narrowly ellipsoid, slightly coiled or not; ribs 4, each with 2 wings; wings light yellow or yellow-brown, submembranous. Fl. Apr-May. Fr. May-Jun. Sand dunes or deserts at 500-1200 m. Distributed in Xinjiang. Also in SW and C Asia.

褐色沙拐枣

Calligonum colubrinum Borsz.

灌木，高1.5-2米。茎多分枝，伸展。老枝灰白色，粗糙，幼枝淡绿色，平滑。叶鳞片状，和托叶鞘合生。花小，2或3朵腋生；花被片果期反折。果深褐色，近球形；瘦果椭球形，扭曲，肋具翅，翅边缘渐撕裂成刺；刺较硬，上部或中部叉状分枝。花期5-6月，果期6-7月。生海拔约600米的半固定或固定沙丘。产新疆东部。哈萨克斯坦和土库曼斯坦亦有。

Shurbs, 1.5-2 m tall. Stems much branched, spreading. Old branches gray-white, scabrous, young branchlets white-green, glabrous. Leaves scalelike, united with ocrea. Flowers small, 2 or 3 at leaf axils; tepals reflexed in fruit. Fruits dark brown, subglobose; achenes ellipsoid, coiled, ribbed with wings, margin of wings gradually splitting into bristles; bristles slightly stiff, forked at middle or above. Fl. May-Jun. Fr. Jun-Jul. Semimobile or stable sand dunes at ca. 600 m. Distributed in E Xinjiang. Also in Kazakhstan and Turkmenistan.

柴达木沙拐枣

Calligonum zaidamense A. Los.

灌木，高0.6-2米，植株近球形。老枝浅灰色或黄灰色；嫩枝草质，灰绿色。花稠密，2-4腋生。果实阔椭球形；瘦果卵球形，扭转或否；肋钝圆；中央生2行刺，刺细弱，易折断。果期7月。生海拔1500-2700米的移动沙丘。产青海和新疆。

Shrubs, 0.6-2 m tall, subglobose. Old branches light gray or yellow-gray; young branchlets herbaceous gray-green. Flowers dense, 2 or 4 at leaf axils. Fruits broadly ellipsoid; achenes ovoid, coiled or not; ribs obtuse; bristles in 2 rows at center of ribs, breakable. Fr. Jul. Mobile sand dunes at 1500-2700 m. Distributed in Qinghai and Xinjiang.

柴达木沙拐枣 *Calligonum zaidamense*

乔木沙拐枣

Calligonum arborescens Litv.

灌木，高2-4米。茎自基部多分枝。老枝黄白色，有裂纹及褐色条纹；嫩枝草质，灰绿色。叶鳞片状，具褐色尖头；托叶鞘和叶合生。花2或3朵腋生；花被片果期反折。果卵球形，幼时黄色或红色，后淡黄色或褐色；瘦果椭球形，强烈扭曲，明显具4肋；每肋具2行刺。花期4-5月。果期5-6月。生海拔500-600米的沙漠沙丘。产甘肃、宁夏和新疆。哈萨克斯坦、乌兹别克斯坦和土库曼斯坦亦有。

Shrubs, 2-4 m tall. Stems much branched from base. Old branches yellow-white, often longitudinally splitting and with brown stripes; young branchlets herbaceous, gray-green. Leaves scalelike, with a brown mucro; ocrea united with leaf. Flowers 2 or 3 at leaf axils; tepals reflexed in fruit. Fruits ovoid, yellow or red when young, becoming light yellow or light brown; achenes ellipsoid, strongly coiled, prominently 4-ribbed; bristles in 2 rows per rib. Fl. Apr-May. Fr. May-Jun. Desert sand dunes at 500-600 m. Distributed in Gansu, Ningxia and Xinjiang. Also in Kazakhstan, Uzbekistan and Turkmenistan.

褐色沙拐枣 *Calligonum colubrinum*

乔木沙拐枣 *Calligonum arborescens*

木蓼 *Atraphaxis frutescens*

木蓼
Atraphaxis frutescens (L.) Eversm.

灌木，高50-100厘米。茎粗壮，树皮灰褐色，纤细状脱落，多分枝，枝顶端无刺。叶蓝绿色至灰绿色，狭披针形、披针形或长圆形，无毛或具乳突状毛；托叶鞘圆柱状，基部褐色，膜质，透明，上部2裂成2锐齿。总状花序顶生，疏松；粉红色，具白色边缘。瘦果狭卵形，具3棱，黑褐色，有光泽。花果期5-8月。生海拔500-3000米的石质或干旱山坡、沙地或沙丘、田边或砾石河岸。产内蒙古、甘肃、宁夏、青海和新疆。哈萨克斯坦、俄罗斯、蒙古和欧洲东部亦有。

Shrubs, much branched, 50-100 cm tall. Stems stout, with gray-brown bark, epidermis splitting fibrously, woody branches spreading, apex without spines. Leaves blue-green to gray-green, narrowly lanceolate, lanceolate, or oblong, glabrous or papillate-hairy; ocrea cylindric, brown at base, membranous, pellucid, upper part cleft into 2 sharp teeth. Flowers fewer in terminal racemes; tepals pink, with white margin. Achenes narrowly ovoid, trigonous, dark brown, shiny. Fl. and fr. May-Aug. Stony or dry slopes, sandy areas or dunes, fieldsides or stony river banks at 500-3000 m. Distributed in Neimenggu, Gansu, Ningxia, Qinghai and Xinjiang. Also in Kazakhstan, Russia, Mongolia and E Europe.

山蓼
Oxyria digyna (L.) Hill

多年生草本。根状茎粗壮；茎直立，单生或数条自根状茎生出，无毛。基生叶肾形或圆肾形，纸质，下面沿脉疏具硬毛；茎生叶通常无；托叶鞘短筒状，膜质，先端偏斜。花序圆锥状，顶生；花两性；花被淡绿色或淡粉色。瘦果卵球形，两侧边缘具宽翅；翅淡红色，边缘具小齿。花期6-10月，果期7-11月。生海拔1300-4900米高山草甸、沼泽地或林中。产中国西南、西北和东北。亚洲、欧洲和北美洲亦有。

Perennial herbs. Rhizomes stout; stems erect, solitary or several from rhizome, usually glabrous. Basal leaves reniform or orbicular-reniform, papery, abaxially sparsely hirtellous along veins; cauline ones often absent; ocrea shortly tubular, membranous, apex oblique. Inflorescences paniculate, terminal; flowers bisexual; tepals 4, greenish or pinkish. Achenes ovoid, broadly winged at margin; wings pink, denticulate. Fl. Jun-Oct. Fr. Jul-Nov. Alpine meadows, marsh lands or forests at 1300-4900 m. Distributed in SW, NW and NE China. Also in Asia, Europe and North America.

戟叶酸模
Rumex hastatus D. Don

灌木。老枝木质，紫褐色；小枝绿色，光滑无毛。叶单生或成簇，戟形，近革质；中裂片线形或狭三角形，侧裂片向上弯曲；托叶鞘先落。花序顶生，圆锥状，疏松；雌花：外轮花被片果期反折；内轮花被片果期增大，圆形或肾状圆形，淡红色。瘦果卵球形，具三棱，褐色，有光泽。花期4-5月，果期5-6月。生海拔600-3200米的阳坡或岩隙。产云南、四川和西藏东南部。印度、尼泊尔、不丹、巴基斯坦和阿富汗亦有。

Shrubs. Old branches lignous, purple-brown; branchlets green, glabrous. Leaves solitary or fascicled, hastate, subleathery; central lobe linear, or narrowly triangular, basal lobes curved upward; ocrea fugacious. Inflorescences

山蓼 *Oxyria digyna*

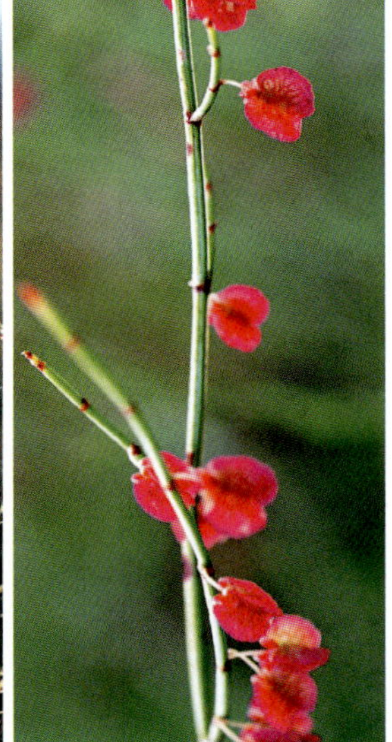
戟叶酸模 *Rumex hastatus*

terminal, paniculate, lax; female flowers: outer tepals reflexed in fruit; inner tepals enlarged in fruit, orbicular or reniform, pinkish. Achenes ovoid, trigonous, brown, shiny. Fl. Apr-May. Fr. May-Jun. Sunny slopes or rocky fissures at 600-3200 m. Distributed in Yunnan, Sichuan and SE Xizang. Also in India, Nepal, Bhutan, Pakistan and Afghanistan.

酸模

Rumex acetosa L.

多年生草本，雌雄异株。茎直立，常不分枝。基部叶卵状披针形至披针形，基生叶和茎下部叶箭形，茎上部叶较小；托叶鞘白色，膜质，先落。花序狭圆锥形，顶生，稀疏；花单性；雄花外花被片小而直立，内花被片椭圆形，雄蕊6；雌花外花被片果期反折，内花被片果期增大，近圆形。瘦果椭球形，具3锐棱，黑褐色，光亮。花期5-7月，果期6-8月。生海拔400-4100米山坡、林缘或潮湿的山谷。产中国大部分地区，尤其是北方。西南亚、东北亚、欧洲和北美洲亦有。

Perennial herbs, dioecious. Stems erect, usually not branched. Basal leaves and cauline ones lower ovate-lanceolate to lanceolate, base sagittate, upper small; ocrea fugacious, white, membranous. Inflorescences terminal, narrowly paniculate, lax; flowers unisexual; male flowers: outer tepals small, erect, inner ones elliptic, stamens 6; female flowers: outer tepals reflexed in fruit, inner ones enlarged in fruit, suborbicular. Achenes ellipsoid, sharply trigonous, blackish brown, shiny. Fl. May-Jul. Fr. Jun-Aug. Mountain slopes, forest edges or moist valleys at 400-4100 m. Distributed in most parts of China, especially the northern parts. Also in SW and NE Asia, Europe and North America.

酸模 *Rumex acetosa*

羊蹄

Rumex japonicus Houtt.

多年生草本。茎直立，上部分枝。基生叶长圆形或披针状长圆形，基部圆形或心形，下面沿脉微具乳突；茎生叶狭长圆形；托叶鞘白色，膜质，先落。花序圆锥状，多花轮生；花两性，花梗纤细，中下部有关节；内花被片果期增大，宽心形，边缘具不整齐小齿，全部具小瘤。瘦果阔卵球形，具3锐棱，深褐色，有光泽。花期5-6月，果期6-7月。生海拔3400米以下的田边、河岸或潮湿山谷。产中国大部分地区。俄罗斯(远东地区)、朝鲜半岛和日本亦有。

Perennial herbs. Stems erect, branched above. Basal leaves oblong or lanceolate-oblong, base orbicular or cordate, abaxially minutely papillate along veins; ocrea fugacious, white, membranous. Inflorescences paniculate; flowers bisexual, pedicels slender, articular below middle; inter tepals enlarged in fruit, broadly cordate, margin irregularly denticulate, all with narrowly ovate tubercles. Achenes broadly ovoid, sharply trigonous, dark brown, shiny. Fl. May-Jun. Fr. Jun-Jul. By fields, stream banks, or wet valleys below 3400 m. Distributed in most parts of China. Also in Russia (Far East), Korean Peninsula and Japan.

羊蹄 *Rumex japonicus*

皱叶酸模 *Rumex crispus*

皱叶酸模
Rumex crispus L.

多年生草本。根粗壮。茎直立，不分枝或上部分枝。基生叶披针形或狭披针形，无毛或沿下面脉具不明显乳突，基部常楔形，边缘皱波状；茎生叶较小，狭披针形。托叶鞘白色，膜质，先落。花序狭圆锥，顶生，分枝近直立或斜升；花两性；内花被片果期增大，宽卵形，基部近截形，边缘全缘，全部具小瘤。瘦果卵球形，具3锐棱，深褐色，有光泽。花期5-6月，果期6-7月。生海拔2500米以下的田边、溪边或荒地。产中国大部分地区。世界广分布。

Perennial herbs. Roots stout. Stems erect, not branched or branched above. Basal leaves lanceolate or narrowly lanceolate, glabrous or indistinctly papillose along veins below, base usually cuneate, margin strongly crisped and undulate. Inflorescences terminal, narrowly paniculate, branches suberect or ascending; flowers bisexual; inter tepals enlarged in fruit, broadly ovate, base subtruncate, margin entire, all with narrowly ovate tubercles. Achenes ovoid, sharply trigonous, dark brown, shiny. Fl. May-Jun. Fr. Jun-Jul. Field edges, streamsides or waste areas below 2500 m. Distributed in most parts of China. Also worldwide.

尼泊尔酸模
Rumex nepalensis Spreng.

多年生草本。根粗壮。茎直立，上部分枝。基生叶宽卵形，两面无毛或下面沿脉微具乳突，基部心形，边缘全缘；茎生叶卵状披针形；托叶鞘膜质，早落。花序圆锥状，顶生；花两性，内花被片果期增大，宽卵形，基部截形，边缘具7-8齿，顶端钩状，一或二部或全部具小瘤。瘦果卵球形，具3锐棱，褐色，有光泽。花期4-5月，果期6-7月。生海拔1000-4300米的山坡、湿润沟谷或草地。产中国西南、华中和华西。南亚、东南亚和西南亚亦有。

Perennial herbs. Roots stout. Stems erect, branched above. Basal leaves broadly ovate, both surfaces glabrous or abaxially minutely papillate along veins, base cordate, margin entire; cauline leaves ovate-lanceolate; ocrea membranous, caudous. Inflorescences paniculate, terminal; flowers bisexual, inter tepals enlarged in fruit, broadly ovate, base truncate, margin with 7-8 teeth, apex acute and hooked, all or 1 or 2 with tubercles. Achenes ovoid, sharply trigonous, brown, shiny. Fl. Apr-May. Fr. Jun-Jul. Slopes, moist valleys or grasslands at 1000-4300 m. Distributed in SW, C and W China. Also in S, SE and SW Asia.

长刺酸模
Rumex trisetifer Stokes

一年生草本。根粗壮，红褐色。茎直立，分枝开展。下部叶长圆形或披针状长圆形，两面无毛，基部楔形，边缘波状；托叶鞘脱落，膜质。总状花序顶生或腋生；花两性；外花被片披针形，较小，内花被片在果期增大，狭三角形，边缘每侧具1枚针刺。瘦果椭球形，具3锐棱，黄褐色，有光泽。花期5-6月，果期6-7月。生海拔1300米以下的田边、潮湿山谷或河边。产中国西南、华南、东南、华中至华东。印度、不丹、缅甸、老挝、泰国和越南亦有。

尼泊尔酸模 *Rumex nepalensis*

长刺酸模 *Rumex trisetifer*

Annual herbs. Roots stout, red-brown. Stems erect, branches spreading. Lower leaves oblong or lanceolate-oblong, both surfaces glabrous, base cuneate, margin undulate; ocrea fugacious, membranous. Inflorescences terminal or axillary, racemose; flowers bisexual; outer tapals lanceolate, small, inner tepals enlarged in fruit, narrowly ovate, margin with 1 tooth, apex narrowly acute. Achenes ellipsoid, sharply trigonous, yellow-brown, shiny. Fl. May-Jun. Fr. Jun-Jul. Field edges, moist valleys or by waters below 1300 m. Distributed in SW, S, SE, C to E China. Also in India, Bhutan, Myanmar, Laos, Thailand and Vietnam.

波叶大黄

Rheum rhabarbarum L.

大型草本，高50-150米。茎粗壮，中空，直立。基生叶叶柄粗壮，绿色；叶片三角状卵形，边缘强皱波，上部叶渐小；托叶鞘膜质，白色，抱茎。圆锥花序大型；花白绿色，5-8簇生；花被片6，外3，稍小、狭窄，内3，稍大、椭圆形；花柱短；柱头头状。瘦果较小，卵状状椭球形，具三窄翅。花期6月，果期7月以后。生海拔1000-1600米的山坡或林缘。产湖北、河北、山西、内蒙古、黑龙江和吉林。俄罗斯(东西伯利亚)和蒙古亦有；欧洲有栽培。

Large herbs, 50-150 m tall. Stems stout, hollow, erect. Basal leaves petioles robust, green; blade large, triangular ovate, margin crispate or sinuate, upper leaves becoming small; ocrea membranous, white, amplexicaul. Panicles large; flowers white-green or yellow-white, 5-8 fascicled; tepals 6, outer 3 smaller and narrow, inner 3, large and elliptic; styles short; stigmas inflated. Achenes smaller, ovate-ellipsoid, trigonous, narrowly 3-winged. Fl. Jun. Fr. after Jul. Slopes or forest edges at 1000-1600 m. Distributed in Hubei, Hebei, Shanxi, Neimenggu, Heilongjiang and Jilin. Also in Russia (E Siberia) and Mongolia; cultivated in Europe.

丽江大黄

Rheum likiangense Sam.

多年生草本。茎直立，密被白色硬毛。基生叶2-4枚；叶阔卵形至圆形，近革质，背面深紫色，密生白毛，基部心形，边缘全缘。圆锥花序1-2分枝，具白色硬毛；花簇生；花被片6，白绿色。果实卵球形，具3翅。花期7月，果期8-9月。生海拔2500-4000米的林中或草地。产云南西北部、四川西南部和西藏东部。

Perennial herbs. Stems erect, densely white hispid. Basal leaves 2-4; broadly ovate to orbicular, nearly leathery, abaxially dark purple, with dense white hairs, base cordate, margin entire. Panicles 1-2-branched, white hispid; flowers fascicled; tepals 6, white-green. Fruits ovoid, 3 winged. Fl. Jul. Fr. Aug-Sep. Forests or grasslands at 2500-4000 m. Distributed in NW Yunnan, SW Sichuan and E Xizang.

波叶大黄 *Rheum rhabarbarum*

丽江大黄 *Rheum likiangense*

药用大黄 *Rheum officinale*

药用大黄

Rheum officinale Baill.

多年生高大草本，高1.5-2米。根及根状茎粗壮。茎粗壮，直立，中空。基生叶大型，近圆形或宽卵圆形，掌状浅裂，裂片三角形，下面具柔毛，基生脉5-7，基部心形；茎生叶向上变小。大型圆锥花序，分枝开展；花4或5簇生；花被片6，绿色至黄白色。果实矩圆状椭球形，具3翅。花期5-6月，果期8-9月。生海拔1200-4000米丘陵、林下，或栽培。产中国西南、东南和华西。

Perennial large herbs, 1.5-2 m tall. Roots and rhizomes robust. Stems stout, erect, hollow. Basal leaves large, suborbicular or broadly ovate, palmately lobed, lobes triangular, abaxially pubescent, basal veins 5-7, base subcordate. Panicles large, branches spreading; flowers 4 or 5fascicled; tepals 6, green to yellow-white. Fruits oblong-ellipsoid, 3-winged. Fl. May-Jun. Fr. Aug-Sep. Hills, forests, or cultivated at 1200-4000 m. Distributed in SW, SE and W China.

掌叶大黄

Rheum palmatum L.

多年生高大草本，高1.5-2米。根及根状茎木质粗壮。茎直立，中空。基生叶大型，近圆形或卵圆形，掌状半裂，基生脉5条，基部心形，先端渐尖或窄急尖；茎生叶向上变小。大型圆锥花序，分枝聚拢；花小，紫红色，有时黄白色.果实椭球形至长球形。花期6月，果期8月。生海拔1500-4400米的山坡或山谷湿地。产中国西南、华北和华西。俄罗斯有栽培。

Perenniak large herbs, 1.5-2 m tall. Roots and rhizomes woody, robust. Stems erect, hollow. Basal leaves large, nearly orbiculate or ovate-orbiculate, palmately divided into pinnatisect lobes, basal veins 5, base cordate, apex acuminate or narrowly acute. Panicles large, branches connivent; flowers small, purple-red, sometimes yellow-white. Achenes oblong-ellipsoid to oblong. Fl. Jun. Fr. Aug. Slopes or moist places in valleys at 1500-4400 m. Distributed in SW, N and W China. Cultivated in Russia.

矮大黄

Rheum nanum Siev. ex Pallas

矮小粗壮草本，高20-35厘米。无茎。叶基生，叶片肾形或圆形，革质，基出脉3-5条，基部圆形或近心形，边缘近全缘。花序自根状茎顶端生出，上部分枝，成宽圆锥状；花被片近肉质，淡黄色。瘦果肾状球形，红色。花期5-6月，果期7-9月。生海拔700-2000米的山坡或山谷。产内蒙古、甘肃和新疆。哈萨克斯坦、俄罗斯和蒙古亦有。

Herbs short, stout, 20-35 cm tall. Stems absent. Leaves basal, blades reniform or orbicular, leathery, basal veins 3-5, base rounded or subcordate, margin nearly entire. Inflorescences raised from apex of rhizome, branched above middle to forming broad panicles; tepals nearly fleshy, yellow-white. Achenes reniform, red,. Fl. May-Jun. Fr. Jul-Sep. Slopes or valleys at 700-2000 m. Distributed in Neimenggu, Gansu and Xinjiang.

掌叶大黄 *Rheum palmatum*

矮大黄 *Rheum nanum*

圆叶大黄 *Rheum tataricum*

网脉大黄 *Rheum reticulatum*

Also in Kazakhstan, Russia and Mongolia.

圆叶大黄

Rheum tataricum L. f.

多年生草本，高35-50厘米。根粗壮。茎直立，粗壮，中空。基生叶1枚或无，大型，平铺地上；叶片纸质，心状圆形或圆形，背面光滑或下面乳头状，基部心形，边缘具锯齿；茎生叶小，近圆形。圆锥花序自中部3次分枝，形成阔圆球形；花1或2簇生；花被片黄白色。瘦果卵球形，紫红色。花期5月，果期6-7月。生海拔500-1000米草地，荒漠或平原。产新疆西部。阿富汗、哈萨克斯坦和俄罗斯(欧洲部分)亦有。

Perennial herbs, 35-50 cm tall. Roots stout. Stems erect, robust, hollow. Basal leaf 1 or absent, large, procumbent; blades papery, cordate-orbicular or orbicular, abaxially glabrous or papilliferous, base cordate, margin serrulate; cauline leaves small, suborbiculate. Panicles branched 3 times from middle to forming broadly globose; flowers 1- or 2-fascicled; tepals yellow-white. Achenes ovoid, purple-red. Fl. May. Fr. Jun-Jul. Grasslands, deserts or plains at 500-1000 m. Distributed in W Xinjiang. Also in Afghanistan, Kazakhstan and Russia (European parts).

网脉大黄

Rheum reticulatum Losinsk.

矮壮草本，高10-30厘米。根状茎顶端有残存托叶鞘。叶基生；叶片革质，三角状卵形，下面紫红色，基出脉5条，脉网极显著。总状花序穗状，自根状茎顶端生出，多枚，可达10；花黄白色，密集。瘦果宽卵球形，三棱状，具翅。花期6月，果期7-8月。生海拔2900-4200米的山坡。产青海和新疆。哈萨克斯坦、吉尔吉斯斯坦和塔吉克斯坦亦有。

Herbs stout, 10-30 cm tall. Rhizomes with remaining ocrea at apex. Leaves basal; blades leathery, triangular-ovate, abaxially purple-red, basal veins 5, reticulation raised obviously. Racemes spiciform, up to 10 from apex of rhizome; flowers yellow-white, dense. Achenes broadly ovoid, trigonous, winged. Fl. Jun. Fr. Jul-Aug. Slopes at 2900-4200 m. Distributed in Qinghai and Xinjiang. Also in Kazakhstan, Kyrgyzstan and Tajikistan.

苞叶大黄(水黄，大苞大黄)

Rheum alexandrae Batalin

多年生草本，高40-80厘米。下部叶卵形至卵状椭圆形，两面无毛；上部叶及叶状苞片较窄小，长卵形，黄绿色，下垂，秋后变红色；托叶鞘褐色，较大，长约7厘米，抱茎。圆锥花序具2或3分枝；花簇生，较小；花被片绿色，基部贴生成杯状。果实菱状椭球形。花期6-7月，果期9月。生海拔3000-4500米的草坡、高山沼泽、溪边、积水草地。产云南西北部、四川西部和西藏东部。

Herbs perennial, 40-80 cm tall. Lower leaves ovate to ovate-elliptic, both surfaces glabrous; upper ones and leafy bracts narrower, smaller, longe ovate, yellow-green, drooping, red at autumn; ocrea brown, large, ca. 7 cm long, clasping. Panicles 2- or 3-branched; flowers fascicled, small; tepals green, connected at base to a cup. Fruits rhomboid-ellipsoid. Fl. Jun-Jul. Fr. Sep. Grassy slopes, alpine swamps, streamsides and water meadows at 3000-4500 m. Distributed in NW Yunnan, W Sichuan and E Xizang.

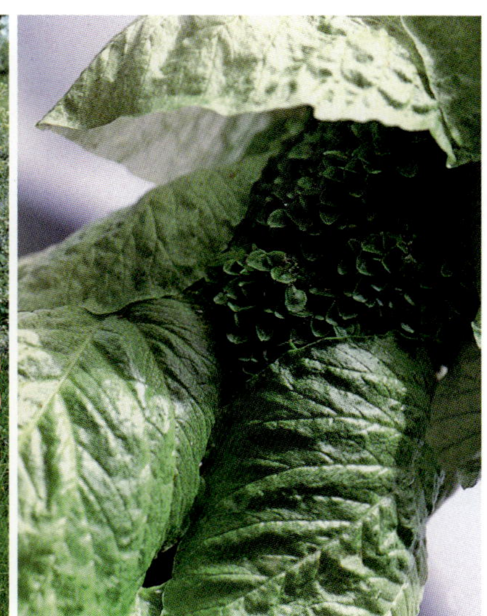
苞叶大黄(水黄、大苞大黄) *Rheum alexandrae*

藜科
Chenopodiaceae

驼绒藜 *Krascheninnikovia ceratoides*

心叶驼绒藜 *Krascheninnikovia ewersmannia*

千针苋
Acroglochin persicarioides (Poir) Moq.

一年生草本。单叶互生，卵形至狭卵形，边缘不整齐羽状浅裂。花序生于几所有叶的叶腋，复二歧聚伞状，末端枝条针状，无花；花被5深裂至近基部，边缘膜质。盖果半球形。花果期6-11月。生田边、路旁、河边或荒地。产中国西南、华中和华西。印度、尼泊尔、不丹和巴基斯坦亦有。

Annual herbs. Leaves simple, alternate, ovate to narrowly so, margin irregularly pinnatilobate. Inflorescences borne in axils of almost all leaves, compoundly dichasium-like, ultimate branches needle-like with no flowers; perianth 5-parted to near base, margin membranous. Pyxidia hemispheric. Fl. and fr. Jun-Nov. Field edges, roadsides, river banks or wastelands. Distributed in SW, C and W China. Also in India, Nepal, Bhutan and Pakistan.

驼绒藜
Krascheninnikovia ceratoides (L.) Gueldenst.

灌木。多分枝；分枝斜展或平展。叶条形至披针形，基部渐狭、楔形或圆形，先端急尖或钝；主脉外凸。雄花序长达4厘米，紧密；雌花管椭圆形，长3-4毫米，宽约2毫米，长为2个角状裂片的1-2倍。胞果椭圆形，被毛。花果期6-9月。生戈壁、半荒漠、干旱山坡。产甘肃、内蒙古、青海、新疆和西藏。亚洲、欧洲东南部和非洲北部的干旱地区亦有。

Shrubs. Much branched; branches spreading. Leaves linear to lanceolate, base attenuate, cuneate, or rounded, apex acute or obtuse; midvein prominent. Male inflorescence to 4 cm, dense; female floral tube ellipsoid, 3-4 mm long, ca. 2 mm wide, 1-2 times as long as 2-cornute free part. Utricle ellipsoid, hairy. Fl. and fr. Jun-Sep. Gobi desert, semideserts, dry slopes. Distributed in Gansu, Neimenggu, Qinghai, Xinjiang and Xizang. Also in Asia, SE Europe and arid regions of N Africa.

千针苋 *Acroglochin persicarioides*

心叶驼绒藜
Krascheninnikovia ewersmannia (Stschegl. ex Losinsk.) Grubov

灌木雌雄同株，全株密被星状毛，上部多分枝。叶柄短；叶片卵形或卵状长圆形，先端急

毛果绳虫实 *Corispermum tylocarpum*

尖或圆形，基部心形。花单性；雄花纤细，具4花被片，无苞片；雌花具2苞片，雌花管椭圆形，裂片略向后弯，果时管外具4束长毛。胞果椭圆形，密被毛。种子直生。花果期7-9月。生半荒漠、沙丘或荒地。产新疆。蒙古和中亚亦有。

Shrubs, monoecious, covered with stellate hairs, much branched above. Petiole short; leaf blade ovate or ovate-oblong, apex acute or rounded, base cordate. Flowers unisexual; male flowers slender, with 4 perianth segments, without bracts; female flowers with 2 bracts, floral tube ellipsoid, with slightly recurved lobes, abaxially with 4 fascicles of long hairs at base in fruit. Utricle ellipsoid, hairy. Seed vertical. Fl. and fr. Jul-Sep. Sandy deserts, dunes, wastelands. Distributed in Xinjiang. Also in Mongolia and C Asia.

毛果绳虫实

Corispermum tylocarpum
Hance

一年生草本。茎直立，多分枝。叶线形至线状披针形，具单脉。穗状花序伸长，线状圆柱形，间断至稍紧密；花被裂片1(或3)。胞果倒卵球状长圆形，无毛或被树枝状或星状毛；翅极狭。花果期6-9月。生沙质荒地、河边、田缘或路边。产华北、华西和西北。蒙古亦有。

Annual herbs. Stems erect, many-branched. Leaves linear to linear-lanceolate, 1-veined. Spikelike inflorescences elongate, linear-cylindric, interrupted to ± dense; perianth segments 1(or 3). Utricles obovate-oblong, glabrous or covered with dendritic or stellate hairs; wings very narrow. Fl. and fr. Jun-Sep. Sandy wastelands, riversides, field edges or roadsides. Distributed in N, W and NW China. Also in Mongolia.

兴安虫实

Corispermum chinganicum
Iljin

一年生草本。茎绿色或红紫色，自基部分枝。穗状花序顶生或侧生。胞果近光亮，具少量棕色斑点，长圆状倒卵形或阔椭圆形，无毛、近无毛或两面覆有星状毛；翅明显，浅黄色，全缘。花果期6-8月。生沙丘、湖畔或草甸。产华北、华西和东北。俄罗斯(西伯利亚东南部)和蒙古亦有。

Annual herbs. Stems green or reddish purple, branched from base. Spikes terminal or lateral. Utricles sublustrous, with a few brown spots, oblong-obovate or broadly elliptic, glabrous, subglabrous, or covered on both sides with stellate hairs; wings distinct, light yellow, margin entire. Fl. and fr. Jun-Aug. Dunes, lake shores or meadows. Distributed in N, W and NE China. Also in Russia (SE Siberia) and Mongolia.

兴安虫实 *Corispermum chinganicum*

刺藜 *Dysphania aristata*

菊叶香藜 *Dysphania schraderiana*

刺藜
Dysphania aristata (L.) Mosyakin et Clemants

一年生草本，常带紫红色，常圆锥状，无毛。茎无毛或稍具腺毛，多分枝。叶线形至狭披针形。复二歧聚伞花序顶生或腋生；花无梗，两性；花被裂片5，果期开展。胞果球形，果皮透明。花期8-9月，果期10月。生田间、山坡或荒地。产中国西南、华北、华西、西北和东北。亚洲和欧洲亦有。

Annual herbs, often tinged purple-red, usually appearing conic, glabrous. Stems glabrous or slightly glandular pubescent, much branched. Leaves linear to narrowly lanceolate. Compound dichasium terminal or axillary; flowers not pedunculate, bisexual; perianth segments 5, spreading in fruit. Utricles globose, pericarp transparent. Fl. Aug-Sep. Fr. Oct. Fields, slopes or wastelands. Distributed in SW, N, W, NW and NE China. Also in Asia and Europe.

菊叶香藜
Dysphania schraderiana (Roem. et Schult.) Mosyakin et Clemants

一年生草本，具强烈香味，被具关节的腺毛和无柄腺体。茎常分枝。叶长圆形，边缘羽状浅裂至深裂。复二歧聚伞花序腋生；花两性。胞果扁球形；外果皮膜质。种子横生，红褐色或黑色，近光亮。花期7-9月，果期9-10月。生草地、林缘、草丛、沟边、村边或农田。产中国西南、华北和华西。亚洲、非洲和欧洲亦有。

Annual herbs, with a strong odor, covered with articulated glandular hairs and sessile glands. Stems usually branched. Leaves oblong, margin pinnately lobed to parted. Compound dichasia axillary; flowers bisexual. Utricles depressed globose; pericarps membranous. Seeds horizontal, red-brown or black, sublustrous. Fl. Jul-Sep. Fr. Sep-Oct. Grassy places, forest edges, meadows, along streams, near villages or fields. Distributed in SW, N and W China. Also in Asia, Africa and Europe.

土荆芥
Dysphania ambrosioides (L.) Mosyakin et Clemants

一年生或多年生草本，具强烈香味。茎直立，多分枝。叶长圆状披针形至披针形，下面具散生腺点，沿脉稍具柔毛，边

土荆芥 *Dysphania ambrosioides*

缘具糙锯齿。花生于上部叶腋，每个团伞花序常3-5花；花被裂片(3或)5。胞果扁球形，包于花被内。花果期较长。生村边、路旁和河岸等处。产长江以南；亦有栽培。亦广布世界的热带和亚热带地区。

Annual herbs or perennial, with strong odor. Stems erect, much branched. Leaves oblong-lanceolate to lanceolate, abaxially with scattered glands, slightly hairy around veins, margin coarsely serrate. Flowers borne in upper leaf axils, usually 3-5 per glomerule; perianth segments (3 or) 5. Utricles depressed globose, enclosed by perianth. Fl. and fr. over a lengthy period. Village edges, roadsides or river banks. Distributed in the south of Yangtze River; also cultivated. Also widespread in tropical and subtropical regions of the world.

球花藜 *Chenopodium foliosum*

灰绿藜 *Chenopodium glaucum*

球花藜
Chenopodium foliosum (Moench) Asch.

一年生草本。茎多自基部分枝。叶狭三角状卵形。花两性和雌性；花被片绿色，通常3深裂，果期红色，肉质；雄蕊1-3；花柱很短，柱头2。胞果扁球形；果皮膜质。花期6-7月。果期8-9月。生林缘、山谷或山坡。产甘肃西部和新疆东部、北部。亚洲中部和西南部、非洲北部和欧洲亦有；偶归化于其他地区。

Annual herbs. Stems mostly branched from base. Leaves narrowly triangular-ovate. Flowers bisexual and female; perianth light green, usually 3-parted, becoming red and succulent in fruit; stamens 1-3; styles very short, stigmas 2. Utricles compressed globose; pericarps membranous. Fl. Jun-Jul. Fr. Aug-Sep. Forest edges, valleys or slopes. Distributed in W Gansu, E and N Xinjiang. Also in C and SW Asia, N Africa and Europe; occasionally naturalized in other regions.

灰绿藜
Chenopodium glaucum L.

一年生草本。茎匍匐或铺散。叶长圆状卵形至披针形，肉质，下面具灰白色粉末，有时稍紫红色。花两性兼雌性；花被裂片3-4，浅绿色；雄蕊1或2；花丝不伸出花被；柱头2，极短。胞果顶端露出花被外。花果期5-10月。生农田、菜园、村边或水边等轻度盐碱的土壤上。产除云南、贵州、广西、广东、福建和江西外的中国各地。全球广布。

Annual herbs. Stems decumbent or diffuse. Leaves oblong-ovate to lanceolate, fleshy, abaxially gray-white farinose, sometimes slightly reddish purple. Flowers bisexual and female; perianth segments 3-4, light green; stamens 1 or 2; filaments not exserted from perianth; stigmas 2, very short. Utricles protruding from perianth. Fl. and fr. May-Oct. Slightly saline-alkaline soil of fields, vegetable gardens, by villages, or by streams. Distributed throughout China, not including Yunnan, Guizhou, Guangxi, Guangdong, Fujian and Jiangxi. Also worldwide.

尖头叶藜
Chenopodium acuminatum Willd.

一年生草本。茎直立，分枝多。叶阔至狭卵形，披针形或长圆形，下面具灰白色粉末，边缘全缘，透明，先端短渐尖，有短尖头。团伞花序在上部枝上排成穗状或穗状圆锥状。胞果球状或卵状，扁平型。花期6-7月，果期8-9月。生河岸、湖畔、海滩、田边或荒地。产华南、东南、华北、华西、华东、西北和东北。中亚和东北亚亦有。

Annual herbs. Stems erect, many-branched. Leaves broadly to narrowly ovate, lanceolate or oblong, abaxially ± gray-white farinose, margin entire, pellucid, apex shortly acuminate, mucronate. Glomerules arranged in spikes or spikelike panicles on upper part of branches. Utricles globose or ovoid, depressed. Fl. Jun-Jul. Fr. Aug-Sep. River banks, lake shores, beaches, field edges or wastelands. Distributed in S, SE, N, W, E, NW and NE China. Also in C and NE Asia.

尖头叶藜 *Chenopodium acuminatum*

杂配藜 *Chenopodium hybridum*

杂配藜
Chenopodium hybridum L.

一年生草本。茎在上部有少分枝，强壮。叶阔卵形至卵状三角形，无毛或稍具粉末，边缘掌状裂至深不规则齿裂。花两性兼有雌性，通常数个团集，在上部分枝上排成开展的圆锥状花序。胞果凸镜状。花果期7-9月。生林缘、灌丛、山谷或山坡。产中国西南、东南、华北、华西和西北。中亚、东北亚和欧洲亦有。

Annual herbs. Stems sparsely branched above, stout. Leaves broadly ovate to ovate-triangular, glabrous or slightly farinose, margin palmately lobed to deeply dentate. Flowers bisexual and female, usually several per glomerule, these arranged in spreading panicles on upper branches. Utricles lenticular. Fl. and fr. Jul-Sep. Forest edges, scrubby areas, valleys or slopes. Distributed in SW, SE, N, W and NW China. Also in C and NE Asia, and Europe.

小藜
Chenopodium ficifolium Sm.

一年生草本。叶片卵状长圆形或长圆形，通常3浅裂。花两性，每个团伞花序具数花，在上部枝条排成顶生的开展的圆锥状；花被近球形，5深裂。胞果包在花被内。花期4-5月。通常为田间杂草，也生荒地或路边。产中国各地。世界各地亦有。

Annual herbs. Leaves ovate-oblong or oblong, usually 3-lobed. Flowers bisexual, several per glomerule, these arranged in spreading, terminal panicles on upper branches; perianth subglobose, 5-parted. Utricles included in perianth. Fl. Apr-May. Common weeds in fields, also wastelands or roadsides. Distributed throughout China. Also worldwide.

藜
Chenopodium album L.

一年生草本。茎直立，多分枝。叶菱状卵形至阔披针形，下面具粉末，边缘具不规则锯齿。团伞花序在上部枝排成大或小的圆锥状或穗状圆锥花序；花两性；花被裂片5。花果期5-10月。生路边、荒地或田间。产中国各地。全球温带至热带亦有。

Annual herbs. Stems erect, much branched. Leaves rhom-

小藜 *Chenopodium ficifolium*

藜 *Chenopodium album*

bic-ovate to broadly lanceolate, abaxially ± farinose, margin irregularly dentate. Glomerules arranged into large or small panicles or spikelike panicles on upper part of branches; flowers bisexual; perianth segments 5. Fl. and fr. May-Oct. roadsides, wastelands or fields. Widespread in China. Also in temperate to tropical regions of the world.

地肤

Kochia scoparia (L.) Schrader

一年生草本。叶披针形或条状披针形，通常具3条分明的主脉。花两性或雌性，每个团伞花序常1-3个生于上部叶腋，且形成疏松的穗状圆锥花序；花被片浅绿色。胞果扁球形。花期6-9月，果期7-10月。生田边、河岸、路旁或荒地。产中国各地。亚洲和欧洲亦有；亦广泛归化于澳大利亚、非洲和美洲。

Annual herbs. Leaves lanceolate or linear-lanceolate, usually with 3 distinct main veins. Flowers bisexual or female, usually 1-3 per glomerule in axils of upper leaves and forming sparse, spike-like panicles; perianth light green. Utricles depressed-globose. Fl. Jun-Sep. Fr. Jul-Oct. Field edges, river banks, roadsides or wastelands. Widespread in China. Also in Asia and Europe; widely naturalized in Australia, Africa and America.

雾冰藜

Bassia dasyphylla (Fisch. et Mey.) Kuntze

一年生草本。植株分枝显著，近球状，密具长柔毛。叶互生，圆柱状或近圆柱状，肉质。花两性，单生或成对，通常仅一花发育；花被5浅裂，具长柔毛。种子近扁球形，光滑。花果期7-9月。生荒漠、盐碱地、沙丘、草原、河滩或台地。产中国西南、华北、西北和东北。西南亚、中亚和西北亚亦有。

Annual herbs. Plants extremely branched, subglobose, densely villous. Leaves alternate, terete or semiterete, fleshy. Flowers bisexual, solitary or paired, usually only 1 flower developing; perianth 5-lobed, villous. Seeds depressed subglobose, smooth. Fl. and fr. Jul-Sep. Deserts, saline-alkaline places, dunes, steppes, river banks or terraces. Distributed in SW, N, NW and NE China. Also in SW, C and NW Asia.

地肤 *Kochia scoparia*

雾冰藜 *Bassia dasyphylla*

碱蓬 *Suaeda glauca*

碱蓬

Suaeda glauca (Bunge) Bunge

一年生草本。茎上部多分枝。叶灰绿色，丝状，稍上弯，近圆柱状，无毛。团伞花序多生于叶基，具1-5花；花两性或有时具雌性；花被片黄绿色，杯状。胞果包袱于花被内；外果皮膜质。花果期7-9月。生海滨、荒地、渠岸或田边等含盐碱的土壤上。产华北、华西、华东、西北和东北。俄罗斯(西伯利亚东南部、远东地区)、蒙古、朝鲜半岛和日本亦有。

Annual herbs. Stems much branched above. Leaves gray-green, filiform-linear, slightly upcurved, semiterete, glabrous. Glomerules mostly inserted near base of leaves, 1-5-flowered; flowers bisexual or sometimes some female; perianth yellow-green, cupular. Utricles enclosed in perianth; pericarps membranous. Fl. and fr. Jul-Sep. Saline-alkaline soils of beaches, wastelands, canal banks or field edges. Distributed in N, W, E, NW and NE China. Also in Russia (SE Siberia, Far East), Mongolia, Korean Peninsula and Japan.

刺毛碱蓬

Suaeda acuminata (C. A. Mey.) Moq.

一年生草本。茎直立，通常多分枝。叶无柄，条形，半圆柱状，先端钝或微尖并具刺毛；刺毛长约3毫米。团伞花序腋生，常具3花；中央的1花较大，两性，侧花雌性；小苞片卵形或卵状披针形，先端渐尖，边缘有微锯齿。胞果包于花被内。花果期7-8月。生盐碱土荒漠、山坡和沙丘。产新疆北部。蒙古、俄罗斯、中亚、西南亚和欧洲东南部亦有。

Annual herbs. Stems erect, usually much branched. Leaves sessile, linear, semiterete, apex obtuse or subacute and with a bristle ca. 3 mm long. Glomerules axillary, usually 3-flowered; central flower larger, bisexual, lateral flowers female; bractlets ovate or ovate-lanceolate, margin slightly serrate, apex acuminate. Utricles enclosed in perianth. Fl. and fr. Jul-Aug. Saline-alkaline deserts, slopes and dunes. Distributed in N Xinjiang. Also in Mongolia, Russia, C and SW Asia, SE Europe.

角果碱蓬

Suaeda corniculata (C. A. Mey.) Bunge

一年生草本，无毛。茎匍匐、平卧或直立。叶无柄，线形，近圆柱状或压扁。团伞花序在上部枝排成穗状，常具3-6花；花两性兼有雌性；花被压扁，5深裂。胞果扁球形。花果期8-9月。生盐碱荒漠、湖畔或河边。产中国西南、华北、华西、西北和东北。中亚、东北亚和欧洲东南部亦有。

刺毛碱蓬 *Suaeda acuminata*

角果碱蓬 *Suaeda corniculata*

Annual herbs, glabrous. Stems prostrate, decumbent, or erect. Leaves sessile, linear, semiterete or compressed. Glomerules arranged into spikes on upper branches, usually 3-6-flowered; flowers bisexual and female; perianth depressed, 5-parted. Utricles depressed globose. Fl. and fr. Aug-Sep. Saline-alkaline deserts, lake shores or riversides. Distributed in SW, N, W, NW and NE China. Also in C and NE Asia, and SE Europe.

白琐琐 *Haloxylon persicum*

盐地碱蓬
Suaeda salsa (L.) Pallas

一年生草本，绿色或紫红色。茎直立，多上部分枝。叶无柄，线形，近圆柱状。团伞花序腋生，常具3-5花，在枝上排成间断的穗状花序；花两性及雌性；花被半球形。胞果被包在花被片中。花果期7-10月。生盐碱滩地或湖畔。产华北、华西、华东、西北和东北。亚洲和欧洲亦有。

Annual herbs, green or purple-red. Stems erect, mostly branched above. Leaves sessile, linear, semiterete. Glomerules axillary, usually 3-5-flowered, arranged into interrupted spikes on branches; flowers bisexual and sometimes female; perianth hemispheric. Utricles enclosed in perianth. Fl. and fr. Jul-Oct. Saline-alkaline soils on beaches or lake shores. Distributed in N, W, E, NW and NE China. Also in Asia and Europe.

白琐琐
Haloxylon persicum Bunge ex Boisss. et Buhse

小乔木。树皮灰白色。叶鳞片状，三角形，先端具芒；叶腋具绵毛。花生于二年生枝条的侧生短枝上；小苞片舟状卵形，与花被片等长，边缘膜质；花盘不明显。胞果淡黄褐色；果皮不与种子贴生。花期5-6月。果期9-10月。生沙丘上。产新疆北部。亚洲中南部和西南部和非洲东北部亦有。

Small trees. Bark gray-white. Leaves scalelike, triangular, apex awned; leaf axils cottony. Flowers borne on dwarf, lateral spurs of previous year's branches; bracteoles navicular-ovate, equaling perianth, margin membranous; disk obscure. Utricles light yellow-brown; pericarps not adnate to seed. Fl. May-Jun. Fr. Sep-Oct. Dunes. Distributed in N Xinjiang. Also in SC and SW Asia, and NE Africa.

无叶假木贼
Anabasis aphylla L.

亚灌木。木质茎多分枝。叶不明显或微鳞片状，阔三角形。花1-3，腋生，在小枝上部形成穗状；外轮3枚花被片基部下面具一横向的翅。胞果垂直，近球形；果皮暗红色，肉质，平滑。花期8-9月，果期10月。生戈壁、沙丘间或砾石洪积扇，有时也见于旱山坡。产甘肃西部和新疆。俄罗斯(西伯利亚西南部)、中亚和欧洲亦有。

Subshurbs. Woody stems much branched. Leaves obscure or slightly scalelike, broadly triangular. Flowers 1-3 in leaf axils, forming spikes on upper parts of branches; outer 3 perianth segments proximally with a transverse wing abaxially. Utricles vertical, subglobose; pericarps dark red, fleshy, smooth. Fl. Aug-Sep. Fr. Oct. Gobi deserts, inter-dunes or gravelly alluvial fans, sometimes on arid slopes. Distributed in W Gansu and Xinjiang. Also in Russia (SW Siberia), C Asia and Europe.

盐地碱蓬 *Suaeda salsa*

无叶假木贼 *Anabasis aphylla*

盐生假木贼 *Anabasis salsa*

盐生草 *Halogeton glomeratus*

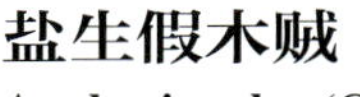

盐生假木贼

Anabasis salsa (C. A. Mey.) Benth. ex Volkens

亚灌木。木质茎多分枝，灰褐色至灰白色。下部和中部叶线形，半圆柱状；上部叶鳞片状。花单生叶腋，于枝端集成短穗状花序；小苞片背面肥厚，边缘膜质。胞果阔卵球形，顶端露出花被；果皮黄褐色，肉质带红色。花果期8-9月。生戈壁或盐碱荒漠地。产新疆北部。哈萨克斯坦、俄罗斯(西伯利亚西南部)、蒙古和高加索地区东部亦有。

Subshurbs. Woody stems much branched, gray-brown to gray-white. Lower and middle leaves linear, semiterete; upper leaves scalelike. Flowers solitary in leaf axils, forming short spikes on upper parts of branches; bracteoles abaxially fleshy, margin membranous. Utricles broadly ovoid, apex protruding from perianth; pericarp yellow-brown or slightly reddish fleshy. Fl. and fr. Aug-Sep. Gobi deserts or saline-alkaline deserts. Distributed in N Xinjiang. Also in Kazakhstan, Russia (SW Siberia), Mongolia and E Caucasus.

盐生草

Halogeton glomeratus (Bieb.) C. A. Mey.

一年生草本。茎与枝光滑，无乳头状小凸起。叶互生，无柄，叶片圆柱形，肉质，顶端有长刺毛，有时长刺毛脱落。花通常4-6朵聚集成腋生的团伞花序；花被片披针形，膜质，背面有1条粗脉；果时自背面近顶部生翅；翅半圆形，大小近相等，膜质，有多数明显的脉，有时翅不发育而花被增厚成革质；雄蕊通常为2。种子直立。花果期7-9月。生戈壁滩、山脚。产甘肃西部、青海、新疆和西藏。蒙古、俄罗斯(西伯利亚)和中亚亦有。

Annual herbs. Stems and branches smooth, not papillate. Leaves alternate, sessile, terete, fleshy, apex aristate awned, awn sometimes deciduous. Flowers usually 4-6 per axillary glomerule; perianth segments lanceolate, membranous, abaxially 1-veined; abaxial wing semiorbicular, subequal, membranous, distinctly veined, sometimes not developed and then perianth thickened, becoming leathery; stamens usually 2. Seed vertical. Fl. and fr. Jul-Sep. Gobi deserts, foothills. Distributed in W Gansu, Qinghai, Xinjiang and Xizang. Also in Mongolia, Russia (Siberia) and C Asia.

蒿叶猪毛菜

Salsola abrotanoides Bunge

亚灌木。木质枝灰棕色，具纵向裂缝；当年生枝密集，黄绿色，草质。叶互生，簇生于老枝的短枝上，半圆柱状。穗状花序纤细，疏松；花被裂片卵形，下面自中部肉质，具翅，边缘膜质。花期7-8月，果期8-9月。生干山坡、冲积扇或岩石河边。产甘肃、青海和新疆。蒙古亦有。

Subshrubs. Woody branches gray-brown, longitudinally fissured; annual branches crowded, yellow-green, herbaceous. Leaves alternate, fascicled on dwarf branches of older branches, semiterete. Inflorescences spicate, slender, loose; perianth segments ovate, abaxially fleshy and winged from middle, margin membranous. Fl. Jul-Aug. Fr. Aug-Sep. Arid slopes, alluvial fans or rocky riversides. Distributed in Gansu, Qinghai and Xinjiang. Also in Mongolia.

蒿叶猪毛菜 *Salsola abrotanoides*

猪毛菜

Salsola collina Pall.

一年生草本，高20-100厘米。茎自基部分枝。叶片开展或稍弯曲，丝状圆柱形，边缘膜质，顶端有刺尖头。花序穗状；柱头丝状，长为花柱的1.5-2倍；花被裂片膜质，果期变硬，下面鸡冠状。花期7-9月，果期9-10月。生村边、路边或荒地。产中国西南、华北、华西、华东、西北和东北。巴基斯坦、俄罗斯、蒙古、朝鲜半岛和中亚亦有；逸生欧洲中部和西部、北美洲。

Annual herbs, 20-100 cm tall. Stems branched from base. Leaves spreading or slightly curved, filiform-terete, margin membranous, apex spinose mucronate. Inflorescences spikelike; stigmas filiform, 1.5-2 × as long as styles; perianth segments membranous, hardened in fruit, abaxially crested. Fl. Jul-Sep. Fr. Sep-Oct. By villages, roadsides or waste places. Distributed in SW, N, W, E, NW and NE China. Also in Pakistan, Russia, Mongolia, Korean Peninsula and C Asia; naturalized in C and W Europe, and North America.

猪毛菜 *Salsola collina*

蒙古猪毛菜

Salsola ikonnikovii Iljin

一年生草本。叶半圆柱形，顶端有刺状尖。花序穗状；花单生；苞片窄卵形，长于小苞片，顶端伸长有刺尖；花药长约1毫米，附属物很小，柱头与花柱近等长。胞果。花期7-8月，果期8-9月。生沙丘或沙地。产内蒙古。蒙古亦有。

Annual herbs. Leaves semiterete, apex spinose mucronate. Inflorescences spicate; flowers solitary; bracts narrowly ovate, longer than bracteoles, apex elongate, spinose mucronate; anthers ca. 1 mm long, appendages extremely small; stigmas equaling styles. Fl. Jul-Aug. Fr. Aug-Sep. Dunes or sandy places. Distributed in Neimenggu. Also in Mongolia.

刺沙蓬

Salsola tragus L.

一年生草本。茎自基部分枝，具白色或紫红色条纹，密具硬毛或近无毛。叶半圆柱形或圆柱形。花序穗状；花被裂片狭卵形，膜质，果时变硬；3翅稍大，有时浅紫红色，2翅较狭窄。花期8-9月，果期9-10月。生沙丘、沙地、砾质戈壁滩、山谷或海滩。产中国西南、华北、华西、华东、西北和东北。原生亚洲中部和西南部、欧洲东南部；现逸生于亚洲、大洋洲、非洲南部、欧洲和美洲。

Annual herbs. Stems branched from base, white, or purple-red striate, densely hispid or subglabrous. Leaves semiterete or terete. Inflorescences spikelike; perianth segments narrowly ovate, membranous, hardened in fruit; 3 wings larger, sometimes light purple-red, other 2 wings narrower. Fl. Aug-Sep. Fr. Sep-Oct. Dunes, sandy places, rocky places in Gobi deserts, valleys or seashores. Distributed in SW, N, W, E, NW and NE China. Native to C and SW Asia, SE Europe; now naturalized in Asia, Oceania, S Africa, Europe and America.

蒙古猪毛菜 *Salsola ikonnikovii*

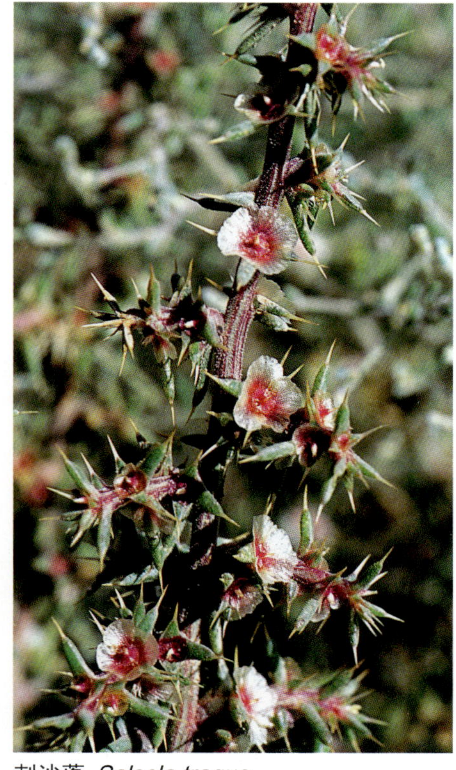

刺沙蓬 *Salsola tragus*

苋科 Amaranthaceae

浆果苋
Deeringia amaranthoides (Lam.) Merr.

攀援灌木。茎长2-6米，幼枝具毛。叶卵形或卵状披针形，先端渐尖或尾状。花组成顶生或腋生的序总状或圆锥状；花被片开展并于花后反折，浅绿色或有时黄色，果期带红色。浆果近球形，红色。花果期10月至翌年3月。生海拔100-2200米的山坡林下或灌丛中。产中国西南和华南。南亚、东南亚和大洋洲亦有。

Climbing shrubs. Stems 2-6 m long, hairy when young. Leaves ovate or ovate-lanceolate, apex acuminate or caudate. Flowers in terminal or axillary racemes or panicles; tepals spreading and reflexed at anthesis, light green or somewhat yellowish, tinged red in fruit. Berries red, globose. Fl. and fr. Oct to next Mar. Forests or thickets on slopes at 100-2200 m. Distributed in SW and S China. Also in S and SE Asia, and Oceania.

白浆果苋
Deeringia polysperma (Roxb.) Moq.

直立草本或亚灌木，高1-2米。叶卵状披针形或长圆状披针形，长8-15厘米，先端渐尖或急尖，有小突尖。穗状花序腋生，单一或成对，长3-12厘米；花被片与果贴生；雄蕊花丝基部扩大，连合成杯状。浆果直径4-5毫米，白色。花果期6月至12月。生中低海拔山坡林中。产台湾和海南。越南、泰国、马来西亚和菲律宾亦产。

Erect herbs or subshrubs, 1-2 m tall. Leaf ovate-lanceolate or oblong-lanceolate, 8-15 cm long, apex acuminate or acute, mucronate. Spikes axillary, solitary or paired, 3-12 cm long; tepals adnate to fruit; filaments connate into a cup at base. Berry 4-5 mm diam, white. Fl. and fr. Jun-Dec. Forests, mountain slopes at low to medium elevations. Distributed in Taiwan and Hainan. Also in Vietnam, Thailand, Malaysia and the Philippines.

白浆果苋 *Deeringia polysperma*

青葙
Celosia argentea L.

一年生草本。叶卵状长圆形至条状披针形，绿色或变红色。穗状花序圆柱形或先端圆锥状；苞片和小苞片白色，光亮；花被片5，白色至粉色；花药和花柱紫红色。胞果卵球形，包于宿存花被中，周裂。花期5-8月，果期6-10月。生海拔1500米以下的田间、山坡或村边，或栽培。产中国除西北和东北以外大部分地区。南亚、东亚和热带非洲亦有。

Annual herbs. Leaves ovate-oblong to linear-lanceolate, green or tinged red. Spikes cylindric or with a conic apex, not branched; flowers dense; bracts and bracteoles white, shiny; tepals 5, white to pink; anthers and styles purple. Utricles ovoid, enveloped in persistent perianth, circumscissile. Fl. May-Aug. Fr. Jun-Oct.

浆果苋 *Deeringia amaranthoides*

青葙 *Celosia argentea*

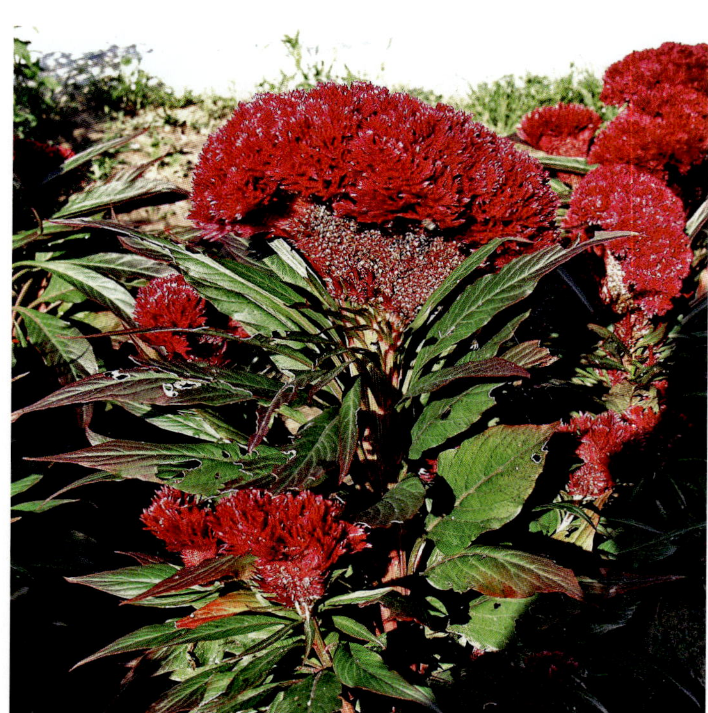
鸡冠花 *Celosia cristata*

Fields, slopes, or by villages below 1500 m, or cultivated. Distributed in most parts of China, except NW and NE China. Also in S and E Asia, and tropical Africa.

鸡冠花
Celosia cristata L.

一年生草本。叶卵形、卵状披针形或披针形。穗状花序鸡冠状、卷冠状或羽毛状，常有小分枝，呈金字塔状长圆形；花被片红色、紫色或橙黄色。胞果卵球形，包被在宿存花被内，周裂。花果期7-9月。广泛栽培或逸生在中国。世界各地亦广泛栽培或逸生。

Annual herbs. Leaves ovate, ovate-lanceolate, or lanceolate. Spikes often cristate, convolute or feathery, with some small branches and pyramidal-oblong in shape; tepals red, purple, yellow or orange. Utricles ovoid, enveloped in persistent perianth, circumscissile. Fl. and fr. Jul-Sep. Widely cultivated or naturalized in China. Also worldwide.

尾穗苋
Amaranthus caudatus L.

一年生草本。茎直立，幼时被短柔毛。叶菱状卵形或菱状披针形，无毛。圆锥花序顶生，下垂，由多数穗状分枝组成；花被片红色，短于果实，透明。胞果上部红色，近球形，长于花被片，周裂。花期7-8月。果期9-10月。栽培或逸生于全国各地。原产南美洲。世界各地栽培。

Annual herbs. Stems erect, pubescent when young. Leaves rhombic-ovate or rhombic-lanceolate, glabrous. Panicles terminal, pendulous, composed of many spike-like branches; tepals red, shorter than fruit, transparent. Utricles red above, subglobose, longer than perianth, circumscissile. Fl. Jul-Aug. Fr. Sep-Oct. Widely cultivated or naturalized in China. Native to South America. Cultivated worldwide.

尾穗苋 *Amaranthus caudatus*

千穗谷
Amaranthus hypochondriacus L.

一年生草本。茎直立，具分枝，绿至红色。叶菱状卵形至长圆状披针形，长3-15cm，先端急尖或渐尖。圆锥花序顶生，具多数斜升的穗状分枝，顶枝常短于侧枝；苞片针状，长3.5-5毫米，长为花被的1.5-2倍；花被片5，先端急尖或渐尖。胞果近菱状卵球形，具粗的花柱基，周裂。花果期7-9月。栽培偶有逃逸。产中国各地。原产墨西哥。世界各地引种。

Annual herbs. Stems erect, branched, green to red. Leaves rhombic-ovate to oblong-lanceolate, 3-15 cm long, apex acute or acuminate. Panicles terminal with numerous, ascending spike-like branches, terminal branch usually shorter than lateral branches; bracts needle-shaped, 3.5-5 mm long, ca. 1.5-2 × as long as the perianth; tepals 5, apex acute or acuminate. Utricles nearly rhombic-ovoid, with thick style bases, circumscissile. Fl. and fr. Jul-Sep. Cultivated, occasionaly escaped. Distributed throughout China. Native to Mexico. Cultivated worldwide.

千穗谷 *Amaranthus hypochondriacus*

老鸦谷 *Amaranthus cruentus*

反枝苋 *Amaranthus retroflexus*

老鸦谷
Amaranthus cruentus L.

一年生草本。茎直立。叶狭椭圆形至宽披针形，无毛。圆锥花序顶生和腋生，直立，由多数上升或平展的穗状分枝组成，顶枝与侧枝近相等；雌花的苞片约与花被片等长。胞果平滑或微皱，周裂。花期7-9月。果期9-10月。中国各地栽培或归化。原产墨西哥。全世界各地栽培。

Annual herbs. Stems erect. Leaves narrowly elliptic to broadly lanceolate, glabrous. Complex thyrsoid structures terminal and axillary, erect, composed of numerous ascending or patent spike-like branches, terminal spike subequal to lateral branches; bracts of pistillate flowers as long as tepals. Utricles smooth or rugose, circumscissile. Fl. Jul-Sep. Fr. Sep-Oct. Widely cultivated or naturalized in China. Native to Mexio. Cultivated worldwide.

反枝苋
Amaranthus retroflexus L.

一年生草本。茎直立，上部密被绵毛。叶卵状菱形或椭圆形。圆锥花序顶生，直立，由多数穗状分枝组成；花被片5，白色，膜质，具一条绿色中脉。胞果淡绿色，卵球形，短于花被，周裂。花期7-8月。果期8-9月。生荒地、田边或路边。产除华南外中国各地。原产北美洲和中美洲。全世界广布。

Annual herbs. Stems erect, upper parts densely woolly. Leaves ovate-rhombic or elliptic. Panicles terminal, erect, composed of many spike-like branches; tepals 5, white, membranous, with a green midvein. Utricles light green, ovoid, shorter than perianth, circumscissile. Fl. Jul-Aug. Fr. Aug-Sep. Waste places, field edges or roadsides. Distributed almost throughout China except S. Native to North and Central America. Now cosmopolitan.

绿穗苋
Amaranthus hybridus L.

一年生草本。茎直立，上部疏被绵毛。叶卵形或菱状卵形，下面疏生柔毛，先端具小尖头。圆锥花序顶生；花被片5，长圆状披针形，具一芒尖。胞果卵球形，微皱，长于宿存花被，周裂。花期7-8月。果期9-10月。生海拔5-1100米的田野、旷地或山坡。产辽宁、华北和长江以南。原产美国至南美洲北部。现在已成为世界性杂草。

Annual herbs. Stems erect, sparsely wooly in the upper part. Leaves ovate or rhombic-ovate, abaxially sparsely pubescent, apex mucronate. Panicles terminal; tepals 5, oblong-lanceolate, with a mucro. Utricles ovoid, longer than persistent tepals, circumscissile. Fl. Jul-Aug. Fr. Sep-Oct. Fields, open places or hill slopes at 5-1100 m. Distributed in Liaoning, N China and throughout the south of Yangtze River. Native to USA to N South America. Now a cosmopolitan weed.

刺苋
Amaranthus spinosus L.

一年生草本。茎直立。叶柄基部具一对刺；叶卵状菱形或卵状披针形。圆锥花序顶生及腋生；花被片5，绿色。胞果包于花被内，长圆形，中部稍下周裂。花果期7-11月。生旷地或园圃。产中国西南、华南、东南、华中和华北。原产热带美洲，目前在暖温带至热带地区广泛分布。

Annual herbs. Stems erect. Petiole with a pair of spines at base; leaf blade ovate-rhombic or ovate-lanceolate. Panicles terminal and axillary; tepals 5, green. Utricles included in perianth, oblong, circumscissile slightly below middle. Fl. and fr. Jul-Nov. Open places or vegetable plots. Distributed in SW, S, SE, C and N China. Native to tropical

绿穗苋 *Amaranthus hybridus*

刺苋 *Amaranthus spinosus*

白苋 *Amaranthus albus*

America, now cosmopolitan in warm-temperate to tropical regions.

苋

Amaranthus tricolor L.

一年生草本。茎直立，绿色或红色。叶片卵形、菱状卵形或披针形，红色、紫色或黄色，有时混有其他颜色。花密簇生于叶腋或在顶端形成穗状；花被片3，长圆形，绿色或黄色。胞果约与花被片等长，周裂。花期5-8月。果期7-9月。中国各地广泛栽培，偶有归化。南亚、中亚和日本亦有。

Annual herbs. Stems erect, green or red. Leaves ovate or rhombic-ovate or lanceolate, red, purple or yellow, sometimes mixed with other colors. Flowers in dense clusters at leaf axils or in spike at apex; tepals 3, oblong, green or yellow. Utricles about as long as tepals, circumscissile. Fl. May-Aug. Fr. Jul-Sep. Widely cultivated throughout China, occasionally naturalized. Also in S and C Asia, and Japan.

白苋

Amaranthus albus L.

一年生草本。茎直立或上升，多分枝，白色，稀具红晕。叶长圆形至匙形，先端钝至微缺，具小尖突，边缘波状。聚伞花簇腋生；苞片钻形，先端刺状，长约为花被的2倍；花被片3。胞果球形，具皱纹，周裂。花果期6-10月。生海拔10-1500米路边、草地、荒地和山坡。产中国东北、华北和西北。原产北美洲，现广布世界温带地区。

Annual herbs. Stems erect or ascending, much branched, white, rarely tinged with red. Leaves oblong to spathulate, apex obtuse to emarginate, mucronate, margin undulate. Cymose clusters axillary; bracts subulate, apex spine-like, ca. twice as long as the perianth; tepals 3. Utricles globose, rugose, circumscissile. Fl. and fr. Jun-Oct. Roadsides, grasslands, waste places and slopes at 10-1500m. Distributed in NE, N and NW China. Native to North America, naturalized in world temperate regions.

苋 *Amaranthus tricolor*

北美苋 *Amaranthus blitoides*

北美苋
Amaranthus blitoides S. Watson

一年生草本。茎平卧和斜升，基部具开展的分枝。叶长圆形至匙形，长1.5-3厘米，先端钝或近急尖。聚伞花簇腋生；苞片短于花被片；雌花花被片4-5，不相等，膜质，具1脉，先端钝。胞果近球形，短于花被，周裂。花果期6-10月。生田间、路边和草地。产中国东北、华北、安徽、上海、四川和新疆。原产美国西部；在欧洲和亚洲温带归化。

Annual herbs. Stems procumbent and ascending. Leaves oblong to spatulate, 1.5-3 cm long, apex obtuse to subacute. Inflorescences axillary; bracts shorter than tepals; tepals of pistillate flowers 4-5, very unequal, membranous, 1 vein, apex obtuse. Utricles subglobose, shorter than tepals, circumscissile. Fl. and fr. Jun-Oct. Fields, roadsides and grassy places. Distributed in NE and N China, Anhui, Shanghai, Sichuan and Xinjiang. Native to W USA; naturalized in Europe and temperate Asia.

合被苋
Amaranthus polygonoides L.

一年生草本。茎平卧和斜升，基部多分枝，上部被短柔毛。叶菱状卵形至倒卵状披针形，长1-2厘米，先端钝至微凹，具小尖突，无毛。花序腋生，近半球状；雌花花被片5，具3脉，下部1/3合生成筒状。胞果狭椭圆体形，具皱纹，不开裂。花果期6-9月。生低海拔草地和路旁。产辽宁、北京、山东和安徽。原产美国和中美洲；在欧洲归化。

Annual herbs. Stems procumbent and ascending, upper parts pubescent. Leaves rhombic-ovate to obovate-lancelate, 1-2 cm long, apex obtuse to emarginate, mucronate, glabrous. Inflorescences axillary, nearly semiglobose; tepals of pistillate flowers 5, with 3 veins, connate to a tube in the lower 1/3. Utricles narrowly ellipsoid, rugose, indehiscent. Fl. and fr. Jun-Sep. Grassy places and roadsides at low elevations. Distributed in Liaoning, Beijing, Shandong and Anhui. Native to USA and Central America; naturalized in Europe.

皱果苋
Amaranthus viridis L.

一年生草本。茎直立，稀上升。叶卵形，卵状长圆形，先端凹缺或圆钝，有一小尖头。圆锥花序顶生，由穗状花序组成；花被片3，长圆形或宽倒披针形。胞果绿色，长于花被，球形，稍扁，极皱，不开裂。花期6-8月。果期8-10月。生村边杂草地或田边。产中国除西北和西藏以外的各地。亦广布热带、亚热带和温带地区。

Annual herbs. Stems erect, rarely ascending. Leaves ovate or ovate-oblong, apex emarginate or rounded, mucronate. Panicles terminal, composed of spikes; tepals 3, oblong or broadly oblanceolate. Utricles green, longer than perianth, globose, slightly compressed, very rugose, indehiscent. Fl. Jun-Aug. Fr. Aug-Oct. Weedy places near villages or by fields. Distributed in all provinces except NW China and Xizang. Also widespread in tropical, subtropical and temperate regions of the world.

合被苋 *Amaranthus polygonoides*

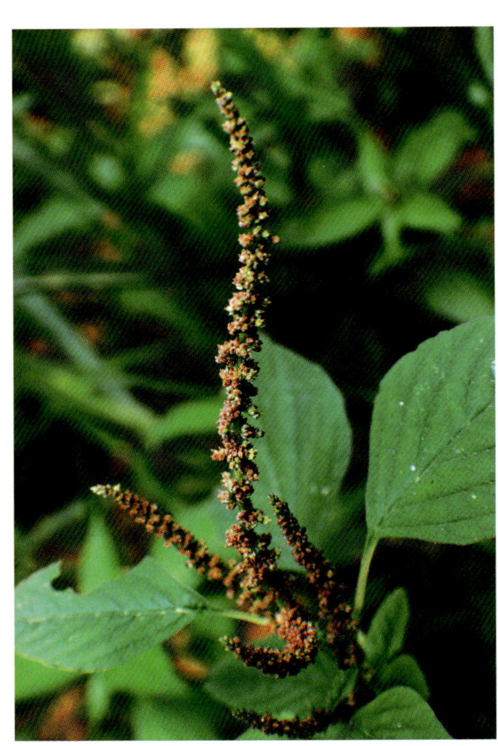
皱果苋 *Amaranthus viridis*

凹头苋
Amaranthus blitum L.

一年生草本，无毛。茎自基部分枝。叶卵形或菱状卵形，顶端有凹缺，有一芒尖。穗状和圆锥花序顶生或腋生；花被片3，浅绿色，长圆形或披针形。胞果超出花被，扁卵球形，近平滑，不开裂。花期7-8月。果期8-9月。生海拔2000米以下的田野、杂草地或村边。产除西藏、内蒙古、宁夏和青海以外的中国各地。全世界广布。

Annual herbs, glabrous. Stems branched from base. Leaves ovate or rhombic-ovate, apex emarginate, mucronate. Spikes and panicles terminal or axillary; tepals 3, light green, oblong or lanceolate. Utricles exceeding perianth, compressed-ovoid, nearly smooth, indehiscent. Fl. Jul-Aug. Fr. Aug-Sep. Fields, weedy places or near villages below 2000 m. Widespread in China except Xizang, Neimenggu, Ningxia and Qinghai. Throughout all the world.

凹头苋 *Amaranthus blitum*

薄叶苋 *Amaranthus tenuifolius*

薄叶苋
Amaranthus tenuifolius Willd.

一年生草本，无毛。茎直立或斜升。叶椭圆状倒披针形，长1-4.5cm，先端凹缺，具小突尖。顶生穗状花序直立；腋生花序团簇状；雄花花被片2，披针形；雌花花被片缺失。胞果倒卵形，具5棱，不开裂。花果期7-9月。生湖边湿地。产山东。原产印度、孟加拉国和巴基斯坦；在欧洲归化。

Annual herbs, glabrous. Stems erect or ascending. Leaves elliptic-oblanceolate, 1-4.5cm long, apex emarginate, mucronate. Spikes terminal, erect; flowers at leaf axils in dense clusters; staminate flower with 2 lanceolate tepals; pistillate flowers without tepals. Utricles obovoid, 5-ribbed, indehiscent. Fl. and fr. Jul-Sep. Wetlands beside lakes. Distributed in Shandong. Native to India, Bangladesh and Pakistan; naturalized in Europe.

长芒苋
Amaranthus palmeri S. Waston

一年生草本，雌雄异株，常无毛。茎直立。叶卵形至披针形，先端钝，急尖或微缺，常具小突尖，基部渐狭至楔形。穗状花序顶生，直立或稍弯；腋生花呈团簇状；苞片常4-6毫米，雌株果期苞片刺状；花被片5，不相等。胞果周裂。花期7-9月，果期8-10月。生海拔30-100米路边和荒地。产辽宁、河北、北京、天津和山东。原产美国；在旧世界归化。

Annual herbs, dioecious, usually glabrous. Stems erect. Leaves ovate to lanceolate, apex obtuse, acute or emarginate, often mucronate, base attenuate to cuneate. Spikes terminal, erect or slightly flexuous; flowers at leaf axils in dense clusters; bracts 4-6mm long, those of fruiting female plants spinelike; tepals 5, unequal. Utricles almost circumscissile. Fl. Jul-Sep. Fr. Aug-Oct. Roadsides and wasteplaces at 30-100m. Distributed in Liaoning, Hebei, Beijing, Tianjin and Shandong. Native to USA; naturalized in the Old World.

长芒苋 *Amaranthus palmeri*

长序苋 *Digera muricata*

长序苋

Digera muricata (L.) Mart.

一年生草本。茎常分枝。叶宽卵形至卵状披针形。先端急尖或渐尖。穗状花序腋生，具长的花序梗；花白色，具粉红色至洋红色晕，稀绿白色；花被片(4-)5；花柱细长，柱头2，反曲。小坚果近球形，被宿存花被所包藏，不开裂。花果期7-10月。生低海拔路边、荒地。在河南和安徽归化。原产非洲东部和西亚至印度，在东南亚归化。

Annual herbs. Stems usually branched. Leaves broadly ovate to ovate-lanceolate, apex acute or acuminate. Spikes axillary, with long peduncles; flowers white tinged with pink to carmine, rarely greenish-white; tepals (4-)5; style long and slender, stigmas 2, recurved. Nutlets subglobose, enclosed by the persistent perianth, indehiscent. Fl. and fr. Jul-Oct. Roadsides or waste grounds at low elevations. Naturalized in Henan and Anhui. Native to E Africa and W Asia to India, naturalized in SE Asia.

头花杯苋

Cyathula capitata (Wall.) Moq.

多年生草本。茎直立，疏生长柔毛。叶纸质，宽卵形或倒卵状长圆形，长5-14厘米，先端尾尖，两面疏生长柔毛。花簇球形或椭圆体形，直径2-4厘米，近单生或组成短穗状花序；假退化雄蕊长方形，长0.6-1毫米，顶端深裂或流苏状。花期8月，果期10月。生海拔1700-2300米山坡杂木林下。产四川、云南和西藏。印度、不丹、尼泊尔和越南亦有。

Perennial herbs. Stems erect, sparsely pilose. Leaves papery, broadly ovate or obovate-oblong, 5-14 cm long, apex caudate, both surfaces sparsely pilose. Flower clusters globose or ellipsoid, 2-4 cm diam, simple or clustered in spikes; pseudostaminodes rectangular, 0.6-1 mm long, parted or fimbriate at apex. Fl. Aug. Fr. Oct. Mixed forests on slopes at 1700-2300 m. Distributed in Sichuan, Yunnan and Xizang. Also in India, Bhutan, Nepal and Vietnam.

川牛膝

Cyathula officinalis K. C. Kuan

多年生草本。单叶，对生，全缘。花小，浅绿色，多数，组成球形的多歧聚伞花序；完全花花被片披针形；退化雄蕊矩形。胞果浅黄色，椭球形或倒卵球形。花期6-7月，果期8-9月。生海拔1500米以上的荒地。产云南、四川和贵州；浙江、河北和山西栽培。尼泊尔亦有。

Perennial herbs. Leaves simple, opposite, entire. Flowers small, light green, numerous in globose multiparous cymes; tepals of perfect flowers lanceolate; pseudostaminodes rectangular. Utricles light yellow, ellipsoid or

头花杯苋 *Cyathula capitata*

川牛膝 *Cyathula officinalis*

白花苋 *Aerva sanguinolenta*

土牛膝 *Achyranthes aspera*

obovoid. Fl. Jun-Jul. Fr. Aug-Sep. Waste places at above 1500 m. Distributed in Yunnan, Sichuan and Guizhou; cultivated in Zhejiang, Hebei and Shanxi. Also in Nepal.

白花苋

Aerva sanguinolenta (L.) Blume

多年生草本，茎直立或稍匍匐，不分枝或分枝。叶卵状椭圆形、长圆形或披针形。花序被白色或紫色丝状毛；花被片白色或粉红色；退化雄蕊三角形。胞果卵球形；种子肾形。花期4-6月，果期8-10月。生海拔1100-2300米的丘陵。产中国西南和华南。南亚和东南亚亦有。

Perennial herbs. Stems erect or slightly creeping, simple or branched. Leaves ovate-elliptic, oblong, or lanceolate. Inflorescences white or purple sericeous; tepals white or pink; pseudostaminodes triangular. Utricles ovate; seeds reniform. Fl. Apr-Jun. Fr. Aug-Oct. Hills at 1100-2300 m. Distributed in SW and S China. Also in S and SE Asia.

土牛膝

Achyranthes aspera L.

多年生草本。单叶，宽倒卵形或椭圆状长圆形，对生，纸质。穗状花序顶生，长10-20厘米，花后反折；小苞片刺状，基部有2翅；退化雄蕊先端有流苏状长缘毛。胞果卵球形。花期6-8月，果期10月。生海拔800-2300米山坡疏林中或林边空旷地。产中国西南、华南、东南和华东。南亚、东南亚、非洲和欧洲亦有。

Perennial herbs. Leaves simple, broadly obovate or elliptic-oblong, opposite, papery. Spikes terminal, 10-20 cm long, reflexed after anthesis; bractlets spiny, base 2-winged; staminode apex with fimbriate long cilia. Utricles ovoid. Fl. Jun-Aug. Fr. Oct. Open forests on slopes or open places near forests at 800-2300 m. Distributed in SW, S, SE and E China. Also in S and SE Asia, Africa and Europe.

牛膝

Achyranthes bidentata Blume

多年生草本。单叶，椭圆形或椭圆状披针形，对生，先端尾尖。穗状花序顶生或腋生，长3-5厘米；花密生；小苞片刺状，基部2深裂；退化雄蕊先端具细锯齿。胞果黄褐色，光亮。花期7-9月，果期9-10月。生海拔200-1800米的山坡林下。产中国除东北和西北以外各地。南亚、东南亚和东北亚亦有。

Perennial herbs. Leaves simple, elliptic or elliptic-lanceolate, opposite, apex caudate. Spikes terminal or axillary, 3-5 cm long; flowers dense; bractlets spiny, base 2-parted; pseudostaminodes serrulate at apex. Utricles yellowish brown, shiny. Fl. Jul-Sep. Fr. Sep-Oct. Forests on slopes at 200-1800 m. Widespread in China, except NE and NW China. Also in S, SE and NE Asia.

牛膝 *Achyranthes bidentata*

柳叶牛膝 *Achyranthes longifolia*

莲子草 *Alternanthera sessilis*

柳叶牛膝

Achyranthes longifolia (Makino) Makino

多年生草本。叶披针形或宽披针形。穗状花序顶生或腋生；花密生；小苞片刺状，基部2深裂；退化雄蕊先端有不明显牙齿。胞果黄褐色，矩圆状。生海拔1200米以下的山丘。产中国西南、华南、东南和华东。老挝、泰国、越南和日本亦有。

Perennial herbs. Leaves lanceolate or broadly lanceolate. Spikes terminal or axillary. Flowers dense; bractlets spiny, base 2-parted; pseudostaminodes indistinctly dentate at apex. Utricles yellowish brown. Hillsides below 1200 m. Distributed in SW, S, SE and E China. Also in Laos, Thailand, Vietnam and Japan.

莲子草

Alternanthera sessilis (L.) R. Br. ex DC.

多年生草本。叶椭圆形，倒卵状椭圆形至倒卵状长圆形。头状花序1-4腋生，无柄，球形至圆柱形，直径3-6毫米；苞片和小苞片白色；花被片白色，卵形；退化雄蕊近钻形。胞果包于花被片中，深褐色，倒卵球形。花期5-7月，果期7-9月。生海拔1800-2100米的草地、田地或沼泽地。产中国西南、华南、东南、华中和华东。南亚和东南亚亦有。

Perennial herbs. Leaves elliptic, obovate-elliptic to obovate-oblong. Heads 1-4, axillary, sessile, globose to cylindric, 3-6 mm diam; bracts and bracteoles white; tepals white, ovate; pseudostaminodes subulate. Utricles enclosed in perianth, dark brown, obovoid. Fl. May-Jul. Fr. Jul-Sep. Grasslands, fields or marshes at 1800-2100 m. Distributed in SW, S, SE, C and E China. Also in S and SE Asia.

空心莲子草

Alternanthera philoxeroides (Mart.) Griseb.

多年生草本。茎基部匍匐，多分枝，上部上升。叶长圆形、长圆状倒卵形或卵状披针形。头状花序单生于叶腋，球状，具花序梗；花被片白色，具光泽；退化雄蕊长圆状线形。花期5-10月。生荒地。栽培或野化于中国西南、东南、华中、华北和华东。原产南美洲。

Perennial herbs. Stems creeping at base, much branched, ascending at upper parts. Leaves oblong, oblong-obovate, or ovate-lanceolate. Heads solitary at leaf axils, globose, pedunculate; tepals white, shiny; pseudostaminodes oblong-linear. Fl. May-Oct. Waste places. Cultivated or naturalized in SW, SE, C, N and E China. Native to South America.

刺花莲子草

Alternanthera pungens Kunth

一年生草本。茎披散，匍匐，多分枝。叶卵形、倒卵形或椭圆状倒卵形。头状花序球状或

空心莲子草 *Alternanthera philoxeroides*

刺花莲子草 *Alternanthera pungens*

矩圆状，1-3个腋生，无柄；花被片不等大，花期后变硬，中脉伸出成刺。胞果褐色，宽椭球形。花期5月。果期7月。生路边。产华东、华南和西南。原产南美洲；世界热带地区归化。

Annual herbs. Stems diffuse, creeping, many-branched. Leaves ovate, obovate, or elliptic-obovate. Heads globose or oblong, 1-3 axillary, sessile; tepals unequal, rigid after anthesis, mid-vein stretching into spines. Utricles brown, broadly ellipsoid. Fl. May. Fr. Jul. Roadsides. Distributed in E, S and SW China. Native to South America; naturalized in tropical regions of the world.

银花苋
Gomphrena celosioides Mart.

一年生草本。茎多分枝。叶片长圆形或长圆状倒卵形，纸质，被白色长毛和缘毛。花簇生成头状，头状花序银色；小苞片紫色；小苞片紫色，三角披针形；花被片于花后变硬；花丝连合成筒，顶端5浅裂；柱头2。胞果近球形，不分裂。花果期2-6月。生路边或荒地。产华南。热带广布。

Annual herbs. Stems much branched. Leaves oblong or oblong-obovate, papery, long white hairy and ciliate. Flowers clustered in heads, heads silvery; bracteoles purple, triangular-lanceolate; tepals rigid after anthesis; filaments connate to a tube, 5-lobed at apex; stigmas 2. Utricles subglobose, indehiscent. Fl. and fr. Feb-Jun. Roadsides or waste places. Distributed in S China. Widely also in tropical areas.

血苋
Iresine herbstii Hook. ex Lindl.

多年生草本，雌雄异株。茎常带红色，粗壮，具分枝。叶紫色，沿脉具浅色条纹。花单性，穗状花序，再排成圆锥花序；花被片浅绿色或浅黄色。花果期9月至翌年的3月。栽培于云南、广西、广东、海南、江苏和上海。原产巴西。

Perennial herbs, dioecius. Stems often tinged red, stout, branched. Leaves purple with lighter bands along main veins. Flowers unisexual, spikes arranged in panicles; tepals greenish white or yellowish white. Fl. and fr. Sep to next Mar. Cultivated in Yunnan, Guangxi, Guangdong, Hainan, Jiangsu and Shanghai. Native to Brazil.

银花苋 *Gomphrena celosioides*

血苋 *Iresine herbstii*

紫茉莉科 Nyctaginaceae

腺果藤
Pisonia aculeate L.

藤状灌木。枝常具粗刺。叶近革质，卵形至椭圆形。聚伞花序腋生；花单性；花梗近顶端具3个卵形小苞片；花被黄色，芳香；雄花花被筒漏斗状；雄蕊6-8，伸出；雌花花被筒卵状圆筒形；花柱伸出。果实棍棒形，5棱，具成行的有柄腺体。花期1-6月。生海岸灌丛和疏林。产台湾和海南。亚洲热带地区、澳大利亚、非洲和美洲亦有。

Vinelike shrubs. Branches usually spiny. Leaf nearly leathery, ovate to elliptic. Cymes axillary; flowers unisexual; pedicel with 3 ovate bracteoles near apex; perianth yellow, fragrant; perianth tube of male flower funnel form; stamens 6-8, exserted; perianth tube of female flower ovate-tubular; style exserted. Fruit clavate, 5-ribbed, with rows of stalked glands. Fl. Jan-Jun. Thickets along seashores and open forests. Distributed in Taiwan and Hainan. Also in tropical Asia, Australia, Africa and America.

抗风桐 *Pisonia grandis*

抗风桐
Pisonia grandis R. Brown

常绿乔木。枝无刺。叶椭圆形、长圆形或卵形。聚伞花序长1-4厘米；花两性；花梗长1-1.5毫米，顶端有2-4长圆形小苞片；花被筒漏斗状。果实棍棒状，长约12毫米，具无柄的腺体，无宿存花被，沿棱具1列有黏液的短皮刺。花期夏季，果期夏末至秋季。生林中。产台湾和西沙群岛。南亚、东南亚、东非、澳大利亚东北部和太平洋岛屿也有。

Evergreen trees. Branches unarmed. Leaves elliptic, oblong or ovate. Cymes 1-4 cm long; flowers bisexual; pedicel 1-1.5 mm long, apex with 2-4 oblong bracteoles; perianth tube funnel form. Fruit clavate, ca. 12 mm long, with sessile glands, without persistent perianth, rib with a row of viscid prickles. Fl. Summer. Fr. late summer to autumn. Forests. Distributed in Taiwan and Xisha Islands. Also in S and SE Asia, E Africa, NE Australia and Pacific Islands.

腺果藤 *Pisonia aculeate*

胶果木 *Pisonia umbellifera*

胶果木
Pisonia umbellifera (J. R. Forster et G. Forster) Seemann

乔木。枝无刺。叶椭圆形、长圆形或卵状披针形。聚伞花序长5-12厘米；花杂性；花梗长1.5-6毫米，近基部有1-3小苞片；花被筒钟形。果实近圆柱状，长2.5-4厘米，平滑，具黏性，果梗顶端有扩展的宿存花被。花果期秋冬。生中低海拔潮湿灌丛和山地疏林中。产台湾和海南。非洲东部、南亚、东南亚、东亚、澳大利亚和太平洋岛屿亦有。

Trees. Branches unarmed. Leaf elliptic, oblong or ovate-lanceolate. Cymes 5-12 cm long; flowers polygamous; pedicel 1.5-6 mm long, with 1-3 bracteoles near base; perianth tube campanu-

光叶子花 *Bougainvillea glabra*

late. Fruit subterete, 2.5-4 cm long, smooth, glutinous, with spreading persistent perianth at apex. Fl. and fr. autumn-winter. Moist thickets, open forests at low to medium elevations. Distributed in Taiwan and Hainan. Also in E Africa, S, SE and E Asia, Australia and Pacific Islands.

光叶子花
Bougainvillea glabra Choisy

藤状灌木。枝无毛或疏生柔毛；刺腋生。叶纸质，卵形或卵状披针形，上面无毛，下面被微柔毛。3个苞片聚生于小枝顶端，叶状，紫色或洋红色，长圆形或椭圆形；花单朵贴生于每个苞片中脉，与苞片约等长；花被管疏生柔毛，顶端5浅裂；雄蕊6-8；花柱侧生，线形。花期冬季至翌年春季间。华南露天栽植，华北栽培于温室。原产巴西。

Vinelike shrubs. Branches glabrous or sparsely pubescent; spines axillary. Leaves chartaceous, ovate or ovate-lanceolate, adaxially glabrous, abaxially sparsely pubescent. 3 bracts clustered at branchlet apex, foliaceous, purple or magenta, oblong or elliptic; single flower adnate to median vein of each bract, about as long as bracts; perianth tube sparsely pubescent, apex 5-lobed lightly; stamens 6-8; style lateral, linear. Fl. winter to next spring. Cultivated in open areas in S China and greenhouse in N China. Native to Brazil.

叶子花
Bougainvillea spectabilis Willd.

藤状灌木。刺下弯。叶椭圆形或卵形，密被短柔毛。苞片深红色或浅紫红色，椭圆状卵形；花被片合生成管状，通常绿色，裂片开展，黄色，密被开展的短柔毛；雄蕊8。果密被毛。花期冬季至翌年春季。华南栽培。原产热带美洲。

Climbing shrubs. Spines recurved. Leaves elliptic or ovate, densely pubescent. Bracts dark red or light purple-red, elliptic-ovate; tepals connatel into a tube, usually green, lobes spreading, yellow, densely spreading-pubescent; stamens 8. Fruits densely hairy. Fl. winter to next spring. Cultivated in S China. Native to tropical America.

紫茉莉
Mirabilis jalapa L.

多年生草本。茎直立，多分枝，节膨大。叶卵形或卵状三角形，基部截形或心形，两面无毛。花常数朵簇生枝顶，芳香；花被高脚碟状，紫红色、黄色、白色或杂色。掺花果黑色，球形，坚硬，具棱及褶。花期6-10月，果期8-11月。中国普遍栽培或逸生。产泛热带；原产热带美洲。

Perennial herbs. Stems erect, much branched, inflated on nodes. Leaves ovate or ovate-triangular, base truncate or cordate, glabrous. Flowers often clustered at branchlet apex, fragrant; perianth salverform, purple-red, yellow, white or variegated. Anthocarps black, globose, hard, ribbed and plicate. Fl. Jun-Oct. Fr. Aug-Nov. Widely cultivated or naturalized in China. Distributed in pantropical; native to tropical America.

叶子花 *Bougainvillea spectabilis*

紫茉莉 *Mirabilis jalapa*

中华山紫茉莉
Oxybaphus himalaicus var. *chinensis*

中华山紫茉莉

Oxybaphus himalaicus Edgew. var. **chinensis** (Heim.) D.Q.Lu

一年生草本。茎平卧，多分枝，疏生腺毛至近无毛。叶卵形，先端渐尖或急尖，基部心形或圆形，下面被毛，边缘具毛或不明显小齿。花顶生或腋生，花梗密被黏腺毛；花被紫红色或粉红色，顶端5裂；雄蕊5。果实椭圆状或卵球形。花果期8-10月。生海拔700-2750(-3400)米干暖河谷的灌丛，草地或石缝中。产陕西、甘肃、四川、重庆、云南和西藏。

Annual herbs. Stems prostrate, many branched, sparsely glandular pubescent to glabrescent. Leaves ovate, apex acuminate or acute, base cordate or rotund, abaxially hairy, margin pubescent or indistinctly denticulate. Flowers terminal or axillary, densely viscid glandular pubescent; perianth purple-red or pink, apex 5-lobed; stamens 5. Fruit ellipsoid or ovoid. Fl. and fr. Aug-Oct. Thickets, grasslands, rock crevices of dry and warm river valleys at 700-2700 (-3400) m. Distributed in Shaanxi, Gansu, Sichuan, Chongqing, Yunnan and Xizang.

黄细心

Boerhavia diffusa L.

多年生蔓性草本。叶对生，卵形，先端钝或急尖，边缘波状，具多细胞毛，两面疏具柔毛，下面灰黄色，干后具皱。头状伞房状圆锥花序顶生；花小。掺花果有黏腺和疏柔毛。花果期春季至秋季。生海拔100-1900米的沿海旷地或干热河谷。产中国西南和华南。南亚、东南亚、琉球群岛、澳大利亚、太平洋岛屿、非洲和美洲亦有。

Perennial trailing herbs. Leaves opposite, ovate, apex obtuse or acute, margin undulate, with muticellular hairs, both surfaces sparsely pubescent, abaxially gray-yellow, wrinkled when dry. Capitate-cymose panicles terminal; flowers small. Anthocarps with viscid glands and sparse pubescences. Fl. and fr. spring to autumn. Coastal open places or dry and hot valleys at 100-1900 m. Distributed in SW and S China. Also in S and SE Asia, Ryukyu Islands, Australia, Pacific Islands, Africa and America.

直立黄细心

Boerhavia erecta L.

草本。茎直立或基部外倾。叶先端急尖，稀钝，基部圆形或楔形。聚伞圆锥花序紧密；花梗长0.5-5毫米，有1-2枚披针形小苞片；花被管状或钟状；雄蕊2-3，稍外伸。果实倒圆锥形，长约3毫米，顶端截形，无毛，5条棱间的沟稍呈波状。花果期夏季。生空旷沙地。产广东和海南。新加坡、马来西亚、印度尼西亚、泰国和太平洋岛屿亦有。

Herbs. Stems erect or decumbent at base. Leaves apex acute, rarely obtuse, base rounded or cuneate. Cymose panicles close together; pedicel 0.5-5 mm long, with 1 or 2 lanceolate bracteoles; perianth tubulose or campanulate; stamens 2-3, slightly exserted. Anthocarp obconic, ca. 3 mm long, apex truncate, glabrous, 5-ribbed, groove between ribs somewhat undulate. Fl. and fr. summer. Open sandy areas. Distributed in Guangdong and Hainan. Also in Singapore, Malaysia, Indonesia, Thailand and Pacific Islands.

中华黏腺果

Commicarpus chinensis (L.) Heim

多年生草本。叶厚纸质，长3-6厘米，宽2.5-5厘米，基部截形或近心形，边缘波状；叶柄长1-3厘米。伞形花序顶生或腋生；花梗长3-7毫米；花粉红色；花被短漏斗状；雄蕊2-4，常伸出；花柱细长，伸出，柱头盾状。果倒圆锥形，具10条纵棱和凸起的小腺体。花果期6-9月。生旷地或丛林中。产海南。南亚和东南亚亦有。

Perennial herbs. Leaves thick chartaceous, 3-6 × 2.5-5 cm, base truncate or subcordate, margin undulate; petiole 1-3 cm long. Umbels axillary or terminal; pedicel 3-7 mm long; perianth pink, short funnelform; stamens

黄细心 *Boerhavia diffusa*

直立黄细心 *Boerhavia erecta*

2-4, usually exserted; style slender, exserted, stigma peltate. Fruit obconic, longitudinally 10-ribbed, with minute glands. Fl. and fr. Jun-Sep. Open places or forests. Distributed in Hainan. Also in S and SE Asia.

澜沧黏腺果
Commicarpus lantsangensis D. Q. Lu

半灌木。叶片稍肉质，长1-2.7厘米，宽1-2.5厘米，基部楔形，边缘全缘；叶柄长0.5-1.3厘米。伞形花序腋生或顶生；花梗长0.5-1.5厘米；花被紫红色，短漏斗状；雄蕊3，外伸；花柱细长，外伸，柱头盾状。果实棍棒状，具10条纵棱，棱上有瘤状腺体。花期6月，果期8月。生海拔2300-3000米干热河谷、路旁，石缝中。产四川、云南北部和西藏东南部。

Subshrubs. Leaves slightly fleshy, 1-2.7 × 1-2.5 cm, base cuneate, margin entire; petiole 0.5-1.3 cm long. Umbels axillary or terminal; peduncle 0.5-1.5 cm long; perianth purple-red, short funnelform; stamens 3, exserted; style slender, exserted, stigma peltate. Fruit clavate, longitudinally 10-ribbed, with wartlike glands on ribs. Fl. Jun. Fr. Aug. Dry and warm river valleys, roadsides, stone crevices at 2300-3000 m. Distributed in Sichuan, N Yunnan and SE Xizang.

中华黏腺果 *Commicarpus chinensis*

澜沧黏腺果 *Commicarpus lantsangensis*

商陆科 Phytolaccaceae

商陆
Phytolacca acinosa Roxb.

多年生草本。叶椭圆形或披针状椭圆形。总状花序直立，常短于叶；花两性；花被片5，通常白色；雄蕊8-10；心皮分离。浆果扁球形，成熟后紫黑色。花期5-8月，果期6-10月。生海拔500-3400米的山谷、山坡、林缘和路边。产中国西南、华南、东南、华北、华西和华东。印度、不丹、缅甸、越南、朝鲜半岛和日本亦有。

Perennial herbs. Leaves elliptic or lanceolate-elliptic. Racemes erect, usually shorter than leaves; flowers bisexual; tepals 5, usually white; stamens 8-10; carpels distinct. Berries oblate, purplish black when mature. Fl. May-Aug. Fr. Jun-Oct. Valleys, slopes, forest edges and roadsides at 500-3400 m. Distributed in SW, S, SE, N, W and E China. Also in India, Bhutan, Myanmar, Vietnam, Korean Peninsula and Japan.

商陆 *Phytolacca acinosa*

多药商陆

Phytolacca polyandra Batalin

草本。叶椭圆状披针形或椭圆形。总状花序直立，圆柱状；花被片5，粉红色；雄蕊12-16，两轮；心皮(6-)8(或9)，合生。浆果后变为膜质，贴生于种子上。花期5-8月，果期6-9月。生海拔1100-3000米的林中、河边或路边。产中国西南、华南和华西。

Herbs. Leaves elliptic-lanceolate or elliptic. Racemes erect, terete; tepals 5, pink; stamens 12-16, bicyclic; carpels (6-)8(or 9), connate. Berries becoming membranous, appressed to seeds. Fl. May-Aug. Fr. Jun-Sep. Forests, riversides or roadsides at 1100-3000 m. Distributed in SW, S and W China.

多药商陆 *Phytolacca polyandra*

日本商陆 *Phytolacca japonica*

日本商陆

Phytolacca japonica Makino

多年生草本。叶长圆形至卵状长圆形。总状花序直立；花被片5，淡红色；雄蕊(10-)12-17；心皮6-10，合生。果序长4.5-11厘米；浆果扁球形，熟时黑色。种子肾圆形，直径约3毫米，具纤细同心条纹。花果期6-8月。生海拔350-1100米山谷水旁林下。产山东、浙江、安徽、江西、福建、台湾、广东和湖南。日本亦有。

Perennial herbs. Leaf oblong to ovate-oblong. Racemes erect; tepals 5, reddish; stamens (10-)12-17; carpels 6-10, connate. Fruiting infructescence 4.5-11 cm long; berry oblate, black when mature. Seeds reniform-orbicular, ca. 3 mm diam, with slender concentric striations. Fl. and fr. Jun-Aug. Valleys, forest understories, riversides at 350-1100 m. Distributed in Shandong, Zhejiang, Anhui, Jiangxi, Fujian, Taiwan, Guangdong and Hunan. Also in Japan.

垂序商陆

Phytolacca americana L.

多年生草本。根厚，倒圆锥状。叶椭圆状卵形或卵状披针形。总状花序顶生或侧生；花被片5，白色；雄蕊、心皮和花柱10；心皮合生。果序下垂；浆果扁，成熟后紫黑色。花期6-8月，果期8-10月。栽培或逸生于中国西南、华南、东南、华中、华北和华东。原产北美洲；世界各地广泛归化。

Perennial herbs. Roots thick, obconic. Leaves elliptic-ovate or ovate-lanceolate. Racemes termi-

垂序商陆 *Phytolacca americana*

数珠珊瑚 *Rivina humilis*

nal or lateral; tepals 5, white, slightly red; stamens, carpels and styles 10; carpels connate. Fruiting infructescences pendent; berries oblate, purple-black when mature. Fl. Jun-Aug. Fr. Aug-Oct. Cultivated or naturalized in SW, S, SE, C, N and E China. Native to North America; widely naturalized in the world.

数珠珊瑚
Rivina humilis L.

半灌木。枝开展，幼时被短柔毛。叶片卵形，先端长渐尖，边缘有微锯齿，托叶细小，早落。总状花序直立或弯曲，被短柔毛；花被片4，白色或粉红色，于花后变绿并反曲；雄蕊4；单心皮。浆果球形，红色。种子双凸镜状。栽培于浙江、福建、广东和广西，逃逸到园圃和路边。原产热带美洲。

Subshrubs. Branches spreading, pubescent when young. Leaves ovate or ovate-lanceolate, slightly serrulate, apex long acuminate. Racemes erect or curved, pubescent; tepals 4, white or pink, green and reflexed after anthesis; stamens 4; carpel 1. Berry globose, red. Seeds lenticular. Cultivated in Zhejiang, Fujian, Guangdong and Guangxi, escaped in nurseries and roadsides. Native to tropical America.

蒜味草
Petiveria alliacea L.

多年生草本，具蒜味。分枝幼时被短柔毛。叶狭椭圆形至卵形，基部楔形，先端急尖至长渐尖。穗状花序纤细；花被片4，开展，白色，花后变绿；雄蕊4-8；柱头侧生。瘦果棍棒状，密被短柔毛，顶端4浅裂，每裂片具1根反折的刺。花期4-11月。生海拔约5米的路边。云南栽培，2001年引自巴西；在福建(厦门)归化；原产热带美洲。

Perennial herbs, alliaceous. Branches pubescent when young. Leaves narrowly elliptic to ovate, base cuneate, argute to long acuminate at apex. Spikes tenuous; tepals 4, white, turning green after anthesis; stamens 4-8; stigma lateral. Achene clavate, densely pubescent, 4-lobed at apex, each lobe with one deflexed spine. Fl. Apr-Nov. Roadsides at ca. 5 m. Cultivated in Yunnan, introduced from Brazil in 2001; naturalized in Fujian (Xiamen); native to tropical America.

蒜味草 *Petiveria alliacea*

番杏科 Aizoaceae

无茎粟米草 *Mollugo nudicaulis*

线叶粟米草 *Mollugo cerviana*

星粟草
Glinus lotoides L.

一年生草本，密被星状柔毛。茎外倾，多分枝。基生叶莲座状，早落；茎生叶轮生或对生，倒卵形至长圆状匙形，长6-24毫米，基部下延。花数朵簇生；花被片5；雄蕊(3-)5(-15)；花柱5。蒴果卵形，与宿存花被等长，5瓣裂。花果期春夏。生海拔500米以下空旷沙滩、河边、稻田或荒地上。产海南、台湾和云南南部。广布于热带地区。

Annual herbs, densely stellate tomentose. Stems decumbent, much branched. Basal leaves in a rosette, drying soon; upper leaves verticillate or opposite, obovate to oblong-spatulate, 6-24 mm long, base decurrent. Flowers several in groups; tepals 5; stamens (3-)5(-15); styles 5. Capsule ovoid, ca. as long as persistent tepals, 5-valved. Fl. and fr. spring to summer. Open sands, riversides, rice fields or waste places at near sea level to 500 m. Distributed in Hainan, Taiwan and S Yunnan. Widely distributed in tropical regions.

无茎粟米草
Mollugo nudicaulis Lam.

一年生草本。叶均基生呈莲座状，椭圆状匙形或倒卵状匙形，长1-5厘米。二歧聚伞花序开展；花被片5，黄白色；雄蕊3-5；花柱3。蒴果近卵球形或椭圆体形，与宿存花被几等长。种子近肾形，具多数颗粒状凸起。花果期几乎全年。多生海岸沙地或旷地。产海南和广东。热带非洲、阿富汗、印度、巴基斯坦、新喀里多尼亚和古巴亦有。

Annual herbs. Leaves all in a basal rosette, elliptic-spatulate or obovate-spatulate, 1-5 cm long. Inflorescence a dichasium, spreading; tepals 5, yellowish white; stamens 3-5; stigmas 3. Capsule subovoid or ellipsoid, ca. as long as persistent tepals. Seed subreinform, granulose. Fl. and fr. almost year-round. Sandy seashores or open places. Distributed in Hainan and Guangdong. Also in tropical Africa, Afghanistan, India, Pakistan, New Caledonia and Cuba.

线叶粟米草
Mollugo cerviana (L.) Ser.

一年生草本。茎多数，斜升。叶无柄；基生叶呈莲座状；茎生叶3-10假轮生，狭条形，长5-10毫米。伞状聚伞花序顶生或腋生；花被片5；中肋绿色，边缘白色；雄蕊3-5；花柱3。蒴果宽椭圆体形。种子近半圆形，有细网纹。花期6-7月。生海拔400-1200米沙地、路边或田间。产新疆和河北。南欧，非洲，亚洲中部、东部和东南部，澳大利亚亦有。

Annual herbs. Stems numerous, ascending. Leaves sessile, basal leaves in a rosette; cauline leaves in pseudowhorls of 3-10, narrowly linear, 5-10 mm long. Umbellate cymes terminal or axillary; tepals 5, midvein green, margin white; Stamens 3-5; styles 3. Capsules broadly ellipsoid. Seeds semi-orbicular, reticulate. Fl. Jun-Jul. Sands, roadsides, or fields at 400-1200 m.

星粟草 *Glinus lotoides*

粟米草 *Mollugo stricta*

Distributed in Xinjiang and Hebei. Also in S Europe, Africa, C, E and SE Asia, and Australia.

粟米草
Mollugo stricta L.

铺散草本。叶披针形或线状披针形。花序顶生或似与叶对生成疏松聚伞状；花被片5，淡绿色，边缘膜质；子房3室。蒴果近球形，3裂。种子多数，栗褐色，肾形，具瘤。花期6-8月，果期8-10月。生海拔100-1800米荒地，农田或海岸沙地。产中国西南、华南、东南、华中和华东。亚洲热带和亚热带地区亦有。

Diffuse herbs. Leaves lanceolate or linear-lanceolate. Inflorescences terminal or in seemingly leaf-opposed lax cymes; tepals 5, greenish, margin membranous; ovary 3-loculed. Capsules subglobose, 3-valved. Seeds numerous, chestnut-colored, reniform, tuberculate. Fl. Jun-Aug. Fr. Aug-Oct. Waste places, farmlands or coastal sandy beaches at 100-1800 m. Distributed in SW, S, SE, C and E China. Also in tropical and subtropical Asia.

种棱粟米草
Mollugo verticillata L.

一年生草本。茎直立或铺散。基生叶莲座状；茎生叶3-7假轮生或2-3生于节侧，倒披针形或条状倒披针形，长1-3厘米；叶柄短。伞状聚伞花序腋生；花被片(4-)5，白或绿白色；雄蕊(2-)3(-5)；花柱3。蒴果椭圆体形或近球形。种子肾形，脊具3-5棱。花果期秋冬。生草地或旱田。产山东和华南。热带美洲、南欧和日本亦有。

Annual herbs. Stem erect or diffuse. Basal leaves in a rosette; cauline leaves in pseudowhorls of 3-7, or 2-3 in groups on one side of node, oblanceolate or linear-oblanceolate, 1-3 cm long; petiole short. Umbellate cymes axillary; tepals (4-)5, pale or greenish white; stamens (2-)3(-5); styles 3. Capsule ellipsoid or subglobose. Seeds reniform, raphe with 3-5 ribs. Fl. and fr. autumn to winter. Grasslands or dry farmlands. Distributed in Shandong and S China. Also in tropical America, S Europe and Japan.

海马齿 *Sesuvium portulacastrum*

海马齿
Sesuvium portulacastrum (L.) L.

多年生草本。茎平卧或匍匐生根，绿色或红色，具白色囊状细胞。叶对生。花两性，单生；雄蕊15-20；子房倒卵球形，无毛；柱头3-5。蒴果倒卵球形，短于花盖。种子亮黑色。花期4-7月。生海滨沙地。产广东、海南、台湾和福建。世界热带和亚热带地区亦有。

Perennial herbs. Stems prostrate or creeping, green or red, with white bladder cells. Leaves opposite. Flowers bisexual, solitary; stamens 15-20; ovary obovoid, glabrous; stigmas 3-5. Capsules ovoid, shorter than perigone. Seeds shiny black. Fl. Apr-Jul. Sands of seashores. Distributed in Guangdong, Hainan, Taiwan and Fujian. Also in tropical and subtropical regions of the world.

种棱粟米草 *Mollugo verticillata*

假海马齿 *Trianthema portulacastrum*

假海马齿
Trianthema portulacastrum L.

多年生草本。茎匍匐。叶对生，无毛，椭圆形至倒卵形，基部楔形；叶柄基部膨大成短鞘。花单生叶腋，无梗；花被5裂，粉红色，稀白色；花被筒贴生叶柄基部；雄蕊10-25；花柱1。蒴果顶端截形，2裂，种子肾形，黑色，具皱纹。花期夏季。生海边空旷干沙地或为田间杂草。产台湾、广东和海南。广布热带地区。

Perennial herbs. Stems procumbent. Leaves opposite, glabrous, elliptic to obovate, base cuneate; petiole base expanded into short sheath. Flowers solitary axillary, sessile; perigone 5-lobed, pink, rarely white; perigone tube fused with basal sheath of pedicel; stamens 10-25; stigma 1. Capsule truncate at apex, 2-lobed. Seeds reniform, black, with low crests. Fl. Summer. Open sunny dry sands or as weeds in farmlands usually near seashore. Distributed in Taiwan, Guangdong and Hainan. Widely also in tropical areas.

番杏
Tetragonia tetragonioides (Pall.) Kuntze

一年生肉质草本，无毛。茎初直立，后平卧上升。叶卵状菱形或卵状三角形，长4-10厘米。花1-3朵簇生叶腋；花被筒长2-3毫米，裂片(3-)4(-5)，内面黄绿色；雄蕊10-13。坚果陀螺形，有4-5角。花果期8-10月。野生于海滩，也有栽培。产江苏、浙江、福建、台湾、广东和云南。非洲、东亚、澳大利亚和南美洲亦有。

Annual fleshy herbs, glabrous. Stems erect when young, becoming decumbent. Leaves rhomboid-ovate or deltoid-ovate, 4-10cm long. Flowers 1-3; perigone tube 2-3 mm, lobes (3-)4(-5), inside yellowish green; stamens 10-13. Fruit turbinate, 4-5-angular. Fl. and fr. Aug-Oct. Sandy shores, also cultivated. Distributed in Jiangsu, Zhejiang, Fujian, Taiwan, Guangdong and Yunnan. Also in Africa, E Asia, Australia and South America.

番杏 *Tetragonia tetragonioides*

马齿苋科 Portulacaceae

马齿苋
Portulaca oleracea L.

一年生草本。植株全体无毛。叶片扁平，倒卵形，肥厚，长1-3厘米。花3-5枚簇生，由2-6个苞片形成的总苞所包被；花萼绿色，盔状；花瓣5，黄色，倒卵形。蒴果卵球形。花期5-8月，果期6-9月。生海拔210-3000米荒地上。产中国各地。全世界广布。

Annual herbs. Whole plant glabrous. Leaves flat, obovate, thick, 1-3 cm long. Flowers in clusters of 3-5, surrounded by involucre of 2-6 bracts; sepals green, helmeted; petals 5, yellow, obovate. Capsules ovoid. Fl. May-Aug. Fr. Jun-Sep. Waste lands at 210-3000 m. Distributed throughout China. Also worldwide.

毛马齿苋
Portulaca pilosa L.

一年生或多年生草本。叶腋密具柔毛；叶圆柱状线形或近钻状披针形。花直径大于2厘米，

马齿苋 *Portulaca oleracea*

毛马齿苋 *Portulaca pilosa*

由6-9个苞片形成的总苞包被且密具绵毛；花瓣5，紫红色，阔倒卵形。蒴果顶端光亮禾秆色，圆柱状卵球形。花果期5-8月。生海拔500-700米的海边干旱开阔地或沙质地上。产中国西南和华南。缅甸、老挝、泰国、越南、马来西亚、印度尼西亚、菲律宾和热带美洲亦有。

Annual or perennial herbs. Leaf axils densely pilose; leaves terete-linear or subulate-lanceolate. Flowers more than 2 cm diam, surrounded by involucre of 6-9 bracts with dense wool; petals 5, red-purple, broadly obovate. Capsules glossy straw colored apically, cylindric-ovoid. Fl. and fr. May-Aug. Dry open places near seashores or sandy lands at 500-700 m. Distributed in SW and S China. Also in Myanmar, Laos, Thailand, Vietnam, Malaysia, Indonesia, the Philippines and tropical America.

大花马齿苋

Portulaca grandiflora Hook.

一年生草本。茎稍带紫红色，平卧或斜升，多分枝。叶不规则互生，圆柱形，肉质。花单生或数朵簇生枝端；花瓣红色、粉红色、紫色或黄白色。蒴果近椭圆体形。花果期7-10月。中国各地及其他国家广栽培。原产巴西。

Annual herbs. Stems slightly violet, prostrate or ascending, multi-branched. Leaves irregularly alternate on stems, terete, fleshy. Flowers solitary or clustered at branch apices; petals red, pinkish, purple or yellowish white. Capsules subellipsoid. Fl. and fr. Jul-Oct. Cultivated throughout China and other countries. Native to Brazil.

土人参

Talinum paniculatum (Jacq.) Gaertn.

一年生或多年生草本。根少分枝，倒钟形，粗，外皮黑褐色，肉质乳白。叶互生或近对生，倒卵形或倒卵状披针形。花萼紫红色；花瓣粉红色或淡紫色。蒴果近球形，纸质。花期6-8月，果期9-11月。生海拔3000米以下阴湿地。华南至华中有栽培或逸生。原产热带美洲；东南亚栽培或归化。

Annual herbs or perennial. Roots few branched, obcampanulate, thick, epidermis black-brown, flesh milky white. Leaves alternate or subopposite, obovate or obovate-lanceolate. Sepals purple-red; petals pink or purplish. Capsules subglobose, papery. Fl. Jun-Aug. Fr. Sep-Nov. Shaded moist places below 3000 m. Cultivated or escaped from S to C China. Native to tropical America; cultivated or naturalized in SE Asia.

大花马齿苋 *Portulaca grandiflora*

土人参 *Talinum paniculatum*

落葵科 Basellaceae

落葵
Basella alba L.

一年生草本。茎缠绕，绿色或红色，肉质，无毛。叶宽卵形或圆形，全缘。穗状花序长3-15(-20)厘米；花无梗；花被片红色至紫黑色；花丝白色。胞果球形，被肉质的花被及小苞片所包被。花期5-9月，果期7-10月。广泛栽培和归化于华南。原产热带亚洲。

Annual herbs. Stems twining, green or red, fleshy, glabrous. Leaves broadly ovate or rotund, margin entire. Spikes 3-15(-20) cm long; flowers sessile; perianth reddish to purple-black; filaments white. Utricles globose, surrounded by persistent fleshy perianth and bracteoles. Fl. May-Sep. Fr. Jul-Oct. Widely cultivated and naturalized in S China. Native to tropical Asia.

落葵薯
Anredera cordifolia (Tenore) Steenis

一年生草本，具粗的根状茎。茎缠绕，无毛。叶宽卵形至近圆形，稍肉质，腋部通常具小块茎(珠芽)。总状或圆锥花序纤细，下垂，长7-25厘米；花梗长2-3毫米；花芳香；花被片白色，于花期开展，花后变黑色。花期6-10月。华东、华南、华中和云南有栽培和归化。原产南美洲热带地区。

落葵 *Basella alba*

落葵薯 *Anredera cordifolia*

Annual herbs, with thick rhizome. Stems twining, glabrous. Leaves broadly ovate to subrotund, thinly fleshy, usually with tubercles (bulbils) in the axils. Racemes or panicles tenuous, pendent, 7-25 cm long; pedicel 2-3 mm long; flowers fragrant; perianth white, patent at anthesis, blackened after anthesis. Fl. Jun-Oct. Cultivated and naturalized in E, S, C China and Yunnan. Native to tropical South America.

石竹科
Caryophyllaceae

荷莲豆草 *Drymaria cordata*

治疝草
Herniaria glabra L.

一年生或多年生草本。茎多分枝，铺散，高5-18(-35)厘米。叶对生，无柄，狭倒卵形，长3-7毫米，无毛。团伞花序与叶对生，具6-10花。萼片5，长1.5毫米；花瓣缺；雄蕊5；子房1室，花柱2深裂。瘦果长2毫米。花期7月。生海拔900-2400米的草甸或山谷中。产四川西部和新疆。蒙古、阿富汗、俄罗斯和欧洲亦有。

Annual or perennial herbs. Stems ramose, diffuse, 5-18(-35) cm tall. Leaves opposite, sessile, narrowly obovate, 3-7 mm long, glabrous. Glomerules leaf-opposed, 6-10-flowered. Sepals 5, 1.5 mm long; petals absent; stamens 5; ovary l-locular and style 2-parted. Achenes 2 mm long. Fl. Jul. Meadows or in valley at 900-2400 m. Distributed in W Sichuan and Xinjiang. Also in Mongolia, Afghanistan, Russia and Europe.

治疝草 *Herniaria glabra*

荷莲豆草
Drymaria cordata (L.) Willd. ex Schult.

一年生草本。茎缠绕，有时近蔓生，茎和叶无毛，有微小乳突。叶片卵状心形，自基部具明显3-5脉；托叶膜质，撕裂成少量白色刚毛。花瓣白色，倒卵状楔形，深2裂。蒴果卵球形，3瓣裂。花期4-10月，果期6-12月。生海拔200-1900(-2400)米的阴湿地，常在河边或灌丛下。产中国西南、华南和东南。热带和亚热带亚洲、大洋洲、非洲和美洲亦有。

Annual herbs. Stems straggling, sometimes subscandent, with leaves glabrous or minutely papillose. Leaves ovate-cordate, prominently 3-5-veined from base; stipules membranous, splitting into few whitish setae. Petals white, obovate-cuneate, deeply 2-cleft. Capsules ovoid, 3-valved. Fl. Apr-Oct. Fr. Jun-Dec. Damp shady sites, often by streams or under thickets at 200-1900 (-2400) m. Distributed in SW, S and SE China. Also in tropical and subtropical Asia, Oceania, Africa and America.

细叶孩儿参 *Pseudostellaria sylvatica*

孩儿参 *Pseudostellaria heterophylla*

细叶孩儿参

Pseudostellaria sylvatica (Maxim.) Pax

多年生小草本。块根通常数个串生，长卵形或短纺锤形。茎近4棱，被2列柔毛。叶无柄；叶片线状或披针状线形，边缘近基部有缘毛。花二型：开花受精花单生茎顶或成二歧聚伞花序；萼片披针形，边缘膜质，外面被柔毛；花瓣白色，倒卵形，顶端浅2裂；花柱2-3；闭花受精花着生下部叶腋或短枝顶端；萼片狭披针形，外面被柔毛；无花瓣。蒴果卵圆形，3瓣裂。花果期6-7月。生松林或混交林下。产甘肃、贵州、河北、黑龙江、河南、湖北、吉林、辽宁、青海、陕西、四川、新疆、西藏和云南。不丹、俄罗斯、日本和朝鲜半岛亦有。

Small perennial herbs. Tuberoids usually several in a row, narrowly ovoid or shortly fusiform. Stems 4-angled, with 2 lines of hairs. Leaves sessile, linear or lanceolate-linear, margin ciliate at base. Flowers of two types: chasmogamic flowers solitary or in dichasium; sepals lanceolate, abaxially pubescent, margin membranous; petals white, obovate, apex 2-lobed; styles 2-3; cleistogamic flowers axillary on stem or terminal on a dwarf shoot; sepals narrowly lanceolate, abaxially pubescent; petals absent. Capsule ovoid, 3-valved. Fl. and fr. Jun-Jul. *Pinus* forests or mixed forests. Distributed in Gansu, Guizhou, Hebei, Heilongjiang, Henan, Hubei, Jilin, Liaoning, Qinghai, Shaanxi, Sichuan, Xinjiang, Xizang and Yunnan. Also in Bhutan, Russia, Japan and Korean Peninsula.

孩儿参

Pseudostellaria heterophylla (Miq.) Pax

多年生小草本。块根长纺锤状。茎单一，直立，具2列毛。基部叶常1或2对，匙形或倒披针形；中部叶披针形。开放花腋生，单生或形成聚伞状；闭锁花腋生，无花瓣。蒴果卵球形，不开裂或3瓣裂。种子具瘤。花期4-7月，果期7-8月。生海拔800-2700米的山谷或阴林中。产中国西南、华中、华北、华西和华东。朝鲜半岛和日本亦有。

Small perennial herbs. Root tubers long fusiform. Stems solitary, erect, with 2 lines of hairs. Proximal leaves usually 1 or 2 pairs, spatulate or oblanceolate; middle leaves lanceolate. Chasmogamic flowers axillary, solitary or in cymes; cleistogamic flowers axillary, petals absent. Capsules ovoid, indehiscent or 3-valved. Seeds tuberculate. Fl. Apr-Jul. Fr. Jul-Aug. Valleys or shaded forests at 800-2700 m. Distributed in SW, C, N, W and E China. Also in Korean Peninsula and Japan.

蔓孩儿参

Pseudostellaria davidii (Franch.) Pax

多年生小草本。块根纺锤状。茎匍匐或蔓生，少分枝。叶卵形，近无柄。开放花腋生，花瓣5，全缘；闭锁花腋生。蒴果卵球形，4瓣裂。种子表面具尖

蔓孩儿参 *Pseudostellaria davidii*

瘤状凸起。花期5-7月，果期7-8月。生海拔1000-3800米的林中、小溪边、石质丘陵或林缘。产中国西南、华北、华西、华东、西北和东北。俄罗斯、蒙古和朝鲜半岛亦有。

Perennial small herbs. Root tubers fusiform. Stems repent or decumbent, sparsely branched. Leaves ovate, subsessile. Chasmogamic flowers axillary; petals 5, margin entire; cleistogamic flowers axillary. Capsules ovoid, 4-valved. Seeds mammillate with pointed projections. Fl. May-Jul. Fr. Jul-Aug. Forests, brooksides, stony hillsides or forest edges at 1000-3800 m. Distributed in SW, N, W, E, NW and NE China. Also in Russia, Mongolia and Korean Peninsula.

球序卷耳 *Cerastium glomeratum*

鹅肠菜

Myosoton aquaticum (L.) Moench

二年生或多年生草本。茎柔弱，顶端具腺毛。叶卵形，托叶不存在。花顶生或腋生；苞片叶状，边缘有腺毛；花梗细，长1-2厘米，密被腺毛；花瓣2裂至基部；雄蕊10。蒴果卵球形，下垂，长于宿存萼片。花期5-6月，果期6-8月。生海拔300-2700米山坡、山谷、林中或田边。产中国各地。北半球广布。

Biennial or perennial herbs. Stems weak, apically glandular hairy. Leaves ovate, exstipulate. Flowers terminal or axillary; bracts leaflike, margin glandular hairy; pedicels 1-2 cm long, slender, densely glandular-hairy; petals bifid to base; stamens 10. Capsules ovoid, pendent, exceeding persistent sepals. Fl. May-Jun. Fr. Jun-Aug. Mountain slopes, valleys, forests or field edges at 300-2700 m. Distributed throughout China. Widely distributed in Northern Hemisphere.

鹅肠菜 *Myosoton aquaticum*

球序卷耳

Cerastium glomeratum Thuill.

一年生草本。茎密被长柔毛，上部混生腺毛。茎下部叶叶片匙形；茎上部叶叶片倒卵状椭圆形，顶端急尖，基部渐狭，两面被长柔毛，边缘具缘毛。聚伞花序呈簇生状或呈头状；花序轴密被腺柔毛；苞片卵状椭圆形，密被柔毛；花梗长1-3毫米；萼片披针形，顶端尖，外面密被长腺毛，边缘狭膜质；花瓣白色，线状长圆形，顶端2浅裂，基部被疏柔毛；花柱5。蒴果长圆柱形，顶端10齿裂。花期3-4月，果期5-6月。生林缘、山坡草地和河畔沙地。产福建、广西、贵州、河南、湖北、湖南、江苏、江西、辽宁、山东、西藏、云南和浙江。全世界广泛分布。

Annual herbs. Stems densely villous, distally glandular pubescent. Proximal leaves spatulate; distal leaves obovate-elliptic, apex acute, base attenuate, both surfaces villous, margin ciliate. Inflorescence of compact, cymose clusters (glomerules); rachis densely glandular pubescent; bracts ovate-elliptic, densely pubescent; pedicel 1-3 mm long; sepals lanceolate, apex acute, abaxially densely long glandular pubescent, margin narrowly membranous; petals white, oblong, apex 2-lobed, base pilose; styles 5. Capsule cylindric, 10-toothed. Fl. Mar-Apr. Fr. May-Jun. Forest margins, mountain slope grasslands and sandy riverside. Distributed in Fujian, Guangxi, Guizhou, Henan, Hubei, Hunan, Jiangsu, Jiangxi, Liaoning, Shandong, Xizang, Yunnan and Zhejiang. Also widespread in the world.

缘毛卷耳

Cerastium furcatum Cham. et Schlecht.

多年生草本。茎高15-55厘米，近顶部被长柔毛和腺毛。茎生叶无柄，狭卵形，长1-3厘米。聚伞花序具5-11花；苞片叶状；萼片5，长5毫米；花瓣5，白色，长椭圆形，长约1厘米，顶端2浅裂；雄蕊10，花丝下部被长柔毛；花柱5。花期5-8月。生海拔1200-3800米的林边或山谷中。产西藏、云南西北部、四川西部、甘肃、陕西南部、山西、宁夏和吉林。俄罗斯(西伯利亚)和朝鲜半岛北部亦有。

Perennial herbs. Stems 15-55 cm tall, near apex villous and glandular pubescent. Cauline leaves sessile, narrowly ovate, 1-3 cm long. Cymes 5-11-flowered; bracts leaflike; sepals 5, 5 mm long; petals 5, white, long elliptic, ca. 1 cm long, apex 2-lobed; stamens 10, filaments below villous; styles 5. Fl. May-Aug. Forest margins or in valley at 1200-3800 m. Distributed in Xizang, NW Yunnan, W Sichuan, Gansu, S Shaanxi, Shanxi, Ningxia and Jilin. Also in Russia (Siberia) and N Korean Peninsula.

山卷耳

Cerastium pusillum Ser.

多年生小草本。茎丛生，高5-15厘米，密被短柔毛。下部叶匙形，上部叶长圆形，长5-15毫米，被柔毛。聚伞花序顶生，具2-7花；萼片5，长5-6毫米；花瓣5，白色，长圆形，长约1厘米，顶端2浅裂；雄蕊10，无毛；花柱5。花期7-8月。生海拔2800-3200米的高山草地。产云南西北部、青海、甘肃、宁夏、内蒙古西南部和新疆。蒙古、哈萨克斯坦和俄罗斯亦有。

Small perennial herbs. Stems caespitose, 5-15 cm tall, densely puberulous. Lower leaves spatulate, upper leaves oblong, 5-15 mm long, puberulous. Cymes terminal, 2-7-flowered; sepals 5, 5-6 mm long; petals 5, white, oblong, ca. 1 cm long, apex 2-lobed; stamens 10, glabrous; styles 5. Fl. Jul-Aug. Alpine meadows at 2800-3200 m. Distributed in NW Yunnan, Qinghai, Gansu, Ningxia, SW Neimenggu and Xinjiang. Also in Mongolia, Kazakhstan and Russia.

卷耳 *Cerastium arvense* subsp. *strictum*

卷耳

Cerastium arvense L. subsp. **strictum** Gaudin

多年生草本。茎疏丛生，下部匍匐，上部直立。叶线形至狭披针形，具柔毛或无毛，半抱茎。聚伞花序顶生，具3-7花；苞片叶状；花瓣倒卵形，2裂至其长度的1/4-1/3。蒴果圆柱状。花期5-8月，果期7-9月。生海拔600-4300米的林下、林缘、灌丛、多草山谷、山谷或沟壑中。产中国西南、华北、华西、西北和东北。哈萨克斯坦、俄罗斯、蒙古、朝鲜半岛、日本、欧洲和美洲亦有。

Perennial herbs. Stems sparsely caespitose, creeping proximally, erect distally. Leaves linear to narrowly lanceolate, pilose or glabrous, amplexicaul. Cymes terminal, 3-7-flowered; bracts leaflike; petals obovate, 2-lobed for 1/4-1/3 their length. Capsules cylindric. Fl. May-Aug. Fr. Jul-Sep. Under forests, forest edges, bushes, grassy mountain valleys, valleys or ditches at 600-4300 m. Distributed in SW, N, W, NW and NE China. Also in Kazakhstan, Russia, Mongolia, Korean Peninsula, Japan, Europe and America.

缘毛卷耳 *Cerastium furcatum*

山卷耳 *Cerastium pusillum*

藏南卷耳 *Cerastium thomsonii*

藏南卷耳

Cerastium thomsonii Hook. f.

多年生草本。茎丛生，密被腺毛。叶椭圆形或长圆形，两面被柔毛，基部渐狭，顶端近尖。聚伞花序近伞形，具少数花；苞片边缘狭膜质；花梗密被腺毛；萼片长圆状披针形，外面疏生腺毛，边缘宽膜质，顶端近尖；花瓣白色，宽倒卵形，顶端2浅裂；花柱5。蒴果长圆形；种子肾形，表面具瘤。花期6-7月。生林下、灌丛、山坡草地和草甸。产西藏。阿富汗、印度西北部、克什米尔地区、尼泊尔和巴基斯坦亦有。

Herbs perennial. Stems caespitose, densely glandular pubescent. Leaves elliptic or oblong, both surfaces long pilose, base attenuate, apex subacute. Cyme subumbellate, few flowered; bracts with narrowly scarious margin; pedicel densely glandular pubescent; sepals oblong-lanceolate, abaxially sparsely glandular pubescent, margin broadly membranous, apex subacute; petals white, broadly obovate, apex 2-lobed; styles 5. Capsule cylindric; seeds reniform, tuberculate. Forests, scrubs, mountain slope grasslands and mire meadows. Distributed in Xizang. Also in Afghanistan, NW India, Kashmir, Nepal and Pakistan.

繁缕

Stellaria media (L.) Vill

一年、二年生或多年生草本。茎匍匐或斜升，浅紫色。基部叶具长柄；叶片卵形至卵状圆形。花疏顶生或腋生为聚伞状；花梗花后伸长且下垂，细弱；花瓣长圆形，2裂至近基部。蒴果卵球形，6瓣裂。花期6-7月，果期7-8月。生海拔540-3700米的田间、路旁、山坡或林下。产中国大部分地区。印度、不丹、巴基斯坦、阿富汗、俄罗斯、朝鲜半岛、日本和欧洲亦有。

Annual, biennial or perennial herbs. Stems decumbent or ascending, pale purplish. Basal leaves long petiolate; blades ovate to ovate-orbicular. Flowers in sparse terminal or axillary cymes; pedicels elongate and nodding after anthesis, slender; petals oblong, 2-cleft nearly to base. Capsules ovoid, 6-valved. Fl. Jun-Jul. Fr. Jul-Aug. Fields, roadsides, slopes or forests at 540-3700 m. Distributed in most parts of China. Also in India, Bhutan, Pakistan, Afghanistan, Russia, Korean Peninsula, Japan and Europe.

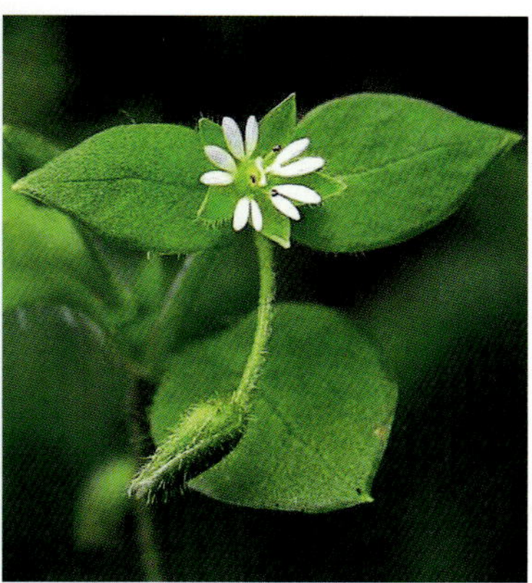
繁缕 *Stellaria media*

箐姑草
Stellaria vestita Kurz

多年生草本，被星状毛。茎疏丛生，铺散或匍匐。叶卵形或椭圆形至长圆状披针形，两面具星状毛。花疏聚伞状；花梗长，密具星状柔毛；花瓣5，2裂至近基部；雄蕊10。蒴果卵球形，6瓣裂。花期4-6月，果期6-8月。生海拔600-3600米的多石地、草坡或林中。产中国西南、东南、华中、华北、华西和华东。南亚和东南亚亦有。

中国繁缕 *Stellaria chinensis*

Perennial herbs, with stellate hairs. Stems sparsely tufted, diffuse or decumbent. Leaves ovate or elliptic to oblong-lanceolate, both surfaces stellate hairy. Flowers in sparse cymes; peduncles long, densely stellate pubescent; petals 5, 2-cleft nearly to base; stamens 10. Capsules ovoid, 6-valved. Fl. Apr-Jun. Fr. Jun-Aug. Stony places, grassy slopes or forests at 600-3600 m. Distributed in SW, SE, C, N, W and E China. Also in S and SE Asia.

中国繁缕
Stellaria chinensis Regel

多年生草本。茎铺散或匍匐，四棱状，无毛。叶卵形至卵状披针形，两面无毛。花疏聚伞状；花瓣5，与花萼近等长，2裂至近基部。蒴果卵球形，6瓣裂。花期5-6月，果期7-8月。生海拔(160-)500-1300(-2500)米的灌丛、林中或湿地。产中国西南、东南、华中、华北、华西和华东。

Perennial herbs. Stems diffuse or decumbent, quadrangular, glabrous. Leaves ovate to ovate-lanceolate, both surfaces glabrous. Flowers in sparse cymes; petals 5, nearly as long as sepals, 2-cleft nearly to base. Capsules ovoid, 6-valved. Fl. May-Jun. Fr. Jul-Aug. Scrubby areas, forests or wetlands at (160-)500-1300(-2500) m. Distributed in SW, SE, C, N, W and E China.

叉歧繁缕
Stellaria dichotoma L.

多年生草本，具腺毛。茎簇生，圆柱形，多二歧分枝。叶卵圆形或卵圆披针形。花多数，形成顶生聚伞状；花瓣5，倒披针形，2裂至近1/3或1/2。蒴果宽卵球形，6瓣裂。花期5-7月，果期7-8月。生海拔200-3100米的山坡、多石丘或沙丘。产华北、华西、西北和东北。俄罗斯(西伯利亚)和蒙古

箐姑草 *Stellaria vestita*

叉歧繁缕 *Stellaria dichotoma*

银柴胡 *Stellaria dichotoma* var. *lanceolata*

亦有。

Perennial herbs, glandular hairy. Stems tufted, terete, with numerous dichotomous branches. Leaves ovate or ovate-lanceolate. Flowers numerous, in terminal cymes; petals 5, oblanceolate, 2-cleft nearly to 1/3 or 1/2. Capsules broadly ovoid, 6-valved. Fl. May-Jul. Fr. Jul-Aug. Slopes, stony or sandy hills at 200-3100 m. Distributed in N, W, NW and NE China. Also in Russia (Siberia) and Mongolia.

银柴胡
Stellaria dichotoma var. **lanceolata** Bunge

多年生草本，具腺毛。茎簇生，圆柱形，多二歧分枝。叶线状披针形、披针形或长圆状披针形，顶端渐尖。花多数，形成顶生聚伞状；花瓣5，倒披针形，2裂至近1/3或1/2。蒴果宽卵球形，6瓣裂。花期6-7月，果期7-8月。生海拔1250-3100米的石质山坡或石质草原。产内蒙古、辽宁、陕西、甘肃和宁夏。蒙古和俄罗斯亦有。

Perennial herbs, glandular hairy. Stems tufted, terete, with numerous dichotomous branches. Leaves linear-lanceolate, lanceolate or oblong-lanceolate, apex acuminate. Flowers numerous, in terminal cymes; petals 5, oblanceolate, 2-cleft nearly to 1/3 or 1/2. Capsules broadly ovoid, 6-valved. Fl. Jun-Jul. Fr. Jul-Aug. Stony mountain slopes or stony grasslands at 1250-3100 m. Distributed in Neimenggu, Liaoning, Shaanxi, Gansu and Ningxia. Also in Mongolia and Russia.

沼生繁缕
Stellaria palustris Ehrh. ex Retz.

多年生草本。茎直立，四棱形，细弱，具乳突。叶无柄，粉绿色，线形至线状披针形，两面无毛。聚伞花序二歧分枝；苞片白色；花瓣5，2裂至近基部。蒴果卵球状圆柱形。花期6-7月，果期7-8月。生海拔1000-3600米的草坡或山谷疏林中。产中国西南、华北、华西和东北。阿富汗、哈萨克斯坦、俄罗斯、蒙古、日本和欧洲亦有。

Perennial herbs. Stems erect, quadrangular, slender, papillose. Leaves sessile, pinkish green, linear to linear-lanceolate, both surfaces glabrous. Flowers in dichotomous cymes; bracts white; petals 5, 2-cleft nearly to base. Capsules ovoid-cylindric. Fl. Jun-Jul. Fr. Jul-Aug. Grassy slopes or sparse valley forests at 1000-3600 m. Distributed in SW, N, W and NE China. Also in Afghanistan, Kazakhstan, Russia, Mongolia, Japan and Europe.

沼生繁缕 *Stellaria palustris*

千针万线草
Stellaria yunnanensis Franch.

多年生草本。茎直立，圆柱状。叶披针形或线状披针形，边有疏缘毛。花形成二歧聚伞状，无毛；花瓣5，2裂至近基部。蒴果卵球状圆形，稍短于宿存花萼，6瓣裂。花期7-8月，果期9-10月。生海拔1800-3300米的林下或林缘。产云南和四川。

Perennial herbs. Stems erect, terete. Leaves lanceolate or linear-lanceolate, margin sparsely ciliate. Flowers in dichotomous cymes, glabrous; petals 5, 2-cleft nearly to base. Capsules ovoid-orbicular, slightly shorter than persistent sepals, 6-valved. Fl. Jul-Aug. Fr. Sep-Oct. Forests or forest edges at 1800-3300 m. Distributed in Yunnan and Sichuan.

兴安繁缕
Stellaria cherleriae (Fisch. ex Ser.) Williams

多年生小草本。茎高5-12(-18)厘米，被短柔毛。叶条形或条状倒披针形，长(0.7-)1-2(-2.5)厘米，两面无毛，被缘毛。聚伞花序顶生；萼片5，狭卵形，长5毫米；花瓣5，白色，长约2毫米，2深裂；雄蕊10。花期6-7月。生海拔2800-3400米的砾石草坡、草原或疏林下。产陕西、山西、河北和内蒙古。蒙古和俄罗斯(西伯利亚)亦有。

千针万线草 *Stellaria yunnanensis*

兴安繁缕 *Stellaria cherleriae*

Small perennial herbs. Stems 5-12(-18) cm long, puberulous. Leaves linear or linear-oblanceolate, (0.7-)1-2(-2.5) cm tall, on both surfaces glabrous, ciliate. Cymes terminal; sepals 5, narrowly ovate, 5 mm long; petals 5, white, ca. 2 mm long, 2-parted; stamens 10. Fl. Jun-Jul. Stony slopes, steppes or sparse forests at 2800-3400 m. Distributed in Shaanxi, Shanxi, Hebei and Neimenggu. Also in Mongolia and Russia (Siberia).

无心菜 (蚤缀)
Arenaria serpyllifolia L.

一年生或二年生草本。茎密生白色短柔毛。叶无柄；叶片卵形，基部狭，边缘具缘毛，顶端急尖。聚伞花序，具多花；苞片草质，卵形，通常密生柔毛；花梗密生柔毛或腺毛；萼片披针形，长于花瓣和雄蕊，边缘膜质，顶端尖，外面被柔毛；花瓣白色，倒卵形；花柱3。蒴果卵圆形，与宿存萼等长。种子肾形，淡褐色，具瘤状凸起和乳突。花期6-8月，果期8-9月。生沙质或石质荒地、田野、园圃、山坡草地。产全国各地。欧洲、北非、亚洲和北美洲也广泛分布。

Annual or biennial herbs. Stems densely white villous. Leaves sessile; leaf blade ovate, base attenuate, margin ciliate, apex acute. Cymes many flowered; bracts ovate, herbaceous, usually densely villous; pedicel densely villous or glandular pubescent; sepals lanceolate, longer than petals and stamens, villous abaxially, margin membranous, apex acute; petals

无心菜(蚤缀) *Arenaria serpyllifolia*

white, obovate; styles 3. Capsule ovoid, equaling persistent sepals. Seeds pale brown, reniform, tuberculate with raised papillae. Fl. Jun-Aug. Fr. Aug-Sep. Sandy or stony barrens, fields, gardens, mountain grassland slopes. Widespread in China. Also in Europe, N Africa, Asia and North America.

老牛筋

Arenaria juncea M. Bieb.

多年生草本。根肉质。叶狭线形，具单脉，基部变宽，抱茎且具鞘，边缘疏具牙齿，具短缘毛，常反卷或平展。聚伞花序少至多数花；花瓣5，白色。蒴果卵球形，先端3瓣裂。花果期7-9月。生海拔800-2200米的林缘、草地、荒漠草原或岩石缝。产华北、华西和东北。俄罗斯、蒙古和朝鲜半岛亦有。

Perennial herbs. Roots fleshy. Leaves narrowly linear, 1-veined, base broadened, clasping and sheathing, margin sparsely dentate, shortly ciliate, often involute or flat. Cymes few- to many-flowered; petals 5, white. Capsules ovoid, apex 3-valved. Fl. and fr. Jul-Sep. Forest edges, grasslands, desert steppes or rock crevices at 800-2200 m. Distributed in N, W and NE China. Also in Russia, Mongolia and Korean Peninsula.

裸茎老牛筋

Arenaria griffithii Boiss.

多年生垫状草本。茎疏散，无毛，基部具枯萎叶基。叶密集；叶片线状钻形或钻形，顶端尖，边缘具缘毛。聚伞花序有1-3花；苞片卵形或卵状钻形；花梗无毛，稀被腺毛；萼片长圆状披针形，顶端尖，边缘膜质，无毛或被腺毛；花瓣白色，有时粉红色，倒卵形或卵状椭圆形，长为萼片的1.5-2倍，顶端钝圆；花盘具明显的5个腺体，花柱3。蒴果长圆形，3瓣裂，裂瓣顶端2裂。花期6-7月。生海拔2200-3000米的山地。产新疆。阿富汗和哈萨克斯坦亦有。

Perennial cashionlike herbs. Stems sparsely clustered, pulvinate, glabrous, with withered persistent leaves at base. Leaves crowded; leaf blade linear-subulate or subulate, apex acute, margin hairy. Cymes 1-3-flowered; bracts ovate or ovate-subulate; pedicel glabrous, rarely glandular hairy; sepals oblong-lanceolate, apex acute, margin membranous, glabrous or glandular hairy; petals white, sometimes pink, obovate or ovate-elliptic, 1.5-2 × as long as sepals, apex obtuse; floral disc with 5 impressed glands; styles 3. Capsule globose, 3-valved; valves 2-cleft at apex. Fl. Jun-Jul. Mountains at 2200-3000 m. Distributed in Xinjiang. Also in Afghanistan and Kazakhstan.

老牛筋 *Arenaria juncea*

裸茎老牛筋 *Arenaria griffithii*

雪灵芝

Arenaria brevipetala Y. W. Tsui et L. H. Zhou

多年生草本。主根强壮，木质。茎垫状，基部具多数枯萎、宿存的叶。花单生或成对，顶生；花梗具腺状长柔毛，先端下垂；花瓣5，白色，卵形；花盘杯状，具5腺体；子房球形。花期6-8月。生海拔3400-4600米的高山草甸或砾石地。产四川、西藏和青海。

Perennial herbs. Principal roots robust, woody. Stems pulvinate, with numerous withered, persistent leaves at base. Flowers solitary or paired, terminal; pedicels glandular villous, apex nutant; petals 5, white, ovate; floral disc cupular, with 5 glands; ovary globose. Fl. Jun-Aug. Alpine meadows or gravel patches at 3400-4600 m. Distributed in Sichuan, Xizang and Qinghai.

青藏雪灵芝

Arenaria roborowskii Maxim.

多年生垫状草本。茎紧密丛生，下部密集枯叶。叶针状线形，基部膜质，抱茎，边缘狭膜质，疏生缘毛，顶端急尖。花单生枝端；苞片线状披针形；花梗长5-10毫米，无毛；萼片披针形，长约5毫米，具1-3脉，边缘狭膜质；花瓣白色，椭圆形，基部楔形，顶端尖；花盘碟状，具大而明显的5个长圆形腺体；花药黄色；花柱3。花期7-8月。生高山草甸和流石滩。产青海、四川和西藏。

青藏雪灵芝 *Arenaria roborowskii*

Perennial cashionlike herbs. Stems densely clustered, pulvinate, with crowded, withered, persistent leaves at base. Leaves linear, base clasping, membranous, margin narrowly membranous, sparsely hairy, apex acute. Flower solitary, terminal; bracts linear-lanceolate; pedicel 5-10 mm long, glabrous; sepals lanceolate, ca. 5 mm long, 1-3-veined, margin narrowly membranous; petals white, elliptic, base cuneate, apex acute; floral disc patelliform, with 5 large, impressed glands; anthers yellow; styles 3. Fl. Jul-Aug. Alpine meadows and shifting screes. Distributed in Qinghai, Sichuan and Xizang.

雪灵芝 *Arenaria brevipetala*

藓状雪灵芝

Arenaria bryophylla Fernald

多年生垫状草本。茎密丛生，基部木质化，下部密集枯叶。叶片针状线形，呈三棱状，基部较宽，抱茎，边缘狭膜质，疏生缘毛，顶端急尖。花单生，无梗；苞片披针形，具1脉；萼片椭圆状披针形；花瓣白色，狭倒卵形，稍长于萼片；花盘碟状，具5个圆形腺体；雄蕊10，等长；花柱3。花期6-7月。生河滩石砾沙地、高山草甸和高山碎石带。产西藏和青海。印度北部和尼泊尔亦有。

Perennial cashionlike herbs. Stems clustered, pulvinate, base woody, below with crowded, withered, persistent leaves. Leaf blade linear, triangular in cross section, base clasping, broadened, margin narrowly membranous, sparsely ciliate, apex acute. Flower solitary, sessile; bracts

藓状雪灵芝 *Arenaria bryophylla*

山生福禄草 *Arenaria oreophila*

lanceolate, 1-veined; sepals elliptic-lanceolate; petals white, narrowly obovate, slightly longer than sepals; floral disc patellate, with 5 orbicular glands; stamens 10, equal; styles 3. Fl. Jun-Jul. Gravelly sands along rivers, alpine meadows and stony slopes. Distributed in Xizang and Qinghai. Also in N India and Nepal.

山生福禄草

Arenaria oreophila Hook. f. ex Edgew. et Hook. f.

多年生草本。茎密垫状，高4-9厘米，被腺毛。基部叶线形，膜质，边缘白色，质硬。单花顶生；花梗密具腺毛；花瓣5，白色。蒴果卵球形，与宿存花萼等长，3瓣裂。花期6-7月，果期7-8月。生海拔3500-5000米的高山草甸或流石滩。产云南、四川、西藏和青海。印度北部亦有。

Perennial herbs. Stems densely pulvinate, 4-9 cm tall, densely glandular hairy. Basal leaves linear, membranous, margin white, hard. Flowers solitary, terminal; pedicels densely glandular hairy; petals 5, white. Capsules ovoid, equaling persistent sepals, 3-valved. Fl. Jun-Jul. Fr. Jul-Aug. Alpine meadows or scree areas at 3500-5000 m. Distributed in Yunnan, Sichuan, Xizang and Qinghai. Also in N India.

密生福禄草

Arenaria densissima Wall. ex Edgew. et Hook. f.

多年生垫状草本。茎高4-5厘米，分枝多而密集。叶极密集，钻形，长5-10毫米，上面具槽。花单生枝端；萼片5，卵形，长2.5毫米；花瓣5，白色，宽匙形，长约5毫米；雄蕊10；花柱3。花期6-8月。生海拔3600-5250米的高山草甸或流石滩上。产西藏南部和东南部、四川西部和青海东南部。尼泊尔和印度北部亦有。

Perennial cashionlike herbs. Stems 4-5 cm tall, strongly and densely ramose. Leaves very dense, subulate, 5-10 mm long, adaxially sulcate. Flowers on branch apex, solitary; sepals 5, ovate, 2.5 mm long; petals 5, white, broadly spatulate, ca. 5 mm long; stamens 10; styles 3. Fl. Jun-Aug. Alpine meadows or shifting screes at 3600-5250 m. Distributed in S and SE Xizang, W Sichuan and SE Qinghai. Also in Nepal and N India.

密生福禄草 *Arenaria densissima*

团状福禄草 *Arenaria polytrichoides*

小腺无心菜 *Arenaria glanduligera*

团状福禄草

Arenaria polytrichoides Edgew. ex Edgew. et Hook. f.

多年生草本。茎密簇生，形成半球形垫状。叶近钻形，近基部疏具缘毛，先端具刺。单花顶生，无柄；花瓣5，白色，稍长于萼片；花药黄色。蒴果卵球形，3瓣裂。花期6-7月，果期8-9月。生海拔3500-5300米的高山草甸或流石滩。产四川、西藏和青海。印度北部亦有。

Perennial herbs. Stems densely clustered, forming hemispheric cushions. Leaves subulate, sparsely ciliate near base, apex spinose. Flowers solitary, terminal, sessile; petals 5, white, slightly longer than sepals; anthers yellow. Capsules ovoid, 3-valved. Fl. Jun-Jul. Fr. Aug-Sep. Alpine meadows or scree areas at 3500-5300 m. Distributed in Sichuan, Xizang and Qinghai. Also in N India.

小腺无心菜

Arenaria glanduligera Edgew.

多年生草本。茎丛生，上部被白色腺柔毛。叶片卵形、椭圆状长圆形或卵状披针形，顶端急尖或渐尖，基部渐狭呈柄状，边缘具缘毛，两面被白色腺柔毛。花1-2朵，生于茎顶端；苞片与叶同形而小；花梗疏被柔毛；萼片披针形或卵状披针形，顶端急尖或钝，外面被腺柔毛；花瓣紫红色，倒卵形或卵状椭圆形，顶端钝；花柱3。花期6-7月。生高山草甸或流石滩。产西藏。克什米尔地区、印度西北部和北部、尼泊尔亦有。

Perennial herbs. Stems clustered, above white glandular hairy. Leaf blade ovate, elliptic-oblong or ovate-lanceolate, apex acute or acuminate, base attenuate into petiole, margin ciliate, both surfaces white glandular pubescent. Flowers solitary or paired, terminal; bracts similar to leaves, smaller; pedicel sparsely villous; sepals lanceolate or ovate-lanceolate, apex obtuse or acute, glandular hairy abaxially; petals violet, obovate or ovate-elliptic, apex obtuse; styles 3. Fl. Jun-Jul. Alpine meadows or shifting screes. Distributed in Xizang. Also in Kashmir, NW and N India, and Nepal.

西南无心菜

Arenaria forrestii Diels

多年生草本。茎簇生，高1-15厘米。近轴叶鳞片状；远轴叶圆披针形或圆卵形，革质，无毛。花单生枝顶；萼片5，黄色；花瓣白色或粉红色；胚珠6-8。花期7-8月。生海拔2900-5300米的山坡、高山草甸或岩石缝。产云南、四川、西藏、甘肃和青海。尼泊尔亦有。

Perennial herbs. Stems clustered, 1-15 cm tall. Proximal cauline leaves scalelike; distal cauline leaves orbicular-lanceolate or orbicular-ovate, leathery, glabrous. Flowers solitary, terminal; sepals 5, yellow; petals white or pink; ovules 6-8. Fl. Jul-Aug. Slopes, alpine meadows or rock crevices at 2900-5300 m. Distributed in Yunnan, Sichuan, Xizang, Gansu and Qinghai. Also in Nepal.

滇藏无心菜

Arenaria napuligera Franch.

一年生草本。根芜菁状。茎基部分枝或不分枝，细弱，高

西南无心菜 *Arenaria forrestii*

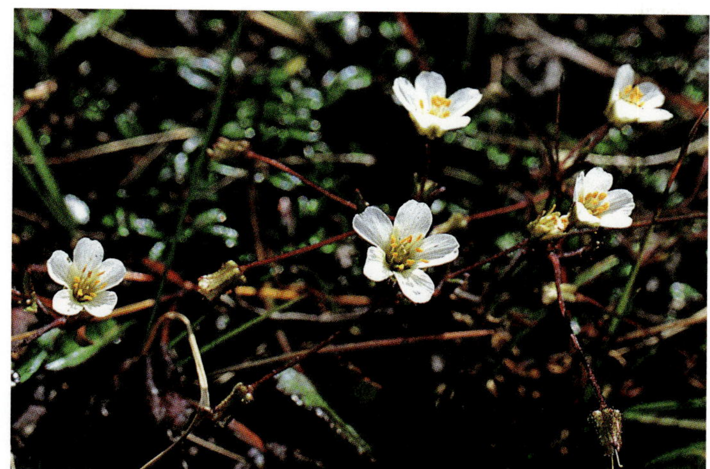
滇藏无心菜 *Arenaria napuligera*

5-15厘米，具白色或紫色腺毛。叶圆形、近圆形或线形。聚伞花序圆锥形，少至多花，或有时花单生；花瓣5，白色或粉红色；子房倒卵球形。花期6-7月。生海拔3000-5100米的高山草地或草甸。产云南、四川和西藏。

Annual herbs. Roots napiform. Stems branched proximally or unbranched, slender, 5-15 cm tall, white or violet glandular villous. Leaves orbicular, narrowly so, or linear. Cymes conic, few- to many-flowered, or sometimes flowers solitary; petals 5, white or pink; ovary obovoid. Fl. Jun-Jul. Alpine grasslands or meadows at 3000-5100 m. Distributed in Yunnan, Sichuan and Xizang.

真齿无心菜

Arenaria euodonta W. W. Sm.

多年生草本。茎直立或铺散，基部分枝，高10-35厘米，疏被长柔毛。花单生或聚伞花序具3-5花；花瓣白色或粉红色，倒披针形或倒卵形，顶端具2-3小齿；花药红色；子房球形。花果期7-8月。生海拔3000-4200米的高山草甸。产云南西北部。

Perennial herbs. Stems erect or diffuse, branched at base, 10-35 cm tall, sparsely villous. Flowers solitary or cymes 3-5-flowered; petals white or pink, oblanceolate or obovate, apex 2-3-toothed; anthers red; ovary globose. Fl. and fr. Jul-Aug. Alpine meadows at 3000-4200 m. Distributed in NW Yunnan.

真齿无心菜 *Arenaria euodonta*

长柱无心菜

Arenaria longistyla Franch.

矮小草本。茎纤细，高4-10厘米，被2行柔毛或褐色腺柔毛。叶在茎上部密聚；叶片宽线形或线状披针形，基部连合呈鞘状，边缘被稀疏的缘毛，顶端具短尖头。花腋生；花梗被腺柔毛；萼片披针形，基部在花期后呈囊状，边缘为白色膜质，顶端急尖；花瓣白色，有时淡红色，倒卵状长圆形，顶端钝圆；雄蕊稍短于花瓣；花柱2，长5-7毫米，钻形。花期6-7月。生林缘、高山草甸、山坡草地和流石滩。产云南西北部、四川西部和西藏东南部。

Small herbs. Stems slender, 4-10 cm tall, villous in 2 lines or brown glandular villous. Leaves aggregated at stem apex; leaf blade broadly linear or linear-lanceolate, base connate into a sheath, margin sparsely hairy, apex mucronate. Flowers axillary; pedicel glandular pubescent; sepals lanceolate, base saccate after anthesis, margin broadly white membranous, apex acute; petals white, sometimes pale red, obovate-oblong, apex obtuse; stamens slightly shorter than petals; styles 2, subulate, 5-7 mm long. Fl. Jun-Jul. Forest margins, alpine meadows, mountain grasslands and shifting screes. Distributed in NW Yunnan, W Sichuan and SE Xizang.

长柱无心菜 *Arenaria longistyla*

四齿无心菜

Arenaria quadridentata (Maxim.) Williams

草本植物。根纺锤形。茎丛生，高10-40厘米。下部叶匙形，上部叶卵状椭圆形或披针形，长1-2厘米。聚伞花序具少数花；萼片5，长圆形，长4-5毫米；花瓣5，白色，倒卵形或楔形，长约8毫米，顶端4齿裂；雄蕊10；花柱2，长5毫米。花期7-9月。生海拔3000-3500米的草坡上。产四川北部和甘肃南部。

Herbs. Roots fusiform. Stems caespitose, 10-40 cm tall. Lower leaves spatulate, upper leaves elliptic or lanceolate, 1-2 cm long. Cymes few-flowered; sepals 5, oblong, 4-5 mm long; petals 5, white, obovate or cuneiform, ca. 8 mm long, apex 4-dentate; stamens 10; styles 2, 5 mm long. Fl. Jul-Sep. Grassy slopes at 3000-3500 m. Distributed in N Sichuan and S Gansu.

滇蜀无心菜

Arenaria dimorphitricha C. Y. Wu ex L. H. Zhou

一年生或二年生草本。茎劲直，少分枝，褐色，下部被长毛，上部被具节腺毛，并疏生长柔毛。叶具短柄；叶片卵状长圆形、长椭圆形或长圆状披针形，稀椭圆形，两面被长柔毛，基部渐狭，顶端钝。聚伞花序具多花；苞片卵形；花梗密被腺柔毛；萼片披针形，顶端尖，边缘膜质，外面密生腺柔毛；花瓣白色，倒卵形，长为萼片的1.5倍，顶端具不整齐的齿裂；花丝稍长于萼片，花药黄绿色或黄褐色；花柱2。蒴果稍短于宿存萼，顶端4裂。种子黑褐色，具皱纹。花果期8-9月。生亚高山混交林、灌丛或山坡草地。产云南西北部和四川西南部。

Annual or biennial herbs. Stems erect, sparingly branched, brown, below long hairy, above nodose hairy and sparsely villous. Leaves stipitate; leaf blade ovate-oblong, narrowly elliptic or oblong-lanceolate, rarely elliptic, both surfaces villous, base attenuate, apex obtuse. Cymes many flowered; bracts ovate; pedicel densely glandular pubescent; sepals lanceolate, glandular pubescent abaxially, margin membranous, apex acute; petals white, obovate, ca. 1.5 × as long as sepals, apex irregularly toothed; filaments slightly longer than sepals, anthers blue-green to yellow; styles 2. Capsule slightly shorter than persistent sepals, apex 4-lobed. Seeds black, rugose. Fl. and fr. Aug-Sep. Subalpine forests, scrubs or mountain grasslands. Distributed in NW Yunnan and SW Sichuan.

髯毛无心菜 *Arenaria barbata*

髯毛无心菜

Arenaria barbata Franch.

多年生草本，全株被长节毛和短腺毛。根簇生。茎常单生，中、下部分枝。叶片长圆状倒卵形或长圆形，基部渐狭，顶端钝或急尖，边缘具白色长缘毛，两面密被腺毛。二歧状聚伞花序；苞片与叶同形而小；花梗密被腺柔毛；萼片披针形，具3脉，外面密被腺柔毛；花瓣5，白色或粉红色，长为萼片的2倍以上，顶端流苏状；雄蕊10，其中5枚的花丝基部扩大；花药紫黑色或黄褐色；花柱2。蒴果4裂。花果期7-9月。生高山草甸、流石滩、林间草地或灌丛。产云南和四川。

Perennial herbs, long nodose hairy and shortly glandular hairy. Roots clustered. Stems usually solitary, branched below middle. Leaf blade oblong-obovate or oblong, base attenuate, apex obtuse or acute, margin long white ciliate, both surfaces glandular pubescent. Cymes dichotomously branched; bracts leaflike, small; pedicel densely glandular pubescent; sepals lanceolate, 3-veined, densely glandular pubescent abaxially; petals 5, white or pink, more than 2 × as long as sepals, apex fimbriate; stamens 10, 5 filaments inflated at base; anthers dark violet or yellow-brown; styles 2. Capsule 4-valved. Fl. and fr. Jul-Sep. Alpine meadows, shifting screes, grasslands or scrubs. Distributed in Yunnan and Sichuan.

四齿无心菜 *Arenaria quadridentata*

滇蜀无心菜 *Arenaria dimorphitricha*

种阜草 *Moehringia lateriflora*

新疆种阜草 *Moehringia umbrosa*

种阜草

Moehringia lateriflora (L.) Fenzl

多年生草本。茎高10-20厘米，被短毛。叶椭圆形或长圆形，长1-2.5厘米。聚伞花序顶生或侧生，萼片5，长2毫米；花瓣5，白色，椭圆状倒卵形，长4-5毫米；雄蕊10；花柱3。种子长1毫米，具白色种阜。花期6月。生海拔780-2300米的林缘。产河北、山西、宁夏、内蒙古、辽宁、吉林和黑龙江。朝鲜半岛、日本、蒙古、俄罗斯(西伯利亚)、哈萨克斯坦和欧洲亦有。

Perennial herbs. Stems 10-20 cm tall, puberulous. Leaves elliptic or oblong, 1-2.5 cm long. Cymes terminal or lateral; petals 5, white. 4-5 mm long; stamens 10; styles 3. Seed 1 mm long, with a white strophiole. Fl. Jun. Forest margin at 780-2300 m. Distributed in Hebei, Shanxi, Ningxia, Neimenggu, Liaoning, Jilin and Heilongjiang. Also in Korean Peninsula, Japan, Mongolia, Russia (Siberia), Kazakhstan and Europe.

新疆种阜草

Moehringia umbrosa (Bunge) Fenzl

多年生小草本，茎丛生，高5-18厘米。叶长圆状披针形或披针形，长1-3厘米。花单生叶腋或茎顶端；萼片5，长2-3毫米；花瓣5，白色，长圆状倒卵形，比萼片长2-2.5倍；雄蕊10；花柱3。花期6-7月。生海拔2000米的草坡上。产新疆。哈萨克斯坦和俄罗斯(西伯利亚)亦有。

Small perennial herbs. Stems caespitose, 5-18 cm tall. Leaves oblong-lanceolate or lanceolate, 1-3 cm long. Flowers solitary, terminal or axillary; sepals 5, 2-3 mm long; petals 5, white, oblong-obovate, 2-2.5 × longer than sepals; stamens 10; styles 3. Fl. Jun-Jul. Grassy slopes at 2000 m. Distributed in Xinjiang. Also in Kazakhstan and Russia (Siberia).

短瓣花

Brachystemma calycinum D. Don

一年生草本。茎铺散或在灌木上攀爬，长达6米，4棱。叶先端渐尖。聚伞圆锥花序疏松，多花；萼片和花瓣均5；花瓣白色，长为花萼的1/3-1/2，边缘全缘。蒴果具1个种子。花期4-7月，果期8-12月。生海拔500-2700米的疏林或草坡。产云南、四川、西藏、贵州和广西。南亚亦有。

Annual herbs. Stems diffuse or climbing among shrubs, up to 6 m long, 4-angled. Leaves apex acuminate. Thyrses lax, many-flowered; sepals and petals 5; petals white, 1/3-1/2 as long as sepals, margin entire. Capsules 1-seeded. Fl. Apr-Jul. Fr. Aug-Dec. Open forests or grassy slopes at 500-2700 m. Distributed in Yunnan, Sichuan, Xizang, Guizhou and Guangxi. Also in S Asia.

短瓣花 *Brachystemma calycinum*

漆姑草 *Sagina japonica*

漆姑草
Sagina japonica (Swartz) Ohwi

一年生或二年生草本。茎丛生，近直立或匍匐，基部分枝，顶端具腺毛。叶线形，无毛，基部贴生。花单生，顶生或腋生；花梗和萼片外面被腺柔毛；花瓣5，白色；花柱5。蒴果球形，5瓣裂。花期4-5月，果期5-6月。生海拔(100-)600-1900(-4000)米的沙质土河边、冲积地、路边草地或林中。产中国大部分地区。印度、尼泊尔、不丹、俄罗斯、朝鲜半岛和日本亦有。

Annual or biennial herbs. Stems Caespitose, suberect or creeping, basally branched, apically glandular hairy. Leaves linear, glabrous, connate at base. Flowers solitary, terminal or axillary; pedicels and sepals abaxially glandular-pubescent; petals 5, white; styles 5. Capsules globose, 5-valved. Fl. Apr-May. Fr. May-Jun. Sandy riversides, floodlands, grasslands by roadsides, or forests at (100-)600-1900(-4000) m. Distributed in most parts of China. Also in India, Nepal, Bhutan, Russia, Korean Peninsula and Japan.

长白米努草
Minuartia macrocarpa (Pursh) var. **koreana** (Nakai) Hara

多年生小草本。茎高4-13厘米，具密集分枝。下部叶密集，条形或钻状条形，上部叶狭卵形，长0.5-1.5厘米，具3脉。花多单朵顶生；萼片5，长4-5毫米，顶端钝；花瓣5，白色，倒卵状长圆形，长7-9毫米；雄蕊10；花柱3。花期6-8月。生海拔2400米的石砾山坡或石上苔藓丛中。产吉林东南部。朝鲜半岛北部亦有。

Small perennial herbs. Stems 4-13 cm tall, densely branched. Lower leaves dense, linear or subulate-linear, upper leaves narrowly ovate, 0.5-1.5 cm long, 3-nerved. Flowers usually solitary, terminal; sepals 5, 4-5 mm long, apex obtuse; petals 5, white, obovate-oblong, 7-9 mm long; stamens 10; styles 3. Fl. Jun-Aug. Stony slopes or in mosses on stones at 2400 m. Distributed in SE Jilin. Also in N Korean Peninsula.

新疆米努草
Minuartia kryloviana Schischk.

多年生小草本。茎高5-18(-20)厘米，分枝，无毛。叶条形，长0.5-15厘米，具3脉。花组成聚伞花序；花梗长1.5-3厘米；萼片5，长3.5-5毫米，顶端急尖；花瓣5，白色，长圆状倒卵形，长约7毫米；雄蕊10；花柱3。花期5-6月。生海拔2150-3400米的山坡上或岩石上。产新疆。阿富汗、哈萨克斯坦和俄罗斯(西西伯利亚)亦有。

Small perennial herbs. Stems 5-18(-20) cm tall, branched, glabrous. Leaves linear, 0.5-1.5 cm long, 3-nerved. Flowers in cymes; pedicels 1.5-3 cm long; sepals 5, 3.5-5 mm long, apex acute; petals 5, white, oblong-obovate, ca. 7 mm long; stamens 10; styles 3. Fl. May-Jun. Slopes or on rocks at 2150-3400 m. Distributed in Xinjiang. Also in Afghanistan, Kazakhstan and Russia (W Siberia).

长白米努草 *Minuartia macrocarpa* var. *koreana*

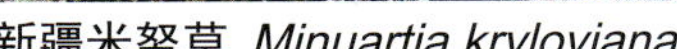
新疆米努草 *Minuartia kryloviana*

麦仙翁 *Agrostemma githago*

剪秋罗 *Lychnis fulgens*

麦仙翁

Agrostemma githago L.

一年生草本，具长平伏浅灰色毛。茎不分枝。叶线形或线状披针形。花单生；花梗极长；花瓣檐部粉色，短于萼齿，爪白色，冠檐深红色，凹缺。蒴果卵球形。花期6-8月，果期7-9月。生田中或路边。产内蒙古、新疆、黑龙江和吉林。原生地中海地区。北亚、北非、欧洲和北美洲亦有。

Annual herbs, with long appressed grayish hairs. Stems unbranched. Leaves linear or linear-lanceolate. Flowers solitary; pedicels very long; petal limbs pink, shorter than calyx teeth, claws white, limbs dark red, emarginate. Capsules ovoid. Fl. Jun-Aug. Fr. Jul-Sep. Fields or roadsides. Distributed in Neimenggu, Xinjiang, Heilongjiang and Jilin. Native to Mediterranean regions. Also in N Asia, N Africa, Europe and North America.

剪秋罗

Lychnis fulgens Fisch. ex Spreng.

多年生草本，具柔毛混生疏多室无腺毛。根丛生，纺锤状，稍肉质。叶卵形或卵状披针形。花大，直径3.5-5厘米；花瓣檐部绯红色，2裂到中部，爪窄披针形，不伸出花萼外；雌雄蕊柄长约5毫米。花期6-7月，果期8-9月。生低山疏林下或潮湿灌丛草甸。产中国西南、华北和东北。俄罗斯(西伯利亚、远东地区)、朝鲜半岛和日本亦有。

Perennial herbs, pilose with sparse multicellular eglandular hairs. Roots caespitose, fusiform, slightly fleshy. Leaves ovate or ovate-lanceolate. Flowers large, 3.5-5 cm diam; petal limbs crimson-red, 2-lobed to the middle, claws narrowly lanceolate, not exceeding calyx; androgynophore ca. 5 mm long. Fl. Jun-Jul. Fr. Aug-Sep. Sparse forests on low hills or wet shady scrub meadows. Distributed in SW, N and NE China. Also in Russia(Siberia, Far East), Korean Peninsula and Japan.

剪红纱花

Lychnis senno Sieb et Zucc.

多年生草本，具柔毛混生多室无腺毛。根丛生，灰黄色，狭圆柱状。叶椭圆状披针形，两面具柔毛。二歧聚伞花序，具多花；花萼管状至狭漏斗形；花瓣檐部深红色，不规则深裂为数个锯齿状裂片；花药黑紫色。蒴果椭圆状卵球形。花期7-8月，果期8-9月。生海拔100-2000米的疏林下或灌丛草地。产中国西南、华中、华北和华东。日本亦有。

Perennial herbs, pilose with sparse multicellular eglandular hairs. Roots caespitose, pale yellow, narrowly cylindric. Leaves elliptic-lanceolate, both surfaces pubescent. Dichasium many-flowered; calyx tubular to narrowly funnel-shaped; petal limbs deep red, irregularly parted into numerous incised-dentate lobes; anthers dark purple. Capsules ellipsoid-ovoid. Fl. Jul-Aug. Fr. Aug-Sep. Sparse forests or scrub grasslands at 100-2000 m. Distributed in SW, C, N and E China. Also in Japan.

剪红纱花 *Lychnis senno*

剪春罗 *Lychnis coronata*

蔓茎蝇子草 *Silene repens*

剪春罗

Lychnis coronata Thunb.

多年生草本，近无毛。根丛生，浅黄色，狭纺锤形。叶卵状披针形，两面近无毛。二歧聚伞状花序，花数朵；花瓣片橙红色，顶端有不规则缺刻状齿，爪不伸出花萼；雌雄蕊柄长1-1.5厘米。蒴果狭卵球形。花期6-7月，果期8-9月。生疏林中或灌丛草地，也常栽培。产中国西南、东南和华东。日本亦有。

Perennial herbs, subglabrous. Roots caespitose, pale yellow, narrowly fusiform. Leaves ovate-lanceolate, both surfaces subglabrous. Dichasium several-flowered; petal limbs salmon-pink, apex irregularly incised-dentate, claws not exceeding calyx; androgynophore 1-1.5 cm long. Capsules narrowly ellipsoid. Fl. Jun-Jul. Fr. Aug-Sep. Sparse forests or scrub grasslands, also cultivated. Distributed in SW, SE and E China. Also in Japan.

鹤草

Silene fortunei Vis.

多年生草本。茎丛生，直立，多分枝。基生叶倒披针形或披针形，花期枯萎。花直立，呈疏而少花的聚伞圆锥状；花萼窄筒状，无毛，在果期上部膨大成棍棒状；花柱3。蒴果短于或几等于花萼。花期6-8月，果期7-9月。生灌丛、高原或低山草地。产中国西南、东南、华北、华西和华东。

Perennial herbs. Stems caespitose, erect, multibranched. Basal leaves oblanceolate or lanceolate, withered at anthesis. Flowers erect, in a lax few-flowered thyrse; calyx narrowly tubular, glabrous, inflated above and clavate in fruit; styles 3. Capsules shorter than or equaling calyx. Fl. Jun-Aug. Fr. Jul-Sep. Scrubby areas, plateaus or scrub grasslands in hill. Distributed in SW, SE, N, W and E China.

蔓茎蝇子草

Silene repens Patr.

多年生草本。根状茎细长，分叉。茎基部常木质化，有时具腋生的不育短枝。叶片线状披针形、披针形、倒披针形或长圆状披针形，基部楔形，两面被柔毛，边缘具缘毛。总状圆锥花序。小聚伞花序常具1-3花；苞片披针形；花萼筒形至棒形，被柔毛；雌雄蕊柄长4-8毫米，被短柔毛；花瓣白色，爪倒披针形，无耳，瓣片浅2裂或深达其中部；副花冠片长圆状；雄蕊微外露；花柱3。蒴果卵形，比宿存萼短。花期6-8月，果期7-9月。生林下、草地、溪岸和沙丘。产甘肃、河北、吉林、内蒙古、陕西、四川和西藏。日本、朝鲜半岛、蒙古、俄罗斯和北美洲西北部也有。

Perennial herbs. Rhizome slender, branched. Stems basally usually lignified, sometimes with short sterile branches at axil. Leaves linear-lanceolate, lanceolate, oblanceolate, or oblong-lanceolate, sparsely pubescent, margin ciliate, base cuneate. Flowers in a racemiform thyrse, cymules usually 1-3-flowered; bracts lanceolate; calyx tubular to clavate, pubescent; androgynophore 4-8 mm long, shortly pubescent; petal white, claws oblanceolate, without auricles, limbs bifid to middle or less; coronal scales oblong; stamens slightly exserted; styles 3.

鹤草 *Silene fortunei*

Capsule ovoid, shorter than calyx. Fl. Jun-Aug. Fr. Jul-Sep. Forests, grasslands, stream edges and sand dunes. Distributed in Gansu, Hebei, Jilin, Neimenggu, Shaanxi, Sichuan and Xizang. Also in Japan, Korean Peninsula, Mongolia, Russia and NW North America.

山蚂蚱草

Silene jenisseensis Willd.

多年生草本。基生叶多数，狭倒披针形或披针状线形。总状聚伞花序；花萼狭钟形，果期稍膨大；花瓣片白色或淡绿色；雌雄蕊柄长约2毫米，被短毛；花柱3。蒴果卵球形。花期7-8月，果期8-9月。生海拔200-1000米林缘、林中草地、高山岩峭、高山草地、砾石草地或沙丘边缘。产华北和东北。俄罗斯、蒙古和朝鲜半岛亦有。

Perennial herbs. Basal leaves numerous, narrowly oblanceolate or lanceolate-linear. Flowers in a racemiform thyrse; calyx narrowly campanulate, slightly inflated in fruit; petal limbs white or pale greenish; androgynophore ca. 2 mm long, shortly hairy; styles 3. Capsules ovoid. Fl. Jul-Aug. Fr. Aug-Sep. Forest edges, forested grasslands, alpine screes, mountain grasslands, gravelly grasslands or dune edges at 200-1000 m. Distributed in N and NE China. Also in Russia, Mongolia and Korean Peninsula.

细蝇子草 *Silene gracilicaulis*

山蚂蚱草 *Silene jenisseensis*

细蝇子草

Silene gracilicaulis C. L. Tang

多年生草本。基部叶多数，线状倒披针形，两面无毛。总状聚伞圆锥花序；花瓣爪倒披针形，无毛，耳呈三角状，瓣片露出花萼，白色，下部紫色或粉红色，2裂至中或下部；花柱3。花期7-8月，果期8-9月。生海拔3000-4000米的沙砾草地或高山。产云南、四川、西藏、青海和内蒙古。

Perennial herbs. Basal leaves numerous, linear-oblanceolate, both surfaces glabrous. Flowers in a racemiform thyrse; petal claws oblanceolate, glabrous, auricles triangular, limbs exserted beyond calyx, white, violet or pink below, bifid to middle or below; styles 3. Fl. Jul-Aug. Fr. Aug-Sep. Gravelly grasslands or mountains at 3000-4000 m. Distributed in Yunnan, Sichuan, Xizang, Qinghai and Neimenggu.

长梗蝇子草 *Silene pterosperma*

Perennial herbs. Basal leaves caespitose, linear or narrowly oblanceolate, 5-10 mm wide. Flowers solitary, rarely 2-3, slightly nutant, densely glandular villous; calyx globose, saccate, membranous, slightly contracted at mouth, with prominent dark violet veins, densely glandular villous, loose in fruit; petals blackish violet. Capsules subglobose. Fl. Jul-Aug. Fr. Aug-Sep. Alpine meadows or gravel grasslands at 3000-4500 m. Distributed in NW Yunnan and SW Sichuan.

林芝蝇子草

Silene wardii (Marq.) Bocquet

多年生草本。茎紫色，基部匍匐，疏丛生，常不分枝，密被腺柔毛。茎生叶无柄，椭圆形，顶端急尖，两面和边缘具腺毛。花单生，稀2-3朵；花梗密被腺毛；花萼紫色，卵状宽钟形，基部脐形，沿脉密被腺毛，萼齿长5-7毫米，顶端钝，边缘具腺缘毛；雌雄蕊柄被长绵毛状柔毛；花瓣紫红色，爪长约1.7厘米，瓣片深2裂，裂片具齿或细圆齿；副花冠片楔形，顶端啮蚀状；花丝下部被绵毛。蒴果卵形，5齿裂。种子具翅。花期7-8月。生流石滩和冰河淤积扇。产西藏东南部。

长梗蝇子草

Silene pterosperma Maxim.

多年生草本。茎高20-50厘米，丛生。基生叶簇生，倒披针状条形，长1.5-3厘米；茎生叶1-2对，较小。总状花序常具对生的花；花梗丝形，比花萼长；花萼狭钟状，长8-9毫米；雌蕊柄长2毫米；花瓣黄白色，爪倒披针形，瓣片伸出，狭长圆形，2深裂；雄蕊10，内藏；花柱3。花期7月。生海拔2100-3700(-4000)米的林缘或灌丛中。产四川、青海、甘肃、陕西和内蒙古。

Perennial herbs. Stems 20-50 cm tall, caespitose. Basal leaves fasciculate, oblanceolate-linear 1.5-3 cm long; cauline leaves 1-2 pairs, smaller. Racemes with often opposite flowers; pedicels filiform, longer than calyx; calyx narrowly campanulate, 8-9 mm long; androgynophore 2 mm long; petals yellow-white, claws oblanceolate, limbs exserted, narrowly oblong, 2-parted; stamens 10, included; styles 3. Fl. Jul. Forest margin or thickets at 2100-3700(-4000) m. Distributed in Sichuan, Qinghai, Gansu, Shaanxi and Neimenggu.

宽叶变黑蝇子草

Silene nigrescens (Edgew. et Hook. f.) Majumdar subsp. **latifolia** Bocquet

多年生草本。基部叶丛生，线形或狭倒披针形，宽5-10毫米。花单生，稀2-3朵，稍下垂，密具腺状长柔毛；花萼球形，囊状，膜质，口部稍收缩，具突出的明显的紫色脉纹，密具腺状长柔毛，果时变疏松；花瓣黑紫色。蒴果近球状。花期7-8月，果期8-9月。生海拔3000-4500米的高山草甸或砾石草地。产云南西北部和四川西南部。

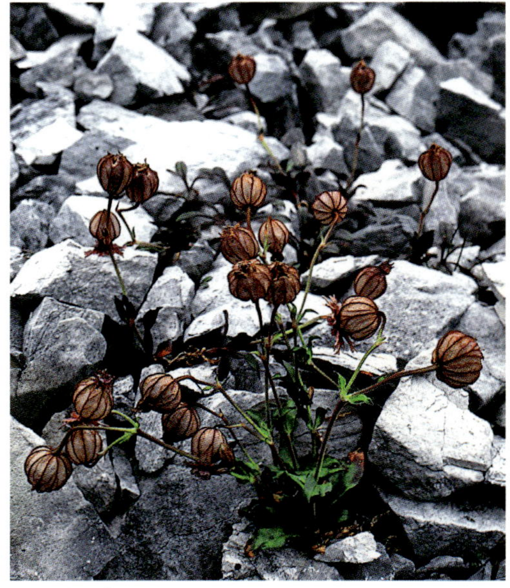
宽叶变黑蝇子草 *Silene nigrescens* subsp. *latifolia*

林芝蝇子草 *Silene wardii*

Perennial herbs. Stems purple, base procumbent, sparsely clustered, usually simple, densely glandular hairy. Cauline leaves sessile, elliptic, both surfaces and margin glandular hairy, apex acute. Flowers solitary, rarely 2-3; pedicel densely glandular hairy; calyx purple, ovoid-campanulate, base umbilicate, densely glandular hairy at veins, calyx teeth 5-7 mm long, margin glandular ciliate, apex obtuse; androgynophore long lanate-villous; petals purple, claws ca. 1.7 cm long, limbs deeply bifid, lobes dentate or with small round teeth; coronal scales cuneate, apex erose; filaments woolly-hairy at base. Capsule ovoid, 5-toothed. Seeds winged. Fl. Jul-Aug. Shifting screes and sandy gravelly deltas of glacier streams. Distributed in SE Xizang.

垫状蝇子草

Silene kantzeensis C. L. Tang

多年生垫状草本。茎紧密丛生，极短。基生叶倒披针状线形，无毛，基部渐狭，顶端渐尖或急尖，边缘具粗短缘毛；茎生叶1-2对或无。花单生，直立，直径1.5-2厘米；花梗密被短柔毛；花萼狭钟形或筒状钟形，被紫色腺毛，纵脉紫色，萼齿三角状卵形，顶端钝，边缘膜质，具缘毛；雌雄蕊柄无毛，长约1毫米；花瓣淡紫色或淡红色，爪狭楔形，耳卵形，瓣片露出花萼，叉状深2裂达瓣片中部；副花冠片倒卵形，全缘或具缺刻。蒴果圆柱形或圆锥形，微长于宿存萼。花期7-8月，果期9-10月。生高山草甸。产青海、四川、西藏和云南。

Perennial cashionlike herbs. Stems densely clustered, very short. Basal leaves oblanceolate-linear, glabrous, base attenuate, apex acuminate or acute, margin coarsely and shortly ciliate; cauline leaves 1-2 pairs or absent. Flower solitary, erect, 1.5-2 cm diam; pedicel densely pubescent; calyx narrowly campanulate or cylindric-campanulate, purple glandular hairy, longitudinal veins purple, calyx teeth triangular-ovate, margin membranous, ciliate, apex obtuse; androgynophore ca. 1 mm long, glabrous; petals lilac or pale red, claws narrowly cuneate, auricles ovoid, limbs exserted beyond calyx, bifid deeply to middle; coronal scales obovoid, margin entire or laciniate. Capsule cylindric or conical, slightly longer than calyx. Fl. Jul-Aug. Fr. Sep-Oct. Alpine meadows. Distributed in Qinghai, Sichuan, Xizang and Yunnan.

湖北蝇子草

Silene hupehensis C. L. Tang

多年生草本。茎高10-30厘米，丛生，无毛。基生叶条形，长5-8厘米；茎生叶少数，较小。聚伞花序具2-5花；花梗长2-5厘米；花萼钟状；长1.2-1.5厘米，无毛；雌雄蕊柄长3-4毫米；花瓣淡红色，长1.5-2厘米，2浅裂；雄蕊稍伸出；花柱3。花期7月。生海拔1200-2700米的草坡上或多石处。产四川、甘肃、陕西、湖北和河南。

Perennial herbs. Stems 10-30 cm tall, caespitose, glabrous. Basal leaves linear, 5-8 cm long; cauline leaves few, smaller. Cymes 2-5-flowered; pedicels 2-5 cm long; calyx campanulate, 1.2-1.5 cm long, glabrous; androgynophore 3-4 mm long; petals reddish, 1.5-2 cm long, 2-lobed; stamens slightly exserted; styles 3. Fl. Jul. Grassy slopes or on rocks at 1200-2700 m. Distributed in Sichuan, Gansu, Shaanxi, Hubei and Henan.

垫状蝇子草 *Silene kantzeensis*

湖北蝇子草 *Silene hupehensis*

女娄菜 *Silene aprica*

坚硬女娄菜(粗壮女娄菜) *Silene firma*

女娄菜

Silene aprica Turcz. ex Fisch. et C. A. Mey.

一年生或二年生草本，全株密被灰柔毛。基部叶倒披针形或狭匙形。小聚伞花序具柄；花梗直立；花萼卵球状钟形，近草质，密具柔毛；花瓣爪下面具缘毛，冠檐白色或粉色。蒴果卵球形。花期5-7月，果期6-8月。生平原、丘陵或山地。产中国各地。俄罗斯、蒙古、朝鲜半岛和日本亦有。

Annual or biennial herbs, densely gray pubescent throughout. Basal leaves oblanceolate or narrowly spatulate. Cymules stalked; pedicels erect; calyx ovoid-campanulate, nearly herbaceous, densely pubescent; petal claws ciliate below, limbs white or pink. Capsules ovoid. Fl. May-Jul. Fr. Jun-Aug. Plains, hills or mountains. Distributed throughout China. Also in Russia, Mongolia, Korean Peninsula and Japan.

坚硬女娄菜(粗壮女娄菜)

Silene firma Sieb. et Zucc.

一年生或二年生草本。茎高50-100厘米，无毛。叶椭圆状披针形，长4-10厘米。花组成不规则聚伞圆锥花序；花梗长5-18(-30)毫米；花萼卵状钟形，长7-9毫米；雌雄蕊柄极短；花瓣5，白色，不伸出，爪倒披针形，瓣片倒卵形，2裂；雄蕊内藏；花柱3。花期6-7月。生海拔300-2500米的草坡上或灌丛中。广布中国北部和长江流域。俄罗斯(远东地区)，朝鲜半岛和日本亦有。

Annual or biennial herbs. Stems 50-100 cm tall, glabrous. Leaves elliptic-lanceolate, 4-10 cm long. Flowers in irregular thyrses; pedicels 5-18(-30) mm long; calyx ovate-campanulate, 7-9 mm long ; androgynophore very short; petals 5, white, not exserted, claws oblanceolate, limbs obovate, 2-lobed; stamens included; styles 3. Fl. Jun-Jul. Grassy slopes or in thickets at 300-2500m. Widespread in N China and the Yangtze River baasin. Also in Russia (Far East), Korean Peninsula and Japan.

毛萼蝇子草

Silene pubicalycina C. Y. Wu

多年生草本。茎不分枝，被长柔毛，上部被腺毛。基生叶匙状倒披针形，基部渐狭成长柄状，顶端急尖，稀圆形，具凸尖，两面和边缘具长柔毛；茎生叶倒披针形或披针形。花序具数花；花梗被柔毛和腺毛；苞片卵状披针形，密被柔毛；花萼钟形，直径4-6毫米，密被

毛萼蝇子草 *Silene pubicalycina*

柔毛，纵脉淡紫色，脉端连结，萼齿三角状卵形，具缘毛；雌雄蕊柄长2-3毫米，无毛；花瓣淡红色，爪倒披针形，无毛，具耳，瓣片露出花萼，深2裂几达瓣片中部，裂片狭卵形，全缘；副花冠片长约1.5毫米，具数缺刻状齿。蒴果卵形，比宿存萼短。花期7-8月，果期8-9月。生林下。产云南、四川和西藏。

Perennial herbs. Stems simple, villous, above glandular hairy. Basal leaves spatulate-oblanceolate, base attenuate into long petiole, apex acute, rarely rounded, mucronulate, both surfaces and margin villous; cauline leaves oblanceolate or lanceolate. Cymes few flowered; pedicel villous and glandular hairy; bracts ovate-lanceolate, densely villous; calyx campanulate, 4-6 mm diam, densely villous, longitudinal veins lilac, cohering at apex, calyx teeth triangular-ovate, ciliate; androgynophore 2-3 mm long, glabrous; petals reddish, claws oblanceolate, glabrous, with auricles, limbs exserted beyond calyx, deeply bifid to middle, lobes narrowly ovate, entire; coronal scales ca. 1.5 mm long, with few indented teeth. Capsule ovoid, shorter than calyx. Fl. Jul-Aug. Fr. Aug-Sep. Forests. Distributed in Yunnan, Sichuan and Xizang.

西南蝇子草

Silene delavayi Franch.

多年生草本。基部叶椭圆状披针形，微具柔毛。聚伞花序具多花；花稍下垂；花梗短于花萼，密具紫色腺毛；花瓣红色或深紫色，浅2裂或微凹缺，上缘啮蚀状。蒴果球状卵球形。花期7-8月，果期9-10月。生海拔3800米以下的高山草地。产云南西北部。

Perennial herbs. Basal leaves elliptic-lanceolate, minutely pubescent. Cymes many flowered; flowers slightly nutant; pedicels shorter than calyx, densely violet glandular hairy; petals red or dark violet, shallowly bifid or emarginate, apex erose. Capsules globose-ovoid. Fl. Jul-Aug. Fr. Sep-Oct. Mountain grasslands below 3800 m. Distributed in NW Yunnan.

石生蝇子草

Silene tatarinowii Regel

多年生草本。根圆柱状或刺状。茎分枝。叶披针形或卵状披针形。聚伞花序二歧分枝；花萼筒状棒形，纵脉绿色；花瓣白色，倒披针形，先端2浅裂；副花冠片椭圆状，全缘。蒴果卵球形或狭卵球形。花期7-8月，果期8-10月。生海拔800-2900米的灌丛、砾石质山坡或岩石缝中。产中国西南、华北和华西。

Perennial herbs. Roots cylindric or spiniform. Stems branched. Leaves lanceolate or ovate-lanceolate. Cymes dichotomous; calyx tubular-clavate, longitudinal veins green; petals white, oblanceolate, apex shallowly bifid; coronal scales elliptic, margin entire. Capsules ovoid or narrowly ovoid. Fl. Jul-Aug. Fr. Aug-Oct. Thickets, stony slopes or rock fissures at 800-2900 m. Distributed in SW, N and W China.

西南蝇子草 *Silene delavayi*

石生蝇子草 *Silene tatarinowii*

大花蝇子草 *Silene grandiflora*

大花蝇子草
Silene grandiflora Franch.

多年生草本。叶披针形或狭披针形。二歧聚伞花序疏松；苞片披针形，草质，被长柔毛；花萼浅绿色或红色，筒状棒形，基部脐状，萼齿边缘膜质；花瓣红色，长2-2.5厘米，爪窄倒披针形，无毛；雄蕊伸出；花柱伸出。蒴果短于花萼。花期7-8月。生海拔约2000米的灌丛草地。产云南。

Perennial herbs. Leaves lanceolate or narrowly lanceolate. Dichasial cymes lax and broad; bracts lanceolate, herbaceous, villous; calyx pale green or red, tubular-clavate, umbilicate at base, calyx teeth margin membranous; petals red, 2-2.5 cm long, claws narrowly oblanceolate, glabrous; stamens exserted; styles exserted. Capsules shorter than calyx. Fl. Jul-Aug. Scrub grasslands at ca. 2000 m. Distributed in Yunnan.

云南蝇子草
Silene yunnanensis Franch.

多年生草本。茎铺散，多分枝，具柔毛。叶窄披针形。聚伞花序具少花，密集至极疏松；花萼筒状棒形，沿纵脉密被刺状毛；花瓣淡粉色到白色，深2裂至中部。蒴果狭卵球形。花期6-7月，果期7-9月。生海拔(2400-)2700-3400(-3900)米的林下或田边。产云南西北部。

Perennial herbs. Stems diffuse, much branched, pubescent. Leaves narrowly lanceolate. Cymes few flowered, dense to rather lax; calyx tubular-clavate, longitudinal veins densely spinose hairy; petal pinkish to white, deeply bifid to middle. Capsules narrowly ovoid. Fl. Jun-Jul. Fr. Jul-Sep. Forests or fields at (2400-)2700-3400(-3900) m. Distributed in NW Yunnan.

心瓣蝇子草
Silene cardiopetala Franch.

多年生草本。根密集，圆柱状或刺状。茎铺散，平卧，纤细，具柔毛。叶椭圆形，两面近无毛。聚伞花序二歧分枝；花萼筒状棒形；花瓣浅红色，冠檐先端稍凹缺或浅2裂。蒴果卵状球形，短于花萼。花期7-9月，果期8-10月。生海拔700-3200米的灌丛中或林缘。产云南、四川和西藏。

Perennial herbs. Roots clustered, cylindric or spiniform. Stems diffuse, prostrate, slender, pubescent. Leaves elliptic, both surfaces subglabrous. Cymes dichotomous; calyx tubular-clavate; petals pale red, limbs slightly emarginate or shallowly bifid at apex. Capsules ovoid-globose, shorter than calyx. Fl. Jul-Sep. Fr. Aug-Oct. Scrubby areas or forest edges at 700-3200 m. Distributed in Yunnan, Sichuan and Xizang.

麦瓶草
Silene conoidea L.

一年生草本。基生叶长圆形至线状披针形，具柔毛。二歧聚伞花序；雌雄蕊柄长达1毫米，近无毛；花瓣长2.5-3.5厘米，爪不伸出花萼，窄披针形，冠檐粉红色，全缘或稍凹缺。蒴果梨形。花期5-6月，果期6-7月。生麦田、荒地或草坡。产西藏和新疆。非洲、亚洲和欧洲亦有。

云南蝇子草 *Silene yunnanensis*

心瓣蝇子草 *Silene cardiopetala*

麦瓶草 *Silene conoidea*

Annual herbs. Basal leaves oblong to linear-lanceolate, pubescent. Dichasial cymes; androgynophore to 1 mm long, subglabrous; petals 2.5-3.5 cm long, claws included in calyx, narrowly lanceolate, limbs pink, margin entire or slightly emarginate. Capsules pyriform. Fl. May-Jun. Fr. Jun-Jul. Wheat fields, waste fields or grassy slopes. Distributed in Xizang and Xinjiang. Also in Africa, Asia and Europe.

狗筋蔓 *Silene baccifera*

狗筋蔓

Silene baccifera (L.) Roth

多年生草本。茎和枝开展。叶卵形、卵状披针形或狭椭圆形，纸质，两面沿脉具毛。花稍下垂；花瓣倒披针形，先端2浅裂。果黑色，球形，肉质，不规则开裂。花期6-8月，果期7-10月。生海拔1000-3600米林缘、灌丛或草地。产中国大部分地区。印度北部、尼泊尔、不丹、克什米尔地区、哈萨克斯坦、俄罗斯、朝鲜半岛、日本和欧洲亦有。

Perennial herbs. Stems and branches spreading. Leaves ovate, ovate-lanceolate or narrowly elliptic, papery, both surfaces hairy at veins. Flowers slightly nutant; petals oblanceolate, apex 2-lobed. Fruits black, globose, fleshy, irregularly dehiscent. Fl. Jun-Aug. Fr. Jul-Oct. Forest edges, thickets or grasslands at 1000-3600 m. Distributed in most parts of China. Also in N India, Nepal, Bhutan, Kashmir, Kazakhstan, Russia, Korean Peninsula, Japan and Europe.

麦蓝菜

Vaccaria hispanica (Mill.) Rauschert

一年生草本。茎顶端分枝，无毛。叶卵状披针形或披针形，先端锐尖。花萼绿色，具5棱，果期呈球形，萼齿三角形，边缘膜质；花瓣冠檐粉色，爪浅绿色，具凹缺，有时缺刻。种子具小瘤。花期4-7月，果期5-8月。生麦田中。产中国西南、华中、华北、华西、华东和西北。亚洲和欧洲亦有。

Annual herbs. Stems apically branched, glabrous. Leaves ovate-lanceolate or lanceolate, apex acute. Calyx green, 5-angled, globose in fruit, calyx teeth triangular, margin scarious; petal limbs pink, claws greenish, emarginate, sometimes erose. Seeds granulate. Fl. Apr-Jul. Fr. May-Aug. Wheat fields. Distributed in SW, C, N, W, E and NW China. Also in Asia and Europe.

麦蓝菜 *Vaccaria hispanica*

石竹 *Dianthus chinensis*

石竹
Dianthus chinensis L.

多年生草本。茎疏丛生，直立。叶线状披针形。花单生或数个成聚伞状；花萼圆柱状，萼齿披针形；花瓣的瓣片鲜红、紫红、粉红或白色，喉部具斑点和疏具须毛，先端具不规则齿。蒴果圆筒形。花期5-6月，果期7-9月。生多沙地林缘、林中草地、山坡矮灌丛、丘陵草地、干旱山地、山谷、沟边或海岸地。并广泛栽培。产华北、华西、西北和东北。哈萨克斯坦、俄罗斯(西伯利亚、远东地区)、蒙古、朝鲜半岛和欧洲亦有。

Perennial herbs. Stems laxly caespitose, erect. Leaves linear-lanceolate. Flowers solitary or several in cymes; calyx cylindric, teeth lanceolate; petal limbs bright red, purple-red, pink or white, throat spotted and sparsely bearded, apex irregularly toothed. Capsules cylindric. Fl. May-Jun. Fr. Jul-Sep. Sandy forest edges, forest grasslands, scrubby areas mountain slopes, hillside grasslands, dry hillsides, valleys, streamsides or seashores. Also widely cultivated. Distributed in N, W, NW and NE China. Also in Kazakhstan, Russia (Siberia, Far East), Mongolia, Korean Peninsula and Europe.

瞿麦
Dianthus superbus L.

多年生草本，全株绿色至灰绿色。叶线状披针形。花1或2，顶生，有时腋生；花萼常紫红色，圆柱状；花瓣淡红色，稀白色，瓣片长约2厘米，边缘裂至近1/2处，喉部具须毛。花期6-9月，果期8-10月。生海拔400-3700米林边、草甸或山谷河边。产中国大部分地区。哈萨克斯坦、俄罗斯(远东地区、西伯利亚)、蒙古、朝鲜半岛、日本和欧洲亦有。

Perennial herbs, green to glaucous. Leaves linear-lanceolate. Flowers 1 or 2, terminal, sometimes axillary; calyx usually reddish purple, cylindric; petals pink or rarely white, petal limbs ca. 2 cm long, margin fimbriate for at least 1/2 its length, throat bearded. Fl. Jun-Sep. Fr. Aug-Oct. Forest edges, meadows or streamsides in valleys at 400-3700 m. Distributed in most parts of China. Also in Kazakhstan, Russia (Far East, Siberia), Mongolia, Korean Peninsula, Japan and Europe.

高山瞿麦
Dianthus superbus subsp. **alpestris** Kablik ex Celakov.

多年生草本，灰白色。叶线状披针形，有时渐无毛。花1或2，顶生，有时腋生；花萼常紫红色，圆柱状；花瓣淡红色或稀白色；瓣片长约3厘米，边缘裂至近1/2处，喉部具须毛。花期6-9月，果期8-10月。生海拔400-3700米的开阔林中、高山林缘、山脚草地或河岸。产中国大部分地区。亚洲和欧洲高山地区亦有。

Perennial herbs, glaucous. Leaves linear-lanceolate, sometimes glabrescent. Flowers 1 or 2, terminal, sometimes axillary; calyx usually reddish purple, cylindric; petals reddish or rarely white, petal limbs ca. 3 cm long, margin fimbriate for at least 1/2 its length, throat bearded. Fl. Jun-Sep. Fr. Aug-Oct. Open forests, forest edges on high mountain,

瞿麦 *Dianthus superbus*

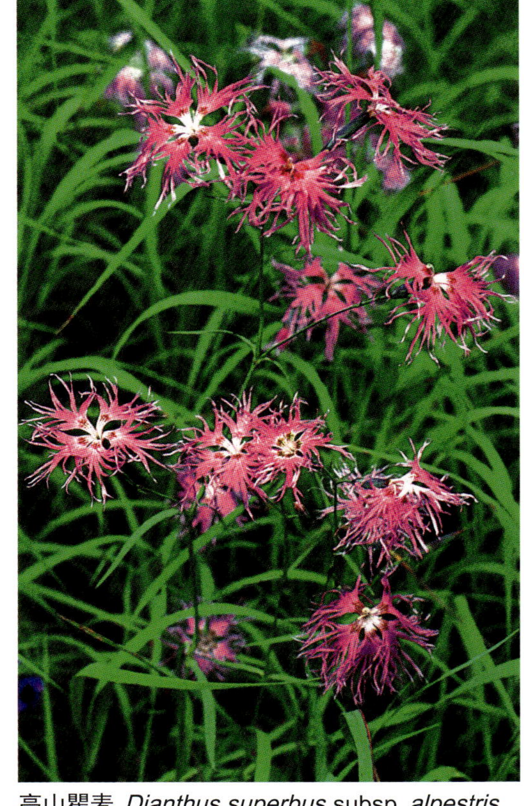

高山瞿麦 *Dianthus superbus* subsp. *alpestris*

长萼瞿麦 *Dianthus longicalyx*

肥皂草 *Saponaria officinalis*

grassy hillsides or river banks at 400-3700 m. Distributed in most parts of China. Also in alpine regions of Asia and Europe.

长萼瞿麦

Dianthus longicalyx Miq.

多年生草本。茎直立，基部分枝，无毛。基生叶数个，花期枯萎。花2至多个，组成疏散的聚伞花序；花萼绿色，长管状，具条纹，萼齿披针形，先端具锐齿尖；花瓣粉红色。蒴果狭圆柱状。花期6-8月，果期8-9月。生海拔800-2100米的林中、山坡、草地、岩地、河床或固定沙丘。产中国西南、华南、东南、华中、华北、华西和华东。朝鲜半岛和日本亦有。

Perennial herbs. Stems erect, basally branched, glabrous. Basal leaves several, withered at anthesis. Flowers 2 to several in lax cymes; calyx green, long tubular, striate, calyx teeth lanceolate, apex sharply pointed; petals pink. Capsules narrowly cylindric. Fl. Jun-Aug. Fr. Aug-Sep. Forests, slopes, grasslands, rocky places, river beds or fixed dunes at 800-2100 m. Distributed in SW, S, SE, C, N, W and E China. Also in Korean Peninsula and Japan.

肥皂草

Saponaria officinalis L.

多年生草本。叶卵形或卵状披针形，半抱茎。花序为聚伞圆锥状，小聚伞花序具3-7花；花大；花萼绿色，有时深紫色，管状，萼齿阔卵形；冠檐白色或粉色，先端凹缺；冠状鳞片线形。蒴果圆锥状卵球形。花期6-9月。西亚和欧洲原产。栽培在公园。

Perennial herbs. Leaves ovate or ovate-lanceolate, semiclasping. Inflorescences a thyrse, cymules 3-7-flowered; flowers large; calyx green, sometimes dark purple, tubular, calyx teeth broadly ovate; petal limbs white or pink, apex emarginate; coronal scales linear. Capsules cylindric-ovoid. Fl. Jun-Sep. Native to W Asia and Europe. Cultivated in park.

刺叶

Acanthophyllum pungens (Bunge) Boiss.

草本亚灌木状。茎丛生，高15-35厘米，被短绒毛。叶钻状针形，长2-4厘米，疏被短绒毛。伞房花序或头状花序顶生；花梗很短；花萼筒状，长6-7毫米，被短柔毛；花瓣5，红色或淡红色，椭圆状倒披针形，长约12毫米；雄蕊10，伸出，长14毫米；花柱2，伸出。花期5-8月。生石砾山坡或沙地，海拔400-1300米。产新疆西北部。蒙古和哈萨克斯坦亦有。

Herbs subshrublike. Stems caespitose, 15-35 cm tall, velutinous. Leaves subulate-needle-shaped, 2-4 cm long, sparsely velutinous. Corymb or capitulum terminal; pedicels very short; calyx tubular, 6-7 mm long, puberulous; petals 5, red or reddish, elliptic-oblanceolate, ca. 12 mm long; stamens 10, exserted, 14 mm long; styles 2, exserted. Fl. May-Aug. Gravelly slopes or sandy places at 400-1300 m. Distributed in NW Xinjiang. Also in Mongolia and Kazakhstan.

刺叶 *Acanthophyllum pungens*

长蕊石头花(霞草)
Gypsophila oldhamiana Miq.

多年生草本。根木质。叶长圆形，质厚，近肉质。伞房状聚伞花序顶生或腋生，密集，无毛；花萼钟形或漏斗形，裂片卵状三角形，边缘白色；花瓣粉红色；雄蕊长于花瓣。蒴果卵球形。花期6-9月，果期8-10月。生海拔2000米以下的灌丛、草坡、岩石地或沿海沙地。产华中、华北和华东。朝鲜半岛亦有。

Perennial herbs. Roots woody. Leaves oblong, thick, subfleshy. Corymbose cymes terminal or axillary, dense, glabrous; calyx campanulate or funnelform, lobes ovate-triangular, margin white; petals pink; stamens longer than petals. Capsules ovoid. Fl. Jun-Sep. Fr. Aug-Oct. Scrubby areas, grassy slopes, rocky places or maritime sands below 2000 m. Distributed in C, N and E China. Also in Korean Peninsula.

草原石头花
Gypsophila davurica Turcz. ex Fenzl

多年生草本，全株无毛。根木质。叶无柄，线状披针形，先端具长尾尖。聚伞花序疏散；花萼钟形，5裂至其长的1/3-1/2，裂片卵状三角形，边缘白色；花瓣淡粉红色或近白色；蒴果卵状。花期6-9月，果期7-10月。生草原、丘陵、固定沙丘或干岩石山坡。产河北、内蒙古、山西、黑龙江、吉林和辽宁。俄罗斯东部和蒙古亦有。

Perennial herbs, glabrous throughout. Roots woody. Leaves sessile, linear-lanceolate, apex long acuminate. Cymes lax; calyx campanulate, 5-lobed for 1/3-1/2 its length, lobes ovate-triangular, margin white; petals pinkish or almost white. Capsules ovoid. Fl. Jun-Sep. Fr. Jul-Oct. Steppes, hills, fixed dunes or dry rocky slopes. Distributed in Hebei, Neimenggu, Shanxi, Heilongjiang, Jilin and Liaoning. Also in E Russia and Mongolia.

紫萼石头花
Gypsophila patrinii Ser.

多年生草本，全株无毛。基生叶簇生，茎生叶稀疏；叶片线形，顶端急尖，基部连合成短鞘状。聚伞花序顶生，花少，疏散；苞片披针形或三角形，顶端渐尖，边缘膜质，具缘毛；花萼钟形，萼脉宽，绿色或带紫色，脉间膜质，淡紫色，萼齿卵形，顶端急尖，边缘膜质，疏生缘毛；花瓣紫红色，倒卵形，顶端微凹，基部楔形。蒴果卵球形，长于宿存萼。花期6-9月，果期7-10月。生山坡草地、石质山坡、戈壁、岩缝和沙地。产宁夏、甘肃、青海和新疆。哈萨克斯坦、俄罗斯(西伯利亚)和蒙古北部也有。

Perennial herbs, glabrous. Basal leaves fascicled, cauline leaves few; blades linear, base connate into a short sheath, apex acute. Cymes terminal, sparsely few flowered; bracts lanceolate or triangular, apex acuminate, margin membranous, ciliate; calyx campanulate, lilac, membranous between green or purplish broad veins, calyx lobes ovate, margin membranous, sparsely ciliate, apex acute; petals purple-red, obovate, apex retuse, base cuneate. Capsule ovoid, longer than calyx. Fl. Jun-Sep. Fr. Jul-Oct. Grassy mountain slopes, rocky slopes, Gobi deserts, rock crevices and sands. Distributed in Ningxia, Gansu, Qinghai and Xinjiang. Also in Kazakhstan, Russia (Siberia) and N Mongolia.

长蕊石头花(霞草) *Gypsophila oldhamiana*

草原石头花 *Gypsophila davurica*

紫萼石头花 *Gypsophila patrinii*

睡莲科
Nymphaeaceae

莲

Nelumbo nucifera Gaertn.

多年生水生草本。叶柄及花梗常有刺；叶下蓝绿色，圆形，直径25-90厘米，纸质。花直径10-23厘米；萼片4-5，花瓣多数；花托鸡冠状凸起，螺旋状。果实长圆形至卵球形。花期6-8月，果期8-10月。生池沼或河流中。中国大部分地区有栽培。全世界大部分地区有栽培。

Perennial aquatic herbs. Petioles and pedicels usually with spines; leaves abaxially blue-green, orbicular, 25-90 cm diam, papery. Flowers 10-23 cm diam; sepals 4-5, petals numerous; receptacles accrescent, turbinate. Fruits oblong to ovoid. Fl. Jun-Aug. Fr. Aug-Oct. Ponds or rivers. Cultivated in most parts of China. Also cultivated in most parts of the world.

莲 *Nelumbo nucifera*

莲 *Nelumbo nucifera*

水盾草 *Cabomba caroliniana*

莼菜 *Brasenia schreberi*

水盾草
Cabomba caroliniana A. Gray

多年生水生草本。沉水叶对生，长2.5-3.8厘米，掌状分裂，裂片3-4次二叉分裂成线形小裂片；浮水叶互生，狭椭圆形，长1-1.6厘米，全缘或基部2浅裂。花单生；花梗长1-1.5厘米，被短柔毛；萼片无毛，椭圆形，长7-8毫米；花瓣绿白色，基部具爪，近基部具1对黄色腺体。花期10月。生河流、湖泊、运河或渠道。产浙江、江苏和北京。原产美国。

Perennial aquatic herbs. Submerged leaves opposite, 2.5-3.8 cm long, palmately parted, lobes 3-4 × dichotomously branched into linear lobules; floating leaves alternate, narrowly elliptic, 1-1.6 cm long, margin entire or 2-lobed at base. Flowers solitary; pedicels 1-1.5 cm long, pubescent; sepals glabrous, elliptic, 7-8 mm long; petals green-white, clawed at base, 1-paired yellow glands near base. Fl. Oct. River, lakes, canals or channels. Distributed in Zhejiang, Jiangsu and Beijing. Native to USA.

莼菜
Brasenia schreberi J. F. Gmelin

多年生水生草本。茎无毛，基部具根状茎。叶椭圆状长圆形，无毛；叶柄长25-40厘米。花暗紫色，直径1-2厘米；萼片及花瓣条形，先端圆钝；雄蕊长为花瓣的1/2；花药条形，长约4毫米。坚果长圆状卵形，长6-10毫米。花期6月；果期10月。生池塘、湖或沼泽。产江苏、安徽、浙江、江西、湖南、四川、云南和台湾。俄罗斯、亚洲、美国、澳大利亚和非洲亦有。

Perennial aquatic herbs. Stems glabrous, base rhizomatous. Leaves elliptic-oblong, glabrous; petiole 25-40 cm. Flowers dull purple, 1-2 cm diam; sepals and petals linear, apex obtuse; stamens 1/2 as long as petals; anthers linear, ca. 4 mm long. Nuts oblong-ovate, 6-10 mm long. Fl. Jun. Fr. Oct. Ponds, lakes or swamps. Distributed in Jiangsu, Anhui, Zhejiang, Jiangxi, Hunan, Sichuan, Yunnan and Taiwan. Also in Russia, Asia, USA, Australia and Africa.

白睡莲
Nymphaea alba L.

多年生水生草本。根状茎匍匐。叶近圆形，纸质，全缘。花浮水，白色；柱头有延伸的圆柱形附属体；内轮雄蕊花丝与花药等宽；柱头具(8-)14-20(-25)辐射线。果实半球形。种子椭圆体形。花期6-8月，果期8-10月。生池沼中。中国大多数地区均有栽培。亚洲和欧洲亦有。引自欧洲。

Perennial aquatic herbs. Rhizomes repent. Leaves suborbicular, papery, margin entire. Flower floating, white; stigmas with elongated cylindrical appendage; filaments of inner stamens ± as wide as anthers; stigma rays (8-)14-20(-25). Fruits semiglobose. Seeds ellipsoid. Fl. Jun-Aug. Fr. Aug-Oct. Ponds. Cultivated in most parts of China. Also in Asia and Europe. Introduced from Europe.

白睡莲 *Nymphaea alba*

睡莲
Nymphaea tetragona Georgi

多年生水生草本。根状茎直立，不分枝。叶全缘，无毛，成熟叶多短于10厘米。花白色，完全开放时直径3-6厘米；花瓣8-15(-17)，有时边缘淡红色；心皮附属物卵球形。果实球形。花期6-8月，果期8-10月。生池沼或河流中。栽培于中国大部分地区。南亚、东亚和美洲亦有。

Perennial aquatic herbs. Rhizomes erect, unbranched. Leaves entire, glabrous, mature leaves mostly less than 10 cm long. Flowers white, 3-6 cm diam when fully open; petals 8-15(-17), margin sometimes pale red; carpellary appendages ovoid. Fruits globose. Fl. Jun-Aug. Fr. Aug-Oct. Ponds or rivers. Cultivated in most parts of China. Also in S and E Asia, and America.

睡莲 *Nymphaea tetragona*

芡实
Euryale ferox Salisb.

多年生水生草本。沉水叶戟形或椭圆形，直径4-10厘米，基部深心形；浮水叶叶柄及沿脉具刺；叶近革质，脉下面强烈具棱；主脉两面具刺。花直径5厘米；花瓣外面深紫色渐褪至内面白色；子房7-16室。果实深紫色，球形，海绵质，密具刺。花期6-8月。生湖中或池塘。中国广泛分布。印度、孟加拉国、俄罗斯(远东地区)、日本和朝鲜半岛亦有。

Perennial aquatic herbs. Submerged leaves sagittate or elliptic, 4-10 cm diam, base deeply cordate; floating leaves prickly on petioles and along veins; blades subleathery, abaxially sparsely pubescent; veins abaxially strongly ribbed; primary veins prickly on both surfaces. Flower 5 cm diam; petals purple-violet outside, fading to white inside; ovary 7-16-loculed. Fruits dark purple, globose,

睡莲属一种 *Nymphaea* sp.

芡实 *Euryale ferox*

spongy, densely prickly. Fl. Jun-Aug. Lakes or ponds. Widely distributed in China. Also in India, Bangladesh, Russia (Far East), Japan and Korean Peninsula.

萍蓬草

Nuphar pumila (Timm) DC.

多年生水生植物。叶片纸质，长6-17厘米，上面无毛，下面无毛至密被柔毛。花直径1-2.5厘米；花梗长40-50厘米；萼片黄色；花瓣狭楔形至阔条形；柱头盘深裂，直径4-7.5毫米。花期5-7月，果期7-9月。生湖泊或池塘。产中国东南、华北和东北。蒙古、俄罗斯、朝鲜半岛、日本和北欧亦有。

Perennial aquatic herbs. Floating leaves 6-17 cm long, papery, adaxially glabrous, abaxially glabrous to densely pubescent. Flowers 1-2.5 cm diam; pedicels 40-50 cm long; sepals yellow; petals narrowly cuneate to broadly linear; stigmatic disc deeply lobed, 4-7.5 mm diam. Fl. May-Jul. Fr. Jul-Sep. Lakes or ponds. Distributed in SE, N and NE China. Also in Mongolia, Russia, Korean Peninsula, Japan and N Europe.

欧亚萍蓬草

Nuphar lutea (L.) Sm.

多年生水生草本。根状茎粗壮。浮水叶椭圆形，15-30 × 10-22厘米，革质；叶柄长约50厘米。花直径4-5厘米；萼片黄色，宽卵形或球形；花瓣条形；柱头盘全缘，直径7-19毫米。花期7-8月，果期9-10月。生湖泊或池塘中。产新疆。俄罗斯(西伯利亚)、哈萨克斯坦、非洲、西南亚和欧洲亦有。

Perennial aquatic herbs. Rhizomes stout. Floating leaves elliptic, 15-30 × 10-22 cm, leathery; petioles ca. 50 cm long. Flower 4-5 cm diam; sepals yellow, broadly ovate to orbicular; petals linear; stigmatic disc entire, 7-19 mm diam. Fl. Jul-Aug. Fr. Sep-Oct. Lakes or ponds. Distributed in Xinjiang. Also in Russia (Siberia) and Kazakhstan, Africa, SW Asia and Europe.

萍蓬草 *Nuphar pumila*

欧亚萍蓬草 *Nuphar lutea*

金鱼藻科
Ceratophyllaceae

五刺金鱼藻 *Ceratophyllum platyacanthum* subsp. *oryzetorum*

金鱼藻
Ceratophyllum demersum L.

多年生沉水草本。叶亮绿色，质地粗糙，轮生(每轮直径1.5-6厘米)，1-2二叉状分枝，裂片线状或丝状。花直径1-3毫米。瘦果深绿色至红褐色，边缘无翅且无刺。花期6-7月，果期8-10月。生池塘、河沟或湖内。中国广泛分布。全世界亦广布。

Perennial herbs, submersed. Leaves bright green, coarse textured, whorled (whorls 1.5-6 cm diam), 1-2 times dichotomous branching, segments linear or filiform. Flowers 1-3 mm diam. Achenes dark green to reddish brown, margins wingless and spineless. Fl. Jun-Jul. Fr. Aug-Oct. Ponds, streams or lakes. Widely distributed in China. Also Widely in the world.

五刺金鱼藻
Ceratophyllum platyacanthum V. Komarov subsp. **oryzetorum** Chamisso

多年生沉水草本。叶深绿色，1-2歧分枝，每轮直径2.5-4厘米，裂片条形。瘦果椭圆体形，褐色或深绿色，平滑，具5尖刺，先端具2刺，不下延，基部具2刺，顶端刺(宿存刺)长2-12.5毫米。花果期6-9月。生池塘或溪中。产华南、华东、华中、西北、华北和东北。俄罗斯(远东地区)、朝鲜半岛和日本亦有。

Perennial herbs, submersed. Leaves deep green, 1-2 times dichotomously divided, whorls 2.5-4 cm diam, segments linear. Achenes ellipsoid, brown or dark green, smooth, with 5 prickles, facial spines 2, not decurrent, basal spines 2, terminal spine (persistent style) 2-12.5 mm long. Fl. and fr. Jun-Sep. Ponds or streams. Distributed in S, E, C, NW, N and NE China. Also in Russia (Far East), Korean Peninsula and Japan.

金鱼藻 *Ceratophyllum demersum*

昆栏树科 Trochodendraceae

昆栏树 *Trochodendron aralioides*

昆栏树

Trochodendron aralioides
Sieb. et Zucc.

灌木或小乔木。叶革质，螺旋状着生；叶片阔卵形，长5-12厘米。花序顶生，总状或圆锥状；花两性，无花被；雄蕊40-70，排成数轮，心皮(4-)6-17，排成一轮。蓇葖轮由数个侧面合生蓇葖果而成。花期5-6月，果期10-11月。生海拔300-2700米混交林中。产台湾。

Shrubs or small trees. Leaves leathery, spirally arranged; leaf blade broadly ovate, 5-12 cm long. Inflorescences terminal, racemes or panicles; flowers bisexual, perianth absent; stamens 40-70, in several series; carpels(4-)6-17, in one series. Fruit composed of laterally fused follicles. Fl. May-Jun. Fr. Oct-Nov. Mixed forests at 300-2700 m. Distributed in Taiwan.

连香树科 Cercidiphyllaceae

Large deciduous trees. Leaves dimorphic, almost rounded, oval or cordate on short shoots, elliptic or triangular on long shoots. Staminate flowers with pink anthers, pistillate flowers with red stigmas. Follicles 2-4, brown to black, oblong, with short persistent styles. Fl. Apr. Fr. Aug. Forests, forest edges or by streams at 600-2700 m. Distributed in SW, SE and C China. Also in Japan.

连香树

Cercidiphyllum japonicum
Sieb. et Zucc.

落叶大乔木。叶二型，生短枝上的叶近圆形、宽卵形或心形，生长枝上的叶椭圆形或三角形。雄花花药粉色，雌花柱头红色。蓇葖果2-4，褐色至黑色，长圆形，具有宿存的短花柱。花期4月，果期8月。生海拔600-2700米的林中、林缘或溪边。产中国西南、东南和华中。日本亦有。

连香树 *Cercidiphyllum japonicum*

星叶草科 Circaeasteraceae

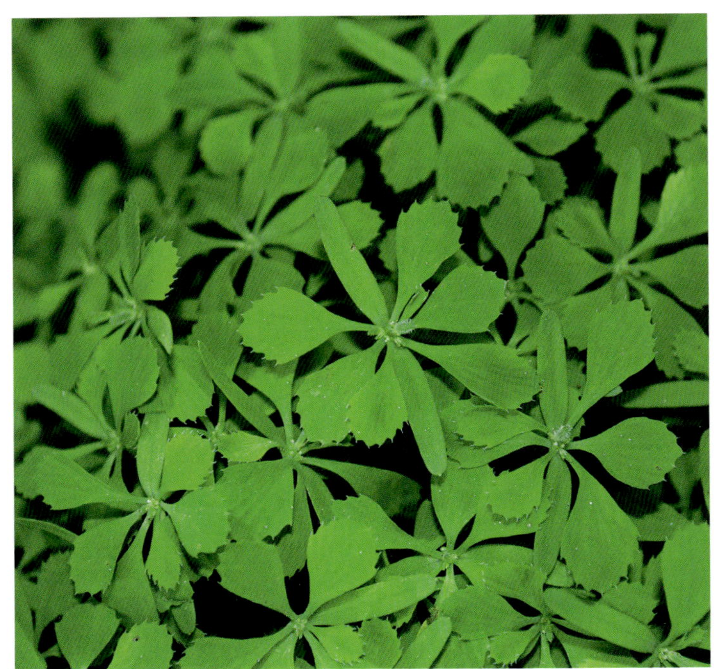
星叶草 *Circaeaster agrestis*

星叶草

Circaeaster agrestis Maxim.

一年生小草本，高3-10厘米。子叶宿存，条形，长4-11毫米。叶莲座状，具柄，菱形或楔形，具小齿，叶脉二歧状分枝。花簇生叶腋；萼片2-3，长约0.5毫米；雄蕊1-2(-3)；心皮1-3，子房有1胚珠。瘦果纺锤形，长约3毫米，具钩状毛。花期4-7月。生海拔2100-5000米的草地上，常生树、灌木或石下阴处。产西藏、云南西北部、四川西部、陕西南部、甘肃南部、青海和新疆西部。不丹、尼泊尔和印度北部亦有。

Small annual herbs, 3-10 cm tall. Cotyledons persistent, linear, 4-11 mm long. Leaves rosulate, petiolate, rhombic or cuneiform, denticulate, with dichotomous veins. Flowers axillary, fascicled; sepals 2-3, ca. 0.5 mm long; stamens 1-2(-3); carpels 1-3, ovule 1 per ovary. Achenes fusiform, ca. 3 mm long, with hooked hairs. Fl. Apr-Jul. Grasslands, or shade of trees, shrubs or rock ledges at 2100-5000 m. Distributed in Xizang, NW Yunnan, W Sichuan, S Shaanxi, S Gansu, Qinghai and W Xinjiang. Also in Bhutan, Nepal and N India.

毛茛科 Ranunculaceae

驴蹄草

Caltha palustris L.

多年生草本，无毛。茎高达48厘米。基生叶3-7，圆卵形，边缘有密小牙齿，茎生叶较小。单歧聚伞花序具心形苞片；萼片5，黄色；雄蕊多数；心皮5-12。花期5-9月。生海拔600-4000米的溪边、草甸上或林下。产云南西北部、四川、甘肃南部、新疆、陕西、河南西部、山西、河北和内蒙古。广布北温带。

Perennial herbs, glabrous. Stems up to 48 cm tall. Basal leaves 3-7, orbicular-ovate, margin densely denticulate, cauline leaves smaller. Monochasia with cordate bracts; sepals 5, yellow; stamens numerous; carpels 5-12. Fl. May-Sep. Streams, alpine meadows or forests at 600-4000 m. Distributed in NW Yunnan, Sichuan, S Gansu, Xingjiang, Shaanxi, W Henan, Shanxi, Hebei and Neimenggu. Widespread in the north temperate zone.

驴蹄草 *Caltha palustris*

三角叶驴蹄草 *Caltha palustris* var. *sibirica*

红花空茎驴蹄草 *Caltha palustris* var. *barthei* f. *atrorubra*

三角叶驴蹄草

Caltha palustris L. var. **sibirica** Regel

草本。茎于花期后外倾。叶基生或茎生，花序下部之叶小于基生叶；叶三角肾形，基部宽心形，边缘基部具齿，顶端波状或近全缘。单歧聚伞花序常单生，具(2或)3-5花；萼片5，黄色；宿存花柱长1-3毫米。花期5-9月。生海拔600-4000米的山谷潮湿地或水边。产中国东北。俄罗斯(远东地区)、蒙古和朝鲜半岛亦有。

Herbs. Stems decumbent after anthesis. Leaves both basal and cauline, below inflorescence usually smaller than basal leaves; triangular-reniform, base broadly cordate, margin dentate at base, apically repand or subentire. Monochasium often solitary, (2 or)3-5-flowered; sepals 5, yellow; persistent styles 1-3 mm long. Fl. May-Sep. Wet places in valleys or by waters at 600-4000 m. Distributed in NE China. Also in Russia (Far East), Mongolia and Korean Peninsula.

掌裂驴蹄草

Caltha palustris L. var. **umbrosa** Diels

草本。茎直立。最上部茎生叶和花序下部苞片掌状分裂；叶圆形至圆肾形或心形。单歧聚伞花序常单生，具(2或)3-5花；萼片5，黄色。蓇葖果(5-)7-25。花期5-9月。生海拔600-4000米的山谷草地区。产云南和四川。

Herbs. Stems erect. Uppermost cauline leaves and bracts palmatipartite; blades orbicular to orbicular-reniform or cordate. Monochasium often solitary, (2 or)3-5-flowered; sepals 5, yellow. Follicles (5-)7-25. Fl. May-Sep. Grassy valley areas at 600-4000 m. Distributed in Yunnan and Sichuan.

掌裂驴蹄草 *Caltha palustris* var. *umbrosa*

红花空茎驴蹄草

Caltha palustris L. var. **barthei** Hance f. **atrorubra** (W. T. Wang) W. T. Wang

草本，具多数肉质根。茎直立、中空。花序下部的叶几乎等大于基生叶；叶圆形至圆肾形或心形。单歧聚伞花序常单生，具(2或)3-5花；萼片5，红色。蓇葖果(5-)7-25。花期5-9月。生海拔3000-3500米的小河边。产四川西北部。

Herbs, with numerous fleshy roots. Stems erect, hollow. Cauline leaves below inflorescences almost equal to basal blades in size; blades orbicular to orbicular-reniform or cordate. Monochasium often solitary, (2 or)3-5-flowered; sepals 5, red. Follicles (5-)7-25. Fl. May-Sep. By streams at 3000-3500 m. Distributed in NW Sichuan.

花葶驴蹄草 *Caltha scaposa*

花葶驴蹄草
Caltha scaposa Hook. f. et Thoms.

多年生小草本。茎单一或分枝。基生叶3-10；叶柄长2.5-10(-15)厘米，具膜质鞘。花单生、顶生或呈单歧聚伞花序；心皮有短柄。蓇葖果(5-)6-8(-11)。花期6-9月，果期7月。生海拔2800-4100米的高山草甸或潮湿草地。产云南、四川、西藏、甘肃和青海。印度北部、尼泊尔和不丹亦有。

Small perennial herbs. Stems simple or branched. Basal leaves 3-10; petioles 2.5-10(-15) cm long, with membranous sheath. Flowers solitary, terminal or in monochasium; carpels shortly stipitate. Follicles (5-)6-8(-11). Fl. Jun-Sep. Fr. Jul. Alpine meadows or wet grasslands at 2800-4100 m. Distributed in Yunnan, Sichuan, Xizang, Gansu and Qinghai. Also in N India, Nepal and Bhutan.

白花驴蹄草
Caltha natans Pall.

沉水或在沼泽匍匐草本。茎长20-50厘米，分枝。茎生叶具长柄，心形或肾形，宽1.5-2.4厘米，有浅齿。单歧聚伞花序有2-5花；萼片5，白色，花瓣状，长约3毫米；雄蕊长2毫米；心皮(10-)20-30。蓇葖果长约5毫米。花期7月。生湿草甸或沼泽中，或水中。产内蒙古和黑龙江。蒙古、俄罗斯(西伯利亚)和北美洲亦有。

Submerged or creeping herbs. Stems 20-50 cm long, branched. Cauline leaves long petiolate, cordate or reniform, 1.5-2.4 cm broad, crenate. Monochasia 2-5-flowered; sepals 5, white, petaloid, ca. 3 mm long; stamens 2 mm long; carpels (10-)20-30. Follicles ca. 5 mm long. Fl. Jul. Wet meadows or marshes, or in water. Distributed in Neimenggu and Heilongjiang. Also in Mongolia, Russia (Siberia) and North America.

鸡爪草
Calathodes oxycarpa Sprague

多年生草本。茎高达45厘米，无毛。基生叶约3，无毛，3全裂；茎生叶较小。花单朵顶生；萼片5，白色；雄蕊多数；心皮7-15。蓇葖果背缝线有1正三角形凸起。花期5-7月。生海拔2400-3200米的山谷林下或草坡阴处。产云南西北部、四川、重庆和湖北西部。

Perennial herbs. Stems up to 45 cm tall, glabrous. Basal leaves ca. 3, glabrous, 3-sect; cauline leaves smaller. Flower solitary, terminal; sepals 5, white; stamens numerous; carpels 7-15. Follicle on dorsal suture with a deltoid projection. Fl. May-Jul. Forests or shady places on grassy slopes at 2400-3200 m. Distributed in NW Yunnan, Sichuan, Chongqing and W Hubei.

白花驴蹄草 *Caltha natans*

鸡爪草 *Calathodes oxycarpa*

云南金莲花 *Trollius yunnanensis*

矮金莲花 *Trollius farreri*

云南金莲花

Trollius yunnanensis (Franch.) Ulbr.

多年生草本。叶基生并茎生，3深裂，五角形。单花顶生，或2或3花形成聚伞状；萼片5(-7)，黄色，干时多少呈绿色；花瓣条形，比雄蕊稍短，长7-8毫米。蓇葖果7-25；宿存花柱长约1毫米。花期6-9月，果期9-10月。生海拔1900-3900米山地草坡或溪边。产云南、四川和甘肃。

Perennial herbs. Leaves basal and cauline, 3-parted, pentagonal. Flowers solitary, terminal, or 2 or 3 in a cyme; sepals 5(-7), yellow, ± green when dried; petals linear, slightly shorter than stamens, 7-8 mm long.Follicles 7-25; persistent styles ca. 1 mm long. Fl. Jun-Sep. Fr. Sep-Oct. Grassy slopes or by streams at 1900-3900 m. Distributed in Yunnan, Sichuan and Gansu.

矮金莲花

Trollius farreri Stapf

多年生小草本，无毛。茎高5-17厘米。叶全部基生或近基生，具长柄，五角形，宽1.4-2.6厘米，3全裂。花单朵顶生，直径1.8-3.4厘米；萼片5，黄色；花瓣比雄蕊稍短。花期6-7月。生海拔3500-4700米的山地草坡。产云南西北部、西藏东北部、四川西部、青海南部和东部、甘肃南部和陕西南部。

Small perennial herbs, glabrous. Stems 5-17 cm tall. Leaves basal or nearly all basal, long petiolate, pentagonal, 1.4-2.6 cm broad, 3-sect. Flower solitary and terminal, 1.8-3.4 cm diam; sepals 5, yellow; petals slightly shorter than stamens. Fl. Jun-Jul. Alpine meadows at 3500-4700 m. Distributed in NW Yunnan, NE Xizang, W Sichuan, S and E Qinghai, S Gansu and S Shaanxi.

毛茛状金莲花

Trollius ranunculoides Hemsl.

多年生草本，无毛。茎高6-30厘米。基生叶数枚，茎生叶较小，五角形，宽1.4-4厘米，3全裂。花单朵顶生，直径2.2-3.2(-4)厘米；萼片5，黄色；花瓣比雄蕊稍短。花期5-7月。生海拔2900-4100米的山地草坡、溪边或疏林中。产西藏东部、云南西北部、四川西部、青海南部和东部、甘肃南部。

Perennial herbs, glabrous. Stems 6-30 cm tall. Basal leaves serveral, cauline leaves smaller, pentagonal, 1.4-4 cm broad, 3-sect. Flower solitary and terminal, 2.2-3.2(-4) cm diam; sepals 5, yellow; petals slightly shorter than stamens. Fl. May-Jul. Grassy slopes, by streams or under sparse forests at 2900-4100 m. Distributed in E Xizang, NW Yunnan, W Sichuan, S and E Qinghai, and S Gansu.

毛茛状金莲花 *Trollius ranunculoides*

鞘柄金莲花 *Trollius vaginatus*

鞘柄金莲花
Trollius vaginatus Hand.-Mazz.

多年生草本。茎高4厘米，果期达11厘米，单一。基生叶1或2(-3)；叶片五角状肾形，3全裂；叶柄基部具鞘。花单生；萼片5，黄色，下面常紫褐色；花瓣约12，匙状线形，近等长于花丝或稍短。蓇葖果4-6。花期6月，果期6月。生海拔3000-4200米的草坡。产云南西北部和四川西南部。

Perennial herbs. Stems 4 cm tall, up to 11 cm at fruiting, simple. Basal leaves 1 or 2(-3); leaf blades pentagonal-reniform, 3-sect; petioles base sheathed. Flowers solitary; sepals 5, yellow, abaxially usually purple-brown; petals ca. 12, spathulate-linear, subequaling filaments or slightly shorter. Follicles 4-6. Fl. Jun. Fr. Jun. Grassy slopes at 3000-4200 m. Distributed in NW Yunnan and SW Sichuan.

长白金莲花
Trollius japonicus Miq.

多年生草本。基生叶3-5，3全裂。花单生；萼片5-7，黄色；花瓣9，条形，近等长于雄蕊。蓇葖果7-15，长1.1厘米，宽3毫米；花柱宿存，长1.5-4毫米。花期7-8月，果期9月。生海拔1200-2300米草坡。产吉林(长白山)。日本亦有。

Perennial herbs. Basal leaves 3-5, 3-sect.Flowers solitary; sepals 5-7, yellow; petals 9, linear, subequaling stamens.Follicles 7-15, 1.1 cm long, 3 mm broad; styles persistent, 1.5-4 mm long. Fl. Jul-Aug. Fr. Sep. Grassy slopes at 1200-2300 m. Distributed in Jilin (Changbai Mountain). Also in Japan.

短瓣金莲花
Trollius ledebouri Reichb.

多年生草本，无毛。茎高60-100厘米。基生叶2-3，具长柄，心状五角形，宽8-13厘米，3全裂。花单朵顶生或2-3朵组成聚伞花序；萼片5-8，黄色，花瓣状，长1.2-2.8厘米；花瓣10-22，条形，长1.3-1.6厘米，宽1毫米；雄蕊长9毫米；心皮20-28。花期6-7月。生海拔100-900米的湿草地上或沟边。产辽宁、黑龙江和内蒙古东北部。蒙古和俄罗斯(远东地区、西伯利亚)亦有。

Perennial herbs, glabrous. Stems 60-100 cm tall. Basal leaves 2-3, long petiolate, cordate-pentagonal, 8-13 cm broad, 3-sect. Flowers singly terminal or 2-3 in cyme; sepals 5-8, yellow, petaloid, 1.2-2.8 cm long; petals 10-22, linear, 1.3-1.6 cm long, 1 mm broad; stamens 9 mm long; carpels 20-28. Fl. Jun-Jul. Wet grasslands or by streams at 100-900 m. Distributed in Liaoning, Heilongjiang and NE Neimenggu. Also in Mongolia and Russia (Far East, Siberia).

阿尔泰金莲花
Trollius altaicus C. A. Mey.

多年生草本。基生叶2-5，五角形，3全裂。花单生，直径3-5厘米；萼片(10-)15-18，橙黄色；花瓣条形，短于或近等于雄蕊；花柱暗紫色或黑色。蓇葖果约16；宿存花柱长1毫米。花期5-7月，果期8月。生海拔1200-2700米草坡或山谷潮湿地。产内蒙古西部和新疆北部。哈萨克斯坦东部、乌兹别克斯坦、吉尔吉斯斯坦、塔吉

长白金莲花 *Trollius japonicus*

短瓣金莲花 *Trollius ledebouri*

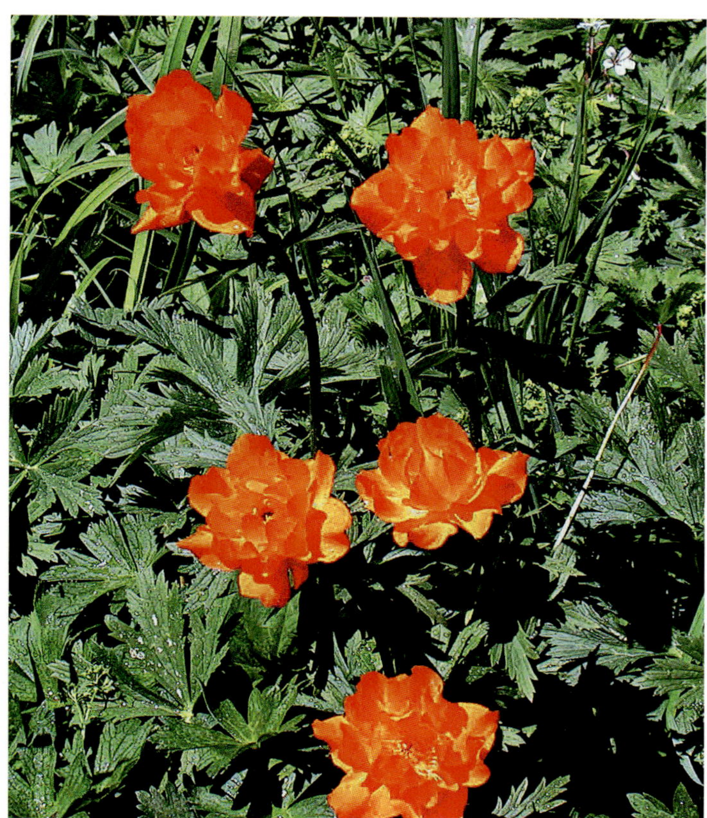
阿尔泰金莲花 *Trollius altaicus*

克斯坦、俄罗斯(西伯利亚)和蒙古亦有。

Perennial herbs. Basal leaves 2-5, pentagonal, 3-sect. Flowers solitary, 3-5 cm diam; sepals (10-) 15-18, orange; petals linear, shorter than or subequaling stamens; styles dark-purple or black.Follicles ca. 16; persistent styles 1 mm long. Fl. May-Jul. Fr. Aug. Grassy slopes or wet places in valleys at 1200-2700 m. Distributed in W Neimenggu and N Xinjiang. Also in E Kazakhstan, Uzbekistan, Kyrgyzstan, Tajikistan, Russia (Siberia) and Mongolia.

金莲花

Trollius chinensis Bunge

多年生草本。叶五角形。花单生或2-3花组成顶生聚伞花序；萼片(6-)10-15(-19)，金黄色；花瓣18-21，狭条形，近等长或稍长于萼片。蓇葖果20-30；宿存花柱长1毫米。花期6-7月，果期8-9月。生海拔1000-2200米的草坡。产华北和东北。

Perennial herbs. Leaves pentagonal.Flowers solitary, or 2-3 in terminal cymes; sepals (6-)10-15(-19), golden yellow; petals 18-21, narrowly linear, subequaling or longer than sepals.Follicles 20-30; persistent styles 1 mm long. Fl. Jun-Jul. Fr. Aug-Sep. Grassy slopes at 1000-2200 m. Distributed in N and NE China.

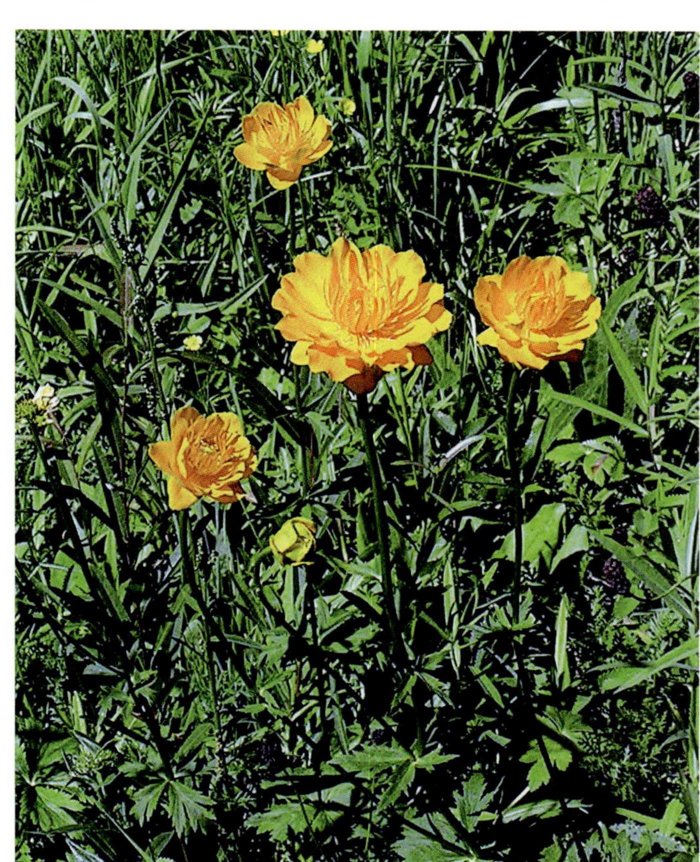
金莲花 *Trollius chinensis*

长瓣金莲花

Trollius macropetalus (Regel) Fr. Schmidt

多年生草本。基生叶3-6，五角形，3全裂。花单生；萼片5-7，金黄色，干后橘黄色；花瓣14-22，窄条形，长于萼片。蓇葖果20-40，长1.3厘米，宽4毫米；花柱宿存，长3.5-4毫米。花期6-9月，果期7月。生海拔400-600米湿草地。产中国东北。俄罗斯(远东地区)和朝鲜半岛北部亦有。

Perennial herbs. Basal leaves 3-6, pentagonal, 3-sect.Flowers solitary;sepals 5-7, golden yellow, orange-yellow when dried; petals 14-22, narrowly linear, longer than sepals. Follicles 20-40, 1.3 cm long, 4 mm wide; styles persistent, 3.5-4 mm long. Fl. Jun-Sep. Fr. Jul. Wet grasslands at 400-600 m. Distributed in NE China. Also in Russia (Far East) and N Korean Peninsula.

长瓣金莲花 *Trollius macropetalus*

铁破锣 *Beesia calthifolia*

铁破锣

Beesia calthifolia (Maxim.) Ulbr.

多年生草本。叶为不分裂的单叶。花序聚伞状；萼片5，白色或带粉红色，狭卵形或椭圆形；花瓣和退化雄蕊均不存在。蓇葖果扁平，披针状条形，具约8个不明显横向脉纹。花期6-8月，果期7-9月。生海拔1400-3500米的林下湿地。产中国西南、华西和华中。缅甸北部亦有。

Perennial herbs. Leaves simple, unlobed. Inflorescences cymose; sepals 5, white or pinkish, narrowly ovate or elliptic; petals and staminodes wanting. Follicles flat, lanceolate-linear, with ca. 8 obliquely transverse veins. Fl. Jun-Aug. Fr. Jul-Sep. Wet places under forests at 1400-3500 m. Distributed in SW, W and C China. Also in N Myanmar.

黄三七

Souliea vaginata (Maxim.) Franch.

多年生草本。叶具柄，二至三回三出全裂，三角形。总状花序具4-6花；花直径1.2-1.4厘米。蓇葖果1或2(或3)，宽条形；种子12-16，成熟时黑色。花期5-6月，果期7-9月。生海拔2800-4000米林中、林缘或草坡上。产中国西南和华西。印度北部、不丹和缅甸北部亦有。

Perennial herbs. Leaves petiolate, 2-3-ternate-sect, blades triangular. Racemes 4-6-flowered; flowers 1.2-1.4 cm diam. Follicles 1 or 2(or 3), broadly linear; seeds 12-16, black when ripe. Fl. May-Jun. Fr. Jul-Sep. Forests, forest edges or grassy slopes at 2800-4000 m. Distributed in SW and W China. Also in N India, Bhutan and N Myanmar.

黄三七 *Souliea vaginata*

小升麻

Cimicifuga japonica (Thunb.) Spreng.

多年生草本。茎高达1米。叶1-2，近基生，为三出复叶；小叶正三角形，掌状浅裂。总状花序狭长，长10-25厘米，单一或有1-3分枝；萼片长3-5毫米；退化雄蕊圆卵形；雄蕊多数；心皮1-2。花期8-9月。生海拔800-2600米的林下或林缘。产四川、湖北、贵州、湖南、广东、浙江、安徽、河南、山西南部、陕西南部和甘肃南部。日本亦有。

Perennial herbs. Stems up to 1 m tall. Leaves 1-2, subbasal, ternate; leaflets deltoid, palmately lobed. Racemes elongate, 10-25 cm long, simple or 1-3-branched; sepals 3-5 mm long; staminodes orbicular-ovate; stamens numerous;

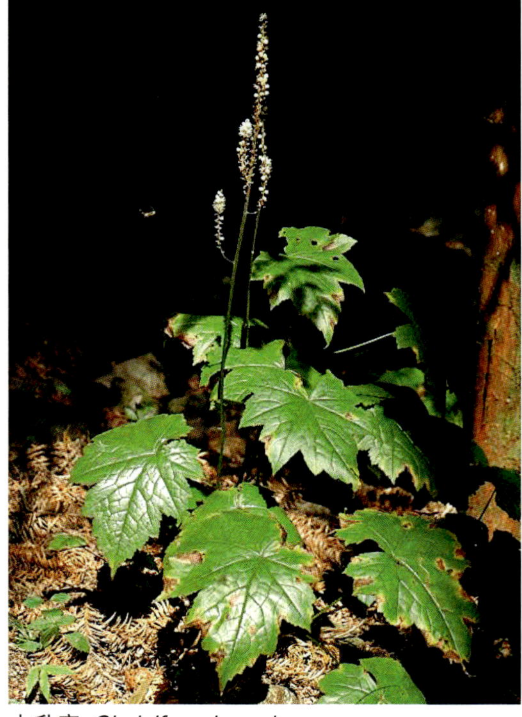

小升麻 *Cimicifuga japonica*

carpels 1-2. Fl. Aug-Sep. Forests or forest edges at 800-2600 m. Distributed in Sichuan, Hubei, Guizhou, Hunan, Guangdong, Zhejiang, Anhui, Henan, S Shanxi, S Shaanxi and S Gansu. Also in Japan.

单穗升麻

Cimicifuga simplex Wormsk.

多年生草本。茎高达1.5米，无毛。叶为一至三回三出羽状复叶；小叶狭卵形或菱形。总状花序长达35厘米，不分枝或近基部有少数短分枝；萼片长4毫米；退化雄蕊近圆形，顶端2浅裂；心皮2-7。花期8-9月。生海拔300-2300米的草丛、草甸或灌丛中。产四川、甘肃、陕西、河北、内蒙古、辽宁、吉林、黑龙江和台湾。俄罗斯(东西伯利亚)、蒙古、朝鲜半岛和日本亦有。

Perennial herbs. Stems up to 1.5 m tall, glabrous. Leaves 1-3-ternate-pinnate; leaflets narrow-ovate or rhombic. Racemes up to 35 cm long, simple or near base with a few short branches; sepals 4 mm long; staminodes suborbicular, apex 2-lobed; carpels 2-7. Fl. Aug-Sep. Grasses, meadows or bushes at 300-2300 m. Distributed in Sichuan, Gansu, Shaanxi, Hebei, Neimenggu, Liaoning, Jilin, Heilongjiang and Taiwan. Also in Russia (E Siberia), Mongolia, Korean Peninsula and Japan.

大三叶升麻

Cimicifuga heracleifolia Kom.

多年生草本。茎高约1米，无毛。叶为一至二回三出复叶；小叶卵形或倒卵形，长达12厘米。总状花序具2-9条分枝；萼片长3-4毫米；退化雄蕊顶端全缘；心皮3-5。蓇葖果长圆形。种子四周有鳞状翅。花期8-9月。生海拔达1000米的草坡或灌丛中。产内蒙古、辽宁、吉林和黑龙江。朝鲜半岛和俄罗斯(远东地区)亦有。

Perennial herbs. Stems ca. 1 m tall, glabrous. Leaves 1-2-ternate; leaflets ovate or obovate, up to 12 cm long. Raceme with 2-9 branches; sepals 3-4 mm long; staminodes at apex entire; carpels 3-5. Follicles oblong. Seeds scaly winged all around. Fl. Aug-Sep. Grassy slopes or bushes at up to 1000 m. Distributed in Neimenggu, Liaoning, Jilin and Heilongjiang. Also in Korean Peninsula and Russia (Far East).

大三叶升麻 *Cimicifuga heracleifolia*

短果升麻

Cimicifuga brachycarpa Hsiao

多年生草本。茎高约1.5米。叶为一至二回三出复叶；小叶菱形或狭卵形，长5-8厘米。总状花序具4-6分枝；花直径约4毫米；心皮(1-)2-3(-4)。蓇葖果近圆形，长约3毫米，无毛。种子周围无鳞状翅。花期7-8月。生海拔1310-2000米的林中或山坡阴处。产云南东北部、四川、重庆、湖北西部和陕西南部。

Perennial herbs. Stems ca. 1.5 m tall. Leaves 1-2-ternate; leaflets rhombic or narrow-ovate, 5-8 cm long. Raceme with 4-6 branches; flower ca. 4 mm diam; carpels (1-)2-3(-4). Follicles suborbicular, ca. 3 mm diam, glabrous. Seeds without scaly wings all around. Fl. Jul-Aug. Forests or shady places on slops at 1310-2000 m. Distributed in NE Yunnan, Sichuan, Chongqing, W Hubei and S Shaanxi.

单穗升麻 *Cimicifuga simplex*

短果升麻 *Cimicifuga brachycarpa*

升麻 *Cimicifuga foetida*

升麻

Cimicifuga foetida L.

草本。叶三角形，为二至三回三出羽状复叶。总状花序具3-20分枝；萼片白色或绿白色；退化雄蕊阔椭圆形，全缘或2浅裂或2裂至中部；心皮2-5，密被灰色柔毛。蓇葖果长圆形；种子周围具鳞质翅。花期7-9月，果期8-10月。生海拔1700-3600米林缘、林中、山地或草坡上。产中国西南、华北和华西。印度、不丹、缅甸、哈萨克斯坦、俄罗斯(西伯利亚)和蒙古亦有。

Herbs. Leaves triangular, 2-3-ternate-pinnate. Inflorescences 3-20-branched; sepals white or greenish white; staminodes broadly elliptic, entire or 2-lobed or 2-parted to middle; carpels 2-5, densely gray pubescent. Follicles oblong; seeds with scaly wings all around. Fl. Jul-Sep. Fr. Aug-Oct. Forest edges, forests, mountains or grassy slopes at 1700-3600 m. Distributed in SW, N and W China. Also in India, Bhutan, Myanmar, Kazakhstan, Russia (Siberia) and Mongolia.

兴安升麻

Cimicifuga dahurica (Turcz.) Maxim.

草本，雌雄异株。叶三角形。花序为复总状，具7-20分枝；退化雄蕊分叉，2裂，具2个不育花药；心皮4-7。蓇葖果顶端具外弯的喙。种子3或4，周围具膜质鳞质翅。花期7-8月，果期8-9月。生海拔300-1200米灌丛、林中、草地或山坡疏林中。产华北和东北。俄罗斯(远东地区、西伯利亚)、蒙古和朝鲜半岛亦有。

Herbs, dioecious. Leaves triangular. Inflorescences compound racemose, 7-20-branched; staminode forked, 2-parted, with 2 sterile anthers; carpels 4-7. Follicles reflexed-beaked at apex. Seeds 3 or 4, with membranous scaly wings all around. Fl. Jul-Aug. Fr. Aug-Sep. Thickets, forests, grasslands or sparse forests on slopes at 300-1200 m. Distributed in N and NE China. Also in Russia (Far East, Siberia), Mongolia and Korean Peninsula.

类叶升麻

Actaea asiatica Hara

草本。叶为二至三回三出复叶或二至三回羽状复叶。花序总状；花梗在果期增粗；萼片倒卵形；花瓣黄色，匙形；心皮1，子房椭圆体形。浆果紫黑色。花期5-6月，果期7-9月。生海拔300-3100米的林下或沟边阴处。产中国西南、华北、华西和东北。俄罗斯(远东地区)、

兴安升麻 *Cimicifuga dahurica*

类叶升麻 *Actaea asiatica*

红果类叶升麻 *Actaea erythrocarpa*

朝鲜半岛和日本亦有。

Herbs. Leaves 2-3-ternate or 2-3-pinnate. Inflorescens racemose; pedicels thickened at fruiting; sepals obovate; petals yellow, spathulate; carpels 1, ovary ellipsoid. Berries purple-black. Fl. May-Jun. Fr. Jul-Sep. Forests or shady places by streams at 300-3100 m. Distributed in SW, N, W and NE China. Also in Russia (Far East), Korean Peninsula and Japan.

红果类叶升麻
Actaea erythrocarpa Fisch.

多年生草本。茎高60-70厘米。下部茎生叶为三回三出羽状复叶，具长柄；小叶卵形，长6-10厘米，3裂，边缘具锯齿。总状花序长约6厘米；花梗果期粗0.6毫米；萼片4，长约2.5毫米；花瓣匙形，长2.5毫米；雄蕊长约5毫米；心皮1，无花柱。果实浆果状，红色，直径约6毫米。花期5-6月。生海拔700-1500米的林中。产山西、河北、内蒙古、辽宁、吉林和黑龙江。蒙古、俄罗斯(远东地区、西伯利亚)、日本和欧洲亦有。

Perennial herbs. Stems 60-70 cm tall. Lower cauline leaves 2-3 times ternately pinnate, long petiolate; leaflets ovate, 6-10 cm long, 3-lobed, serrate. Racemes ca. 6 cm long; pedicels at fruiting time 0.6 mm across; sepals 4, ca. 2.5 mm long; petals spatulate, 2.5 mm long; stamens ca. 5 mm long; carpel 1, without style. Fruits berry-like, red, ca. 6 mm diam. Fl. May-Jun. Forests at 700-1500 m. Distributed in Shanxi, Hebei, Neimenggu, Liaoning, Jilin and Heilongjiang. Also in Mongolia, Russia (Far East, Siberia), Japan and Europe.

铁筷子
Helleborus thibetanus Franch.

多年生草本。茎高30-50厘米，无毛。基生叶1(-2)，具长柄，鸡足状3全裂，无毛；茎生叶较小。花1(-2)朵顶生；萼片5，粉红色，果期变绿色；花瓣8-10，筒状；雄蕊多数；心皮2-3。花期4月。生海拔1100-3700米的山地林中或灌丛中。产四川西北部、甘肃南部、陕西南部和湖北西部。

Perennial herbs. Stems 30-50 cm tall, glabrous. Basal leaves 1(-2), long petiolate, pedately 3-sect, glabrous; cauline leaves smaller. Flowers 1(-2) terminal; sepals 5, pink, at fruiting time turning green; petals 8-10, tubular; stamens numerous; carpels 2-3. Fl. Apr. Forests or bushes at 1100-3700 m. Distributed in NW Sichuan, S Gansu, S Shaanxi and W Hubei.

菟葵
Eranthis stellate Maxim.

根状茎球形。基生叶1或不存在，有长柄；叶片圆肾形，3全裂，无毛。花葶高达20厘米，无毛；总苞苞片细裂；花单朵顶生；萼片5，黄色；花瓣约10，顶端2裂；雄蕊多数；心皮6-9。花期3-4月。生林下或林边。产辽宁和吉林。朝鲜半岛北部和俄罗斯(远东地区)亦有。

Rhizomes globose. Basal leaf 1 or wanting, long petiolate; blades orbicular-reniform, 3-sect, glabrous. Scapes up to 20 cm tall, glabrous; involucral bracts dissected; flower solitary, terminal; sepals 5, yellow; petals ca. 10, at apex 2-lobulate; stamens numerous; carpels 6-9. Fl. Mar-Apr. Forests or forest edges. Distributed in Liaoning and Jilin. Also in N Korean Peninsula and Russia (Far East).

铁筷子 *Helleborus thibetanus*

菟葵 *Eranthis stellate*

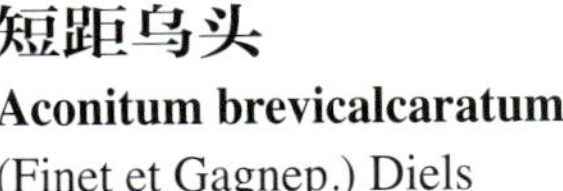

短距乌头 *Aconitum brevicalcaratum*

短距乌头

Aconitum brevicalcaratum (Finet et Gagnep.) Diels

根状茎直径约1厘米。茎高达1米，被贴伏毛。基生叶3-4，具长柄；叶片肾形，宽7.5-13厘米，3中裂；茎生叶较小。总状花序长达40厘米；花梗基部具2小苞片；上萼片圆筒形，长1.3-1.5厘米；心皮3。花期8-10月。生海拔3500米的山地草坡。产云南西北部和四川西南部。

Rhizomes ca. 1 cm diam. Stems up to 1 m tall, appressed-puberulous. Basal leaves 3-4, long petiolate; blades reniform, 7.5-13 cm broad, 3-fid; cauline leaves smaller. Racemes up to 40 cm long; pedicels at base 2-bracteolate; upper sepal cylindric, 1.3-1.5 cm long; carpels 3. Fl. Aug-Oct. Grassy slopes at 3500 m. Distributed in NW Yunnan and SW Sichuan.

花葶乌头 *Aconitum scaposum*

花葶乌头

Aconitum scaposum Franch.

根状茎直径约0.8厘米。茎高约50厘米。基生叶3-4，具长柄，叶片肾状五角形，宽8.5-20厘米；茎生叶小。总状花序长达40厘米；花梗基部具2小苞片；上萼片蓝紫色，圆筒形；花瓣具长距；心皮3。花期8-9月。生海拔1200-2000米的山谷林中或阴处。产四川、贵州北部、湖北、江西、陕西南部和河南西南部。

Rhizomes ca. 0.8 cm diam. Stems ca. 50 cm tall. Basal leaves 3-4, long petiolate, blades reniform-pentagonal, 8.5-20 cm broad; cauline leaves small. Racemes up to 40 cm long; pedicels at base 2-bracteolate; upper sepal blue-purple, cylindric; petals long spurred; carpels 3. Fl. Aug-Sep. Forests or shady places in valley at 1200-2000 m. Distributed in Sichuan, N Guizhou, Hubei, Jiangxi, S Shaanxi and SW Henan.

粗花乌头

Aconitum crassiflorum Hand.-Mazz.

根状茎直径约1厘米。茎高达1米。基生叶2-3，具长柄，叶片圆肾形，宽7.6-12厘米，3深裂；茎生叶较小。总状花序长14-36厘米；花梗在基部或中部具2小苞片；上萼片蓝紫色，筒形，长1.6-2.3厘米；心皮3。花期7-8月。生海拔3200-4200米的草坡、灌丛或林中。产云南西北部和四川西南部。

Rhizomes ca. 1 cm diam. Stems up to 1 m tall. Basal

粗花乌头 *Aconitum crassiflorum*

赣皖乌头 *Aconitum finetianum*

两色乌头 *Aconitum alboviolaceum*

leaves 2-3, long petiolate, blades orbicular-reniform, 7.6-12 cm broad, 3-parted; cauline leaves smaller. Racemes 14-36 cm long; pedicels at base or near the middle 2-bracteolate; upper sepal blue-purple, tubular, 1.6-2.3 cm tall; carpels 3. Fl. Jul-Aug. Grassy slopes, bushes or forests at 3200-4200 m. Distributed in NW Yunnan and SW Sichuan.

赣皖乌头

Aconitum finetianum Hand.-Mazz.

根状茎直径3-4毫米。茎缠绕，长约1米，疏被短柔毛。下部茎生叶具长柄，叶片五角状肾形，宽10-18厘米，3深裂；上部茎生叶较小。总状花序具4-9花；轴和花梗被反曲短柔毛；萼片白色带淡紫色，上萼片圆筒形；心皮3。花期8-9月。生海拔850-1600米的山地阴湿处。产湖南、江西、浙江西部和安徽南部。

Rhizomes 3-4 mm diam. Stems twining, ca. 1 m long, sparsely puberulous. Lower cauline leaves long petiolate, blades pentagonal-reniform, 10-18 cm broad, 3-parted; upper cauline leaves smaller. Racemes 4-9-flowered; rachis and pedicels retrorsely puberulous; sepals white, tinged with purplish, upper sepal cylindric; carpels 3. Fl. Aug-Sep. Shady damp places at 850-1600 m. Distributed in Hunan, Jiangxi, W Zhejiang and S Anhui.

两色乌头

Aconitum alboviolaceum Kom.

多年生草本。茎缠绕。叶3裂超过中部。总状花序长6-14厘米，具3-8花；小苞片和苞片条形；萼片淡紫色或近白色，上萼片圆筒状，高1.3-2厘米，具短喙；心皮3。蓇葖果直立。花期8-9月。生海拔300-1400米林中、山谷灌丛或山地。产河北、辽宁、吉林和黑龙江。俄罗斯(远东地区)和朝鲜半岛亦有。

Perennial herbs. Stems twining. Leaves 3-parted ± beyond middle. Racemes 6-14 cm long, 3-8-flowered; bracts and bracteoles linear; sepals purplish or nearly white, upper sepals cylindric, 1.3-2 cm tall, shortly beaked; carpels 3. Follicles erect. Fl. Aug-Sep. Forests, thickets in valleys, or mountains at 300-1400 m. Distributed in Hebei, Liaoning, Jilin and Heilongjiang. Also in Russia (Far East) and Korean Peninsula.

高乌头

Aconitum sinomontanum Nakai

根状茎长达20厘米。茎高达1.5米。叶五角形，宽达28厘米，3深裂，基部心形。总状花序长达50厘米，具多数花；轴和花梗被短柔毛；萼片蓝紫色，上萼片筒状，高约2厘米；心皮3。花期6-9月。生海拔1000-3700米的山坡草地或林中。产贵州、湖北西部、四川、青海东部、陕西南部、山西和河北。

Rhizomes up to 20 cm long. Stems up to 1.5 m tall. Leaves pentagonal, up to 28 cm broad, 3-parted, base cordate. Racemes up to 50 cm long, multi-flowered; rachis and pedicels puberulous; sepals blue-purple, upper sepal tubular, ca. 2 cm tall; carpels 3. Fl. Jun-Sep. Grassy slopes or forests at 1000-3700 m. Distributed in Guizhou, W Hubei, Sichuan, E Qinghai, S Shaanxi, Shanxi and Hebei.

高乌头 *Aconitum sinomontanum*

白喉乌头
Aconitum leucostomum Worosch.

草本。叶片肾形或圆肾形，3深裂。总状花序长20-45厘米，多花，花轴和花梗密具开展的淡黄色短腺毛；萼片淡蓝紫色，下部白色，上萼片筒状；花瓣距稍拳卷；心皮3。蓇葖果长1-1.2厘米。花期7-8月。生海拔900-2600米草坡或山谷、森林、林缘。产甘肃北部和新疆。哈萨克斯坦、吉尔吉斯斯坦、俄罗斯(西西伯利亚)和蒙古亦有。

Herbs. Leaves reniform or orbicular-reniform, 3-parted. Racemes 20-45 cm long, many flowered; rachis and pedicels densely spreading yellowish shortly glandular pubescent; sepals pale blue-purple, lower part white, upper sepals tubular; petal spurs slightly circinate; carpels 3. Follicles 1-1.2 cm long. Fl. Jul-Aug. Grassy slopes or valleys, forests, forest edges at 900-2600 m. Distributed in N Gansu and Xinjiang. Also in Kazakhstan, Kyrgyzstan, Russia (W Siberia) and Mongolia.

白喉乌头 *Aconitum leucostomum*

山地乌头
Aconitum monticola Steinb.

多年生草本。茎高约1.2米，上部被开展短柔毛。叶圆肾形，宽达22厘米，3深裂。总状花序长约25厘米，具密集花；轴和花梗被开展黄色短毛；上萼片黄色，筒状，高约1.5厘米；心皮3，无毛。花期8月。生海拔2300米的山地阴湿处。产新疆。哈萨克斯坦亦有。

Perennial herbs. Stems ca. 1.2 m tall, above spreading-puberulous. Leaves orbicular-reniform, up to 22 cm broad, 3-parted. Racemes ca. 25 cm long, with dense flowers; rachis and pedicels spreading-yellow-puberulous; upper sepal yellow, tubular, ca. 1.5 cm tall; carpels 3, glabrous. Fl. Aug. Shady places on slopes at 2300 m. Distributed in Xinjiang. Also in Kazakhstan.

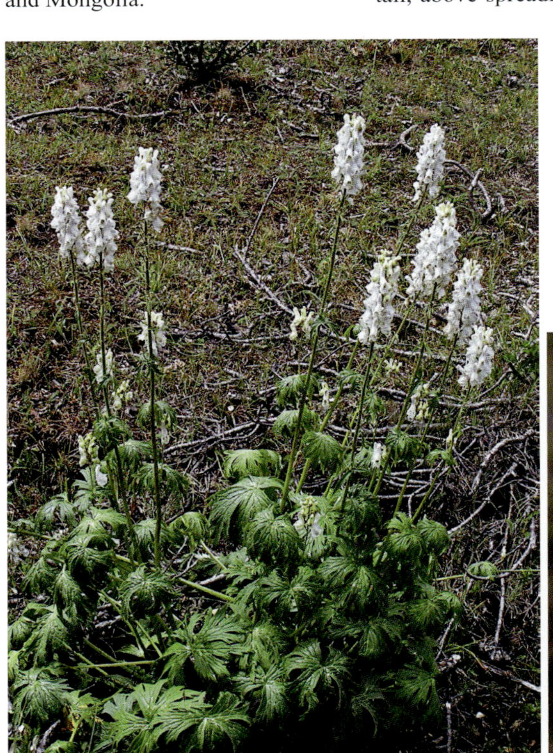

山地乌头 *Aconitum monticola*

华山乌头
Aconitum kirinense Nakai var. **australe** W. T. Wang

多年生草本，具根状茎。茎与叶柄被长硬毛。基生叶肾状五角形，3深裂。总状花序长约20厘米；上萼片黄色，筒形，高2厘米；心皮3，有毛。花期8-9月。生海拔780-2000米的林中或山坡石上。产湖北西部、陕西东南部、山西南部和河南西部。

Perennial herbs, with rhizomes. Stem with petioles hirsute. Basal leaves reniform-pentagonal, 3-parted. Racemes ca. 20 cm long; upper sepal yellow, tubular, 2 cm tall; carpels 3, hairy. Fl. Aug-Sep. Forest or on rocks of slopes at 780-2000 m. Distributed in W Hubei, SE Shaanxi, S Shanxi and W Henan.

细叶黄乌头
Aconitum barbatum Pers.

根状茎圆柱形，长达15厘米。茎高55-90厘米，下部被开展短柔毛。基生叶2-4，与茎下部叶具长柄，心状五角形，宽7-20厘

华山乌头 *Aconitum kirinense* var. *australe*

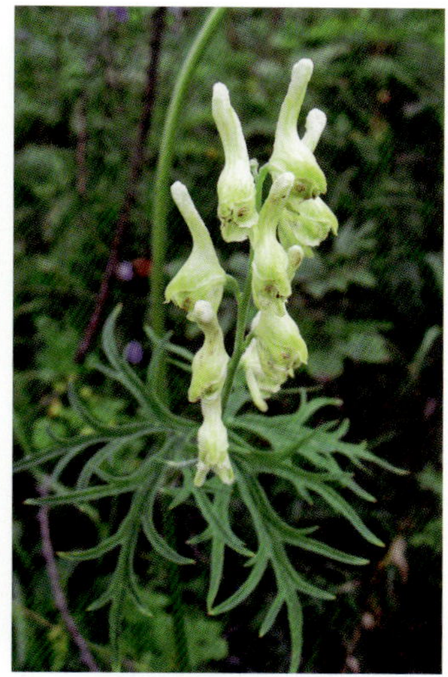

细叶黄乌头 *Aconitum barbatum*

米，3全裂，全裂片细裂。总状花序长约20厘米；轴和花梗密被贴伏短柔毛；萼片黄色，上萼片筒形，高1.3-1.7厘米，下缘长1-1.2厘米；心皮3。花期7-8月。生海拔400-900米的草坡上、林中或林边。产内蒙古北部。俄罗斯(西伯利亚、远东地区)亦有。

Rhizome terete, up to 15 cm long. Stems 55-90 cm tall, below spreading-puberulous. Basal leaves 2-4, with lower cauline leaves long petiolate, cordate-pentagonal, 7-20 cm broad, 3-sect, segments dissected. Raceme ca. 20 cm long; rachis and pedicels densely appressed-puberulous; sepals yellow, upper sepal tubular, 1.3-1.7 cm tall, lower margin 1-1.2 cm long; carpels 3. Fl. Jul-Aug. Grassy slopes, forests or forest margins at 400-900 m. Distributed in N Neimenggu. Also in Russia (Siberia, Far East)

牛扁 *Aconitum barbatum* var. *puberulum*

牛扁

Aconitum barbatum Pers. var. **puberulum** Ledeb.

多年生草本。茎高达1.1米，被贴伏短柔毛。叶肾形，宽达22厘米，3全裂，全裂片细裂。总状花序长达24厘米，具密集的花；上萼片黄色，筒状，高约2厘米；心皮3，被短柔毛。花期6-8月。生海拔400-2700米的疏林下或阴湿处。产新疆东部、内蒙古、山西和河北。俄罗斯(西伯利亚)亦有。

Perennial herbs. Stems up to 1.1 m tall, apressed-puberulous. Leaves reniform, up to 22 cm broad, 3-sect, segments dissected. Racemes up to 24 cm long, with dense flowers; upper sepal yellow, tubular, ca. 2 cm tall; carpels 3, puberulous. Fl. Jun-Aug. Sparse forests or shady places at 400-2700 m. Distributed in E Xinjiang, Neimenggu, Shanxi and Hebei. Also in Russia (Siberia).

甘青乌头

Aconitum tanguticum (Maxim.) Stapf

块根小，长约2厘米。茎高达50厘米。基生叶7-9，具长柄，叶片肾形，宽达6.8厘米，3中裂。总状花序具3-5花；上萼片船形，长约2厘米；花瓣的距极小；心皮5。花期7-8月。生海拔3200-4800米的山地草坡或沼泽草地。产西藏东部、云南西北部、四川西部、青海东部、甘肃南部和陕西南部。

Tubers small, ca. 2 cm long. Stems up to 50 cm tall. Basal leaves 7-9, long petiolate, blades reniform, up to 6.8 cm broad, 3-fid. Racemes 3-5-flowered; upper sepal navicular, ca. 2 cm long; petals with very small spurs; carpels 5. Fl. Jul-Aug. Grassy slopes or grassy places of swamps at 3200-4800 m. Distributed in E Xizang, NW Yunnan, W Sichuan, E Qinghai, S Gansu and S Shaanxi.

甘青乌头 *Aconitum tanguticum*

船盔乌头 *Aconitum naviculare*

船盔乌头
Aconitum naviculare (Bruhl) Stapf

块根小，长0.8-1.5厘米。茎高5-30(-45)厘米。叶基生并茎生，叶片五角形，宽约3厘米，3中裂。总状花序有1-5花；萼片堇色或紫色，上萼片船形，长约1.5厘米；花瓣的距小，向前弯；心皮5。花期9月。生海拔3200-5000米的草坡或灌丛中。产西藏南部。不丹和印度北部亦有。

Tubers small, 0.8-1.5 cm long. Stems 5-30(-45) cm tall. Leaves basal and cauline; blades pentagonal, ca. 3 cm broad, 3-fid. Racemes 1-5-flowered; sepals violet or purple, upper sepal navicular, ca. 1.5 cm long; petaline spur small, forward curved; carpels 5. Fl. Sep. Grassy slopes or bushes at 3200-5000 m. Distributed in S Xizang. Also in Bhutan and N India.

美丽乌头
Aconitum pulchellum Hand.-Mazz.

块根小，长约7毫米。茎高6.5-30(-50)厘米。叶基生并茎生；叶片五角形，宽约3厘米，3全裂或3深裂。总状花序具1-4花；萼片蓝色，上萼片盔状船形或盔形；花瓣的距向内弯；心皮5。花期8-9月。生海拔3500-4500米的草坡或石处。产西藏东南部、云南西北部和四川南部。印度北部、不丹和缅甸北部亦有。

Tubers small, ca. 7 mm long. Stems 6.5-30(-50) cm tall. Leaves basal and cauline; blades pentagonal, ca. 3 cm broad, 3-sect or 3-parted. Racemes 1-4-flowered; sepals blue, upper sepal galeate-navicular or galeate; petaline spurs incurved; carpels 5. Fl. Aug-Sep. Grassy slopes or rocky places at 3500-4500 m. Distributed in SE Xizang, NW Yunnan and S Sichuan. Also in N India, Bhutan and N Myanmar.

保山乌头
Aconitum nagarum Stapf

多年生草本。叶片五角状肾形，纸质或革质，3全裂。总状花序狭长，长12-30厘米，具6-25花；轴和花梗密具白色弯曲和平伏柔毛；萼片蓝紫色，上萼片船状盔形；心皮5，密被白色柔毛。花期10月。生海拔1800-3800米的林缘、矮灌丛或草坡上。产云南。印度东北部和缅甸北部亦有。

Perennial herbs. Leaves pentagonal-reniform, papery or coriaceous, 3-sect. Racemes elongate, 12-30 cm long, 6-25-flowered; rachis and pedicels densely curved and appressed white pubescent; sepals blue-purple, upper sepal navicular-galeate; carpels 5, densely white pubescent. Fl. Oct. Forest edges, thickets or grassy slopes at 1800-3800 m. Distributed in Yunnan. Also in NE India and N Myanmar.

褐紫乌头
Aconitum brunneum Hand.-Mazz.

块根长1.5-3.5厘米。茎高85-110厘米。叶基生并茎生；叶片肾形五角形，宽约10厘米，3深裂。总状花序长达50厘米；萼片褐紫色，上萼片船形或船状盔形，长约1厘米；花瓣无距；心皮3。花期8-9月。生海拔3000-4250米的山坡阴处或

美丽乌头 *Aconitum pulchellum*

保山乌头 *Aconitum nagarum*

褐紫乌头 *Aconitum brunneum*

膝瓣乌头 *Aconitum geniculatum*

冷杉林中。产四川西北部和甘肃西南部。

Tubers 1.5-3.5 cm long. Stems 85-110 cm tall. Leaves basal and cauline; blades reniform-pentagonal, ca. 10 cm broad, 3-parted. Racemes up to 50 cm long; sepals brown-purple, upper sepal navicular or navicular-galeate, ca. 1 cm long; petals not spurred; carpels 3. Fl. Aug-Sep. Shady places on slopes or *Abies* forests at 3000-4250 m. Distributed in NW Sichuan and SW Gansu.

膝瓣乌头

Aconitum geniculatum Fletcher et Lauener

草本。叶圆状五角形，3深裂。花序总状或伞房状，长3-8.5厘米，具2-8花；花序梗和花梗无毛；上萼片高盔状；花瓣无毛，爪顶端膝状；心皮5，无毛。花期7月。生海拔3200-3600米山地、草坡、湿草地或河边。产云南和四川。

Herbs. Leaves orbicular-pentagonal, 3-parted. Inflorescences racemose or corymbose, 3-8.5 cm long, 2-8-flowered; rachis and pedicels glabrous; upper sepal highly galeate; petals glabrous, claws geniculate at apex; carpels 5, glabrous. Fl. Jul. Mountains, grassy slopes, wet grassy places or by streams at 3200-3600 m. Distributed in Yunnan and Sichuan.

普格乌头

Aconitum pukeense W. T. Wang

茎高约1米，无毛。叶心状五角形，宽约10厘米，3深裂。总状花序约有10花；下部花梗长5-8厘米，无毛，其小苞片叶状，3深裂；萼片紫色，上萼片船状盔形，高2厘米；花瓣和雄蕊均无毛；心皮5，无毛。花期8月。生海拔3400-3500米的山谷林边或沟边。产四川(米易县、普格县)。

Stems ca. 1 m tall, glabrous. Leaves cordate-pentagonal, ca. 10 cm broad, 3-parted. Raceme ca. 10-flowered; lower pedicels 5-8 cm long, glabrous, with bracteoles foliaceous, 3-parted; sepals purple, upper sepal navicular-galeate, 2 cm tall; petals and stamens glabrous; carpels 5, glabrous. Fl. Aug. Forest edges in valley or by streams at 3400-3500 m. Distributed in Sichuan (Miyi County, Puge County).

普格乌头 *Aconitum pukeense*

聂拉木乌头 *Aconitum nielamuense*

聂拉木乌头

Aconitum nielamuense W. T. Wang

块根长3.5-6.5厘米。茎高60-150厘米。中部茎生叶有柄；叶片肾状五角形，宽约9.5厘米，3深裂。总状花序长6.5-30厘米；萼片暗紫色，上萼片盔状船形，长1.8-2.4厘米；心皮5，密被柔毛。花期8-9月。生林边或灌丛中，海拔3400-3900米。特产西藏(聂拉木县)。

Tubers 3.5-6.5 cm long. Stems 60-150 cm tall. Middle cauline leaves petiolate; blades reniform-pentagonal, ca. 9.5 cm broad, 3-parted. Racemes 6.5-30 cm long; sepals dark-purple, upper sepal galeate-navicular, 1.8-2.4 cm long; carpels 5, densely pubescent. Fl. Aug-Sep. Forest edges or bushes at 3400-3900 m. Endemic to Xizang (Nyalam County).

展毛大渡乌头

Aconitum franchetii Finet et Gagnep. var. **villosulum** W. T. Wang

块根胡萝卜形，长6厘米。茎高达1.2米。中部茎生叶具稍长柄，心状五角形，宽约8厘米，3深裂。总状花序长10-35厘米；轴和花梗被开展的短柔毛；萼片蓝色，外面无毛。上萼片盔形，高1.8-2.4厘米；花瓣无毛；心皮3或5。花期7-9月。生海拔3400-4000米的草坡或林中。产四川西部。

Tubers carrot-shaped, 6 cm long. Stems up to 1.2 m tall. Middle cauline leaves slightly long petiolate, cordate-pentagonal, ca. 8 cm broad, 3-parted. Racemes 10-35 cm long; rachis and pedicels spreading puberulous; sepals blue, abaxially glabrous, upper sepal galeate, 1.8-2.4 cm tall; petals glabrous; carpels 3 or 5. Fl. Jul-Sep. Grassy slopes or forests at 3400-4000 m. Distributed in W Sichuan.

丽江乌头

Aconitum forrestii Stapf

多年生草本。茎中部以上的叶有短柄或无柄，叶片3浅裂。总状花序顶生，较长，多花，有淡黄色开展的毛；萼片蓝紫色，上萼片盔形；花瓣无毛或被柔毛，唇长约5毫米，2裂，距稍内卷；心皮5，无毛或被柔毛。花期9月。生海拔约3100米山地草坡或林缘。产云南和四川。

Perennial herbs. Upper cauline leaves with short petioles or sessile; blades 3-parted. Racemes terminal, long, many flowered, yellowish spreading pubescent; sepals purple-blue, upper sepal galeate; petals glabrous or pubescent, lips ca. 5 mm long, 2-lobed, spurs slightly incurved; carpels 5, glabrous or pubescent. Fl. Sep. Grassy slopes or forest edges at ca. 3100 m. Distributed in Yunnan and Sichuan.

独龙乌头

Aconitum taronense (Hand.-Mazz.) Fletcher et Lauener

块根长约4厘米。茎高80-110厘米。茎生叶心状五角形，3深裂。总状花序长15-37厘米；轴被短柔毛；上萼片紫色，盔形，心皮4或5。花期8-9月。生海拔2600-4100米的草坡上。产云南西北部和西藏东南部(察隅县)。

Tubers ca. 4 cm long. Stems 80-110 cm tall. Cauline leaves cordate-pentagonal, 3-parted. Racemes 15-37 cm long; rachis puberulous; upper sepal purple, galeate; carpels 4 or 5. Fl. Aug-Sep. Grassy slopes at 2600-4100 m. Distributed in NW Yunnan and

展毛大渡乌头 *Aconitum franchetii* var. *villosulum*

丽江乌头 *Aconitum forrestii*

独龙乌头 *Aconitum taronense*

腺毛加查乌头

Aconitum chiachaense W. T. Wang var. **glandulosum** W. T. Wang

块根长约4厘米。茎高约30厘米。茎生叶肾状五角形，3深裂。总状花序长12厘米；轴被腺毛；上萼片紫色，盔形或船状盔形，高1.5厘米；心皮5。花期9月。生海拔4400米的短圆柏丛中。特产西藏(加查县)。

Tubers ca. 4 cm long. Stems ca. 30 cm tall. Cauline leaves reniform-pentagonal, 3-parted. Racemes ca. 12 cm long; rachis glandular-hairy; upper sepal purple, galeate or navicular-galeate, ca. 1.5 cm tall; carpels 5. Fl. Sep. Low *Sabina* bushes at 4400 m. Endemic to Xizang (Gyaca County) .

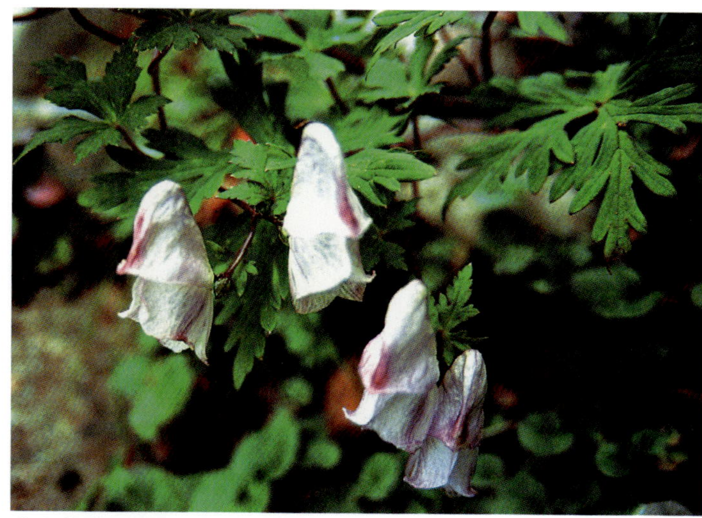

苍山乌头 *Aconitum contortum*

苍山乌头

Aconitum contortum Finet et Gagnep.

多年生草本。叶五角形，3全裂。花序具2-5花；轴具反曲柔毛；苞片叶状；花萼蓝紫色，上萼片高盔形；花瓣距内曲，球形；心皮3-5。蓇葖果长1.4-1.8厘米。花期9-10月。生海拔约3400米的高山草地。产云南西部。

Perennial herbs. Leaves pentagonal, 3-sect. Racemes 2-5-flowered; rachis retrorse pubescent; bracts leaflike; sepals blue-purple, upper sepal highly galeate; petal spurs incurved, globose; carpels 3-5. Follicles 1.4-1.8 cm long. Fl. Sep-Oct. Alpine grasslands at ca. 3400 m. Distibuted in W Yunnan.

SE Xizang (Zayü County).

短梗墨脱乌头

Aconitum elliotii var. **doshongense** (Lauener) W. T. Wang

块根长约5厘米。茎高约50厘米。茎生叶肾状五角形，3深裂。总状花序长约20厘米；花梗长达4.5厘米，被短柔毛；上萼片蓝紫色，盔形；心皮5。花期7-8月。生海拔3700-4100米的草坡上。特产西藏(墨脱县)。

Tubers ca. 5 cm long. Stems ca. 50 cm tall. Cauline leaves reniform-pentagonal, 3-parted. Racemes ca. 20 cm long; pedicels to 4.5 cm long, puberulous; upper sepal blue-purple, galeate; carpels 5. Fl. Jul-Aug. Grassy slopes at 3700-4100 m. Endemic to Xizang (Mêdog County).

短梗墨脱乌头 *Aconitum elliotii* var. *doshongense*

腺毛加查乌头 *Aconitum chiachaense* var. *glandulosum*

太白乌头 *Aconitum taipaicum*

太白乌头

Aconitum taipaicum Hand.-Mazz.

块根长1.5-3厘米。茎高35-60厘米。中部茎生叶具长柄；叶片五角形，宽约6厘米，3深裂。总状花序具2-4花；轴和花梗被反曲短柔毛；上萼片蓝色，盔形，高1.7厘米，无毛；心皮5。花期9月。生海拔2600-3400米的高山草地。产陕西南部和河南西部。

Tubers 1.5-3 cm long. Stems 35-60 cm tall. Middle cauline leaves long petiolate; blades pentagonal, ca. 6 cm broad, 3-parted. Racemes 2-4-flowered; rachis and pedicels puberulous; upper sepal blue, galeate, 1.7 cm tall, glabrous; carpels 5. Fl. Sep. Alpine meadows at 2600-3400 m. Distributed in S Shaanxi and W Henan.

瓜叶乌头

Aconitum hemsleyanum E. Pritz. ex Diels

草质藤本。叶五角形或卵状五角形，3深裂。总状花序顶生，具2-6(-12)花；花萼深蓝色，上萼片高盔或圆柱状盔状，具爪或不明显；花瓣距内曲；心皮5。蓇葖果直立。花期8-10月。生海拔1700-3500米山坡林中或灌丛中。产华中和华东。缅甸亦有。

Herbaceous vines. Leaves pentagonal or ovate-pentagonal, 3-parted. Racemes terminal, 2-6(-12)-flowered; sepals dark blue, upper sepal high galeate or cylindric-galeate, clawed or indistinctly so; petal spurs incurved; carpels 5. Follicles erect. Fl. Aug-Oct. Forests or bushes on slopes at 1700-3500 m. Distributed in C and E China. Also in Myanmar.

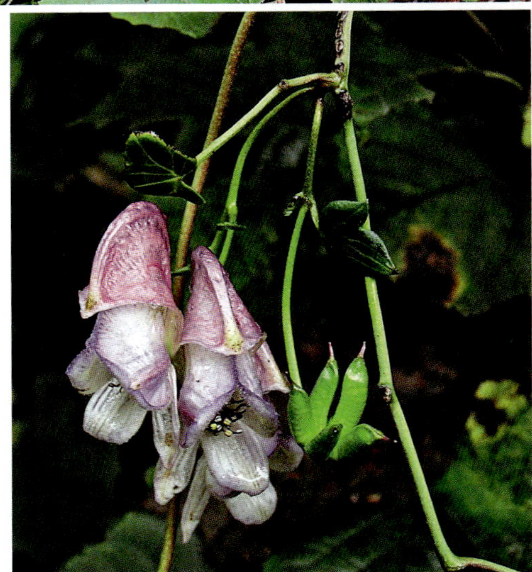

瓜叶乌头 *Aconitum hemsleyanum*

黄草乌

Aconitum vilmorinianum Kom.

草质藤本。块根椭圆状球形或胡萝卜状，直径约1厘米。茎缠绕。叶3全裂。花序具3-6花；花轴和花梗密具淡黄色反曲柔毛；上萼片高盔形；心皮5。蓇葖果直立。花期8-10月。生海拔2100-3000米的矮灌丛中。产云南中部、四川西南部和贵州西部。

Herbaceous vines. Tubers ellipsoid-globose or carrot-shaped, ca. 1 cm diam. Stems twining. Leaves 3-sect. Inflorescences 3-6-flowered; rachis and pedicels densely retrorse yellowish pubescent; upper sepal high galeate; carpels 5. Follicles erect. Fl. Aug-Oct. Thickets at 2100-3000 m. Distributed in C Yunnan, SW Sichuan and W Guizhou.

黄草乌 *Aconitum vilmorinianum*

川鄂乌头

Aconitum henryi Pritz.

块根长1.5-3.8厘米。茎缠绕，无毛。叶五角形，3全裂。总状

川鄂乌头 *Aconitum henryi*

花序有少数花；轴和花梗几无毛；上萼片蓝色，高盔形，高2-2.5厘米；花瓣的距内弯，长4-5毫米；心皮3。花期8-10月。生海拔1000-2000米的山地丛林中。产四川、重庆、湖北西部和陕西西南部。

Tubers 1.5-3.8 cm long. Stems twining, glabrous. Leaves pentagonal, 3-sect. Racemse few-flowered; rachis and pedicels subglabrous; upper sepal blue, high-galeate, 2-2.5 cm tall; petaline spurs incurved, 4-5 mm long; carpels 3. Fl. Aug-Oct. Thickets at 1000-2000 m. Distributed in Sichuan, Chongqing, W Hubei and S Shaanxi.

蔓乌头

Aconitum volubile Pall. ex Koelle

草本。茎缠绕，分枝。叶具柄，五角形，宽8-10厘米，3全裂，全裂片细裂。总状花序顶生和腋生，有3-5花；轴和花梗密被开展短柔毛；萼片蓝紫色，长萼片高盔形，高1.8-2.7厘米，下缘长1-1.5厘米；花瓣距长2-3毫米；心皮5。花期8-9月。生海拔200-1000米的草坡或林中。产辽宁、吉林和黑龙江。俄罗斯(西伯利亚)亦有。

Herbs. Stems twining, branched. Leaves petiolate, pentagonal, 8-10 cm broad, 3-sect, segments dissected. Racemes terminal and axillary, 3-5-flowered; rachis and pedicels densely spreading-puberulous; sepals blue-purple, upper sepal high galeate, 1.8-2.7 cm tall, lower margin 1-1.5 cm long; petal spurs 2-3 mm long; carpels 5. Fl. Aug-Sep. Grassy slopes or forests at 200-1000 m. Distributed in Liaoning, Jilin and Heilongjiang. Also in Russia (Siberia).

薄叶乌头

Aconitum fischeri Reichb.

块根倒圆锥形。茎高1-1.6米，被短柔毛。叶五角形，宽8-15厘米，3深裂，中裂片菱形，3浅裂，侧裂片斜扇形，不等2深裂。顶生总状花序有4-6花；萼片淡蓝紫色，上萼片高盔形，高约2厘米，下缘长1.2-1.7厘米；心皮3。花期8月。生海拔400-800米的林中或草地上。产内蒙古东北部和黑龙江北部。俄罗斯(远东地区)亦有。

Tuber obconical. Stems 1-1.6 m tall, puberulous. Leaves pentagonal, 8-15 cm broad, 3-parted, central lobe rhombic, 3-lobed, lateral lobes obliquely flabellate, unequally 2-parted. Terminal raceme 4-6-flowered; sepals blue-purplish, upper sepal high galeate, ca. 2 cm tall, lower margin 1.2-1.7 cm long; carpels 3. Fl. Aug. Forests or grasslands at 400-800 m. Distributed in NE Neimenggu and N Heilongjiang. Also in Russia (Far East).

蔓乌头 *Aconitum volubile*

薄叶乌头 *Aconitum fischeri*

大苞乌头 *Aconitum raddeanum*

大苞乌头
Aconitum raddeanum Regel

草本。茎无毛。叶五角形，3全裂，中央裂片菱形。花序具数花；花轴和花梗无毛；苞片叶状；花萼蓝紫色，上萼片高盔形；花瓣距拳卷；心皮3-5，密被棕色柔毛。花期7月。生山坡。产黑龙江和吉林。俄罗斯(远东地区)亦有。

Herbs. Stems glabrous. Leaves pentagonal, 3-sect, central segments rhombic. Inflorescences several flowered; rachis and pedicels glabrous; bracts leaflike; sepals purple-blue, upper sepal high galeate; petal spurs incurved; carpels 3-5, densely brown pubescent. Fl. Jul. Mountain slopes. Distributed in Heilongjiang and Jilin. Also in Russia (Far East).

展毛乌头 *Aconitum carmichaelii* var. *truppelianum*

乌头 *Aconitum carmichaelii*

乌头
Aconitum carmichaelii Debx.

草本。叶五角形，3全裂至近基部。总状花序顶生，多花；花梗和花梗密具反曲或平伏柔毛；花萼蓝紫色，上萼片高盔形；花瓣距常拳卷；心皮3-5。蓇葖果长1.5-1.8厘米。花期9-10月。生海拔100-2200米的林边、灌丛、草坡。自云南东部和四川向东至华东广布。越南亦有。

Herbs. Leaves pentagonal, 3-sect or nearly to base. Racemes terminal, many flowered; rachis and pedicels ± densely retrorse and appressed pubescent; sepals blue-purple, upper sepal high galeate; petal spurs usually circinate; carpels 3-5. Follicles 1.5-1.8 cm long. Fl. Sep-Oct. Forest edges, thickets, grassy slopes at 100-2200 m. Widespread from E Yunnan and Sichuan eastward to E China. Also in Vietnam.

展毛乌头
Aconitum carmichaelii Debx. var. **truppelianum** (Ulbr.) W. T. Wang et Hsiao

本变种与乌头的区别在于本变种的花序轴和花梗被开展的毛。花期7-9月。生海拔280-400米的草坡、林边或灌丛中。产浙江北部、江苏、河南南部、山东和辽宁南部。

This variety differs from the typical variety in its spreading-pubescent raceme rachis and pedicels. Fl. Jul-Sep. Grassy slopes, forest margins or bushes 280-400 m. Distributed in N Zhejiang, Jiangsu, S Henan, Shandong and S Liaoning.

黄山乌头
Aconitum carmichaelii Debx. var. **hwangshanicum** (W. T. Wang et Hsiao) W. T. Wang et Hsiao

本变种与模式变种的区别在于叶较薄，草质，中央全裂片顶端渐尖或长渐尖，总状花序的轴很短。花期8-9月。生海拔1000米的山坡林边草地。产浙江西北部、安徽南部和江西东北部。

This variety differs from var. *carmichaeli* in its thinner herbaceous leaves with central segments acuminate or long acuminate at apex and in its short raceme rachises. Fl. Aug-Sep. Grassy places near forests on slopes at 1000 m. Distributed in NW

黄山乌头 *Aconitum carmichaelii* var. *hwangshanicum*

北乌头 *Aconitum kusnezoffii*

长白乌头 *Aconitum tschangbaischanense*

Zhejiang, S Anhui and NE Jiangxi.

北乌头
Aconitum kusnezoffii Reichb.

草本。叶五角形，3全裂，中央裂片菱形。花序顶生，具9-22花；花轴和花梗无毛；萼片蓝紫色，上萼片盔状或高盔状，高1.5-2.5厘米；心皮(4或)5。蓇葖果直立，长(0.8-)1.2-2厘米。花期7-9月。生海拔2000-2400米的草坡或林缘。产河北、山西、内蒙古、黑龙江、吉林和辽宁。朝鲜半岛、俄罗斯(远东地区、西伯利亚)和日本北部亦有。

Herbs. Leaves pentagonal, 3-sect, central segments rhombic. Racemes terminal, 9-22-flowered; rachis and pedicels glabrous; sepals purple-blue, upper sepal galeate or high galeate, 1.5-2.5 cm tall; carpels (4 or)5. Follicles erect, (0.8-)1.2-2 cm long. Fl. Jul-Sep. Grassy slopes or forest edges at 2000-2400 m. Distributed in Hebei, Shanxi, Neimenggu, Heilongjiang, Jilin and Liaoning. Also in Korean Peninsula, Russia (Far East, Siberia) and N Japan.

长白乌头
Aconitum tschangbaischanense S.H. Li et Y.H. Huang

草本。叶片肾状五角形，3全裂，全裂片细裂。总状花序顶生或腋生，长11-14.5厘米，具7-14花；上萼片高盔状或盔状，高1.5-2厘米；花瓣无毛，唇长约6毫米，距内弯；心皮5。生海拔1000-1700米的草坡或林缘。产吉林(长白山)。

Herbs. Leaves reniform-pentagonal, 3-sect, segments dissected. Racemes terminal or axillary, 11-14.5 cm long, 7-14-flowered; upper sepals high galeate or galeate, 1.5-2 cm tall; petals glabrous, lips ca. 6 mm long, spurs incurved; carpels 5. Grassy slopes or forest edges at 1000-1700 m. Distributed in Jilin (Changbai Mountain).

细叶乌头
Aconitum macrorhynchum Turcz.

块根长1.2-2.8厘米。茎高68-100厘米。叶圆卵形，宽达10厘米，3全裂，全裂片细裂。总状花序有5-15花；上萼片紫蓝色，高盔形，高1.5-1.9厘米，外面疏被短柔毛；心皮5(-8)。花期8-9月。生海拔200-500米的山地草甸上或山坡上。产吉林和黑龙江。俄罗斯(西伯利亚、远东地区)亦有。

Tubers 1.2-2.8 cm long. Stems 68-100 cm tall. Leaves orbicular-ovate, up to 10 cm broad, 3-sect, segments dissected. Racemes 5-15-flowered; upper sepal blue-purple, high galeate, 1.5-1.9 cm tall, abaxially with sparse hairs; carpels 5(-8). Fl. Aug-Sep. Meadows or slopes at 200-500 m. Distributed in Jilin and Heilongjiang. Also in Russia (Siberia, Far East).

细叶乌头 *Aconitum macrorhynchum*

缺刻乌头 *Aconitum incisofidum*

新都桥乌头 *Aconitum tongolense*

冕宁乌头 *Aconitum legendrei*

缺刻乌头

Aconitum incisofidum W. T. Wang

草本。块根胡萝卜状。叶圆五角形，3全裂至近基部，裂片细裂。总状花序顶生，长38厘米，多花；轴和花梗密具反折柔毛；花萼紫堇色，上萼片盔形；花瓣唇微凹，距内曲；心皮3。蓇葖果长约1.1厘米。花期8-9月。生海拔3700-4000米的冷杉林下。产云南和四川。

Herbs. Tubers carrot-like. Leaves orbicular-pentagonal, 3-sect nearly to base, segments dissected. Racemes terminal, to 38 cm long, many flowered; rachis and pedicels densely retrorse pubescent; sepals violet-blue, upper sepal galeate; petal lips emarginate, spurs incurved; carpels 3. Follicles ca. 1.1 cm long. Fl. Aug-Sep. *Abies* forests at 3700-4000 m. Distributed in Yunnan and Sichuan.

新都桥乌头

Aconitum tongolense Ulbr.

块根胡萝卜形。茎高达1米。叶心状五角形，宽达8厘米，3全裂，全裂片细裂。总状花序长达50厘米；轴和花梗被贴伏短柔毛；萼片蓝色，上萼片船形，下缘凹，长达1.5厘米；心皮4，无毛。花期7-8月。生海拔3500米的草坡或溪边灌丛。产四川西部。

Tubers carrot-shaped. Stems up to 1 m tall. Leaves cordate-pentagonal, to 8 cm broad, 3-sect, segments dissected. Raceme to 50 cm long; rachis and pedicels appressed-puberulous; sepals blue, upper sepal navicular, lower margin concave, up to 1.5 cm long; carpels 4, glabrous. Fl. Jul-Aug. Grassy slopes or bushes by streams at 3500 m. Distributed in W Sichuan.

冕宁乌头

Aconitum legendrei Hand.-Mazz.

块根长约3.5厘米。茎高达1.2米，上部被反曲短柔毛。中部茎生叶五角形，3全裂，末回裂片狭卵形或披针状条形。总状花序狭长；轴和花梗被开展短毛；上萼片蓝色，盔形，高1.5-2厘米，外面被短柔毛；花瓣距长1毫米；心皮3-5。花期8-9月。生海拔2500-2800米的山地草坡。产四川西南部。

Tubers ca. 3.5 cm long. Stems up to 1.2 m tall, above retrorsely puberulous. Middle cauline leaves pentagonal, 3-sect, ultimate lobules narrow-ovate or lanceolate-linear. Racemes elongate; rachis and pedicels spreading-puberulous; upper sepal blue, galeate, 1.5-2 cm tall, abaxially puberulous; petaline spurs 1 mm long; carpels 3-5. Fl. Aug-Sep. Grassy slopes at 2500-2800 m. Distributed in SW Sichuan.

兴安乌头 *Aconitum ambiguum*

兴安乌头

Aconitum ambiguum Reichb.

草本无毛。茎高50-100厘米。叶心状五角形，宽5-13厘米，3全裂，全裂片细裂。总状花序顶生，有1-5花；萼片蓝紫色，上萼片盔形，高1.3-1.5厘米，下缘长1.5厘米；花瓣距小，约长1毫米；心皮3-5。花期8月。生海拔400-500米的林下或林缘。产内蒙古东北部。蒙古和俄罗斯(东西伯利亚)亦有。

Herbs glabrous. Stems 50-100 cm tall. Leaves cordate-pentagonal, 5-13 cm broad, 3-sect, segments dissected. Racemes terminal, 1-5-flowered; sepals blue-purple, upper sepal galeate, 1.3-1.5 cm tall, lower margin 1.5 cm long; petal spurs small, ca. 1 mm diam; carpels 3-5. Fl. Aug. Forests or forest margins at 400-500 m. Distribtued in NE Neimenggu. Also in Mongolia and Russia (E Siberia).

阿尔泰乌头

Aconitum smirnovii Steinb.

块根倒圆锥形，长约3.5厘米。茎高70-100厘米。叶五角形，宽达6.5厘米，3全裂，全裂片细裂。总状花序长达15厘米；轴和花梗被开展短柔毛；萼片蓝紫色，上萼片船形，下缘长1.4厘米；花瓣和雄蕊无毛；心皮3。花期8-9月。生海拔1800米的草坡上。产新疆(布尔津县)。蒙古和俄罗斯(中西伯利亚)亦有。

Tubers obconical, ca. 3.5 cm long. Stems 70-100 cm tall. Leaves pentagonal, to 6.5 cm broad, 3-sect, segments dissected. Raceme to 15 cm long; rachis and pedicels spreading-puberulous; sepals blue-purple, upper sepal navicular, lower margin 1.4 cm long; petals and stamens glabrous; carpels 3. Fl. Aug-Sep. Grassy slopes at 1800 m. Distributed in Xinjiang (Buerjin County) Also in Mongolia and Russia (C Siberia).

托里乌头

Aconitum tuoliense W. T. Wang

草本。茎高约50厘米，无毛。叶无毛，3全裂，全裂片细裂，小裂片条形。总状花序长约15厘米，花极密集；萼片深蓝色，上萼片船状盔形，高约2厘米，侧萼片圆倒卵形，下萼片椭圆形；花瓣2，具短距；雄蕊花丝多有1-2小齿；心皮3。花期8月。生海拔1800米的石砾草坡。特产新疆(托里县)。

Herbs. Stems ca. 50 cm tall, glabrous. Leaves glabrous, 3-sect, segments dissected into linear lobules. Racemes ca. 15 cm long, with very dense flowers; sepals deeply blue, upper sepal navicular-galeate, ca. 2 cm tall, lateral sepals orbicular-obovate, lower sepals elliptic; petals 2, with short spurs; stamen filaments mostly 1-2-denticulate; carpels 3. Fl. Aug. Gravelly grassy slopes at 1800 m. Endemic to Xinjiang (Toli County).

阿尔泰乌头 *Aconitum smirnovii*

托里乌头 *Aconitum tuoliense*

工布乌头
Aconitum kongboense

工布乌头

Aconitum kongboense Lauener

草本。块根近圆柱形。叶五角形，3全裂，裂片细裂。总状花序有多数花；近轴苞片叶状；花萼白色带紫色或淡紫色，上萼片盔形或船形，具短爪；心皮3-11。花期7-8月。生海拔3000-4000米的山坡草地或灌丛中。产云南、四川和西藏。

Herbs. Tubers almost cylindric. Leaves pentagonal, 3-sect, segments dissected. Racemes many-flowered; proximal bracts leaflike; sepals white-purple or purplish, upper sepals galeate or navicular, shortly clawed; carpels 3-11. Fl. Jul-Aug. Grasslands on mountain slopes or bushes at 3000-4000 m. Distributed in Yunnan, Sichuan and Xizang.

华北乌头

Aconitum jeholense Nakai et Kitag. var. **angustius** (W. T. Wang) Y. Z. Zhao

草本。叶3全裂，裂片细裂。总状花序长10-40厘米，具10-35花；花序轴和花梗疏生倒向的柔毛或近无毛；花萼蓝紫色，上萼片船状盔形；心皮3-5。蓇葖果椭圆体形，长约1厘米。花期8月，果期9月。生海拔2000-3000米林中、林缘或草地。产华北。俄罗斯(西伯利亚)亦有。

Herbs. Leaves 3-sect, segments dissected. Racemes 10-40 cm long, 10-35-flowered; rachis and pedicels sparsely retrorse pubescent or subglabrous; sepals blue-purple, upper sepal navicular-galeate; carpels 3-5. Follicles ellipsoid, ca. 1 cm long. Fl. Aug. Fr. Sep. Forests, forest edges or grasslands at 2000-3000 m. Distributed in N China. Also in Russia (Siberia).

类乌齐乌头

Aconitum leiwuqiense W. T. Wang

块根长3.5-7厘米。茎高约45厘米，无毛。中部茎生叶具稍长柄，肾状五角形，宽约6厘米，3全裂，全裂片细裂。总状花序长6-11厘米；轴和花梗被反曲短毛；小苞片条形；上萼片蓝色，船形，长约2.2厘米，无毛；心皮5。花期8月。生海拔4500米山地。特产西藏(类乌齐县)。

Tubers 3.5-7 cm long. Stems ca. 45 cm tall, glabrous. Middle cauline leaves slightly long petiolate, reniform-pentagonal, ca. 6 cm broad, 3-sect, segments dissected. Racemes 6-11 cm long; rachis and pedicels retrorsely puberulous; bracteoles linear; upper sepal blue, navicular, ca. 2.2 cm long, glabrous; carpels 5. Fl. Aug. Montane regions at 4500 m. Endemic to Xizang (Riwoqě County).

华北乌头 *Aconitum jeholense* var. *angustius*

类乌齐乌头 *Aconitum leiwuqiense*

伏毛铁棒锤 *Aconitum flavum*

多根乌头
Aconitum karakolicum

多根乌头

Aconitum karakolicum
Rapaics

块根数个呈链状。茎高约1米。茎生叶五角形，3全裂，全裂片细裂。总状花序顶生；上萼片紫色，盔形，高约2厘米；心皮3-5，无毛。花期7-8月。生海拔1900米的草坡上。产新疆。哈萨克斯坦亦有。

Tubers several chain-like. Stems ca. 1 m tall. Cauline leaves pentagonal, 3-sect, segments dissected. Racemes terminal; upper sepal purple, galeate, ca. 2 cm tall; carpels 3-5, glabrous. Fl. Jul-Aug. Grassy slopes at 1900 m. Distributed in Xinjiang. Also in Kazakhstan.

伏毛铁棒锤

Aconitum flavum Hand.-Mazz.

块根长约3.5厘米。茎高35-100厘米。中部茎生叶密集，具短柄或近无柄，宽3.6-4.5厘米，3全裂，全裂片细裂。总状花序狭长，有12-25花；轴和花梗被贴伏短毛；上萼片紫色或绿黄色，盔状船形，外面被短毛；心皮5。花期8月。生海拔2000-3700米的山地草坡或疏林下。产西藏东部、四川西部、青海、甘肃、宁夏南部和内蒙古南部。

Tubers ca. 3.5 m long. Stems 35-100 cm tall. Middle cauline leaves dense, shortly petiolate or subsessile, 3.6-4.5 cm broad, 3-sect, segments dissected. Racemes elongate, 12-25-flowered; rachis and pedicels appressed-puberulous; upper sepal purple or greenish-yellow, galeate-navicular, abaxially puberulous; carpels 5. Fl. Aug. Grassy slopes or sparse forests at 2000-3700 m. Distributed in E Xizang, W Sichuan, Qinghai, Gansu, S Ningxia and S Neimenggu.

莲花山乌头

Aconitum lianhuashanicum
W. T. Wang

茎高约135厘米，无毛。叶均茎生，密集，五角形，宽5-10厘米，3全裂，全裂片二至三回细裂，末回裂片狭条形。总状花序顶生，长约48厘米；轴无毛；花梗被短柔毛；萼片淡紫色，上萼片有长爪，还有喙；心皮5。花8-9月。生海拔2100米的田间沟中。特产甘肃(莲花山)。

Stem ca. 135 cm tall, glabrous. Leaves all cauline, dense, pentagonal, 5-10 cm broad, 3-sect, segments 2-3 times dissected, ultimate lobes narrow-linear. Raceme terminal, ca. 48 cm long; rachis glabrous; pedicels puberulous; sepals purplish, upper sepal long clawed, and rostrate; carpels 5. Fl. Aug-Sep. Ditch between fields at 2100 m. Endemic to Gansu (Lianhua Mountain).

莲花山乌头 *Aconitum lianhuashanicum*

铁棒锤 *Aconitum pendulum*

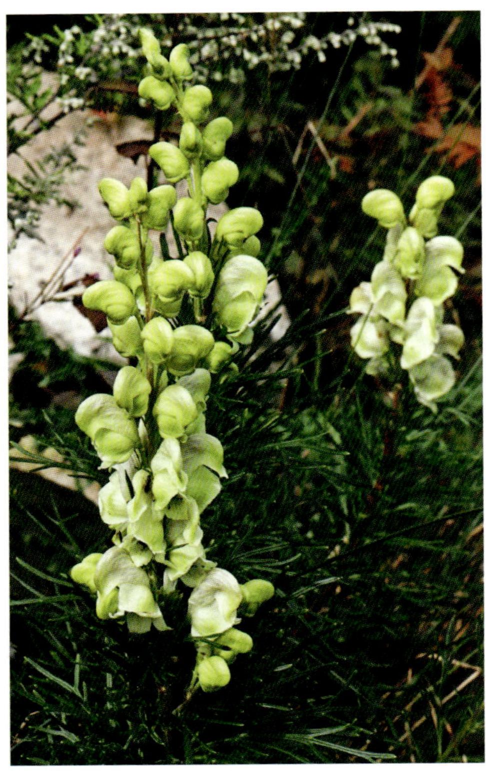
栾川乌头 *Aconitum luanchuanense*

黄花乌头

Aconitum coreanum (Lévl.) Rapaics

块根长约2.8厘米。茎高30-100厘米。中部茎生叶具稍长柄，宽3.6-6.4厘米，3全裂，全裂片细裂。总状花序有2-7花；上萼片淡黄色，船状盔形或盔形，外面被短毛；心皮3。花期8-9月。生海拔200-900米的山地草坡或疏林下。产河北北部、辽宁、吉林和黑龙江东部。朝鲜半岛和俄罗斯(远东地区)亦有。

Tubers ca. 2.8 cm long. Stems 30-100 cm tall. Middle cauline leaves slightly long petiolate, 3.6-6.4 cm broad, 3-sect, segments dissected. Racemes 2-7-flowered; upper sepal yellowish, navicular-galeate or galeate, abaxially puberulous; carpels 3. Fl. Aug-Sep. Grassy slopes or sparse forests at 200-900 m. Distributed in N Hebei, Liaoning, Jilin and E Heilongjiang. Also in Korean Peninsula and Russia (Far East).

铁棒锤

Aconitum pendulum Busch

块根倒圆锥形。茎高26-100厘米。中部茎生叶密集，具短柄，宽4.5-5.5厘米，3全裂，全裂片细裂。总状花序狭长，有8-35朵花；轴和花梗被黄色开展短毛；上萼片蓝色或绿黄色，船状镰刀形，长1.6-2厘米，外面有短毛；心皮5。花期7-9月。生海拔2800-4500米的山地草坡或林边。产西藏、云南西北部、四川西部、青海、甘肃南部、陕西南部和河南西部。

Tubers obconical. Stems 26-100 cm tall. Middle cauline leaves dense, shortly petiolate, 4.5-5.5 cm broad, 3-sect, segments dissected. Racemes elongate, 8-35-flowered; rachis and pedicels spreading-yellow-puberulous; upper sepal blue or greenish-yellow, navicular-falcate, 1.6-2 cm long, abaxially puberulous; carpels 5. Fl. Jul-Sep. Grassy slopes or forest edges at 2800-4500 m. Distributed in Xizang, NW Yunnan, W Sichuan, Qinghai, S Gansu, S Shaanxi and W Henan.

合作乌头

Aconitum hezuoense W. T. Wang

茎高约68厘米，上部被短柔毛。叶均茎生，密集，五角形，宽2.4-6.5厘米，3全裂，全裂片一至三回细裂，末回裂片条形。总状花序顶生，长8厘米；轴和花梗被短柔毛；小苞片3裂；萼片淡黄色，上萼片船状盔形，长达2.2厘米；心皮5，无毛。花期8-9月。生海拔2850米的山坡灌木和草丛中。特产甘肃(合作)。

Stems ca. 68 cm tall, above puberulous. Leaves all cauline, dense, pentagonal, 2.4-6.5 cm broad, 3-sect, segments 1-3 times dissected, ultimate lobes linear. Raceme terminal, 8 cm long; rachis and pedicels puberulous; bracteoles 3-parted; sepals yellowish, upper sepal navicular-galeate, up to 2.2 cm long; carpels 5, glabrous. Fl. Aug-Sep. Shrubs and grasses on slopes at 2850 m. Endemic to Gansu (Hezuo).

栾川乌头

Aconitum luanchuanense W. T. Wang

块根细胡萝卜形，长约6厘米，粗6毫米。茎高约54厘米，上部被短柔毛。叶均茎生，密集，宽卵形，3全裂，全裂片二至三回细裂，末回裂片条形。总状花序长约10厘米；萼片黄色，上萼片船状盔形，高2-2.2厘米；心皮5，只在背缝线上疏被柔毛。花期8月。生海拔2200米的悬崖上。特产河南(栾川县)。

Tuber thinly carrot-shaped, ca. 6 cm long, 6 mm across. Stem ca. 54 cm tall, above puberulous. Leaves all cauline, dense, broadly ovate, 3-sect, segments 2-3 times dissected, ultimate lobes linear. Raceme ca. 10 cm long; sepals yellow, upper sepal navicular-galeate, 2-2.2 cm tall; carpels 5, only on dorsal suture sparsely puberulous. Fl. Aug. Cliff at 2200 m. Endemic to Henan (Luanchuan County).

拟黄花乌头

Aconitum anthoroideum DC.

块根长1-7厘米。茎高30-100厘米。中部茎生叶具短柄，五角

合作乌头 *Aconitum hezuoense*

黄花乌头 *Aconitum coreanum*

拟黄花乌头 *Aconitum anthoroideum*

形，3全裂，全裂片细裂。总状花序有2-12花；轴和花梗密被开展黄色短毛；上萼片淡黄色，盔形，高1.2-1.7厘米，具喙，外面密被短毛；心皮4-5。花期8-9月。生海拔1400-1950米的山地草坡或灌丛中。产新疆北部。俄罗斯(西伯利亚)和欧洲东部亦有。

Tubers 1-7 cm long. Stems 30-100 cm tall. Middle cauline leaves shortly petiolate, pentagonal, 3-sect, segments dissected. Racemes 2-12-flowered; rachis and pedicels densely spreading-yellow-puberulous; upper sepal yellowish, galeate, 1.2-1.7 cm tall, rostrate, abaxially puberulous; carpels 4-5. Fl. Aug-Sep. Grassy slopes or bushes at 1400-1950 m. Distributed in N Xinjiang. Also in Russia (Siberia) and E Europe.

露蕊乌头

Aconitum gymnandrum Maxim.

一年生草本。叶3全裂，裂片二至三回羽状分裂。总状花序有6-16花；基部苞片叶状；小苞片叶状至条形；萼片蓝紫色，有长爪，上萼片船形；花瓣唇扇形；心皮6-13，被毛。蓇葖果0.8-1.2厘米。花期6-8月。生海拔500-3800米草坡，草地或沟边。产云南、四川、西藏、甘肃和青海。

Annual herbs. Leaves 3-sect, segments 2- to 3-pinnately parted. Racemes 6-16- flowered; basal bracts leaf-like; bracteoles leaf-like to linear; sepals blue-purple, long clawed, upper sepal navicular; petal lips flabellate; carpels 6-13, hairy. Follicles 0.8-1.2 cm long. Fl. Jun-Aug. Grassy slopes, grasslands or by streams at 500-3800 m. Distributed in Yunnan, Sichuan, Xizang, Gansu and Qinghai.

露蕊乌头 *Aconitum gymnandrum*

短距翠雀花

Delphinium forrestii Diels

多年生草本。茎高约35厘米。基生叶和下部茎生叶有长柄，圆肾形，宽达8厘米，3深裂。总状花序有密集的花；萼片淡灰蓝色，萼距比萼片稍短，圆锥状钻形；退化雄蕊黑色，无毛，瓣片比爪稍宽，顶端微2裂；心皮3。花期8-10月。生海拔3800-4100米的山地多石山坡，陡崖或杜鹃花灌丛。产云南西北部和四川西南部。

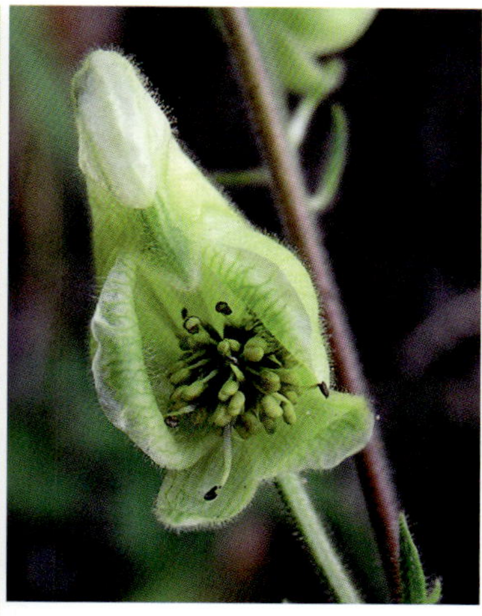

Perennial herbs. Stems ca. 35 cm tall. Basal and lower cauline leaves long petiolate, orbicular-reniform, up to 8 cm broad, 3-parted. Racemes with dense flowers; sepals grey-bluish, spur shorter than sepals, conic-subulate; staminodes black, glabrous, with limbs slightly broader than claws; carpels 3. Fl. Aug-Oct. Rock slopes, cliffs or *Rhododendron* bushes at 3800-4100 m. Distributed in NW Yunnan and SW Sichuan.

短距翠雀花 *Delphinium forrestii*

光茎短距翠雀花 *Delphinium forrestii* var. *viride*

毛翠雀花 *Delphinium trichophorum*

光茎短距翠雀花

Delphinium forrestii Diels var. **viride** (W. T. Wang) W. T. Wang

与短距翠雀花的区别在于本变种的茎疏被短柔毛，变无毛，萼片淡绿色或白色，外面被短硬毛。花期8-9月。生海拔3100-4100米的草坡上或林边。产云南西北部(德钦县)和西藏东南部(察隅县)。

This variety differs from the typical variety in its sparsely puberulous and glabrescent stems and greenish or whitish sepals abaxially with short rigid hairs. Fl. Aug-Sep. Grassy slopes or forest edges at 3100-4100 m. Distributed in NW Yunnan (Dêqên County) and SE Xizang (Zayü County).

毛翠雀花

Delphinium trichophorum Franch.

多年生草本。茎高达65厘米。基生叶3-5，具长柄，圆肾形，宽达15厘米，3深裂。茎生叶1-2，小。总状花序狭长；萼片淡蓝色或紫色，被糙毛，距下垂，圆筒状钻形，比萼片稍长；退化雄蕊黑色，无毛；心皮3。花期8-10月。生海拔2100-4600米的草坡。产西藏东部、四川西部、青海东南部和东部、甘肃南部。

Perennial herbs. Stems up to 65 cm tall. Basal leaves 3-5, long petiolate, orbicular-reniform, up to 15 cm broad, 3-parted. Cauline leaves 1-2, small. Racems elongate; sepals bluish or purple, strigose, spurs pendulous, cylindric-subulate, slightly longer than sepals; staminodes black, glabrous; carpels 3. Fl. Aug-Oct. Grass slopes at 2100-4600 m. Distributed in E Xizang, W Sichuan, SE and E Qinghai, and S Gansu.

狭序翠雀花

Delphinium wrightii Chen

多年生草本。茎高40-70厘米，被糙硬毛，不分枝。基生叶约4，具长柄，肾形，宽6-10厘米，3深裂，深裂片浅裂。总状花序狭长，有10-13花，萼片紫色，上萼片长1.8-2.3厘米，距水平开展，钻形，长1.6-2厘米；退化雄蕊瓣片卵形，2浅裂，爪比瓣片长；心皮3。花期8-9月。生海拔3400米的草坡上。特产四川(木里县)。

Perennial herbs. Stems 40-70 cm tall, hispid, simple. Basal leaves ca. 4, long petiolate, reniform, 6-10 cm broad, 3-parted, lobes lobulated. Racemes elongate, 10-13-flowered; sepals purple, upper sepal 1.8-2.3 cm long, spur horizontally spreading, subulate, 1.6-2 cm long; staminode limb ovate, 2-lobed, claw longer than limb; carpels 3. Fl. Aug-Sep. Grassy slopes at 3400 m. Endemic to Sichuan (Muli County).

拟毛翠雀花

Delphinium trichophoroides W. T. Wang

多年生草本。叶3，均基生，具长柄，五角形，宽6-12厘米，3全裂。花葶长约1厘米。圆锥花序长约50厘米，顶生总状花序狭长，长约45厘米；萼片蓝白色，距下垂，圆锥状钻形，比萼片稍短；退化雄蕊黑色，无毛；心皮3。花期8月。生海拔4400米的灌丛草甸。特产四川(理塘县)。

Perennial herbs. Leaves 3, all basal, long petiolate, pentagonal, 6-12 cm broad, 3-sect. Scape ca. 1 cm long. Panicle ca. 50 cm

狭序翠雀花 *Delphinium wrightii*

拟毛翠雀花 *Delphinium trichophoroides*

密花翠雀花 *Delphinium densiflorum*

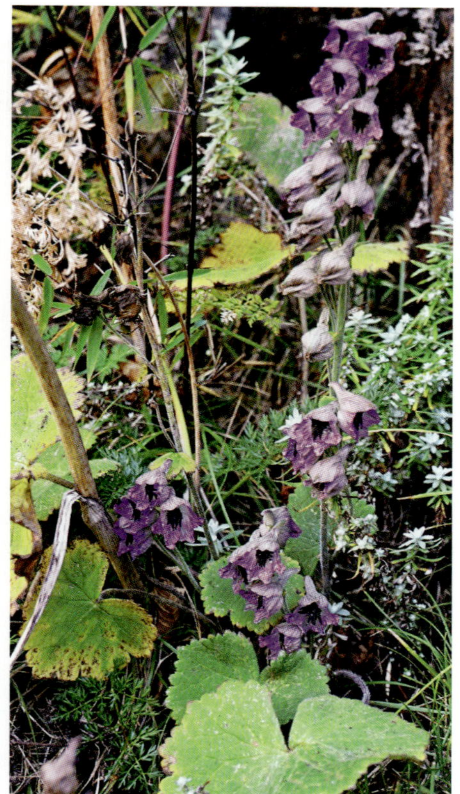
浅裂翠雀花 *Delphinium vestitum*

long, terminal raceme elongate, ca. 45 cm long; sepals bluish-whitish, spur pendulous, conic-subulate, slightly shorter than sepals; staminodes black, glabrous; carpels 3. Fl. Aug. Bush-meadows at 4400 m. Endemic to Sichuan (Litang County).

密花翠雀花

Delphinium densiflorum Duthie ex Huth

多年生草本。茎高约40厘米。叶基生并茎生，肾形，3深裂。总状花序有30-40朵密集的花；萼片淡灰蓝色，距圆锥状，比萼片短，长约1.4厘米；退化雄蕊近黑色，瓣片卵形，2深裂，腹面中央密被长柔毛；心皮3。花期7-8月。生海拔3300-4500米的灌丛中、河滩或冲积扇上。产青海和甘肃中部。尼泊尔和印度北部亦有。

Perennial herbs. Stems ca. 40 cm tall. Leaves basal and cauline, reniform, 3-parted. Racems with 30-40 dense flowers; sepals grey-bluish, spur conical, shorter than sepals, ca. 1.4 cm long; staminodes nearly black, limbs ovate, 2-parted, adaxially on the middle densely villous; carpels 3. Fl. Jul-Aug. Bushes, river banks or alluvial fans at 3300-4500 m. Distributed in Qinghai and C Gansu. Also in Nepal and N India.

浅裂翠雀花

Delphinium vestitum Wall. ex Royle

多年生草本。茎高约35厘米。基生叶约2，与下部茎生叶有长柄，圆肾形，3浅裂。总状花序有4-9花；萼片淡蓝紫色，距筒状圆锥形，比萼片短，长1.1-1.3厘米；退化雄蕊瓣片卵形，2中裂，腹面中央被黄色髯毛；心皮3。花期8-9月。生海拔3400米的山地林边。产西藏南部。克什米尔地区、尼泊尔和印度西北部亦有。

Perennial herbs. Stems ca. 35 cm tall. Basal leaves ca. 2, with lower cauline leaf long petiolate, orbicular-reniform, 3-lobed. Racemes 4-9-flowered; sepals bluish-purple, spur tubular-conical, shorter than sepals, 1.1-1.3 cm long; staminode limbs ovate, 2-fid, adaxially on the middle yellow-barbate; carpels 3. Fl. Aug-Sep. Forest edges at 3400 m. Distributed in S Xizang. Also in Kashmir, Nepal and NW India.

朗县翠雀花

Delphinium langxianense W. T. Wang

多年生草本。茎高15-45厘米。基生叶与茎生叶具长柄，肾形，宽1.5-3.6厘米，3浅裂。总状花序长14-24厘米，有7-8花；花梗长3.5-12厘米；萼片长1.6-2厘米，距钻形，长1.6-1.8厘米；退化雄蕊黑色，瓣片2中裂，腹面疏被长柔毛；心皮3。花期8-9月。生海拔4620米的杜鹃花灌丛中。特产西藏(朗县)。

Perennial herbs. Stems 15-45 cm tall. Basal leaves and cauline leaves long petiolate, reniform, 1.5-3.6 cm broad, 3-lobed. Racemes 14-24 cm long, 7-8-flowered; pedicels 3.5-12 cm long; sepals 1.6-2 cm long, spur subulate, 1.6-1.8 cm long; staminodes black, limbs 2-fid, adaxially sparsely villous; carpels 3. Fl. Aug-Sep. *Rhododendron* bushes at 4620 m. Endemic to Xizang (Nang County).

朗县翠雀花 *Delphinium langxianense*

囊距翠雀花 *Delphinium brunnonianum*

黄毛翠雀花 *Delphinium chrysotrichum*

囊距翠雀花
Delphinium brunnonianum Royle

多年生草本。茎高10-34厘米。基生叶和下部茎生叶有长柄，肾形，宽5-8.5厘米，3深裂。伞房花序有2-4花；萼片蓝紫色，长1.8-3厘米，距圆锥状，长6-10毫米；退化雄蕊瓣片2深裂，腹面中央有黄色髯毛；心皮3。花期8月。生海拔4500-6000米的草地或多石处。产西藏南部。尼泊尔、印度北部、克什米尔地区和阿富汗亦有。

Perennial herbs. Stems 10-34 cm tall. Basal leaves and lower cauline leaves long petiolate, reniform, 5-8.5 cm broad, 3-parted. Corymbs 2-4-flowered; sepals blue-purple, 1.8-3 cm long, spur conical, 6-10 mm long; staminode limbs 2-parted, adaxially on the middle yellow-barbate; carpels 3. Fl. Aug. Grassy or rocky places at 4500-6000 m. Distributed in S Xizang. Also in Nepal, N India, Kashmir and Afghanistan.

黄毛翠雀花
Delphinium chrysotrichum Finet et Gagnep.

多年生草本。茎高10-20厘米。叶肾形，宽3-5厘米，3深裂。伞房花序有2(-4)花；花梗被开展柔毛和腺毛；萼片紫色，长2-3厘米，距圆锥状钻形，长1.5-2.4厘米；退化雄蕊瓣片黑色，2深裂，有黄色髯毛，心皮3。花期8-9月。生多石砾山坡或流石滩，海拔4200-5000米。产西藏东南部和四川西部。

Perennial herbs. Stems 10-20 cm tall. Leaves reniform, 3-5 cm broad, 3-parted. Corymbs 2(-4)-flowered; pedicels spreading-puberulous and glandular-puberulous; sepals purple, 2-3 cm long, spur conic-subulate, 1.5-2.4 cm long; staminode limbs black, 2-parted, yellow-barbate; carpels 3. Fl. Aug-Sep. Gravelly slopes or screes at 4200-5000 m. Distributed in SE Xizang and W Sichuan.

察瓦龙翠雀花
Delphinium chrysotrichum var. **tsarongense** (Hand.-Mazz.) W. T. Wang

本变种与黄花翠雀花的区别是本变种的花梗被反曲并贴伏的短柔毛，无腺毛。生海拔3000-4600米的草坡或多石砾山坡。产西藏东南部(察隅县)和云南西北部(德钦县)。

This variety differs from the typical variety in its retrorse-appressed-puberulous pedicels lacking glandular hairs. Grassy or gravelly slopes at 3000-4600 m. Distributed in SE Xizang (Zayü County) and NW Yunnan (Dêqên County).

灰花翠雀花
Delphinium tephranthum W. T. Wang

多年生小草本。茎高8-16厘米，无毛。茎生叶2-3，具长柄，肾形，宽达4.6厘米，3深裂。伞房花序有2-3花；萼片灰色，有蓝色脉，长约2.2厘米，距圆锥状钻形，长1.5-2厘米；退化雄蕊瓣片黑色，2浅裂，腹面疏被柔毛；心皮3。花期7-8月。生海拔4800米的灌丛草甸上。特产西藏(昌都)。

Small perennial herbs. Stems 8-16 cm tall, glabrous. Cauline leaves 2-3, long petiolate, reniform, up to 4.6 cm broad, 3-parted. Corymbs 2-3-flowered; sepals greyish, blue-nerved, ca. 2.2 cm long, spur conic-subulate, 1.5-2 cm long; staminode limbs black, 2-lobed, adaxially sparsely pubescent; carpels 3. Fl. Jul-Aug. Bush-meadows at 4800 m. Endemic to Xizang (Changdu).

察瓦龙翠雀花 *Delphinium chrysotrichum* var. *tsarongense*

灰花翠雀花 *Delphinium tephranthum*

唐古拉翠雀花 *Delphinium tangkulaense*

唐古拉翠雀花

Delphinium tangkulaense W. T. Wang

多年生小草本。茎高4.8-10厘米。基生叶和下部茎生叶有长柄，圆肾形，宽达3厘米，3全裂，全裂片互相覆压。花单朵顶生；萼片蓝紫色，长1.8-2.7厘米，距钻状圆筒形，长1-1.3厘米；退化雄蕊瓣片黑色，2中裂，有黄色髯毛；心皮3。花期7-8月。生海拔4700-4900米的山坡上或湖边沙地。产西藏中部和青海西南部。

Small perennial herbs. Stems 4.8-10 cm tall. Basal leaves and lower cauline leaves long petiolate, orbicular-reniform, up to 3 cm long, 3-sect, with segments imbricate among themselves. Flowers solitary, terminal; sepals blue-purple, 1.8-2.7 cm long, spur subulate-cylindric, 1-1.3 cm long; staminode limbs black, 2-fid, yellow-barbate; carpels 3. Fl. Jul-Aug. Slopes or sandy banks of lakes at 4700-4900 m. Distributed in C Xizang and SW Qinghai.

白缘翠雀花

Delphinium chenii W. T. Wang

草本。叶掌状3深裂，一回裂片浅裂。花序伞房状，具2-4花；萼片宿存，上萼片淡蓝色，萼距近钻形，其他萼片边缘白色；退化雄蕊长圆形，2半裂，黑色，腹面有黄色髯毛；心皮3。花期8月。生海拔3900-5000米的林缘。产云南(香格里拉)和四川(稻城县)。

Herbs. Leaves palmately 3-parted, primary lobes lobed. Inflorescences corymbose, 2-4-flowered; sepals persistent, upper sepal bluish, spur subulate, other sepals white at margin; staminodes oblong, 2-cleft, black, adaxially yellow-barbate; carpels 3. Fl. Aug. Forest edges at 3900-5000 m. Distributed in Yunnan (Shangri-La) and Sichuan (Daocheng County).

白缘翠雀花 *Delphinium chenii*

宽萼翠雀花

Delphinium pseudopulcherrimum W. T. Wang

多年生草本。茎高20-50厘米。中部茎生叶具稍长柄，五角形，3深裂近基部。伞房花序有2-6花；萼片蓝紫色，长1.8-2.4厘米，距钻形，长1.8-2厘米；退化雄蕊瓣片黑色，2浅裂，有黄色髯毛；心皮3。花期8-9月。生海拔4000-5000米的山地草坡。特产西藏拉萨一带。

Perennial herbs. Stems 20-50 cm tall. Middle cauline leaves slightly long petiolate, pentagonal, 3-parted to near base. Corymbs 2-6-flowered; sepals blue-purple, 1.8-2.4 cm long, spur subulate, 1.8-2 cm long; staminode limbs black, 2-lobed, yellow-barbate; carpels 3. Fl. Aug-Sep. Grassy slopes at 4000-5000 m. Endemic to areas around the city of Lasa, Xizang.

宽萼翠雀花 *Delphinium pseudopulcherrimum*

三果大通翠雀花 *Delphinium pylzowii* var. *trigynum*

三果大通翠雀花

Delphinium pylzowii Maxim. var. **trigynum** W. T. Wang

多年生草本。茎高约35厘米。下部茎生叶具长柄，圆五角形，宽达5厘米，3全裂，全裂片细裂。伞房花序约有4花；萼片淡灰蓝色，距钻形，比萼片长；退化雄蕊瓣片黑褐色，2浅裂，有黄色髯毛；心皮3。花期7-8月。生海拔3500-4500米的高山草甸。产西藏东部、四川西北部、青海南部和东南部、甘肃西南部。

Perennial herbs. Stems ca. 35 cm tall. Lower cauline leaves long petiolate, orbicular-pentagonal, up to 5 cm broad 3-sect, segments dissected. Corymbs ca. 4-flowered; sepals greyish-bluish, spur subulate, longer than sepals; staminode limbs black-brown, 2-lobed, yellow-barbate; capels 3. Fl. Jul-Aug. Alpine meadows at 3500-4500 m. Distributed in E Xizang, NW Sichuan, S and SE Qinghai, and SW Gansu.

单花翠雀花

Delphinium candelabrum Ostf. var. **monanthum** Hand.-Mazz.

草本。叶肾状五角形，3全裂，末回裂片卵形。伞房花序单花，或偶有2花；苞片叶状；花萼蓝紫色，下面具柔毛，萼距近钻形，近直伸或稍弯曲；花瓣不分裂；退化雄蕊圆卵形，2裂，蓝色或下部暗褐色。花期8-10月，果期11月至翌年1月。生海拔4100-5000米的高山石坡或草甸。产四川西北部、西藏东北部、甘肃西南部、青海东部和东南部。

Herbs. Leaves reniform-pentagonous, 3-sect, ultimate lobes ovate. Corymb 1- or occasionally 2-flowered; bracts leaflike; sepals blue-purple, abaxially puberulent, spur subulate, nearly straight or slightly recurved; petals undivided; staminodes orbicular-ovate, 2-lobed, blue or below dark brown. Fl. Aug-Oct. Fr. Nov to next Jan. Alpine rocky slopes or meadows at 4100-5000 m. Distributed in NW Sichuan, NE Xizang, SW Gansu, E

单花翠雀花 *Delphinium candelabrum* var. *monanthum*

白蓝翠雀花 *Delphinium albocoeruleum*

文采翠雀花 *Delphinium wentsaii*

and SE Qinghai.

白蓝翠雀花
Delphinium albocoeruleum Maxim.

多年生草本。茎高40-60(-100)厘米。茎生叶具长或短柄，五角形，3深裂。伞房花序有3-7花；萼片蓝紫色或蓝白色，长2-2.5(-3)厘米，距钻形，长1.7-2.5(-3.3)厘米；退化雄蕊瓣片黑褐色，2浅裂，有黄色髯毛；心皮3。花期7-9月。生海拔3600-4700米的山地草坡或圆柏林下。产西藏东北部、四川西北部、青海东部和甘肃。

Perennial herbs. Stems 40-60 (-100) cm tall. Cauline leaves long to shortly petiolate, pentagonal, 3-parted. Corymbs 3-7-flowered; sepals blue-purple or bluish-white, 2-2.5(-3) cm long, spur subulate, 1.7-2.5(-3.3) cm long; staminode limbs black-brown, 2-lobed, yellow-barbate; carpels 3. Fl. Jul-Sep. Grassy slopes or *Sabina* forests at 3600-4700 m. Distributed in NE Xizang, NW Sichuan, E Qinghai and Gansu.

文采翠雀花
Delphinium wentsaii Y. Z. Zhao

多年生草本。茎高约20厘米，密被硬毛。茎生叶心状五角形，3深裂。总状花序6花，密被硬毛；萼片蓝紫色，背面密被硬毛，距钻形，长1.2厘米；退化雄蕊瓣片黑色，被白色髯毛；心皮3。花期8月。生高山草地。特产新疆西部。

Perennial herbs. Stems ca. 20 cm tall, densely hirsute. Cauline leaves cordate-pentagonal, 3-parted. Racemes 6-flowered, densely hirsute; sepals blue-purple, spur subulate, 1.2 cm long; staminode limbs black, white-barbate; carpels 3. Fl. Aug. Alpine meadows. Endemic to W Xinjiang.

东北高翠雀花
Delphinium korshinskyanum Nevski

多年生草本。茎高50-120厘米，下部被开展长毛。叶心状五角形，宽达13厘米，3深裂。总状花序狭长；轴和花梗无毛；萼片蓝紫色，长1-1.3厘米，距钻形，长1.2-1.6厘米；退化雄蕊黑褐色，被黄髯毛；心皮3，无毛。花期7-8月。生海拔370-750米的草甸上。产内蒙古东北部和黑龙江。俄罗斯(东西伯利亚、远东地区)亦有。

Perennial herbs. Stems 50-120 cm tall, below spreading-hirsute. Leaves cordate-pentagonal, to 13 cm broad, 3-parted. Raceme elongate; rachis and pedicels glabrous; sepals blue-purple, 1-1.3 cm long, spur subulate, 1.2-1.6 cm long; staminodes black-brown, yellow-barbate; carpels 3, glabrous. Fl. Jul-Aug. meadows at 370-750 m. Distributed in NE Neimenggu and Heilongjiang. Also in Russia (E Siberia, Far East).

东北高翠雀花 *Delphinium korshinskyanum*

全裂翠雀花 *Delphinium trisectum*

粗距翠雀花 *Delphinium pachycentrum*

全裂翠雀花

Delphinium trisectum W. T. Wang

多年生草本。茎高约45厘米。下部茎生叶有长柄，圆肾形，宽7-12厘米，3全裂。总状花序有10-14花；萼片蓝紫色，长1.4-1.7厘米，距钻形，长1.6-2.1厘米；退化雄蕊瓣片黑褐色，2浅裂，有淡黄色短髯毛；心皮3。花期4-5月。生海拔400-800米的山地。产湖北北部和河南东南部。

Perennial herbs. Stems ca. 45 cm tall. Lower cauline leaves long petiolate, orbicular-reniform, 7-12 cm broad, 3-sect. Racemes 10-14-flowered; sepals blue-purple, 1.4-1.7 cm long, spur subulate, 1.6-2.1 cm long; staminode limbs black-brown, 2-lobed, shortly yellowish-barbate; carpels 3. Fl. Apr-May. Montane regions at 400-800 m. Distributed in N Hubei and SE Henan.

粗距翠雀花

Delphinium pachycentrum Hemsl.

多年生草本。茎高20-50厘米。中部茎生叶有稍长柄，五角形，宽5.5-7.5厘米，3深裂。总状花序有5-12朵密集的花；萼片紫蓝色，长1.7-2.2厘米，两面被短柔毛，距圆筒形，与萼片近等长；退化雄蕊瓣片蓝色，具黄色髯毛；心皮3。花期7-8月。生海拔4000-4500米的多石砾草坡。产四川西部和青海东南部。

Perennial herbs. Stems 20-50 cm tall. Middle cauline leaves slightly long petiolate, pentagonal, 5.5-7.5 cm, 3-parted. Racemes densely 5-12-flowered; sepals purple-blue, 1.7-2.2 cm long, on both surfaces puberulous, spur cynlidric, nearly as long as sepals; staminode limbs blue, yellow-barbate; carpels 3. Fl. Jul-Aug. Gravelly and grassy slopes at 4000-4500 m. Distributed in W Sichuan and SE Qinghai.

拉萨翠雀花

Delphinium gyalanum Marq. et Shaw

多年生草本。茎高55-110厘米。基生叶和下部茎生叶有长柄，肾状五角形，宽9-22厘米，3深裂。总状花序长20-25厘米；萼片蓝紫色，长1.4-1.7厘米，距圆筒状钻形，长1.2-1.8厘米；退化雄蕊瓣片蓝色，2浅裂，有黄色髯毛；心皮3。花期7-9月。生海拔3000-4500米的草坡或灌丛中。特产西藏南部和东南部。

Perennial herbs. Stems 55-110 cm tall. Basal and lower cauline leaves long petiolate, reniform-pentagonal, 9-22 cm broad, 3-parted. Racemes 20-25 cm long; sepals blue-purple, 1.4-1.7 cm long, spur cylindric-subulate, 1.2-1.8 cm long; staminode limbs blue, 2-lobed, yellow-barbate; carpels 3. Fl. Jul-Sep. Grassy slopes or bushes at 3000-4500 m. Endemic to S and SE Xizang.

拉萨翠雀花 *Delphinium gyalanum*

兴安翠雀花 *Delphinium hsinganense*

滇川翠雀花 *Delphinium delavayi*

兴安翠雀花

Delphinium hsinganense S. H. Li et Z. F. Fang

多年生草本。茎高75-95厘米，近无毛。叶具柄，心状五角形，宽6-10厘米，3深裂。总状花序长约20厘米；轴和花梗被反曲短柔毛；萼片蓝色，长1.4-1.7厘米，距钻形，长1.6-1.8厘米；退化雄蕊瓣片宽椭圆形，顶端微凹，腹面有黄色髯毛；心皮3，子房被开展短柔毛。花期6-7月。生林缘。产内蒙古东北部。

Perennial herbs. Stems 75-95 cm tall, subglabrous. Leaves petiolate, cordate-pentagonal, 6-10 cm broad, 3-parted. Raceme ca. 20 cm long; rachis and pedicels retrorsely puberulous; sepals blue, 1.4-1.7 cm long, spur subulate, 1.6-1.8 cm long; staminode limbs broadly elliptic, at apex mucronate, adaxially yellow-barbate; carpels 3, ovaries spreading-puberulous. Fl. Jun-Jul. Forest margins. Distributed in NE Neimenggu.

滇川翠雀花

Delphinium delavayi Franch.

草本。茎和叶柄密被反曲短硬毛，毛长0.5-2毫米。叶3深裂，中央裂片菱形。总状花序具5-15花，具白色糙伏毛和黄色短腺毛；小苞片狭披针形，与花邻接；退化雄蕊长圆形，2裂；花丝无毛；心皮3。花期7-11月。生海拔2600-3600米的斜坡、林下或溪边。产云南、四川和贵州。

Herbs. Stems and petioles densely retrovsely hispidulous, hairs 0.5-2 mm long. Leaves 3-parted, central lobes rhombic. Racemes 5-15-flowered, white strigose and yellow glandular puberulent; bracteoles narrow-lanceolate, contiguous to flower; staminode limbs oblong, 2-lobed; filaments glabrous; carpels 3. Fl. Jul-Nov. Slopes, forests or by streams at 2600-3600 m. Distributed in Yunnan, Sichuan and Guizhou.

须花翠雀花

Delphinium delavayi Franch. var. **pogonanthum** (Hand.-Mazz.) W. T. Wang

多年生草本。茎和叶柄密被开展硬毛，毛长2-3毫米。叶片3深裂，中裂片菱形。总状花序具硬毛和黄色短腺毛；退化雄蕊长圆形，2裂；花丝无毛；心皮3。生海拔2600-3600米的灌丛、疏林或草坡。产云南北部、四川西南部和贵州西部。

Perennial herbs. Stems and petioles densely covered with spreading hairs, 2-3 mm long. Leaves 3-parted, central lobe rhombic. Racemes strigose and yellow glandular puberulent; staminode limbs oblong, 2-lobed; filaments glabrous; carpels 3. Thickets, sparse forests or grassy slopes at 2600-3600 m. Distributed in N Yunnan, SW Sichuan and W Guizhou.

峨眉翠雀花

Delphinium omeiense W. T. Wang

多年生草本。茎高60-95厘米，与叶柄均被硬毛。下部茎生叶有长柄，五角形，3深裂。总状花序长12-30厘米；花梗长1-4厘米，近中部具2小苞片；萼片蓝紫色，长1.2-1.6厘米，距钻形，长2-2.6厘米；退化雄蕊瓣片紫色，2中裂，有黄色髯毛；心皮3。花期7-8月。生海拔2500-3300米的山地草坡或林中。产四川西部和云南东北部。

Perennial herbs. Stmes 60-95 cm tall, with petioles hirsute. Lower cauline leaves long petiolate, pentagonal, 3-parted. Racemes 12-30 cm long; pedicels 1-4 cm long, near the middle 2-bracteolate; sepals blue-purple, 1.2-1.6 cm long, spur subulate, 2-2.6 cm long; staminode limbs purple, 2-fid, yellow-barbate; carpels 3. Fl. Jul-Aug. Grassy slopes or forests at 2500-3300 m. Distributed in W Sichuan and NE Yunnan.

须花翠雀花 *Delphinium delavayi* var. *pogonanthum*

峨眉翠雀花 *Delphinium omeiense*

川西翠雀花 *Delphinium tongolense*

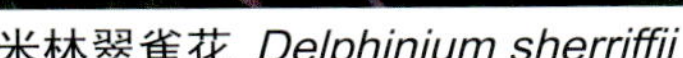

米林翠雀花 *Delphinium sherriffii*

川西翠雀花

Delphinium tongolense Franch.

草本。茎无毛。叶3深裂，中裂片3浅裂。总状花序长12-38厘米，具8-25花，被黄色腺毛，下部苞片叶状，上部者条形或丝状；萼片蓝紫色，萼距钻形；退化雄蕊近长圆形，2裂；花丝无毛；心皮3。花期7-8月。生海拔2200-3900米的林下或草坡。产云南北部和四川西部。

Herbs. Stems glabrous. Leaves 3-parted, central lobe 3-lobed. Racemes 12-38 cm long, 8-25-flowered, yellow glandular-puberulent; proximal bracts leaf-like, distal ones linear or filiform; sepals blue-purple, spur subulate; staminode limbs suboblong, 2-cleft; filaments glabrous; carpels 3. Fl. Jul-Aug. Forests or grassy slopes at 2200-3900 m. Distributed in N Yunnan and W Sichuan.

米林翠雀花

Delphinium sherriffii Munz

多年生草本。茎高达1.6米。中部茎生叶有稍长柄，五角形，3深裂。圆锥花序长约40厘米；轴被黄色腺毛；小苞片丝形，长2-5毫米；萼片紫色，长1-1.2厘米，距钻形，长2-2.2厘米；退化雄蕊瓣片紫色，2深裂，有淡黄色髯毛；心皮3。花期6-7月。生海拔3000-3500米的沟边或林下。特产西藏(米林县)。

Perennial herbs. Stems up to 1.6 m tall. Middle cauline leaves slightly long petiolate, pentagonal, 3-parted. Panicles ca. 40 cm long; rachis yellow-glandular-puberulous; bracteoles filiform, 2-5 mm long; sepals purple, 1-1.2 cm long, spur subulate, 2-2.2 cm long; steminode limbs purple, 2-parted, yellowish-barbate; carpels 3. Fl. Jun-Jul. By streams or under forests at 3000-3500 m. Endemic to Xizang (Mainling County).

毛茎翠雀花

Delphinium hirticaule Franch.

多年生草本。茎高约70厘米，下部疏被长硬毛。茎生叶约5，五角形，3深裂。总状花序狭长，有5-10花；花梗长1.5-3厘米，无毛；小苞片长2.5-4毫米；萼片蓝紫色，长1.2-2厘米，距钻形，长1.7-2厘米；退化雄蕊瓣片蓝色，有黄色髯毛；心皮3。花期8月。生海拔1400-2900米的山地草坡。产重庆、湖北西部和陕西南部。

Perennial herbs. Stems ca. 70 cm tall, below sparsely hirsute. Cauline leaves ca. 5, pentagonal, 3-parted. Racemes elongate, 5-10-flowered; pedicels 1.5-3 cm long, glabrous; bracteoles 2.5-4 mm long; sepals blue-purple, 1.2-2 cm long, spur subulate, 1.7-2 cm long; staminode limbs blue, yellow-barbate; carpels 3. Fl. Aug. Grassy slopes at 1400-2900 m. Distributed in Chongqing,

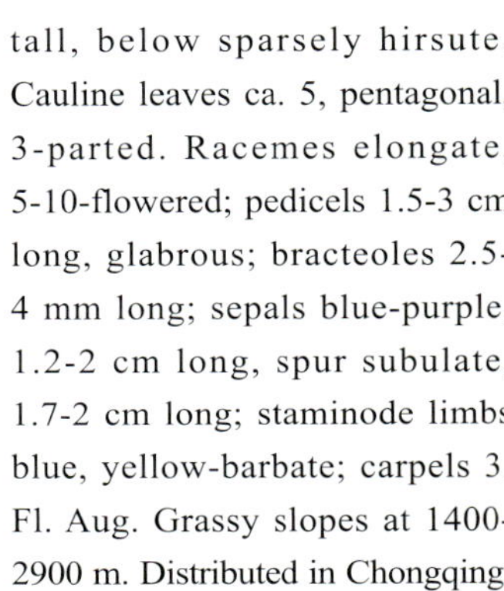

毛茎翠雀花 *Delphinium hirticaule*

毛梗翠雀花 *Delphinium eriostylum*

螺距黑水翠雀花 *Delphinium potaninii* var. *bonvalotii*

W Hubei and S Shaanxi.

毛梗翠雀花

Delphinium eriostylum Lévl.

多年生草本。茎高40-60厘米。叶片心状五角形，宽7-16厘米，3深裂，被糙伏毛。伞房花序具3-5花；苞片披针形；花梗长2-3.5厘米，近中部有2小苞片；萼片5，紫蓝色，长约1.7厘米；距钻形，长1.7-2厘米，稍向下弯；花瓣2，无毛；退化雄蕊瓣片长圆形，2裂，被黄色髯毛；心皮3，被疏毛或无毛。花期3-6月。生海拔500-2000米的山谷草坡或溪边。产贵州、四川东南部和重庆南部。

黑水翠雀花 *Delphinium potaninii*

Perennial herbs. Stems 40-60 cm tall. Leaf blades cordate-pentagonal, 7-16 cm broad, deeply 3-lobed, strigose. Corymb 3-5-flowered; bracts lanceolate; pedicels 2-3.5 cm long, near the middle 2-bracteolate; sepals 5, purple-blue, ca. 1.7 cm long, spur subulate, 1.7-2 cm long, slightly decurved; petals 2, glabrous; staminode limbs oblong, 2-lobed, yellow-barbate; carpels 3, pilose or glabrous. Fl. Mar-Jun. Grassy slopes or by streams in valley at 500-2000 m. Distributed in Guizhou, SE Sichuan and S Chongqing.

黑水翠雀花

Delphinium potaninii Huth

多年生草本。茎高60-120厘米，无毛。中部茎生叶有稍长柄，五角形，3深裂。顶生总状花序长20-30厘米；轴和花梗无毛；萼片蓝紫色，长1-1.8厘米，距钻形，长2-2.5(-3)厘米；退化雄蕊瓣片紫色，2中裂，有黄色髯毛；心皮3，无毛。花期8-9月。生海拔1800-3300米的山坡上或林中。产四川西北部、重庆北部、湖北西北部、甘肃南部和陕西南部。

Perennial herbs. Stems 60-120 cm tall, glabrous. Middle cauline leaves slightly long petiolate, pentagonal, 3-parted. Terminal raceme 20-30 cm long; rachis and pedicels glabrous; sepals blue-purple, 1-1.8 cm long, spur subulate, 2-2.5 (-3) cm long; steminode limbs purple, 2-fid, yellow-barbate; carpels 3, glabrous. Fl. Aug-Sep. Slopes or forests at 1800-3300 m. Distributed in NW Sichuan, N Chongqing, NW Hubei, S Gansu and S Shaanxi.

螺距黑水翠雀花

Delphinium potaninii var. **bonvalotii** (Franch.) W. T. Wang

多年生草本。茎及花序轴、花梗均无毛。叶掌状3深裂。总状花序长6-14厘米，具3-15花，呈伞房状；花左右对称；小苞片窄条形或丝状；萼片蓝紫色，萼距“U”字形或螺旋状扭曲。花期6-9月。生海拔1100-3800米的林缘。产四川西部和贵州。

Perennial herbs. Stems with corymb rachis and pedicels glabrous. Leaves palmately 3-parted. Racemes 6-14 cm long, 3-15-flowered, often corymbiform; flowers zygomorphic; bracteoles narrowly linear or filiform; sepals blue-purple, spur U-shaped or spirally recurved. Fl. Jun-Sep. Forest edges at 1100-3800 m. Distributed in W Sichuan and Guizhou.

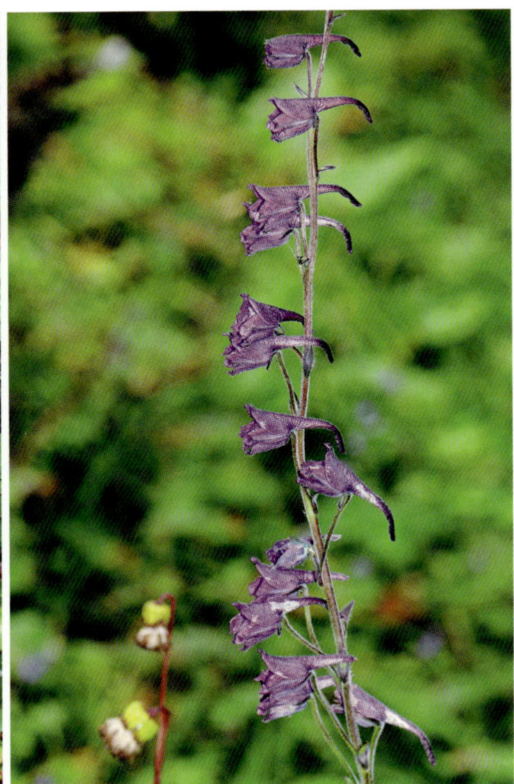

拟螺距翠雀花 *Delphinium bulleyanum*

错那翠雀花

Delphinium conaense W. T. Wang

多年生草本。茎高约60厘米，被贴伏短柔毛。茎生叶具稍长柄，五角形，宽4.8-6厘米，3全裂。总状花序有2-9花；萼片淡蓝色，长约9毫米，距钻形，长15毫米；退化雄蕊瓣片淡蓝色，2浅裂，有淡黄色髯毛；心皮3。花期8-9月。生海拔3400米的林下。特产西藏(错那县)。

Perennial herbs. Stems ca. 60 cm tall, appressed-puberulous. Cauline leaves slightly long petiolate, pentagonal, 4.8-6 cm broad, 3-sect. Racemes 2-9-flowered; sepals bluish, ca. 9 mm long, spur subulate, 15 mm long; staminode limbs bluish, 2-lobed, yellowish-barbate; carpels 3. Fl. Aug-Sep. Forests at 3400 m. Endemic to Xizang (Cona County).

拟螺距翠雀花

Delphinium bulleyanum Forrest ex Diels

草本。茎无毛。叶掌状3深裂，一回裂片顶端渐尖。花序总状或复总状，具6-16花，被反曲柔毛和黄色短腺毛；萼片蓝色，下面密被短柔毛，萼距常“U”字形弯曲；退化雄蕊狭倒卵形，2深裂；心皮3。花期8-9月。生海拔3100-4800米的间草坡或灌丛中。产云南西北部和四川西南部。

Herbs. Stems glabrous. Leaves palmately 3-parted, primary lobes acuminate. Inflorescences racemose or compoundly racemose, 6-16-flowered, retrorsely puberulent and yellow glandular puberulent, blue, abaxially densely puberulent, sepal spur usually U-shaped recurved; staminode limbs narrowly obovate, 2-parted; carpels 3. Fl. Aug-Sep. Grassy slopes or bushes at 3100-4800 m. Distributed in NW Yunnan and SW Sichuan.

弯距翠雀花

Delphinium campylocentrum Maxim.

多年生草本。茎高75-100厘米，无毛。茎中部叶有稍长柄，五角形，3全裂。圆锥花序有多数花；轴和花梗被黄色腺毛；萼片蓝紫色，长1-1.4厘米，外面顶端之下有2裂的角状凸起，距呈“U”字形下弯；心皮3。花期7-9月。生海拔3400-3900米的云杉林下或林边草坡。产四川西北部和甘肃南部。

Perennial herbs. Stems 75-100 cm tall, glabrous. Middle cauline leaves slightly long petiolate, pentagonal, 3-sect. Panicle multi-flowered; rachis and pedicels yellow-glandular-puberulous, septals blue-purple, 1-1.4 cm long, abaxially below apex with 2-lobed horn-like projection, spur U-form decurved; carpels 3. Fl. Jul-Sep. *Picea* forests or grassy slopes near forests at 3400-3900 m. Distributed in NW Sichuan and S Gansu.

秦岭翠雀花

Delphinium giraldii Diels

多年生草本。茎高55-110厘米，无毛。下部茎生叶有稍长柄，五角形，3全裂。圆锥花序顶生；轴和花梗无毛；萼片蓝紫色，长1-1.3厘米，距钻形，长1.6-2厘米；退化雄蕊瓣片蓝色，2深裂，有黄色髯毛；心皮3，无毛。花期7-8月。生海拔960-2000米的草坡或林中。产四川西北部、甘肃东南部、宁夏南部、陕西南部、湖北西部、河南西部和山西南部。

弯距翠雀花 *Delphinium campylocentrum*

错那翠雀花 *Delphinium conaense*

秦岭翠雀花 *Delphinium giraldii*

大理翠雀花 *Delphinium taliense*

Perennial herbs. Stems 55-110 cm tall, glabrous. Lower cauline leaves slightly long petiolate, pentagonal, 3-sect. Panicle terminal; rachis and pedicels glabrous; sepals blue-purple, 1-1.3 cm long, spur subulate, 1.6-2 cm long; staminode limbs blue, 2-parted, yellow-barbate; carpels 3, glabrous. Fl. Jul-Aug. Grassy slopes or forests at 960-2000 m. Distributed in NW Sichuan, SE Gansu, S Ningxia, S Shaanxi, W Hubei, W Henan and S Shanxi.

大理翠雀花

Delphinium taliense Franch.

多年生草本。茎高60-130厘米。基生叶和下部茎生叶有长柄，五角形，3深裂。总状花序有6-17花；轴无毛；萼片紫蓝色，长约1.4厘米，距钻形，长1.8-2.1厘米；退化雄蕊瓣片蓝色，2裂，有黄色髯毛；心皮3。花期8-11月。生海拔2800-3500米的松林边草坡或林中。产云南北部和西北部、四川西南部。

Perennial herbs. Stems 60-130 cm tall. Basal and lower cauline leaves long petiolate, pentagonal, 3-parted. Racemes 6-17-flowered; rachis glabrous; sepals purple-blue, ca. 1.4 cm long, spur subulate, 1.8-2.1 cm long; staminode limbs blue, 2-lobed, yellow-barbate; carpels 3. Fl. Aug-Nov. Grassy slopes near *Pinus* forests or forests at 2800-3500 m. Distributed in N and NW Yunnan, and SW Sichuan.

角萼翠雀花

Delphinium ceratophorum Franch.

草本。叶约5，基生，1-3茎生，掌状3深裂，一回裂片浅裂。总状花序具5-10(-17)花；萼片紫蓝色，顶端具角状凸起，萼距筒状钻形，长1.9-2.3厘米，微弯曲或呈"U"字形；退化雄蕊倒卵形，2浅裂，2深裂或凹缺；心皮3。花期8-10月。生海拔2800-4200米的山地草坡、林缘或草甸。产云南西北部。

Herbs. Leaves basal, ca. 5, cauline 1-3, palmately 3-parted, primary lobes lobed. Racemes 5-10(-17)-flowered; sepals purple-blue, corniculate near apex, spur cylindric-subulate, 1.9-2.3 cm long, slightly recurved or U-shaped; staminode limbs obovate, 2-lobed, 2-cleft, or emarginate; carpels 3. Fl. Aug-Oct. Grassy slopes, forest edges or meadows at 2800-4200 m. Distributed in NW Yunnan.

角萼翠雀花 *Delphinium ceratophorum*

长距翠雀花 *Delphinium tenii*

澜沧翠雀花 *Delphinium thibeticum*

锐裂翠雀花 *Delphinium thibeticum* var. *laceratilobum*

长距翠雀花
Delphinium tenii Lévl.

草本。茎无毛。叶基生并茎生，掌状三深裂近基部，一回裂片细裂。总状花序少花；轴和花梗均无毛；萼片蓝色，长1-1.4厘米，萼距钻形，长1.8-2.5cm；退化雄蕊倒卵形，2浅裂；花丝无毛；心皮3。花期7-10月。生海拔1900-3400米的草坡或林缘。产云南西北部、四川西南部和西藏东南部。

Herbs. Stems glabrous. Leaves basal and cauline, palmately 3-parted nearly to base, with dissected primary lobes. Racemes few-flowered; rachis and pedicels glabrous; sepals blue, 1-1.4 cm long, spur subulate, 1.8-2.5 cm long; staminodes obovate, 2-lobed; filaments glabrous; carpels 3. Fl. Jul-Oct. Grassy slopes or forest edges at 1900-3400 m. Distributed in NW Yunnan, SW Sichuan and SE Xizang.

太白翠雀花 *Delphinium taipaicum*

澜沧翠雀花
Delphinium thibeticum Finet et Gagnep.

多年生草本。茎高28-80厘米。基生叶约3，有长柄，肾形，宽8-12(-20)厘米，3全裂，全裂片稍细裂。总状花序狭长；轴和花梗被贴伏短柔毛；萼片蓝紫色，长1.2-1.4厘米，距钻形，长1.9-2.2厘米；退化雄蕊瓣片蓝色，2中裂，有黄色髯毛；心皮3。花期8-9月。生海拔2800-3800米的山地草坡或多石砾山坡或疏林中。产西藏东南部、云南西北部和四川西部。

Perennial herbs. Stems 28-80 cm tal. Basal leaves ca. 3, long petiolate, reniform, 8-12(-20) cm broad, 3-sect, segments slightly dissected. Racemes elongate; rachis and pedicels appressed-puberulous; sepals blue-purple, 1.2-1.4 cm long, spur subulate, 1.9-2.2 cm long; staminode limbs blue, 2-fid, yellow-barbate; carpels 3. Fl. Aug-Sep. Grassy or gravelly slopes or sparse forests at 2800-3800 m. Distributed in SE Xizang, NW Yunnan and W Sichuan.

锐裂翠雀花
Delphinium thibeticum var. **laceratilobum** W.T. Wang

本变种与澜沧翠雀花的区别在于本变种的叶有多数密集的狭三角形至披针状条形、顶端锐尖的小裂片；花序轴和花梗被开展短柔毛。生海拔3700米的山谷岩石上。产西藏东部(江达县)和四川西部(德格县)。

This variety differs from the typical variety in its multi-lobulate leaves with lobules narrow-triangular to lanceolate-linear and at apex pungent, and in its spreading-puberulous rachis and pedicels. Rocks in valley at 3700 m. Distributed in E Xizang (Jomda County) and W Sichuan (Dege County).

太白翠雀花
Delphinium taipaicum W. T. Wang

多年生草本。茎高22-30厘米。基生叶约4，有长柄，五角状肾形，3全裂。总状花序长4-10厘米；萼片紫色，长1.5-1.8厘米，距圆筒状钻形，长1.4-2厘米；退化雄蕊瓣片紫色，2浅裂，有黄色髯毛；心皮3。花期9月。生海拔3600-3900米的山地草坡。特产陕西(太白山)。

Perennial herbs. Stems 22-30 cm tall. Basal leaves ca. 4, long petiolate, pentagonal-reniform, 3-sect. Racemes 4-10 cm long;

金沙翠雀花 *Delphinium majus*

翠雀 *Delphinium grandiflorum*

sepals purple, 1.5-1.8 cm long, spur cylindric-subulate, 1.4-2 cm long; staminode limbs purple, 2-lobed, yellow-barbate; carpels 3. Fl. Sep. Grassy slopes at 3600-3900 m. Endemic to Shaanxi (Taibai Mountain).

金沙翠雀花
Delphinium majus (W. T. Wang) W. T. Wang

多年生草本。茎高26-65厘米。基生叶和下部茎生叶有长柄，五角状肾形，3全裂，末回小裂片多为狭三角形。总状花序有4-7花；萼片紫蓝色，长1.1-1.6厘米，距钻形，长1.6-2.4厘米；退化雄蕊瓣片宽倒卵形，微凹，有黄色髯毛；心皮3。花期8-9月。生海拔1600-1800米的草坡上或灌丛中。产云南西北部和四川西南部。

Perennial herbs. Stems 26-65 cm tall. Basal and lower cauline leaves long petiolate, pentagonal-reniform, 3-sect, ultimate lobules mostly narrow-triangular. Racemes 4-7-flowered; sepals purple-blue, 1.1-1.6 cm broad, spur subulate, 1.6-2.4 cm long; staminode limbs broad-obovate, emarginate, yellow-barbate; carpels 3. Fl. Aug-Sep. Grassy slopes or bushes at 1600-1800 m. Distributed in NW Yunnan and SW Sichuan.

翠雀
Delphinium grandiflorum L.

草本。茎被贴伏短柔毛。叶3全裂，全裂片细裂，末回裂片条形。总状花序长5-20厘米，具3-15花；萼片紫蓝色或蓝色，外面被短柔毛，萼距钻状或筒状钻形，直或微下弯；花瓣不裂；退化雄蕊近圆形或宽倒卵形；心皮3。花期5-10月。生海拔100-3500米的疏林、灌丛或草坡。产中国西南、华北和东北。俄罗斯(西伯利亚)和蒙古亦有。

Herbs. Stems appressed-puberulous. Leaves 3-sect, segments dissected, ultimate lobules linear. Racemes 5-20 cm long, 3-15-flowered; sepals purple-blue or blue, abaxially puberulent, spur subulate or cylindric-subulate, straighten decurved; petals undivided; staminode limbs suborbicular or broadly obovate; carpels 3. Fl. May-Oct. Sparse forests, thickets or grassy slopes at 100-3500 m. Distributed in SW, N and NE China. Also in Russia (Siberia) and Mongolia.

光序翠雀花
Delphinium kamaonense Huth

多年生草本。茎疏被反曲短柔毛。叶掌状3全裂，全裂片细裂。花序通常复总状；轴和花梗近无毛；退化雄蕊宽倒卵形；萼片深或亮蓝色，萼距与萼片等长，长1.2-1.6厘米。花期6-8月。生海拔2800-4200米的高山草坡。产西藏南部。尼泊尔和印度西北部亦有。

Perennial herbs. Stems sparsely retrorse-pubescent. Leaves palmately 3-sect, with dissected segments. Inflorescences usually compound racemes; rachis and pedicels subglabrous; staminodes broadly obovate; sepals deep or light-blue; sepal spurs as long as sepals, 1.2-1.6 cm long. Fl. Jun-Aug. Alpine grassy slopes at 2800-4200 m. Distributed in S Xizang. Also in Nepal and NW India.

光序翠雀花 *Delphinium kamaonense*

展毛翠雀花
Delphinium kamaonense var. *glabrescens*

宽距翠雀花 *Delphinium beesianum*

展毛翠雀花
Delphinium kamaonense Huth var. **glabrescens** (W. T. Wang) W. T. Wang

本变种与光序翠雀花的区别在于本变种的萼距比萼片长，稍向下弯；雄蕊无毛。生海拔2500-4200米的高山草地。产西藏东部、四川西部、青海南部和东南部、甘肃西南部。

This variety differs from the typical variety in its sepal spurs which are longer than sepals, and slightly decurved, and in its glabrous stamens. Alpine meadows at 2500-4200 m. Distributed in E Xizang, W Sichuan, S and SE Qinghai, and SW Gansu.

宽距翠雀花
Delphinium beesianum W. W. Sm.

多年生草本。茎高8-28厘米。基生叶有长柄，五角形，3全裂，末回裂片狭三角形或条形。伞房花序有1-5花；萼片蓝色，长2-2.5厘米，距钻状圆筒形，长2.4-2.8厘米；心皮3。花期9-11月。生海拔3500-4600米的草坡或多石砾处。产云南西北部和四川西南部。

Perennial herbs. Stems 8-28 cm tall. Basal leaves long petiolate, pentagonal, 3-sect, ultimate lobules narrow-triangular or linear. Corymbs 1-5-flowered; sepals blue, 2-2.5 cm long, spur subulate-cylindric, 2.4-2.8 cm long; carpels 3. Fl. Sep-Nov. Grassy slopes or gravelly places at 3500-4600 m. Distributed in NW Yunnan and SW Sichuan.

蓝翠雀花
Delphinium caeruleum Jacq. ex Camb.

多年生草本。茎高8-60厘米。叶基生并茎生，近圆形，3全裂，全裂片细裂。伞房花序有1-7花；萼片紫蓝色，长1.5-2厘米，距钻形；退化雄蕊瓣片圆倒卵形，有黄色髯毛；心皮5。花期7-9月。生海拔2100-4000米的草坡或多石砾山坡。产西藏、四川西部、青海和甘肃。不丹和尼泊尔亦有。

Perennial herbs. Stems 8-60 cm tall. Leaves basal and cauline, suborbicular, 3-sect, segments dissescted. Corymbs 1-7-flowered; sepals purple-blue, 1.5-2 cm long, spur subulate; staminode limbs orbicular-obovate, yellow-barbate; carpels 5. Fl. Jul-Sep. Grassy or gravelly slopes at 2100-4000 m. Distributed in Xizang, W Sichuan, Qinghai and Gansu. Also in Bhutan and Nepal.

蓝翠雀花 *Delphinium caeruleum*

细距翠雀花
Delphinium lagarocentrum W. T. Wang

多年生草本。茎高30-35厘米。茎生叶具长或短柄，近圆形，3全裂，全裂片细裂，末回裂片狭条形。伞房花序有2-5花；萼片蓝色，长1.1-1.5厘米，距细钻形，长2.1-2.4厘米；退化雄蕊瓣片卵形，2中裂，有黄色髯毛；心皮5。花期7-8月。生海拔3682米的云杉林岩壁上。特产西藏(类乌齐县)。

Perennial herbs. Stems 30-35 cm tall. Cauline leaves long or shortly petiolate, suborbicular, 3-sect, segments dissected, ultimate lobes narrow-linear. Corymbs 2-5-flowered; sepals blue, 1.1-1.5 cm long, spur thin-subulate, 2.1-2.4 cm long; staminode limbs ovate, 2-fid, yellow-barbate; car-

细距翠雀花 *Delphinium lagarocentrum*

pels 5. Fl. Jul-Aug. Rocky cliff in *Picea* forests at 3682 m. Endemic to Xizang (Riwoqê County).

康定翠雀花

Delphinium tatsienense Franch.

多年生草本。叶掌状3全裂，裂片细裂。花序伞房状，有3-12花，密被反曲短柔毛，并常混有黄色腺毛；萼片紫蓝色，萼距钻形；退化雄蕊宽倒卵形，顶端不裂或2浅裂；心皮3，子房密被短柔毛。花期7-9月。生海拔2300-4000米的草坡或高山草甸。产云南北部和四川西部。

Perennial herbs. Leaves palmately 3-sect, with dissected segments. Inflorescences corymbous, 3-12-flowered, densely retrorsely puberulent, often mixed with yellow glandular hairs; sepals purple-blue, spur subulate; staminodes broadly obovate, apex undivided or 2-lobulate; carpels 3, ovary densely puberulent. Fl. Jul-Sep. Grassy slopes or alpine meadows at 2300-4000 m. Distributed in N Yunnan and W Sichuan.

还亮草

Delphinium anthriscifolium Hance

一年生草本。二或三回羽状复叶。总状花序长2-12厘米，具2-10花，被反曲短柔毛；萼片堇色或紫色，萼距近钻形，长0.5-1.5厘米；退化雄蕊瓣片斧形或卵形，无毛，2深裂；心皮3。种子近球形，有横膜翅。花期3-5月。生海拔200-1200米山坡草地、丘陵地河边或低山区。产中国西南、华南、华中和华东。越南北部亦有。

还亮草 *Delphinium anthriscifolium*

康定翠雀花 *Delphinium tatsienense*

Annual herbs. Leaves 2-or 3-pinnate. Racemes 2-12 cm long, 2-10-flowered, retrorsely puberulent; sepals violet or purple, spur subulate, 0.5-1.5 cm long; staminode limbs dolabriform or ovate, glabrous, 2-parted; carpels 3. Seeds subglobose, transversely lamellate. Fl. Mar-May. Grassy slopes, by streams of hilly places, or low-elevation mountains at 200-1200 m. Distributed in SW, S, C and E China. Also in N Vietnam.

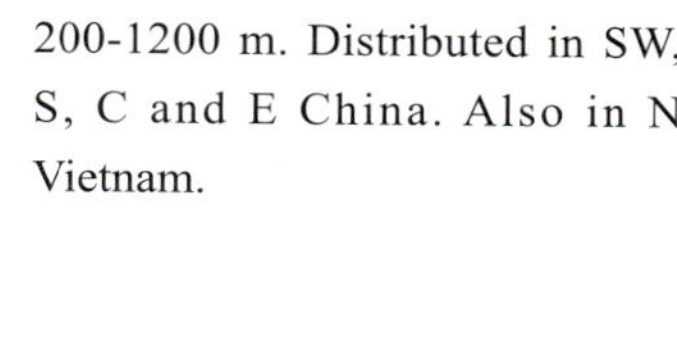

飞燕草 *Consolida ajacis*

拟扁果草 *Enemion raddeanum*

飞燕草
Consolida ajacis (L.) Schur

一年生草本。茎高约40厘米。中部茎生叶具短柄，宽约4厘米，3全裂，全裂片细裂，末回裂片狭条形。总状花序顶生；萼片紫色、粉红色或白色，长约1.2厘米，距长1.6厘米；花瓣2，合生；退化雄蕊不存在；心皮1。原产欧洲南部和亚洲西南部；在中国各城市有栽培。

Annual herbs. Stems ca. 40 cm tall. Middle cauline leaves shortly petiolate, ca. 4 cm broad, 3-sect, segments dissected, ultimate lobutes narrow-linear. Racemes terminal; sepals purple, pink or white, ca. 1.2 cm long, spur 1.6 cm long; petals 2, connate; staminodes wanting; carpel 1. Native to S Europe and SW Asia; cultivated in several cities of China as an ornamental.

拟扁果草
Enemion raddeanum Regel

多年生草本。茎高20-40厘米。基生叶1，为二回三出复叶，茎生叶也1，小叶3全裂，全裂片3浅裂或不等2深裂。伞形花序有1-8花；萼片5，白色，椭圆形，长4-6毫米；雄蕊多数；心皮约5。花期5月。生山地林下。产辽宁、吉林和黑龙江。俄罗斯(远东地区)、朝鲜半岛和日本亦有。

Perennial herbs. Stems 20-40 cm tall. Basal leaf 1, biternate; cauline leaf also 1, ternate; leaflets 3-sect, segments 3-lobed or unequally 2-parted. Umbels 1-8-flowered; sepals 5, white, elliptic, 4-6 mm long; stamens numerous; carpels ca. 5. Fl. May. Forests in montane regions. Distributed in Liaoning, Jilin and Heilongjiang. Also in Russia (Far East), Korean Peninsula and Japan.

东北扁果草
Isopyrum manshuricum Kom.

根状茎细长，具多数须根和块根。茎高10-18厘米，无毛。基生叶约4，为二回三出复叶。花序有2-3花；苞片叶状；萼片5，白色，长6.5-7.5毫米；花瓣5，长3毫米，基部浅囊状；雄蕊多数；心皮(1-)2。花期5月。生海拔800米的山地林下。产辽宁、吉林和黑龙江。

Rhizome slender, with numerous fibrous roots and tuberoids. Stems 10-18 cm tall, glabrous. Basal leaves ca. 4, 2-ternate. Inflorescences 2-3-flowered; bracts foliaceous; sepals 5, white, 6.5-7.5 mm long; petals 5, 3 mm long, at base shallowly scrotiform; stamens numerous; carpels (1-)2. F1.May. Forests montane regions at 800 m. Distributed in

东北扁果草 *Isopyrum manshuricum*

蓝堇草 *Leptopyrum fumarioides*

Liaoning, Jilin and Heilongjiang.

蓝堇草

Leptopyrum fumarioides (L.) Reichb.

一年生草本。茎数条，渐升，高达30厘米。基生叶具长柄，为二回三出复叶，无毛；茎生叶1，小。单歧聚伞花序有数花；萼片5，黄色，长3-4.5毫米；花瓣2-3，长约1毫米，二唇形；雄蕊10-15；心皮6-20。花期5-6月。生海拔100-1400米的林边、草地或田边。广布于中国西北、华北和东北。朝鲜半岛、蒙古、俄罗斯(西伯利亚)和欧洲亦有。

Annual herbs. Stems several, ascending, up to 30 cm tall. Basal leaves long petiolate, 2-ternate, glabrous; cauline leaf 1, small. Monochasia few-flowered; sepals 5, yellow, 3-4.5 mm long; petals 2-3, ca. 1 mm long, bilabiate; stamens 10-15; carpels 6-20. Fl. May-Jun. Forest edges, grassy places or by fields at 100-1400 m. Widespread in NW, N and NE China. Also in Korean Peninsula, Mongolia, Russia (Siberia) and Europe.

蕨叶人字果

Dichocarpum dalzielii (Drumm. et Hutch.) W. T. Wang et Hsiao

草本，无毛。叶3-11，基生，具小叶7-15，纸质，无毛。花葶3-11，高20-28厘米，具3-8花；苞片无柄，3深裂；萼片白色。蓇葖果2，倒披针形；宿存花柱约长2毫米。花期4-5月，果期5-6月。生海拔700-1600米的林下或河岸阴而潮湿地。产中国西南、华南和华东。

Herbs,glabrous. Leaves 3-11, basal, 7-15-foliolate, papery, glabrous. Scapes 3-11, 20-28 cm tall, 3-8-flowered; bracts sessile, 3-sect; sepals white. Follicles 2, oblanceolate; persistent styles ca. 2 mm long. Fl. Apr-May. Fr. May-Jun. Forests or shady and wet places by streams at 700-1600 m. Distributed in SW, S and E China.

耳状人字果

Dichocarpum auriculatum (Franch.) W. T. Wang et Hsiao

草本，无毛。叶少数，基生，5小叶；小叶不等大；茎生叶2(-4)，似基生叶。花序长7-19厘米，(1-)3-7花；萼片白色；瓣片宽倒卵形，顶端缺或全缘。蓇葖果2，倒披针形，长1.1-1.5厘米；宿存花柱长约2毫米。花期4-5月，果期4-6月。生海拔600-1500米林下石旁或阴湿的地方。产云南东北部、四川和湖北西部。

Herbs, glabrous. Leaves few, 5-foliolate, leaflets unequal in size; stem leaves 2(-4), similar to basal leaves. Inflorescences 7-19 cm long, (1-)3-7-flowered; sepals white; petal limbs broadly obovate, apex retuse or entire. Follicles 2, oblanceolate, 1.1-1.5 cm long; persistent styles ca. 2 mm long. Fl. Apr-May. Fr. Apr-Jun. By rocks in forests or shady wet places at 600-1500 m. Distributed in NE Yunnan, Sichuan and W Hubei.

蕨叶人字果 *Dichocarpum dalzielii*

耳状人字果 *Dichocarpum auriculatum*

裂瓣人字果 *Dichocarpum lobatipetalum*

裂瓣人字果

Dichocarpum lobatipetalum W. T. Wang et Bing Liu

多年生草本，无毛。根状茎长约4.5厘米。基生叶1，具长柄，为三回鸟足状复叶，具9-12小叶；较大小叶披针形长约9厘米，有牙齿。花葶1，高约38厘米；聚伞花序2-3回分枝，有9-14花；苞片3-4，轮生；萼片5，淡粉红色，长椭圆形，长约1.5厘米；花瓣4-5，橙黄色，具细爪，瓣片2浅裂，基部盾形；雄蕊多数；心皮2。花期3-4月。生海拔1534米的石灰山林下。特产云南(麻栗坡县)。

Perennial herbs, glabrous. Rhizome ca. 4.5 cm long. Basal leaf solitary, long petiolate, 3 times pedately compound, 9-12-foliolate; larger leaflets lanceolate, ca. 9 cm long, dentate. Scape solitary, ca. 38 cm tall; cyme 2-3 times branched, 9-14-flowered; bracts 3-4, verticillate; sepals 5, pinkish, long elliptic, ca. 1.5 cm long; petals 4-5, orange, slenderly clawed, limbs 2-lobed, at base peltate; stamens numerous; carpels 2 Fl. Mar-Apr. Forest in limestone hills at 1534 m. Endemic to Yunnan (Malipo County).

人字果

Dichocarpum sutchuenense (Franch.) W. T. Wang et Hsiao

多年生草本，无毛。茎高8-30厘米。基生叶少数，为鸟足状复叶；小叶5-13，圆倒卵形或宽卵形。复单歧聚伞花序有3-8花；苞片叶状；萼片5，白色，长6-11毫米；花瓣黄色，长3毫米；雄蕊多数；心皮2。花期4-5月。生海拔1450-2150米的林下或溪边。产云南、四川、湖北和浙江。

Perennial herbs, glabrous. Stems 8-30 cm tall. Basal leaves few, pedately compound; leaflets 5-13. Compound monochasia 3-8-flowered; bracts foliaceous; sepals 5, white, 6-11 mm long; petals 5, yellow, 3 mm long; stamens numerous; carpels 2. Fl. Apr-May. Forests or by streams at 1450-2150 m. Distributed in Yunnan, Sichuan, Hubei and Zhejiang.

小花人字果

Dichocarpum franchetii (Finet et Gagnep.) W. T. Wang et Hsiao

多年生草本，无毛。茎高9-26厘米。基生叶少数，为鸟足状复叶；小叶7-8，近圆形或圆卵形。复单歧聚伞花序有3-7花；苞片叶状；萼片5，白色，长3.5-4.5毫米；花瓣5，黄色，长1-1.2毫米；雄蕊多数；心皮2。花期4-5月。生海拔1300-3200米的林中或溪边。产云南、四川、贵州、湖南和广西。

Perennial herbs, glabrous. Stems 9-26 cm tall. Basal leaves few, pedately compound; leaflets 7-8, suborbicular or orbicular-ovate. Compound monochasia 3-7-flowered; bracts foliaceous; sepals 5, white, 3.5-4.5 mm long; petals 5, yellow, 1-1.2 mm long; stamens numerous; carpels 2. Fl. Apr-May. Forests or by streams at 1300-3200m. Distributed in Yunnan, Sichuan, Guizhou, Hunan and Guangxi.

人字果 *Dichocarpum sutchuenense*

小花人字果 *Dichocarpum franchetii*

纵肋人字果 *Dichocarpum fargesii*

拟耧斗菜 *Paraquilegia microphylla*

纵肋人字果

Dichocarpum fargesii (Franch.) W. T. Wang et Hsiao

多年生草本，无毛。茎高14-35厘米。叶基生和茎生，为鸟足状复叶，小叶约5。复单歧聚伞花序有2-4花；萼片5，白色，长4-5毫米；花瓣漏斗状，长约2毫米；雄蕊10；心皮2。种子有纵肋。花期5-6月。生海拔1300-1600米的山谷阴湿处。产重庆、贵州、湖北西部、河南西南部、陕西南部和甘肃南部。

Perennial herbs, glabrous. Stems 14-35 cm tall. Leaves basal and cauline, pedately compound, leaflets ca. 5. Compound monochasia 2-4-flowered; sepals 5, white, 4-5 mm long; petals funnelform, ca. 2 mm long; stamens 10; carpels 2. Seeds longitudinally ribbed. Fl. May-Jun. shady, damp places of valley at 1300-1600 m. Distributed in Chongqing, Guizhou, W Hubei, SW Henan, S Shaanxi and S Gansu.

拟耧斗菜

Paraquilegia microphylla (Royle) Drumm. et Hutch.

多年生草本。叶均基生，三角状卵形，二回三出分裂。花葶高3-18厘米；花单朵顶生；萼片5，淡紫色至紫红色；花瓣比萼片小得多；雄蕊多数；心皮5。种子褐色，具狭翅。花期6-8月，果期8-9月。生海拔2700-4300米的高山岩石上或裂隙中。产四川西部、云南西北部、西藏、甘肃西南部、青海和新疆。印度北部、尼泊尔、不丹、巴基斯坦北部、哈萨克斯坦和塔吉克斯坦亦有。

Perennial herbs. Leaves all basal, deltoid-ovate, 2-ternate. Scapes 3-18 cm tall; flower solitary, terminal; sepals 5, purplish to purplish red; petals quite smaller than sepals; stamens numerous; carpels 5. Seeds brown, narrowly winged. Fl. Jun-Aug. Fr. Aug-Sep. Alpine rocks or cliff crevices at 2700-4300 m. Distributed in W Sichuan, NW Yunnan, Xizang, SW Gansu, Qinghai and Xinjiang. Also in N India, Nepal, Bhutan, N Pakistan, Kazakhstan and Tajikistan.

天葵

Semiaquilegia adoxoides (DC.) Makino

多年生草本。块根长1-2厘米。三出复叶，小叶片卵形、近圆形或肾形。单歧聚伞花序有2-3花；萼片5，白色，常带淡紫色，长4-6毫米；花瓣5，匙形，长3毫米；雄蕊8-14；退化雄蕊约2，白色，条状披针形，膜质，无毛。蓇葖果3，长椭圆体形，有横向脉纹。花期3-4月，果期4-5月。生海拔100-1100米的林下或路边。产中国西南、东南、华中和华东。朝鲜半岛和日本亦有。

Perennial herbs. Root tubers 1-2 cm long. Leaves ternate; leaflets ovate, suborbicular or reniform. Monochasia 2-3-flowered; sepals 5, white, usually purple-tinged, 4-6 mm long; petals 5, spathulate, 3 mm long; staminodes ca. 2, white, linear-lanceolate, membranous, glabrous. Follicles 3, long ellipsoid, transversely nerved. Fl. Mar-Apr. Fr. Apr-May. Forests or roadsides at 100-1100 m. Distributed in SW, SE, C and E China. Also in Korean Peninsula and Japan.

天葵 *Semiaquilegia adoxoides*

尾囊草 *Urophysa henryi*

距囊尾囊草 *Urophysa rockii*

尾囊草

Urophysa henryi (Oliv.) Ulbr.

多年生小草本。叶约10，均基生，具长柄，3全裂。花葶与叶近等长；聚伞花序有3花；苞片倒卵形；萼片5，蓝色或粉白色，长10-14毫米；花瓣5，长5毫米，基部囊状；雄蕊多数；退化雄蕊数枚，狭披针形；心皮5(-8)。花期3-4月。生山地石旁或陡崖上。产重庆、湖北西部、湖南北部和贵州。

Small perennial herbs. Leaves ca. 10, all basal, long petiolate, 3-sect. Scape nearly as long as leaves; cymes 3-flowered; bracts obovate; sepals 5, blue or pinkish-white, 10-14 mm long; petals 5, 5 mm long, base saccate; stamens numerous; staminodes several, narrow-lanceolate; carpels 5(-8). Fl. Mar-Apr. Rocks or cliffs. Distributed in Chongqing, W Hubei, N Hunan and Guizhou.

距囊尾囊草

Urophysa rockii Ulbr.

多年生小草本。叶约10，均基生，有长柄，宽2.6-7厘米，3全裂。花葶高7-12厘米；聚伞花序具1花；苞片1-2，条形；萼片5，蓝色，长约2厘米；花瓣5，长6毫米，距长2毫米；雄蕊多数；退化雄蕊披针形；心皮5。花期3月。生溪边潮湿处。特产四川涪江上游。

Small perennial herbs. Leaves ca. 10, all basal, long petiolate, 2.6-7 cm broad, 3-sect. Scape 7-12 cm tall; cymes 1-flowered; bracts 1-2, linear; sepals 5, blue, ca. 2 cm long; petals 5, 6 mm long, with spurs 2 mm long; stamens numerous; staminodes lanceolate; carpels 5. Fl. Mar. Shady, damp places by streams. Endemic to upper reaches of Fujiang River, Sichuan.

无距耧斗菜

Aquilegia ecalcarata Maxim.

多年生草本。茎高20-60厘米。基生叶数枚，为二回三出复叶，具长柄；茎生叶小。单歧聚伞花序有2-6花；花梗被短柔毛；萼片5，紫色，长1-1.4厘米；花瓣5，无距；雄蕊多数；心皮4-5。花期5-8月。生海拔1800-3500米的山地林下或路边。产西藏东部、四川、贵州北部、湖北西部、河南西部、陕西南部、甘肃和青海。

Perennial herbs. Stems 20-60 cm tall. Basal leaves several, 2-ternate, long petiolate; cauline leaves small. Monochasium 2-6-flowered; pedicels puberulous; sepals 5, purple, 1-1.4 cm long; petals 5, without spurs; stamens numerous; carpels 4-5. Fl. May-Aug. Forests or roadsides at 1800-3500 m. Distributed in E Xizang, Sichuan, N Guizhou, W Hubei, W Henan, S Shaanxi, Gansu and Qinghai.

无距耧斗菜 *Aquilegia ecalcarata*

直距耧斗菜

Aquilegia rockii Munz

多年生草本。叶基生并茎生，为二回三出复叶。单歧聚伞花序具1-3花；花下垂；萼片和花瓣紫色或蓝色；花瓣距直伸或顶端稍弯曲；花药黑色；心皮

直距耧斗菜 *Aquilegia rockii*

细距耧斗菜 *Aquilegia leptoceras*

5，直立，密具腺毛。花期6-8月，果期7-9月。生海拔2500-3500米的林中、山坡草地或路边。产云南东北部、四川西南部和西藏东南部。

Perennial herbs. Basal and cauline leaves 2-ternat. Monochasium 1-3-flowered; flowers pendulous; sepals and petals purple or blue; petal spurs straight or apically slightly incurved; anthers black; carpels 5, erect, densely glandular hairy. Fl. Jun-Aug. Fr. Jul-Sep. Forests, grassy slopes or alongside roads at 2500-3500 m. Distributed in NE Yunnan, SW Sichuan and SE Xizang.

小花耧斗菜

Aquilegia parviflora Ledeb.

多年生草本。茎高15-45厘米，无毛。基生叶少数，为二回三出复叶，具长柄；小叶长1.6-3.5厘米，2-3浅裂。单歧聚伞花序有3-6花；萼片5，淡蓝色，长1.2-2厘米；花瓣瓣片近白色，长3-5毫米，距蓝紫色，长3-5毫米；雄蕊比花瓣短；心皮5。花期6-7月。生海拔2500-3500米的林中或林边。产内蒙古北部和黑龙江。蒙古和俄罗斯(东西伯利亚)亦有。

Perennial herbs. Stems 15-45 cm tall, glabrous. Basal leaves few, 2-ternate, long petiolate; leaflets 1.6-3.5 cm long, 2-3-lobed. Monochasium 3-6-flowered; sepals 5, bluish, 1.2-2 cm long; petal limbs nearly white, 3-5 mm long, spurs blue-purple, 3-5 mm long; stamens shorter than petals; carpels 5. Fl. Jun-Jul. Forests or forest margins at 2500-3500 m. Distributed in N Neimenggu and Heilongjiang. Also in Mongolia and Russia (E Siberia).

细距耧斗菜

Aquilegia leptoceras Fisch. et Mey.

多年生草本。茎高20-30厘米。基生叶具长柄，与茎生叶均为三出复叶。单歧聚伞花序有3-4花；花具长梗，直径3-5厘米，淡紫色；萼片5，椭圆状卵形；花瓣5，瓣片扇状倒卵形，长约1.2厘米，距暗紫色，直，钻形；雄蕊稍伸出。花期5月。生山地林边。产内蒙古东北部。俄罗斯(东西伯利亚)亦有。

Perennial herbs. Stems 20-30 cm tall. Basal leaves long petiolate, with cauline leaves all ternate. Monochasium 3-4-flowered; flower long pedicellate, 3-5 cm diam, purplish; sepals 5, elliptic-ovate; petals 5, limbs flabellate-obovate, ca. 1.2 cm long, spurs dark-purple, straight, subulate; stamens slightly exserted. Fl. May. Margins of Montane forest. Distributed in NE Neimenggu. Also in Russia (E Siberia).

小花耧斗菜 *Aquilegia parviflora*

耧斗菜 *Aquilegia viridiflora*

耧斗菜
Aquilegia viridiflora Pall.

多年生草本。茎高15-50厘米。基生叶为二回三出复叶，具长柄。单歧聚伞花序有3-7花；花梗被短柔毛和腺毛；萼片5，淡绿色，长1.2-1.5厘米；花瓣也呈淡绿色，距长1.2-1.8厘米；雄蕊伸出花外；心皮5。花期5-7月。生海拔200-2300米的山地路旁、溪边或草坡。产青海东部、甘肃、宁夏、陕西、华北和东北。俄罗斯(远东地区)亦有。

Perennial herbs. Stems 15-50 cm tall. Basal leaves 2-ternate, long petiolate. Monochasia 3-7-flowered; pedicels puberulous and glandular-puberulous; sepals 5, greenish, 1.2-1.5 cm long; petals also greenish, spurs 1.2-1.8 cm long; stamens exserted; carpels 5. Fl. May-Jul. Roadside, by streams, or grassy slopes at 200-2300 m. Distributed in E Qinghai, Gansu, Ningxia, Shaanxi, N and NE China. Also in Russia (Far East).

紫花耧斗菜
Aquilegia viridiflora Pall. f. **atropurpurea** (Willd.) Kitag.

多年生草本。茎被柔毛和腺毛。单歧聚伞花序具3-7花；苞片3全裂；花下垂；萼片和花瓣紫色；花瓣距直伸或先端稍弯曲；花药黄色；心皮5，密具腺毛。蓇葖果具有宿存花柱。花期5-7月，果期7-8月。生林中、山谷或溪边多石处。产青海东部、山西、山东东部、河北、内蒙古和辽宁南部。俄罗斯(西伯利亚)、蒙古和日本亦有。

Perennial Herbs. Stems pubescent and glandular hairy. Monochasia 3-7-flowered; bracts 3-sect; flowers pendulous; sepals and petals purple; petal spurs straight or apically slightly incurved; anthers yellow; carpels 5, densely glandular hairy. Follicles with persistent styles. Fl. May-Jul. Fr. Jul-Aug. Forests, valleys or rocky places by streams. Distributed in E Qinghai, Shanxi, E Shandong, Hebei, Neimenggu and S Liaoning. Also in Russia (Siberia), Mongolia and Japan.

尖萼耧斗菜
Aquilegia oxysepala Trautv. et C. A. Mey.

多年生草本。单歧聚伞花序具3-5花；苞片3全裂，顶端圆钝；花微下垂；萼片紫色，窄卵形，先端尖；花瓣瓣片淡黄色，距紫色；心皮5，具白色柔毛；宿存花柱长3-10毫米。花期5-6月，果期7-8月。生海拔400-2700米的林缘或草坡。产中国东北。俄罗斯(远东地区)和朝鲜半岛亦有。

Perennial herbs. Monochasia 3-5-flowered; bracts 3-sect, apex obtuse; flowers slightly pendulous; sepals purple, narrowly ovate, apex acute; petal limbs yellowish, spurs purple; carpels 5, white pubescent; styles persistent, 3-10 mm long. Fl. May-Jun. Fr. Jul-Aug. Forest edges or grassy slopes at 400-2700 m. Distributed in NE China. Also in Russia (Far East) and Korean Peninsula.

阿穆尔耧斗菜
Aquilegia amurensis Kom.

多年生草本。茎高20-40厘米。叶均基生，为二回三出复叶，具长柄；小叶倒卵形，3裂。聚伞花序有1-3花；萼片5，蓝紫色，长2-2.5厘米；花瓣5，瓣片长1厘米，上部黄白色，距蓝紫色，长1.5厘米，末端钩状；雄蕊伸出；心皮5。花期6月。生海拔1000米的山顶石缝。产内蒙古北部。俄罗斯(远东地区、东西伯利亚)和朝鲜半岛亦有。

Perennial herbs. Stems 20-40 cm tall. Leaves all basal, 2-ternate, long petiolate; leaflets obovate, 3-lobed. Cyme 1-3-flowered; sepals 5, blue-purple, 2-2.5 cm long; petals 5, limbs 1 cm long, above yellow-white, spurs blue-purple, 1.5 cm long, basally

紫花耧斗菜 *Aquilegia viridiflora* f. *atropurpurea*

尖萼耧斗菜 *Aquilegia oxysepala*

阿穆尔耧斗菜 *Aquilegia amurensis*

hooked; stamens exserted; carpels 5. Fl. Jun. Rock fissures on mountain top at 1000 m. Distributed in N Neimenggu. Also in Russia (Far East, E Siberia) and Korean Peninsula.

华北耧斗菜

Aquilegia yabeana Kitag.

多年生草本。茎疏被柔毛和腺毛，顶端分枝。单歧聚伞花序具少数花，密被短腺毛；萼片紫色，狭卵形；花瓣紫色，近直立，距向内钩状弯曲。蓇葖果明显具脉网，具宿存花柱。花期5-6月。生林缘或草坡。产华中和华北。

Perennial herbs. Stems sparsely pubescent and glandular hairy, apically branched. Monochasia few-flowered, densely glandular puberulent; sepals purple, narrowly ovate; petals purple, suberect, spurs hooked-incurved. Follicles conspicuously reticulate, with persistent styles. Fl. May-Jun. Forest edges or grassy slopes. Distributed in C and N China.

长白耧斗菜

Aquilegia japonica Nakai et Hara

草本。叶全部基生，为二回三出复叶。单歧聚伞花序具1-3花；苞片条状披针形；萼片蓝紫色，花瓣状，椭圆状倒卵形；花瓣黄白色或白色，距紫色，长1-1.6厘米，末端弯曲呈钩状；退化雄蕊膜质，白色；心皮5，无毛。花期7月。生海拔1400-2500米的山坡草地。产黑龙江南部和吉林(长白山)。朝鲜半岛和日本亦有。

Herbs. Leaves basal, 2-ternate. Monochasia 1-3-flowered; bracts linear-lanceolate; sepals blue-purple, petaloid, elliptic-obovate; petals yellow-white or white, spurs purple, 1-1.6 cm long, basally incurved in a hook; staminodes membranous, white; carpels 5, glabrous. Fl. Jul. Grassy slopes at 1400-2500 m. Distributed in S Heilongjiang and Jilin (Changbai Mountain). Also in Korean Peninsula and Japan.

华北耧斗菜 *Aquilegia yabeana*

长白耧斗菜 *Aquilegia japonica*

大花耧斗菜 *Aquilegia glandulosa*

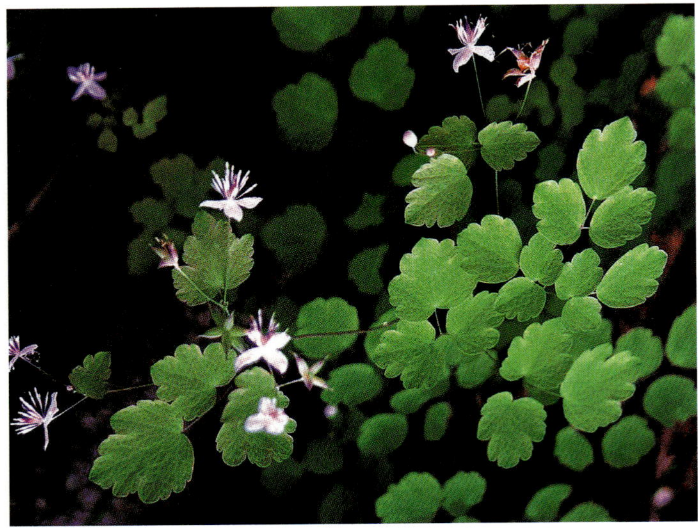

华东唐松草 *Thalictrum fortunei*

大花耧斗菜

Aquilegia glandulosa Fisch. ex Link

草本。花序1(-3)花；苞片1-3，披针形至长圆形；花直径6-9厘米；萼片蓝色，开展，卵形至长圆状卵形；花瓣蓝色或白色；退化雄蕊条形；心皮8-10，密被开展长柔毛。花期6-8月。生海拔1900-2700米的针叶林下、草坡或山谷河边。产新疆。俄罗斯(西伯利亚)和蒙古亦有。

Herbs. Inflorescences 1(-3)-flowered; bracts 1-3, lanceolate to oblong; flowers 6-9 cm diam; sepals blue, spreading, ovate to oblong-ovate; petals blue or white; staminodes linear; carpels 8-10, densely spreading villous. Fl. Jun-Aug. Coniferous forests, grassy slopes or by streams in valleys at 1900-2700 m. Distributed in Xinjiang. Also in Russia (Siberia) and Mongolia.

爪哇唐松草

Thalictrum javanicum Blume

多年生草本，无毛。叶为三或四回三出复叶。花序多歧聚伞状，少花或多花；花丝基部丝状，顶端倒披针形，宽于花药；心皮8-15；花柱拳卷状。瘦果狭卵球形。花期4-7月。生海拔1500-3400米的林中、灌丛、山坡或潮湿而多石脊处。产中国西南、华南、华中和华东。不丹、印度、尼泊尔、斯里兰卡和印度尼西亚亦有。

Perennial herbs, glabrous. Leaves 3- or 4-ternate. Inflorescences pleiochasial, few- or many-flowered; filament base filiform, apex oblanceolate, broader than anther; carpels 8-15, styles circinate. Achenes narrowly ovoid. Fl. Apr-Jul. Forests, thickets, slopes or damp rocky ledges at 1500-3400 m. Distributed in SW, S, C and E China. Also in, Bhutan, India, Nepal, Sri Lanka and Indonesia.

华东唐松草

Thalictrum fortunei S. Moore

多年生草本。叶为二回三出复叶；托叶半圆形，膜质；小叶圆形。单歧聚伞花序；萼片4，白色，带紫色，倒卵形，长3-4.5毫米；心皮3-6，花柱直，短。瘦果纺锤形，具6-8条纵肋。花期3-5月。生海拔100-1500米的林中、山坡或潮湿而多石脊处。产浙江、安徽、江苏和江西。

Perennial herbs. Leaves 2-ternate; stipules semiorbicular, membranous; leaflets orbicular. Inflorescences monochasial; sepals 4, white, purplish-tinged, obovate, 3-4.5 mm long; carpels 3-6, styles straight, short.

爪哇唐松草 *Thalictrum javanicum*

长喙唐松草 *Thalictrum macrorhynchum*

粗壮唐松草 *Thalictrum robustum*

Achenes fusiform, 6-8-ribbed. Fl. Mar-May. Forests, slopes or damp rocky ledges at 100-1500 m. Distributed in Zhejiang, Anhui, Jiangsu and Jiangxi.

长喙唐松草

Thalictrum macrorhynchum Franch.

多年生草本，无毛。茎高45-65厘米。叶基生并茎生，为二至三回三出复叶。圆锥花序有稀疏分枝；萼片白色，早落；雄蕊长4毫米；心皮10-20。瘦果狭卵球形，有8条纵肋，基部有短柄，宿存花柱拳卷。花期6月。生海拔850-2900米的林中或灌丛中。产重庆北部、湖北西部、甘肃南部、陕西南部、山西和河北西部。

Perennial herbs, glabrous. Stems 45-65 cm tall. Leaves basal and cauline, 2-3-ternate. Panicles sparsely branched; sepals white, caducous; stamens 4 mm long; carpels 10-20. Achenses narrow-ovoid, longitudinally 8-ribbed, base shortly stipitate, persistent styles circinate. Fl. Jun. Forests or bushes at 850-2900 m. Distributed in N Chongqing, W Hubei, S Gansu, S Shaanxi, Shanxi and W Hebei.

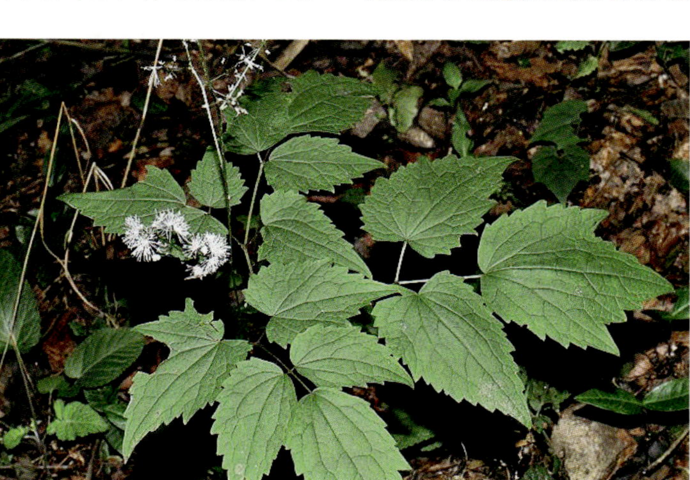
大叶唐松草 *Thalictrum faberi*

大叶唐松草

Thalictrum faberi Ulbr.

多年生草本，无毛。茎高45-110厘米。叶为二至三回三出复叶；小叶卵形，长达10厘米。圆锥花序长达40厘米；雄蕊花丝上部狭倒披针形；心皮3-6。瘦果长5毫米，宿存花柱钩状，长1毫米。花期7-8月。生海拔800-1300米的山地林下。产湖南东部、江西、福建、浙江、江苏南部、安徽、河南南部和陕西南部。

Perennial herbs, glabrous. Stems 45-110 cm tall. Leaves 2-3-ternate; leaflets ovate, up to 10 cm long. Panicles up to 40 cm long; filaments above narrow-oblanceolate; carpels 3-6. Achenes 5 mm long, persistent styles hook-like, 1 mm long. Fl. Jul-Aug. Forests at 800-1300 m. Distributed in E Hunan, Jiangxi, Fujian, Zhejiang, S Jiangsu, Anhui, S Henan and S Shaanxi.

粗壮唐松草

Thalictrum robustum Maxim.

多年生草本。茎高80-150厘米。叶为二至三回三出复叶；小叶卵形，长3-8.5厘米，背面被短柔毛。圆锥花序有多数花；雄蕊花丝条状倒披针形；心皮6-16。瘦果长约2毫米，宿存花柱钩状，长0.6-0.8毫米。花期6-7月。生海拔940-2100米的山地林中。产四川北部、重庆、湖北西北、甘肃南部、陕西南部和河南西部。

Perennial herbs. Stems 80-150 cm tall. Leaves 2-3-ternate; leaflets ovate, 3-8.5 cm long, abaxially puberulous. Panicles multi-flowered; filaments narrow-oblanceolate; carpels 6-16. Achenes ca. 2 mm long, persistent styles hook-like, 0.6-0.8 mm long. Fl. Jun-Jul. Forests at 940-2100 m. Distributed in N Sichuan, Chongqing, NW Hubei, S Gansu, S Shaanxi and W Henan.

钩柱唐松草 *Thalictrum uncatum*

西南唐松草 *Thalictrum fargesii*

钩柱唐松草

Thalictrum uncatum Maxim.

多年生草本，无毛。茎高45-90厘米。叶为四至五回三出复叶；小叶宽达7-10毫米。圆锥花序狭，似总状花序；萼片淡紫色，长3毫米；花丝上部狭条形。瘦果新月形，长4-5毫米，有8条纵肋，有短柄，宿存花柱钩状。花期5-7月。生海拔2700-3200米的草坡或灌丛边。产西藏东部、云南西北部、四川西部、青海东部和甘肃南部。

Perennial herbs, glabrous. Stems 45-90 cm tall. Leaves 4-5-ternate; leaflets up to 7-10 mm broad. Panicles narrow, racemiform; sepals purplish, 3 mm long; filaments above narrow-linear. Achenes lunate, 4-5 mm long, 8-ribbed, shortly stipitate, persistent styles hook-like. Fl. May-Jul. Grassy slopes or bush edges at 2700-3200 m. Distributed in E Xizang, NW Yunnan, W Sichuan, E Qinghai and S Gansu.

狭序唐松草 *Thalictrum atriplex*

狭序唐松草

Thalictrum atriplex Finet et Gagnep.

多年生草本，无毛。茎高40-80厘米。下部茎生叶为四回三出复叶；小叶宽0.8-3厘米。圆锥花序狭，似总状花序；萼片长约3毫米；花丝上部棒状，下部丝形。瘦果卵球形，长2.5毫米，宿存花柱钩状。花期6-7月。生海拔2300-3600米的草坡、林边或疏林中。产西藏东部、云南西北部和四川西部。

Perennial herbs, glabrous. Stems 40-80 cm tall. Lower cauline leaves 4-ternate; leaflets 0.8-3 cm broad. Panicles narrow, racemiform; sepals ca. 3 mm long; filaments above clavate, below filiform. Achenes ovoid, 2.5 mm long, persistent styles hook-like. Fl. Jun-Jul. Grassy slopes, forest edges or sparse forests at 2300-3600 m. Distributed in E Xizang, NW Yunnan and W Sichuan.

西南唐松草

Thalictrum fargesii Franch.

多年生草本，无毛。茎高约40厘米。茎生叶为三至四回三出复叶；小叶宽达2.5厘米。单歧聚伞花序顶生；萼片4，白色或粉红色；花丝上部倒披针形。瘦果狭卵球形，长约4.5毫米，宿存花柱和柱头长3毫米。花期5-6月。生海拔1300-2400米的林中、草坡或溪边。产四川、贵州北部、重庆、湖北、河南西南部、陕西南部和甘肃南部。

Perennial herbs, glabrous. Stems ca. 40 cm tall. Cauline leaves 3-4-ternate; leaflets up to 2.5 cm broad. Monochasia terminal; sepals 4, white or pink; filaments above oblanceolate. Achenes narrow-ovoid, ca. 4.5 mm long, persistent style and stigma 3 mm long. Fl. May-Jun. Forests, grassy slopes or by streams at 1300-2400 m. Distributed in Sichuan, N Guizhou, Chongqing, Hubei, SW Henan, S Shaanxi and S Gansu.

武夷唐松草

Thalictrum wuyishanicum W. T. Wang et S. H. Wang

多年生草本，无毛。茎高17-29厘米。基生叶有长柄，为二回三出复叶；小叶宽1.5-2.5厘米。单歧聚伞花序顶生；萼片淡粉红色，长约3.5毫米；花丝上部倒披针形。瘦果长约3.5毫米，有8纵肋，无花柱，宿存柱头扁头形。花期4月。生海拔2000米的山地石上。产江西东

武夷唐松草 *Thalictrum wuyishanicum*

部和福建西北部。

Perennial herbs, glabrous. Stems 17-29 cm tall. Basal leaves 2-ternate, long petiolate; leaflets 1.5-2.5 cm broad. Monochasia terminal; sepals pinkish, ca. 3.5 mm long; filaments above oblanceolate. Achenes ca. 3.5 mm long, 8-ribbed, without styles, persistent stigmas depressed-capitate. Fl. Apr. Rocks at 2000 m. Distributed in E Jiangxi and NW Fujian.

贝加尔唐松草

Thalictrum baicalense Turcz.

多年生草本，无毛。茎高45-80厘米。中部茎生叶为三回三出复叶；小叶宽2-5厘米。圆锥花序伞状；萼片长约2毫米；花丝上部倒披针形。瘦果近球形，长3毫米，有8纵肋和短柄，宿存花柱短。花期5-6月。生海拔900-2800米的林下或湿草坡。产西藏东南部、青海东部、甘肃南部、陕西南部、河南、山西、河北、内蒙古和东北。朝鲜半岛和俄罗斯(西伯利亚、远东地区)亦有。

Perennial herbs, glabrous. Stems 45-80 cm tall. Middle cauline leaves 3-ternate; leaflets 2-5 cm broad. Panicles umbelliform; sepals ca. 2 mm long; filaments above oblanceolate. Achenes subglobose, 3 mm long, 8-ribbed, shortly stipitate, persistent styles short. Fl. May-Jun. Forests or damp grassy slopes at 900-2800 m. Distributed in SE Xizang, E Qinghai, S Gansu, S Shaanxi, Henan, Shanxi, Hebei, Neimenggu and NE China. Also in Korean Peninsula and Russia (Siberia, Far East).

瓣蕊唐松草 *Thalictrum petaloideum*

瓣蕊唐松草

Thalictrum petaloideum L.

多年生草本，全株无毛。叶为三至四回三出复叶。复单歧聚伞花序伞房状，具少花或多花；花丝白色，狭倒披针形，明显比花药宽；心皮4-13。瘦果无柄，卵球形，具8条纵肋。花期6-7月。生海拔700-3000米的山坡或草甸。产中国西南、华北、华西和东北。俄罗斯(西伯利亚)、蒙古和朝鲜半岛亦有。

Perennial herbs, glabrous. Leaves 3-4-ternate. Compound monochasia corymbiform, few- or many-flowered; filaments white, narrowly oblanceolate, much broader than anthers; carpels 4-13. Achenes sessile, ovoid, longitudinally 8-ribbed. Fl. Jun-Jul. Slopes or meadows at 700-3000 m. Distributed in SW, N, W and NE China. Also in Russia (Siberia), Mongolia and Korean Peninsula.

贝加尔唐松草 *Thalictrum baicalense*

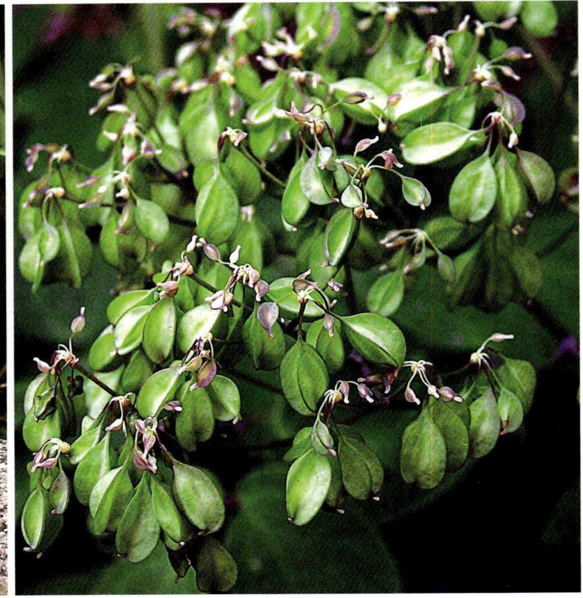

唐松草 *Thalictrum aquilegiifolium* var. *sibiricum*

唐松草

Thalictrum aquilegiifolium L. var. **sibiricum** Regel

多年生草本。叶为三或四回三出复叶；小叶倒卵形或阔圆形。复单歧聚伞花序伞房状，多花；雄蕊多数；心皮6-8，花柱宿存。瘦果倒卵形，长4-7毫米，有3条宽翅；心皮柄长3-5毫米。花期7月。生海拔500-1800米林中、林缘或草坡。产华东、华北和东北。俄罗斯(西伯利亚)、蒙古、朝鲜半岛和日本亦有。

Perennial herbs. Leaves 3-4-ternate; leaflets obovate or broadly orbicular. Compound monochasia corymbiform, many-flowered; stamens numerous; carpels 6-8, styles persistent. Achenes obovoid, 4-7 mm long, longitudinally 3-winged; carpophores 3-5 mm long. Fl. Jul. Forests, forest edges, grassy slopes or meadows at 500-1800 m. Distributed in E, N and NE China. Also in Russia (Siberia), Mongolia, Korean Peninsula and Japan.

长柄唐松草 *Thalictrum przewalskii*

长柄唐松草

Thalictrum przewalskii Maxim.

多年生草本。茎高50-120厘米，无毛。下部茎生叶为四回三出复叶，有长柄；小叶宽0.9-3.5厘米，下面有短毛或无毛。圆锥花序具多数花；萼片白色或淡黄绿色；花丝上部条状倒披针形。瘦果扁，长6-12毫米，有8条纵肋，心皮柄长0.8-3毫米。花期6-8月。生海拔750-3500米的灌丛边、草坡或林下。产西藏东部、四川西部、青海东部、甘肃、陕西、湖北、河南西部、山西、河北西部和内蒙古南部。

Perennial herbs. Stems 50-120 cm tall, glabrous. Lower cauline leaves 4-ternate; leaflets 0.9-3.5 cm broad, abaxially with short hairs or glabrous. Panicles multi-flowered; sepals white or pale yellowish-green; filaments above linear-oblanceolate. Achenes compressed, 6-12 mm long, 8-ribbed, carpophore 0.8-3 mm long. Fl. Jun-Aug. Bush edges, grassy slopes or forests at 750-3500 m. Distributed in E Xizang, W Sichuan, E Qinghai, Gansu, Shaanxi, Hubei, W Henan, Shanxi, W Hebei and S Neimenggu.

阴地唐松草

Thalictrum umbricola Ulbr.

多年生草本，无毛。茎高15-50厘米。基生叶为二至三回三出复叶，具长柄；小叶宽1-2.5厘米。单歧聚伞花序有少数花；萼片白色，长2毫米；花丝上部狭倒披针形。瘦果长3毫米，有8条纵肋，无花柱，心皮柄长1-2毫米。花期3-4月。生海拔1000-1300米的林中、沟边或陡崖上。产广西东部、广东、湖南东南部和江西南部。

Perennial herbs, glabrous. Stems 15-50 cm tall. Basal leaves 2-3-ternate, long petiolate; leaflets 1-2.5 cm broad. Monochasia few-flowered; sepals white, 2 mm long; filaments narrow-oblanceolate. Achenes 3 mm long, 8-ribbed, without styles, carpophores 1-2 mm long. Fl. Mar-Apr. Forests, by steams or cliffs at 1000-1300 m. Distributed in E Guangxi, Guangdong, SE Hunan and S Jiangxi.

阴地唐松草 *Thalictrum umbricola*

尖叶唐松草 *Thalictrum acutifolia*

尖叶唐松草

Thalictrum acutifolia (Hand.-Mazz.) B. Boivin

多年生草本。叶1-2基生，为二回三出复叶；小叶卵形，草质，边缘具少数牙齿。单歧聚伞花序，少花；萼片4，早落，白色或粉色；心皮6-12，花柱短。瘦果扁压，长椭圆形，有8条纵肋；心皮柄比瘦果短。花期4-7月。生海拔600-2000米的林缘、山坡或潮湿而多石脊处。产中国西南、华南、东南和华东。

Perennial hebs. Leaves 1-2 basal, 2-ternate; leaflets ovate, herbaceous, margin sparsely toothed. Monochasia few flowered; sepals 4, caducous, white or pinkish; carpels 6-12, styles short. Achenes compressed, long elliptic, longitudinally 8-ribbed; carpophore shorter than achene body. Fl. Apr-Jul. Forest edges, slopes or damp rocky ledges at 600-2000 m. Distributed in SW, S, SE and E China.

盾叶唐松草

Thalictrum ichangense Lecoy. ex Oliv.

多年生草本。叶为一至三回三出复叶；小叶盾形。单歧聚伞花序；萼片早落，白色，卵形；心皮5-12(-16)。瘦果镰刀形，长约4.5毫米，有8条纵肋；心皮柄细，长约1.5毫米。花期5-7月。生海拔600-1900米林中、灌丛或潮湿而多石脊处。产中国西南、浙江、湖北、陕西南部和辽宁。

Perennial herbs. Leaves 1-3-ternate; leaflets peltate. Inflorescences monochasial; sepals caducous, white, ovate; carpels 5-12(-16). Achenes falcate, ca. 4.5 mm long, longitudinally 8-ribbed; carpophores slender, ca. 1.5 mm long. Fl. May-Jul. Forests, thickets or damp rocky ledges at 600-1900 m. Distributed in SW China, Zhejiang, Hubei, S Shaanxi and Liaoning.

小果唐松草

Thalictrum microgynum Lecoy. ex Oliv.

多年生草本，无毛。茎高20-42厘米。基生叶1，为二至三回三出复叶；小叶宽1.5-4.8厘米。单歧聚伞花序似伞形花序；萼片白色，长1.5毫米；花丝上部倒披针形。瘦果长1.8毫米，有6条纵肋，无花柱，心皮柄长1.2毫米。花期4-7月。生海拔700-2800米的林下、草坡或悬崖上。产云南西北部、四川、重庆、湖南西北部、湖北西部和陕西南部。

Perennial herbs, glabrous. Stems 20-42 cm tall. Basal leaf 1, 2-3-ternate; leaflets 1.5-4.8 cm broad. Monohasia umbelliform; sepals white, 1.5 mm long; filaments above oblanceolate. Achenes 1.8 mm long, 6-ribbed, without styles, carpophores 1.2 mm long. Fl. Apr-Jul. Forests, grassy slopes or cliffs at 700-2800 m. Distributed in NW Yunnan, Sichuan, Chongqing, NW Hunan, W Hubei and S Shaanxi.

盾叶唐松草 *Thalictrum ichangense*

小果唐松草 *Thalictrum microgynum*

芸香叶唐松草 *Thalictrum rutifolium*

帚枝唐松草 *Thalictrum virgatum*

芸香叶唐松草

Thalictrum rutifolium Hook. f. et Thoms.

多年生草本，无毛。茎高11-50厘米。叶基生并茎生，为三至四回羽状复叶；小叶长3-8毫米。复单歧聚伞花序似总状花序；萼片淡黄色，长1.5毫米；花丝丝形。瘦果下垂，新月形，长4-6毫米，有8条纵肋，心皮柄长1毫米，宿存花柱长0.3毫米。花期6月。生海拔2280-4300米的草坡或河滩上。产西藏、云南西北部、四川西部、青海、甘肃中部和南部。印度北部亦有。

Perennial herbs, glabrous. Stems 11-50 cm tall. Leaves basal and cauline, 3-4-ternate; leaflets 3-8 mm long. Compound monochasia racemiform; sepals yellowish, 1.5 mm long; filaments filiform. Achenes pendulous, lunate, 4-6 mm long, 8-ribbed, carpophores 1 mm long, persistent styles 0.3 mm long. Fl. Jun. Grassy slopes or river banks at 2280-4300 m. Distributed in Xizang, NW Yunnan, W Sichuan, Qinghai, C and S Gansu. Also in N India.

帚枝唐松草

Thalictrum virgatum Hook. f. et Thoms.

多年生草本，无毛。茎高16-65厘米。叶均茎生，为三出复叶，具短柄或无柄。单歧聚伞花序顶生；萼片4-5，白色或淡粉色，长4-8毫米；花丝狭条形。瘦果扁，长3毫米，有8条纵肋，无花柱，心皮柄长0.4毫米。花期6-7月。生海拔2300-3500米的林下或林边石上。产西藏南部、云南北部和西部、四川西部。不丹和尼泊尔亦有。

Perennial herbs, glabrous. Stems 16-65 cm tall. Leaves all cauline, ternate, shortly petiolate or sessile. Monochasia terminal; sepals 4-5, white or pinkish, 4-8 mm long; filaments narrow-linear. Achenes compressed, 3 mm long, 8-ribbed, without styles, carpophores 0.4 mm long. Fl. Jun-Jul. Forests or rocks at forest edges at 2300-3500 m. Distributed in S Xizang, N and W Yunnan, and W Sichuan. Also in Bhutan and Nepal.

丝叶唐松草

Thalictrum foeniculaceum Bunge

多年生草本，无毛。茎高11-78厘米。叶基生并茎生，为二至四回三出复叶；小叶狭条形，宽0.5-1.5毫米。聚伞花序伞房状；萼片4，粉红色或白色；花丝丝形。瘦果长约4毫米，有8-10条纵肋，宿存花柱极短。花期6-7月。生海拔590-1000米的干燥草坡、多石砾地或草丛。产甘肃北部、陕西北部、山西、山东(博山)、河北北部和辽宁西部。

Perennial herbs, glabrous. Stems 11-78 cm tall. Leaves basal and cauline, 2-4-ternate; leaflets narrow-linear, 0.5-1.5 mm broad. Cymes corymbiform; sepals 4, pink or white; filaments filiform. Achenes ca. 4 mm long, 8-10-ribbed, persistent style very short. Fl. Jun-Jul. Dry grassy slopes, gravelly or grassy places at 590-1000 m. Distributed in N Gansu, N Shaanxi, Shanxi, Shandong (Boshan), N Hebei and W Liaoning.

偏翅唐松草

Thalictrum delavayi Franch.

多年生草本，无毛。三或四回羽状复叶；小叶圆卵形、倒卵形或椭圆形。花序圆锥状，长15-40厘米；花梗长8-25毫米；萼片长圆形、长椭圆形或卵形，长6-11毫米；花丝丝形；心皮15-22。瘦果斜倒卵形，长5-8毫米，沿二缝线有狭翅，基部具长1-3毫米之柄。花期6-9月。生海拔1800-3400米的灌丛、林下、草坡、阴处或潮湿而多石脊处。产云南、四川西部和西藏东部。

Perennial herbs, glabrous. Leaves 3- or 4-pinnate; leaflets orbicular-ovate, obovate or elliptic. Inflorescences paniculate, 15-40 cm long; pedicels 8-25 mm long; sepals oblong, long elliptic or ovate, 6-11 mm long; filaments filiform; carpels 15-22. Achenes

丝叶唐松草 *Thalictrum foeniculaceum*

偏翅唐松草 *Thalictrum delavayi*

美丽唐松草 *Thalictrum reniforme*

obliquely obovoid, 5-8 mm long, along both sutures narrowly winged, base with 1-3 mm long stipes. Fl. Jun-Sep. Shrubs, forests, grassy slopes, shady places or damp rocky edges at 1800-3400 m. Distributed in Yunnan, W Sichuan, E Xizang.

美丽唐松草

Thalictrum reniforme Wall.

多年生草本，植株被短腺毛。叶为三回羽状复叶，长约20厘米。花序圆锥形，长20-30厘米；花萼4片，粉红色；心皮15-20，被短腺毛。瘦果斜狭倒卵状。花期7-10月。生海拔3100-3700米的云杉林中、灌丛中或山坡。产西藏南部。印度北部、尼泊尔和不丹亦有。

Perennial herbs, glandular puberulent. Leaves 3-pinnate, ca. 20 cm long. Inflorescences paniculate, 20-30 cm long; sepals 4, reddish pink; carpels 15-20, glandular puberulent. Achenes obliquely narrowly obovoid. Fl. Jul-Oct. *Picea* forests, thickets or slopes at 3100-3700 m. Distributed in S Xizang. Also in N India, Nepal and Bhutan.

美花唐松草

Thalictrum callianthum W. T. Wang

多年生草本，无毛。茎高约1.5米。中部茎生叶为三至四回三出复叶；小叶宽5-9毫米。圆锥花序具多数花；萼片4(-5)，紫色，卵形，长11-16毫米；花丝丝形。瘦果新月形，长约5毫米，宿存花柱长2-2.2毫米，顶端具钩。花期7-8月。生海拔3400米的山地灌丛中。特产西藏(米林县)。

Perennial herbs, glabrous. Stems ca. 1.5 m tall. Middle cauline leaves 3-4-ternate; leaflets 5-9 mm broad. Panicles multi-flowered; sepals 4(-5), purple, ovate, 11-16 mm long; filaments filiform. Achenes lunate, ca. 5 mm long, persistent styles 2-2.2 mm long, apex hooked. Fl. Jul-Aug. Bushes at 3400 m. Endemic to Xizang (Mainling County).

美花唐松草 *Thalictrum callianthum*

珠芽唐松草 *Thalictrum chelidonii*

珠芽唐松草

Thalictrum chelidonii DC.

多年生草本，无毛。茎高约1米，在上部叶腋有珠芽。叶为三回羽状复叶；小叶长1-3厘米。聚伞圆锥花序有多数花；萼片4，粉红色，长5-8毫米。瘦果扁平，半倒卵形，长4毫米，有长约0.6毫米的细柄。花期8月。生海拔2600米的山地林中。产西藏南部。不丹和尼泊尔亦有。

Perennial herbs, glabrous. Stems ca. 1 m tall, in upper leaf axils with bulbils. Leaves 3-pinnate; leaflets 1-3 cm long. Thyrse many-flowered; sepals 4, pink, 5-8 mm long. Achene flat, semiobovate, ca. 4 mm long; stipe ca. 0.6 mm long. Fl. Aug. Montane forests at 2600 m. Distributed in S Xizang. Also in Bhutan and Nepal.

毛发唐松草

Thalictrum trichopus Franch.

多年生草本，无毛。茎高达120厘米。下部茎生叶为三回羽状复叶，长约30厘米；小叶宽0.6-1.3厘米。圆锥花序稀疏，花梗丝形，长1.4-3.5厘米；萼片4，白色；花丝丝形。瘦果长约3.5毫米，有8-9条纵肋，宿存柱头长0.4毫米。花期6-7月。生海拔2000-2500米的灌丛中。产云南西部和西北部、四川西南部。

Perennial herbs, glabrous. Stems up to 120 cm tall. Lower cauline leaves 3-pinnate, ca. 30 cm long; leaflets 0.6-1.3 cm broad. Panicles sparse-flowered; pedicels filiform, 1.4-3.5 cm long; sepals 4, white; filaments filiform. Achenes ca. 3.5 mm long, 8-9-ribbed, persistent stigmas 0.4 mm long. Fl. Jun-Jul. Bushes at 2000-2500 m. Distributed in W and NW Yunnan, and SW Sichuan.

毛发唐松草 *Thalictrum trichopus*

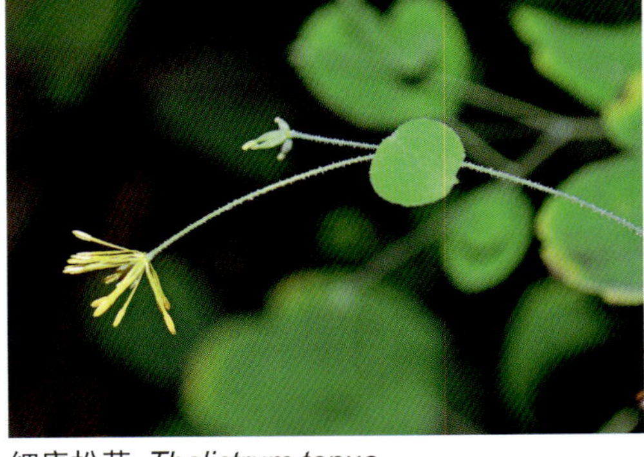

细唐松草 *Thalictrum tenue*

细唐松草

Thalictrum tenue Franch.

多年生草本，全株无毛，有白粉。叶为三或四回羽状复叶；小叶小，卵形，全缘。单歧聚伞花序具少花；萼片4，早落，淡黄绿色；心皮4-6。瘦果两侧压扁，狭倒卵状。花期6月。生于旱山坡或田边。产华北和华西。

Perennial herbs, glabrous, glaucous. Leaves 3- or 4-pinnate; leaflets small, ovate, entire. Inflorescences monochasial few-flowered; sepals 4, caducous, pale yellowish-green; carpels 4-6. Achenes laterally compressed, narrowly obovoid. Fl. Jun. Dry slopes or field edges. Distributed in N and W China.

多叶唐松草

Thalictrum foliolosum DC.

多年生草本，无毛。叶为三回三出或羽状复叶；小叶长1-2.5厘米。花序顶生或腋生，圆锥状，多花；萼片4，早落，淡黄绿色；心皮4-6。瘦果纺锤形，有8条纵肋。花期8-9月。生海拔1500-3200米山地林中或草坡上。产云南、四川西南部和西藏南部。印度、尼泊尔、缅甸和泰国亦有。

Perennial herbs, glabrous. Leaves 3-ternate or 3-pinnate; leaflets 1-2.5 cm long. Inflorescences terminal or axillary, paniculate, many flowered; sepals 4, caducous, pale yellowish-green; carpels 4-6. Achenes fusiform, longitudinally 8-ribbed. Fl. Aug-Sep. Forests or grassy slopes at 1500-3200 m. Distributed in Yunnan, SW Sichuan and S Xizang. Also in India, Nepal, Myanmar and Thailand.

多叶唐松草 *Thalictrum foliolosum*

高原唐松草 *Thalictrum cultratum*

高原唐松草

Thalictrum cultratum Wall.

多年生草本。茎高50-120厘米。中部茎生叶为三至四回羽状复叶；小叶宽3-10(-14)毫米，背面有白粉，有稀疏短毛或无毛。圆锥花序长10-24厘米；萼片绿白色；花丝丝形；心皮4-9，柱头三角形。瘦果长3.5毫米，有8条纵肋，宿存花柱长1毫米。花期6-7月。生海拔1700-3800米的草坡或灌丛中。产西藏南部、云南西北部、四川西部和甘肃南部。尼泊尔和印度北部亦有。

Perennial herbs. Stems 50-120 cm tall. Middle cauline leaves 3-4-pinnate; leaflets 3-10 (-14) mm long, abaxially glaucous, with sparse short hairs or glabrous. Panicles 10-24 cm long; sepals greenish-white; filaments filiform; carpels 4-9, stigmas triangular. Achenes 3.5 mm long, 8-ribbed, persistent stigmas 1 mm long. Fl. Jun-Jul. Grassy slopes or bushes at 1700-3800 m. Distributed in S Xizang, NW Yunnan, W Sichuan and S Gansu. Also in Nepal and N India.

腺毛唐松草 *Thalictrum foetidum*

腺毛唐松草
Thalictrum foetidum L.

多年生草本。茎高15-100厘米。中部茎生叶为三回羽状复叶；小叶宽3.5-15毫米，背面脉上有短毛和腺毛，稀无毛。圆锥花序顶生；花梗通常有短毛和极短的腺毛；萼片5，淡黄绿色；花丝近丝形；心皮4-8，子房有疏毛，柱头箭头形。瘦果扁，长3-5毫米，有8条纵肋。花期5-7月。生海拔350-4500米的草坡或多石砾处。广布中国西南、西北和华北。亚洲西部和欧洲亦有。

Perennial herbs. Stems 15-100 cm tall. Middle cauline leaves 3-pinnate; leaflets 3.5-15 mm broad, abaxially on nerves with short hairs and glandular hairs, rarely glabrous. Panicules terminal; pedicels with short hairs and very short glandular hairs; sepals 5, yellowish-green; carpels 4-8, ovaries hairy, stigmas sagittate. Achenes flattened, 3-5 mm long, 8-ribbed. Fl. May-Jul. Grassy slopes or gravelly places at 350-4500 m. Widespread in SW, NW and N China. Also in W Asia and Europe.

东亚唐松草
Thalictrum minus L. var. **hypoleucum** (Sieb. et Zucc.) Miq.

多年生草本，全株无毛。叶为四回三出羽状复叶；小叶背面被白粉，背面脉隆起。花序圆锥状；萼片4，早落，带淡绿色；花丝丝形，花药狭长圆形；心皮3-5，柱头具翅，三角形。瘦果无柄，狭椭圆状。花期6-7月。生长海拔1400-2700米的生林缘或潮湿岩石边。产中国大部分地区。朝鲜半岛和日本亦有。

Perennial herbs, glabrous. Leaves 4-ternate-pinnate; leaflets abaxially glaucous; veins prominent abaxially. Inflorescences paniculate; sepals 4, caducous, greenish tinged; filiments filiform, anthers narrowly oblong; carpels 3-5, stigmas triangular, winged. Achenes sessile, narrowly ellipsoid. Fl. Jun-Jul. Forest edges or damp rocky ledges at 1400-2700 m. Distributed in most parts of China. Also in Korean Peninsula and Japan.

东亚唐松草 *Thalictrum minus* var. *hypoleucum*

展枝唐松草
Thalictrum squarrosum Steph. ex Willd.

多年生草本，无毛。茎高60-100厘米，中部之上二歧状分枝。茎生叶为二至三回羽状复叶；小叶宽0.6-1.5(-2.6)厘米。圆锥花序近二歧状分枝；萼片4，淡黄绿色；花丝丝形；心皮1-3(-5)，柱头箭头状。花期7-8月。生海拔200-1900米的草坡、平原草地或田边。产陕西北部、华北和东北。蒙古和俄罗斯(西伯利亚)亦有。

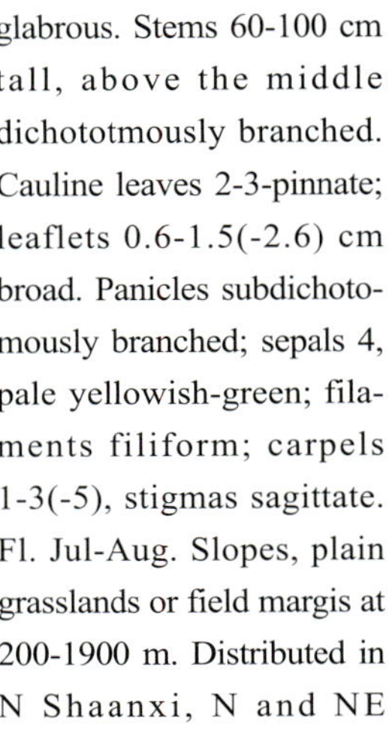

Perennial herbs, glabrous. Stems 60-100 cm tall, above the middle dichototmously branched. Cauline leaves 2-3-pinnate; leaflets 0.6-1.5(-2.6) cm broad. Panicles subdichotomously branched; sepals 4, pale yellowish-green; filaments filiform; carpels 1-3(-5), stigmas sagittate. Fl. Jul-Aug. Slopes, plain grasslands or field margis at 200-1900 m. Distributed in N Shaanxi, N and NE China. Also in Mongolia and Russia (Siberia).

直梗高山唐松草
Thalictrum alpinum L. var. **elatum** Ulbr.

多年生小草本，通常无毛。叶均基生，为二回羽状三出复叶；小叶宽3-5(-20)毫米。花葶高约10厘米；花梗向上直展；萼片4，早落；花丝丝形；心皮3-5。柱头箭头状。瘦果长约3毫米，有8条纵肋。花期6-8月。生海拔2400-4600米的高山草甸。产西藏东部、云南北部、四川西部、青海南部和东

展枝唐松草 *Thalictrum squarrosum*

部、甘肃南部、宁夏、陕西南部、山西和河北西北部。

Small perennial herbs, usually glabrous. Leaves all basal, 2-pinnate-ternate; leaflets 3-5(-20) mm broad. Scapes ca. 10 cm tall; pedicels erect, straight; sepals 4, caduceus; filaments filiform; carpels 3-5, stigmas sagittate. Achenes ca. 3 mm long, 8-ribbed. Fl. Jun-Aug. Alpine meadows at 2400-4600 m. Distributed in E Xizang, N Yunnan, W Sichuan, S and E Qinghai, S Gansu, Ningxia, S Shaanxi, Shanxi and NW Hebei.

直梗高山唐松草 *Thalictrum alpinum* var. *elatum*

石砾唐松草

Thalictrum squamiferum Lecoy.

多年生小草本，无毛。茎渐升或直立，长6-20厘米，下部被围以石砾，在节上有鳞片。茎生叶为三至四回羽状复叶；小叶宽约0.6毫米。花单生叶腋，向顶发育；萼片4，淡黄绿色；花丝丝形；心皮4-6，柱头箭头状。瘦果长3毫米，有8条纵肋。花期7月。生海拔3600-5000米的碎石滩、河岸石砾地或林边。产西藏东南至西南部、云南西北部、四川西部和青海南部。印度北部亦有。

Small perennial herbs, glabrous. Stems ascending or erect, 6-20 cm long, lower parts surrounded by gravels and on nodes with scales. Cauline leaves 3-4-pinnate; leaflets ca. 0.6 mm broad. Flowers singly axillary, acropetal; sepals 4, pale yellowish-green; filaments filiform; carpels 4-6, stigmas sagittate. Achenes 3 mm long, 8-ribbed. Fl. Jul. Screes, river gravelly banks or forest margins at 3600-5000 m. Distributed in SE to SW Xizang, NW Yunnan, W Sichuan and S Qinghai. Also in N India.

石砾唐松草 *Thalictrum squamiferum*

鞭柱唐松草 *Thalictrum smithii*

峨眉黄连 *Coptis omeiensis*

鞭柱唐松草

Thalictrum smithii Boivin

多年生草本，无毛。茎高30-90厘米。下部茎生叶为三回羽状复叶；小叶宽0.6-1.5厘米。圆锥花序长10-35厘米；花两性或单性；萼片4，淡黄绿色；花丝丝形；心皮6-20，柱头细钻形，比子房长，长1.2毫米。瘦果长1.6毫米，有8条纵肋。花期5-7月。生海拔2300-3000米的林边、草坡或田边。产西藏东部和四川西部。

Perennial herbs, glabrous. Stems 30-90 cm tall. Lower cauline leaves 3-pinnate; leaflets 0.6-1.5 cm broad. Panicles 10-35 cm long; flowers bisexual or unisexual; sepals 4, pale-yellowish-green; filaments filiform; carpels 6-20, stigmas slenderly subulate, longer than ovaries, 1.2 mm long. Achenes 1.6 mm long, 8-ribbed. Fl. May-Jul. Forest, grassy slopes or field margins at 2300-3000 m. Distributed in E Xizang and W Sichuan.

黄连

Coptis chinensis Franch.

多年生草本。叶基生，近革质，三全裂，中央裂片深3-5裂。花葶1或2，高12-25厘米；花序具3-8花；萼片5，淡黄绿色；花瓣条状披针形；心皮8-12。花期2-3月，果期4-7月。生海拔500-2000米的山地林中或山谷阴处，或栽培。产中国西南和华中。

Perennial herbs. Leaves basal, subcoriaceous, 3-sect, central segments deeply 3-5 lobed. Scapes 1 or 2, 12-25 cm tall; inflorescences 3-8-flowered; sepals 5, pale-yellowish-green; petals linear-lanceolate; carpels 8-12. Fl. Feb-Mar. Fr. Apr-Jul. Forests or shady places in valleys at 500-2000 m, or cultivated. Distributed in SW and C China.

峨眉黄连

Coptis omeiensis (Chen) C. Y. Cheng

状茎圆柱形，黄色。叶均基生，具长柄；叶片三全裂，中全裂片菱状披针形，侧全裂片

黄连 *Coptis chinensis*

较小，不等2深裂。花葶高15-27厘米；聚伞花序顶生；萼片黄绿色，长约9毫米；花瓣9-12，条状披针形，为萼片的1/2；雄蕊多数；心皮9-14，有柄。花期2-3月。生海拔1000-1700米的悬崖或石上。特产四川峨眉、峨边和洪雅一带。

Rhizomes terete, yellow. Leaves all basal, long petiolate; blades 3-sect, central segment rhombic-lanceolate, lateral segments smaller, unequally 2-parted. Scapes 15-27 cm tall; cymes terminal; sepals yellow-green, ca. 9 mm long; petals 9-12, linear-lanceolate, 2 times shorter than sepals; stamens numerous; carpels 9-14, stipitate. Fl. Feb-Mar. Cliffs or rocks at 1000-1700 m. Endemic to Sichuan (Emei, Ebian and Hongya).

裂叶星果草

Asteropyrum cavaleriei (Lévl. et Vaniot) Drumm. et Hutch.

多年生草本。基生叶2-7，五角形，3-5裂，裂片三角形，顶端急尖。花单生花葶顶端；萼片5，花瓣状，白色。蓇葖果5-8，成熟时星状开展。花期5-6月，果期6-7月。生海拔1000-1100米的林中或河边阴湿处。产云南、四川、贵州、广西北部和湖南西部。

Perennial Herbs. Leaves basal, 2-7, pentagonal in outline, 3-5-lobed, lobes triangular, apex acute. Flower solitary, terminal to scape; sepals 5, petaloid, white. Follicles 5-8, radiately spreading when mature. Fl. May-Jun. Fr. Jun-Jul. Forests or shady places by streams at 1000-1100 m. Distributed in Yunnan, Sichuan, Guizhou, N Guangxi and W Hunan.

星果草

Asteropyrum peltatum (Franch.) Drumm. et Hutch.

多年生小草本。叶均基生，具长柄；叶片圆形或圆五角形，宽2-3厘米，基部盾形。花葶6-10厘米高；苞片2，长约3毫米；花单朵顶生，直径1.2-1.5厘米；萼片5，白色，倒卵形；花瓣黄色，具爪；雄蕊11-18；心皮5-8。花期5-6月。生海拔2000-4000米的山地林下。产云南西北部、四川和湖北西部。缅甸北部和不丹亦有。

Small perennial herbs. Leaves all basal, long petiolate; blades orbicular or orbicular-pentagonal, 2-3 cm broad, base peltate. Scapes 6-10 cm tall; bracts 2, ca. 3 mm long; flower solitary, terminal, 1.2-1.5 cm diam; sepals 5, white, obovate; petals yellow, clawed; stamens 11-18; carpels 5-8. Fl. May-Jun. Forests at 2000-4000 m. Distributed in NW Yunnan, Sichuan and W Hubei. Also in N Myanmar and Bhutan.

西南银莲花

Anemone davidii Franch.

多年生草本。基生叶2-5，具长柄，为三出复叶，宽5-15厘米。花葶高20-40厘米；聚伞花序有1-3花；总苞苞片3-5，叶状，具柄；萼片5(-6)，白色，长1.5-2厘米；花丝丝形；心皮45-70，花柱短，柱头头形。花期5-6月。生海拔1000-3500米的林中、溪边或石上。产西藏东南(墨脱县)、云南西北部、四川、重庆、湖北西部、贵州、湖南西部。

Perennial herbs. Basal leaves 2-5, long petiolate, ternate, 5-15 cm broad. Scapes 20-40 cm tall; cymes 1-3-flowered; bracts 3-5, foliaceous, petiolate; sepals 5(-6), white, 1.5-2 cm long; filaments filiform; carpels 45-70, styles short, stigmas capitate. Fl. May-Jun. Forests, by streams or rocks at 1000-3500 m. Distributed in SE Xizang (Mêdog County), NW Yunnan, Sichuan, Chongqing, W Hubei, Guizhou and W Hunan.

裂叶星果草 *Asteropyrum cavaleriei*

星果草 *Asteropyrum peltatum*

西南银莲花 *Anemone davidii*

小银莲花 *Anemone exigua*

多被银莲花 *Anemone raddeana*

小银莲花

Anemone exigua Maxim.

多年生小草本。基生叶2-5，为三出复叶，具长柄。花葶高5-15(-25)厘米；聚伞花序具1花；苞片3，具柄，3深裂或3浅裂；萼片5(-6)，白色，长5-8毫米；花丝丝形；心皮5-10，子房被短柔毛，花柱短，顶端弯曲。花期6-8月。生海拔2000-3500米的云杉林中或灌丛中。产云南西北部、四川西部、青海东部、甘肃南部、陕西南部、河南西部、山西南部和台湾。

Small perennial herbs. Basal leaves 2-5, ternate, long petiolate. Scapes 5-15(-25) cm tall; cyme 1-flowered; bract 3, petiolate, 3-parted or 3-lobed; sepals 5(-6), white, 5-8 mm long; filaments filiform; carpels 5-10, ovaries puberulous, styles short, apex curved. Fl. Jun-Aug. *Picea* forests or bushes at 2000-3500 m. Distributed in NW Yunnan, W Sichuan, E Qinghai, S Gansu, S Shaanxi, W Henan, S Shanxi and Taiwan.

阴地银莲花 *Anemone umbrosa*

阴地银莲花

Anemone umbrosa C. A. Mey.

多年生草本。基生叶1，为三出复叶，具长柄。花葶高8-20(-30)厘米；花序具1花；苞片3，叶状，具柄；萼片5，白色，长7-14毫米；花丝丝形；心皮约11，子房狭卵球形，密被柔毛，花柱直或稍弯。花期5-6月。生海拔200-500米的林中草地或阴坡上。产辽宁、吉林和黑龙江。朝鲜半岛北部和俄罗斯(远东地区)亦有。

Perennial herbs. Basal leaf 1, ternate, long petiolate. Scapes 8-20(-30) cm tall; cyme 1-flowered; bracts 3, foliaceous, petiolate; sepals 5, white, 7-14 mm long; filaments filiform; carpels ca. 11, ovaries narrow-ovoid, densely puberulous, styles straight or slightly curved. Fl. May-Jun. Grassy places under forests or shady slopes at 200-500 m. Distributed in Liaoning, Jilin and Heilongjiang. Also in N Korean Peninsula and Russia (Far East).

多被银莲花

Anemone raddeana Regel

多年生草本。叶三出，中央小叶倒卵形，先端3微裂或具3齿。聚伞花序具1花；萼片9-15，白色或白带紫色，条状长圆形；心皮约30，子房卵球形，具柔毛。瘦果狭卵球形，稍压扁，具狭肋。花期4-5月。生海拔500-1000米的林中、阴湿的山谷或阴草地。产山东东部、辽宁、吉林和黑龙江。俄罗斯(远东地区)、朝鲜半岛和日本亦有。

Perennial herbs. Leaves ternate, central leaflet obovate, apex 3-lobulate or 3-dentate. Cymes 1-flowered; sepals 9-15, white or white-tinged-purple, linear-oblong; carpels ca. 30, ovary ovoid, pilose. Achenes narrowly ovoid, slightly compressed, narrowly ribbed. Fl. Apr-May. Forests, shady places in valleys, or shady grasslands at 500-1000 m. Distributed in E Shandong, Liaoning, Jilin and Heilongjiang. Also in Russia (Far East), Korean Peninsula and Japan.

阿尔泰银莲花

Anemone altaica Fisch. ex C. A. Mey.

多年生草本。基生叶1，为三出复叶，具长柄。花葶高10-20(-30)厘米；聚伞花序有1花；苞片3，具柄，叶片3全裂；萼片8-12，白色或淡蓝色，长10-20毫米；花丝丝形；心皮20-30，子房狭卵球形，密被柔毛，花柱直。花期3-5月。生海拔1200-1800米的林中或灌丛中或沟边。产湖北西北部、河南西部、山西南部、陕西南部和新疆北部。俄罗斯和欧洲东南部(罗马尼亚)亦有。

Perennial herbs. Basal leaf 1, ternate, long petiolate. Scapes 10-20(-30) cm tall; cyme 1-flowered; bracts 3, petiolate, 3-sect; sepals 8-12, white or bluish, 10-20 mm long; filaments fili-

阿尔泰银莲花 *Anemone altaica*

序有2-3花；苞片3，无柄；萼片(4-)5(-8)，白色，长7-10毫米；花丝丝形；心皮约8，子房卵球形，密被柔毛，花柱不明显，柱头球形。花期4-6月。生海拔1100-3000米的林中或阴湿草地。产云南西北部、四川、湖北西部、贵州、湖南、江西、安徽南部、江苏南部和浙江。日本、俄罗斯(远东地区)和萨哈林岛(库页岛)亦有。

Perennial herbs. Basal leaves 1-3, long petiolate; blades reniform-pentagonal, 3-sect. Scapes 15-25(-40) cm tall; cyme 2-3-flowered; bracts 3, sessile; sepals (4-) 5(-8), white, 7-10 mm long; filaments filiform; carpels ca. 8, ovary ovoid, densely puberulous, style inconspicuous, stigma globose. Fl. Apr-Jun. Forests or shady grassy places at 1100-3000 m. Distributed in NW Yunnan, Sichuan, W Hubei, Guizhou, Hunan, Jiangxi, S Anhui, S Jiangsu and Zhejiang. Also in Japan, Russia (Far East), and Sakhalin.

form; carpels 20-30, ovary narrow-ovoid, densely puberulous, styles straight. Fl. Mar-May. Forests or bushes or by streams at 1200-1800 m. Distributed in NW Hubei, W Henan, S Shanxi, S Shaanxi and N Xinjiang. Also in Russia and SE Europe (Romania).

反萼银莲花

Anemone reflexa Steph. ex Willd.

多年生草本。基生叶1，为三出复叶，具长柄。花葶高15-25(-30)厘米；聚伞花序有1-3花；苞片3，具柄，3全裂；萼片5-6(-7)，反折，白色，条状披针形，长5-7毫米；花丝狭条形；心皮约12，子房狭卵球形，密被柔毛，花柱直。花期4-5月。生林中或灌丛中。产陕西南部、河南西部和吉林东部。朝鲜半岛北部、蒙古和俄罗斯(西伯利亚)亦有。

Perennial herbs. Basal leaf 1, ternate, long petiolate. Scapes 15-25(-30) cm tall; cyme 1-3-flowered; bracts 3, petiolate, 3-sect; sepals 5-6(-7), reflexed, white, linear-lanceolate, 5-7 mm long; filaments narrow-linear; carpels ca. 12, ovaries narrow-ovoid, densely puberulous, styles straight. Fl. Apr-May. Forests or bushes. Distributed in S Shaanxi, W Henan and E Jilin. Also in N Korean Peninsula, Mongolia and Russia (Siberia).

鹅掌草

Anemone flaccida F. Schmidt

多年生草本。基生叶1-3，具长柄；叶片肾状五角形，3全裂。花葶高15-25(-40)厘米；聚伞花

反萼银莲花 *Anemone reflexa*

鹅掌草 *Anemone flaccida*

毛果银莲花 *Anemone baicalensis*

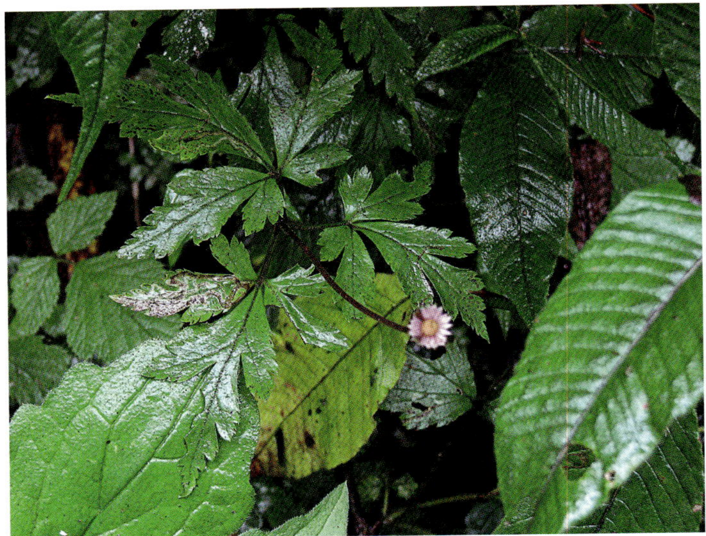
滇川银莲花 *Anemone delavayi*

毛果银莲花

Anemone baicalensis Turcz.

多年生草本。基生叶约2，具长柄，心状五角形，宽5-10厘米，3全裂。花葶高15-28厘米；苞片3，无柄，3裂；萼片5(-6)，白色，长1-1.5厘米；雄蕊花丝丝形；心皮6-16，子房密被短柔毛，柱头近无柄，头形。花期5-9月。生海拔2800-3100米的林下，灌丛中或草地。产四川西部、甘肃南部、陕西南部、辽宁、吉林和黑龙江。朝鲜半岛和俄罗斯(西伯利亚)亦有。

Perennial herbs. Basal leaves ca. 2, long petiolate, cordate-pentagonal, 5-10 cm broad, 3-sect. Scape 15-28 cm tall; involucral bracts 3, sessile, 3-fid; sepals 5(-6), white, 1-1.5 cm long; filaments filiform; carpels 6-16, ovaries densely puberulous, stigmas subsessile, capitate. Fl. May-Sep. Forests or bushes or grasslands at 2800-3100 m. Distributed in W Sichuan, S Gansu, S Shaanxi, Liaoning, Jilin and Heilongjiang. Also in Korean Peninsula and Russia (Siberia).

滇川银莲花

Anemone delavayi Franch.

多年生草本。根状茎细且长；鳞片状叶2或3枚。叶单生，心状五角形，3全裂。总苞片3；萼片5，白色稍带淡红色或蓝色。瘦果卵球形。花期5-7月。生海拔2400-3000米山地林下或林缘。产云南西北部和四川西南部。

Perennial herbs. Rhizomes slender, elongate; scalelike leaves 2 or 3. Leaves solitary, cordate-pentagonal 3-sect. Involucral bracts 3; sepals 5, white and reddish tinged or blue. Achenes ovoid. Fl. May-Jul. Montane forests or forest edges at 2400-3000 m. Distributed in NW Yunnan and SW Sichuan.

川西银莲花 *Anemone pratii*

川西银莲花

Anemone pratii Huth ex Ulbr.

根状茎念珠状至近圆柱形，包含2-10个块茎。基生叶1-2(-3)，具长柄，肾形，3全裂。花葶高约15厘米，无毛；花序具1(-2)花；苞片3，无柄；萼片4-5，白色，长8-12毫米；花丝丝形；心皮4-7。花期4-5月。生海拔1420-2400米的山谷林下。产四川、重庆和云南东北部。

Rhizome moniliform to subterete, containing 2-10 tubers. Basal leaves 1-2(-3), long petiolate, reniform, 3-sect. Scapes ca. 15 cm tall, glabrous; cyme 1-2-flowered; bracts 3, sessile; sepals 4-5, white, 8-12 mm long; filaments

草玉梅 *Anemone rivularis*

filiform; carpels 4-7. Fl. Apr-May. Forests in valleys at 1420-2400m. Distributed in Sichuan, Chongqing and NE Yunnan.

草玉梅

Anemone rivularis Buch.-Ham. ex DC.

草本。叶3-5，肾状五角形，3全裂。花葶1-3；复聚伞花序，具2或3分枝，多花；总苞片3或4；萼片5-10，白色、蓝色、紫色或淡紫色；心皮30-60，无毛，子房狭卵球形，花柱具钩。瘦果卵球形或纺锤形，稍压扁。花期5-8月。生海拔800-4900米的林缘、草坡、溪边或湖边。产中国西南和华西。印度北部、尼泊尔、不丹、斯里兰卡和印度尼西亚(苏门答腊)亦有。

Herbs. Leaves 3-5, reniform-pentagonal, 3-sect. Scapes 1-3; cymes compound, 2- or 3-branched, many-flowered; involucral bracts 3 or 4; sepals 5-10, white, blue, purple, or mauve; carpels 30-60, glabrous, ovary narrowly ovoid, style uncinate. Achenes ovoid or fusiform, slightly compressed. Fl. May-Aug. Forest edges, grassy slopes, streamsides or lakesides at 800-4900 m. Distributed in SW and W China. Also in N India, Nepal, Bhutan, Sri Lanka and Indonesia (Pulau Sumatera).

卵叶银莲花 *Anemone begoniifolia*

米林银莲花

Anemone milinense W. T. Wang

米林银莲花 *Anemone milinense*

多年生小草本，无毛。基生叶数枚，具长柄；叶片肾状五角形，3全裂。花葶高12-16厘米；花序有1-2花；苞片3，无柄；萼片5，淡蓝紫色，长约4.5毫米；雄蕊约18；心皮约16，子房长1-2毫米，花柱长0.5-1毫米，顶端具钩。花期7-8月。生海拔4200米的草地上。特产西藏(米林县)。

Small perennial herbs, glabrous. Basal leaves several, long petiolate; blades reniform-pentagonal, 3-sect. Scapes 12-16 cm tall; cyme 1-2-flowered; bracts 3, sessile; sepals 5, pale bulish-purple, ca. 4.5 mm long; stamens ca. 18; carpels ca. 16, ovaries 1-2 mm long, styles 0.5-1 mm long, apex hooked. Fl. Jul-Aug. Alpine meadows at 4200 m. Endemic to Xizang (Mainling County).

卵叶银莲花

Anemone begoniifolia Lévl. et Vant.

多年生草本。基生叶3-9，有长柄，卵形，长1.5-8.8厘米，不明显3-5浅裂。花葶高15-39厘米；伞形花序有3-7花；苞片3，无柄；萼片5，白色，长约1厘米；花丝丝形；心皮约40，无毛，花柱短。花期2-4月。生海拔650-1000米的山谷林中、溪边或石上。产云南东南部、广西西部、贵州和重庆南部。

Perennial herbs. Basal leaves 3-9, long petiolate, ovate, 1.5-8.8 cm long, inconspicuously 3-5-lobed. Scapes 15-39 cm tall; umbel 3-7-flowered; bracts 3, sessile; sepals 5, white, ca. 1 cm long; filaments filiform; carpels ca. 40, glabrous, styles short. Fl. Feb-Apr. Forests, by streams or rocks at 650-1000 m. Distributed in SE Yunnan, W Guangxi, Guizhou and S Chongqing.

打破碗花花 *Anemone hupehensis*

打破碗花花

Anemone hupehensis (Lem.) Lem.

草本。叶三出，疏具糙毛。聚伞花序2或3分枝，多花；总苞片3，具柄；萼片5，紫红色；心皮约400，具长柄，子房具绒毛。瘦果卵球形，具绵毛。花期7-10月。生海拔400-1800米的矮灌丛中、草坡或丘陵地河边。产中国西南、华南、华中和华东。

Herbs. Leaves ternate, sparsely strigose. Cymes 2- or 3-branched, many-flowered; involucral bracts 3; sepals 5, purple-red; carpels ca. 400, long stipitate, ovary velutinous. Achenes ovoid, lanate. Fl. Jul-Oct. Thickets, grassy slopes or streamsides in hilly regions at 400-1800 m. Distributed in SW, S, C and E China.

水棉花

Anemone hupehensis (Lem.) Lem. f. **alba** W. T. Wang

本变型与打破碗花花的区别在于本变型的萼片呈白色或带粉红色。花期7-10月。生海拔1200-3500米的草坡、沟边或路边。产云南、四川西部和贵州西部。

This form differs from the typical form in its white or pinkish sepals. Fl. Jul-Oct. Grassy slopes, by streams or at roadsides at 1200-3500 m. Distributed in Yunnan, W Sichuan and W Guizhou.

大火草 *Anemone tomentosa*

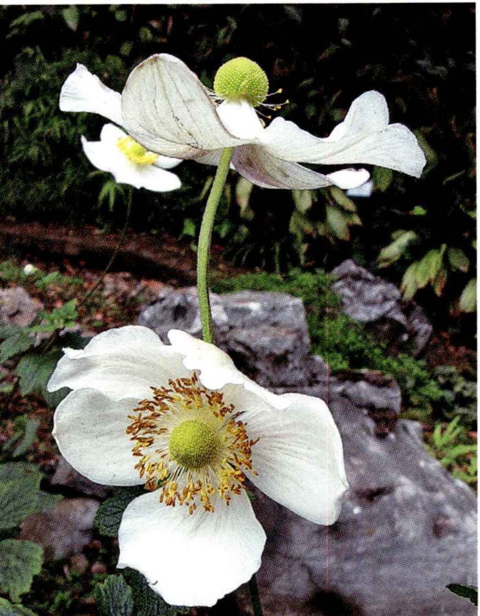

水棉花 *Anemone hupehensis* f. *alba*

大火草

Anemone tomentosa (Maxim.) C. Péi

多年生草本。茎基木质，直立，分枝。叶三出，小叶下面被白色绒毛。聚伞花序二或三回分枝，多花；总苞片3，类似于叶；萼片5，白色或淡粉红色；子房被绒毛。瘦果纺锤形。花期7-10月。生海拔700-3400米草坡。产四川西部、青海东部、甘肃、陕西、湖北西部、河南西部、山西和河北西部。

Perennial herbs. Caudex woody, erect, branched. Leaves ternate; leaflets abaxially white tomentose. Cymes 2- or 3-branched, many-flowered; involucral bracts 3, similar to leaves; sepals 5, white or pinkish; ovary tomentose. Achenes fusiform. Fl. Jul-Oct. Grassy slopes at 700-3400 m. Distributed in W Sichuan, E Qinghai, Gansu, Shaanxi, W Hubei, W Henan, Shanxi and W Hebei.

野棉花

Anemone vitifolia Buch.-Ham. ex DC.

多年生草本。基生叶2-5，均为单叶，心状卵形，长10-20厘米，3-5浅裂，背面被绒毛。花葶高60-100厘米；花序长30-60厘米；苞片3，有柄；萼片5，白色，长1.4-1.8厘米；花丝丝形；心皮约400，有柄，子房被绵毛。花期7-10月。生海拔1200-2700米的山地草坡，沟边或疏林中。产西藏南部和东南部、云南和四川西南部。缅甸北部、印度北部、不丹和尼泊尔亦有。

野棉花 *Anemone vitifolia*

岩生银莲花 *Anemone rupicola*

Perennial herbs. Basal leaves 2-5, all simple, cordate-ovate, 10-20 cm long, 3-5-lobed, abaxially tomentose. Scapes 60-100 cm tall; cymes 30-60 cm long; bracts 3, petiolate; sepals 5, white, 1.4-1.8 cm broad; filaments filiform; carpels ca. 400, stipitate, ovaries lanate. Fl. Jul-Oct. Grassy slopes, by streams or sparse forests at 1200-2700 m. Distributed in S and SE Xizang, Yunnan and SW Sichuan. Also in N Myanmar, N India, Bhutan and Nepal.

岩生银莲花

Anemone rupicola Camb.

多年生草本。基生叶有长柄，心状五角形，3全裂。花葶高6-20(-30)厘米；花序有1(-2)花；苞片2(-3)，无柄，3深裂；萼片5，白色，长2-3厘米；花丝丝形；心皮90-120，子房密被绵毛，花柱短。花期6-8月。生海拔2400-4200米的石崖上、多石砾山坡、沟边或林下。产西藏南部、云南西北部和四川西部。不丹、尼泊尔和印度北部亦有。

Perennial herbs. Basal leaves long petiolate, cordate-pentagonal, 3-sect. Scapes 6-20(-30) cm tall; cyme 1(-2)-flowered; bracts 2(-3), sessile; sepals 5, white, 2-3 cm long; filaments filiform; carpels 90-120, ovaries densely lanate, styles short. Fl. Jun-Aug. Rocky cliffs, gravelly slopes, by streams or forests at 2400-4200 m. Distributed in S Xizang, NW Yunnan and W Sichuan. Also in Bhutan, Nepal and N India.

大花银莲花

Anemone sylvestris L.

多年生草本。基生叶3-9，具长柄，心状五角形，3全裂。花葶高20-50厘米；花序有1花；苞片3，有柄；萼片5(-6)，白色，长1.5-2厘米；花丝丝形；心皮180-240，子房密被柔毛，柱头球形。瘦果密被绵毛。花期5-6月。生海拔1280-3400米的草坡或多沙山坡，桦树林边或平原上。产新疆北部、内蒙古、河北北部和东北。亚洲北部和欧洲亦有。

Perennial herbs. Basal leaves 3-9, long petiolate, cordate-pentagonal, 3-sect. Scapes 20-50 cm tall; cyme 1-flowered; bracts 3, petiolate; sepals 5(-6), white, 1.5-2 cm long; filaments filiform; carpels 180-240, ovaries densely pubescent, stigmas globose. Achenes densely lanate. Fl. May-Jun. Grassy, sandy slopes or *Betula* forest margins at 1280-3400 m. Distributed in N Xinjiang, Neimenggu, N Hebei and NE China. Also in N Asia and Europe.

大花银莲花 *Anemone sylvestris*

西藏银莲花 *Anemone tibetica*

西藏银莲花

Anemone tibetica W. T. Wang

多年生草本。基生叶约8，有长柄，心状五角形，3深裂。花葶高约40厘米；聚伞花序有2花；苞片3，无柄，3浅裂；萼片5，白色，长1.8-2厘米；花丝丝形；心皮约50，长1.2毫米，子房密被柔毛，花柱钻形。花期3月。生海拔3100米的山谷沟边。特产西藏朗县。

Perennial herbs. Basal leaves ca. 8, long petiolate, cordate-pentagonal, 3-parted. Scapes ca. 40 cm tall; cyme 2-flowered; bracts 3, sessile, 3-lobed; sepals 5, white, 1.8-2 cm long; filaments filiform; carpels ca. 50, 1.2 mm long, ovaries densely pubescent, styles subulate. Fl. Mar. By stream in valley at 3100 m. Endemic to Xizang (Nang County).

路边青银莲花

Anemone geum Lévl.

多年生小草本。基生叶5-15，具长柄，卵形，三全裂。花葶高5-25厘米，被长柔毛；聚伞花序具1花；苞片长1-2厘米，无柄；萼片5(-8)，白色，蓝色或黄色，长5-12毫米；花丝条形；子房通常被毛。花期5-8月。生海拔1900-5000米的高山草地或灌丛中。产西藏、云南西北部、四川西部、青海、新疆南部、甘肃、宁夏南部、陕西、山西和河北西部。尼泊尔和印度北部亦有。

Small perennial herbs. Basal leaves 5-15, long petiolate, ovate, 3-sect. Scapes 5-25 cm tall, villous; cyme 1-flowered; bracts 1-2 cm long, sessile; sepals 5(-8), white, blue or yellow, 5-12 mm long; filaments linear; ovaries usually hairy. Fl. May-Aug. Alpine meadows or bushes at 1900-5000 m. Distributed in Xizang, NW Yunnan, W Sichuan, Qinghai, S Xinjiang, Gansu, S Ningxia, Shaanxi, Shanxi and W Hebei. Also in Nepal and N India.

湿地银莲花

Anemone rupestris Hook. f. et Thoms.

多年生小草本。基生叶4-6，具长柄；叶片卵形，长1-2.5厘米，近二回三全裂。花葶1-3，高达18厘米；聚伞花序有1花；苞片3，无柄；萼片5(-7)，白或紫色，长7-11毫米；花丝条形；心皮5-12，有柔毛或无毛，花柱短。花期6-8月。生海拔2500-4200米的山地草甸、草坡或沟边。产西藏东南部，云南西北部和东北部。不丹、尼泊尔和印度北部亦有。

Small perennial herbs. Basal leaves 4-6, long petiolate; blades ovate, 1-2.5 cm long, nearly two times 3-sect. Scapes 1-3, up to 18 cm tall; cyme 1-flowered; bracts 3, sessile; sepals 5(-7), white or purple, 7-11 mm long; filaments linear; carpels 5-12, pubescent or glabrous, styles short. Fl. Jun-Aug. Alpine meadows, grassy slopes, or by streams at 2500-4200 m. Distributed in SE Xizang, and NW and NE Yunnan. Also in Bhutan, Nepal and N India.

条叶银莲花

Anemone coelestina Franch. var. **linearis** (Brühl) Ziman et B.E. Dutton

草本。基生叶无柄，条状倒披针形，常不分裂，顶端具3牙齿。花葶2-8；聚伞花序具1花；总苞片不分裂，条状披针形；萼片常白色或蓝色；退化雄蕊不存在。花期6-9月。生海拔3500-5000米的高山草地或灌丛中。产云南西北部、四川西部、西藏东部和南部、甘肃西南部和青海南部。不丹亦有。

Herbs. Basal leaves sessile, linear-oblanceolate, undivided, apex 3-dentate. Scapes 2-8;

路边青银莲花 *Anemone geum*

湿地银莲花 *Anemone rupestris*

条叶银莲花 *Anemone coelestina* var. *linearis*

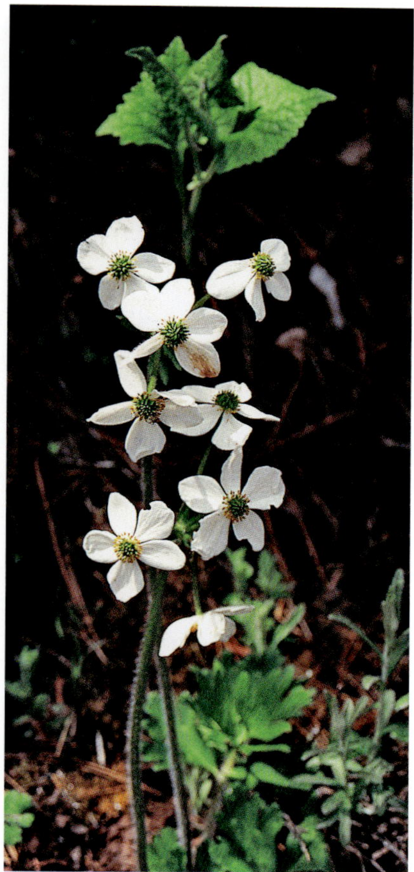
展毛银莲花 *Anemone demissa*

cymes 1-flowered; involucral bracts undivided, linear-lanceolate; sepals usually white or blue; staminodes absent. Fl. Jun-Sep. Alpine meadows or bushes at 3500-5000 m. Distributed in NW Yunnan, W Sichuan, E and S Xizang, SW Gansu and S Qinghai. Also in Bhutan.

小五台银莲花

Anemone xiaowutaishanica W. T. Wang et Bing Liu

多年生小草本。基生叶约12，具长柄，心状五角形，3全裂。花葶高约7厘米；花序有1花；苞片3，无柄；萼片5-6，白色，长4-6毫米；花丝条形；心皮3，无毛，子房两侧扁，无边，花柱钻形，不具柱头。花期9月。生海拔1692米的山谷石边草丛中。特产河北(小五台山)。

小五台银莲花 *Anemone xiaowutaishanica*

Small perennial herbs. Basal leaves ca. 12, long petiolate, cordate-pentagonal, 3-sect. Scapes ca. 7 cm tall; cyme 1-flowered; bracts 3, sessile; sepals 5-6, white, 4-6 mm long; filaments linear; carpels 3, glabrous, ovaries bilaterally compressed, not marginate, styles subulate, adaxially not stigmatose. Fl. Sep. Grassy places by stones in valley at 1692 m. Endemic to Hebei (Xiaowutai Mountain).

展毛银莲花

Anemone demissa Hook. f. et Thoms.

草本。叶片卵形，宽达5厘米；侧裂片极小于中央裂片；叶柄具疏毛。花葶长10-20厘米；聚伞花序具少数花；总苞片3或4；萼片5-7，蓝色、紫色或白色；退化雄蕊有时存在。瘦果宽椭圆体形或倒卵球形。花期6-7月。生海拔3200-4600米的疏林、草坡或高山灌丛。产西藏东部和南部、四川西部、青海南部和甘肃西南部。印度北部、尼泊尔、不丹和巴基斯坦亦有。

Herbs. Blades ovate, up to 5 cm wide; lateral segments much smaller than central one; petioles sparsely villous. Scapes 10-20 cm long; cymes few flowered; involucral bracts 3 or 4; sepals 5-7, blue, purple or white; staminodes sometimes present. Achenes broadly ellipsoid or obovoid. Fl. Jun-Jul. Sparse forests, grassy slopes or alpine shrubs at 3200-4600 m. Distributed in E and S Xizang, W Sichuan, S Qinghai and SW Gansu. Also in N India, Nepal, Bhutan and Pakistan.

宽叶展毛银莲花

Anemone demissa Hook. f. et Thoms. var. **major** W.T. Wang

草本。叶片常五角形，宽达12厘米；侧裂片稍小于中央裂片；叶柄具疏或密生柔毛。花葶长20-45厘米；聚伞花序具少数花；萼片5-6，白色或蓝紫色；退化雄蕊有时存在。瘦果宽椭圆体形或倒卵球形。花期6-7月。生海拔3200-4100米的疏林、灌丛或草坡。产云南西北部、四川西部和西藏南部。不丹亦有。

Herbs. Leaf blades usually pentagonal, up to 12 cm broad; lateral segments smaller than central one; petioles sparsely or densely villous. Scapes 20-45 cm long; cymes few flowered; sepals 5-6, white or purple; staminodes sometimes present. Achenes broadly ellipsoid or obovoid. Fl Jun-Jul. Sparse forests, thickets or grassy slopes at 3200-4100 m. Distributed in NW Yunnan, W Sichuan and S Xizang. Also in Bhutan.

长毛银莲花

Anemone narcissiflora L. var. **crinita** (Juz.) Tamura

多年生草本。基生叶有长柄，心状五角形，宽4-9厘米，3全裂，全裂片细裂，两面被长柔毛。花葶高30-60厘米，与叶柄被开展长柔毛；伞形花序有2-6花；苞片掌状深裂，被长柔毛；萼片5，白色，长约1.5厘米，无毛；雄蕊多数；心皮无毛。瘦果扁，宽倒卵形，长5-7毫米。花期5-6月。生海拔2100-4000米的草坡上、林中或灌丛中。产河北北部、内蒙古、宁夏和新疆东部。蒙古和俄罗斯(西伯利亚)亦有。

Perennial herbs. Basal leaves long petiolate, cordate-pentagonal, 4-9 cm broad, 3-sect, segments dissected, on both surfaces villous. Scapes 30-60 cm tall, with petioles spreading-villous; umbel 2-6-flowered; involucral bracts palmately parted, villous; sepals 5, white, ca. 1.5 cm long, glabrous; stamens numerous; carpels glabrous. Achenes bilaterally compressed, broadly obovate, 5-7 mm long. Fl. May-Jun. Grassy slopes or forests or bushes at 2100-4000 m. Distributed in N Hebei, Neimenggu, Ningxia and E Xinjiang. Also in Mongolia and Russia (Siberia).

长毛银莲花 *Anemone narcissiflora* var. *crinita*

银莲花

Anemone cathayensis Kitag.

多年生草本。叶片3全裂，圆肾形；叶柄长6-25厘米，疏被长柔毛或无毛。花葶2-6；聚伞花序具2-5花；总苞片3-5；萼片5-8(-10)，白色或淡粉红色；子房具边缘，花柱钻形，腹面具狭条形柱头。瘦果扁，宽椭圆形，具边缘。花期4-7月。生海拔1000-2800米的草坡、砾石坡或溪边。产河北和山西。朝鲜半岛亦有。

Perennial herbs. Blades 3-sect, orbicular-reniform; petioles 6-25 cm long, sparsely villous or glabrous. Scapes 2-6; cymes 2-5-flowered; involucral bracts 3-5; sepals 5-8(-10), white or pinkish; ovaries magninate, styles subulate, adaxially with narrow-linear stigmas. Achenes compressed, broadly elliptic marginate. Fl. Apr-Jul. Grassy slopes, gravelly slopes, or streamsides at

宽叶展毛银莲花 *Anemone demissa* var. *major*

银莲花 *Anemone cathayensis*

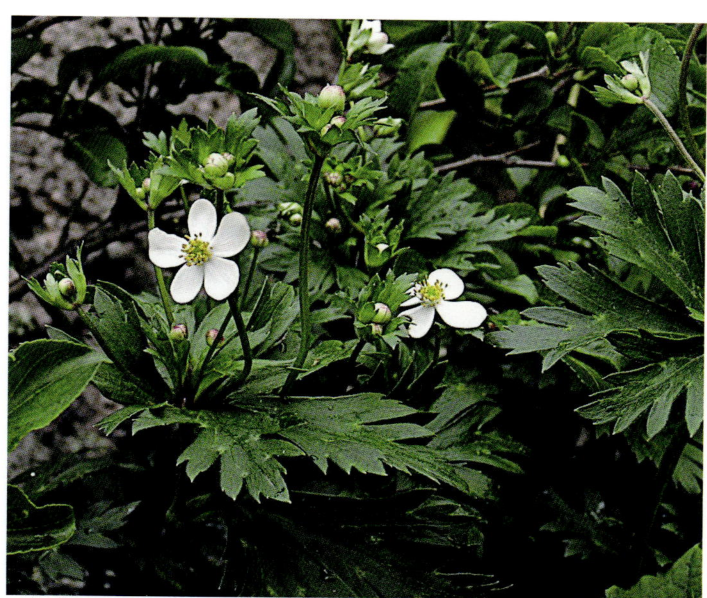
山东银莲花 *Anemone shikokiana*

1000-2800 m. Distributed in Hebei and Shanxi. Also in Korean Peninsula.

山东银莲花
Anemone shikokiana (Maxim.) Makino

多年生草本。基生叶6-8，有长柄，圆肾形，宽5-14厘米，3全裂。花葶高10-55厘米；复伞形花序长4-12厘米；苞片2-3，无柄；萼片4(-5)，白色，长6-10毫米；花丝披针状条形；心皮2-5。瘦果扁平，宽6毫米，有边缘。花期6-7月。生海拔600-1100米的草地或多石山坡。产山东东部。朝鲜半岛和日本亦有。

Perennial herbs. Basal leaves 6-8, long petiolate, orbicular-reniform, 5-14 cm broad, 3-sect. Scapes 10-55 cm tall; compound umbels 4-12 cm long; bracts 2-3, sessile; sepals 4(-5), white, 6-10 mm long; filaments lanceolate-linear; carpels 2-5. Achenes flattened, 6 mm broad, marginate. Fl. Jun-Jul. Meadows or rocky slopes at 600-1100 m. Distributed in E Shandong. Also in Korean Peninsula and Japan.

叠裂银莲花 *Anemone imbricate*

叠裂银莲花
Anemone imbricate Maxim.

多年生草本。基生叶4-7，具长柄；叶片狭卵形，长1.5-2.8厘米，近二回三全裂，背面密被柔毛。花葶1-4，高5-13厘米，密被长柔毛；花序有1花；苞片3，无柄；萼片6-9，白色、紫色或黑紫色，长10-13毫米；花丝条形；心皮约30，无毛。瘦果扁平，宽4毫米，有边缘。花期5-8月。生海拔3200-5300米的草坡或灌丛中。产西藏东部、四川西部、甘肃西南部和中部，青海。

Perennial herbs. Basal leaves 4-7, long petiolate; blades narrow-ovate, 1.5-2.8 cm long, nearly two times 3-sect, abaxially densely villous. Scapes 1-4, 5-13 cm tall, densely villous; cymes 1-flowered; bracts 3, sessile; sepals 6-9, white, purple or black-purple, 10-13 mm long; filaments linear; carpels ca. 30, glabrous. Achenes flattened, 4 mm broad, marginate. Fl. May-Aug. Grassy slopes or bushes at 3200-5300 m. Distributed in E Xizang, W Sichuan, SW and C Gansu, and Qinghai.

二歧银莲花
Anemone dichotoma L.

多年生草本。基生叶1，通常不存在。花葶高35-60厘米；花序二至三回二歧状分枝；总苞和小总苞均具2苞片；萼片5，白色，长7-12毫米；花丝狭条形；心皮约30，无毛。瘦果扁平，长约6毫米，有边缘。花期6月。生湿草地或林中。产吉林和黑龙江。蒙古、俄罗斯和欧洲亦有。

Perennial herbs. Basal leaf 1, usually wanting. Scapes 35-60 cm tall; cymes 2-3 times dichotomously branched; involucre and involucel all 2-bracteate; sepals 5, white, 7-12 mm long; filaments narrow-linear; carpels ca. 30, glabrous. Achenes flattened, ca. 6 mm long, marginate. Fl. Jun. Damp grassy places or forests. Distributed in Jilin and Heilongjiang. Also in Mongolia, Russia and Europe.

二歧银莲花 *Anemone dichotoma*

川鄂獐耳细辛 *Hepatica henryi*

獐耳细辛 *Hepatica nobilis* var. *asiatica*

川鄂獐耳细辛

Hepatica henryi (Oliv.) Steward

草本。叶基生，单叶，不明显3浅裂或分裂，裂片边缘1或2齿，顶端急尖。花葶1或2，被毛；总苞片卵形；萼片6，花瓣状；心皮约10，花柱短，子房疏被毛。花期4-5月。生海拔1300-2500米林中或草坡。产四川、湖北西部、湖南西北部和陕西南部。

Herbs. Leaves basal, simple, inconspicuously 3-lobed or 3-fid, lobe margin 1- or 2-toothed, apex acute. Scapes 1 or 2, hairy; involucral bracts ovate; sepals 6, petaloid; carpels ca. 10, styles short, ovary pilose. Fl. Apr-May. Forests or grassy slopes at 1300-2500 m. Distributed in Sichuan, W Hubei, NW Hunan and S Shaanxi.

獐耳细辛

Hepatica nobilis Schreb. var. **asiatica** (Nakai) Hara

多年生草本。基生叶3-6，具长柄；叶片心状正三角形，宽4.5-7.5厘米，3中裂，全缘。花葶1-6，高8-18厘米；苞片3，无柄，长7-12毫米；萼片6-11，粉红色，长8-14毫米；雄蕊长2-6毫米；子房密被柔毛。花期4-5月。生林下或草坡岩石上。产浙江西北部、安徽、河南、陕西东南部和辽宁东南部。

Perennial herbs. Basal leaves 3-6, long petiolate; blades cordate-deltoid, 4.5-7.5 cm broad, 3-fid, entire. Scapes 1-6, 8-18 cm tall; bracts 3, sessile, 7-12 mm long; sepals 6-11, pink, 8-14 mm long; stamens 2-6 mm long; ovaries densely hairy. Fl. Apr-May. Forests or rocks on grassy slopes. Distributed in NW Zhejiang, Anhui, Henan, SE Shaanxi and SE Liaoning.

罂粟莲花

Anemoclema glaucifolium (Franch.) W. T. Wang

草本。基生叶4-7，具羽状脉，羽状深裂。花序具2-4花；总苞片细羽状深裂；小苞片宽披针形；萼片5，倒卵形，蓝紫色；心皮有长花柱。瘦果长约1.2厘米，近椭圆体形，稍扁。花期7-9月。生海拔1700-3000米山地草坡或松林中草地。产云南西北部和四川西南部。

Herbs. Basal leaves 4-7, penninerved, pinnatipartite. Inflorescences 2-4-flowered; involucral bracts finely pinnatipartite; bracteoles broadly lanceolate; sepals 5, obovate, blue-purple; carpels with long styles. Achenes ca. 1.2 cm long, subellipsoid, slightly compressed. Fl. Jul-Sep. Grassy slopes of mountains or grasslands in pine forests at 1700-3000 m. Distributed in NW Yunnan and SW Sichuan.

白头翁

Pulsatilla chinensis (Bunge) Regel

多年生草本，具根状茎。叶片宽卵形，具3小叶；叶柄密被长柔毛。花葶1（或2）；总苞片基部合生成一个长3-10毫米的管，顶端常掌状3裂，每个裂片条形；花萼紫色。瘦果扁，疏被微柔毛，宿存花柱羽毛状。花期3-5月，果期6-7月。生海拔200-3200米的林缘或山坡。产四川西部、华北、华东和东北。俄罗斯(远东地区)和朝鲜半岛亦有。

Perennial herbs, with rhizomes. Blades broadly ovate, 3-foliolate; petioles densely long pilose. Scapes 1(or 2); involucral bracts basally connate into a 3-10 mm long tube, apically palmately 3-lobed, each lobe linear; sepals purple. Achenes compressed, sparsely puberulent, persistente styles plumose. Fl. Mar-May. Fr. Jun-Jul. Forest edges or slopes at 200-3200 m. Distributed in W Sichuan, N, E and NE China. Also in Russia (Far East) and Korean Peninsula.

掌叶白头翁

Pulsatilla patens (L.) Mill. var. **multifida** (Pritz.) S. H. Li et Y. H. Huang

多年生草本。基生叶具长柄，轮廓五角状心形，宽达7厘米，3全裂，全裂片细裂。花葶高达40厘米；总苞长3-5厘米，密被长柔毛，裂片狭条形；萼片6，

罂粟莲花 *Anemoclema glaucifolium*

白头翁 *Pulsatilla chinensis*

掌叶白头翁
Pulsatilla patens var. *multifida*

直立，蓝紫色，长2.5-4厘米，外面被长柔毛。瘦果长4毫米；宿存花柱羽毛状，3.5-4.8厘米。花期5-6月。生草坡上。产黑龙江、内蒙古北部和新疆北部。蒙古、俄罗斯、欧洲北部和北美洲亦有。

Perennial herbs. Basal leaves long petiolate, pentagonal-cordate, up to 7 cm broad, 3-sect, segments dissected. Scapes up to 40 cm tall; involucre 3-5 cm long, densely villous, lobes narrow-linear; sepals 6, erect, purple-blue, 2.5-4 cm long, abaxially villous. Achenes 4 mm long; persistent styles plumose, 3.5-4.8 cm long. Fl. May-Jun. Grassy slopes. Distributed in Heilongjiang, N Neimenggu and N Xinjiang. Also in Mongolia, Russia, N Europe and North America.

兴安白头翁

Pulsatilla dahurica (Fisch. ex DC.) Spreng.

多年生草本。叶7-9，卵形，3小叶。花葶2-4，约高7.5厘米，在果期伸长，被毛；总苞片长4-5厘米，基部合生成长1.2-1.4毫米的管，顶端裂片卵形；萼片紫色至淡蓝紫色；雄蕊约2倍短于萼片。瘦果密被短柔毛。花期5-6月。生草坡。产中国东北和内蒙古。俄罗斯(远东地区，东西伯利亚)和朝鲜半岛亦有。

Perennial herbs. Leaves 7-9, ovate, 3-foliolate. Scapes 2-4, ca. 7.5 cm tall, elongated in fruiting stage, hairy; involucral bracts 4-5 cm long, basally connate into a 1.2-1.4 mm long tube, apical lobes ovate; sepals purple to bluish violet; stamens ca. 2 times shorter than sepals. Achenes densely puberulent. Fl. May-Jun. Grassy slopes. Distributed in NE China and Neimenggu. Also in Russia (Far East, E Siberia) and Korean Peninsula.

蒙古白头翁

Pulsatilla ambigua Turcz.

多年生草本。叶6-8，具3小叶；小叶细裂。花葶1或2，直立，高约4厘米，果期伸长达16厘米；总苞片3，长1.5-2.8厘米，基部合生成长约2毫米的管；萼片暗堇色；雄蕊约为萼片的1/2。瘦果被毛；花柱宿存，长2.5-3厘米。花果期4-7月。生海拔2000-3400米的草坡或林缘。产内蒙古、甘肃北部、宁夏、青海北部、新疆和黑龙江。俄罗斯(西西伯利亚)和蒙古亦有。

Perennial herbs. Leaves 6-8, 3-foliolate; leaflets dissected. Scapes 1 or 2, erect, ca. 4 cm tall, elongated to 16 cm in fruiting stage; involucral bracts 3, 1.5-2.8 cm long, basally connate into a ca. 2 mm long tube; sepals dark-violet; stamens ca. 2 times shorter than sepals. Achenes hairy; styles persistent, 2.5-3 cm long. Fl. and fr. Apr-Jul. Grassy slopes or forest edges at 2000-3400 m. Distributed in Neimenggu, N Gansu, Ningxia, N Qinghai, Xinjiang and Heilongjiang. Also in Russia (W Siberia) and Mongolia.

兴安白头翁 *Pulsatilla dahurica*

蒙古白头翁 *Pulsatilla ambigua*

细叶白头翁
Pulsatilla turczaninovii

细叶白头翁

Pulsatilla turczaninovii Krylov et Serg.

多年生草本。叶4或5，狭椭圆形，有时卵形，三回羽状分裂。花葶果期伸长至15厘米；总苞片基部合生成长5-6毫米的管，细裂；花萼蓝紫色；花药黄色。瘦果密被微柔毛。花期5-6月，果期6-7月。生草坡。产华北、西北和东北。俄罗斯(远东地区、西伯利亚)和蒙古亦有。

Perennial herbs. Leaves 4 or 5, narrowly elliptical, sometimes ovate, 3 times pinnately divided. Scapes elongated to 15 cm in fruiting stage; involucral bracts basally connate into a 5-6 mm long tube, finely divided; sepals blue-violet; anthers yellow. Achenes densely puberulent. Fl. May-Jun. Fr. Jun-Jul. Grassy slopes. Distributed in N, NW and NE China. Also in Russia (Far East, Siberia) and Mongolia.

美花铁线莲 *Clematis potaninii*

美花铁线莲

Clematis potaninii Maxim.

木质藤本。枝被短柔毛。叶对生，为(一至)二回羽状复叶；小叶长1.5-6厘米，3裂或不分裂，有小齿。聚伞花序腋生，有1-3花；萼片5-6(-7)，开展，白色，长1.8-3.8厘米；雄蕊无毛；子房被柔毛。花期6-8月。生海拔1400-3000米的林下或林边。产西藏东部、云南西北部、四川西部、甘肃南部和陕西南部。

Woody vines. Branches puberulous. Leaves opposite, (1-)2-pinnate; leaflets 1.5-6 cm long, 3-lobed or undivided, denticulate. Cymes axillary, 1-3-flowered; sepals 5-6(-7), spreading, white, 1.8-3.8 cm long; stamens glabrous; ovaries hairy. Fl. Jun-Aug. Forests or forest margins at 1400-3000 m. Distributed in E Xizang, NW Yunnan, W Sichuan, S Gansu and S Shaanxi.

槭叶铁线莲

Clematis acerifolia Maxim.

小灌木。枝圆柱状。叶五角形，掌状5裂，纸质，边缘具疏锯齿。花2-4朵与叶同生于茎顶；萼片5-8，白色或粉红色，开展，两面无毛；宿存花柱长约2.5厘米。花期4-5月，果期5-6月。生海拔约200米的岩壁或沙坡。产北京。

槭叶铁线莲 *Clematis acerifolia*

绣球藤 *Clematis montana*

薄叶铁线莲 *Clematis gracilifolia*

Shrublets. Branches subterete. Leaves pentagonal, palmately 5-lobed, papery, margin sparsely dentate. Flowers 2-4 borne together with leaves from a bud at apex of stem; sepals 5-8, white or pinkish, spreading, both surfaces glabrous; persistent styles ca. 2.5 cm long. Fl. Apr-May. Fr. May-Jun. Rocky cliffs or soil slopes at ca. 200 m. Distributed in Beijing.

绣球藤

Clematis montana Buch.-Ham. ex DC.

木质藤本。叶为三出复叶；小叶纸质至草质，边缘具疏齿。花(1或)2-4(-6)与数叶自老枝腋芽生出，直径3-5厘米；萼片4，白色或有时带粉色，开展；宿存花柱长2-6(-7)厘米，羽毛状。花期4-9月，果期7-9月。生海拔1000-4000米的山坡、灌丛、林缘或沟边。产中国西南、东南、华中、华西和华东。印度北部、尼泊尔、不丹、巴基斯坦北部、缅甸北部和阿富汗亦有。

Vines woody. Leaves ternate; leaflets papery to herbaceous, margin sparsely dentate. Flowers (1 or) 2-4(-6) borne together with several leaves from axillary bud of old branch, 3-5 cm diam; sepals 4, white or sometimes tinged pink, spreading; persistent styles 2-6(-7) cm long, plumose. Fl. Apr-Sep. Fr. Jul-Sep. Slopes, bushes, forests edges, or by streams at 1000-4000 m. Distributed in SW, SE, C, W and E China. Also in N India, Nepal, Bhutan, N Pakistan, N Myanmar and Afghanistan

红木通 *Clematis montana* var. *rubens*

红木通

Clematis montana Buch.-Ham. ex DC. var. **rubens** Wils.

本变种与绣球藤的区别在于本变种的花较大，直径5-11厘米，萼片呈粉红色，长2.5-5.5厘米。花期4-8月。生海拔1300米的山谷林缘。产重庆中部和湖北西部。

This variety differs from the typical variety in its larger flowers 5-11 cm diam. and in its pink, larger sepals 2.5-5.5 cm long. Fl. Apr-Aug. Forest margins in valleys at 1300 m. Distributed in C Chongqing and W Hubei.

薄叶铁线莲

Clematis gracilifolia Rehd. et Wils.

木质藤本。叶通常为羽状复叶，具5小叶；小叶卵形或狭卵形，长约3.5厘米，渐尖，边缘有牙齿。花1-5朵与数叶簇生；萼片4，白色，长1-2厘米。瘦果扁，长约4毫米，无毛；宿存花柱羽毛状，长约2厘米。花期4-6月。生海拔2000-3800米的林中或沟边。产西藏东部、云南西北部、四川西部和甘肃南部。

Woody vines. Leaves usually pinnate, 5-foliolate; leaflets elliptic or narrowly elliptic, ca. 3.5 cm long, acuminate, margin dentate. Flowers 1-5 with several leaves fascicled; sepals 4, white, 1-2 cm long. Achenes compressed, ca. 4 mm long, glabrous; persistent styles plumose, ca. 2 cm long. Fl. Apr-Jun. Forests or by streams at 2000-3800 m. Distributed in E Xizang, NW Yunnan, W Sichuan and S Gansu.

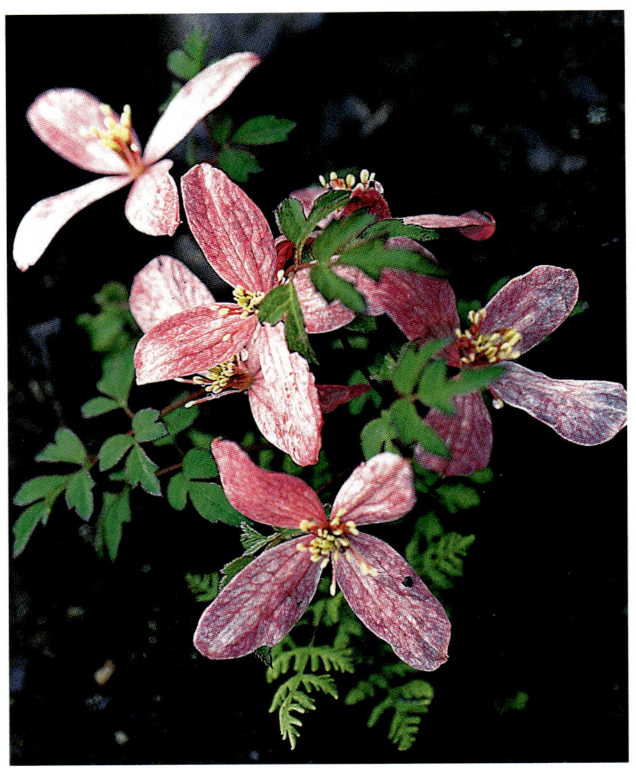

狭裂薄叶铁线莲 *Clematis gracilifolia* var. *dissectifolia*

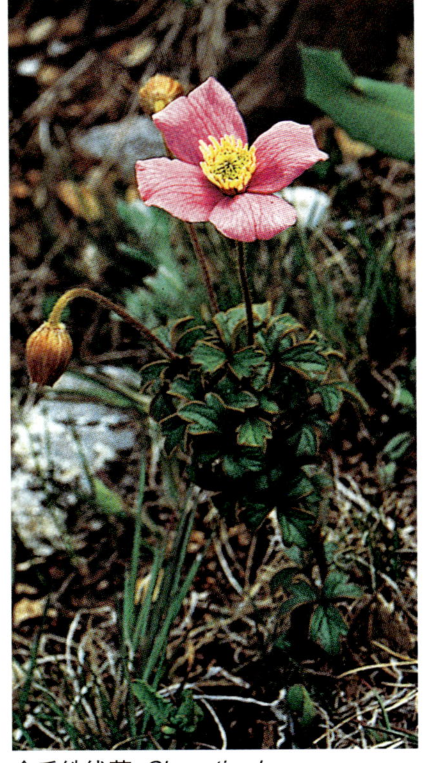

金毛铁线莲 *Clematis chrysocoma*

狭裂薄叶铁线莲

Clematis gracilifolia Rehd. et Wils. var. **dissectifolia** W. T. Wang et M. C. Chang

木质藤本。叶为羽状复叶，具5小叶；小叶2-3深裂或全裂，边缘具疏细齿至锯齿。花1-5朵与数片叶生自老枝的腋芽，直径1.5-2.3厘米；萼片4，白色下面带粉红色；子房和瘦果无毛，宿存花柱长1.2-2.2厘米，羽毛状。花期5-6月，果期9月。生海拔2800-4000米山坡或矮灌丛中。产四川西部和西藏东部。

Vines woody. Leaves pinnate, 5-foliolate; leaflets 2-3-parted or -sect, margin sparsely serrate to denticulate. Flowers 1-5 together with several leaves arising from axillary bud of old branch, 1.5-2.3 cm diam; sepals 4, white or abaxially tinged pinkish; ovary and achenes glabrous, persistent styles 1.2-2.2 cm long, plumose. Fl. May-Jun. Fr. Sep. Slopes or thickets at 2800-4000 m. Distributed in W Sichuan and E Xizang.

丽叶铁线莲

Clematis venusta M. C. Chang

木质藤本。叶三出；小叶片纸质，披针形至狭卵形，全缘或具2小牙齿。1-3花与数叶同自一腋芽生出；萼片4，白色，开展，下面具平伏毛；花柱长约7毫米，被长柔毛。花期5月。生海拔2300-2700米林下或矮灌丛下。产云南(丽江、香格里拉)。

Vines woody. Leaves ternate; leaflets papery, lanceolate to narrowly ovate, margin entire or 2-denticulate. Flowers 1-3 together with several leaves arising from a axillary bud; sepals 4, white, spreading, abaxially sparsely appressed pubescent; styles ca. 7 mm long, densely villous. Fl. May. Forests or thickets at 2300-2700 m. Distributed in Yunnan (Lijiang, Shangri-La).

金毛铁线莲

Clematis chrysocoma Franch.

木质藤本。小枝、叶柄、花梗和小叶背面密被黄色短柔毛。三出复叶；小叶菱状倒卵形或菱状卵形，纸质，边缘具少量锯齿。花1-6朵与数片叶生于老枝腋芽处或单生于当年生枝叶腋；萼片4，白色或粉红色，开展；宿存花柱长2.2-2.7厘米，棕色羽毛状。花期4-7月，果期7-10月。生海拔1000-3000米的草坡或灌丛中。产云南、四川西部和贵州西部。

Vines woody. Branchlets, petioles, pedicels and lower surfaces of leaflets densely yellow-pubescent. Leaves ternate; leaflets rhombic-obovate to rhombic-ovate, papery, margin few dentate. Flowers 1-6 borne together with several leaves from axillary buds of old branches or solitary in leaf axils of current year's branches; sepals 4, white or pink, spreading; persistent styles 2.2-2.7 cm long, brownish plumose. Fl. Apr-Jul. Fr. Jul-Oct. Grassy slopes or thickets along streams at 1000-3000 m. Distributed in Yunnan, W Sichuan and W Guizhou.

丽叶铁线莲 *Clematis venusta*

文山铁线莲 *Clematis wenshanensis*

女萎 *Clematis apiifolia*

文山铁线莲
Clematis wenshanensis W. T. Wang

木质藤本。叶三出；小叶卵形或狭卵形，纸质，边缘具锯齿。花2与数叶自老枝腋芽生出；萼片4，白色，开展，狭卵形，下面近无毛；子房无毛，花柱长8-10毫米，密被长柔毛。花期5月。生海拔2400-2700米的林下或山坡竹林中。产云南（文山县）。

Vines woody. Leaves ternate; leaflets ovate or narrowly so, papery, margin dentate. Flowers 2, with several leaves together arising from a axillary bud of old branch; sepals 4, white, spreading, narrowly ovate, abaxially subglabrous; ovary glabrous, styles 8-10 mm long, densely villous. Fl. May. Forests or bamboo thickets on slopes at 2400-2700 m. Distributed in Yunnan (Wenshan County).

滑叶藤
Clematis fascieuliflora Franch.

木质藤本。枝疏被短柔毛。叶对生或数叶与花簇生，为三出复叶；小叶狭卵形，长达10厘米，全缘。花1-9簇生；萼片4，白色，近直立，长约2厘米；雄蕊无毛；子房无毛。花期12月至翌年3月。生海拔1000-2700米的林下或灌丛中。产云南、四川西部和广西西部。缅甸北部和越南北部亦有。

Woody vines. Branches sparsely puberulous. Leaves opposite or several fasciculate, ternate; leaflets narrow-ovate, up to 10 cm long, entire. Flowers 1-9 fasciculate; sepals 4, white, nearly erect, ca. 2 cm long; stamens glabrous; ovaries glabrous. Fl. Dec to next Mar. Forests or bushes at 1000-2700 m. Distributed in Yunnan, W Sichuan and W Guangxi. Also in N Myanmar and N Vietnam.

女萎
Clematis apiifolia DC.

木质藤本。小枝密被短柔毛。叶对生，为三出复叶；小叶卵形，长2.5-8厘米，常不明显3浅裂，有小牙齿或牙齿，下面有疏毛。圆锥花序；萼片4，白色，长约8毫米；雄蕊无毛。瘦果狭卵形，长3-5毫米。花期7-9月。生海拔150-1000米的林边或草坡上。产江西东北部、福建北部、浙江、江苏南部和安徽南部。朝鲜半岛和日本亦有。

Woody vines. Branches densely puberulous. Leaves opposite, ternate; leaflets ovate, 2.5-8 cm long, often indistinctly 3-lobed, denticulate or dentate, abaxially sparsely puberulous. Panicles axiallary; sepals 4, white, ca. 8 mm long; stamnes glabrous. Achenes narrow-ovate, 3-5 mm long. Fl. Jul-Sep. Forest margins or grassy slopes at 150-1000 m. Distributed in NE Jiangxi, N Fujian, Zhejiang, S Jiangsu and S Anhui. Also in Korean Peninsula and Japan.

滑叶藤 *Clematis fascieuliflora*

钝齿铁线莲 *Clematis apiifolia* var. *argentilucida*

钝齿铁线莲

Clematis apiifolia DC. var. **argentilucida** (Lévl. et Vant.) W. T. Wang

本变种与女萎的区别：本变种的小叶较大，长达13厘米，宽达9厘米，边缘具粗牙齿，下面被密柔毛或绒毛，宿存花柱长达2.7厘米。生海拔400-2300米的林中或沟边。产云南东部、四川、甘肃南部、陕西南部、重庆、贵州、广西北部、广东北部、湖南、湖北、江西、浙江、江苏南部和安徽南部。

This variety differs from the typical variety in its leaflets which are larger, up to 13 cm long and 9 cm broad, at margin coarsely obtusely dentate, and abaxially usually densely pubescent or tomentose, and in its persistent styles up to 2.7 cm long. Forests or by streams at 400-2300 m. Distributed in E Yunnan, Sichuan, S Gansu, S Shaanxi, Chongqing, Guizhou, N Guangxi, N Guangdong, Hunan, Hubei, Zhejiang, S Jiangsu and S Anhui.

粗齿铁线莲 *Clematis grandidentata*

金佛铁线莲

Clematis gratopsis W. T. Wang

木质藤本。叶纸质，羽状，(3-)5小叶；小叶阔卵形至三角状卵形，纸质，边缘疏具不等大锯齿，两面密被短柔毛。聚伞花序具3-14花；萼片4，白色，开展，下面密具柔毛至绢毛；宿存花柱长2.3-4.5厘米，羽毛状。花期8-10月，果期10-11月。生海拔200-1700米山坡灌丛下或河边。产四川东北部和东南部、湖北西部、湖南西北部、陕西南部和甘肃南部。

Vines woody. Leaves papery, pinnate, (3-)5-foliolate; leaflets broadly ovate to deltoid-ovate, papery, margin sparsely unequally dentate, on both surfaces densely puberulous. Cymes 3-14-flowered; sepals 4, white, spreading, abaxially densely puberulous to velutinous; persistent styles 2.3-4.5 cm long, plumose. Fl. Aug-Oct. Fr. Oct-Nov. Thickets on slopes or along streams at 200-1700 m. Distributed in NE and SE Sichuan, W Hubei, NW Hunan, S Shaanxi and S Gansu.

粗齿铁线莲

Clematis grandidentata (Rehd. et Wils.) W. T. Wang

木质藤本。枝被短柔毛。叶对生，为羽状复叶；小叶(3-)5，卵形，长达8厘米，边缘有大牙齿，下面常密被短柔毛。聚伞花序腋生并顶生，腋生花序常具3-7花，似总状花序；萼片4-5，白色，开展，长10-15毫米；雄蕊无毛；子房有毛。花期5-8月。生海拔400-3200米的山坡灌丛中。产云南北部、贵州、湖南西部、四川、重庆、湖北、安徽西部、甘肃南部、陕西南部、河南西部、山西南部和河北西南部。

Woody vines. Branches puberulous. Leaves opposite, pinnate; leaflets (3-)5, ovate, up to 8 cm long, at margin coarsely dentate, abaxially often densely puberulous. Cymes axillary and terminal, axillary ones often 3-7-flowered, racemiform; sepals 4-5, white, spreading, 10-15 mm long; stamens glabrous; ovaries hairy. Fl. May-Aug. Bushes on slopes at 400-3200 m. Distributed in N Yunnan, Guizhou, W Hunan, Sichuan, Chongqing, Hubei, W Anhui, S Gansu, S Shaanxi, W Henan, S Shanxi and SW Hebei.

金佛铁线莲 *Clematis gratopsis*

黄荆铁线莲 *Clematis huangjingensis*

毛果铁线莲 *Clematis peterae* var. *trichocarpa*

黄荆铁线莲

Clematis huangjingensis W. T. Wang et L. Q. Li

木质藤本。叶对生，为具5小叶的羽状复叶；小叶卵形，长6-9厘米，上面无毛，下面被短柔毛，边缘上部有少数牙齿。聚伞花序单生叶腋，有1花；苞片2，对生，长2毫米；花梗长1.5厘米。瘦果狭倒卵球形，长3.5毫米，被疏毛；宿存花柱长约2.6厘米，羽毛状，毛白色。果期9月。生海拔1338-1513米的山坡常绿阔叶林中。特产四川(古蔺县)。

Woody vines. Leaves opposite, 5-foliolately pinnate; leaflets ovate, 6-9 cm long, adaxially glabrous, abaxially puberulous, margin above few-dentate. Cymes singly axillary, 1-flowered; bracts 2, opposite, 2 mm long; pedicel 1.5 cm long. Achenes narrowly obovoid, ca. 3.5 mm long, slightly pilose; persistent styles ca. 2.6 cm long, white-plumose. Fr. Sep. Evergreen broad-leaved forests on slopes at 1338-1513 m. Endemic to Sichuan(Gulin County).

钝萼铁线莲

Clematis peterae Hand.-Mazz.

木质藤本。叶为羽状复叶，有5小叶；小叶椭圆状卵形，长3-7厘米，近无毛，边缘有疏齿或全缘。聚伞花序顶生或腋生；萼片4，白色，长约8毫米；雄蕊无毛。瘦果长4毫米，无毛；宿存花柱羽毛状，长达3厘米。花期6-9月。生海拔1650-3400米的灌丛或草坡上。产云南北部和东部、贵州、四川、湖北、甘肃、陕西、河南、山西南部和河北西南部。

Woody vines. Leaves pinnate, 5-foliolate; leaflets elliptic-ovate, 3-7 cm long, subglabrous, margin with few teeth or entire. Cymes terminal or axillary; sepals 4, white, ca. 8 mm long; stamens glabrous. Achenes 4 mm long, glabrous; persistent styles plumose, up to 3 cm long. Fl. Jun-Sep. Bushes or grassy slopes at 1650-3400 m. Distributed in N and E Yunnan, Guizhou, Sichuan, Hubei, Gansu, Shaanxi, Henan, S Shanxi and SW Hebei.

毛果铁线莲

Clematis peterae Hand.-Mazz. var. **trichocarpa** W. T. Wang

本变种与原变种的区别在于本变种的心皮子房和瘦果均被短柔毛。花期1-9月，果期8-11月。生海拔600-1900米的山坡灌丛中或溪边。产四川、甘肃南部、陕西南部、湖北、贵州、湖南、江西北部、浙江北部、江苏南部、安徽南部和河南南部。

This variety differs from var. *peterae* in its puberulous ovaries and achenes. Fl. Jan-Sep. Fr. Aug-Nov. Bushes on slopes or by streams at 600-1900 m. distributed in Sichuan, S Gansu, S Shaanxi, Hubei, Guizhou, Hunan, N Jiangxi, N Zhejiang, S Jiangsu, S Anhui and S Henan.

钝萼铁线莲 *Clematis peterae*

小蓑衣藤 *Clematis gouriana*

短尾铁线莲 *Clematis brevicaudata*

小蓑衣藤
Clematis gouriana Roxb. ex DC.

木质藤本。枝被短柔毛。叶对生，为羽状复叶；小叶5，狭卵形，长达10厘米，通常全缘，近无毛。聚伞圆锥花序具多数花；萼片4，开展，白色，长5-9毫米；雄蕊无毛；子房有毛。瘦果狭卵形或纺锤形，长约4毫米。花期9-10月。生海拔100-1800米的山坡上，山谷灌丛中或沟边。产云南、广西、广东、湖南西部、贵州、湖北西部、重庆和四川。不丹、尼泊尔、印度、缅甸、菲律宾和新几内亚岛亦有。

Woody vines. Branches puberulous. Leaves opposite, pinnate; leaflets 5, narrow-ovate, up to 10 cm long, usually entire, subglabrous. Thyrses multi-flowered; sepals 4, spreading, white, 5-9 mm long; stamens glabrous; ovaries hairy. Achenes narrow-ovate or fusiform, ca. 4 mm long. Fl. Sep-Oct. slopes, bushes in valleys or by streams at 100-1800 m. Distributed in Yunnan, Guangxi, Guangdong, W Hunan, Guizhou, W Hubei, Chongqing and Sichuan. Also in Bhutan, Nepal, India, Myanmar, the Philippines and New Guinea.

短尾铁线莲
Clematis brevicaudata DC.

藤本。叶为二回羽状复叶；小叶不裂或3浅裂，纸质，边缘各具1-3(-6)个锯齿。聚伞花序常短于叶，具4-25花；萼片4，白色，开展，下面具平伏柔毛；宿存花柱长1.2-2厘米。花期7-9月，果期9-10月。生海拔400-2800米灌丛或疏林中。产中国西南、华中、华北、华西和东北。俄罗斯(远东地区)、蒙古和朝鲜半岛北部亦有。

细木通 *Clematis subumbellata*

Vines. Leaves bi-pinnate; leaflets undivided or 3-lobed, papery, margin 1-3(-6)-dentate on each side. Cymes usually shorter than leaves, 4-25-flowered; sepals 4, white, spreading, abaxially appressed puberulous; persistent styles 1.2-2 cm long. Fl. Jul-Sep. Fr. Sep-Oct. Bushes or sparse forests at 400-2800 m. Distributed in SW, C, N, W and NE China. Also in Russia (Far East), Mongolia and N Korean Peninsula.

细木通
Clematis subumbellata Kurz.

木质藤本。一或二回羽状复叶。有小叶5-21；小叶纸质至近革质，全缘，背面密被贴伏短柔毛。聚伞花序腋生或顶生，具8至多花；萼片4，白色，开展；花柱宿存，长2.5-4厘米，羽毛状。花期12月至翌年1月，果期2-4月。生海拔400-1900米的林缘、山坡或灌丛。产云南南部和西南部。老挝、缅甸北部、泰国北部和越南北部亦有。

Vines woody. Leaves 2-pinnate or 1-pinnate, 5-21-foliolate; leaflets papery to subleathery, margin entire, abaxialy densely appressed

裂叶铁线莲 *Clematis parviloba*

puberulous. Cymes axillary or terminal, 8- to many-flowered; sepals 4, white, spreading; persistent styles 2.5-4 cm long plumose. Fl. Dec to next Jan. Fr. Feb-Apr. Forest edges, slopes or thickets at 400-1900 m. Distributed in S and SW Yunnan. Also in Laos, N Myanmar, N Thailand and N Vietnam.

裂叶铁线莲

Clematis parviloba Gardn et Champ.

木质藤本。叶二回羽状，小叶全缘。腋生聚伞花序具3花；萼片4，白色，开展，披针形或长圆披针形；花药窄卵形或窄椭圆形。瘦果窄卵球形或窄椭圆形；宿存花柱长2-3.2厘米，羽毛状。花期4-10月。生海拔800-2500米的林中，灌丛或河边。产中国西南、华南和东南部。

Vines woody. Leaves 2 pinnate, leaflets entire at margin. Cymes 3-flowered, axillary; sepals 4, white, spreading, lanceolate or oblong-lanceolate; anthers narrowly-ovate or narrowly-elliptic. Achenes narrowly ovoid or narrowly ellipsoid; persistent styles 2-3.2 cm long, plumose. Fl. Apr-Oct. Forests, thickets or by streams at 800-2500 m. Distributed in SW, S and SE China.

毛果扬子铁线莲

Clematis puberula Hook. f. et Thoms. var. **tenuisepala** (Maxim.) W.T. Wang

藤本。二回羽状复叶或二回三出复叶；小叶卵形或披针形，边缘具不规则锯齿或全缘，背面近无毛。圆锥状聚伞花序腋生或顶生，具 (3-)9至多花；萼片4，白色，开展；子房有毛；宿存花柱长2-3.5厘米，羽毛状。瘦果有毛。花期7-10月。生海拔200-3300米的林下、山谷或草坡中。产中国西南、东南、华中、华北和华东。

Vines. Leaves 2-pinnate or 2-ternate; leaflets ovate or lanceolate, margin irregularly dentate or entire, abaxially subglabrous. Cymes both axillary and terminal, (3-)9- to many-flowered; sepals 4, white, spreading; ovary hairy; persistent styles 2-3.5 cm long, plumose.Achenes hairy. Fl Jul-Oct. Forests, valleys or grassy slopes at 200-3300 m. Distributed in SW, SE, C, N and E China.

扬子铁线莲

Clematis puberula Hook. f. et Thoms. var. **ganpiniana** (Lévl. et Vant.) W. T. Wang

木质藤本，干时变黑色。枝疏被短柔毛。叶对生，为一至二回羽状复叶或二回三出复叶。聚伞花序腋生并顶生；萼片4，开展，白色，长9-14毫米；雄蕊和子房无毛。瘦果无毛。花期7-9月。生海拔400-3300米的山坡上、灌丛中或林中。产西藏东部、云南北部、四川、重庆、陕西南部、河南南部、湖北、贵州、广西北部、广东西部、湖南、江西、福建北部、浙江和安徽南部。

Woody vines, turning black when drying. Branches sparsely puberulous. Leaves opposite, 1-2-pinnate or 2-ternate. Cymes axillary and terminal; sepals 4, spreading, white, 9-14 mm long; stamens and ovaries glabrous. Achenes glabrous. Fl. Jul-Sep. Slopes, bushes or forests at 400-3300 m. Distributed in E Xizang, N Yunnan, Sichuan, Chongqing, S Shaanxi, S Henan, Hubei, Guizhou, N Guangxi, W Guangdong, Hunan, Jiangxi, N Fujian, Zhejiang and S Anhui.

毛果扬子铁线莲 *Clematis puberula* var. *tenuisepala*

扬子铁线莲 *Clematis puberula* var. *ganpiniana*

厚叶铁线莲 *Clematis crassifolia*

厚叶铁线莲

Clematis crassifolia Benth.

木质藤本。叶三出；小叶近革质，全缘。聚伞花序多花，顶生或腋生；萼片4，白色或带粉色，开展，下面近无毛至具柔毛；花丝皱缩；宿存花柱长2.4-4厘米。花期12月至翌年1月，果期2月。生海拔300-2300米的林下、山坡或沿小河边。产广西、广东、海南、台湾、福建和湖南。日本南部亦有。

Woody vines. Leaves ternate; leaflets subleathery, margin entire. Cymes axillary or terminal, many flowered; sepals 4, white or tinged ± pink, spreading, abaxially subglabrous to puberulous; filaments rugose; persistent styles 2.4-4 cm long. Fl. Dec to next Jan. Fr. Feb. Forests, slopes or along streams at 300-2300 m. Distributed in Guangxi, Guangdong, Hainan, Taiwan, Fujian and Hunan. Also in S Japan.

披针叶铁线莲

Clematis lancifolia Bur. et Franch.

小灌木。茎高35-60厘米，近无毛。叶对生，披针形，长5.5-10厘米，全缘，近无毛。聚伞花序顶生，有3-5花；萼片4-6，白色，开展，长约10毫米；雄蕊无毛；子房有毛。花期7-8月。生海拔1000-1900米的林下或草地。产云南中北部和东北部、四川西部和西南部。

Small shrubs. Stems 35-60 cm tall, subglabrous. Leaves opposite, lanceolate, 5.5-10 cm long, entire, subglabrous. Cymes terminal, 3-5-flowered; sepals 4-6, white, spreading, ca. 10 mm long; stamens glabrous; ovaries hairy. Fl. Jul-Aug. Forests or grassy areas at 1000-1900 m. Distributed in NC and NE Yunnan, W and SW Sichuan.

披针叶铁线莲 *Clematis lancifolia*

准噶尔铁线莲

Clematis songorica Bunge

直立小灌木或多年生草本。叶全部不裂，条状披针形。聚伞花序顶生或腋生，具1至数花；萼片4-6，白色，开展；宿存花柱长1.4-2.6厘米，羽毛状。花期6-8月，果期7-9月。生海拔400-2500米的草地、砾石地、多石山坡或溪边。产甘肃北部、内蒙古西南部和新疆。阿富汗、哈萨克斯坦、吉尔吉斯斯坦、塔吉克斯坦和蒙古亦有。

Shrubs small, erect or perennial herbs. All leaves undivided, linear-lanceolate. Cymes terminal or axillary, (1-)few- to many-flowered; sepals 4-6, white, spreading; persistent styles 1.4-2.6 cm long, plumose. Fl. Jun-Aug. Fr. Jul-Sep. Grasslands, gravelly, rocky slopes or along streams at 400-2500 m. Distributed in N Gansu, SW Neimenggu and Xinjiang. Also in Afghanistan, Kazakhstan, Kyrgyzstan, Tajikistan and Mongolia.

蕨叶铁线莲

Clematis songarica Bunge var. **asplenifolia** (Schrenk) Trautv.

本变种与准噶尔铁线莲的区别在于本变种的上部茎生叶羽状全裂。花期7-8月，生海拔500-2500米的山谷草地、溪边或河滩多石砾处。产新疆西部。哈萨克斯坦和阿富汗亦有。

This variety differs from the typical variety in its upper pinnatisect cauline leaves. Fl. Jul-Aug. Grassy places by streams in valleys, or gravelly river banks at 500-2500 m. Distributed in W Xinjiang. Also in Kazakhstan and

准噶尔铁线莲 *Clematis songorica*

蕨叶铁线莲 *Clematis songarica* var. *asplenifolia*

疏毛银叶铁线莲 *Clematis delavayi* var. *calvescens*

Afghanistan.

银叶铁线莲
Clematis delavayi Franch.

小灌木。茎高达1.5米，被短柔毛。叶为羽状复叶，有7-13小叶；小叶长0.8-4厘米，背面密被银白色毡毛。聚伞花序顶生；萼片4-6，白色，平展，长8-14毫米；雄蕊无毛。瘦果长4毫米；宿存花柱羽毛状，长2.5厘米。花期7-8月。生海拔1800-3000米的山谷灌丛或林中，或山坡上。产云南西北部和四川西南部。

Small shrubs. Stems up to 1.5 m tall, puberulous. Leaves pinnate, 7-13-foliolate; leaflets 0.8-4 cm long, abaxially densely silvery-pannose. Cymes terminal; sepals 4-6, white, spreading, 8-14 mm long; stamens glabrous. Achenes 4 mm long; persistent styles plumose, 2.5 cm long. Fl. Jul-Aug. Bushes or forests, or slopes in valley at 1800-3000 m. Distributed in NW Yunnan and SW Sichuan.

疏毛银叶铁线莲
Clematis delavayi Franch. var. **calvescens** Schneid.

小灌木。茎高约1米，与花梗和小叶疏被短柔毛。叶为羽状复叶，具5小叶；小叶背面疏被柔毛侧生小叶长椭圆形，长5-14毫米，全缘，顶生小叶菱形，长7-25毫米，常3浅裂。聚伞花序有3-5花；萼片4，白色，卵形，长6-8毫米。瘦果长4毫米；宿存花柱细，羽毛状，长2厘米。花期8月。生海拔1950米的河边草丛。产云南西北和四川西南部。

Small shrubs. Stems ca. 1 m tall, with pedicels sparsely puberulous. Leaves 5-foliolately pinnate; leaflets abaxially sparsely pubescent, lateral ones long elliptic, 5-14 mm long, entire, terminal ones rhombic, 7-25 mm long, usually 3-lobed. Cymes 3-5-flowered; sepals 4, white, ovate, 6-8 mm long. Achenes 4 mm long; persistent styles slender, plumose, 2 cm long. Fl. Aug. Grassy river banks at 1950 m. Distributed in NW Yunnan and SW Sichuan.

银叶铁线莲 *Clematis delavayi*

长冬草 *Clematis hexapetala* var. *tchefouensis*

棉团铁线莲 *Clematis hexapetala*

棉团铁线莲

Clematis hexapetala Pall.

多年生草本。叶一至二回羽状全裂；末回裂片条状披针形、椭圆形或条形，革质，基部楔形，全缘。聚伞花序顶生或腋生，具(1-)3至多花；萼片(4-)5-6(-8)，白色，开展。瘦果倒卵球形，具柔毛；花柱宿存，长1.5-3厘米，羽毛状。花期6-8月，果期7-10月。生海拔100-1300米草地、干燥山坡或沙丘。产华北、华西和东北。俄罗斯(东西伯利亚)、蒙古和朝鲜半岛亦有。

Perennial herbs. Leaves 1- or 2-pinnatisect; ultimate lobe linear-lanceolate, elliptic or linear, leathery, base cuneate, margin entire. Cymes termial or axillary, (1-)3- to many-flowered; sepals (4-)5-6 (-8), white, spreading. Achenes obovoid, pubescent; persistent styles 1.5-3 cm long, plumose. Fl. Jun-Aug. Fr. Jul-Oct. Grasslands, dry slopes or dunes at 100-1300 m. Distributed in N, W and NE China. Also in Russia (E Siberia), Mongolia and Korean Peninsula.

长冬草

Clematis hexapetala var. **tchefouensis** (Debeaux) S. Y. Hu

与棉团铁线莲的主要区别为萼片除外面边缘有绒毛外，其余无毛。花期6-8月，果期7-9月。生海拔100-900米的草坡。产山东东部和江苏北部。

This variety differs from var. *hexapetala* in its sepals which are abaxially on margin velutinous and glabrous on both surfaces. Fl. Jun-Aug. Fr. Jul-Sep. Grassy slopes at 100-900 m. Distributed in E Shandong and N Jiangsu.

毛柱铁线莲

Clematis meyeniana Walp.

木质藤本。叶三出；小叶近革质至纸质，全缘。聚伞花序腋生或顶生，常圆锥状，多花；萼片4，白色，开展，下面无毛；花柱密具柔毛。宿存花柱长2-4厘米，黄色，羽毛状。花期6-8月，果期8-10月。生海拔300-1800米的林下、林缘、矮灌丛中或沿河岸。产中国西南、华南、东南和华中。缅甸、老挝、越南、菲律宾和琉球群岛亦有。

Vines woody. Leaves ternate; leaflets subleathery to papery, margin entire. Cymes axillary or terminal, usually paniclelike and many-flowered; sepals 4, white, spreading, abaxially glabrous; styles densely villous. Persistent styles 2-4 cm long, yellow, plumose. Fl. Jun-Aug. Fr. Aug-Oct. Forests, forest edges, thickets or along streams at 300-1800 m. Distributed in SW, S, SE and C China. Also in Myanmar, Laos, Vietnam, the Philippines and Ryukyu Islands.

山木通

Clematis finetiana Lévl. et Vaniot

木质藤本。叶为三出复叶；小叶卵状披针形、狭卵形或卵

毛柱铁线莲 *Clematis meyeniana*

山木通 *Clematis finetiana*

形，近革质至革质，全缘。花序有1-7花；萼片4(-6)，白色，开展，下面除边缘具绢毛外无毛；花柱宿存，长1.5-2.5厘米，黄褐色羽毛状。花期4-6月，果期7-11月。生海拔100-1200米的疏林、山坡、灌丛或河岸。产中国西南、东南、华中和华东。

Vines woody. Leaves ternate; leaflets ovate-lanceolate narrowly ovate or ovate, subleathery to leathery, margin entire. Cymes 1-7-flowered; sepals 4(-6), white, spreading, abaxially glabrous except for velutinous margin; persistent styles 1.5-2.5 cm long, yellow-brown plumose. Fl. Apr-Jun. Fr. Jul-Nov. Sparse forests, slopes, thickets or along streams at 100-1200 m. Distributed in SW, SE, C and E China.

滇北铁线莲

Clematis jialasaensis W. T. Wang var. **macrantha** W. T. Wang

木质藤本。叶对生，为三出复叶；小叶狭卵形，长6-9厘米，全缘。圆锥花序具多数花；花直径2.5-3.7厘米；萼片4，白色，开展，倒披针形，长1.4-2厘米，宽4-6毫米；雄蕊无毛；花药条形，顶端钝；子房被短柔毛，花柱密被长柔毛。花期4月。生海拔900-1000米的山坡灌丛。产云南北部和四川东南部。

Woody vines. Leaves opposite, ternate; leaflets narrowly ovate, 6-9 cm long, entire. Panicle many-flowered. Flower 2.5-3.7 cm diam; sepals 4, white, spreading, oblanceolate, 1.4-2 cm long, 4-6 mm broad; stamens glabrous; anthers linear, apex obtuse; ovaries puberulous, styles densely villous. Fl. Apr. Bushes on slopes at 900-1000 m. Distributed in N Yunnan and SE Sichuan.

滇北铁线莲 *Clematis jialasaensis* var. *macrantha*

小木通

Clematis armandii Franch.

木质藤本。枝疏被短柔毛。叶对生，为三出复叶；小叶革质，狭卵形，长达15厘米，全缘，无毛。聚伞圆锥花序顶生并腋生，腋生花序自腋芽生出；萼片4(-5)，白色，开展，长1.2-2.4厘米；雄蕊无毛；子房有毛。瘦果狭卵形。花期3-4月。生海拔100-2400米的山坡灌丛、林边或沟边。产西藏东部、云南、贵州、四川、甘肃南部、陕西南部、重庆、湖北、湖南、广西、广东和福建西南部。越南、缅甸和印度东北部亦有。

Woody vines. Branches sparsely puberulous. Leaves opposite, ternate; leaflets coriaceous, narrow-ovate, up to 15 cm long, entire, glabrous. Thyrses terminal and axillary, axillary ones arising from axillary buds; sepals 4(-5), white, spreading, 1.2-2.4 cm long; stamens glabrous; ovaries hairy. Achenes narrow-ovate. Fl. Mar-Apr. Bushes, forest margins or by streams at 100-2400 m. Distributed in E Xizang, Yunnan, Guizhou, Sichuan, S Gansu, S Shaanxi, Chongqing, Hubei, Hunan, Guangxi, Guangdong and SW Fujian. Also in Vietnam, Myanmar and NE India.

小木通 *Clematis armandii*

大花小木通 *Clematis armandii* var. *farquhariana*

威灵仙 *Clematis chinensis*

大花小木通

Clematis armandii Franch. var. **farquhariana** (Rehd. et Wils.) W. T. Wang

木质藤本。三出复叶；小叶狭卵形、披针形或卵形，革质，全缘。聚伞花序具3花；萼片4(-6)，白色或淡粉色，开展，长2.1-4厘米；宿存花柱长1.6-4.8厘米。花期3-4月，果期4-7月。生海拔500-1500米的林中、灌丛和河边。产重庆、湖北西部和湖南西北部。

Vines woody. Leaves ternate; leaflets narrowly ovate, lanceolate or ovate, leathery, margin entire. Cymes 3-flowered; sepals 4(-6), white or pinkish, spreading, 2.1-4 cm long; persistent styles 1.6-4.8 cm long. Fl. Mar-Apr. Fr. Apr-Jul. Forests, scrubs and along streams at 500-1500 m. Distributed in Chongqing, W Hubei and NW Hunan.

威灵仙

Clematis chinensis Osbeck

木质藤本。一回羽状复叶，通常有5小叶；小叶卵形或披针形，纸质，全缘。圆锥状聚伞花序通常多花；萼片4，白色，开展，下面近顶部具柔毛；宿存花柱长1.8-4厘米。花期6-9月，果期8-11月。生海拔100-1500米的灌丛、山坡或河边。产中国西南、华南、东南、华中和华东。越南和日本南部亦有。

Vines woody. Leaves pinnate, usually 5-foliolate; leaflets ovate or lanceolate, papery, margin entire. Cymes usually paniclelike and many-flowered; sepals 4, white, spreading, abaxially puberulous near apex; persistent styles 1.8-4 cm long. Fl. Jun-Sep. Fr. Aug-Nov. Thickets, slopes or along streams at 100-1500 m. Distributed in SW, S, SE, C and E China. Also in Vietnam and S Japan.

太行铁线莲

Clematis kirilowii Maxim.

木质藤本。一至二回羽状复叶；小叶不分裂或2或3浅裂，革质，全缘。聚伞花序顶生或腋生；萼片4(-6)，白色，下面具平伏柔毛；宿存花柱长约1.8厘米，羽毛状。花期4-8月，果期8-10月。生海拔100-1700米的灌丛或山坡。产华北和华东。

Woody vines. Leaves 1- or 2-pinnate; leaflets undivided or 2- or 3-fid, leathery, margin entire. Cymes both terminal and axillary; sepals 4(-6), white, abaxially appressed puberulous; persistent styles ca. 1.8 cm long, plumose. Fl. Apr-Aug. Fr. Aug-Oct. Thickets or slopes at 100-1700 m. Distributed in N and E China.

秦岭铁线莲

Clematis obscura Maxim.

木质藤本，干时变黑色。枝疏被短柔毛。叶对生，为一至二回羽状复叶；小叶5-11，卵形，长1.2-7.8厘米，全缘，近无毛。聚伞花序腋生并顶生，有1-3(-5)花；萼片(4-)5-6(-7)，白色，开展，长1.2-2.6厘米；雄蕊无毛；子房有毛。花期4-6月。生海拔400-2600米的灌丛中或山坡上。产四川西北部、湖北西北部、甘肃南部、陕西南部、河南西部和山西南部。

Woody vines, turning black when drying. Branches sparsely puberulous. Leaves opposite, 1-2-pinnate; leaflets 5-11, ovate, 1.2-7.8 cm long, margin entire, sub-

太行铁线莲 *Clematis kirilowii*

秦岭铁线莲 *Clematis obscura*

辣蓼铁线莲
Clematis terniflora var. *mandshurica*

glabrous. Cymes axillary and terminal, 1-3(-5)-flowered; sepals (4-)5-6(-7), white, spreading, 1.2-2.6 cm long; stamens glabrous; ovaries hairy. Fl. Apr-Jun. Bushes or slopes at 400-2600 m. Distributed in NW Sichuan, NW Hubei, S Gansu, S Shaanxi, W Henan and S Shanxi.

圆锥铁线莲
Clematis terniflora DC.

木质藤本。茎和枝条被短柔毛。叶对生，为羽状复叶；小叶5(-7)，卵形，长2.5-8厘米，全缘。聚伞圆锥花序腋生或顶生，长5-19厘米；萼片4，白色，开展，长6-13毫米；雄蕊无毛。瘦果扁，橙黄色，长6-9毫米，宽3-6毫米，宿存花柱长达4厘米。花期6-8月。生海拔400米以下的林边或草地。产陕西东南部、河南南部、湖北、江西、浙江、江苏和安徽南部。朝鲜半岛和日本亦有。

Woody vines. Stems and branches puberulous. Leaves opposite, pinnate; leaflets 5(-7), ovate, 2.5-8 cm long, entire. Thyrses axillary or terminal, 5-19 cm long; sepals 4, white, spreading, 6-13 mm long; stamens glabrous. Achenes compressed, orange-yellow, 6-9 mm long, 3-6 mm broad, persistent styles up to 4 mm long. Fl. Jun-Aug. Forest margins or grassy places below 400 m. Distributed in SE Shaanxi, S Henan, Hubei, Jiangxi, Zhejiang, Jiangsu and S Anhui. Also in Korean Peninsula and Japan.

辣蓼铁线莲
Clematis terniflora DC. var. **mandshurica** (Rupr.) Ohwi

木质藤本。一回羽状复叶，有小叶5(-7)；小叶纸质至近革质，全缘。聚伞花序腋生或顶生，常具多花；萼片4，白色，开展；花药顶端有尖头。瘦果橘黄色，长4-6毫米，宽2.5-4毫米；宿存花柱长1.2-4厘米，羽毛状。花期6-8月，果期8-11月。生海拔200-800米的林缘或山坡灌丛中。产内蒙古和东北。俄罗斯(西伯利亚)、蒙古和朝鲜半岛北部亦有。

Vines woody. Leaves pinnate, 5(-7)-foliolate; leaflets papery to subleathery, margin entire. Cymes axillary or terminal, usually many-flowered; sepals 4, white, spreading; anther apex minutely apiculate. Achenes orange-yellow, 4-6 mm long, 2.5-4 mm broad; styles persistent, 1.2-4 cm long, plumose. Fl. Jun-Aug. Fr. Aug-Nov. Forest edges or thickets on slopes at 200-800 m. Distributed in Neimenggu and NE China. Also in Russia (Siberia), Mongolia and N Korean Peninsula.

圆锥铁线莲 *Clematis terniflora*

柱果铁线莲 *Clematis uncinata*

柱果铁线莲

Clematis uncinata Champ. ex Benth.

木质藤本。茎无毛。叶对生，为一至二回羽状复叶，无毛；小叶5-15，卵形，全缘。聚伞圆锥花序腋生或顶生；萼片4，白色，开展，长10-15毫米；雄蕊和子房无毛。瘦果近圆柱形，长约6毫米，宿存花柱长1.5-2厘米。花期6-7月。生海拔100-2500米的灌丛中或林边。产云南东南部、广西、广东、台湾、福建、浙江、江苏南部、安徽南部、江西、湖南、贵州、湖北、重庆、四川、甘肃南部和陕西南部。越南亦有。

Woody vines. Stems glabrous. Leaves opposite, 1-2-pinnate, glabrous; leaflets 5-15, ovate, margin entire. Thyrses axillary or terminal; sepals 4, white, spreading, 10-15 mm long; stamens and ovaries glabrous. Achenes subterete, ca. 6 mm long, persistent styles 1.5-2 cm long. Fl. Jun-Jul. Bushes or forest margins at 100-2500 m. Distributed in SE Yunnan, Guangxi, Guangdong, Taiwan, Fujian, Zhejiang, S Jiangsu, S Anhui, Jiangxi, Hunan, Guizhou, Hubei, Chongqing, Sichuan, S Gansu and S Shaanxi. Also in Vietnam.

甘青铁线莲

Clematis tangutica (Maxim.) Korsh.

落叶藤本。一至二回羽状复叶；小叶菱状卵形至狭卵形，纸质，边缘具牙齿。单花顶生，有时1-3花呈腋生聚伞状；萼片4，黄色；花丝有柔毛；子房被柔毛，花柱密被长柔毛。瘦果菱状倒卵球形，被毛；宿存花柱长达5厘米，羽毛状。花期6-9月，果期9-10月。生海拔300-4900米灌丛、草地或多砾石河岸。产新疆、青海、四川西部、甘肃和陕西。哈萨克斯坦亦有。

Decidious vines. Leaves 1- or 2-pinnate; leaflets rhombic-ovate to narrowly ovate, papery, margin dentate. Flowers solitary, terminal, or sometimes in axillary 1-3-flowered cymes; sepals 4, yellow; filaments hairy; ovary puberulous, styles densely villose. Achenes rhomboid-obovoid, puberulous; persistent styles up to 5 cm long, plumose. Fl. Jun-Sep. Fr. Sep-Oct. Thickets, grasslands or gravelly river banks at 300-4900 m. Distributed in Xinjiang, Qinghai, W Sichuan, Gansu and Shaanxi. Also in Kazakhstan.

甘川铁线莲

Clematis akebioides (Maxim.) Hort. ex Veitch.

木质藤本。叶为一至二回羽状复叶；小叶卵形、椭圆形或长圆形，薄纸质至草质，边缘具浅圆齿。聚伞花序腋生，具1-3花；萼片4，黄色，淡黄绿色或有时带紫色，下面除边缘具绒毛外无毛；花丝有毛；宿存花柱长约3厘米。花期7-9月，果期9-10月。生海拔1200-3600米的高山草地、溪边或灌丛。产西藏东部、云南西北、四川西部、青海南部、甘肃、陕西、山西和内蒙古西南部。

Vines woody. Leaves 1- or 2-pinnate; leaflets ovate, elliptic or oblong, thin papery to herbaceous, margin crenate. Cymes axillary, 1-3-flowered; sepals 4, yellow, pale greenish-yellow or sometimes tinged purple, abaxi-

甘青铁线莲 *Clematis tangutica*

甘川铁线莲 *Clematis akebioides*

黄花铁线莲 *Clematis intricata*

厚萼中印铁线莲 *Clematis tibetana* var. *vernayi*

ally glabrous except for velutinous margin; filaments hairy; persistent styles ca. 3 cm long. Fl. Jul-Sep. Fr. Sep-Oct. Alpine grasslands, along streams or shrubs at 1200-3600 m. Distributed in E Xizang, NW Yunnan, W Sichuan, S Qinghai, Gansu, Shaanxi, Shanxi and SW Neimenggu.

黄花铁线莲

Clematis intricata Bunge

木质藤本。叶为二回羽状复叶，有时为一回羽状；小叶纸质，披针形或条状披针形，全缘或具2锯齿。聚伞花序腋生，具(1-)3(-5)花；萼片4，黄色，斜升，下面无毛；花丝有毛；宿存花柱长2.5-4厘米，羽毛状。花期6-7月，果期8-9月。生海拔400-2600米的林中、山坡或灌丛中。产华北和华西。蒙古亦有。

Woody vines. Leaves mostly 2-pinnate, sometimes 1-pinnate; leaflets papery, lanceolate or linear-lanceolate, margin entire or 2-denticulate. Cymes axillary, (1-)3(-5)-flowered; sepals 4, yellow, ascending, abaxially glabrous; filaments hairy; persistent styles 2.5-4 cm long, plumose. Fl. Jun-Jul. Fr. Aug-Sep. Forests, slopes or thickets at 400-2600 m. Distributed in N and W China. Also in Mongolia.

厚萼中印铁线莲

Clematis tibetana Kuntze var. **vernayi** (C. E. C. Fisch.) W. T. Wang

木质藤本。叶通常二回羽状，有时一回；顶端小叶裂片披针形至窄卵形。单花顶生，或腋生，为1-3花组成的聚伞状；萼片4，革质，黄色、黄褐色或棕紫色，斜升；花丝有毛；花柱长0.5-1.5厘米，密被长柔毛。花期5-7月。生海拔2200-4800米的山坡、灌丛、草地或多石河岸。产西藏东部至南部。印度北部和尼泊尔西部亦有。

Vines woody. Leaves usually 2-pinnate, sometimes 1-pinnate; terminal lobes of leaflets lanceolate to narrowly ovate. Flowers solitary and terminal or in axillary 1-3-flowered cymes; sepals 4, leathery, yellow, yellow-brown or brown-purple, ascending; filaments hairy; styles 0.5-1.5 cm long, densely villous. Fl. May-Jul. Slopes, thickets, grasslands or gravelly river banks at 2200-4800 m. Distributed in E to S Xizang. Also in N India and W Nepal.

粉绿铁线莲

Clematis glauca Willd.

木质藤本。枝疏被短柔毛。叶对生，为一至二回羽状复叶；小叶狭卵形，长2-5厘米，全缘。聚伞花序腋生，通常有1-7花；萼片4，黄色，渐升，狭长圆形，长约2厘米，内面无毛或有疏毛；花丝条状披针形，被缘毛。花期7-9月。生海拔500-2000米的山坡上或溪边。产甘肃中部和新疆。蒙古、俄罗斯(西伯利亚)和哈萨克斯坦亦有。

Woody vines. Branches sparsely puberulous. Leaves opposite, 1-2-pinnate; leaflets narrow-ovate, 2-5 cm long, margin entire. Cymes axillary, usually 1-7-flowered; sepals 4, yellow, ascending, narrow-oblong, ca. 2 cm long, adaxially glabrous or with a few hairs; filaments linear-lanceolate, ciliate. Fl. Jul-Sep. Slopes or by streams at 500-2000 m. Distributed in C Gansu and Xinjiang. Also in Mongolia, Russia (Siberia) and Kazakhstan.

粉绿铁线莲 *Clematis glauca*

齿叶铁线莲 *Clematis serratifolia*

东方铁线莲 *Clematis orientalis*

齿叶铁线莲
Clematis serratifolia Rehd.

木质藤本。枝疏被短柔毛或无毛。叶对生，为二回三出复叶；小叶狭卵形，有锯齿，无毛。聚伞花序腋生，有1-3花；萼片4，黄色，渐升，披针形，长1.5-2.5厘米，内面有短柔毛；花丝狭条形，被短柔毛；子房有毛。花期8月。生海拔400米的山地林下或河岸多石砾地。产辽宁和吉林东部。朝鲜半岛、俄罗斯(远东地区)和日本亦有。

Woody vines. Branches sparsely puberulous or glabrous. Leaves opposite, 2-ternate; leaflets narrow-ovate, serrate, glabrous. Cymes axillary, 1-3-flowered; sepals 4, yellow, ascending, lanceolate, 1.5-2.5 cm long, adaxially puberulous; filaments narrow-linear, puberulous; ovaries hairy. Fl. Aug. Forests or gravelly river banks at 400 m. Distributed in Liaoning and E Jilin. Also in Korean Peninsula, Russia (Far East) and Japan.

东方铁线莲
Clematis orientalis L.

木质藤本。茎被短柔毛。叶对生，为一至二回三出复叶；小叶灰绿色，通常3深裂或3浅裂，中央裂片狭卵形或披针形，侧裂片较小。聚伞花序有少数至多数花；萼片4，黄色，披针形，长1.2-2厘米；花丝条状披针形，有缘毛。花期6-7月。生海拔1000-2000米的沟边或山坡上。产新疆。亚洲西部和欧洲东南部亦有。

Woody vines. Stems puberulous. Leaves opposite, 1-2-ternate; leaflets grey-green, usually 3-parted or 3-lobed, central lobe narrow-ovate or lanceolate, lateral lobes smaller. Cymes few- to multi-flowered; sepals 4, yellow, lanceolate, 1.2-2 cm long; filaments linar-lanceolate, ciliate. Fl. Jun-Jul. By streams or slopes at 1000-2000 m. Distributed in Xinjiang. Also in W Asia and SE Europe.

灌木铁线莲
Clematis fruticosa Turcz.

小灌木。茎高约1米。叶对生，狭披针形，长1.5-4.5厘米，边缘有牙齿，近基部常深裂。聚伞花序顶生或腋生，有1-3花；萼片4，黄色，斜上展，卵形，长1-2厘米；雄蕊无毛，花丝狭披针形。花期7-8月。生海拔800-1800米的灌丛中或路边。产陕西北部、山西北部、河北北部和内蒙古。蒙古亦有。

Small shrubs. Stems ca. 1 m tall. Leaves opposite, narrowly lanceolate, 1.5-4.5 cm long, margin dentate, near base often parted. Cymes terminal or axillary, 1-3-flowered; sepals 4, yellow, ascending, ovate, 1-2 cm long; stamens glabrous, filaments narrow-lanceolate. Fl. Jul-Aug. Bushes or roadsides at 800-1800 m. Distributed in N Shaanxi, N Shanxi, N Hebei and Neimenggu. Also in Mongolia.

小叶铁线莲
Clematis nannophylla Maxim.

小灌木。茎高30-100厘米。叶对生，革质，卵形，长5-14(-20)毫米，羽状全裂。聚伞花序顶生，有1-3(-7)花；萼片4，黄色，渐升，狭卵形，长8-16毫米；雄蕊无毛；花丝狭披针形；子房有毛。花期7-8月。生海拔1200-3200米的山坡上。产青海东部、甘肃、陕西、宁夏和内蒙古西南部。

灌木铁线莲 *Clematis fruticosa*

小叶铁线莲 *Clematis nannophylla*

菝葜叶铁线莲 *Clematis smilacifolia*

Small shrubs. Stems 30-100 cm tall. Leaves opposite, coriaceous, ovate, 5-14(-20) mm long, pinnatisect. Cymes terminal, 1-3 (-7)-flowered; sepals 4, yellow, ascending, narrow-ovate, 8-16 mm long; stamens glabrous; filaments narrow-lanceolate; ovaries hairy. Fl. Jul-Aug. Dry slopes at 1200-3200 m. Distributed in E Qinghai, Gansu, Shaanxi, Ningxia and SW Neimenggu.

菝葜叶铁线莲

Clematis smilacifolia Wall.

木质藤本。茎无毛。叶对生，卵形，长10-16厘米，全缘，无毛。圆锥花序腋生；花梗密被锈色短柔毛；萼片4-5，蓝紫色，开展，长1.6-1.8厘米；雄蕊无毛；花丝狭条形；花药顶端具1钻形尖头；子房有毛。花期11-12月。生海拔500-2300米的林中或沟边。产西藏东南部、云南南部、贵州西部、广西西部和海南。印度、越南、印度尼西亚和菲律宾亦有。

Woody vines. Stems glabrous. Leaves opposite, ovate, 10-16 cm long, margin entire, glabrous. Panicles axillary; pedicels densely rusty-puberulous; sepals 4-5, blue-purple, spreading, 1.6-1.8 cm long; stamens glabrous; filaments narrow-linear; anther apex with a subulate apicula; ovaries hairy. Fl. Nov-Dec. Forests or by streams at 500-2300 m. Distributed in SE Xizang, S Yunnan, W Guizhou, W Guangxi and Hainan. Also in India, Vietnam, Indonesia and the Philippines.

大花威灵仙 *Clematis courtoissii*

大花威灵仙

Clematis courtoissii Hand.-Mazz.

短柱铁线莲 *Clematis cadmia*

多年生草质藤本。茎疏被柔毛。叶对生，为一至二回三出复叶；小叶狭卵形，全缘。花单生叶腋；苞片2，对生，宽卵形；萼片6，白色，开展，长2.7-5厘米；雄蕊无毛；花丝条形；宿存花柱长1.2-3厘米，羽毛状。花期5-6月。生海拔200-500米的山坡、灌丛或林中。产湖南东部、河南东南部、安徽南部、江苏南部和浙江北部。

Perennial, herbaceous vines. Stems sparsely pubescent. Leaves opposite, 1-2-ternate; leaflets narrow-ovate, margin entire. Flowers singly axillary; bracts 2, opposite, broad-ovate; sepals 6, white, spreading, 2.7-5 cm long; stamens glabrous; filaments linear; persistent styles 1.2-3 cm long, plumose. Fl. May-Jun. slopes, bushes or forests at 200-500 m. Distributed in E Hunan, SE Henan, S Anhui, S Jiangsu and N Zhejiang.

短柱铁线莲

Clematis cadmia Buch.-Ham. ex Hook. f. et Thoms.

多年生草质藤本。茎长约1米。叶对生，为二回三出复叶；小叶狭卵形，全缘。花单生叶腋；苞片2，对生，三角状卵形；萼片6，淡紫色，开展，长2-3厘米；雄蕊无毛；宿存花柱钻形，长1-3毫米。花期4-5月。生海拔100-1600米的溪边或湿草地。产云南、湖南、湖北、江西北部、浙江北部、江苏南部和安徽南部。越南、缅甸和印度东北部亦有。

Perennial, herbaceous vines. Stems ca. 1 m long. Leaves opposite, 2-ternate; leaflets narrow-ovate, entire. Flowers singly axillary; bracts 2, opposite, triangular-ovate; sepals 6, purplish, spreading, 2-3 cm long; stamens glabrous; persistent styles subulate, 1-3 mm long. Fl. Apr-May. Damp grassy places or by streams at 100-1600 m. Distributed in Yunnan, Hunan, Hubei, N Jiangxi, N Zhejiang, S Jiangsu and S Anhui. Also in Vietnam, Myanmar and NE India.

重瓣铁线莲 *Clematis florida* var. *plena*

重瓣铁线莲

Clematis florida Thunb. var. **plena** D. Don

多年生草质藤本。叶一或二回三出；小叶狭卵形至披针形，纸质，全缘。萼片6，白色，下面沿脉密具绢毛；雄蕊全部成花瓣状的退化雄蕊，狭披针形，长1-1.2厘米；宿存花柱长约8毫米，下部有开展短柔毛，上部无毛。花期4-6月。生海拔约1700米的灌丛中或沿河岸。产云南西南部和浙江。

Perennial, herbaceous vines. Leaves 1- or 2-ternate; leaflets narrowly ovate to lanceolate, papery, margin entire. Sepals 6, white, abaxially densely velutinous along midvein; stamens all becoming petaloid staminodes, narrowly lanceolate, 1-1.2 cm long; persistent styles ca. 8 mm long, basally spreading puberulous, apically glabrous. Fl. Apr-Jun. Thickets or along streams at ca. 1700 m. Distributed in SW Yunnan and Zhejiang.

转子莲

Clematis patens Morr. et Decne.

多年生草质藤本。茎疏被柔毛。叶对生，为羽状复叶；小叶3-5，卵形，全缘。花单生茎顶端；萼片8，白色或淡黄色，开展，长3.5-6厘米；雄蕊无毛，花丝条形；子房被柔毛，宿存羽毛状花柱长达3.8厘米。花期5-6月。生海拔200-1000米的草坡上或灌丛中。产山东东部和辽宁东部。朝鲜半岛和日本亦有。

Perennial, herbaceous vines. Stems sparsely pubescent. Leaves opposite, pinnate; leaflets 3-5, ovate, margin entire. Flowers solitary, terminal; sepals 8, white or yellowish, spreading, 3.5-6 cm long; stamens glabrous, filaments linear; ovaries villous, persistent styles plumose, up to 3.8 cm long. Fl. May-Jun. Grassy slopes or bushes at 200-1000 m. Distributed in E Shandong and E Liaoning. Also in Korean Peninsula and Japan.

湖州铁线莲

Clematis huchouensis Tamura

草质藤本。茎被短柔毛。叶对生，为羽状复叶；小叶(3-)5-9，椭圆形或狭卵形，全缘。聚伞花序腋生，有1-3花；萼片4，白色，斜上展，长1.4-2.2厘米；雄蕊无毛；宿存花柱钻形，长8-13毫米，有短毛。花期6-8月。生海拔约100米的草坡上或湖岸。产湖南北部、江西北部、浙江北部、江苏东南部和安徽南部。

Herbaceous vines. Stems puberulous. Leaves opposite, pinnate; leaflets (3-)5-9, elliptic or narrow-ovate, margin entire. Cymes axillary, 1-3-flowered; sepals 4, white, ascending, 1.4-2.2 cm long; stamens glabrous; persistent styles subulate, 8-13 mm long, with short hairs. Fl. Jun-Aug. Grassy slopes or lake banks at ca. 100 m. Distributed in N Hunan, N Jiangxi, N Zhejiang, SE Jiangsu and S Anhui.

大叶铁线莲

Clematis heracleifolia DC.

多年生草本。茎高20-70厘米。叶对生，为三出复叶；小叶宽4-13厘米。聚伞圆锥花序长达30厘米；花梗被短柔毛；萼片4，蓝紫色，直立，狭长圆形，长1.6-2.5厘米；花丝顶端和药隔有疏毛；宿存花柱羽毛状，长约2厘米。花期7-9月。生海拔85-1600米的山坡草丛或疏林下。产河南、山西、山东、河北、内蒙古南部和辽宁。朝鲜半岛北部亦有。

Perennial herbs. Stems 20-70 cm tall. Leaves opposite, ternate; leaflets 4-13 cm broad. Thyrses up to 30 cm long; pedicels puberulous; sepals 4,

转子莲 *Clematis patens*

湖州铁线莲 *Clematis huchouensis*

大叶铁线莲 *Clematis heracleifolia*

狭卷萼铁线莲 *Clematis tubulosa* var. *ichangensis*

blue-purple, erect narrow-oblong, 1.6-2.5 cm long; filament apex and connective with a few hairs; persistent styles plumose, ca. 2 cm long. Fl. Jul-Sep. Grassy places on slopes or sparse forests at 85-1600 m. Distributed in Henan, Shanxi, Shandong, Hebei, S Neimenggu and Liaoning. Also in N Korean Peninsula.

卷萼铁线莲

Clematis tubulosa Turcz.

亚灌木或多年生草本。叶三出；小叶纸质，常3裂，边缘具锐齿至锯齿。复聚伞花序顶生或腋生；花梗粗壮，短，被短绒毛；萼片4，蓝色至紫色，直立，宽达12毫米；宿存花柱长约2.5厘米，羽毛状。花期8-9月，果期10月。生海拔80-1400米的林缘或矮灌丛中。产中国江苏东北部、山东东部、河北北部和辽宁。朝鲜半岛亦有。

Subshrubs or perennial herbs. Leaves ternate; leaflets papery, often 3-lobed, margin incised to dentate. Compound cymes terminal or axillary; pedicels robust, short, velutionous; sepals 4, blue to purple, erect, up to 12 mm broad; persistent styles ca. 2.5 cm long, plumose. Fl. Aug-Sep. Fr. Oct. Forest edges or thickets at 80-1400 m. Distributed in NE Jiangsu, E Shandong, N Hebei and Liaoning. Also in Korean Peninsula.

狭卷萼铁线莲

Clematis tubulosa Turcz. var. **ichangensis** (Rehd. et Wils.) W. T. Wang

本变种与卷萼铁线莲的区别在于本变种的萼片在花开之后只在近顶端处稍微展宽，呈狭长圆形，上部和下部的宽度近相同；花梗被短绒毛；萼片长12-18毫米，宽2.2-7毫米。花期6-9月。生海拔50-2000米的草坡上、灌丛中、溪边或悬崖上。产贵州东部、湖南西部、湖北西部、陕西南部、山西、山东、河南、安徽和浙江西北部。

This variety differs from the typical variety in its sepals which are narrowly oblong in outine, after anthesis near apex slightly dilated and with nearly same width above and below; pedicels velutinous; sepals 12-18 mm long, 2.2-7 mm wide. Fl. Jun-Sep. Grassy slopes, bushes, by streams or cliffs at 50-2000 m. Distributed in E Guizhou, W Hunan, W Hubei, S Shaanxi, Shanxi, Shandong, Henan, Anhui and NW Zhejiang.

卷萼铁线莲 *Clematis tubulosa*

单叶铁线莲 *Clematis henryi*

单叶铁线莲
Clematis henryi Oliv.

木质藤本。主根膨大。单叶，叶片卵状披针形，厚纸质，边缘具小牙齿。聚伞花序腋生，具1(-5)花；萼片4，直立，白色或淡黄色，下面仅先端具柔毛；花丝密被长柔毛。瘦果狭卵形；宿存花柱长达4厘米。花期11-12月，果期翌年3-4月。生海拔200-2500米的林中、林缘、阴湿的坡地、灌丛或溪边。产中国西南、华南、东南、华中和华东。

Woody vines. Roots have tubers. Leaves simple, ovate-lanceolate, thick papery, margin denticulate. Cymes axillary, 1(-5)-flowered; sepals 4, erect, white or yellowish, abaxially puberulous only near apex; filaments densely villous. Achenes narrowly ovate; persistent styles up to 4 cm long. Fl. Nov-Dec. Fr. next Mar-Apr. Forests, forest edges, shady slopes, thickets or along streams at 200-2500 m. Distributed in SW, S, SE, C and E China.

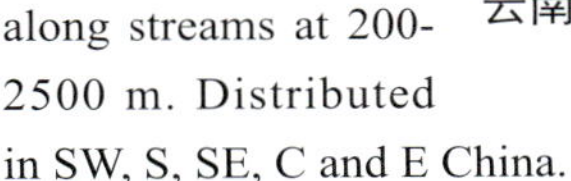

云南铁线莲 *Clematis yunnanensis*

云南铁线莲
Clematis yunnanensis Franch.

木质藤本。茎被短柔毛。叶对生，为三出复叶；小叶条状披针形或狭披针形。聚伞花序腋生，有1-8花；萼片4，白色或淡黄色，直立，长约15毫米；花丝密被长柔毛；宿存花柱羽毛状，长约2厘米。花期11-12月。生海拔1600-2800米的山地林边或林中。产云南和四川西南部。

Woody vines. Stems puberulous. Leaves opposite, ternate; leaflets linear-lanceolate or narrow-lanceolate. Cymes axillary, 1-8-flowered; sepals 4, white or yellowish, erect, ca. 15 mm long; filaments densely villous; persistent styles plumose, ca. 2 cm long. Fl. Nov-Dec. Forest margins or forests at 1600-2800 m. Distributed in Yunnan and SW Sichuan.

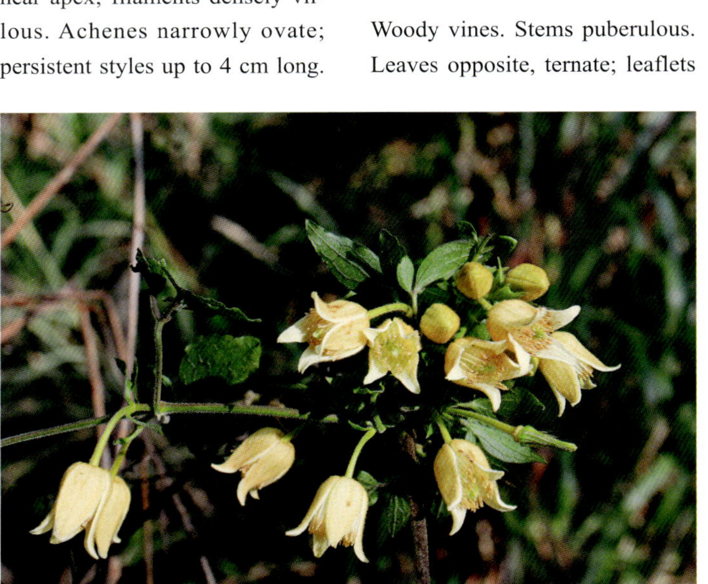
锈毛铁线莲 *Clematis leschenaultiana*

锈毛铁线莲
Clematis leschenaultiana DC.

木质藤本。茎与叶柄和花梗均密被黄褐色柔毛。叶对生，为三出复叶；小叶狭卵形，长5-11厘米，背面密被柔毛。聚伞花序腋生，有3-10花；萼片4，淡黄色，直立，长约2厘米；花丝被长柔毛；宿存羽毛状花柱长3-4厘米。花期1-2月。生海拔500-1200米的山坡灌丛中。产云南、四川、贵州、湖南、广西、广东、福建和台湾。越南、菲律宾和印度尼西亚亦有。

Woody vines. Stems with petioles and pedicels densely fulvous-pubescent. Leaves opposite, ternate; leaflets narrrow-ovate, 5-11 cm long, abaxially densely pubescent. Cymes axillary, 3-10-flowered; sepals 4, yellowish, erect, ca. 2 cm long; filaments villous; persistent styles plumose, 3-4 cm long. Fl. Jan-Feb. Bushes on slopes at 500-2000 m. Distributed in Yunnan, Sichuan, Guizhou, Hunan, Guangxi, Guangdong, Fujian and Taiwan. Also in Vietnam, the Philippines and Indonesia.

毛木通
Clematis buchananiana DC.

木质藤本。小枝密被淡黄色柔毛。一回羽状复叶；小叶宽卵形、卵形或椭圆形，纸质，下面具柔毛，边缘具小牙齿至锯齿。聚伞圆锥花序腋生，具多花；萼片4，黄色，直立，下面密具平伏柔毛；宿存花柱长3.5-5厘米。花期10-12月，果期翌年2-3月。生海拔1200-2800米的山区林缘、开阔地带灌丛中或沟边。产云南、四川西南部和西藏南部。印度、尼泊尔、不丹和缅甸北部亦有。

Woody vines. Branches densely yellowish pubescent. Leaves pinnate; leaflets broadly ovate, ovate or elliptic, papery, abaxially pubescent, margin denticulate to dentate. Thyrses axillary, usually many-flowered; sepals 4, yellow, erect, abaxially densely appressed puberulous; persistent styles 3.5-5 cm long. Fl. Oct-Dec. Fr. next Feb-Mar. Forest edges on mountains, thickets on open places or along streams at 1200-2800 m. Distributed in Yunnan, SW Sichuan and S Xizang. Also in India, Nepal, Bhutan and N Myanmar.

曲柄铁线莲
Clematis repens Finet et Gagnep.

木质藤本。幼枝被柔毛。叶对生，为单叶或三出复叶，无毛，不分裂或三裂。花序有1花；萼片4，黄色，直立，长约2厘米；花丝

毛木通 *Clematis buchananiana*

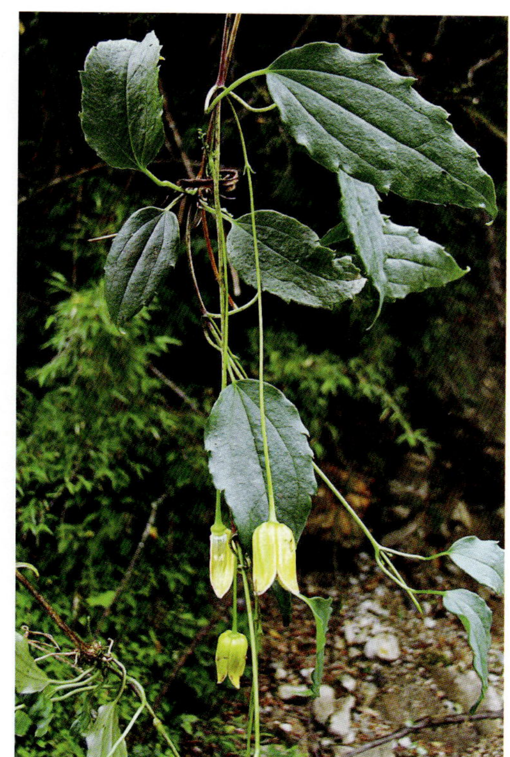
曲柄铁线莲 *Clematis repens*

和花药被短柔毛；宿存羽毛状花柱长达5.5厘米。花期7-8月。生海拔1300-2700米的林中或河边。产云南东北部、四川、贵州、广西北部和广东北部。

Woody vines. Young branches pubescent. Leaves opposite, simple or ternate, glabrous, undivided or 3-lobed. Cymes 1-flowred; sepals 4, yellow, erect, ca. 2 cm long; filaments and anthers puberulous; persistent styles up to 5.5 cm long. Fl. Jul-Aug. Forests or river banks at 1300-2700 m. Distributed in NE Yunnan, Sichuan, Guizhou, N Guangxi and N Guangdong.

华中铁线莲

Clematis pseudootophora M. Y. Fang

木质藤本。茎无毛。叶对生，为三出复叶，无毛；小叶狭卵形，边缘有小锯齿。聚伞花序腋生，有1-3花；萼片4，淡黄色，直立，长2.5-3厘米；花丝和花药被短柔毛；宿存羽毛状花柱长4-5厘米。花期8-9月。生海拔1280-1800米的林下、灌丛中或沟边。产贵州北部、湖南中部、江西、福建北部、浙江西北部、湖北西南部和河南南部。

Woody vines. Stems glabrous. Leaves opposite, ternate, glabrous; leaflets narrow-ovate, serrulate. Cymes axillary, 1-3-flowered; sepals 4, yellowish, erect, 2.5-3 cm long; filaments and anthers puberulous; persistent styles plumose, 4-5 cm long. Fl. Aug-Sep. Forests, bushes or by streams at 1280-1800 m. Distributed in N Guizhou, C Hunan, Jiangxi, N Fujian, NW Zhejiang, SW Hubei and S Henan.

须蕊铁线莲

Clematis pogonandra Maxim.

藤本。茎在节上被短柔毛。叶对生，为三出复叶；小叶狭卵形，全缘。单花腋生，有细花梗，无花序梗和苞片；萼片4，淡黄色，直立，长2.5-3厘米；花丝和花药被柔毛；宿存羽毛状花柱长达3厘米。花期6-7月。生海拔2200-3400米的山坡林边或灌丛中。产四川、甘肃南部、陕西南部、重庆、湖北西部和湖南西北部。

Vines. Stems on nodes puberulous. Leaves opposite, ternate; leaflets narrow-ovate, entire. Flowers solitary, axillary, slenderly pedicellate, without peduncle and bracts; sepals 4, yellowish, erect, 2.5-3 cm long; filaments and anthers pubescent; persistent styles plumose, up to 3 cm long. Fl. Jun-Jul. Forest margins or bushes at 2200-3400 m. Distributed in Sichuan, S Gansu, S Shaanxi, Chongqing, W Hubei and NW Hunan.

华中铁线莲 *Clematis pseudootophora*

须蕊铁线莲 *Clematis pogonandra*

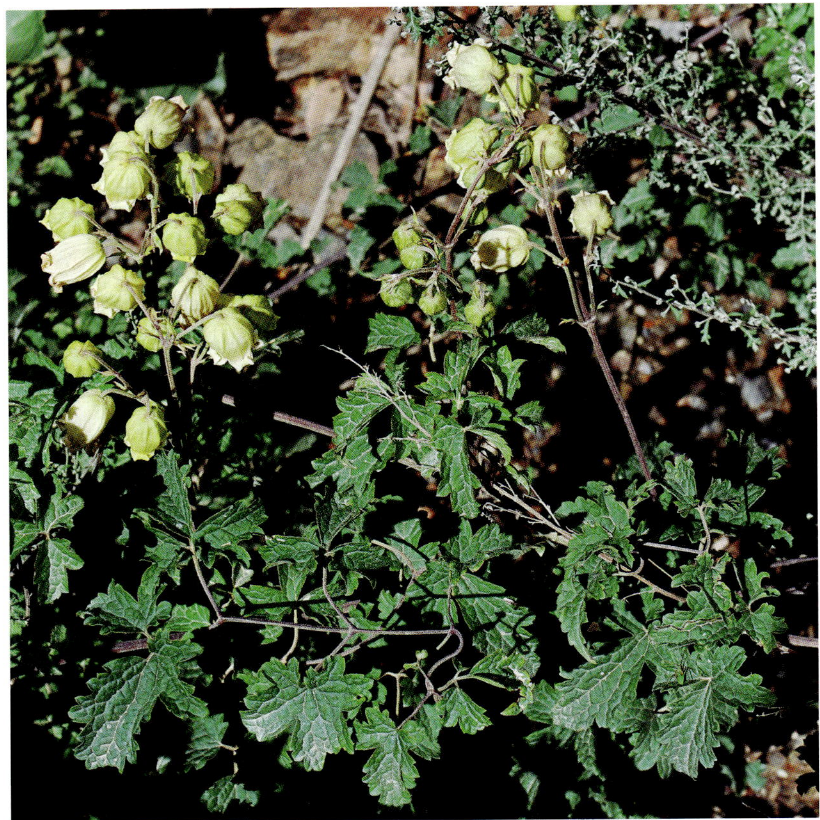
长花铁线莲 *Clemaitis rehderiana*

长花铁线莲
Clematis rehderiana Craib

木质藤本。二回或有时为一回羽状复叶。聚伞花序腋生，具4至多花，常圆锥状；花直径1.4-1.8厘米；萼片4，淡黄色，长圆形；花柱宿存，长2-2.5厘米，羽毛状。瘦果卵圆形至近圆形，被柔毛。花期7-8月，果期9月。生海拔2000-2500米的山坡上、灌丛或溪边。产云南西北、四川西部、西藏东部和青海南部。尼泊尔亦有。

Vines woody. Leaves 2- or sometimes 1-pinnate. Cymes axillary, 4- to many flowered, often panicle-like; flowers 1.4-1.8 cm diam; sepals 4, yellowish, oblong; styles persistent, 2-2.5 cm long, plumose. Achenes ovoid to suborbicular, pubescent. Fl. Jul-Aug. Fr. Sep. Slopes, thickets or by streams at 2000-2500 m. Distributed in NW Yunnan, W Sichuan, E Xizang and S Qinghai. Also in Nepal.

柄铁线莲
Clematis connata DC.

木质藤本。叶为一回羽状复叶；叶柄基部稍膨大，与对生叶柄的基部合生，形成一个碟形的构造；小叶卵形、宽卵形或狭卵形，纸质，边缘具锯齿。聚伞花序腋生，具(5-)11至多花；萼片4，淡黄色，直立，下面密具平伏柔毛；宿存花柱长约4厘米。全株有毒。花期8-10月，果期9-11月。生海拔2000-3400米的林下或溪边灌丛。产云南西北部、四川西部和西藏南部。印度北部、尼泊尔、不丹和巴基斯坦亦有。

Woody vines. Leaves pinnate; petioles base dilated and connate to that of opposite petiole forming a disclike structure; leaflets ovate, broadly ovate or narrowly ovate, papery, margin dentate.Cymes axillary, (5-)11- to many flowered, often panicle-like; sepals 4, yellowish, erect, abaxially appressed puberulous; persistent styles ca. 4 cm long. Whole plant poisonous. Fl. Aug-Oct. Fr. Sep-Nov. Forests or shrubs by streams at 2000-3400 m. Distributed in NW Yunnan, W Sichuan and S Xizang, Also in N India, Nepal, Bhutan and Pakistan.

毛蕊铁线莲
Clematis lasiandra Maxim.

多年生草质藤本。叶二回羽状或二出；小叶草质至薄纸质，边缘具细齿、小牙齿或锯齿。萼片4，紫红色，直立，下面除边缘具绒毛外无毛；花丝基部具柔毛，顶端密具柔毛；花柱宿存，长2-3厘米，羽毛状。花期8-10月，果期11月。生海拔500-2800米的山坡、灌丛或河边。产中国西南、华南、华中和华东。日本亦有。

Vines perennial, herbaceous. Leaves 2-pinnate or 2-ternate; leaflets herbaceous to thinly papery, margin serrate, denticulate, or dentate. Sepals 4, purple red, erect, abaxially glabrous except for velutinous margin; filaments basally pubescent, apically densely villous; persistent styles 2-3 cm long, plumose. Fl. Aug-Oct. Fr. Nov. Slopes, thickets

柄铁线莲 *Clematis connata*

毛蕊铁线莲 *Clematis lasiandra*

or along streams at 500-2800 m. Distributed in SW, S, C and E China. Also in Japan.

毛茛铁线莲

Clematis ranunculoides Franch.

多年生草质藤本。上部叶一回羽状或二回三出；小叶卵形或五角形，纸质，边缘具不规则牙齿至小牙齿。聚伞花序腋生，常具1-3花，顶生聚伞花序具3-7花；萼片4，紫红色，直立，背面具2或3条纵向狭翅；宿存花柱长1.5-2.5厘米。花期9-10月，果期10-11月。生海拔500-3000米的林下、灌丛或草地。产云南、四川西南部、贵州西部和广西西北部。

Vines perennial, herbaceous. Upper leaves 1-pinnate or 2-ternate; leaflets ovate or pentagonal, papery, margin irregularly dentate to denticulate. Cymes axillary, usually 1-3-flowered, terminal cymes 3-7-flowered; sepals 4, purple-red, erect, abaxially with 2 or 3 narrow longitudinal wings; persistent styles 1.5-2.5 cm long. Fl. Sep-Oct. Fr. Oct-Nov. Forests, thickets or grasslands at 500-3000 m. Distributed in Yunnan. SW Sichuan, W Guizhou and NW Guangxi.

毛茛铁线莲 *Clematis ranunculoides*

芹叶铁线莲

Clematis aethusifolia Turcz.

多年生草质藤本。叶二至四回羽状全裂；羽片4或5对，细裂。聚伞花序腋生或顶生，具1-5花；萼片4，淡黄色，直立，两面近无毛；宿存花柱长1.6-2.7厘米。花期7-8月，果期9月。生海拔300-3000米的灌丛、山坡或溪边。产华北和华西。俄罗斯(西伯利亚)和蒙古亦有。

Perennial vines, herbaceous. Leaflets 2-4-pinnatisect; pinnae 4 or 5 pairs, dissected. Cymes axillary or terminal, 1-5-flowered; sepals 4, yellowish, erect, both surfaces subglabrous; persistent styles 1.6-2.7 cm long. Fl. Jul-Aug. Fr. Sep. Thickets, slopes or streamsides at 300-3000 m. Distributed in N and W China. Also in Russia (Siberia) and Mongolia.

西南铁线莲

Clematis pseudopogonandra Finet et Gagnep.

木质藤本。二回三出复叶；小叶纸质，边缘疏具小牙齿。花与叶自同一腋芽中生出；萼片4，紫红色或暗紫色；花丝和花药被柔毛；宿存花柱长约3.5厘米，浅褐色至浅黄色，羽毛状。花期6-7月，果期8-9月。生海拔2700-4300米的溪边、山谷、灌丛或林下。产云南西北部、四川西部、西藏东部和东南部。

Vines woody. Leaves 2-ternate; leaflets papery, margin sparsely dentate. Flowers 1-3, with 2 leaves borne from the same axillary bud; sepals 4, purple-red or purple-black; filaments and anthers villous; persistent styles ca. 3.5 cm long, brownish to yellowish, plumose. Fl. Jun-Jul. Fr. Aug-Sep. By streams, valleys, thickets or forests at 2700-4300 m. Distributed in NW Yunnan, W Sichuan, E and SE Xizang.

芹叶铁线莲 *Clematis aethusifolia*

西南铁线莲 *Clematis pseudopogonandra*

褐毛铁线莲 *Clematis fusca*

褐毛铁线莲
Clematis fusca Turcz.

多年生藤本或直立草本。一回羽状复叶，具(5-)7(-9)小叶；小叶阔至狭卵形，纸质，全缘。花单生枝顶或聚伞花序腋生；萼片4，紫色，直立；宿存花柱长3-4厘米，褐色，羽毛状。花期6-7月，果期8-9月。生海拔500-1000米的林中、林缘、山坡或灌丛中。产山东东部、河北东北部、内蒙古、辽宁、吉林和黑龙江。俄罗斯(远东地区)、朝鲜半岛和日本亦有。

Perennial vines or erect herbs. Leaves pinnate, (5-)7(-9)-foliolate; leaflets broadly to narrowly ovate, papery, margin entire. Flowers usually solitary and terminal, or in axillary cymes; sepals 4, purple, erect; persistent styles 3-4 cm long, brown, plumose. Fl. Jun-Jul. Fr. Aug-Sep. Forests, forest edges, slopes or thickets at 500-1000 m. Distributed in E Shandong, NE Hebei, Neimenggu, Liaoning, Jilin and Heilongjiang. Also in Russia (Far East), Korean Peninsula and Japan.

全缘铁线莲
Clematis integrifolia L.

直立草本或半灌木。茎高达1.5米。叶对生，无柄，宽卵形，长5-14厘米，全缘。单花顶生，下垂；萼片4，蓝色、紫色或白色，直立，长3-4.5厘米；花丝和花药被柔毛；宿存花柱羽毛状，长4-5厘米。花期6-7月。生海拔1200-2000米的草坡、河滩或灌丛中。产新疆北部。亚洲西部和欧洲亦有。

Erect herbs or subshrubs. Stems up to 1.5 m tall. Leaves opposite, sessile, broadly ovate, 5-14 cm long, entire. Flowers solitary, pendulous; sepals 4, blue, purple or white, erect, 3-4.5 cm long; filaments and anthers pubescent; persistent styles plumose, 4-5 cm long. Fl. Jun-Jul. Grassy slopes. river banks, or bushes at 1200-2000 m. Distributed in N Xinjiang. Also in W Asia and Europe.

互叶铁线莲
Clematis alternata Kitam. et Tamura

攀援小灌木。单叶互生，心状卵形，背面被短柔毛，边缘具小齿。聚伞花序腋生，有1-3花；花下垂；萼片4，直立，紫红色，长1.8-2.2厘米；花丝被柔毛；子房被柔毛，花柱被长柔毛。花期7月。生海拔2200-2500米的林缘或灌丛中。产西藏南部。尼泊尔亦有。

Scandent shrublets. Leaves simple, alternate, cordate-ovate, abaxially puberulous, margin denticulate. Cymes axillary, 1-3-flowered; flowers pendulous; sepals 4, erect, purple-red, 1.8-2.2 cm long; filaments pubescent; ovaries pubescent, styles villous. Fl. Jul. Forest margins or bushes at 2200-2500 m. Distributed in S Xizang. Also in Nepal.

西伯利亚铁线莲
Clematis alpina (L.) Mill. subsp. **sibirica** (Mill.) Kuntze

木质藤本。茎长达3米，无毛。叶具柄，为二回三出复叶；小叶狭卵形，长1.5-5厘米，边缘具小牙齿。花单生叶腋，具长柄；萼片4，近直立，淡黄或白色，长3-4.5厘米；退化雄蕊条状匙形，长约2厘米；雄蕊和心皮均多数。花期6-7月。生海拔1200-2000米的林中或林缘。产新疆、青海东部、甘肃中部、宁夏、内蒙古和黑龙江。蒙古、俄罗斯和欧洲西北部亦有。

Woody vines. Stems up to 3 m long, glabrous. Leaves petiolate, 2-ternate; leaflets narrowly ovate, 1.5-5 cm long, margin denticulate. Flowers singly axillary, long pedicellate; sepals 4, suberect, yellowish or white, 3-4.5 cm long; staminodes linear-spatulate, ca. 2 cm long; stamens and car-

全缘铁线莲 *Clematis integrifolia*

互叶铁线莲 *Clematis alternata*

西伯利亚铁线莲 *Clematis alpina* subsp. *sibirica*

半钟铁线莲 *Clematis alpina* subsp. *ochotensis*

pels all numerous. Fl. Jun-Jul. Forests or forest margins at 1200-2000 m. Distributed in Xinjiang, E Qinghai, C Gansu, Ningxia, Neimenggu and Heilongjiang. Also in Mongolia, Russia and NW Europe.

半钟铁线莲

Clematis alpina (L.) Mill. subsp. **ochotensis** (Pall.) Kuntze

木质藤本。枝具4-6棱。二回三出复叶；小叶狭卵形或卵形，纸质，边缘具锯齿。花单生，直径3-6厘米；萼片4，蓝色或紫色，斜升，下面密被柔毛，退化雄蕊约比萼片短2倍；宿存花柱长3.5-4厘米，羽毛状。花期7月，果期8月。生海拔600-1200米的林中。产山西北部、河北北部、内蒙古、吉林和黑龙江。俄罗斯(远东地区、西伯利亚)和蒙古亦有。

Woody vines. Branches 4-6-angulate. Leaves 2-ternate; leaflets narrowly ovate or ovate, papery, margin serrate. Flowers solitary, 3-6 cm diam; sepals 4, blue or purple, ascending, abaxially densely puberulous; staminodes ca. 2 times shorter than sepals; persistent styles 3.5-4 cm long, plumose. Fl. Jul. Fr. Aug. Forests at 600-1200 m. Distributed in N Shanxi, N Hebei, Neimenggu, Jilin and Heilongjiang. Also in Russia (Far East, Siberia) and Mongolia.

长瓣铁线莲

Clematis macropetala Ledeb.

木质藤本。幼枝被短柔毛。叶对生，为二回三出复叶；小叶狭卵形，边缘有小齿。花单生于当年生枝顶端；萼片4，蓝色或紫色，斜上展，长3-4.8厘米；外退化雄蕊狭披针形，与萼片近等长，内退化雄蕊较小，条状匙形；雄蕊有毛。花期7月。生海拔2000-2600米的草坡、多石处或林下。产青海东部、甘肃、陕西、宁夏、山西、河北北部、内蒙古和辽宁西南部。蒙古东部和俄罗斯(西伯利亚)亦有。

Woody vines. Young branches puberulous. Leaves opposite, 2-ternate; leaflets narrow-ovate, serrulate. Flowers singly terminal to the hornotinous branches; sepals 4, blue or purple, ascending, 3-4.8 cm long; outer staminodes narrow-lanceolate, nearly as long as sepals, inner staminodes smaller, linear-spatulate; stamens hairy. Fl. Jul. Grassy slopes, gravelly places or forests at 2000-2600 m. Distributed in E Qinghai, Gansu, Shaanxi, Ningxia, Shanxi, N Hebei, Neimenggu and SW Liaoning. Also in E Mongolia and Russia (Siberia).

长瓣铁线莲 *Clematis macropetala*

锡兰莲 *Naravelia laurifolia*

两广锡兰莲 *Naravelia pilulifera*

锡兰莲

Naravelia laurifolia Wall. ex Hook. f. et Thoms.

木质藤本。茎无毛，有纵沟。叶为羽状复叶，有5小叶，近无毛；基部2小叶宽卵形，长6-11厘米，顶生3小叶变成3卷须。圆锥花序具多数花；萼片4，长5-7毫米；花瓣8-10，狭匙形，长5-7毫米；雄蕊条形。瘦果纺锤形，宿存花柱羽毛状，长约2.5厘米。花期10月。生海拔1000米的林下。产云南南部和海南。印度东北部、缅甸、泰国、马来半岛、菲律宾和印度尼西亚亦有。

Woody vines. Stems glabrous, sulcate. Leaves pinnate, 5-foliolate, subglabrous; basal 2 leaflets broadly ovate, 6-11 cm long, distal 3 leaflets transformed into 3 tendrils. Panicle many-flowered; sepals 4, 5-7 mm long; petals 8-10, narrowly spatulate, 5-7 mm long; stamens linear. Achenes fusiform; persistent styles plumose, ca. 2.5 cm long. Fl. Oct. Forests at 1000 m. Distributed in S Yunnan and Hainan. Also in NE India, Myanmar, Thailand, Malay Pennisula, the Philippines and Indonesia.

两广锡兰莲

Naravelia pilulifera Hance

本质藤本。茎长2-3米。叶对生，为羽状复叶，具5小叶，基部2小叶宽卵形，长7-11厘米，具5出脉，上部3小叶转变为卷须。花序圆锥状，腋生；花直径约1.5厘米；萼片4，长约7毫米；花瓣8-12，具圆形瓣片；雄蕊多数。瘦果长约5毫米，宿存羽毛状花柱长约2厘米。花期9月。生海拔100-1000米的林中或灌丛中。产广东、海南、广西和云南南部。

Woody vinces. Stem 2-3 m long. Leaves opposite, 5-foliolately pinnate; basal 2 leaflets broadly ovate, 7-11 cm long, 5-nerved, apical 3 leaflets transformed into tendrils. Panicles axillary; flower ca. 1.5 cm diam; sepals 4, ca. 7 mm long; petals 8-12, with orbicular limbs; stamens numerous. Achenes ca. 5 mm long, with plumose persistent styles ca. 2 cm long. Fl. Sep. In forests or scrubs at 100-1000 m. Distributed in Guangdong, Hainan, Guangxi and S Yunnan.

独叶草

Kingdonia uniflora Balf. f. et W. W. Sm.

多年生小草本。芽鳞约3，膜质。叶通常1，基生，具长柄，柄长5-11厘米；叶片5全裂，具二叉分枝脉。花葶高7-12厘米；花顶生，单生；萼片浅绿色，(4-)5(-7)；花瓣不存在；雄蕊5-8；退化雄蕊8-11；心皮3-7。瘦果窄倒披针形。花期5-6月。生海拔2700-3900米的林中。产云南西北部、四川西部、甘肃南部和陕西南部。

Perennial small herbs. Bud scales ca. 3, membranous. Leaves usually 1, basal, with long petioles, petioles 5-11 cm long; blades 5-sect, with dichotomous veins. Scapes 7-12 cm tall; flowers terminal, solitary; sepals pale green, (4-)5(-7); petals absent; stamens 5-8; staminodes 8-11; carpels 3-7. Achenes narrowly oblanceolate. Fl. May-Jun. Forests at 2700-3900 m. Distributed in NW Yunnan, W Sichuan, S Gansu and S Shaanxi.

独叶草 *Kingdonia uniflora*

太白美花草 *Callianthemum taipaicum*

短柱侧金盏花 *Adonis brevistyla*

太白美花草

Callianthemum taipaicum W. T. Wang

草本，无毛。叶窄卵形；小叶片2或3对；2深裂。花单朵顶生；萼片5，蓝紫色；花瓣9-13，白色，倒披针形至窄倒卵形，长11-14毫米，顶端截形；雄蕊多数；心皮18-22。花期6月。生海拔3400-3600米的草坡。产陕西(太白山)。

Herbs, glabrous. Leaves narrowly ovate; leaflets 2 or 3 paris, 2-parted. Flowers singly terminal; sepals 5, blue-purple; petals 9-13, white, oblanceolate to narrowly obovate, 11-14 mm long, apex truncate; stamens numerous; carpels 18-22. Fl. Jun. Grassy slopes at 3400-3600 m. Distributed in Shaanxi (Taibai Mountain).

美花草

Callianthemum pimpinelloides (D. Don) Hook. f. et Thoms.

草本。叶卵形至狭卵形，羽状全裂。萼片5；花瓣5-7(-9)，条状倒披针形，白色、粉色或浅紫色，倒卵状长圆形至宽条形，长5-10毫米；雄蕊多数；心皮8-14。瘦果卵球形，具皱。花期4-6月。生海拔3200-5600米的高山草地。产云南、四川、西藏和青海。印度北部、尼泊尔、不丹、巴基斯坦北部和阿富汗亦有。

Herbs. Leaves ovate to narrowly ovate, pinnatisect. Sepals 5; petals 5-7(-9), white, pink or pale purple, obovate-oblong to broadly linear, 5-10 mm long; stamens numerous; carpels 8-14. Achenes ovoid, rugose. Fl. Apr-Jun. Alpine meadows at 3200-5600 m. Distributed in Yunnan, Sichuan, Xizang and Qinghai. Also in N India, Nepal, Bhutan, N Pakistan and Afghanistan.

美花草 *Callianthemum pimpinelloides*

短柱侧金盏花

Adonis brevistyla Franch.

多年生草本。茎下部叶有长柄，叶片3全裂。花单朵顶生；萼片5-7，椭圆形；花瓣7-14，白色，有时带紫色；雄蕊与萼片近等长；花柱极短，柱头球形。瘦果倒卵形，有宿存花柱。花期4-8月。生海拔1900-3500米的林中、山地草坡或溪岸。产中国西南、华中和华西。不丹亦有。

Perennial herbs. Lower leaves with long petioles, blades 3-sect. Flower solitary, terminal; sepals 5-7, elliptic; petals 7-14, white, sometimes tinged with purple; stamens as long as sepals; styles very short, stigmas globose. Achenes obovate, styles persistent. Fl. Apr-Aug. Forests, grassy slopes or river banks at 1900-3500 m. Distributed in SW, C and W China. Also in Bhutan.

蜀侧金盏花 *Adonis sutchuenensis*

甘青侧金盏花 *Adonis bobroviana*

蜀侧金盏花

Adonis sutchuenensis Franch.

多年生草本。茎高达40厘米，无毛。叶无毛，具长或短柄，三全裂，全裂片二回羽状全裂或深裂。花单生茎顶端；萼片约6，淡绿色；花瓣8-12，黄色，长1.5-2厘米；雄蕊和心皮均多数。花期4-6月。生海拔1100-3300米的林中或草坡上。产四川北部、湖北西部和陕西南部。

Perennial herbs. Stems up to 40 cm tall, glabrous. Leaves glabrous long or shortly petiolate, 3-sect; segments 2-pinnatisect or 2-pinnatipartite. Flowers solitary, terminal; sepals ca. 6, greenish; petals 8-12, yellow, 1.5-2 cm long; stamens and carpels numerous. Fl. Apr-Jun. Forests or grassy slopes at 1100-3300 m. Distributed in N Sichuan, W Hubei and S Shaanxi.

侧金盏花

Adonis amurensis Regel et Radde

多年生草本。叶片三角形，二至三回羽状深裂。萼片约9，淡灰紫色，长圆形至倒卵长圆形；花瓣约10，黄色，倒卵长圆形至窄披针形。瘦果倒卵球形，被柔毛。花期3-4月。生林中或草坡。产黑龙江、吉林和辽宁。俄罗斯、朝鲜半岛和日本亦有。

Perennial herbs. Leaves triangular, 2-3 pinnately divided. Sepals ca. 9, pale-grayish-purple, oblong to obovate-oblong; petals ca. 10, yellow, obovate-oblong to narrowly lanceolate. Achenes obovoid, pubescent. Fl. Mar-Apr. Forests or grassy slopes. Distributed in Heilongjiang, Jilin and Liaoning. Also in Russia, Korean Peninsula and Japan.

甘青侧金盏花

Adonis bobroviana Sim.

多年生草本。茎高达30厘米，被极短腺毛。叶长4-9厘米，二至三回羽状细裂，末回裂片条形。花单朵顶生；萼片5，淡绿色，长0.5-1.7厘米；花瓣9-13(-16)，黄色，长1-2厘米；雄蕊和心皮均多数。花期4-7月。生海拔1900-3200米的干草坡。产青海东北部、甘肃中部和宁夏。

Perennial herbs. Stems up to 30 cm tall, with leaves very shortly glandular-puberulous. Leaves 4-9 cm long, 2-3-pinnately dissected, ultimate lobules linear. Flowers singly terminal; sepals 5, greenish, 0.5-1.7 cm long; petals 9-13(-16), yellow, 1-2 cm long; stamens and carpels numerous. Fl. Apr-Jul. Dry grassy slopes at 1900-3200 m. Distributed in NE Qinghai, C Gansu and Ningxia.

北侧金盏花

Adonis sibirica Patr. ex Ledeb.

多年生草本。茎高约40厘米，无毛。叶无柄，无毛，卵形或三角形，长约6厘米，二至三回羽状细裂，末回裂片狭条形。花直径4-5.5厘米；萼片5，长约1厘米，无毛；花瓣10-18，黄色，长2-2.3厘米。瘦果长4毫米，疏被短柔毛。花期5-6月。生草坡上，海拔1900米。产内蒙古东北部和新疆北部。蒙古、俄罗斯和欧洲亦有。

Perennial herbs. Stems ca. 40 cm tall, glabrous. Leaves sessile, glabrous, ovate or triangular, 2-3

侧金盏花 *Adonis amurensis*

北侧金盏花 *Adonis sibirica*

夏侧金盏花 *Adonis aetivalis*

times finely pinnately divided, ultimate lobes narrow-linear. Flower 4-5.5 cm diam; sepals 5, ca. 1 cm long, glabrous; petals 10-18, yellow, 2-2.3 cm long. Achenes 4 mm long, sparsely pubescent. Fl. May-Jun. Grassy slopes at 1900 m. Distributed in NE Neimenggu and N Xinjiang. Also in Mongolia, Russia and Europe.

蓝侧金盏花

Adonis coerulea Maxim.

多年生小草本。茎高3-15厘米，无毛。叶长达8厘米，无毛，二至三回羽状细裂。花单朵顶生；萼片5-7，长4-6毫米；花瓣约8，淡蓝色或淡紫色，长6-11毫米；心皮多数。花期4-7月。生草坡上或灌丛中，海拔2300-5000米。产西藏东北部、四川西北部、青海和甘肃。

Small perennial herbs. Stems 3-15 cm tall, glabrous. Leaves up to 8 cm long, glabrous, 2-3-pinnately dissected. Flowers singly terminal; sepals 5-7, 4-6 mm long; petals ca. 8, bluish or purplish, 6-11 mm long; carpels numerous. Fl. Apr-Jul. Grassy slopes or bushes at 2300-5000 m. Distributed in NE Xizang, NW Sichuan, Qinghai and Gansu.

夏侧金盏花

Adonis aetivalis L.

一年生草本。茎高10-30厘米。叶长达6厘米，二至三回羽状细裂，无毛或有疏毛。花单朵顶生；萼片5，长约8毫米；花瓣约8，橙黄色，下部黑紫色，长约10毫米；心皮多数。花期5-6月。生海拔1300米的田边草地。产新疆西部。广布亚洲西部和欧洲。

Annual herbs. Stems 10-30 cm tall. Leaves up to 6 cm long, 2-3-pinnately dissected, glabrous or wth sparse hairs. Flowers singly terminal; sepals 5, ca. 8 mm long; petals ca. 8, orange-yellow, below black-purple, ca. 10 mm long; carpels numerous. Fl. May-Jun. Grassland by fields at 1300 m. Distributed in W Xinjiang. Widespread in W Asia and Europe.

天山毛茛

Ranunculus popovii Ovcz.

多年生草本。茎高5-25厘米，被柔毛。基生叶具长柄，叶片形状多样，宽卵形，椭圆形，宽菱形等，3-11浅裂，有时深裂，下面被柔毛。茎生叶通常3全裂。花1朵生茎或枝的顶端；萼片5，长约4.5毫米；花瓣5，黄色，长6-8毫米；雄蕊和心皮多数。花期6月。生海拔2400-2900米的山谷草甸。产内蒙古西南部和新疆。哈萨克斯坦亦有。

Perennial herbs. Stems 5-25 cm tall, pubescent. Basal leaves long petiolate; leaf blades with variable shapes, broadly ovate, elliptic, broadly rhombic, 3-11-lobed or 3-parted, abaxially pubescent. Cauline leaves 3-sect. Flowers solitary, terminal; sepals 5, ca. 4.5 mm long; petals 5, yellow, 6-8 mm long; stamens and carpels numerous. Fl. Jun. Meadows in valleys at 2400-2900 m. Distributed in SW Neimenggu and Xinjiang. Also in Kazakhstan.

蓝侧金盏花 *Adonis coerulea*

天山毛茛 *Ranunculus popovii*

深山毛茛 *Ranunculus franchetii*

矮毛茛 *Ranunculus pseudopygmaeus*

深山毛茛

Ranunculus franchetii H. Boiss.

多年生草本。茎高15-20厘米。基生叶具长柄；叶片肾形，宽2.5-4厘米，3深裂。上部茎生叶无柄，3全裂。花序具1-2花；花瓣5-7，黄色，长7-8毫米。瘦果宽约1.5毫米，被短柔毛。花期5-7月。生海拔300-1300米的林边、灌丛中或溪边。产辽宁、吉林和黑龙江。朝鲜半岛、俄罗斯(远东地区)和日本亦有。

Perennial herbs. Stems 15-20 cm tall. Basal leaves long petiolate; blades reniform, 2.5-4 cm broad, 3-parted. Upper cauline leaves sessile, 3-sect. Monochasia 1-2-flowered; petals 5-7, yellow, 7-8 mm long. Achenes ca. 1.5 mm wide, densely puberulous. Fl. May-Jul. Forest margins, bushes or by streams at 300-1300 m. Distributed in Liaoning, Jilin and Heilongjiang. Also in Korean Peninsula, Russia (Far East) and Japan.

矮毛茛

Ranunculus pseudopygmaeus Hand.-Mazz.

多年生小草本。茎高约5厘米。基生叶约4，具长柄；叶片肾状五角形，宽6-15毫米，3深裂。茎生叶1-2。花单生茎顶端；花瓣5，黄色，长2-2.5毫米。瘦果卵球形，长1毫米，无毛。花期7-8月。生海拔3000-4000米的草坡或砾石地。产西藏东南部(墨脱县)和云南西北部。尼泊尔亦有。

Small perennial herbs. Stems ca. 5 cm tall. Basal leaves ca. 4, long petiolate; blades reniform-pentagonal, 6-15 mm broad, 3-parted. Cauline leaves 1-2. Flowers singly terminal; petals 5, yellow, 2-2.5 mm long. Achenes ovoid, 1 mm long, glabrous. Fl. Jul-Aug. Grassy slopes or gravelly places at 3000-4000 m. Distributed in SE Xizang (Mêdog County) and NW Yunnan. Also in Nepal.

多雄拉毛茛

Ranunculus duoxionglashanicus W. T. Wang

多年生小草本。茎高12-18厘米。基生叶约4，具长柄；叶片五角形，宽约1.8厘米，3深裂。花单朵顶生或2朵单歧聚伞花序；花瓣5，黄色，宽椭圆形，长3-4毫米；心皮无毛。花期7-8月。生海拔4200米的高山草地。特产西藏(米林县)。

Small perennial herbs. Stems 12-18 cm tall. Basal leaves ca. 4, long petiolate; blades pentagonal, ca. 1.8 cm broad, 3-parted. Flowers singly terminal or 2 in monochasium; petals 5, yellow, broad-elliptic, 3-4 mm long; carpels glabrous. Fl. Jul-Aug. Alpine meadow at 4200 m. Endemic to Xizang (Mainling County).

太白山毛茛

Ranunculus petrogeiton Ulbr.

多年生小草本。茎渐升，长3-12(-20)厘米，无毛。基生叶具柄，宽卵形，3全裂；茎生叶小。单花顶生；萼片5，长3-6毫米；花瓣5，黄色，长4.5-9毫米；心皮约10。花期6-7月。生海拔3000-4800米的高山草地。产四川西部、甘肃南部和陕西南部。

Small perennial herbs. Stems ascending, 3-12(-20) cm long, glabrous. Basal leaves petiolate, broadly ovate, 3-sect, glabrous; cauline leaves small. Flowers singly terminal; sepals 5, 3-6 mm

多雄拉毛茛 *Ranunculus duoxionglashanicus*

太白山毛茛 *Ranunculus petrogeiton*

爬地毛茛 *Ranunculus pegaeus*

三裂毛茛 *Ranunculus hirtellus* var. *orientalis*

long; petals 5, yellow, 4.5-9 mm long; carpels ca. 10. Fl. Jun-Jul. Alpine meadows at 3000-4800 m. Distributed in W Sichuan, S Gansu and S Shaanxi.

爬地毛茛

Ranunculus pegaeus Hand.-Mazz.

多年生小草本。茎纤细，匍匐地面，长约20厘米，在节上生根。叶具柄，心状五角形，宽3-10毫米，3深裂或3全裂。花单生，具长柄；萼片5，长约2.5毫米；花瓣5-7，黄色，长圆形，长约3毫米；雄蕊8-10；心皮无毛。花期6-8月。生海拔3600-4100米的沟边或冷杉林下。产云南西北部和西藏南部。尼泊尔和印度北部亦有。

Small perennial herbs. Stems slender, creeping, ca. 20 cm long, from nodes rooting. Leaves long petiolate, cordate-pentagonal, 3-10 mm long, 3-parted or 3-sect. Flowers solitary, long pedicellate; sepals 5, ca. 2.5 mm long; petals 5-7, yellow, oblong, ca. 3 mm long; stamens 8-10; carpels glabrous. Fl. Jun-Aug. By streams or *Abies* forests at 3600-4100 m. Distributed in NW Yunnan and S Xizang. Also in Nepal and N India.

三裂毛茛

Ranunculus hirtellus Royle var. **orientalis** W. T. Wang

多年生小草本。茎高5-15(-20)厘米。基生叶数枚，有长柄；叶片肾状五角形，3深裂或3全裂；茎生叶较小。花单朵顶生；花瓣5，黄色，长约6毫米。瘦果长约1.5毫米，无毛。花期6-8月。生海拔3000-5000米的高山草地或多石砾地。产西藏东部、云南西北部、四川西部和青海东部。

Small perennial herbs. Stems 5-15(-20) cm tall. Basal leaves several, long petiolate; blades reniform-pentagonal, 3-parted or 3-sect; cauline leaves smaller. Flowers singly terminal; petals 5, yellow, ca. 6 mm long. Achenes ca. 1.5 mm long, glabrous. Fl. Jun-Aug. Alpine meadows or gravelly slopes at 3000-5000 m. Distributed in E Xizang, NW Yunnan, W Sichuan and E Qinghai.

高原毛茛

Ranunculus tanguticus (Maxim.) Ovcz.

多年生草本。基生叶的裂片一至三回细裂，末回裂片披针状条形。单歧聚伞花序顶生，具2或3花；花托具柔毛；花瓣5，蜜槽无鳞片。聚合果狭卵球形；瘦果稍两侧压扁，倒卵球形。花果期6-10月。生海拔2200-4200米的山坡、沼泽地或潮湿地。产中国西南、华北和华西。尼泊尔亦有。

Perennial herbs. Segments of basal leaves 1-3 times dissected, ultimate lobes lanceolate-linear. Monochasium terminal, 2- or 3-flowered; receptacles puberulent; petals 5, nectary pit without a scale. Aggregate fruits narrowly ovoid; achenes slightly bilaterally compressed, obovoid. Fl. and fr. Jun-Oct. Slopes, marshes or wet places at 2200-4200 m. Distributed in SW, N and W China. Also in Nepal.

高原毛茛 *Ranunculus tanguticus*

云生毛茛 *Ranunculus nephelogenes*

单叶毛茛 *Ranunculus monophyllus*

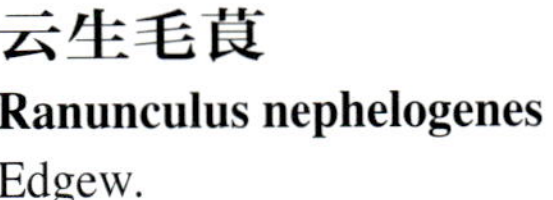

云生毛茛

Ranunculus nephelogenes Edgew.

多年生草本。茎直立。基生叶4-9，叶片卵形至披针形，有时3浅裂；上部茎生叶无柄，叶片披针状条形，不分裂或有时3深裂。花单生茎顶；花瓣5(-7),有短爪，黄色，蜜槽呈杯状或袋穴。聚合果卵球形；瘦果斜倒卵球形。花果期3-8月。生海拔1700-5200米的高山草甸、砾坡、河边或沼泽地。产中国西南和西北。尼泊尔和巴基斯坦亦有。

Perennial herbs. Stems erect. Basal leaves 4-9, ovate to lanceolate, sometimes 3-lobed; upper stem leaves sessile, lanceolate-linear, undivided or rarely 3-lobed. Flowers solitary, terminal; petals 5(-7), with short claw, yellow, nectary pits cup or sack shaped. Aggregate fruits ovoid; achenes obliquely obovoid. Fl. and fr. Mar-Aug. Alpine meadows, gravelly slopes, by streams or swamps at 1700-5200 m. Distributed in SW and NW China. Also in Nepal and Pakistan.

单叶毛茛

Ranunculus monophyllus Ovcz.

多年生草本。茎高20-30厘米，无毛。基生叶1(-3)，具长柄；叶片肾形，3浅裂或3深裂；茎生叶无柄，5全裂。单花顶生；花瓣5，黄色，长6-7毫米。瘦果长约2毫米，被短柔毛。花期5-7月。生海拔1700-2000米的灌丛中或溪边湿草甸上。产山西、河北北部、黑龙江、内蒙古和新疆。蒙古、哈萨克斯坦和俄罗斯亦有。

Perennial herbs. Stems 20-30 cm tall, glabrous. Basal leaves 1(-3), long petiolate; blades reniform, 3-lobed or 3-parted; cauline leaves sessile, 5-sect. Flowers singly terminal; petals 5, yellow, 6-7 mm long. Achenes ca. 2 mm long, puberulous. Fl. May-Jul. Bushes or meadows by stream at 1700-2000 m. Distribtued in Shanxi, N Hebei, Heilongjiang, Neimenggu and Xinjiang. Also in Mongolia, Kazakhstan and Russia.

宽瓣毛茛 *Ranunculus albertii*

宽瓣毛茛

Ranunculus albertii Regel et Schmalh.

多年生小草本。茎高8-30厘米。基生叶具柄，肾形，边缘有圆齿。花单朵顶生；花瓣5-8，黄色，宽8-14毫米。瘦果卵球形，长约1.6毫米。花期6-8月。生海拔1800-3300米的草坡上。产新疆。哈萨克斯坦亦有。

Small perennial herbs. Stems 8-30 cm tall. Basal leaves petiolate, reniform, margin rounded-dentate. Flowers singly terminal; petals 5-8, yellow, 8-14 mm wide. Achenes ovoid, ca. 1.6 mm long. Fl. Jun-Aug. Grassy slopes at 1800-3300 m. Distributed in Xinjiang. Also in Kazakhstan.

云南毛茛

Ranunculus yunnanensis Franch.

多年生草本。基生叶3-7，有柄，倒卵形或匙形，不裂；上部茎生叶三裂。单歧聚伞花序顶生，具2或3花；花托无毛；花瓣5(-8)，蜜槽无鳞片。聚合果卵球形；瘦果稍两侧压扁，斜倒卵球形。花期6-9月。生海拔2800-4800米的林缘、溪边或草坡上。产云南北部和四川西南部。

Perennial herbs. Basal leaves 3-7, petiolate, obovate or spathulate, undivided; upper cauline leaves 3-partite. Monochasium terminal, 2- or 3-flowered; receptacles glabrous; petals 5(-8), nectary pit without a scale. Aggregate fruits ovoid; achenes slightly bilaterally compressed, obliquely obovoid. Fl. Jun-Sep. Forest edges, by streams or grassy slopes at 2800-4800 m. Distributed in N Yunnan and SW Sichuan.

阿尔泰毛茛

Ranunculus altaicus Laxm.

多年生草本。茎高5-20厘米。基生叶具长柄；叶片楔形，长

阿尔泰毛茛 *Ranunculus altaicus*

1.4-4厘米，顶端有3-5齿；茎生叶小，3中裂或3深裂。单花顶生；花瓣5(-6-8)，黄色，长1.2-1.6厘米。瘦果长2-2.5毫米，无毛或有少数毛。花期6-8月。生高山草地或沼泽边，海拔2500-2700米。产新疆北部。哈萨克斯坦、蒙古和俄罗斯(西伯利亚)亦有。

Perennial herbs.Stems 5-20 cm tall. Basal leaves long petiolate; blades cuneate, 1.4-4 cm long, apex 3-5-dentate; cauline leaves small, 3-fid or 3-parted. Flowers singly terminal; petals 5(-6-8), yellow, 1.2-1.6 cm long. Achenes 2-2.5 mm long, glabrous or with a few hairs. Fl. Jun-Aug. Alpine meadows or swamp margins at 2500-2700 m. Distributed in N Xinjiang. Also in Kazakhstan, Mongolia and Russia (Siberia).

丽江毛茛

Ranunculus dielsianus Ulbr. var. **suprasericeus** Hand.-Mazz.

多年生小草本。茎高7-12厘米。基生叶具长柄，肾状五角形，5微裂，上面被贴伏柔毛，边缘有圆齿。花序约有2花；花瓣5，黄色，长约5毫米。瘦果有柔毛。花期6-7月。生海拔3500-3700米的溪边、石上或灌丛中。产云南西北和东北、四川西南部。

Small perennial herbs. Stems 7-12 cm tall. Basal leaves long petiolate, reniform-pentagonal, 5-lobulate, adaxially appressed-pubescent, margin rounded-dentate. Monochasia 2-flowered; petals 5, yellow, ca. 5 mm long. Achenes hairy. Fl. Jun-Jul. By streams, on rocks or in bushes at 3500-3700 m. Distributed in NW and NE Yunnan, and SW Sichuan.

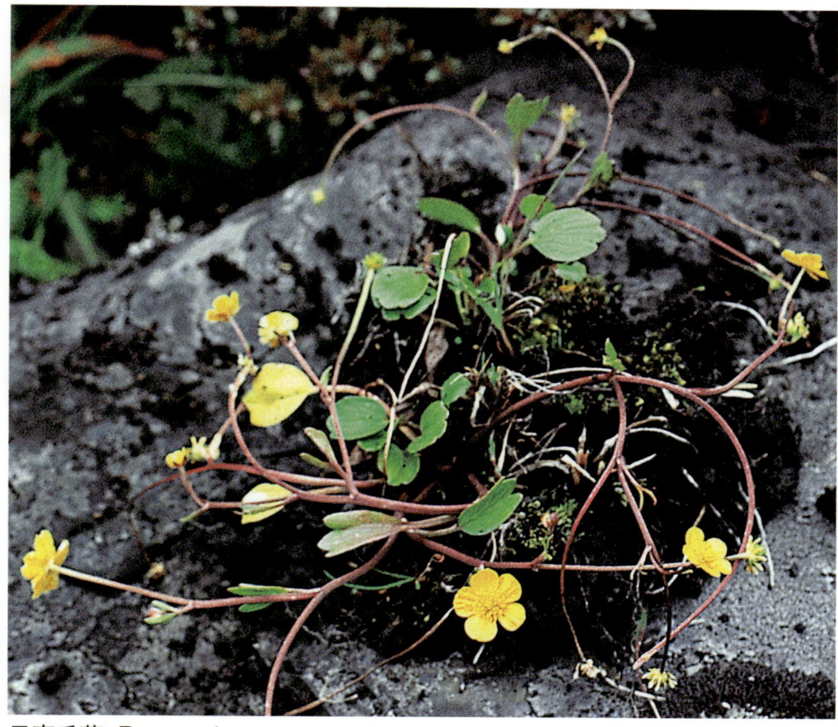
云南毛茛 *Ranunculus yunnanensis*

苞毛茛

Ranunculus similis Hemsl.

多年生小草本。茎高3-8厘米，无毛。基生叶2-4，具长柄，无毛；叶片肾形，3-5微裂；茎生叶2-3，无柄，与花邻接，3裂。单花顶生；花瓣5，黄色，长8-12毫米。瘦果长2毫米，无毛。花期5-8月。生海拔4000-5700米的砾石山坡或碎石河岸。产西藏、青海西南部和南部、新疆东南部。

Small perennial herbs. Stems 3-8 cm tall, glabrous. Basal leaves 2-4, long petiolate, glabrous; blades reniform, 3-5-lobulate; cauline leaves 2-3, sessile, contiguous to flower, 3-fid. Flowers singly terminal; petals 5, yellow, 8-12 mm long. Achenes 2 mm long, glabrous. Fl. May-Aug. Grassy or gravelly slopes or gravelly river banks at 4000-5700 m. Distributed in Xizang, SW and S Qinghai, and SE Xinjiang.

丽江毛茛 *Ranunculus dielsianus* var. *suprasericeus*

苞毛茛 *Ranunculus similis*

小掌叶毛茛 *Ranunculus gmelinii*

松叶毛茛 *Ranunculus reptans*

小掌叶毛茛
Ranunculus gmelinii DC.

多年生水生小草本。茎细弱，长约30厘米。叶基生并茎生，具柄；叶片圆卵形或肾形，宽1.2-1.5厘米，3深裂或3全裂，一回裂片多少细裂。花序有1-4花；花瓣5，黄色，长约3.5毫米。瘦果无毛。花期6-8月。生沼泽中或溪水中。产内蒙古、吉林和黑龙江。蒙古、俄罗斯(西伯利亚)和欧洲北部亦有。

Small aquatic perennial herbs. Stems slender, ca. 30 cm long. Leaves basal and cauline, petiolate; blades orbicular-ovate or reniform, 1.2-1.5 cm wide, 3-parted or 3-sect, primary lobes more or less dissected. Monochasia 1-4-flowered; petals 5, yellow, ca. 3.5 mm long. Achenes glabrous. Fl. Jun-Aug. Swamps or streams. Distributed in Neimenggu, Jilin and Heilongjiang. Also in Mongolia, Russia (Siberia) and N Europe.

浮毛茛
Ranunculus natans C. A. Mey.

多年生水生草本。茎长20厘米以上，无毛，常节上生根。叶基生并茎生；叶片肾形，宽1.5-2.5厘米，3-5浅裂。花与叶对生；花瓣5，黄色，长2.5-4毫米。瘦果无毛。花期6-8月。生海拔1800-3500米的溪边或沼泽中。产西藏、青海、新疆、内蒙古和黑龙江。哈萨克斯坦、蒙古和俄罗斯(西伯利亚)亦有。

Aquatic perennial herbs. Stems more than 20 cm long, glabrous, often rooting at nodes. Leaves basal and cauline; blades reniform, 1.5-2.5 cm broad, 3-5-lobed. Flowers leaf-opposed; petals 5, yellow, 2.5-4 mm long. Achenes glabrous. Fl. Jun-Aug. By streams or in swamps at 1800-3500 m. Distributed in Xizang, Qinghai, Xinjiang, Neimenggu and Heilongjiang. Also in Kazakhstan, Mongolia and Russia (Siberia).

松叶毛茛
Ranunculus reptans L.

多年生小草本。茎细，具匍匐茎，长8-25厘米，节上生根。叶基生并茎生，近无柄，狭条形或条状倒披针形。单花顶生；花瓣5-7，黄色，长3-4.5毫米。瘦果无毛。花期7-9月。生海拔200-1500米的河岸或湖边。产黑龙江、内蒙古和新疆北部。广布亚洲北部、欧洲和北美洲。

Small perennial herbs. Stems slender, stoloniferous, 8-25 cm long, rooting at nodes. Leaves basal and cauline, subsessile, narrow-linear or linar-oblanceolate. Flowers singly terminal; petals 5-7, yellow, 3-4.5 mm long. Achenes glabrous. Fl. Jul-Sep. River banks or by lakes at 200-1500 m. Distributed in Heilongjiang, Neimenggu and N Xinjiang. Widespread in N Asia, Europe and North America.

浮毛茛 *Ranunculus natans*

猫爪草 *Ranunculus ternatus*

猫爪草
Ranunculus ternatus Thunb.

多年生草本。块根卵球形。叶三出，有时单叶。单花顶生；花托无毛；花瓣5，顶端圆形，蜜槽无鳞片。聚合果卵球形；瘦果卵球形；花柱宿存。花期3-5月。生海拔500米以下的田间、草坡或林中。产中国东南、华中和华东。日本亦有。

Perennial herbs. Root tubers ovoid. Leaves ternate or sometimes simple. Flowers solitary, terminal; receptacles glabrous; petals 5, apex rounded, nectary pit without a scale. Aggregate fruits ovoid; achenes ovoid; styles persistent. Fl. Mar-May. Fields, grassy slopes or forests below 500 m. Distributed in SE, C and E China. Also in Japan.

石龙芮
Ranunculus sceleratus L.

一年生草本。基生叶及茎生叶3深裂。复单歧聚伞花序顶生，聚伞状；花小，直径4-8毫米；花托近圆柱状；花瓣5，蜜槽不具鳞片。聚合果圆柱形；瘦果稍两侧压扁，斜卵球形，长1-1.2毫米。花期5-8月。生海拔50-2300米的湖边、河边或湿地上。产中国各省区。亚洲、欧洲和北美洲亦有。

Annual herbs. Basal and cauline leaves 3-parted. Compound monochasium terminal, corymbose; flowers small, 4-8 mm diam; receptacles subterete; petals 5, nectary pit without a scale. Aggregate fruits cylindric; achenes slightly bilaterally compressed, obliquely obovoid, 1-1.2 mm long. Fl. May-Aug. By rivers, lakes or moist places at 50-2300 m. Widely distributed in China. Also in Asia, Europe and North America.

石龙芮 *Ranunculus sceleratus*

西南毛茛

Ranunculus ficariifolius Lévl. et Vant.

多年生草本。茎高10-30厘米。叶基生并茎生，宽卵形、正三角状卵形或卵形，边缘有少数小齿。花与叶对生；花瓣5，黄色，长4-5毫米。瘦果长1.5毫米，无毛，有小瘤状凸起。花期4-7月。生海拔1000-3200米的林边湿地或溪边。产云南西部和北部、贵州、四川、重庆南部、湖北西部、湖南南部和江西西部。尼泊尔、印度北部和泰国亦有。

Perennial herbs. Stems 10-30 cm tall. Leaves basal and cauline, broad-ovate, deltoid-ovate or ovate, margin few-denticulate. Flowers leaf-opposed; petals 5, yellow, 4-5 mm long. Achenes 1.5 mm long, glabrous, tuberculate. Fl. Apr-Jul. Meadows near forests or by streams at 1000-3200 m. Distributed in W and N Yunnan, Guizhou, Sichuan, S Chongqing, W Hubei, S Hunan and W Jiangxi. Also in Nepal, N India and Thailand.

西南毛茛 *Ranunculus ficariifolius*

毛茛 *Ranunculus japonicus*

毛茛

Ranunculus japonicus Thunb.

多年生草本。基生叶3-6，心状五角形，3深裂。复单歧聚伞花序顶生，具 (1-)3-15花；花托无毛；花瓣5，蜜槽由一鳞片包被。聚合果近球形；瘦果斜阔倒卵形。花果期4-9月。生海拔100-3500米的林缘、路边、山谷旁或湿草地。产中国大部分地区。俄罗斯(远东地区)、蒙古和日本亦有。

Perennial herbs. Basal leaves 3-6, cordate-pentagonal, 3-partite. Compound monochasium terminal, (1-)3-15-flowered; receptacles glabrous; petals 5, nectary pit covered by a scale. Aggregate fruits subglobose; achenes obliquely broadly obovate. Fl. and fr. Apr-Sep. Forest edges, roadsides, valleysides or damp grasslands at 100-3500 m. Distributed in most parts of China. Also in Russia (Far East), Mongolia and Japan.

棱喙毛茛

Ranunculus trigonus Hand.-Mazz.

多年生草本。基生叶和下部叶为3出复叶；叶片五角形至宽卵形。单歧聚伞花序顶生，具2或3花；花托具柔毛；花瓣(3-)5，蜜槽上有鳞片。聚合果近球形；瘦果扁平，斜阔倒卵球形；花柱宿存。花期3-9月。生海拔1300-3300米的草坡、湿草地、林中或沟旁。产云南、四川西南部和西藏东南部。

Perennial herbs. Basal and lower stem leaves ternate, blades pentagonal to broadly ovate. Monochasium terminal, 2- or 3-flowered; receptacles puberulent; petals (3-)5, nectary pit covered by a scale. Aggregate fruits subglobose; achenes flat, obliquely broadly obovoid; styles persistent. Fl. Mar-Sep. Grassy slopes, moist meadows, woods, or by streams at 1300-3300 m.

棱喙毛茛 *Ranunculus trigonus*

褐鞘毛茛 *Ranunculus sinovaginutus*

Distributed in Yunnan, SW Sichuan and SE Xizang.

褐鞘毛茛

Ranunculus sinovaginutus W. T. Wang

多年生草本。茎渐升或直立，高5-28厘米。基生叶具长柄，为三出复叶。单花顶生；萼片5，反折；花瓣5，黄色，长6.5-11毫米。瘦果扁平，无毛。花期4-9月。生海拔1500-3200米的草地、溪边或林中。产云南西北、四川西部、甘肃南部和陕西南部。

Perennial herbs. Stems ascending or erect, 5-28 cm tall. Basal leaves long petiolate, ternate. Flowers singly terminal; sepals 5, reflexed; petals 5, yellow, 6.5-11 mm long. Achenes flat, glabrous. Fl. Apr-Sep. Grassy places, by streams or under forests at 1500-3200 m. Distributed in NW Yunnan, W Sichuan, S Gansu and S Shaanxi.

铺散毛茛

Ranunculus diffuses DC.

多年生草本。茎渐升，高20-40厘米。基生叶和下部茎上叶具长柄，为三出复叶；茎生叶3深裂。花与叶对生；花瓣5，黄色，长5-6毫米。瘦果扁平，长2.5毫米，无毛。花期3-6月。生海拔1000-3000米的草坡、多石处或溪边。产西藏南部和东部、云南西部和北部。缅甸、印度北部、不丹、尼泊尔、巴基斯坦北部和阿富汗亦有。

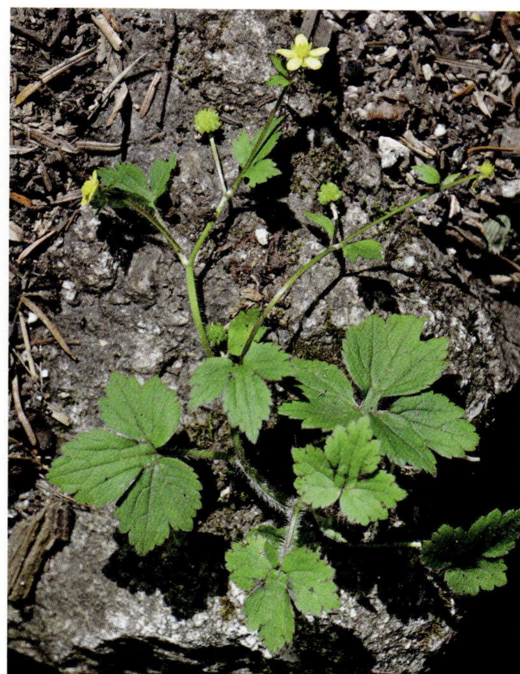
铺散毛茛 *Ranunculus diffuses*

Perennial herbs. Stems ascending, 20-40 cm tall. Basal and lower cauline leaves long petiolate, ternate; other cauline leaves 3-parted. Flowers leaf-opposed; petals 5, yellow, 5-6 mm long. Achenes flattened, 2.5 mm long, glabrous. Fl. Mar-Jun. Grassy slopes, gravelly places or by streams at 1000-3000 m. Distributed in S and E Xizang, W and N Yunnan. Also in Myanmar, N India, Bhutan, Nepal, N Pakistan and Afghanistan.

扬子毛茛

Ranunculus sieboldii Miq.

多年生草本。基生叶3-7，三出，卵形。花与叶对生；花托具柔毛；萼片5，反折，外面有硬毛；花瓣5，蜜槽被鳞片。聚合果近球形；瘦果扁平，斜倒卵球形。花期3-10月。生海拔50-2500米的草地、灌丛或河边。产中国西南、东南、华中和华东。日本亦有。

Perennial herbs. Basal leaves 3-7, ternate, ovate in outline. Flowers leaf-opposed; receptacles puberulent; sepals 5, reflexed, abaxially strigose; petals 5, nectary pits covered by scales. Aggregate fruits subglobose; achenes flat, obliquely obovoid. Fl. Mar-Oct. Grassy places, thickets or by rivers at 50-2500 m. Distributed in SW, SE, C and E China. Also in Japan.

扬子毛茛 *Ranunculus sieboldii*

禺毛茛 *Ranunculus cantoniensis*

钩柱毛茛 *Ranunculus silerifolius*

禺毛茛

Ranunculus cantoniensis DC.

多年生草本。茎高20-65厘米，被长硬毛。基生叶和下部茎生叶具长柄，为三出复叶。花序有4-10花；花瓣5，黄色，长4-7.5毫米。瘦果扁平，长约2.5毫米，宿存花柱长1毫米。花期3-9月。生海拔100-1700米的草坡、溪边或林边。云南东南部、华南至陕西南部和河南广布。朝鲜半岛、日本和尼泊尔亦有。

Perennial herbs. Stems 20-65 cm tall, hirsute. Basal and lower cauline leaves long petiolate, ternate. Compound monochasia 4-10-flowered; petals 5, yellow, 4-7.5 mm long. Achenes flattened, ca. 2.5 mm long, persistent styles 1 mm long. Fl. Mar-Sep. Grassy slopes, by streams or forest margins at 100-1700 m. Wide spread from SE Yunnan, S China to S Shaanxi and Henan. Also in Korean Peninsula, Japan and Nepal.

钩柱毛茛

Ranunculus silerifolius Lévl.

多年生草本。茎高28-95厘米，下部被长硬毛。基生叶和下部茎生叶具长柄，为三出复叶。花序有少数至多数花；花瓣5，长4-10毫米。瘦果扁平，长2-2.8毫米，宿存花柱长达1.2毫米，顶端具钩。花期4-10月。生海拔100-2500米的草坡、溪边或林中。产云南南部和东部、贵州、四川、重庆、湖北西南部、湖南、广西、广东北部、江西南部、福建和台湾。不丹、印度东北、印度尼西亚、朝鲜半岛和日本亦有。

Perennial herbs. Stems 28-95 cm tall, below hirsute. Basal and lower cauline leaves long petiolate, ternate. Compound monochasia few- to many-flowered; petals 5, yellow, 4-10 mm long. Achenes flattened, 2-2.8 mm long, persistent styles to 1.2 mm long, apex hooked. Fl. Apr-Oct. Grassy slopes, by streams or under forests at 100-2500 m. Distributed in S and E Yunnan, Guizhou, Sichuan, Chongqing, SW Hubei, Hunan, Guangxi, N Guangdong, S Jiangxi, Fujian and Taiwan. Also in Bhutan, NE India, Indonesia, Korean Peninsula and Japan.

匍枝毛茛

Ranunculus repens L.

多年生草本，具匍匐茎。茎渐升或近直立，高10-60厘米。基生叶具长柄，为三出复叶；茎生叶较小。单歧聚伞花序有2至数花；花瓣5，黄色，长7-10毫米。瘦果无毛。花期4-8月。生海拔300-3300米的草地或溪边。产云南西北部、山西、辽宁、吉林、黑龙江、内蒙古和新疆。广布亚洲北部和西部、欧洲和北美洲。

Perennial herbs, with stolons. Stems ascending or suberect, 10-60 cm tall. Basal leaves long petiolate, ternate; cauline leaves smaller. Monochasia 2- to several-flowered; petals 5, yellow, 7-10 mm long. Achenes glabrous. Fl. Apr-Aug. Meadows or by streams at 300-3300 m. Distributed in NW Yunnan, Shanxi, Liaoning, Jilin, Heilongjiang, Neimenggu and Xinjiang. Widespread in N and W Asia, Europe and North America.

茴茴蒜

Ranunculus chinensis Bunge

一年生草本。叶为三出复叶，基生叶和下部叶具长柄。复单歧聚伞花序顶生，具3至数花；花

匍枝毛茛 *Ranunculus repens*

茴茴蒜 *Ranunculus chinensis*

托具柔毛；苞片叶状；萼片5，反折；花瓣5，蜜槽由一鳞片包被。聚合果卵状圆柱形；瘦果两侧压扁，斜倒卵球形，无毛。有毒植物。花期4-9月。生海拔3000米以下的草地、草坡或河边。产中国大部分地区。南亚、中亚和东北亚亦有。

Annual herbs. Leaves ternate, basal and lower cauline ones with long petioles. Compound monochasium terminal, 3- to several flowered; receptacles puberulent; bracts leaflike; sepals 5, reflexed; petals 5, nectary pit covered by a scale. Aggregate fruits ovoid-cylindric; achenes bilaterally compressed, obliquely obovoid, glabrous. Poisonous. Fl. Apr-Sep. Grasslands, grassy slopes or by streams below 3000 m. Distributed in most parts of China. Also in S, C and NE Asia.

刺果毛茛

Ranunculus muricatus L.

一年生草本。基生叶6-9，3浅裂。花单生，与叶对生；花瓣5，蜜槽上有小鳞片。聚合果近球形；瘦果扁平，具刺。花期3-4月。生草坡或水田。产浙江、安徽和江苏。原产西亚和欧洲。

Annual herbs. Basal leaves 6-9, 3-lobed. Flowers solitary, leaf-opposed; petals 5, nectary pits covered by small scales. Aggregate fruits subglobose; achenes complanate, spiny. Fl. Mar-Apr. Grassy places or paddy fields. Distributed in Zhejiang, Anhui and Jiangsu. Native to W Asia and Europe.

刺果毛茛 *Ranunculus muricatus*

鸦跖花

Oxygraphis glacialis (Fisch. ex DC.) Bunge

多年生小草本。叶卵形或倒卵形。花葶1-8，长于2厘米，果期伸长至10厘米；花单生，直径1.5-3厘米；萼片5，近革质；花瓣12-19，黄色。聚合果近球形；瘦果狭倒卵形，每侧有1纵肋。花期4-9月，果期6-10月。生海拔2700-5000米的草地或高山草甸、河边。产中国西南和西北。印度北部、尼泊尔、不丹、哈萨克斯坦、俄罗斯(西伯利亚)和蒙古亦有。

Small perennial herbs. Leaves ovate or obovate. Scapes 1-8, more than 2 cm long, elongating to 10 cm in fruit; flowers solitary, 1.5-3 cm diam; sepals 5, subleathery; petals 12-19, yellow. Aggregate fruits subglobose; achenes narrowly obovate, on each side 1-ribbed. Fl. Apr-Sep. Fr. Jun-Oct. Grasslands or alpine meadows, by rivers at 2700-5000 m. Distributed in SW and NW China. Also in N India, Nepal, Bhutan, Kazakhstan, Russia (Siberia) and Mongolia.

鸦跖花 *Oxygraphis glacialis*

脱萼鸦跖花

Oxygraphis delavayi Franch.

草本。基生叶3-5，圆卵形。花葶1-3，超过4厘米，果期伸长至15厘米；花单生，或2-3成单岐聚伞花序；萼片5，纸质，无毛，脱落；花瓣5-10，白色。聚合果宽卵球形。花期4-8月，果期7-8月。生海拔3500-5000米高山草甸、草坡或砾石地。产云南西北部、四川西北部和西藏东南部(波密县)。

Herbs. Basal leaves 3-5, orbicular-ovate. Scapes 1-3, more than 4 cm long, elongating to 15 cm in fruit stage; flowers solitary, or 2-3 in a monochasium; sepals 5, papery, glabrous, deciduous; petals 5-10, white. Aggregate fruits broadly ovoid. Fl. Apr-Aug. Fr. Jul-Aug. Alpine meadows, grassy slopes, or gravelly places at 3500-5000 m. Distributed in NW Yunnan, NW Sichuan and SE Xizang (Bomê County).

脱萼鸦跖花 *Oxygraphis delavayi*

北京水毛茛 *Batrachium pekinense*

碱毛茛 *Halerpestes sarmentosa*

北京水毛茛
Batrachium pekinense L. Liou

多年生沉水草本。茎长30厘米或更长。沉水叶片三至四回细裂，裂片丝形；浮水叶片约三回细裂，末回裂片披针状条形。花直径1-1.3厘米；花瓣5，白色。瘦果长约1毫米，有7条横皱。花期5-8月。生海拔100-500米山地或平原溪水中。产北京和内蒙古南部。

Perennial submerged herbs. Stems 30 cm or more. Submerged leaf blades 3-4 times dissected, with filiform lobules; floating leaf blades ca. 3 times dissected, with lanceolate-linear ultimate lobules. Flower 1-1.3 cm diam; petals 5, white. Achenes ca. 1 mm long, with 7 transverse wrinkles. Fl. May-Aug. Streams on plain or montane regions at 100-500 m. Distributed in Beijing and S Neimenggu.

水毛茛
Batrachium bungei (Steud.) L. Liu

多年生沉水草本。叶片扇形或半圆形，3全裂；裂片4或5回细裂，末回裂片丝形。花托圆锥状；花瓣（4或)5，白色，基部黄色。聚合果近球形或阔卵球形；瘦果斜倒卵球形，具约6个横向皱纹。花期5-8月，果期7-8月。生海拔3800米以下的池塘、湖、溪和湿地中。产中国西南、华中、华北、华东和华西。

Perennial submerged herbs. Leaves flabellate or semicircular, 3-sect; segments 4 or 5 times dissected, ultimate lobules filiform. Receptacles conical; petals (4 or) 5, white with yellow base. Aggregate fruits subglobose or broadly ovoid; achenes obliquely obovoid, with ca. 6 transverse wrinkles. Fl. May-Aug. Fr. Jul-Aug. Ponds, lakes, streams or swamps below 3800 m. Distributed in SW, C, N, E and W China.

碱毛茛
Halerpestes sarmentosa (Adans) Kom. et Aliss.

多年生小草本，具匍匐茎。叶基生，具长柄，圆卵形，长0.5-2.5厘米，具圆齿。花葶高3-16厘米；花单朵顶生或2-4组成花序；花瓣5，黄色。瘦果有3-5条纵肋。花期5-8月。生海拔10-3500米的盐碱性沼泽或湖边。产西藏、四川北部、中国西北部、华北和东北。哈萨克斯坦、蒙古、俄罗斯(西伯利亚)和朝鲜半岛亦有。

Small perennial herbs, with stolons. Leaves basal, long petiolate, orbicular-ovate, 0.5-2.5 cm long, rounded-dentate. Scapes 3-16 cm tall; flowers singly terminal or 2-4 in monochasium; petals 5, yellow. Achenes longitudinally 3-5-ribbed. Fl. May-Aug. Saline swamps or by lakes at 10-3500 m. Distributed in Xizang, N Sichuan, and NW, N, NE China. Also in Kazakhstan, Mongolia, Russia (Siberia) and Korean Peninsula.

长叶碱毛茛
Halerpestes ruthenica (Jacq.) Ovcz.

多年生草本。匍匐茎长15-65厘米。叶片卵状梯形或宽长圆形，顶端有3-6齿。花葶1-4；花1朵，有时2朵组成单歧聚伞花序；萼片绿色，5；花瓣黄色，6-12枚。聚合果卵球形。花期5-8月，果期8月。生海拔1400米以下的湿草地或水边。产华北、华西、西北和东北。

水毛茛 *Batrachium bungei*

变叶三裂碱毛茛 *Halerpestes tricuspis* var. *variifolia*

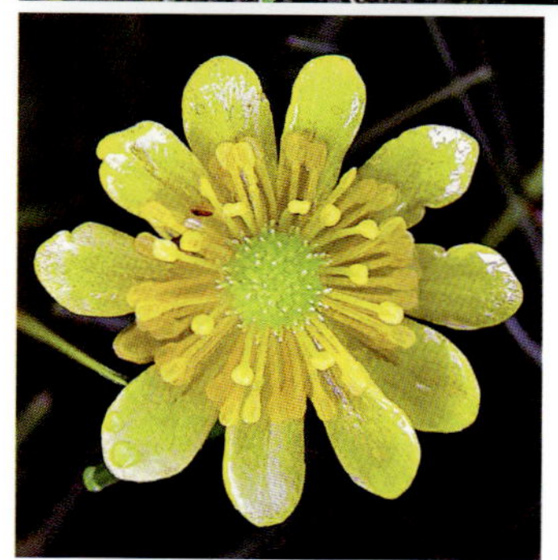

长叶碱毛茛
Halerpestes ruthenica

哈萨克斯坦、俄罗斯(西伯利亚)和蒙古亦有。

Perennial herbs. Stolons 15-65 cm long. Leaves ovate-trapeziform or broadly oblong, apically 3-6-dentate. Scapes 1-4; flowers 1, sometimes 2 in monochasium; sepals green, 5; petals yellow, 6-12. Aggregate fruits ovoid. Fl. May-Aug. Fr. Aug. Damp grasslands or by waters below 1400 m. Distributed in N, W, NW and NE China. Also in Kazakhstan, Russia (Siberia) and Mongolia.

变叶三裂碱毛茛

Halerpestes tricuspis Hand.-Mazz. var. **variifolia** (Tamura) W. T. Wang

多年生小草本，具匍匐茎，基生叶5-11，无毛；叶片五角形，宽达1.2厘米，3深裂或3全裂；叶柄长达6厘米。花葶长约5厘米，具1花；萼片5，狭卵形；花瓣5-6，黄色，长5-7毫米；雄蕊和心皮多数。瘦果长约2毫米，每侧有2-3条纵肋。花期5-7月。生海拔2000-5000米的草甸或沼泽上或沟边。产西藏南部、四川西南部、甘肃西南部和宁夏。尼泊尔亦有。

Small perennial herbs, with long stolons. Basal leaves 5-11, glabrous; blades pentagonal, up to 1.2 cm wide, 3-parted or 3-sect; petioles up to 6 cm long. Scapes ca. 5 cm long, 1-flowered; sepals 5, narrowly ovate; petals 5-6, yellow, 5-7 mm long; stamens and carpels numerous. Achenes ca. 2 mm long, on each side with 2-3 longitudinal ribs. Fl. May-Jul. Meadows or swamps, or by streams at 2000-5000 m. Distributed in S Xizang, SW Sichuan, SW Gansu and Ningxia. Also in Nepal.

角果毛茛

Ceratocephala testiculata (Crantz) Roth

一年生小草本。叶基生，最外4叶无柄，条形，其他叶具柄，一至二回三全裂，末回裂片狭条形。花葶高达5.8厘米；单花顶生；花瓣5，淡黄色，长4-6毫米。瘦果长约1.8毫米，基部有2突起，宿存花柱刺状，长约4毫米。花期4-5月。生海拔600-1600米的山坡、沙漠或河岸上。产新疆。俄罗斯(西西伯利亚)、哈萨克斯坦、巴基斯坦北部、西南亚和欧洲亦有。

Small annual herbs. Leaves basal, outermost ca. 4, sessile, linear, other ones petiolate, 1-2 times 3-sect, ultimate lobules narrow-linear. Scapes to 5.8 cm tall; flowers singly terminal; petals 5, yellowish, 4-6 mm long. Achene ca. 1.8 mm long, at base with 2 protuberances, persistent style spine-like, ca. 4 mm long. Fl. Apr-May. Dry slopes, deserts or river banks at 600-1600 m. Distributed in Xinjiang. Also in Russia (W Siberia), Kazakhstan, N Pakistan, SW Asia and Europe.

角果毛茛 *Ceratocephala testiculata*

芍药科 Paeoniaceae

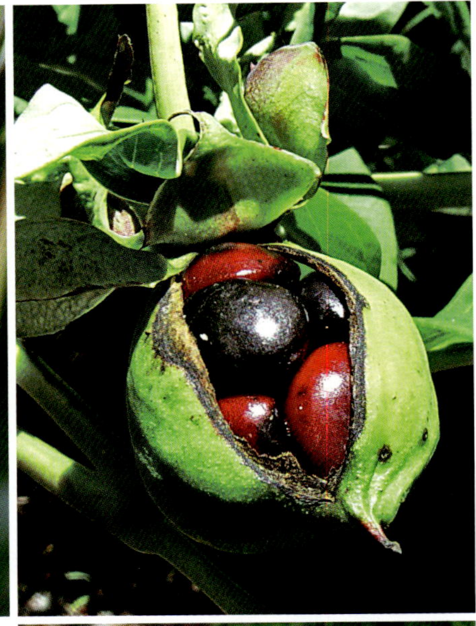

牡丹
Paeonia suffruticosa Andrews

落叶灌木。茎生叶二回三出；顶生小叶深3裂，裂片再次2或3裂。单花顶生，宽10-17厘米；萼片5，绿色，不等大；花瓣5-11，白色、粉色、红色或紫红色；花盘于花期完全包被心皮，紫红色，革质；心皮5。花期5-6月，果期8月。中国广泛栽培。全球亦广泛栽培。原产河南和安徽。

Deciduous shrubs. Proximal leaves 2-ternate; terminal leaflets deeply 3-lobed, lobes again 2- or 3-lobed. Flowers solitary, terminal, 10-17 cm wide; sepals 5, green, unequal; petals 5-11, white, pink, red, or red-purple; disc wholly enveloping carpels at anthesis, purple-red, leathery; carpels 5. Fl. May-Jun. Fr. Aug. Cultivated in most parts of China. Also commonly cultivated in the world. Native to Henan and Anhui.

滇牡丹
Paeonia delavayi Franch.

亚灌木。茎生叶二回三出；小叶常分裂；裂片再分裂。每枝具花1-3，稍下垂，单生，宽6-10厘米；花萼2-9，上面绿色，基部具粉色，或全为紫色或紫红色；花瓣(4-)7-11(-13)；心皮2-5(或6-8)。花期5-6月，果期8-9月。生海拔2000-3600米的石灰山开阔林中或石坡上。产云南、四川和西藏。

Subshrubs. Proximal leaves 2-ternate; leaflets always segmented; segments again segmented. Flowers 1-3 per shoot, ± pendulous, single, 6-10 cm wide; sepals 2-9, green with a pink base adaxially, or wholly purple or purple-red; petals (4-)7-11 (-13); carpels 2-5(or 6-8). Fl. May-Jun. Fr. Aug-Sep. Limestone open forests or rocky slopes at 2000-3600 m. Distributed in Yunnan, Sichuan and Xizang.

滇牡丹 *Paeonia delavayi*

牡丹 *Paeonia suffruticosa*

牡丹 *Paeonia suffruticosa*

大花黄牡丹 *Paeonia ludlowii*

大花黄牡丹

Paeonia ludlowii (Stern et Taylor) D.Y. Hong

丛生灌木，全体无毛。茎生叶二回三出；小叶常3裂至近基部。每枝具3-4花，腋生，宽10-12厘米；花萼3-5，绿色；花瓣开展，纯黄色。蓇葖果圆柱状。花期5月，果期8月。生海拔2900-3500米的疏林或灌丛中。产西藏东南部。

Shrubs caespitose, glabrous throughout. Proximal leaves 2-ternate; leaflets usually 3-segmented almost to base. Flowers 3-4 per shoot, axillary, 10-12 cm wide; sepals 3-5, green; petals spreading, pure yellow. Follicles cylindric. Fl. May. Fr. Aug. Sparse forests or thickets at 2900-3500 m. Distributed in SE Xizang.

草芍药

Paeonia obovata Maxim.

多年生草本。叶二回三出复叶。单花顶生，宽7-12厘米；萼片(2或)3(或4)，不等大；花瓣4-7，白色、玫瑰色、粉红色或红色，基部和边缘带粉红色。蓇葖果渐反曲，椭圆体形。花期5-6月，果期9月。生海拔200-2800米落叶阔叶林、混交林或针叶林下。产中国西南、华中、华北、华西、华东和东北。俄罗斯(远东地区)、朝鲜半岛和日本亦有。

Perennial herbs. Leaves 2-ternate. Flowers solitary, terminal, 7-12 cm wide; sepals (2 or)3(or 4), unequal; petals 4-7, white, rose-colored, pink or red, pinkish at

草芍药 *Paeonia obovata*

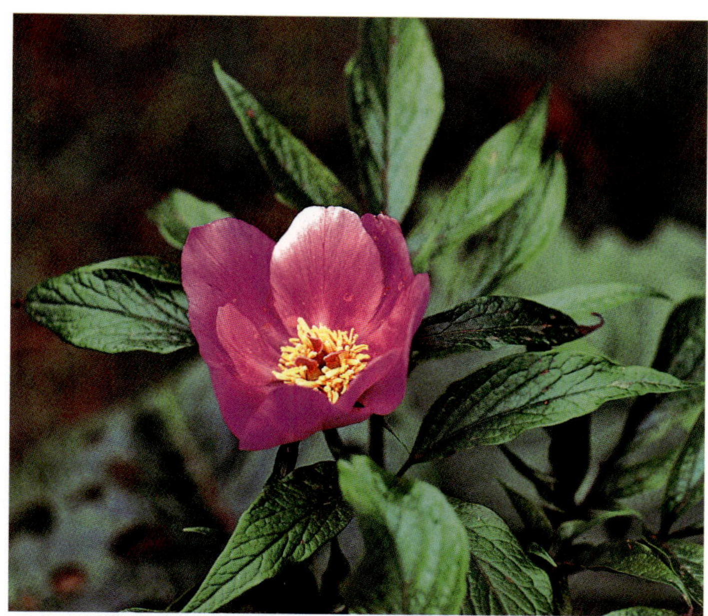

美丽芍药 *Paeonia mairei*

base and margin. Follicles gradually recurved, ellipsoid. Fl. May-Jun. Fr. Sep. Deciduous broad-leaved, mixed or coniferous forests at 200-2800 m. Distributed in SW, C, N, W, E and NE China. Also in Russia (Far East), Korean Peninsula and Japan.

美丽芍药

Paeonia mairei Lévl.

多年生草本。叶二回三出复叶；苞片1-3，叶状或线形，长可达9厘米。单花顶生，宽7.5-14厘米；花瓣7-9，粉红至红色，长3.5-7厘米，宽2-4.5厘米；花丝紫红色；花盘黄色，环形；心皮2或3。花期4-5月，果期8月。生海拔1500-2700米落叶阔叶林中。产云南、四川、贵州、湖北、陕西和甘肃。

Perennial herbs. Leaves 2-ternate; bracts 1-3, leaflike or linear, up to 9 cm long. Flowers solitary, terminal, single, 7.5-14 cm wide; petals 7-9, pink to red, 3.5-7 cm long, 2-4.5 cm wide; filaments purple-red; disc yellow, annular; carpels 2 or 3. Fl. Apr-May. Fr. Aug. Deciduous broad-leaved forests at 1500-2700 m. Distributed in Yunnan, Sichuan, Guizhou, Hubei, Shaanxi and Gansu.

芍药

Paeonia lactiflora Pall.

多年生草本。茎生叶常二回三出；顶生小叶2或3次分裂；小叶和裂片多至15。花通常几朵，顶生或腋生，有时只有顶端1朵发育；单瓣(野生的)，重瓣(栽培的)，宽8-13厘米；苞片4-5；花瓣9-13，白色或粉红色(野生的)或颜色有变化(栽培的)。花期5-6月，果期8月。生海拔400-2300米林中或草地。产华北、华西和东北。俄罗斯(西伯利亚和远东地区)、蒙古、朝鲜半岛和日本亦有。

Perennial herbs. Proximal leaves 2-ternate; terminal ones often 2- or 3-segmented; leaflets and segments up to 15. Flowers usually several per shoot, terminal or axillary, sometimes only terminal one developed, single (wild) or double (cultivated), 8-13 cm wide; bracts 4-5; petals 9-13, white or pink (wild), or varying in color (cultivated). Fl. May-Jun. Fr. Aug. Forests or grasslands at 400-2300 m. Distributed in N, W and NE China. Also in Russia (Siberia and Far East), Mongolia, Korean Peninsula and Japan.

芍药 *Paeonia lactiflora*

新疆芍药 (窄叶芍药)
Paeonia anomala L.

多年生草本。近轴叶二回三出；小叶羽状分裂。单花顶生，花红色，稀于远处叶腋具1或2个未开花芽；萼片3或4；花瓣6-9，玫瑰色至红色；花盘黄色；心皮2-5，疏或密生黄褐色硬毛或柔毛。花期4-7月，果期8-9月。生海拔1200-1800米的沟谷中杨木林或针叶林。产新疆北部。哈萨克斯坦东北部、蒙古北部和俄罗斯(欧洲部分东北部和西伯利亚)亦有。

Perennial herbs. Proximal leaves 2-ternate; leaflets pinnately segmented, . Flowers solitary, terminal, flowers red, both terminal and axillary, rarely 1 or 2 underdeveloped flower buds also present in axils of distal leaves; sepals 3 or 4; petals 6-9, rose to red; disc yellow; carpels 2-5, sparsely to densely brown-yellow hispid or hirsute. Fl. Apr-Jul. Fr. Aug-Sep. *Populus* or coniferous forests in valleys at 1200-1800 m. Distributed in N Xinjiang. Also in NE Kazakhstan, N Mongolia and Russia (NE European parts and Siberia).

川赤芍
Paeonia anomala L. subsp. **veitchii** (Lynch) D.Y. Hong et K.Y. Pan

本亚种与新疆芍药的区别在于本亚种的花每枝（1或)2-4，顶生及腋生，常于远处叶腋具1-3朵未开花芽。花期4-7月，果期8-9月。生海拔1800-3900米的林缘、草地、灌丛或阳坡上。产中国西南、华北和华西。

This subspecies differs from the typical subspecies in its flowers (1 or)2-4 per shoot, both terminal and axillary, usually 1-3 underdeveloped flower buds also present in axils of distal leaves. Fl. Apr-Jul. Fr. Aug-Sep. Forests edges, grasslands, thickets or sunny slopes at 1800-3900 m. Distributed in SW, N and W China.

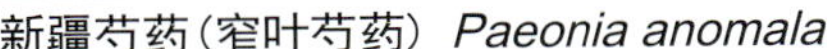
新疆芍药(窄叶芍药) *Paeonia anomala*

川赤芍 *Paeonia anomala* subsp. *veitchii*

木通科 Lardizabalaceae

木通
Akebia quinata (Houtt.) Decne.

木质藤本，攀援状。掌状复叶通常具5小叶。花略芳香，雌花远比雄花大，花被片3，很少为4或5，卵圆形。果矩圆状至椭圆状，肉质果沿腹缝线开裂。花期4-5月，果期6-8月。生海拔300-1500米的林缘、溪边或山坡。产长江中下游地区。朝鲜半岛和日本亦有。

Woody climbers. Leaf blade palmately 5-foliolate. Flowers slightly fragrant, female flowers larger than male flowers, tepals 3, rarely 4 or 5. Fruits oblong to ellipsoid. Fl. Apr-May. Fr. Jun-Aug. Forest edges, streamsides or slopes at 300-1500 m. Distributed in the middle and lower reaches of Yangtze River. Also in Korean Peninsula and Japan.

白木通
Akebia trifoliata (Thunb.) Koidz. subsp. **australis** (Diels) T. Shimizu

木质藤本，攀援。掌状复叶通

木通 *Akebia quinata*

白木通 *Akebia trifoliata* subsp. *australis*

常具3(-5)小叶，小叶革质，全缘或浅波状。雌花远比雄花大，花被片3，很少为4或5，卵圆形。肉质果沿腹缝线开裂。花期4-5月，果期6-9月。生海拔300-2100米的沟谷疏林或灌丛中。产中国西南、华南、华中、华东和华北。

Woody climbers. Leaflets 3(-5), leathery, margin usually entire. Female flowers larger than male flowers, tepals 3, rarely 4 or 5. Fruits fleshy. Fl. Apr-May. Fr. Jun-Sep. Open forests along valleys, or shrubs on slopes at 300-2100 m. Distributed in SW, S, C, E and N China.

三叶木通
Akebia trifoliata (Thunb.) Koidz.

木质藤本，攀援。小叶3，纸质至近革质，边缘深波至浅裂状。花被片3，很少为4或5，卵圆形；雌花远比雄花大；雄花萼片宽椭圆形至椭圆形，淡紫色；雌花萼片3，近圆形，紫褐色或暗紫色；心皮6-9。果成熟时灰白色或微紫色，长圆形，沿腹缝线开裂。花期4-5月，果期7-8月。生海拔200-2100米的峡谷疏林下或山脚灌丛中。产华北、华中和华东。日本亦有。

Woody climbers. Leaflets 3, papery to subleathery, margin sinuate to slightly lobed. Female flowers larger than male flowers; male flowers: sepals 3, pale purple, broadly elliptic to elliptic; famale flowers: sepals 3, purplish brown or dark purple, suborbicular; carpels 6-9. Fruits grayish white or slightly purple at maturity, oblong. Fl. Apr-May. Fr. Jul-Aug. Open forests along valleys or scrubby areas on hillsides at 200-2100 m. Distributed in N, C and E China. Also in Japan.

猫儿屎
Decaisnea insignis (Griff.) Hook. f. et Thoms.

直立灌木，落叶，高2-5米。冬芽大、卵形，具2枚大型鳞片；叶痕大而明显；髓大，白色。羽状复叶着生茎顶，大，长50-90厘米；小叶13-25，对生，叶下表具易脱落的单细胞柔毛。大型圆状花序顶生或腋生，大，长6-30厘米；雌花与雄花等大，淡绿色，萼片6，2轮，披针形，内面被微柔毛；花瓣缺；花丝合生成长筒，药隔角状体显著。浆果圆柱状，稍弯曲，成熟时蓝色或蓝紫色，外覆白粉，果皮表面具颗粒状小突起，开裂或偶有不裂。生海拔900-3600米的山坡灌丛或山谷杂木林中。花期4-6月，果期7-8月。产中国西南至华中。印度东北部、不丹、尼泊尔和缅甸亦有。

Shrubs, stems erect. Leaves odd-pinnate, with 13-25 leaflets. Panicles terminal or axillary, 6-30 cm long; flowers greenish, sepals 6 in two whorls, petals absent. Fruits pendulous, cylindric, bluish black. Fl. Apr-Jun. Fr. Jul-Aug. Scrubby areas on slopes or mixed forests of valleys at 900-3600 m. Distributed in SW to C China. Also in NE India, Bhutan, Nepal and Myanmar.

三叶木通 *Akebia trifoliata*

猫儿屎 *Decaisnea insignis*

五月瓜藤 *Holboellia angustifolia*

五月瓜藤
Holboellia angustifolia Wall.

木质藤本，长2-6(-8)米，落叶。茎有细纵纹。掌状复叶3-7小叶，小叶形态变异较大，通常外狭长圆形或披针形，大小变化也很大，长为宽2倍以上，上面绿色，下面灰绿色，两面侧脉不明显。伞房花序数个簇生叶腋，花芳香，吊钟状；雄花乳白色；雌花紫色，较大，萼片6，外轮较大，退化花瓣鳞片状，极小。果紫红色，长圆形，干后表面常结肠状，不开裂。花期4-5月，果期9-10月。生海拔500-3000米的林缘或灌丛。产中国西南、东南和华中。不丹、尼泊尔、印度北部和缅甸亦有。

Woody climbers. Leaflets 3-7, ovate-elliptic, more than 2 × as long as wide, apex acute. Corymbs clustered and axillary, flowers fragrant; female flowers purple, sepals 6. Fruits purple at maturity. Fl. Apr-May. Fr. Sep-Oct. Forest edges or thickets at 500-3000 m. Distributed in SW, SE and C China. Also in Bhutan, Nepal, N India and Myanmar.

鹰爪枫
Holboellia coriacea Diels

木质藤本。掌状3小叶，叶厚革质，上面深绿色，有光泽，下面粉绿色，基出3脉。伞房状花序具5-8花；雄花：白色或淡紫色，萼片6，两轮；雌花紫红色，心皮3。果实成熟时淡紫色。花期4-5月，果期8-9月。生海拔400-1800米的山谷、溪边和山坡灌丛中。广布于黄河以南。

Woody climbers. Leaf blade palmately 3-foliolate, thick leathery, abaxially farinaceous green, adaxially dark green and shiny, primary veins 3 from base. Racemes corymb-like, flowers 5-8; male flowers: white or purplish, sepals 6 in two whorls; female flowers: purple, carpels 3. Fruits purplish at maturity. Fl. Apr-May. Fr. Aug-Sep. Valleys, streamsides and bushes on mountain slopes at 400-1800 m. Distributed in the south of Yellow River.

八月瓜
Holboellia latifolia Wall.

木质攀援藤本，雌雄同株，常

鹰爪枫 *Holboellia coriacea*

八月瓜 *Holboellia latifolia*

绿，长3-10米。茎圆柱形，干后灰白色或棕色。枝有线纹。掌状3-9小叶，小叶常卵状长圆形或卵圆形，基部宽楔形至钝圆，顶端常渐尖偶为钝圆形或急尖，上面亮绿色，下面淡绿色，革质。伞房花序数个腋生，芳香；雄花吊钟型，花乳白色；雌花较大，卵圆形，淡紫色。果实圆柱形或卵圆形，偶为结肠状，成熟时紫红色。花期4-5月，果期7-9月。生海拔600-2600米的山坡或山谷阔叶林林缘。产云南、西藏东南部和南部、四川南部和贵州西部。尼泊尔、印度东北部、不丹、孟加拉国北部和缅甸北部亦有。

Woody climbers. Leaves palmately compound, 3-9-foliolate; leaflets ovate or oblong, apex acuminate. Flowers several in corymbose raceme. Fruits cylindric, purple at maturity. Fl. Apr-May. Fr. Jul-Sep. Slopes or dense forest edges in valleys at 600-2600 m. Distributed in Yunnan, SE and S Xizang, S Sichuan and W Guizhou. Also in Nepal, NE India, Bhutan, N Bangladesh and N Myanmar.

串果藤

Sinofranchetia chinensis (Franch.) Hemsl.

木质藤本，落叶。幼枝表面有白粉。冬芽具数枚鳞片，覆瓦状排列。掌状3小叶密集生于短枝，幼时淡红色，叶全缘或有时浅波状；顶生小叶菱状倒卵形，叶柄长达20厘米，基部宽楔形；侧生小叶较小，基部偏斜，叶柄极短，叶纸质。总状花序纤细，下垂，长11-29厘米；花小，近无柄，雌雄花等大，扁圆形，直径约5毫米，淡绿色；雌蕊具3心皮。浆果较小，椭圆状球形，长约2厘米，淡紫蓝色，含种子数粒。种子小、卵形、压扁，种皮棕黄色，无光泽。花期5-6月，果期9-10月。生海拔900-2450米的山坡阔叶林内或林缘或山沟灌丛中。产陕西、甘肃、云南、四川、广东、湖南、湖北和江西。

Woody climbers. Leaves usually fascicled, pinnately 3-foliolate. Racemes pendulous; male flowers: petals 6, fleshey, subobcordate, less than 1 mm; female flowers: petals very minute; carpels 3. Fruits pale purplish blue, ellipsoid. Fl. May-Jun. Fr. Sep-Oct. Dense forests along valleys, forest edges, or among shrubs at 900-2450 m. Distributed in Shaanxi, Gansu, Yunnan, Sichuan, Guangdong, Hunan, Hubei and Jiangxi.

串果藤 *Sinofranchetia chinensis*

三叶野木瓜

Parvatia brunoniana (Wall.) Decne.

木质藤本，常绿。幼枝绿色或淡红色，老茎皮灰白色，栓质化，不规则分裂。3小叶，叶背脉较明显。总状花序几个聚生叶腋，花小，黄白色，有香气；雌花比雄花稍大；萼片6，卵状披针形；花瓣6，较大；雄蕊花丝连合成筒，药隔突起比花丝筒长；心皮柱头圆锥形，延长状。果实圆球形，不开裂，成熟时表面粗超，黄褐色或灰白色。花期11月。生海拔900-1500米的山坡、山谷溪边疏林中或林缘和灌丛中。产四川南部和云南西南部。印度东北部、缅甸中部、泰国北部和越南北部亦有。

Woody climbers. Leaves palmately 3-foliolate. Racemes 2-5, many-flowered; female flowers larger than male flowers, sepals 6, petals 6, filaments fused. Fruits obovoid-oblong. Fl. Nov. Forests on mountain slopes at 900-1500 m. Distributed in S Sichuan and SW Yunnan. Also in NE India, C Myanmar, N Thailand and N Vietnam.

三叶野木瓜 *Parvatia brunoniana*

野木瓜

Stauntonia chinensis DC.

木质藤本，常绿。老枝干后灰白色或灰褐色。小叶通常5-7，长圆形，先端长渐尖，干时常歪斜；老叶革质，干后上面有光泽，下面具浅色斑点。通常3-5朵排成伞房花序，总花梗纤细，基部托以大型苞片；花淡黄色或乳白色，内面有紫斑；雌花较大，外轮萼片卵形长尖，长达2.5厘米，宽1厘米。果实椭圆形，熟时橙黄色，不开裂。花期4-5月，果期6-10月。生海拔500-1300米的常绿阔叶林下，山谷、溪边和路边的灌木丛中。产云南南部和东南部、广西东南部、广东北部、海南、福建南部和香港。老挝和越南北部亦有。

Woody climbers. Leaves palmately compound, 5-7-foliolate, leathery, adaxially shiny. Flowers monoecious; petals 6, ligulate, ca. 1.5 mm long. Fruits ellipsoid, orange at maturity. Fl. Apr-May. Fr. Jun-Oct. Dense forests on mountain slopes, open forests by streams along valleys, or in thickets at 500-1300 m. Distributed in S and SE Yunnan, SE Guangxi, N Guangdong, Hainan, S Fujian and Hong Kong. Also in Laos and N Vietnam.

大血藤

Sargentodoxa cuneata (Oliver) Rehd. et Wils.

落叶缠绕藤本，通常近十余米长。当年枝条暗红色，老树皮常常纵裂。三出复叶，稀为单叶，中间小叶近棱状倒卵圆形，侧生小叶斜卵形。总状花序长，雌花比雄花稍大，苞片矩圆形，干膜质；萼片长圆形，花瓣状，长达1厘米，退化花瓣圆形，长约1毫米，蜜腺性；雌蕊多数，螺旋状生于卵状突起的花托上。浆果多数，近球形，直径约1厘米，成熟时黑蓝色。种子卵球形，长约5毫米，黑色，光亮，平滑；种脐显著。花期4-5月，果期6-9月。生海拔400-

野木瓜 *Stauntonia chinensis*

大血藤 *Sargentodoxa cuneata*

倒卵叶野木瓜 *Stauntonia obovata*

1600米的山坡灌丛、疏林和林缘。产华南、华中至东南。老挝和越南北部亦有。

Woody climbers. Leaves ternate, rarely simple, deciduous. Racemes long; female flowers slightly larger than male flowers, sepals 6, petaloid, carpels numerous. Aggregate fruits blackish blue at maturity. Fl. Apr- May. Fr. Jun-Sep. Thickets on mountain slopes, sparse forests or forest edges at 400-1600 m. Distributed in S, C to SE China. Also in Laos and N Vietnam.

倒卵叶野木瓜
Stauntonia obovata Wu

木质藤本。小叶5-7，通常倒卵形，薄革质，下面干后灰白色。花白色或黄白色，小。雌花比雄花略大；萼片披针或线形，长6-10毫米；心皮3。果实椭圆形，成熟时黄色。生海拔300-1500米的山谷、溪边、山坡灌丛中及疏林边缘。产长江以南。

Woody climbers. Leaf blade palmately 5-7, leaflets usually obovate, subleathery, abaxially farinaceous whitish green. Flowers white or yellowish, small; female flowers slightly larger than male flowers; sepals lanceolate or linear, 6-10 mm long; carpels 3. Fruits ellipsoid, yellow at maturity. Valleys, streamsides, bushes on mountain slopes and forest edges at 300-1500 m. Distributed in the south of Yangtze River.

尾叶那藤
Stauntonia obovatifoliola Hayata

木质藤本。5-7(-9)小叶，通常倒卵形，薄革质。伞房花序数个簇生叶腋，苞片大，宽卵形或长圆形。雌花比雄花大；萼片淡黄白色。果实椭圆形，成熟时淡黄色。生山谷、溪旁和山坡杂木林等阴湿处。产江苏南部、浙江、安徽南部、江西、福建、广东、台湾、广西、湖南、湖北、贵州和云南。

Woody climbers. Leaf blade palmately 5-7(-9), leaflets usually obovate, subleathery. Corymbs clustered and axillary, bracts large, ovate or oblong. Female flowers larger than male flowers; sepals yellowish white. Fruit ellipsoid, yellowish at maturity. Valleys, streamsides, forests on mountain slopes. Distributed in S Jiangsu, Zhejiang, S Anhui, Jiangxi, Fujian, Guangdong, Taiwan, Guangxi, Hunan, Hubei, Guizhou and Yunnan.

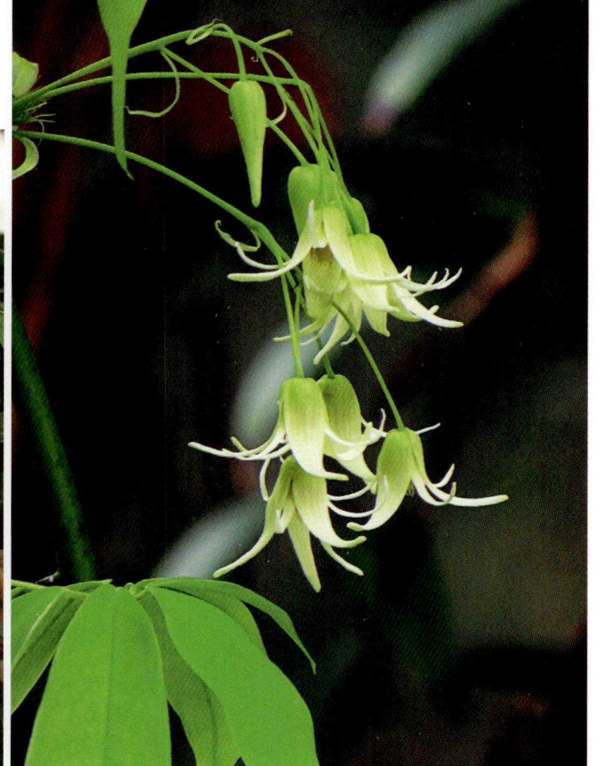
尾叶那藤 *Stauntonia obovatifoliola*

小檗科
Berberidaceae

南天竹 *Nandina domestica*

南天竹
Nandina domestica Thunb.

常绿灌木。叶为二至三回羽状复叶，长30-50厘米；小叶全缘。花序直立，长20-35厘米；花小，白色，具芳香。浆果红色或紫色。花期3-6月，果期5-11月。产中国西南、华南、华东、华中和华北。日本亦有；北美洲东南部有栽培。

Evergreen shrubs. Leaves compound, bipinnate to tripinnate, 30-50 cm long; leaflets entire. Inflorescences erect, 20-35 cm long; flowers small, white, fragrant. Berries red or purplish. Fl. Mar-Jun. Fr. May-Nov. Distributed in SW, S, SE, C and N China. Also in Japan; cultivated in SE North America.

刺红珠
Berberis dictyophylla Franch.

落叶灌木。茎刺3分叉，有时单生。叶革质，全缘，侧脉与网脉两面显著可见。花单生；萼片2轮；花瓣先端全缘；胚珠3或4；浆果红色，具白粉，柱头宿存。花期5-6月，果期7-9月。生海拔2500-4000米的林缘、草坡或山坡灌丛中。产云南、西藏和四川。尼泊尔和印度亦有。

Deciduous shrubs. Spines 3-fid, sometimes single. Leaves coriaceous, entire, veins obvious on both surfaces. Flowers solitary; sepals in 2 whorls; petals apex entire; ovules 3 or 4. Berries red, pruinose, styles persistent. Fl. May-Jun. Fr. Jul-Sep. Forest edges, grassy slopes, or thickets on slopes at 2500-4000 m. Distributed in Yunnan, Xizang and Sichuan. Also in Nepal and India.

刺红珠 *Berberis dictyophylla*

木里小檗
Berberis muliensis Ahrendt

落叶灌木。刺3叉，细弱，有时缺无。叶纸质，边缘平展全缘或具刺齿。花单生；萼片2轮；花瓣先端具缺凹缺；胚珠3或4。浆果较大，红色，卵球形或长圆状卵球形，长10-14毫米，柱头不宿存。花期7-8月，果期9-10月。生海拔2800-4300米的灌丛。产中国西南。缅甸亦有。

Deciduous shrubs. Spines 3-fid, slender, sometimes absent. Leaves papery, margin applanate-entire

木里小檗 *Berberis muliensis*

or spinose-serrate. Flowers solitary; sepals in 2 whorls; petals apex shortly emarginate; ovules 3 or 4. Berries large, red, ovoid or oblong-ovoid, 10-14 mm long, styles not persistent. Fl. Jul-Aug. Fr. Sep-Oct. Thickets at 2800-4300 m. Distributed in SW China. Also in Myanmar.

疣枝小檗

Berberis verruculosa Hemsl. et Wils.

常绿灌木。枝条棕黄色，圆柱形，密生疣点。叶椭圆形至倒卵状椭圆形，背面被白粉。花常单生，黄色。浆果紫黑色，卵形，被白粉。花期5-6月，果期7-12月。生海拔1900-3200米的山地灌丛及林下。产四川、云南和甘肃。

Evergreen shrubs. Branches brownish yellow, terete, densely verruculose. Leaves elliptic to obovate-elliptic, abaxially pruinose. Flowers usually solitary, yellow. Berries purplish black, ovoid, pruinose. Fl. May-Jun. Fr. Jul-Dec. Thickets and forests on mountains at 1900-3200 m. Distributed in Sichuan, Yunnan and Gansu.

西伯利亚小檗

Berberis sibirica Pall.

落叶灌木。刺3-9分叉，细弱；叶纸质，每侧具4-7个硬直刺状牙齿。花单生，花瓣基部有2腺体；萼片2轮，花瓣先端具短凹缺；胚珠5-8。浆果红色，倒卵形，不具白粉，柱头不宿存。花期5-7月，果期8-9月。生海拔1450-3000米高山碎石坡、林下、陡峭山坡或荒漠地区。产华北、东北和西北。蒙古和俄罗斯(西伯利亚)亦有。

Deciduous shrubs. Spines 3-9-fid, slender. Leaves papery, margin coarsely 4-7-aristate-dentate on each side. Flowers solitary: sepals in 2 whorls; petals with 2 glands at base, apex shortly emarginate; ovules 5-8. Berries red, obovoid, not pruinose, styles not persistent. Fl. May-Jul. Fr. Aug-Sep. Rocky slopes, under forests, precipitous slopes, or wilderness at 1450-3000 m. Distributed in N, NE and NW China. Also in Mongolia and Russia (Siberia).

疣枝小檗 *Berberis verruculosa*

西伯利亚小檗 *Berberis sibirica*

卷叶小檗 *Berberis replicata*

卷叶小檗
Berberis replicata W. W. Sm.

常绿灌木。叶长圆状椭圆形，背面苍白色，叶缘向背面反卷，每边具1-8刺齿。花2-10朵簇生，花黄色。浆果长圆形至倒卵状椭圆形，紫黑色。种子1-4。花期3-5月，果期6-12月。生海拔1700-2100米的山地灌丛、松林及林缘。产云南(腾冲县)。

Evergreen shrubs. Leaves oblong-elliptic, abaxially pruinose, margin revolute, with 1-8 spinescent teeth on each side. Flowers 2-10 fascicled, yellow. Berries purplish black, oblong to oboval ellipsoid. Seeds 1-4. Fl. Mar-May. Fr. Jun-Dec. Thickets, Pinus forests and sparse forests on mountains at 1700-2100 m. Distributed in Yunnan (Tengchong County).

金花小檗
Berberis wilsoniae Hemsl.

半常绿灌木。刺3叉，偶生或缺。叶革质，网脉闭锁状，全缘或偶有1-2细刺齿。花4-7朵簇生；萼片2轮；胚珠3-5；浆果粉红色，近球形，稍具白粉，柱头明显宿存。花期7-9月，果期翌年1-2月。生海拔1000-4000米的山坡、灌丛、林缘或路边。产云南、四川、西藏和甘肃。

Semi-evergreen shrubs. Spines 3-fid, sometimes single or absent. Leaves leathery, with closed netted venation, margin entire or occasionally with 1-2 -spinose-serrate on each side. Flowers 4-7 fascicled; sepals in 2 whorls; ovules 3-5. Berries pink, subovoid, slightly pruinose, styles distinctly persistent. Fl. Jul-Sep. Fr. next Jan-Feb. Slopes, thickets, forest edges or roadsides at 1000-4000 m. Distributed in Yunnan, Sichuan, Xizang and Gansu.

西山小檗
Berberis wangii Schneid.

常绿灌木。叶椭圆形或椭圆状披针形，长6-10厘米，边缘每边具5-18细刺齿；叶柄长2-4毫米。花6-17朵簇生；花梗长10-20毫米；萼片3轮，外萼片长三角形；花瓣长椭圆形，先端全缘；胚珠单生。浆果椭圆形，红色，长7-8毫米，顶端具明显宿存花柱，不被白粉。花期3月，果期8-11月。生海拔1600-2300米的山坡灌丛中、混交林中或沙石坡。产云南。

Evergreen shrubs. Leaves elliptic or elliptic-lanceolate, 6-10 cm long, margin 5-18-spinose-serrate on each side; petiole 2-4 mm long. Flowers 6-17 fascicled; pedicels 10-20 mm long; sepals in 3 whorls, outer sepals long triangular; petals long elliptic, apex entire; ovules solitary. Berry elliptic, red, 7-8 mm long, with conspicuously persistent style at apex, not pruinose. Fl. Mar. Fr. Aug-Nov. Thickets on slopes, mixed forests or gravel slopes at 1600-2300 m. Distributed in Yunnan.

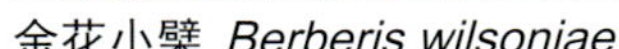

金花小檗 *Berberis wilsoniae*

西山小檗 *Berberis wangii*

显脉小檗 *Berberis phanera*

显脉小檗

Berberis phanera Schneid.

常绿灌木。枝棕灰色，圆柱形。叶长圆状椭圆形或椭圆状披针形，每边具7-12刺齿。花2-6朵簇生，黄色；花梗长2-3厘米。浆果紫黑色，椭圆形，被白粉。花期5-6月，果期7-12月。生海拔2800-4000米的山地灌丛及林下。产四川和云南。

Evergreen shrubs. Branches brownish gray, terete. Leaves oblong-elliptic or elliptic-lanceolate, margin with 7-12 spinescent teeth on each side. Flowers 2-6 fascicled, yellow; pedicels 2-3 cm long. Berries purplish black, ellipsoid, pruinose. Fl. May-Jun. Fr. Jul-Dec. Thickets, forests on mountains at 2800-4000 m. Distributed in Sichuan and Yunnan.

近光滑小檗

Berberis sublevis W. W. Sm.

常绿灌木。叶椭圆状披针形，叶缘具密集的细刺齿。花2-14朵簇生；花黄色；胚珠单生。浆果紫黑色，椭圆形至卵状椭圆形。种子1粒。花期2-3月，果期4-12月。生海拔1600-2500米的山地灌丛及林中。产云南西部。缅甸和印度东北部亦有。

Evergreen shrubs. Leaves elliptic-lanceolate, margin densely spinulose-serrulate. Flowers 2-14 fascicled, yellow; ovule solitary. Berries purplish black, ellipsoid to oval ellipsoid. Seed 1. Fl. Feb-Mar. Fr. Apr-Dec. Thickets and forests on mountains at 1600-2500 m. Distributed in W Yunnan. Also in Myanmar and NE India.

春小檗

Berberis vernalis (Schneid.) Chamberlain et C. M. Hu

常绿灌木。叶椭圆形至椭圆状披针形或长圆状椭圆形，长3-12厘米，边缘每边具10-24刺齿；叶柄长2-4毫米。花8-30朵簇生；萼片2轮，外萼片卵形；花瓣倒卵形，先端全缘，基部缢缩呈爪；药隔先端平截；胚珠单生，近无柄。浆果椭圆形，顶端具明显宿存花柱，不被白粉。花果期6-10月。生海拔1300-2600米的路边，灌丛或林中。产云南和湖南。

Evergreen shrubs. Leaves elliptic to elliptic-lanceolate or oblong-elliptic, 3-12 cm long, margin 10-24-spinose-serrate on each side; petiole 2-4 mm long. Flowers 8-30 fascicled; sepals in 2 whorls, outer sepals ovate; petals obovate, apex entire, base constricted into claw; connectives truncate at apex; ovules solitary, subsessile. Berry elliptic, with conspicuously persistent style at apex, not pruinose. Fl and fr. Jun-Oct. Thickets, roadsides or forests at 1300-2600 m. Distributed in Yunnan and Hunan.

近光滑小檗 *Berberis sublevis*

春小檗 *Berberis vernalis*

豪猪刺 *Berberis julianae*

滑叶小檗 *Berberis liophylla*

豪猪刺

Berberis julianae Schneid.

常绿灌木。刺3叉。叶革质，每侧具10-20个刺齿。花10-25朵簇生；萼片2轮；胚珠单生。浆果蓝黑色，矩圆状，具白粉，柱头宿存。花期3月，果期5-11月。生海拔1100-2100米山坡、林中、灌丛中或溪边。产华中和西南。

Evergreen shrubs. Spines 3-fid. Leaves leathery, margin 10-20-spinose-serrate on each side. Flowers 10-25 fascicled; sepals in 2 whorls; ovules solitary. Berries blue-black, oblong, white pruinose, styles persistent. Fl. Mar. Fr. May-Nov. Mountain slopes, forests, thickets or streamsides at 1100-2100 m. Distributed in C and SW China.

滑叶小檗

Berberis liophylla Schneid.

常绿灌木。叶椭圆形至倒卵状椭圆形，叶缘每边具5-10刺齿。花3-12朵簇生；花黄色；胚珠单生。浆果紫黑色，椭圆形至长圆形，具白粉。种子1粒。花期3-4月，果期5-12月。生海拔2100-2800米的山地灌丛及林中。产四川和云南。

Evergreen shrubs. Leaves elliptic to obovate-elliptic, margin with 5-10 spinescent teeth on each side. Flowers 3-12 fascicled, yellow; ovule solitary. Berries purplish black, ellipsoid to oblong, pruinose. Seed 1. Fl. Mar-Apr. Fr. May-Dec. Thickets and forest edges on mountains at 2100-2800 m. Distributed in Sichuan and Yunnan.

粉叶小檗

Berberis pruinosa Franch.

常绿灌木。刺3叉。叶硬革质，背面有白粉，边缘稍反卷或平展，每侧常具1-6个粗刺状齿。花(8-)10-20簇生；萼片2轮；花瓣先端深缺裂或全缘；胚珠2或3。浆果椭球形或近球形，密具白粉，柱头不宿存。花期3-4月，果期6-8月。生海拔1800-4000米林缘、山谷或石灰岩山坡。产云南、贵州、四川和西藏。

Evergreen shrubs. Spines 3-fid. Leaves stiff coriaceous, white-farinose abaxially, margin slightly revolute or flat, usually coarsely 1-6-aristate-dentate on each side. Flowers (8-)10-20 fascicled; sepals in 2 whorls; petals apex deeply incised or entire; ovules 2 or 3. Berries ellipsoid or subglobose, densely white pruinose, styles not persistent. Fl. Mar-Apr. Fr. Jun-Aug. Forest edges, valleys or limestone slopes at 1800-4000 m.

粉叶小檗 *Berberis pruinosa*

刺黑珠 *Berberis sargentiana*

Distributed in Yunnan, Guizhou, Sichuan and Xizang.

刺黑珠

Berberis sargentiana Schneid.

常绿灌木。叶长圆状椭圆形，长4-15厘米，先端急尖，基部楔形，边缘每边具15-25刺齿；近无柄。花4-10朵簇生；萼片3轮；花瓣倒卵形，具2枚橙色腺体，先端缺裂，具圆形裂片；胚珠1-2枚。浆果长圆形或长圆状椭圆形，顶端不具宿存花柱，不被白粉。花期4-5月，果期6-11月。生海拔700-2100米的灌丛中、路边、竹林或溪边林下。产湖北和四川。

Evergreen shrubs. Leaves oblong-elliptic, 4-15 cm long, apex acute, base cuneate, margin 15-25-spinose-serrate on each side; subsessile. Flowers 4-10 fascicled; sepals in 3 whorls; petals obovate, with 2 orange glands, apex emarginate with rounded lobes; ovules 1-2. Berry oblong or oblong-ellipsoid, style not persistent, not pruinose. Fl. Apr-May. Fr. Jun-Nov. Thickets, roadsides, bamboo forests, streamsides of forest understories at 700-2100 m. Distributed in Hubei and Sichuan.

丽江小檗

Berberis lijiangensis C. Y. Wu ex S. Y. Bao

常绿灌木。老枝棕灰色，近圆柱形。叶长圆状椭圆形或椭圆形，背面被白粉，叶缘每边具3-5刺齿。花3-6朵簇生，黄色。浆果紫黑色，卵状椭圆形，被白粉。花期5-6月，果期7-12月。生海拔2700-3400米的山地灌丛及林下。产云南(玉龙县)。

Evergreen shrubs. Branches brownish gray, subterete. Leaves oblong-elliptic or elliptic, abaxially pruinose, margin with 3-5 spinescent teeth on each side. Flowers 3-6 fascicled, yellow. Berries purplish black, oval ellipsoid, pruinose. Fl. May-Jun. Fr. Jul-Dec. Thickets, forests on mountains at 2700-3400 m. Distributed in Yunnan (Yulong County).

腰果小檗

Berberis johannis Ahrendt

落叶灌木。叶狭倒卵形至倒卵状披针形。近伞形花序，具3-10花；花黄色。浆果红色，狭卵状椭圆形，中部缢缩，略弯曲。花期5-6月，果期7-10月。生海拔3000-4000米的山地灌丛及林下。产西藏东南部。

Deciduous shrubs. Leaves narrowly obovate to obovate- lanceolate. Inflorescences subumbellates, 3-10-flowered; flowers yellow. Berries red, narrowly ovate-ellipsoid, contracted at the middle, slightly bending. Fl. May-Jun. Fr. Jul-Oct. Thickets and forests on mountains at 3000-4000 m. Distributed in SE Xizang.

丽江小檗 *Berberis lijiangensis*

腰果小檗 *Berberis johannis*

日本小檗 *Berberis thunbergii*

日本小檗

Berberis thunbergii DC.

落叶灌木。具刺，刺单生。叶薄纸质，全缘。(1或) 2-5花近簇生，呈伞形状；萼片2轮；花瓣先端具钝凹缺；胚珠1或2。浆果亮红色，椭球形，柱头不宿存。花期4-6月，果期7-10月。中国各地广泛栽培。原产日本。

Deciduous shrubs. Spiny, spine single. Leaves thinly papery, entire. Inflorescences umbels with sub-fascicled flowers, (1 or)2-5-flowered; sepals in 2 whorls; petals apex obtusely emarginate; ovules 1 or 2. Berries shiny, red, ellipsoid, styles not persistent. Fl. Apr-Jun. Fr. Jul-Oct. Widely cultivated in China. Native to Japan.

涝峪小檗

Berberis gilgiana Fedde

落叶灌木。刺单生或3叉。叶纸质，全缘或具2-9个小刺齿。总状花序具10-25花；花萼2轮；花瓣先端凹缺或全缘。浆果红色，矩圆状，稍具白粉，柱头不宿存。花期4-5月，果期8-10月。生海拔800-2000米的山坡或山谷。产河南、湖北和陕西。

Deciduous shrubs. Spines single or 3-fid. Leaves papery, margin entire or 2-9-spinulose-serrulate on each side. Racemes 10-25-flowered; sepals in 2 whorls; petals apex emarginate or entire. Berries red, oblong, slightly pruinose, styles not persistent. Fl. Apr-May. Fr. Aug-Oct. Slopes or valleys at 800-2000 m. Distributed in Henan, Hubei and Shaanxi.

细叶小檗

Berberis poiretii Schneid.

落叶灌木。刺缺乏、3叉或单生。叶纸质，全缘。穗状总状花序，具8-15花；小苞片2，披针形；萼片2轮；花瓣先端具细齿；胚珠1 (或2)。浆果红色，长圆状，不具白粉，柱头不宿存。花期5-6月，果期7-9月。生海拔600-2300米山地灌丛，砾质地，山沟河岸或林下。产中国东北、华北和西北。俄罗斯(远东地区)、蒙古和朝鲜半岛亦有。

Deciduous shrubs. Spines absent, 3-fid or single. Leaves papery, margin entire. Spicate-raceme 8-15-flowered; bracteoles 2, lanceolate; sepals in 2 whorls; petals apex incised; ovules 1(or 2). Berries red, oblong, not pruinose, styles not persistent. Fl. May-Jun. Fr. Jul-Sep. Thickets on mountains, gravel areas, by streams in valleys, or forests at 600-2300 m. Distributed in NE, N and NW China. Also in Russia (Far East), Mongolia and Korean Peninsula.

涝峪小檗 *Berberis gilgiana*

细叶小檗 *Berberis poiretii*

淡色小檗 *Berberis pallens*

淡色小檗
Berberis pallens Franch.

落叶灌木。刺3叉；枝条暗红色，微被白粉。叶厚纸质，具芒。伞形总状花序具3-8花；萼片3轮；花瓣先端钝凹缺；胚珠1或2。浆果红色，长圆状椭球形，具白粉；柱头宿存。花果期5-8月。生海拔3000-3500米的灌丛。产云南。

Deciduous shrubs. Spines 3-fid; branches dull red, slightly white-farinose. Leaves thickly papery, aristate. Inflorescences umbellate racemes, 3-8-flowered; sepals in 3 whorls; petals apex obtusely emarginate; ovules 1 or 2. Berries red, oblong-ellipsoid, pruinose; styles persistent. Fl. and fr. May-Aug. Thickets at 3000-3500 m. Distributed in Yunnan.

拉萨小檗
Berberis hemsleyana Ahrendt

落叶灌木。叶狭倒卵形至倒卵状椭圆形。伞形状总状花序，基部常有数花簇生；花黄色。浆果红色，卵状椭圆形，微被白粉。花期5-6月，果期7-10月。生海拔3600-4400米的山地灌丛。产西藏。

Deciduous shrubs. Leaves narrowly obovate to obovate-elliptic. Inflorescences subumbellate-racemes, 4-8-flowered, sometimes mixed with fascicled flowers; flowers yellow. Berries red, ovate-ellipsoid, slightly pruinose. Fl. May-Jun. Fr. Jul-Oct. Thickets on mountains at 3600-4400 m. Distributed in Xizang.

拉萨小檗 *Berberis hemsleyana*

察瓦龙小檗
Berberis tsarongensis Stapf

落叶灌木。枝光滑无毛；刺单生或3叉。叶薄纸质，全缘或每侧具1-4个刺齿。伞形总状花序具4-6花；萼片2轮；花瓣先端凹缺；胚珠2。浆果红色，长圆状椭球形，不具白粉，柱头不宿存。花期4-5月，果期6-10月。生海拔2900-3880米的林中。产云南和西藏。

Deciduous shrubs. Branches glabrous; spines single or 3-fid. Leaves thinly papery, margin entire or 1-4-spinose-serrate on each side. Inflorescences umbellate racemes, 4-6-flowered; sepals in 2 whorls; petals apex emarginate; ovules 2. Berries red, oblong-ellipsoid, not pruinose, styles not persistent. Fl. Apr-May. Fr. Jun-Oct. Forests at 2900-3880 m. Distributed in Yunnan and Xizang.

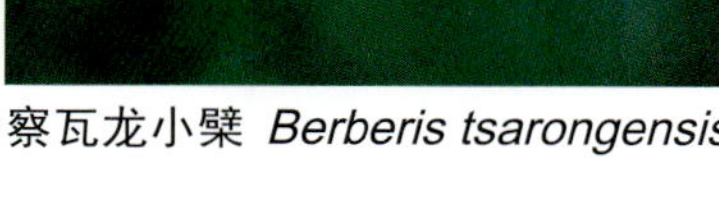
察瓦龙小檗 *Berberis tsarongensis*

川滇小檗 *Berberis jamesiana*

安徽小檗 *Berberis anhweiensis*

川滇小檗

Berberis jamesiana Forrest et W.W. Sm.

落叶灌木。刺单生或3叉，粗壮。叶近革质，全缘或具致密小齿。总状花序常具9-20花；花萼2轮；花瓣黄色，基部有2枚腺体；胚珠2。浆果最初乳白色，后为亮红色，近球形，半透明，不具白粉，柱头不宿存。花期4-5月，果期6-9月。生海拔2100-3600米山坡、林缘、河边、林中或灌丛中。产云南、四川和西藏。

Deciduous shrubs. Spines single or 3-parted, stout. Leaves subleathery, margin entire or closely spinulose. Flowers 9-20 in racemes; sepals in 2 whorls; petals yellow, base with 2 glands; ovules 2. Berries initially creamy white, finally light red, subglobose, translucent, not pruinose, styles not persistent. Fl. Apr-May. Fr. Jun-Sep. Slopes, forest edges, by streams, forests or bushes at 2100-3600 m. Distributed in Yunnan, Sichuan and Xizang.

安徽小檗

Berberis anhweiensis Ahrendt

落叶灌木。叶近圆形，叶缘具密集的细刺齿。总状花序；花黄色。浆果红色，宽椭圆形。种子1-2。花期5-6月，果期7-10月。生海拔700-1800米的山地灌丛、疏林。产安徽、浙江和湖北。

Deciduous shrubs. Leaves rotund, margin densely spinulose-serrulate. Inflorescences racemes; flowers yellow. Berries red, broadly ellipsoid. Seeds 1-2. Fl. May-Jun. Fr. Jul-Oct. Thickets and sparse forests on mountains at 700-1800 m. Distributed in Anhui, Zhejiang and Hubei .

黄芦木

Berberis amurensis Rupr.

落叶灌木。叶倒卵状椭圆形，叶缘每边具40-60细刺齿。花序总状；花黄色；花瓣先端凹缺。浆果红色，卵形至长圆球形。花期4-5月，果期8-9月。生海拔1100-2800米山地灌丛中、沟谷、林缘、疏林。产中国东北和华北。日本、朝鲜半岛和俄罗斯亦有。

Deciduous shrubs. Leaves obovate-elliptic, margin 40-60-spinulose-serrulate on each side. Inflorescences racemes; flowers yellow; petals apex emarginated. Berries red, ovate to oblong. Fl. Apr-May. Fr. Aug-Sep. Bushes on mountains, valleys, forest edges, sparse forests at 1100-2800 m. Distributed in NE and N China. Also in Japan, Korean Peninsula and Russia.

黄芦木 *Berberis amurensis*

异果小檗 *Berberis heteropoda*

异果小檗

Berberis heteropoda Schrenk.

落叶灌木。刺三叉或单生，叶厚纸质，侧脉2-4对，全缘或具不明显刺齿。总状花序或伞形状总状花序，具4-9花；萼片2轮；花瓣先端圆，全缘；胚珠4-6。浆果近球色，黑色，柱头不宿存。花期5-6月，果期7-10月。生海拔950-3200米石质山坡、河滩地、疏林或云杉林下、灌丛或荒草原上。产新疆。俄罗斯亦有。

Deciduous shrubs. Spines 3-fid or single. Leaves thickly papery, lateral veins 2-4 pairs, margin entire or indistinctly spinose-serrate. Racemes or umbellate-raceme, 4-9-flowered; sepals in 2 whorls; petals apex rounded, entire; ovules 4-6. Berries subglobose, black, styles not persistent. Fl. May-Jun. Fr. Jul-Oct. Rocky slopes, flood plains, sparse forests or *Picea* forests, thickets, or grassy wastelands at 950-3200 m. Distributed in Xinjiang. Also in Russia.

喀什小檗

Berberis kaschgarica Rupr.

落叶灌木。刺3叉。叶纸质侧脉2-3对，不明显；总状花序具5-9花；花黄色；小苞片披针形；萼片2轮；花瓣先端凹缺；胚珠5。浆果黑色，卵状球形，不具白粉，柱头宿存。花期5-6月，果期6-8月。生海拔1900-2800米山谷、山坡、林缘或灌丛中。产新疆。俄罗斯亦有。

Deciduous shrubs. Spines 3-fid. Leaves papery, lateral veins 2-3 paired, indistinct. Racemes 5-9-flowered; flowers yellow; bracteoles lanceolate; sepals 2-verticillate; petals apex emarginate; ovules 5. Berries black, ovoid-globose, not pruinose, styles persistent. Fl. May-Jun. Fr. Jun-Aug. Valleys, mountain slopes, forest edges or thickets at 1900-2800 m. Distributed in Xinjiang. Also in Russia.

波密小檗

Berberis gyalaica Ahrendt

落叶灌木。叶宽倒卵形至倒卵状椭圆形。圆锥花序具10-50花；花黄色。浆果蓝紫色，狭卵状椭圆形，中部缢缩，略弯曲，被白粉。花期6-7月，果期8-10月。生海拔2500-3200米的山地灌丛及林下。产西藏东南部。

Deciduous shrubs. Leaves broadly obovate to obovate-elliptic. Inflorescences panicles, 10-50-flowered; flowers yellow. Berries bluish purple, narrowly ovate-ellipsoid, contracted at the middle, slightly bending, pruinose. Fl. Jun-Jul. Fr. Aug-Oct. Thickets and forests on mountains at 2500-3200 m. Distributed in SE Xizang.

喀什小檗 *Berberis kaschgarica*

波密小檗 *Berberis gyalaica*

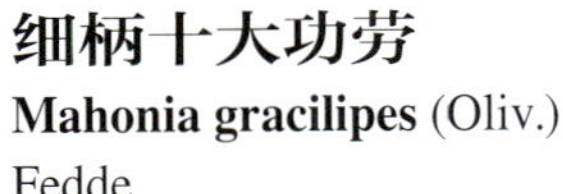
细柄十大功劳 *Mahonia gracilipes*

阿里山十大功劳 *Mahonia oiwakensis*

细柄十大功劳
Mahonia gracilipes (Oliv.) Fedde

灌木。小叶2或3对，每侧具1-5个刺齿。总状花序分枝或不分枝；花梗细，长1.3-2.4厘米；花瓣淡黄色或白色，萼片2，紫红色基部有2腺体。浆果球形，黑色，被白粉。花期4-8月，果期9-11月。生海拔700-2400米阔叶林中、林缘或阴坡。产四川和云南。

Shrubs. Leaflets 2 or 3 pairs, 1-5-spinose-serrate on each side. Racemes branched or not branched; pedicels slender, 1.3-2.4 cm long; petals pale yellow or white; sepals 2-whorled, purplish-red, base with 2 glands. Berries globose, black, glaucescent. Fl. Apr-Aug. Fr. Sep-Nov. Broad-leaved forests, forest edges, or shady slopes at 700-2400 m. Distributed in Sichuan and Yunnan.

阿里山十大功劳
Mahonia oiwakensis Takeda

乔木。叶下黄绿色；小叶12-20对，长圆状披针形，每侧边缘具2-9个刺齿。7-18个总状花序簇生，长9-25厘米；花柄粗壮，长3-5毫米。浆果卵球形，蓝黑色。花期8-11月，果期11月至翌年5月。生海拔650-3800米的阔叶林下、灌丛或山坡。产中国西南至东南、台湾。

Trees. Leaves abaxially yellowish green; leaflets 12-20 pairs, oblong-lanceolate, margin 2-9-spinose-serrate on each side. Inflorescences 7-18 fascicled racemes, 9-25 cm long; pedicels stout, 3-5 mm long. Berries ovoid, bluish black, pruinose. Fl. Aug-Nov. Fr. Nov to next May. Under broad-leaved forests, thickets or slopes at 650-3800 m. Distributed in SW to SE China and Taiwan.

长柱十大功劳
Mahonia duclouxiana Gagnep.

灌木。叶下黄绿色，薄纸质至薄革质；小叶4-9对，每侧具3-12个牙齿；总状花序长8-30厘米，基部常有分枝；花柱长达3毫米。浆果深紫色，具白粉，球形至近球形。花期11月至翌年4月，果期翌年3-6月。生海拔1800-2700米的林中、灌丛、山坡或路边。产云南、四川和广西。印度、泰国和缅甸亦有。

Shrubs. Leaves abaxially yellowish green, thinly papery to thinly leathery; leaflets 4-9 pairs, margin with 3-12 teeth on each side. Racemes 8-30 cm long, often branched at base; styles up to 3 mm long. Berries deep purple, pruinose, globose to subglobose. Fl. Nov to next Apr. Fr. next Mar-Jun. Forests, bushes, slopes or roadsides at 1800-2700 m. Distributed in Yunnan, Sichuan and

长柱十大功劳 *Mahonia duclouxiana*

独龙十大功劳 *Mahonia taronensis*

Guangxi. Also in India, Thailand and Myanmar.

独龙十大功劳

Mahonia taronensis Hand.-Mazz.

灌木。小叶5-10对，先端尾状渐尖，基部楔形，每侧具9-23个牙齿；3-5个总状花序簇生，长5-8厘米；花浅黄绿色。浆果蓝色，具白粉，球形。花期5月，果期6-10月。生海拔2000-2600米的林中。产云南和西藏。

Shrubs. Leaflets 5-10 pairs, apex caudate-acuminate, base cuneate, margin with 9-23 teeth on each side. Inflorescences 3-5 fascicled racemes, 5-8 cm long; flowers pale greenish yellow. Berries blue, pruinose, globose. Fl. May. Fr. Jun-Oct. Forests at 2000-2600 m. Distributed in Yunnan and Xizang.

靖西全缘叶十大功劳

Mahonia jingxiensis J. Y. Wu, M. Ogisu, H. N. Qin et S. N. Lu

灌木。叶狭椭圆形，具5-7对小叶，最下部的小叶卵形，长0.5-3厘米，全缘，稀具1齿；顶生小叶略大于其余小叶，长圆状椭圆形，长4.5-9.5厘米。4-12个总状花序簇生，总状花序有时分枝；花金黄色；中萼片与内萼片近等大；花瓣基部具2枚腺体，先端具缺刻。浆果蓝黑色，梨形。花期9-10月，果期10-11月。生海拔约500米的常绿林中。产广西(靖西县)。

Shrubs. Leaves narrowly elliptic, with 5-7 pairs of leaflets, lowest pair of leaflets ovate, 0.5-3 cm long, entire, rarely 1-dentate; terminal leaflet slightly larger than others, oblong-elliptic, 4.5-9.5 cm long. Inflorescence of 4-12 fascicled racemes, sometimes branched racemes; flowers golden-yellow; median sepals almost equal to inner sepals; petals base with two glands, apex incised. Berry blue-black, pyriform. Fl. Sep-Oct. Fr. Oct-Nov. Evergreen forests at ca. 500 m. Distributed in Guangxi (Jingxi County).

靖西十大功劳

Mahonia subimbricata W. Y. Chun et F. Chun

灌木。叶椭圆形至倒披针形，长12-22厘米；小叶8-13对，邻接或覆瓦状接叠，基出脉3条，节间长1-2厘米；小叶边缘每边具2-7刺锯齿。总状花序9-13个簇生；花瓣黄色，狭椭圆形，基部腺体显著，先端全缘；药隔延伸；胚珠1或2。浆果倒卵形，长约8毫米，黑色，被白粉。花期9-11月，果期11月至翌年5月。生海拔约1900米灌丛或林中。产广西和云南。

Shrubs. Leaves elliptic to oblanceolate, 12-22 cm long; leaflets 8-13 pairs, close or overlapping, basal veins 3, internodes 1-2 cm long; leaflets margin with 2-7 spiny teeth on each side. Inflorescence 9-13 fascicled racemes; petals yellow, narrowly elliptic, base with distinct glands, apex entire; ovules 1 or 2. Berry obovoid, ca. 8 mm long, black, pruinose. Fl. Sep-Nov. Fr. Nov to next May. Thickets or woodlands at ca. 1900 m. Distributed in Guangxi and Yunnan.

靖西全缘叶十大功劳 *Mahonia jingxiensis*

靖西十大功劳 *Mahonia subimbricata*

十大功劳 *Mahonia fortunei*

十大功劳
Mahonia fortunei (Lindl.) Fedde

灌木。叶下浅黄绿色；小叶2-5对，倒卵形至　倒卵形，每侧边缘具5-10对刺齿；4-10个总状花序簇生，3-7厘米长；花黄色。浆果球形，紫黑色，被白粉。花期7-9月，果期9-11月。生海拔300-2000米山坡沟谷林、灌丛、路边或河边。产华中、西南和华东。印度尼西亚、日本、美国和其他地区亦有栽培。

Shrubs. Leaves abaxially pale yellowish green; leaflets 2-5 pairs, obovate to obovate-elliptic, margin with 5-10 spinose teeth on each side. Inflorescences 4-10 fascicled racemes, 3-7 cm long; flowers yellow. Berries globose, purple-black, pruinose. Fl. Jul-Sep. Fr. Sep-Nov. Forests on slopes or by streams, thickets, roadsides or by rivers at 300-2000 m. Distributed in C, SW and E China. Also cultivated in Indonesia, Japan, USA and elsewhere.

安坪十大功劳
Mahonia eurybracteata Fedde subsp. **ganpinensis** (Levl.) Ying et Burff.

灌木。小叶5-7对，椭圆状披针形至狭卵形，宽1.5厘米以下，边缘每边具3-9刺齿；近无柄或长达约3厘米。总状花序4-10个簇生；花梗长1.5-2毫米；外萼片卵形至卵状长圆形；胚珠2枚。浆果倒卵状或长圆状，蓝色或淡红紫色，被白粉，具宿存花柱。花期7-10月，果期11月至翌年5月。生海拔200-1200米的林下、林缘或溪边。产贵州、四川和湖北。

Shrubs. Leaflets 5-7 paired, elliptic-lanceolate to narrowly ovate, 1.5 cm or less wide, margin 3-9-spinose-serrate on each side; sessile or with petiolule to ca. 3 cm. Inflorescence 4-10 fascicled racemes; pedicel 1.5-2 mm long; outer sepals ovate to ovate-oblong; ovules 2. Berry obovoid or oblong, blue or reddish purple, pruinose, with persistent style. Fl. Jul to Oct. Fr. Nov to next May. Under forest, forest margins or streamsides at 200-1200 m. Distributed in Guizhou, Sichuan and Hubei.

阔叶十大功劳
Mahonia bealei (Fort.) Carr.

灌木或小乔木；叶27-51 × 10-20厘米，具4-10对小叶，下面覆白粉；花序直立，3-9个总状花序簇生。浆果深紫色或深蓝色，卵圆状或矩圆状卵球形，1.5 × 1-1.2厘米。花期9月至翌年6月，果期翌年3-5月。生海拔500-2000米的林中、林缘、山坡、溪边、路边或灌丛中。产中国西南、华南、东南和华中。许多国家广泛栽培。

安坪十大功劳 *Mahonia eurybracteata* subsp. *ganpinensis*

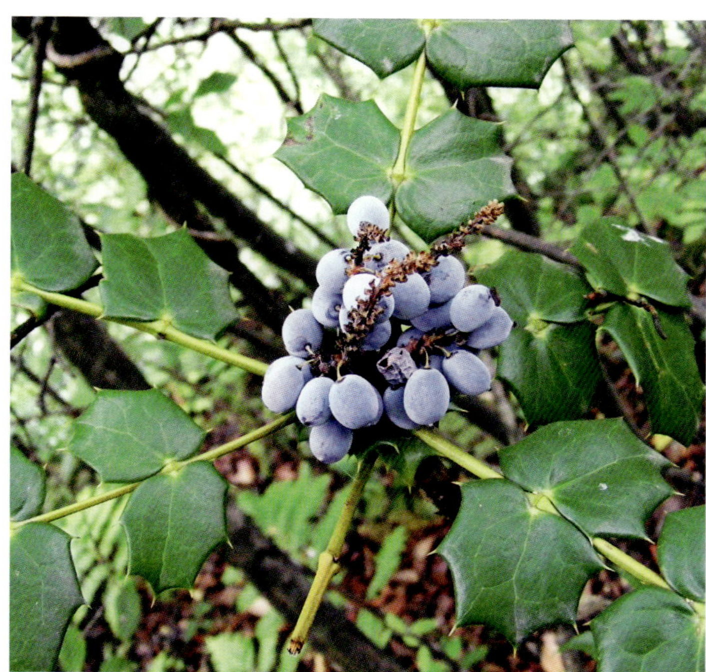
阔叶十大功劳 *Mahonia bealei*

Shrubs or small trees. Leaves 27-51 × 10-20 cm, with 4-10 pairs of leaflets, abaxially glaucous. Inflorescences erect, 3-9 fascicled racemes. Berries dark purple or dark blue, ovoid or oblong-ovoid, 1.5 × 1-1.2 cm. Fl. Sep to next Jun. Fr. next Mar-May. Forests, forest edges, slopes, streamsides, roadsides or thickets at 500-2000 m. Distributed in SW, S, SE and C China. Widely cultivated in many countries.

长苞十大功劳
Mahonia longibracteata Takeda

灌木。叶下浅黄绿色，具4或5对小叶，每侧具（3或)4-7(-11)牙齿；苞片披针形；花黄色；外萼片披针形，中萼片长圆形。浆果长圆形。亮红色，不具白粉。花期4-5月，果期5-10月。生海拔1900-3300米的山坡林下、灌丛中、阴坡或铁杉林下。产云南和四川。

Shrubs. Leaves abaxially pale yellowish green, with 4 or 5 pairs of leaflets, margin with (3 or)4-7(-11) teeth on each side. Racemes 6-9-clustered; bracts lanceolate; flowers yellow; outer sepals broadly lanceolate, central sepals oblong. Berries oblong, brightly red, not pruinose. Fl. Apr-May. Fr. May-Oct. Forests on mountain slopes, thickets, shaded slopes, or Tsuga forests at 1900-3300 m. Distributed in Yunnan and Sichuan.

密叶十大功劳
Mahonia conferta Takeda

灌木或小乔木。叶厚革质，羽状复叶，具8-18对小叶，小叶每边近轴处具2或3个牙齿，远轴处具3-5个牙齿；3-6个总状花序簇生。浆果椭圆形，稍被白粉。花期7-10月，果期10-12月。生海拔1500-2100米林中或山坡阴处。产云南。

Shrubs or small trees. Leaves thick leathery, pinnately compound, with 8-18 pairs of leaflets, leaflet proximal margins with 2 or 3 teeth, distal margins with 3-5 teeth. Inflorescences 3-6 fascicled racemes. Berries ellipsoid, slightly pruinose. Fl. Jul-Oct. Fr. Oct-Dec. Forests or shady places on slopes at 1500-2100 m. Distributed in Yunnan.

长苞十大功劳 *Mahonia longibracteata*

密叶十大功劳 *Mahonia conferta*

桃儿七 *Sinopodophyllum hexandrum*

桃儿七
Sinopodophyllum hexandrum (Royle) Ying

多年生草本。叶2，薄纸质，非盾状，基部心形，3-5深裂；花单生，先叶开放，粉红色；萼片6；雄蕊药隔较窄。浆果红色，卵状球形，肉质。花期5-6月，果期7-9月。生海拔2200-4300米的冷杉林下或松林下，林缘湿地或草丛。产中国西南和西北。尼泊尔、不丹、印度北部、阿富汗东部和巴基斯坦亦有。

Perennial herbs. Leaves 2, papery, never peltate, cordate at base, 3-5-parted. Flowers solitary, appearing before leaves, pink; connective of stamen narrow. Berries red, ovoid-globose, fleshy. Fl. May-Jun. Fr. Jul-Sep. Fir or pine forests, damp places of forest edges, or grasslands at 2200-4300 m. Distributed in SW and NW China. Also in Nepal, Bhutan, N India, E Afghanistan and Pakistan.

贵州八角莲
Dysosma majorensis (Gagnep.) Ying

多年生草本。叶互生，肾状圆形，长10-20厘米，4-6深裂，裂片先端3小裂，背面被微柔毛，边缘具稀疏细齿；叶柄长4-20厘米。花2-5朵着生于近叶基处，紫色；萼片不等大，椭圆形，淡绿色，无毛；花瓣椭圆状披针形，长约9厘米。浆果长圆形，熟时红色。花期4-6月，果期6-9月。生海拔1300-1800米的林下或竹林下。产贵州、四川、湖北、广西和云南。

Perennial herbs. Leaves alternate, reniform-orbicular, 10-20 cm long, deeply 4-6-parted, lobes 3-fid at apex, abaxially puberulent, margin remotely serrulate; petiole 4-20 cm long. Flowers 2-5, attached near base of blade, purple; sepals unequal in size, elliptic, pale green, glabrous; petals elliptic-lanceolate, ca. 9 cm long. Berry oblong, red when mature. Fl. Apr-Jun. Fr. Jun-Sep. Forests or bamboo forests at 1300-1800 m. Distributed in Guizhou, Sichuan, Hubei, Guangxi and Yunnan.

八角莲
Dysosma versipellis (Hance) M. Cheng ex Ying

多年生草本。茎生叶1，偶有2，互生，近圆形，直径达30厘米，薄纸质，下面具柔毛，掌状4-9裂；5-8花簇生；花瓣匙状倒卵形。浆果椭球形或球形。花期3-6月，果期5-9月。生海拔300-2400米的山坡林下、灌丛、潮湿地或石灰山林下。产中国西南、东南、华中和华北。

Perennial herbs. Stem leaves 1, occasionally 2, alternate, suborbicular, up to 30 cm diam, thinly papery, abaxially pubescent, palmately 4-9-lobed. Inflorescences 5-8-fascicled-flowered; petals spathulate-obovate. Berries ellipsoid or ovoid. Fl. Mar-Jun. Fr. May-Sep. Forests on slopes, thickets, damp places, or forests on limestone hills at 300-2400 m. Distributed in SW, SE, C and N China.

贵州八角莲 *Dysosma majorensis*

八角莲 *Dysosma versipellis*

小八角莲 *Dysosma difformis*

小八角莲

Dysosma difformis (Hemsl. et Wils.) T. H. Wang ex Ying

多年生草本。叶互生，不等大，偏盾状着生，两面无毛，全缘或3-5浅裂，边缘疏生不明显细齿。花2-5朵着生于叶基部处；萼片长圆状披针形，外面被柔毛，内面无毛；花瓣淡赭红色，长圆状条带形，长4-5厘米，无毛。浆果圆球形，直径1.7-2.7厘米。花期4-6月，果期6-9月。生海拔700-1800米密林下。产四川、贵州、湖北、湖南和广西。越南亦有。

Perennial herbs. Leaves alternate, unequal in size, obliquely peltate, both surfaces glabrous, undivided or 3-5-lobed, margin sparsely inconspicuously serrate. Flowers 2-5 fascicled near base of blade; sepals oblong-lanceolate, outside pubescent, inside glabrous; petals pale brownish red, oblong-loriform, 4-5 cm long, glabrous. Berry globose, 1.7-2.7 cm diam. Fl. Apr-Jun. Fr. Jun-Sep. Dense forests at 700-1800 m. Distributed in Sichuan, Guizhou, Hubei, Hunan and Guangxi. Also in Vietnam.

六角莲

Dysosma pleiantha (Hance) Woodson

多年生草本。茎单一。叶对生，盾形，直径16-33厘米，纸质，两面无毛，5-9浅裂；5-8花簇生；花瓣6-9，紫红色，倒卵状长圆形，长3-4厘米。浆果红色，倒卵状长圆形或椭圆形。花期3-6月，果期7-9月。生海拔400-1600米的山谷河边或阴湿草地。产华南、东南至华中。

Perennial herbs. Stems solitary. Leaves opposite, large, peltate, 16-33 cm diam, papery, both surfaces glabrous, 5-9-lobed. Inflorescences 5-8-fascicled-flowered; petals 6-9, purplish red, obovate-oblong, 3-4 cm long. Berries red, obovoid-oblong or ellipsoid. Fl. Mar-Jun. Fr. Jul-Sep. Streamsides in valleys or grassy shaded places at 400-1600 m. Distributed in S, SE to C China.

六角莲 *Dysosma pleiantha*

西藏八角莲 *Dysosma tsayuensis*

西藏八角莲
Dysosma tsayuensis Ying

多年生草本。叶对生，圆形或近圆形，直径约30厘米，两面被短伏毛，5-7深裂，几达中部，裂片楔状长圆形，边缘具刺细齿和睫毛。花2-6朵簇生于叶柄基部；花瓣白色，倒卵状椭圆形，长2.7-2.8厘米；胚珠多数。浆果卵形或卵状椭圆形，长约3厘米，红色。花期5月，果期7月。生海拔2500-3500米的松林、冷杉林、云杉林下或林间空地。产西藏。

Perennial herbs. Leaves opposite, orbicular or suborbicular, ca. 30 cm diam, both surfaces strigose, 5-7-parted to ca. midway, lobes cuneate-oblong, margin spinose-serrulate, ciliolate. Flowers 2-6 fascicled at base of petiole; petals white, obovate-elliptic, 2.7-2.8 cm long; ovules numerous. Berry ovoid or ovoid-ellipsoid, ca. 3 cm long, red. Fl. May. Fr. Jul. *Picea*, *Abies*, and *Pinus* forests and openings in forests at 2500-3500 m. Distributed in Xizang.

川八角莲
Dysosma delavayi (Franch) Hu

多年生草本。茎生叶对生，盾状，2枚，裂片先端3裂，叶下中脉具柔毛，渐无毛。伞形花序具2-6花；花瓣长圆状披针形，6瓣，紫红色，长约2厘米。浆果红色，椭球形。花期4-5月，果期6-9月。生海拔1200-2500米的山谷林下或水边。产云南、四川和贵州。

Perennial herbs. Stem leaves opposite, peltate, 2, lobe apex 3-lobed, midvein pubescent and glabrescent abaxially. Umbels 2-6-flowered; petals oblong-lanceolate, 6, purplish red, ca. 2 cm long. Berries red, ellipsoid. Fl. Apr-May. Fr. Jun-Sep. Forests in valleys or by waters at 1200-2500 m. Distributed in Yunnan, Sichuan and Guizhou.

南方山荷叶
Diphylleia sinensis H. L. Li

多年生草本。叶盾状，肾形或肾状圆形至横向长圆形，呈2半裂，每半裂具3-6浅裂或波状，边缘具不规则锯齿，齿端具尖头，背面被柔毛。外轮萼片披针形至线状披针形；胚珠5-11。浆果球形至宽椭圆形，深蓝或紫黑色，微被白粉。花期5-6月，果期7-8月。生海拔1900-3700米的潮湿落针叶林下、灌丛或竹丛下。产湖北、陕西、甘肃、云南和四川。

Perennial herbs. Leaves peltate, reniform or reniform-orbicular to transversely oblong, 2-cleft with divisions undulate or shallowly 3-6-lobed, margin irregularly dentate with teeth apiculate,

川八角莲 *Dysosma delavayi*

南方山荷叶 *Diphylleia sinensis*

abaxially pubescent. Outer sepals lanceolate to linear-lanceolate; ovules 5-11. Berry globose to broadly ellipsoid, dark blue or purplish black, slightly pruinose. Fl. May-Jun. Fr. Jul-Aug. Moist deciduous and coniferous forests, thickets, bamboo thickets at 1900-3700 m. Distributed in Hubei, Shaanxi, Gansu, Yunnan and Sichuan.

光叶淫羊藿

Epimedium sagittatum (Sieb. et Zucc.) Maxim. var. **glabratum** Ying

多年生草本。花茎具2枚对生三出复叶。小叶3，卵形至卵状披针形，长5-19厘米，基部心形，顶生小叶长圆形，背面无毛，边缘密具刺齿；圆锥花序尖塔形；花梗无毛；花黄色，直径约8毫米；萼片2轮；花瓣囊状，先端钝，长1.5-2毫米。蒴果长约1厘米，具宿存花柱。花期4-5月，果期5-7月。生海拔约700米的林下。产贵州和湖北。

Perennial herbs. Flowering stem with 2 opposite trifoliolate leaves. Leaflets 3, ovate to ovate-lanceolate, 5-19 cm long, base cordate, terminal leaflet oblong, abaxially glabrous, margin closely spinulose-subserrulate; panicle pyramidal; pedicel glabrous; flowers yellow, ca. 8 mm diam; sepals in 2 whorls; petals saccate, apex blunt, 1.5-2 mm long. Capsules ca. 1 cm long, with persistent style. Fl. Apr-May. Fr. May-Jul. Forests at ca. 700 m. Distributed in Guizhou and Hubei.

柔毛淫羊藿

Epimedium pubescens Maxim.

多年生草本。叶一回三出复叶，基部深心形或心形，边缘具密刺齿，下面密被绒毛。圆锥花序疏松，具30至多于100花；花瓣淡黄色，囊状。蒴果长圆形。花期4-5月，果期5-7月。生海拔300-2000米林下、灌丛、山坡或山沟阴处。产华西和华中。

Perennial herbs. Leaves 2, opposite, ternate; leaflets at base deeply cordate or cordate, margin minutely setose-serrate, abaxially densely tomentose. Panicles loose, 30- to more than 100-flowered; petals pale yellow, saccate. Capsules oblong. Fl. Apr-May. Fr. May-Jul. Forests, bushes, slopes, or shady places by streams at 300-2000 m. Distributed in W and C China.

柔毛淫羊藿 *Epimedium pubescens*

光叶淫羊藿 *Epimedium sagittatum* var. *glabratum*

朝鲜淫羊藿 *Epimedium koreanum*

朝鲜淫羊藿
Epimedium koreanum Nakai

多年生草本。根状茎匍匐，粗壮。花茎具1枚二回三出复叶。叶纸质，下面无毛或疏具柔毛。总状花序具有花4-16朵；花白色、淡黄色、红色或紫蓝色。蒴果狭纺锤状。花期4-5月，果期5月。生海拔400-1500米的林中或灌丛中。产中国东北和华东。朝鲜半岛和日本亦有。

Perennial herbs. Rhizomes creeping, robust. Flowering stem with 1 biternate leaf. Leaves papery, abaxially glabrous or sparsely puberulent. Racemes 4-16-flowered; flowers white, pale yellow, red or purplish blue. Capsules narrowly fusiform. Fl. Apr-May. Fr. May. Forests or thickets at 400-1500 m. Distributed in NE and E China. Also in Korean Peninsula and Japan.

宝兴淫羊藿
Epimedium davidii Franch.

多年生草本。花茎具2枚对生3小叶复叶。叶下苍白色，革质或纸质，下面具乳突和疏柔毛及短伏毛。圆锥花序具6-24花，长15-25厘米；花淡黄色，直径2-3厘米。蒴果长1.5-2厘米。花期4-5月，果期5-8月。生海拔1400-3000米林下、灌丛、岩石上或河边杂木林中。产云南和四川。

Perennial herbs. Flowering stem with 2 trifoliolate opposite leaves. Leaves abaxially glaucescent, leathery or papery, abaxially papillose and sparsely pubescent with short appressed hairs. Panicles loosely 6-24-flowered, 15-25 cm long; flowers pale yellow, 2-3 cm diam. Capsules 1.5-2 cm long. Fl. Apr-May. Fr. May-Aug. Forests, bushes, rocks, or mixed forests by streams at 1400-3000 m. Distributed in Yunnan and Sichuan.

粗毛淫羊藿
Epimedium acuminatum Franch.

多年生草本。花茎具2枚对生3小叶复叶，有时3枚轮生。小叶3，薄革质，基部脉7条，边缘具细密刺齿，具密或疏的短粗伏毛，有时近无毛。圆锥花序具10-50花；花黄色、白色、红色或淡青色。蒴果。花期4-5月，果期5-7月。生海拔270-2400米草丛、石灰岩陡坡、林下或灌丛中。产中国西南至华中。

Perennial herbs. Flowering stems with 2 trifoliolate opposite leaves, sometimes 3 leaves whorled. Leaflets 3, thinly coriaceous, margin densely setose-dentate, veins 7 at base, with dense or sparse shortly appressed fairly stout bristles, sometimes nearly glabrous. Panicles 10-50-flowered; flowers yellow, white, red or pale-blue. Fruit a capsule. Fl. Apr-May. Fr. May-Jul. Grasslands, limestones on slopes, under forests or bushes at 270-2400 m. Distributed in SW to C China.

淫羊藿
Epimedium brevicornu Maxim.

多年生草本。根状茎短。花茎具2枚对生二回三出复叶。叶纸质或稍厚，下面无毛或具粗糙柔毛。圆锥花序具20-50花；花白色或淡黄色；外层萼片深绿色，内层白色或淡黄色。花期5-6月，果期6-8月。生海拔600-3500米的林缘、灌丛或山坡。产山西、陕西、青海、甘肃、河南、湖北和四川。

宝兴淫羊藿 *Epimedium davidii*

粗毛淫羊藿 *Epimedium acuminatum*

淫羊藿 *Epimedium brevicornu*

Perennial herbs. Rhizomes short. Flowering stems with 2 opposite usually biternate leaves. Leaves papery or thickly so, abaxially glabrous or scarcely pubescent. Panicles 20-50-flowered; flowers white or pale yellow; outer sepals dark green, inner sepals white or pale yellow. Fl. May-Jun. Fr. Jun-Aug. Forest edges, thickets or slopes at 600-3500 m. Distributed in Shanxi, Shaanxi, Qinghai, Gansu, Henan, Hubei and Sichuan.

囊果草 *Liontice incerta*

囊果草

Liontice incerta Pall.

多年生草本。块根卵球形或近球形，直径2-5厘米。茎高5-20厘米，基部有数枚鳞片。茎生叶2，生于近茎顶部，具柄，2-3回三出全裂，裂片椭圆形或倒卵形，全缘。总状花序顶生，长4-6厘米；苞片近圆形；萼片6，黄色，椭圆形；花瓣蜜腺状，倒卵形；雄蕊6；心皮1。瘦果近球形，直径2.5-4.5厘米。花期4月。生海拔600米荒漠低山山坡、沙地。产新疆。哈萨克斯坦亦有。

Perennial herbs. Root tuber ovoid or subglobose, 2-5 cm diam. Stems 5-20 cm tall, at base with several scales. Cauline leaves 2, growing near stem apex, petiolate, 2 to 3 times ternate-sect, with segments elliptic or obovate, entire. Raceme terminal, 4-6 cm long; bracts suborbicular; sepals 6, yellow, elliptic; petals nectary-like, obovate; stamens 6; carpel 1. Achenes subglobose, 2.5-4.5 cm diam. Fl. Apr. On slopes of low hills of desert or in sandy places at 600 m. Distributed in Xinjiang. Also in Kazakhstan.

红毛七

Caulophyllum robustum Maxim.

多年生草本。小叶卵形，长圆形或宽披针形，长4-8厘米，边缘全缘，有时2-3裂，两面无毛。花淡黄色；萼片倒卵形；花瓣远较萼片小，基部缢缩呈爪；胚珠2。花期5-6月，果期7-9月。生海拔950-3500米的林下、竹林、银杉林或山沟阴湿处。产中国东北、华中、山西、甘肃、河北、安徽、浙江、云南、贵州和西藏。朝鲜半岛、日本和俄罗斯亦有。

Perennial herbs. Leaflets ovate, oblong, or broadly lanceolate, 4-8 cm long, margin entire, sometimes 2- or 3-lobed, both surfaces glabrous. Flowers pale yellow; sepals obovate; petals much smaller than sepals, base constricted into claw; ovules 2. Fl. May-Jun. Fr. Jul-Sep. Forests, bamboo thickets, cathaya forests or moist places in valleys at 950-3500 m. Distributed in NE and C China, Shanxi, Gansu, Hebei, Anhui, Zhejiang, Yunnan, Guizhou and Xizang. Also in Korean Peninsula, Japan and Russia.

红毛七 *Caulophyllum robustum*

防己科
Menispermaceae

崖藤 *Albertisia laurifolia*

崖藤
Albertisia laurifolia Yamamoto

木质藤本。叶革质，侧脉每边3-5。聚伞花序；萼片3轮，内面的大，内轮合生成花冠管状；花瓣6，2轮；聚药雄蕊具27雄蕊，排成垂直6列；心皮6。核果椭圆体形。花期夏季，果期秋季。生海拔200-1000米林中。产云南南部、广西南部和海南南部。越南北部亦有。

Woody vines. Leaves coriaceous; lateral veins 3-5 per side. Cymes; sepals 3-whorled, inner ones large, inner whorl connate into corolloid tube; petals 6, 2-whorled; synandrium with 27 anthers in 6 vertical rows; carpels 6. Drupes ellipsoid. Fl. summer. Fr. autumn. Forests at 200-1000 m. Distributed in S Yunnan, S Guangxi and S Hainan. Also in N Vietnam.

大叶藤
Tinomiscium petiolare Hook. f. et. Thoms.

木质藤本。叶阔卵形，薄革质。花序数个生自老茎突出处，总状，常下垂，具淡紫褐色绒毛或柔毛；胎座迹不明显；花被明显分化为萼片和花瓣。核果起初绿色，具白色斑点，后为白色至黄色(或橙色)，具白色蜡，扁椭圆体形。花期春夏季，果期秋季。生海拔750-1400米的林中。产云南南部和东南部、广西南部。泰国、越南中部和北部、马来西亚、印度尼西亚和巴布亚新几内亚亦有。

Woody vines. Leaves broadly ovate, thinly leathery. Inflorescences several arising together from protuberances on old stems, racemose, often pendulous, purplish ferruginous tomentose or puberulent; placentation trace inconspicuous; tepals conspicuously differentiated into sepals and petals. Drupes at first green with white spots, later white to yellow (or orange), with white latex, compressed ellipsoidal. Fl. spring to summer. Fr. autumn. Forests at 750-1400 m. Distributed in S and SE Yunnan, and S Guangxi. Also in Thailand, C and N Vietnam, Malaysia, Indonesia and Papua New Guinea.

天仙藤
Fibraurea recisa Pierre

大型木质藤本。叶不明显盾形，革质，长圆状卵形。花序自无叶老茎伸出，圆锥状；雄花序疏松，长达30厘米；花被片8-12；雄蕊3。核果黄色，长圆状椭圆体形。花期春夏，果期秋。生海拔180-680米热带密林中。产云南、广西和广东。老挝、越南和柬埔寨亦有。

Large woody vines. Leaves inconspicuously peltate, coriaceous, oblong-ovate. Inflore-

大叶藤 *Tinomiscium petiolare*

天仙藤 *Fibraurea recisa*

scences arising from leafless old stems, paniculate; male inflorescences lax, up to 30 cm long; tepals 8-12; stamens 3. Drupes yellow, oblong-ellipsoid. Fl. spring to summer. Fr. autumn. Tropical dense forests at 180-680 m. Distributed in Yunnan, Guangxi and Guangdong. Also in Laos, Vietnam and Cambodia.

波叶青牛胆

Tinospora crispa (L.) Miers ex Hook. f. et Thoms.

落叶藤本。叶宽卵状心形至心状近圆形，稍肉质，掌状5(-7)脉。花序总状，不分枝，先叶生长，花2或3朵簇生；雄花：萼片6排成2轮；花瓣3-6，黄色；雄蕊6；雌花多1朵生节间。核果橘黄色，近球形。花期春季，果期夏季。生海拔200-900米的疏林中或灌丛中。产云南。南亚和东南亚亦有。

Vines deciduous. Leaves broadly ovate to orbicular, slightly fleshy, palmately 5(-7)-veined. Inflorescences racemose, unbranched, appearing before leaves; flowers 2 or 3 fascicled; male flowers: sepals 6 in 2 whorls; petals 3-6, yellow; stamens 6; female flowers mostly 1 per node. Drupes orange, subglobose. Fl. spring. Fr. summer. Open forests or thickets at 200-900 m. Distributed in Yunnan. Also in S and SE Asia.

中华青牛胆 *Tinospora sinensis*

中华青牛胆

Tinospora sinensis (Lour.) Merr.

近肉质藤本。枝叶均被密毛。叶阔卵形至近圆形，纸质，基出5(-7)掌状脉。花序于植株无叶时生出；雄花序单生或少数簇生；雄蕊：萼片6枚2轮；花瓣6；雌花序单生；心皮3。核果红色，近球形。花期4月，果期5-6月。生海拔560-900米的林中。产云南、广西和广东。印度、缅甸、老挝、泰国、越南和马来西亚亦有。

Subfleshy vines. Stems and leaves with dense hairs. Leaves broadly ovate to subrotund, papery, palmately 5(-7)-veined at base. Inflorescences appearing when plant is leafless; male inflorescences solitary or few fascicled; male flower: sepals 6 in 2 whorls; petals 6; female inflorescences solitary; carpels 3. Drupes red, subglobose. Fl. Apr. Fr. May-Jun. Forests at 560-900 m. Distributed in Yunnan, Guangxi and Guangdong. Also in India, Myanmar, Laos, Thailand, Vietnam and Malaysia.

波叶青牛胆 *Tinospora crispa*

青牛胆 *Tinospora sagittata*

夜花藤 *Hypserpa nitida*

青牛胆

Tinospora sagittata (Oliv.) Gagnep.

草质藤本。块根念珠状，黄色。叶披针状箭形或有时披针状戟形，纸质或薄革质。花序腋生，常少数至多数花簇生，聚伞状，有时假圆锥状；雄花：萼片6；花瓣6；心皮3。核果红色，近球形。花期4月，果期秋季。生林中、林缘、竹林下或草地。产中国西南、华南、东南和华中。越南北部亦有。

Vines herbaceous. Root tubers rosary-shaped, yellow. Leaves lanceolate-sagittate or sometimes lanceolate-hastate, papery to thinly leathery. Inflorescences axillary, often a few or many flowers fascicled, cymose, sometimes pseudopanicles; male flower: sepals 6; petals 6; female flower: carpels 3. Drupes red, subglobose. Fl. Apr. Fr. autumn. Forests, forest edges, bamboo forests or grasslands. Distributed in SW, S, SE and C China. Also in N Vietnam.

连蕊藤

Parabaena sagittata Miers

草质藤本，通常被糙硬毛。叶纸质或干后膜质，长8-16(-25)厘米。花序单生或有时成对，聚伞状，具绒毛；雌花：萼片4；花瓣4，2轮；心皮3。核果近球形，稍扁；果核密被短刺。花期4-5月，果期8-9月。生林缘或灌丛中。产中国西南。南亚亦有。

Vines herbaceous, usually strigose. Leaves papery or dry-membranous, 8-16(-25) cm long. Inflorescences solitary or sometimes paired, corymbose, tomentose; female flower: sepals 4; petals 4, 2-whorled; carpels 3. Drupes subglobose, slightly flat; stones densely short-spinescent. Fl. Apr-May. Fr. Aug-Sep. Forest edges or shrublands. Distributed in SW China. Also in S Asia.

连蕊藤 *Parabaena sagittata*

夜花藤

Hypserpa nitida Miers

木质藤本。叶片纸质至革质，掌状3脉。雄花序常具少数花，聚伞状至圆锥状；雄花：萼片7-11，心皮2，子房半球形或近椭圆形。核果黄色或橙黄色，近球形，稍扁。花果期夏季。生林中或林缘。产云南、广西、广东、海南和福建。南亚和东南亚亦有。

Woody vines. Leaves papery to coriaceous, palmately 3-veined.

细圆藤 *Pericampylus glaucus*

苍白秤钩风 *Diploclisia glaucescens*

Male inflorescences usually only few flowered, cymose to paniculate; male flower: sepals 7-11; carpels 2, ovary semiglobose or subellipsoid. Drupes yellow or orange-yellow, subglobose, slightly flat. Fl. and fr. summer. Forests or forest edges. Distributed in Yunnan, Guangxi, Guangdong, Hainan and Fujian. Also in S and SE Asia.

细圆藤

Pericampylus glaucus (Lam.) Merr.

木质藤本。幼茎常长且下垂，常具淡黄色绒毛。叶三角状卵形至三角状长圆形，掌状(3-)5裂。花序伞状聚伞状；雄花：萼片9；柱头2裂。核果红色或紫红色。花期4-6月，果期9-10月。生海拔约700米林中、林缘或灌丛中。产长江流域以南。南亚和东南亚亦有。

Woody vines. Young stems often long and pendulous, usually yellowish tomentose. Leaves triangular-ovate to triangular-oblong, palmately (3-)5-veined. Inflorescences corymbose cymes; male flower: sepals 9; stigma 2-lobed. Drupes red or purple. Fl. Apr-Jun. Fr. Sep-Oct. Forests, forest edges or thickets at ca. 700 m. Distributed in south of Yangtze River. Also in S and SE Asia.

苍白秤钩风

Diploclisia glaucescens (Blume) Diels

木质藤本。叶革质，下面被白霜。圆锥花序生老茎或老枝上，稍下垂；花瓣淡黄色，微香；雄花：萼片带黑色斑纹；雌花之花瓣顶端2裂。核果狭倒卵球形，黄红色，下面微弯。花期4月，果期8月。生海拔540-1200米的林中或灌丛中。产云南、广西、广东和海南。南亚、东南亚和巴布亚新几内亚亦有。

Woody vines. Leaves coriaceous, abaxially glaucescent. Panicles inserted on old stems or branches, ± pendulous; petals pale yellow, slightly fragrant; male flower: sepals marked by a dark reticulum; petals of female flowers 2-lobed at apex. Drupes narrowly obovoid, yellow-red, below slightly recurved. Fl. Apr. Fr. Aug. Forests or thickets at 540-1200 m. Distributed in Yunnan, Guangxi, Guangdong and Hainan. Also in S and SE Asia, and Papua New Guinea.

木防己

Cocculus orbiculatus (L.) DC.

木质藤本。叶纸质或近革质，掌状3 (-5)脉。花序腋生，聚伞状，少花或多花排列成狭顶生或腋生聚伞圆锥状；花瓣6，顶端2裂。核果近球形，红色或紫红色。生海拔1200米以下的灌丛、灌丛或林缘。产中国大部分地区。东南亚、东亚、日本和美国(夏威夷)亦有。

Woody vines. Leaves papery or subcoriaceous, palmately 3(-5)-veined. Inflorescences axillary, cymose, few flowered, or many flowered arranged in a narrow terminal or axillary thyrse; petals 6, apex 2-lobed. Drupes subglobose, red or purple-red. Sparse forests, scrubby areas or forest edges below 1200 m. Distributed in most parts of China. Also in SE and E Asia, Japan and USA (Hawaii).

木防己 *Cocculus orbiculatus*

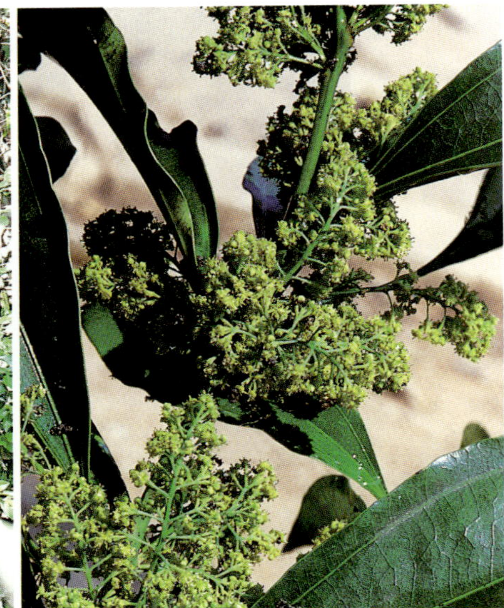

樟叶木防己 *Cocculus laurifolius*

樟叶木防己

Cocculus laurifolius DC.

直立灌木或小乔木。叶椭圆形、卵形或长椭圆形至披针状椭圆形，薄革质，掌状三出脉中侧生的一对几乎伸至叶片顶部。花序腋生，伞形或聚伞圆锥状；雄花：萼片6；花瓣6。核果黑色，球形，稍扁平。花期春末夏初，果期秋季。生海拔240-1600米的密林或灌丛中。产西藏、贵州、台湾和湖南。南亚、东南亚和日本亦有。

Erect shrubs or small trees. Leaves elliptic, ovate, or long elliptic to lanceolate-elliptic, thinly leathery, trinerved, lateral basal nerves nearly extending to leaf apices. Inflorescences axillary, cymose or thyrsoid; male flower: sepals 6; petals 6. Drupes black, rotund, slightly flattened. Fl. from late spring to early summer. Fr. autumn. Dense forests or thickets at 240-1600 m. Distributed in Xizang, Guizhou, Taiwan and Hunan. Also in S and SE Asia, and Japan.

风龙

Sinomenium acutum (Thunb.) Rehd. et Wils.

木质藤本。叶革质至纸质，掌状5(-7)脉。圆锥花序长可达30厘米；雄花：小苞片2，萼片淡黄绿色；雌花：退化雄蕊丝状；心皮无毛。核果红色至暗紫色或蓝黑色。花期夏季，果期秋末。生林中。产中国西南、华北和华东。印度、尼泊尔、泰国和日本亦有。

Woody vines. Leaves coriaceous to papery, palmately 5(-7)-veined. Panicles to 30 cm long; male flower: bracteoles 2, sepals yellowish green; female flowers: staminodes filiform; carpels glabrous. Drupes red to dark purple or blue-black. Fl. summer. Fr. late autumn. Forests. Distributed in SW, N and E China. Also in India, Nepal, Thailand and Japan.

蝙蝠葛

Menispermum dauricum DC.

落叶草质藤本。叶轮廓常心状扁圆形，常3-9浅裂，纸质或近膜质，掌状9-12脉。花序圆锥状，单生或有时成对，具20花，花多无柄，有时伞状簇生；雄花：萼片4-8。核果紫黑色。花期6-7月，果期8-9月。生海拔800米以下的路边或疏林

风龙 *Sinomenium acutum*

蝙蝠葛 *Menispermum dauricum*

一文钱 *Stephania delavayi*

中。产中国西南、华中、华北、华西、华东和东北。俄罗斯(南西伯利亚)、朝鲜半岛和日本亦有。

Herbaceous deciduous vines. Leaves usually cordate-oblate in outline, usually shallowly 3-9-lobed, papery or submembranous, palmately 9-12-veined. Inflorescences paniculate, solitary or paired, 20-flowered with flower in mostly sessile, sometimes umbel-like fascicles; male flower: sepals 4-8. Drupes purplish black. Fl. Jun-Jul. Fr. Aug-Sep. Roadsides or sparse forests below 800 m. Distributed in SW, C, N, W, E and NE China. Also in Russia (S Siberia), Korean Peninsula and Japan.

一文钱

Stephania delavayi Diels

草质藤本。叶明显盾状，三角状圆形，薄纸质，掌状9或10脉，纤细，稍突起。花序腋生或生于短枝，具退化叶，复合伞形聚伞状；雄花：萼片6(或8)，排成2轮；花瓣3或4。核果红色，无毛。生海拔1700-2700米的灌丛、田间或山坡。产云南、四川和贵州。

Vines herbaceous. Leaves conspicuously peltate, triangular-rotund, thinly papery, palmately 9- or 10-veined, veins slender, slightly raised abaxially. Inflorescences axillary or on short axillary branches with reduced leaves, compound umbelliform cymes; male flower: sepals 6 (or 8) in 2 whorls; petals 3 or 4. Drupes red, glabrous. Thickets, fields or slopes at 1700-2700 m. Distributed in Yunnan, Sichuan and Guizhou.

桐叶千金藤

Stephania japonica var. **discolor** (Blume) Forman

藤本。茎具柔毛。叶明显盾形，下面被丛卷柔毛。花序轴具柔毛；花序复伞状聚伞状，常腋生；雄花：萼片倒披针形至匙形，有时狭椭圆形，具柔毛。核果红色，倒卵球状近球形；胎座迹具穿孔。花期夏季，果期秋季。生海拔600-1700米的林下、灌丛或石灰山上。产云南、四川、贵州和广西。南亚、东南亚和澳大利亚亦有。

Vines. Stems pubescent. Leaves conspicuously peltate, with flocci abaxially. Inflorescence compound umbelliform cymes, usually axillary, axes pubescent; male flowers: sepals oblanceolate to spatulate, sometimes narrowly elliptic, pubescent. Drupes red, obovoid-subglobose; condyles perforate. Fl. summer. Fr. autumn. Forests, thickets or limestone hills at 600-1700 m. Distributed in Yunnan, Sichuan, Guizhou and Guangxi. Also in S and SE Asia, and Australia.

桐叶千金藤 *Stephania japonica* var. *discolor*

千金藤 *Stephania japonica*

千金藤
Stephania japonica (Thunb.) Miers

藤本。叶纸质或坚纸质，掌状脉10-11条，盾状着生。复伞形聚伞花序，常腋生，无毛；雄花：萼片6或8，两轮，膜质，淡黄绿色；花瓣黄色，肉质。核果红色；胎座迹不具穿孔。花期春季和夏季，果期秋季和冬季。生村边、石灰山或旷野灌丛中。产中国西南、东南、华中和华东。南亚、东南亚、东北亚、澳大利亚和太平洋岛屿亦有。

Vines. Leaves papery or hard-papery, palmately 10-11-veined, peltate. Inflorescences compound umbelliform cymes, usually axillary, glabrous; male flower: sepals 6 or 8 in 2 whorls, membranous, yellowish green; petals yellow, fleshy. Drupes red; condyles not perforate. Fl. spring and summer. Fr. autumn and winter. By villages, limestone mountainsor thickets of wastelands. Distributed in SW, SE, C and E China. Also in S, SE and NE Asia, Australia and Pacific Islands.

粪箕笃 *Stephania longa*

粪箕笃
Stephania longa Lour.

草质藤本。叶纸质，三角状卵形；掌状10-11脉。复伞形聚伞花序腋生；雄花：萼片8，偶6，排成2轮；花瓣4或3，绿黄色；雌花：萼片和花瓣均为4；子房无毛。核果红色。花期春末夏初，果期秋季。生灌丛或林缘。产中国西南和华南。老挝亦有。

Vines herbaceous. Leaves papery, triangular-ovate, palmately 10-11-veined. Compound-umbellate cymes, axillary; male flower: sepals 8, rarely 6, 2-verticillate, petals 4 or 3, yellow-green; female flower: sepals and petals 4; ovary glabrous. Drupes red. Fl. late spring to early summer. Fr. autumn. Thickets or forest edges. Distributed in SW and S China. Also in Laos.

地不容
Stephania epigaea H. S. Lo

草质落叶藤本，无毛。块根硕大，通常扁球形。叶盾形，扁圆形。花序单生或伞状聚伞花序，腋生，常淡紫红色且具白粉；小聚伞花序少数至10簇生，具2或3(-7)花；雄花：萼片6；花瓣3，紫色或橘黄色，具紫色斑点。核果红色。花期春夏季，果期秋季。生海拔1500-2700米石山坡或灌丛，亦常见栽培。产云南、四川和贵州。

Herbaceous deciduous vines, glabrous. Root tubers large, usually oblate-spheroidal. Leaves peltate, oblate. Inflorescences simple umbelliform cymes, axillary, often purplish red and glaucous; cymelets few to 10, fascicled, 2- or 3(-7)-flowered; male flower: sepals 6; petals 3, purple or orange with purple dots. Drupes red. Fl. spring to summer. Fr. autumn. Rocky slopes or thickets

地不容 *Stephania epigaea*

汝兰 *Stephania sinica*

at 1500-2700 m, also often cultivated. Distributed in Yunnan, Sichuan and Guizhou.

汝兰

Stephania sinica Diels

落叶藤本，全株无毛。叶膜质或近纸质。复伞房花序；花梗和小伞形花序梗肉质，无苞片和小苞片；雄花：萼片6，2轮，微肉质；花瓣3；雌花：萼片1；花瓣2。果梗肉质，干后黑色。花期6月，果期8-9月。生林中或河边。产云南、四川、贵州、湖北和湖南。

Vines deciduous, glabrous. Leaves membranous or subpapery. Compound corymbs; peduncles and umbellet pedicels fleshy, without bract and bracteole; male flower: sepals 6, 2-whorled, slightly fleshy; petals 3; female flower: sepal 1; petals 2. Fruiting pedicels fleshy, black when dry. Fl. Jun. Fr. Aug-Sep. Forests or by streams. Distributed in Yunnan, Sichuan, Guizhou, Hubei and Hunan.

金线吊乌龟

Stephania cepharantha Hayata

草质藤本。叶纸质，盾状，三角状扁圆形或近圆形，掌状7-9脉。雄花序常腋生圆锥状，头状，具碟形花托，花梗丝状；雄花：萼片4或6(或8)；花瓣3或4(-6)。核果红色，阔圆球形。花期4-5月，果期6-7月。生海拔1000米以下灌丛中或林缘。产长江流域以南。

Herbaceous vines. Leaves papery, peltate, triangular-oblate or suborbicular, with palmate veins 7-9. Male inflorescences often in axillary panicles, capitate, with discoid receptacle; peduncles filamentous; male flower: sepals 4 or 6(or 8); petals 3 or 4(-6). Drupes red, broadly rotund. Fl. Apr-May. Fr. Jun-Jul. Thickets or forest edges below 1000 m. Distributed in south of Yangtze River.

荷包地不容

Stephania dicentrinifera H. S. Lo et M. Yang

落叶草质藤本。叶纸质，三角状近圆形，掌状10-11脉。花序复伞状聚伞状，常腋生或有生腋生短枝上；雄花6萼片；雌花1萼片，小；花瓣2；心皮无毛。核果具肉质果梗，成熟时红色。花期6-7月，果期9-10月。常生多砾石的林下。产云南。

Vines deciduous herbaceous. Leaves papery, triangular-suborbioular, palmately 10-11-veined. Inflorescences compound umbelliform cymes, usually axillary or sometimes on short axillary branches;male flower with 6 sepals; female flower with one small sepal; petals 2; carpels glabrous. Drupes with fleshy carpopodium, red when ripe. Fl. Jun-Jul. Fr. Sep-Oct. Often on gravelly patches under forests. Distributed in Yunnan.

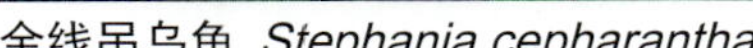

金线吊乌龟 *Stephania cepharantha*

荷包地不容 *Stephania dicentrinifera*

广西地不容 *Stephania kwangsiensis*

Woody vines. Branches slender, striate, usually densely pubescent. Leaves cordate-rotund or rotund, palmately 5-7-veined. Male flowers in a corymbose-cyme; female flowers in a panicle; sepals free. condyles bordered by a horseshoe-shaped ridge. Fl. Apr-May. Fr. Aug-Sep. Dense forests at 200-600 m. Distributed in S Yunnan, Guizhou and Guangxi. Pantropical spreading.

广西地不容

Stephania kwangsiensis Lo

草质、落叶藤本。叶纸质，三角状圆形至圆形，长、宽近相等，5-12厘米或稍长，全缘或有角状粗齿，两面无毛；叶柄长4-9厘米，基部扭曲。复伞形聚伞花序腋生；雄花：萼片6，排成2轮；花瓣3，里面有2个大腺体；雌花：萼片1，偶有2，近卵形；花瓣2，偶有3。核果红色。花期5月。生石灰岩山地。产广西和云南。

Herbaceous and deciduous vines. Leaves papery, triangular-rounded to rotund, length and width subequal, 5-12 cm or slightly longer, margin entire or horned serrate, both surfaces glabrous; petiole 4-9 cm long, twining at base. Inflorescences compound umbelliform cymes, axillary; staminate flower: sepals 6 in 2 whorls; petals 3, with 2 large glands inside; pistillate flower: sepal 1(or 2), subovate; petals 2(or 3). Drupes red. Fl. May. Limestone mountains. Distributed in Guangxi and Yunnan.

锡生藤

Cissampelos pareira L. var. **hirsuta** (Buch. ex DC.) Forman

木质藤本。枝条细弱，具条纹，常密具柔毛。叶心状圆形或圆形，具掌状5-7脉。雄花伞房状聚伞花序；雌花圆锥花序，萼片分离；胎座迹基部马蹄形。花期4-5月，果期8-9月。生海拔200-600米的密林中。产云南南部、贵州和广西。泛热带分布。

粉叶轮环藤

Cyclea hypoglauca (Schauer) Diels

木质藤本。叶纸质，两面无毛或疏具白色柔毛，掌状5-7脉。雄花：萼片4-5，离生；花瓣4-5，连合，杯状；雌花：萼片2；花瓣2，不等大，微肉质。果红色，无毛。生林缘或山地灌丛中。产中国西南、华南和东南。越南北部亦有。

Woody vines. Leaves papery, both surfaces glabrous or sparsely whitish puberulent, palmately 5-7-veined. Male flower: sepals 4-5, free; petals 4-5, connate, cup-like; famale flower: sepals 2; petals 2, unequal, slightly fleshy. Fruits red, glabrous. Forest edges or thickets on mountains. Distributed in SW, S and SE China. Also in N Vietnam.

锡生藤 *Cissampelos pareira* var. *hirsuta*

粉叶轮环藤 *Cyclea hypoglauca*

木兰科
Magnoliaceae

锈毛木莲 *Manglietia rufibarbata*

粗梗木莲 *Manglietia crassipes*

大果木莲
Manglietia grandis Hu et W. C. Cheng

常绿乔木。小枝浅灰，粗壮，无毛。托叶痕约为叶柄的1/4。叶革质，下面具乳突。花大，芳香，花梗粗壮；花被片12，外轮3片较薄，淡红色。聚合果长圆状卵球形，长10-12厘米。花期5月，果期9-10月。生海拔800-1500米的常绿阔叶林中。产云南和广西。

Evergreen trees. Twigs pale gray, thick and strong, glabrous. Stipular scar as long as 1/4 of petiole. Leaves leathery, abaxially papillate. Flowers large, fragrant, peduncles stout; tepals 12, outer 3 thinner, pale red. Aggregate fruits oblong-ovoid, 10-12 cm long. Fl. May. Fr. Sep-Oct. Evergreen broad-leaved forests at 800-1500 m. Distributed in Yunnan and Guangxi.

锈毛木莲
Manglietia rufibarbata Dandy

常绿乔木。幼枝、营养芽、托叶、叶柄、幼叶和果梗密具锈色绒毛。叶革质，倒卵形或倒卵状椭圆形，下面被褐色长毛。花大，乳白色，单生枝头；花被片9-12。聚合果狭长圆柱形，长5-7厘米。花期5月，果期9-10月。生海拔约1500米的常绿阔叶林中。产云南(马关县)。越南北部亦有。

Evergreen trees. Young twigs, vegetative buds, stipules, petioles, young leaves and fruiting peduncles densely rust-colored tomentose. Leaves leathery, obovate or obovate-elliptic, abaxially with brown long hairs. Flowers large, creamy white; solitary and terminal; tepals 9-12. Aggregate fruits narrowly cylindric, 5-7 cm long. Fl. May. Fr. Sep-Oct. Evergreen broad-leaved forests at ca. 1500 m. Distributed in Yunnan (Maguan County). Also in N Vietnam.

粗梗木莲
Manglietia crassipes Y. W. Law

常绿小乔木或灌木。小枝、营养芽和叶下具白粉。托叶痕长约为叶柄长的1/6。叶革质；佛焰苞1，质薄。花梗粗壮，长3.5-4厘米；花被片9，3轮，白色。聚合果卵球形，长5-6厘米；蓇葖果椭圆体，具短喙；果梗粗，直径约1厘米。花期5月，果期9月。生海拔约1300米林中。仅产广西(金秀)。

Small evergreen trees or shrubs. Twigs, vegetative buds and abaxial surfaces of leaves glaucous. Stipular scar as long as 1/6 of petiole. Leaves leathery. Spathe 1, thin; peduncles thick and strong, 3.5-4 cm long; tepals 9, 3-whorled, white. Aggregate fruits ovoid, 5-6 cm long; follicles ellipsoid, with short beak at pex; carpopodia thick, ca. 1 cm diam. Fl. May. Fr. Sep. Forests at ca. 1300 m. Distributed only in Guangxi (Jinxiu).

大果木莲 *Manglietia grandis*

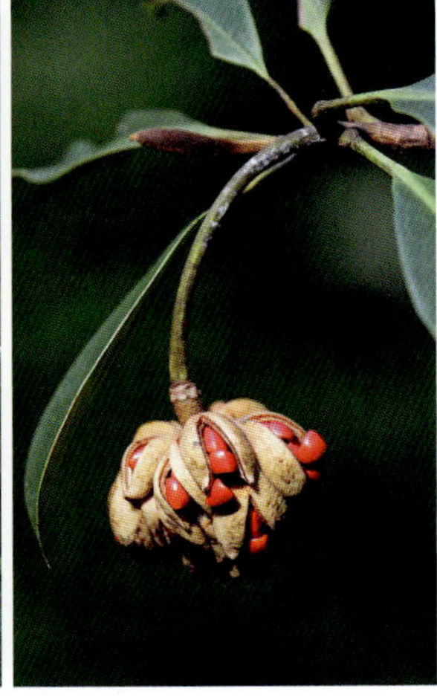

桂南木莲 *Manglietia conifera*

桂南木莲

Manglietia conifera Dandy

常绿乔木。营养芽和幼枝具红褐色柔毛。叶长12-15厘米，革质，下面起初灰绿色，被硬毛或白粉；叶柄长2-3厘米；托叶痕长3-5毫米。花被片9-11，外面3个花被片常绿色，内轮3或4个纯白色；花梗细长，4-7厘米，向下弯垂。聚合果长4-5厘米；种子表面具点。生海拔700-1300米山地林中。花期5-6月，果期9-10月。产云南、贵州、广西、广东和湖南。

Evergreen trees. Vegetative buds and young twigs reddish brown pubescent. Leaves 12-15 cm long, leathery, abaxially at first grayish green and hispidulous or glaucous; petiole 2-3 cm long; stipular scar 3-5 mm long. Tapels 9-11, outer 3 tepals usually green, inner 3 or 4 tepals pure white; pedicels slender, 4-7 cm long, pendulous. Aggregated fruits 4-5 cm long; endotesta dotted. Fl. May-Jun. Fr. Sep-Oct. Montane forests at 700-1300 m. Distributed in Yunnan, Guizhou, Guangxi, Guangdong and Hunan.

香木莲

Manglietia aromatica Dandy

常绿大乔木，除营养芽外无毛。叶薄革质，侧脉12-16对。花梗粗壮，果期长10-15毫米；花芳香；花被片白色，11-12片。果木质，成熟时心皮顶端截平。花期5-6月，果期9-10月。生海拔900-1600米的山地丘陵常绿阔叶林中。产云南、贵州和广西。

Large evergreen trees, glabrous except vegetative buds. Leaves thinly leathery, lateral nerves 12-16 pairs. Peduncles strong, 10-15 mm long in fruit; flowers fragrant; tepals white, 11-12. Fruits ligneous, carpels truncate at apex when mature. Fl. May-Jun. Fr. Sep-Oct. Evergreen broad-leaved forests on hills at 900-1600 m. Distributed in Yunnan, Guizhou and Guangxi.

中缅木莲

Manglietia hookeri Cubitt. et W. W. Sm.

常绿大乔木。新生部分具灰白色至浅褐色平伏毛。叶柄长3-5厘米；托叶痕钝三角形，长2-3厘米。叶革质，披针形、长圆倒卵形或狭倒卵形，侧脉16-20对。花白色，芳香，直径约10厘米，花被片9-12。聚合果卵状长圆形。花期4-5月，果期9月。生海拔1400-3000米的常绿阔叶林中。产云南和贵州。缅

香木莲 *Manglietia aromatica*

中缅木莲 *Manglietia hookeri*

四川木莲 *Manglietia szechuanica*

甸亦有。

Large evergreen trees. Young parts grayish white to pale brown appressed pilose. Petioles 3-5 cm long; stipular scar obtusely triangular, 2-3 cm long; leaf blades leathery, lanceolate, oblong-obovate or narrowly obovate, lateral nerves 16-20 pairs. Flowers white, fragrant, ca. 10 cm diam; tepals 9-12. Aggregate fruits ovoid-oblong. Fl. Apr-May. Fr. Sep. Evergreen broad-leaved forests at 1400-3000 m. Distributed in Yunnan and Guizhou. Also in Myanmar.

四川木莲
Manglietia szechuanica Hu

常绿乔木。幼枝绿色，密具柔毛，毛渐脱落至仅于节上留存。叶革质。花芳香；花被片9，紫红色；雌蕊群卵状椭圆体形；心皮狭椭圆体形，密被褐色短柔毛。聚合果卵球形，长8-10厘米。花期4-5月，果期8-9月。生海拔1300-2000米的常绿阔叶林中。产云南北部、四川中部和南部。

Evergreen trees. Young twigs green, densely villous, hairs gradually deciduous, only residual at nodes. Leaves leathery. Flowers fragrant; tepals 9, purple-red; gynoecia ovoid-ellipsoid; carpels narrowly ellipsoid, densely rusty-pubescent. Aggregate fruits ovoid, 8-10 cm long. Fl. Apr-May. Fr. Aug-Sep. Evergreen broad-leaved forests at 1300-2000 m. Distributed in N Yunnan, and C and S Sichuan.

川滇木莲
Manglietia duclouxii Finet et Gagnep.

常绿乔木。小枝无毛。托叶痕长约为叶柄的1/3。叶薄革质，两面无毛。花梗无毛；花被片9，肉质，外轮3片红色，内轮2片紫红色；雌蕊群狭椭圆体形，长7-8毫米。聚合果卵球状椭圆体形。花期5-6月，果期9-10月。生海拔1300-2000米的常绿阔叶林中。产云南、四川和广西。

Evergreen trees. Twigs glabrous. Stipular scar ca. 1/3 as long as petiole. Leaves thinly leathery, glabrous on both surfaces. Peduncles glabrous; tepals 9, fleshy, outer 3 red, tepals of inner 2 whorls purplish red; gynoecia narrowly ellipsoid, 7-8 mm long. Aggregate fruits ovoid-ellipsoid. Fl. May-Jun. Fr. Sep-Oct. Evergreen broad-leaved forests at 1300-2000 m. Distributed in Yunnan, Sichuan and Guangxi.

川滇木莲 *Manglietia duclouxii*

红花木莲 *Manglietia insignis*

滇桂木莲 *Manglietia forrestii*

红花木莲

Manglietia insignis (Wall.) Blume

常绿乔木。小枝无毛，或节间幼时具锈色至黄褐色柔毛。叶革质，倒披针形，长圆形或长圆状椭圆形。花芳香；花被片9-12，外轮花被片红色或紫红色。聚合果成熟后紫红色，卵球状椭圆体形，长7-12厘米。花期5-6月，果期8-9月。生海拔900-1200米的常绿阔叶林中。产云南、四川、西藏、贵州、广西和湖南。尼泊尔、印度北部、缅甸北部和泰国亦有。

Evergreen trees. Twigs glabrous or nodes ferruginous to yellowish brown pubescent when young. Leaves leathery, oblanceolate, oblong or oblong-elliptic. Flowers fragrant; tepals 9-12, outer tepals red or purple-red. Aggregate fruits purplish red when fresh, ovoid-ellipsoid, 7-12 cm long. Fl. May-Jun. Fr. Aug-Sep. Evergreen broad-leaved forests at 900-1200 m. Distributed in Yunnan, Sichuan, Xizang, Guizhou, Guangxi and Hunan. Also in Nepal, N India, N Myanmar and Thailand.

滇桂木莲

Manglietia forrestii W. W. Sm. ex Dandy

常绿乔木。小枝、花芽、叶柄和外轮花被片背面和花梗具红褐色平伏、有光泽的柔毛。叶革质。花芳香，白色，单生枝顶；花被片9(或10)，白色。聚合果卵球形，长4-6厘米。花期6月，果期9-10月。生海拔1100-2900米的林中。产云南和广西。

Evergreen trees. Young twigs, buds, petioles, abaxial base of outer tepals, and peduncles reddish brown appressed glossy villous. Leaves leathery. Flowers fragrant, white, solitary and terminal; tepals 9(or 10), white. Aggregate fruits ovoid, 4-6 cm long. Fl. Jun. Fr. Sep-Oct. Forests at 1100-2900 m. Distributed in Yunnan and Guangxi.

木莲

Manglietia fordiana Oliv.

常绿乔木。幼枝和营养芽具红褐色柔毛，后渐无毛。叶革质，下面疏具红褐色柔毛。花梗长5-11(-40)毫米；花单生枝顶；花被片9，3片轮生，白色。聚合果卵球形，长2-5厘米。花期4-5月，果期8-10月。生海拔300-1200米的山地、森林或河边。产中国西南、华南、东南、华中和华东。越南亦有。

Evergreen trees. Young twigs and vegetative buds reddish brown pubescent, later glabrescent. Leaves leathery, abaxially sparsely reddish brown pubescent. peduncles 5-11(-40) mm long; flowers solitary and terminal;

木莲 *Manglietia fordiana*

海南木莲 *Manglietia fordiana* var. *hainanensis*

tepals 9, 3-whorled, white. Aggregate fruits ovoid, 2-5 cm long. Fl. Apr- May. Fr. Aug-Oct. Hills, forests or beside rivers at 300-1200 m. Distributed in SW, S, SE, C and E China. Also in Vietnam.

海南木莲

Manglietia fordiana Oliv. var. **hainanensis** (Dandy) N. H. Xia

常绿大乔木。叶薄革质，下面疏生红褐色平伏毛，侧脉12-16对。花序梗无毛；花被片9，外轮3片较薄，阔卵形至倒卵形；外面绿色，内轮6片白色。聚合果褐色。花期4-5月，果期9-10月。生海拔300-1200米的山地、森林和河边。产海南。

Large evergreen trees. Leaves thinly coriacecus, abaxially sparsely rusty-hairy, lateral veins 12-16 pairs. Peduncles glabrous; tepals 9, outer 3 thinner, broadly ovate to obovate, abaxially green, inner 6 white. Aggregate fruits brown. Fl. Apr-May. Fr. Sep-Oct. Hills, forests and beside rivers at 300-1200 m. Distributed in Hainan.

华盖木

Pachylarnax sinica (Law) N. H. Xia et C. Y. Wu

常绿大乔木。托叶与叶柄离生，无托叶痕。叶革质，在芽时不对折，平展紧贴芽。侧脉每边13-16条。花单生，顶生，芳香；花被片9，3轮，外轮3枚外面深红色，内面浅红色，内2轮6枚白色；雄蕊65；心皮13-16。聚合果成熟时绿色，干后暗褐色；蓇葖果成熟时厚木质，沿腹缝线及顶端开裂。花期4月，果期9-10月。生海拔1300-1600米的山谷常绿阔叶林中。产云南(马关县和喜洲)。

Large evergreen trees. Stipules free from petiole. Leaf blades leathery, not folded and firmly adnate to young buds when young; lateral veins 13-16 per side. Flowers solitary and terminal, fragrant; tepals 9, 3-whorled, outer 3 abaxially deep red, adaxially reddish, inner 6 white; stamens 65; carpels 13-16. Aggregate fruits green when mature, dark-brown when dry; follicles thickly woody when mature, completely dehiscing along ventral sutures and at apex. Fl. Apr. Fr. Sep-Oct. Evergreen broad-leaved forests in valleys at 1300-1600 m. Distributed in Yunnan (Maguan County and Xizhou).

华盖木 *Pachylarnax sinica*

大叶木兰
Magnolia henryi Dunn

落叶乔木。托叶痕几达叶柄全长；叶柄长4-11厘米。叶极大型，长20-70厘米，革质，长圆形或倒卵状长圆形。花梗下垂，长约8厘米；花大，芳香，乳白色，肉质；雌蕊群狭椭圆体形；心皮85-95。花期5月，果期8-9月。生海拔500-1500米的林中。产云南南部和东南部。缅甸、老挝和泰国亦有。

Deciduous trees. Stipular scars nearly reaching apex of petioles; petioles 4-11 cm long. Leaf blades very large, 20-70 cm long, leathery, oblong or obovate-oblong. Peduncles pendulous, ca. 8 cm long; flowers large, fragrant, creamy white, fleshy; gynoecium narrowly ellipsoid; carpels 85-95. Fl. May. Fr. Aug- Sep. Forests at 500-1500 m. Distributed in S and SE Yunnan. Also in Myanmar, Laos and Thailand.

山玉兰
Magnolia delavayi Franch.

常绿小乔木。树皮粗糙纵裂。托叶痕几达叶柄全长。叶卵形至卵状长圆形，革质，长10-20(-32)厘米。花梗直立，长3-4厘米；花大，芳香，杯状；花被片9-10，肉质；心皮约100。果实卵状椭圆体形。花期4-6月，果期8-10月。生海拔1500-2800米的林中、湿润坡地及石灰岩地区。产云南、四川和贵州。

Small evergreen trees. Bark coarse and fissured. Stipular scars nearly reaching apex of petioles; leaf blades ovate to ovate-oblong, leathery, 10-20(-32) cm long. Peduncles erect, 3-4 cm long; flowers large, fragrant, cupulate; tepals 9-10, fleshy; carpels ca. 100. Fruits ovoid-ellipsoid. Fl. Apr-Jun. Fr. Aug-Oct. Forests, wet slopes, limestone areas at 1500-2800 m. Distributed in Yunnan, Sichuan and Guizhou.

山玉兰 *Magnolia delavayi*

大叶木兰 *Magnolia henryi*

夜香木兰

Magnolia coco (Lour.) DC.

常绿灌木或小乔木。叶革质，侧脉每边8-10。花梗下垂，具3或4个苞片脱落痕；花顶生单生，大型，圆球形，芳香；花被片9，外轮3片淡绿色，内轮6片白色；心皮约10，狭卵球形，背面有1纵沟至花柱基部。聚合果约长3厘米。花期5-6月，果期9-10月。生海拔600-900米湿润肥沃土壤的林中。产云南、广西、广东、台湾、福建和浙江。越南亦有。广泛栽培于东南亚。

Evergreen shrubs or small trees. Leaves leathery, lateral veins 8-10 per side. Peduncles pendulous, with 3 or 4 bract scars; flowers terminal and solitary, large, globose, fragrant; tepals 9, outer 3 pale green, inner 6 white; carpels ca. 10, narrowly ovoid, abaxially with 1 furrow downward to base of style. Aggregate fruits ca. 3 cm long. Fl. May-Jun. Fr. Sep-Oct. Forests with moist and rich soil at 600-900 m. Distributed in Yunnan, Guangxi, Guangdong, Taiwan, Fujian and Zhejiang. Also in Vietnam. Widely cultivated in SE Asia.

馨香木兰 *Magnolia odoratissima*

厚朴 *Magnolia officinalis*

馨香木兰

Magnolia odoratissima Law et R. Z. Zhou

乔木。叶革质，长8-14(-30)厘米，侧脉每边9-13条，下面被白色弯曲毛。花白色，极芳香；花被片9，内弯，肉质，乳白色；外轮花被片具9脉纹；雄蕊约175，长约3厘米。花期5月，果期9月。生海拔约1100米的林中。产云南东南部。

Trees. Leaves leathery, 8-14 (-30) cm long, lateral veins 9-13 per side, abaxially white-curved-hairy. Flowers white, very fragrant; tepals 9, incurved, fleshy, creamy white, tepals of outer whorl 9-veined; stamens ca. 175, ca. 3 cm long. Fl. May. Fr. Sep. Forests at ca. 1100 m. Distributed in SE Yunnan.

厚朴

Magnolia officinalis Rehder et E. H. Wils.

落叶乔木。叶大，近革质，7-9朵簇生于小枝顶端，嫩叶背被白色长毛。花直径10-15厘米，芳香；花被片9-12(-17)，白色，厚肉质，浅绿色。成熟蓇葖果椭圆状卵球形，具3-4毫米长的喙。花期5-6月，果期8-10月。生海拔300-1500米的林中，多栽培于山麓或村舍附近。产中国西南、华南、东南、华中、华北、华西和华东。

Deciduous trees. Leaves large, nearly leathery, 7-9 clustered on twig apex, young leaves abaxially with white villose. Flowers 10-15 cm diam, fragrant; tepals 9-12(-17), white, thickly fleshy, pale green. Mature follicles ellipsoid-ovoid, with a 3-4 mm long beak. Fl. May-Jun. Fr. Aug-Oct. Forests at 300-1500 m, often cultivated at foothills or near villages. Distributed in SW, S, SE, C, N, W and E China.

夜香木兰 *Magnolia coco*

凹叶厚朴 *Magnolia biloba*

西康天女花 *Magnolia wilsonii*

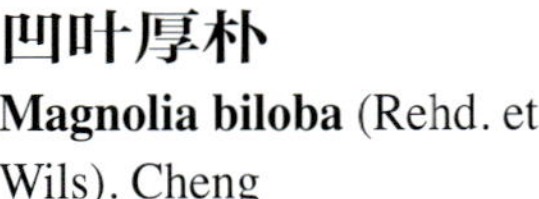

凹叶厚朴

Magnolia biloba (Rehd. et Wils). Cheng

乔木，高达15米。叶大，近革质，狭倒卵形，长15-30厘米，顶端有凹缺，侧脉15-25对。花与叶同时开放，直径10-15厘米；花被片9-12，白色，倒披针形；雄蕊多数；心皮多数。聚合果狭卵球形，长11-16厘米。花期4-5月。生海拔300-1000米的林中。产广西南部、广东南部、湖南、江西、福建、浙江西部和安徽。

Tree, up to 15 m tall. Leaves large, subcoriaceous, narrowly obovate or oblanceolate, 15-30 cm long, at apex deeply retuse and with 2 lobules, lateral nerves 15-25 pairs. Flower opening at the same time with the leaves, 10-15 cm diam; tepals 9-12, white, oblanceolate; stamens numerous; carpels numerous. Aggregate fruits narrowly ovoid, 11-16 cm long. Fl. Apr-May. Forests at 300-1000 m. Distributed in S Guangxi, S Guangdong, Hunan, Jiangxi, Fujian, W Zhejiang and Anhui.

西康天女花

Magnolia wilsonii (Finet et Gagnep.) Rehder

落叶灌木或小乔木。托叶痕长为叶柄的4/5-5/6。叶纸质，长圆状卵形或椭圆状卵形，叶背被灰色平伏长毛。花叶同期，起初杯状，花期为碟形，长10-12厘米；花被片9(-12)，白色。果实红色，成熟后变为紫色，下垂。花期5-6月，果期9-10月。生海拔1900-3000米的林中。产云南、四川和贵州。

Deciduous shrubs or small trees. Stipular scars 4/5-5/6 as long as petioles. Leaf blades papery, oblong-ovate or elliptic-ovate, abaxially densely adpressed villose. Flowers appearing at same time as leaves, cupular at first, plate-shaped at anthesis, 10-12 cm diam, fragrant; tepals 9(-12), white. Fruits red and then becoming purple when mature, terete, pendulous. Fl. May-Jun. Fr. Sep-Oct. Forests at 1900-3000 m. Distributed in Yunnan, Sichuan and Guizhou.

天女花

Magnolia sieboldii K. Koch

乔木。托叶痕长为叶柄的近1/2；叶膜质，下面沿脉被多室

天女花 *Magnolia sieboldii*

绒毛，散生金黄色腺点和白色长丝毛。花叶同期，起初杯状，花期为碟形，直径7-10厘米，直立或稍下垂，芳香；花被片9，白色。聚合果熟时红色，倒卵球形或长圆形。花期3-4月，果期9-10月。生海拔1600-2000米的林地或山地。产中国西南、华南、华北、华东和东北。朝鲜半岛和日本亦有。

Trees. Stipular scars nearly 1/2 as long as petioles; leaves membranous, abaxially covered with multicellular trichomes, scattered golden yellow dots, and white long sericeous hairs along veins. Flowers appearing at same time as leaves, cupular at first but plate-shaped at anthesis, 7-10 cm diam, erect or slightly nodding, fragrant; tepals 9, white. Aggregate fruits red when mature, obovoid or oblong. Fl. Mar-Apr. Fr. Sep-Oct. Forests or mountains at 1600-2000 m. Distributed in SW, S, N, E and NE China. Also in Korean Peninsula and Japan.

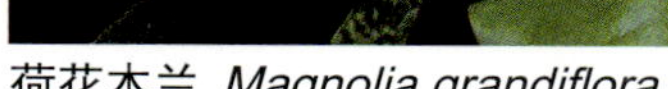
荷花木兰 *Magnolia grandiflora*

荷花木兰
Magnolia grandiflora L.

常绿大乔木。叶厚革质；托叶与叶柄离生，叶柄上无托叶痕。花大，直径15-20厘米，极芳香，乳白色，厚肉质；花被片9-12。果实圆柱形至卵球形；种子卵球形。花期5-6月，果期9-10月。长江以南有栽培。原产美洲东南部。

Large evergreen trees. Leaves thickly leathery; stipules separate from petioles, without stipule scars on petioles. Flowers large, 15-20 cm diam, very fragrant, creamy white, thickly fleshy; tepals 9-12. Fruits terete to ovoid; seeds ovoid. Fl. May-Jun. Fr. Sep-Oct. Cultivated to the south of Yangtze River. Native to SE America.

凹叶玉兰
Magnolia sargentiana Rehder et E. H. Wils.

落叶乔木。叶近革质，倒卵形，先端圆形具凹缺或具短尖，下面具浅绿色和密生银灰色波曲长柔毛。花先叶开放，长15-33(-36)厘米；花被片10-14，淡红色或紫红色。果实圆柱形。花期4-5月，果期9

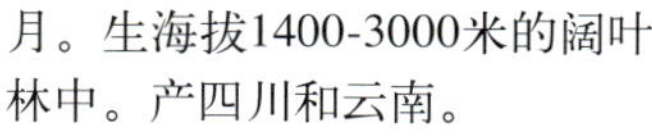
月。生海拔1400-3000米的阔叶林中。产四川和云南。

Deciduous trees. Leaves nearly leathery, obovate, apex rounded and emarginate or mucronate, abaxially pale green and densely silvery gray wavy villous. Flowers appearing before leaves, 15-33 (-36) cm long; tepals 10-14, pink or purple-red. Fruits terete. Fl. Apr-May. Fr. Sep. Broad-leaved forests at 1400-3000 m. Distributed in Sichuan and Yunnan.

凹叶玉兰 *Magnolia sargentiana*

武当玉兰 *Magnolia sprengeri*

武当玉兰

Magnolia sprengeri Pamp.

落叶乔木。叶纸质，倒卵形，长10-18厘米。每花蕾具花1朵；花先叶开放，杯形，芳香；花被片12(-14)，外面玫瑰红色，有深紫色纵纹。果实圆柱形，长6-18厘米。花期3-4月，果期8-9月。生海拔1300-2400米的林地或灌丛中。产中国西南、华中、华北、华西和华东，广泛栽培。

Deciduous trees. Leaves papery, obovate, 10-18 cm long. Each flower bud with 1 flower; flowers appearing before leaves, cupular, fragrant; tepals 12(-14), outside rosy red and dark purple striated. Fruits cylindric, 6-18 cm long. Fl. Mar-Apr. Fr. Aug-Sep. Forests or thickets at 1300-2400 m. Distributed in SW, C, N, W and E China, widely cultivated.

玉兰

Magnolia denudata Desr.

落叶乔木。托叶痕为叶柄长的1/4-1/3；叶纸质，长10-15(-18)厘米。花先叶开放，直径10-16厘米，直立，芳香；花被片9，白色，近等大，基部常粉红色。果实圆柱形。花期2-3月，果期8-9月。生海拔500-1000米的林中。产中国西南、华南、东南、华中、华西和华东。世界各地广泛栽培。

玉兰 *Magnolia denudata*

玉兰 *Magnolia denudata*

Deciduous trees. Stipular scars as long as 1/4-1/3 of petioles; leaf blades papery, 10-15(-18) cm long. Flowers appearing before leaves, 10-16 cm diam, erect, fragrant; tepals 9, white, subequal, base usually pinkish. Fruits cylindric. Fl. Feb-Mar. Fr. Aug-Sep. Forests at 500-1000 m. Distributed in SW, S, SE, C, W and E China. Widely cultivated throughout the world.

二乔玉兰

Magnolia soulangeana Soul. Bod.

乔木。托叶痕长为叶柄的约1/3；叶纸质，倒卵形。花先叶开放；花被片6-9，浅红色至深红色，外轮花被片稍短于内轮；成熟心皮黑色，卵球形至倒卵球形。花期2-3月，果期9-10月。栽培于中国。世界各地栽培。

Trees. Stipular scars ca. 1/3 as long as petioles. Leaf blades papery, obovate. Flowers appearing before leaves; tepals 6-9, pink to dark red, outer whorl of tepals slightly shorter than inner ones; mature carpels black, ovoid to obovoid. Fl. Feb-Mar. Fr. Sep-Oct. Cultivated in China. Also throughout the world.

二乔玉兰 *Magnolia soulangeana*

二乔玉兰 *Magnolia soulangeana*

宝华玉兰 *Magnolia zenii*

天目玉兰 *Magnolia amoena*

宝华玉兰

Magnolia zenii Cheng

落叶乔木。小枝绿色，无毛。叶膜质，侧脉每边8-10条，下面沿脉被长弯曲毛。花先叶开放，芳香；花被片9，白色；雄蕊群圆柱形，长约2厘米。聚合果圆柱形，长5-7厘米。花期3-4月，果期8-9月。生海拔约200米的丘陵地。仅产江苏(宝华山)。

Deciduous trees. Young twigs green, glabrous. Leaves membranous, lateral veins 8-10 per side, abaxially curved-villose along veins. Flowers appearing before leaves, fragrant; tepals 9, white; gynoecia terete, ca. 2 cm long. Aggregate fruits terete, 5-7 cm long. Fl. Mar-Apr. Fr. Aug-Sep. Hilly regions at ca. 200 m. Distributed only in Jiangsu (Baohua Mountain).

天目玉兰

Magnolia amoena Cheng

落叶乔木。托叶痕达叶柄1/5-1/2；叶长10-15厘米，纸质，侧脉10-13对。花先叶开放，芳香；花被片9，红色至淡红色，直径约6厘米；花丝紫红色。聚合果圆柱形，长4-10厘米。花期4-5月，果期9-10月。生海拔700-1000米的林中。产福建、浙江、河北、安徽、江苏和江西。

Deciduous trees. Stipular scars 1/5-1/2 as long as petioles; leaf blades 10-15 cm long, papery, secondary veins 10-13 on each side of midvein. Flowers blossom before leaves, fragrant; tepals 9, red to pale red, ca. 6 cm diam; filaments purplish red. Aggregate fruits cylindric, 4-10 cm long. Fl. Apr-May. Fr. Sep-Oct. Forests at 700-1000 m. Distributed in Fujian, Zhejiang, Hebei, Anhui, Jiangsu and Jiangxi.

紫玉兰

Yulania liliiflora (Desr.) D. C. Fu

落叶灌木。托叶痕为叶柄长的约1/2；叶纸质，椭圆状倒卵形或倒卵形，基部沿叶柄下延。花叶同期，瓶形，直立，稍芳香；花被片9-12，紫色或紫红色。果实深紫褐色，圆柱形。花期3-4月，果期8-9月。生海拔300-1600米的山坡林。产云南、四川、福建、重庆、湖北和陕西。

Deciduous shrubs. Stipular scars ca. 1/2 as long as petioles; leaf blades papery, elliptic-obovate or obovate, base gradually narrowing along petioles. Flowers appearing at same time with leaves, vase-shaped, erect, slightly fragrant; tepals 9-12, purple or purple-red. Fruits dark purplish brown, cylindric. Fl. Mar-Apr. Fr. Aug-Sep. Hillside forests at 300-1600 m. Distributed in Yunnan, Sichuan, Fujian, Chongqing, Hubei and Shaanxi.

紫玉兰 *Yulania liliiflora*

黄山玉兰 *Magnolia cylindrica*

黄山玉兰
Magnolia cylindrica E. H. Wils.

落叶乔木。幼枝、叶柄和叶背均具浅黄色平伏绒毛。叶膜质；托叶痕为叶柄长1/6-1/3。花被片6，外轮3片非常小，膜质，内2轮6片白色，基部稍带红色；雌蕊群绿色。蓇葖果全部愈合成圆柱形聚合果。花期4-6月，果期8-9月。生海拔700-1600米的山地林间。产福建、浙江、湖北、河南、安徽和江西。

Deciduous trees. Young twigs, petioles, and leaf abaxial surfaces with pale yellow appressed trichomes. Leaves membranous; scars of stipules as long as 1/6-1/3 length of petioles. Tepals 6, outer 3 very small, membranous, inner 6 white, slightly reddish at base; gynoecia green. Follicles all connate in terete aggregate fruits. Fl. Apr-Jun. Fr. Aug-Sep. Forests in mountains at 700-1600 m. Distributed in Fujian, Zhejiang, Hubei, Henan, Anhui and Jiangxi.

盖裂木
Talauma hodgsonii Hook. f. et Thoms.

常绿乔木。小枝稍具粉，无毛。叶革质，倒卵状长圆形，长20-50厘米，侧边每边10-20条。花梗粗壮；苞片佛焰苞状，紫色；花被片9，厚肉质，外轮3个草绿色，中部和内部者乳白色。蓇葖果40-80。花期4-5月，果期8月。生海拔800-1500米林间。产云南西南部和西藏南部。印度东北部、尼泊尔、不丹、缅甸北部和泰国北部亦有。

Evergreen trees. Twigs slightly glaucous, glabrous. Leaves leathery, obovate-oblong, 20-50 cm long, laterally 10-20 per side. Peduncles stout; bracts spathe-like, purple; tepals 9, thickly fleshy, outer 3 tepals grassy green, middle and inner whorls milky white. Follicles 40-80. Fl. Apr-May. Fr. Aug. Forests at 800-1500 m. Distributed in SW Yunnan and S Xizang. Also in NE India, Nepal, Bhutan, N Myanmar and N Thailand.

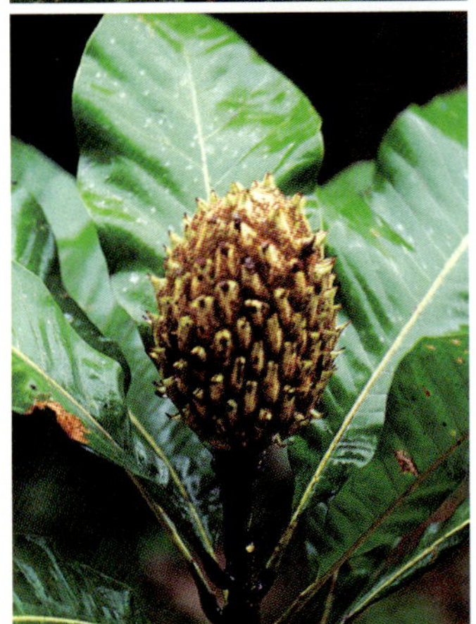
盖裂木 *Talauma hodgsonii*

光叶拟单性木兰
Parakmeria nitida (W. W. Sm.) Y. W. Law

常绿大乔木，具两性花。叶革质，椭圆形、长椭圆形或倒卵状椭圆形，长5.5-9.5厘米；侧脉7-13对。花两性，单生枝顶，芳香；花被约12，浅黄色。聚伞果长5-7.5厘米。花期3-5月，果期9-10月。生海拔1800-2500米的常绿阔叶林中。产云南和西藏。缅甸北部亦有。

Large evergreen trees, with bisexual flowers. Leaves leathery, elliptic, oblong or obovate-elliptic, 5.5-9.5 cm long, secondary veins 7-13 on each side of midvein. Flowers bisexu-

光叶拟单性木兰 *Parakmeria nitida*

云南拟单性木兰 *Parakmeria yunnanensis*

长蕊木兰 *Alcimandra cathcartii*

al, solitary and terminal, fragrant; tepals ca. 12, pale yellow. Aggregate fruits 5- 7.5 cm long. Fl. Mar-May. Fr. Sep-Oct. Evergreen broad-leaved forests at 1800-2500 m. Distributed in Yunnan and Xizang. Also in N Myanmar.

云南拟单性木兰

Parakmeria yunnanensis Hu

常绿大乔木，雄全异株。叶薄革质，卵状长圆形或卵状椭圆形，长 6.5-15(-20)厘米，两面幼时紫红色。花杂性，单生枝顶，芳香，白色；雄性花花被片9-12。聚合果长卵球形，长约6厘米。花期5月，果期9-10月。生海拔1200-1500米的林中。产云南和西藏。缅甸北部亦有。

Large evergreen trees, andro-dioecious. Leaves thinly leathery, ovate-oblong or ovate-elliptic, 6.5-15(-20) cm long, both surfaces purplish red when young. Flowers polygamous, solitary and terminal, fragrant, white; male tepals 9-12. Aggregate fruits long ovoid, ca. 6 cm long. Fl. May. Fr. Sep-Oct. Forests at 1200-1500 m. Distributed in Yunnan and Xizang. Also in N Myanmar.

乐东拟单性木兰

Parakmeria lotungensis (Chun et C. Tsoong) Y. W. Law

常绿乔木，雄全异株。叶狭倒卵状椭圆形、倒卵状椭圆形或狭椭圆形，革质，长6-11厘米，侧脉每边9-13条。雄花被片9-12，外轮3-4片浅黄色，内轮6-8片乳白色；雄蕊10-35；雌蕊群卵圆形。聚合果长3-6厘米。花期4-5月，果期8-9月。生海拔700-1400米的阔叶林中。产中国西南、华南和东南。

Evergreen trees, andro-dioecious. Leaves narrowly obovate-elliptic, obovate-elliptic, or narrowly elliptic, leathery, 6-11 cm long, lateral veins 9-13 per side. Male tepals 9-12, outer 3-4 pale yellow, inner 6-8 creamy white; stamens 10-35; gynoecia ovoid. Aggregate fruits 3-6 cm long. Fl. Apr-May. Fr. Aug-Sep. Broad-leaved forests at 700-1400 m. Distributed in SW, S and SE China.

长蕊木兰

Alcimandra cathcartii (Hook. f. et Thoms.) Dandy

常绿乔木。叶薄革质，卵形或椭圆状卵形，全缘。佛焰苞状苞片绿色，紧接花被片；花被片9，白色，具透明腺体；雄蕊长约4厘米，雌蕊群长约2厘米；成熟心皮扁球形，具白色皮孔。花期5月，果期8-9月。生海拔1800-2700米的林中或混交林中。产云南和西藏。印度北部、不丹、缅甸北部和越南北部亦有。

Evergreen trees. Leaves thinly leathery, ovate or elliptic-ovate, entire. Spathaceous bracts green, just basal to tepals; tepals 9, white, with pellucid glands; stamens ca. 4 cm long; gynoecia ca. 2 cm long. Mature carpels compressed globose, white lenticellate. Fl. May. Fr. Aug-Sep. Forests or mixed forests at 1800-2700 m. Distributed in Yunnan and Xizang. Also in N India, Bhutan, N Myanmar and N Vietnam.

乐东拟单性木兰 *Parakmeria lotungensis*

绒叶含笑 *Michelia velutina*

白兰 *Michelia alba*

绒叶含笑
Michelia velutina DC.

常绿乔木。幼嫩枝和叶背密被灰色长绒毛。托叶痕长为叶柄的约1/2；叶柄长1-2厘米；叶薄革质，狭椭圆形或椭圆形。花芳香；花被片10-12，淡黄色。聚合果长10-13厘米。花期5-6月，果期8-9月。生海拔1500-2400米的林中。产云南和西藏。印度东北部、尼泊尔和不丹亦有。

Evergreen trees. Young branchlets and abaxial leaves densely gray-tomentose. Stipular scars ca. 1/2 as long as petioles; petioles 1-2 cm long; leaf blades thinly leathery, narrowly elliptic or elliptic. Flowers fragrant; tepals 10-12, yellowish. Aggregate fruits 10-13 cm long. Fl. May-Jun. Fr. Aug-Sep. Forests at 1500-2400 m. Distributed in Yunnan and Xizang. Also in NE India, Nepal and Bhutan.

白兰
Michelia alba DC.

常绿大乔木。叶薄革质；托叶痕短于叶柄的一半；叶柄长1.5-2厘米。花白色，极香；花被片10，白色，披针形，长3-4厘米。花期4-9月。栽培于云南、广西、广东、海南、台湾和福建。东南亚广泛栽培。原产印度尼西亚(爪哇)。

Large evergreen trees. Leaves thinly leathery; stipule scars shorter than 1/2 length of petioles; petioles 1.5-2 cm long. Flowers white, very fragrant; tepals 10, white, lanceolate, 3-4 cm long. Fl. Apr-Sep. Cultivated in Yunnan, Guangxi, Guangdong, Hainan, Taiwan and Fujian. Widely cultivated in SE Asia. Native to Indonesia (Java).

南亚含笑
Michelia doltsopa Buch.-Ham. ex DC.

常绿乔木。叶薄革质，椭圆形、长圆状椭圆形或狭椭圆形，网脉密致；托叶痕长为叶柄的1/5。花芳香；花被片12-16，狭倒卵状匙形，白色。聚合果长4-7厘米。蓇葖近倒卵球形。花期4-5月，果期9-10月。生海拔1500-2400米的常绿阔叶林中。产云南和西藏。印度东北部、尼泊尔、不丹和缅甸北部亦有。

南亚含笑 *Michelia doltsopa*

Evergreen trees. Leaves thinly leathery, elliptic, oblong-elliptic or narrowly elliptic, reticulate veins dense; length of stipule scars equal to 1/5 petioles. Flowers fragrant; tepals 12-16, narrowly obovate-spathulate, white. Aggregate fruits 4-7 cm long. Mature carpels nearly obovoid. Fl. Apr-May. Fr. Sep-Oct. Evergreen broad-leaved forests at 1500-2400 m. Distributed in Yunnan and Xizang. Also in NE India, Nepal, Bhutan and N Myanmar.

黄兰
Michelia champaca L.

常绿乔木。托叶痕长为叶柄的0.5-1倍；叶柄长2-4厘米；叶薄革质，椭圆形或卵形。花单生叶腋，黄色，极香；花被片15-20，倒披针形。蓇葖倒卵球状椭圆体形。花期6-7月，果期9-10月。生海拔200-1600米的常绿阔叶林。产中国西南、华南和东南。印度、尼泊尔、缅甸、泰国、越南、马来西亚和印度尼西亚亦有。

Evergreen trees. Stipular scars 0.5-1 × as long as petioles; petioles 2-4 cm long; leaf blades thinly leathery, elliptic or ovate.

黄兰 *Michelia champaca*

Flowers solitary and axillary, yellow, very fragrant; tepals 15-20, oblanceolate. Mature carpels obovoid-ellipsoid. Fl. Jun-Jul. Fr. Sep-Oct. Evergreen broad-leaved forests at 200-1600 m. Distributed in SW, S and SE China. Also in India, Nepal, Myanmar, Thailand, Vietnam, Malaysia and Indonesia.

多花含笑

Michelia floribunda Finet et Gagnep.

常绿乔木。托叶痕长为叶柄之半或过半；叶柄长1-1.5(-2.5)厘米；叶革质，狭卵状披针形、狭卵形或狭倒卵状披针形。花单生叶腋，芳香；花被片11-13，白色，匙形或倒披针形。花期2-4月，果期8-9月。生海拔1300-2700米的林中。产云南、四川、西藏、重庆、湖北和湖南。缅甸、老挝、泰国和越南亦有。

Evergreen trees. Stipule scars about 1/2 of petioles or longer; petioles 1-1.5(-2.5) cm long; leaf blades leathery, narrowly ovate-elliptic, narrowly ovate or narrowly obovate-elliptic. Flowers solitary and axillary, fragrant; tepals 11-13, white, spathulate or oblanceolate. Fl. Feb-Apr. Fr. Aug-Sep. Forests at 1300-2700 m. Distributed in Yunnan, Sichuan, Xizang, Chongqing, Hubei and Hunan. Also in Myanmar, Laos, Thailand and Vietnam.

多花含笑 *Michelia floribunda*

紫花含笑

Michelia crassipes Y. W. Law

常绿小乔木或灌木。叶革质；叶柄短，长约2毫米，密被黄褐色绒毛。花单生叶腋，极芳香；花被片6，紫红色或深紫色；花梗密被黄褐色绒毛。聚合果长2.5-5厘米；蓇葖多于10枚。花期4-5月，果期8-9月。生海拔300-1000米的常绿阔叶林和峡谷中。产广西东北部、广东北部和湖南南部。

Small evergreen trees or shrubs. Leaves leathery; petioles short, ca. 2 mm long, densely yellowish-brown tomentose. Flowers solitary and axillary, very fragrant; tepals 6, purplish red or dark purple; pedicels densely yellowish-brown-tomentose. Aggregate fruits 2.5-5 cm long; mature carpels more than 10. Fl. Apr-May. Fr. Aug-Sep. Evergreen broad-leaved forests, ravines at 300-1000 m. Distributed in NE Guangxi, N Guangdong and S Hunan.

紫花含笑 *Michelia crassipes*

云南含笑 *Michelia yunnanensis*

云南含笑
Michelia yunnanensis Franch. ex Finet et Gagnep.

常绿乔木。幼枝、花芽和幼叶背和叶柄具深红色平伏毛。托叶痕长约为叶柄的2/3或达叶柄顶端；叶柄长4-5毫米；叶革质。花单生叶腋，极香；花被片6-12(-17)，白色；雌蕊群较雄蕊群长。聚合果通常仅5-9枚蓇葖果发育。花期3-4月，果期8-9月。生海拔1100-2300米的丛林中。产云南、四川、西藏和贵州。

Evergreen trees. Young twigs, buds, young leaves adaxial surfaces and petioles with dark red appressed trichomes. Stipular scars ca. 2/3 as long as petiole or reaching petiole apex; petioles 4-5 mm long; leaf blades leathery. Flowers solitary and axillary, very fragrant; tepals 6-12(-17), white; gynoecia longer than stamens. Aggregate fruits usually only with developed 5-9 follicles. Fl. Mar-Apr. Fr. Aug-Sep. Thickets at 1100-2300 m. Distributed in Yunnan, Sichuan, Xizang and Guizhou.

含笑花
Michelia figo (Lour.) Spreng.

常绿灌木。小枝和叶密集；幼枝、花芽和叶柄密具黄褐色绒毛。托叶痕达叶柄顶端；托叶长2-4毫米；叶革质。花单生叶腋，极香；花被片6，稍肉质，淡黄色或乳白色，边缘常染有紫色；雌蕊群及聚合果均无毛。花期3-5月，果期7-8月。原产华南，现广泛栽培。

Evergreen shrubs. Twigs and leaves dense; young twigs, buds and petioles densely yellowish brown tomentose. Stipular scars reaching petiole apex; petioles 2-4 mm long; leaf blades leathery. Flowers solitary and axillary, very fragrant; tepals 6, slightly fleshy, pale yellow or creamy white, margin sometimes purple-tinged; gynoecia and aggregated fruits glabrous. Fl. Mar-May. Fr. Jul-Aug. Native to S China, now widely

cultivated.

野含笑
Michelia skinneriana Dunn

常绿乔木。托叶痕与叶柄等长；叶柄长2-4毫米；叶革质，侧脉每边10-13条。花芳香，淡黄色；花被片6；心皮密被褐色毛。聚合果长4-7厘米；蓇葖果黑色，球形或椭圆形，具短尖喙。花期5-6月，果期8-9月。生海拔1200米以下的山谷、山坡或溪边密林中。产广西、广东、福建、浙江、湖南和江西。

Evergreen trees. Stipular scars as long as petioles; petioles 2-4 mm long; leaf blades leathery, lateral veins 10-13 per side. Flowers fra-

含笑花 *Michelia figo*

野含笑 *Michelia skinneriana*

苦梓含笑 *Michelia balansae*

grant, pale yellow; tepals 6; carpels densely brown-hairy. Aggregate fruits 4-7 cm long; follicles black, globose or ellipsoid, beaked at apex. Fl. May-Jun. Fr. Aug-Sep. Valleys, slopes, or dense forests by streams below 1200 m. Distributed in Guangxi, Guangdong, Fujian, Zhejiang, Hunan and Jiangxi.

苦梓含笑

Michelia balansae (A. DC.) Dandy

常绿乔木。小枝被毛。叶柄长1.5-4厘米，无托叶痕；叶厚革质，长10-20(-28)厘米。花单生叶腋，极芳香；花被片6，白色带淡绿色。蓇葖果长2-6厘米。花期3-7月，果期8-10月。生海拔300-1000米的常绿阔叶林或溪旁。产云南、贵州、广西、广东、海南和福建。越南亦有。

Evergreen trees. Branchlets hairy. Petioles 1.5-4 cm long, without a stipular scar; leaf blades thickly leathery, 10-20(-28) cm long. Flowers solitary and axillary, very fragrant; tepals 6, pale greenish-white. Follicles 2-6 cm long. Fl. Mar-Jul. Fr. Aug-Oct. Evergreen broad-leaved forests or along rivers at 300-1000 m. Distributed in Yunnan, Guizhou, Guangxi, Guangdong, Hainan and Fujian. Also in Vietnam.

黄心夜合

Michelia martinii (Lévl.) Lévl.

常绿乔木。幼枝橄榄绿色，无毛。叶柄长1.5-2厘米，无托叶痕；叶革质，倒披针形或狭倒卵状椭圆形。花芳香，淡黄色；花被片6-8。果实长8-15厘米，具皱。花期2-3月，果期8-10月。生海拔1000-2000米的林中。产中国西南、华南和华中。越南亦有。

Evergreen trees. Young twigs olive green, glabrous. Petioles 1.5-2 cm long, without a stipular scar; leaf blades leathery, oblanceolate or narrowly obovate-elliptic. Flowers fragrant, light yellow; tepals 6-8. Fruits 8-15 cm long, wrinkled. Fl. Feb-Mar. Fr. Aug-Oct. Forests at 1000-2000 m. Distributed in SW, S and C China. Also in Vietnam.

黄心夜合 *Michelia martinii*

乐昌含笑 *Michelia chapensis*

乐昌含笑

Michelia chapensis Dandy

常绿大乔木。小枝无毛或于节间幼时具淡灰色柔毛；叶柄长1-2.5厘米；叶薄革质，倒卵形、狭倒卵形或长圆倒卵形。花单生叶腋，芳香；花被片6，淡黄色。聚合果长约10厘米。花期3-4月，果期8-11月。生海拔500-1700米的常绿阔叶林中。产云南、贵州、广西、广东、湖南和江西。越南亦有。

Large evergreen trees. Twigs glabrous or nodes grayish puberulous when young. Petioles 1-2.5 cm long; leaf blades thinly leathery, obovate, narrowly obovate or oblong-obovate. Flowers solitary and axillary, fragrant; tepals 6, yellowish. Aggregate fruits ca. 10 cm long. Fl. Mar-Apr. Fr. Aug-Nov. Evergreen broad-leaved forests at 500-1700 m. Distributed in Yunnan, Guizhou, Guangxi,

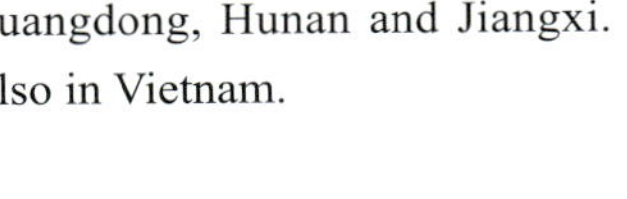

Guangdong, Hunan and Jiangxi. Also in Vietnam.

黄花含笑

Michelia xanthantha C. Y. Wu ex Y. W. Law et Y. F. Wu

常绿大乔木。叶柄长2-2.5厘米，无托叶痕；叶薄革质，长圆形或倒卵状长圆形，两面无毛。花大，芳香；花被片6，2轮，黄色。聚合果下垂，长约21厘米。花期3月，果期9-10月。生海拔1300-1400米的密林中。产云南(西双版纳)。

Large evergreen trees. Petioles 2-2.5 cm long, without a stipular scar; leaf blades thinly leathery, oblong or obovate-oblong, both surfaces glabrous. Flowers large, fragrant; tepals 6, in 2 whorls, yellow. Aggregate fruits pendulous, ca. 21 cm long. Fl. Mar. Fr. Sep-Oct. Dense forests at 1300-1400 m. Distributed in Yunnan (Sipsongpanna).

醉香含笑

Michelia macclurei Dandy

常绿大乔木。叶柄长1.5-4厘米，无托叶痕；叶革质，下面被灰色毛，侧脉每边10-15条。花单生或形成2-3朵的聚伞花序，极香；花被片9，乳白色，3轮。聚合果穗状，长3-7厘米。花期2-4月，果期9-11月。

黄花含笑 *Michelia xanthantha*

醉香含笑 *Michelia macclurei*

生海拔200-1500米密林中。产云南(喜洲镇)、广西、广东和海南。越南北部亦有。

Large evergreen trees. Petioles 1.5-4 cm long, without a stipular scar; leaf blades leathery, abaxially gray-hairy, lateral veins 10-15 per side. Flowers solitary or 2-3 in cyme, very fragrant; tepals 9, creamy white, 3-whorled. Aggregate fruits spicate, 3-7 cm long. Fl. Feb-Apr. Fr. Sep-Nov. Dense forests at 200-1500 m. Distributed in Yunnan (Xizhou Town), Guangxi, Guangdong and Hainan. Also in N Vietnam.

香子含笑
Michelia hypolampra Dandy

常绿乔木。叶有香味，薄革质，侧脉每边8-10条。花芳香；花被片9，白色，3轮，条形，长约1.5厘米；心皮10。蓇葖果灰黑色，椭圆形，长2-4.5厘米，密生皮孔。花期3-4月，果期9-10月。生海拔300-800米的山坡或沟谷林中。产云南、广西和海南。越南亦有。

Evergreen trees. Leaves fragrant, thinly leathery, lateral veins 8-10 per side. Flowers fragrant, tepals 9, white, 3-whorled, linear, ca. 1.5 cm long; carpels 10. Follicles gray-black, ellipsoid, 2-4.5 cm long, densely lenticellate. Fl. Mar-Apr. Fr. Sep-Oct. Slopes or forests in valleys at 300-800 m. Distributed in Yunnan, Guangxi and Hainan. Also in Vietnam.

阔瓣含笑
Michelia cavaleriei Finet et Gagnep. var. **platypetala** (Hand.-Mazz.) N. H. Xia

常绿乔木，高达20米。叶柄长1.5-3厘米，无托叶痕；叶薄革质，侧脉8-14对；嫩枝和芽疏被红色卷曲毛。花芳香，单生叶腋；花被片9，3轮，白色，外轮长5-7厘米。聚合果穗状，长5-15厘米。花期3-4月，果期9-10月。生海拔800-2400米的密林中。产贵州、广西、广东、湖北和湖南。

Evergreen trees, up to 20 m tall. Young branches and buds sparsely reddish-floccose-hairy. Petioles 1.5-3 cm long, without a stipular scar; leaf blades thinly leathery, lateral veins 8-14 per side. Flowers fragrant, solitary and axillary; tepals 9, 3-whorled, white, outer ones 5-7 cm long. Aggregate fruits spicate, 5-15 cm long. Fl. Mar-Apr. Fr. Sep-Oct. Dense forests at 800-2400 m. Distributed in Guizhou, Guangxi, Guangdong, Hubei and Hunan.

香子含笑 *Michelia hypolampra*

阔瓣含笑 *Michelia cavaleriei* var. *platypetala*

棕毛含笑
Michelia fulva Hung T. Chang et B. L. Chen

常绿小乔木。嫩枝和芽密被黄褐色长绒毛。叶革质。花黄色，芳香；花被片9-12(-14)，白色或黄色；雌蕊柄长0.6-2.4厘米，具柔毛；心皮约152，狭卵球形，密具金黄色柔毛。花期3月，果期11月。生海拔600-1700米的石灰岩山地林中。产云南和广西。

Small evergreen trees. Young branchlets and buds long brown-tomentose. Leaves coriaceous. Flowers yellow, fragrant; tepals 9-12(-14), white or yellow; gynophores 0.6-2.4 cm long, pubescent; carpels ca. 152, narrowly ovoid, densely golden yellow pubescent. Fl. Mar. Fr. Nov. Forests on limestone hills at 600-1700 m. Distributed in Yunnan and Guangxi.

棕毛含笑 *Michelia fulva*

深山含笑 *Michelia maudiae*

深山含笑

Michelia maudiae Dunn

常绿乔木。幼枝、花芽、叶下面和苞片具白色粉末。叶柄长1-3厘米，无托叶痕；叶革质，侧脉每边7-12条。花芳香；花被片9，纯白色，基部稍带淡红色。聚合果长7-15厘米。花期2-3月，果期9-10月。生海拔600-1500米的常绿阔叶林中。产中国西南、华南和东南。

Evergreen trees. Young twigs, buds, leaves abaxial surfaces, and bracts white powdery. Petioles 1-3 cm long, without a stipular scar; leaf blades leathery, lateral veins 7-12 per side. Flowers fragrant; tepals 9, white, slightly reddish at base. Aggregate fruits 7-15 cm long. Fl. Feb-Mar. Fr. Sep-Oct. Evergreen broad-leaved forests at 600-1500 m. Distributed in SW, S and SE China.

石碌含笑

Michelia shiluensis Chun et Y. F. Wu

乔木。顶芽狭椭圆形，被橙黄色或灰色有光泽的短柔毛。叶柄长1-3厘米，具宽沟，无托叶痕；叶革质，倒卵状长圆形，长8-14(-20)厘米，先端钝，具短尖，下面被白霜；侧脉8-12对。花被片9，3轮，倒卵形；花丝红色。果长4-5厘米，成熟的果片先端具短喙。花期3-5月，果期6-8月。生海拔200-1500米的常绿阔叶林中、山沟或路旁。产海南。

Trees. Terminal buds narrowly ellipsoid, orangish yellow or gray, glossy pubescent. Petiole 1-3 cm long, broadly furrowed, without a stipular scar; leaf blades leathery, obovate-oblong, 8-14(-20) cm long, apex obtuse and mucronate, abaxially glaucous; secondary veins 8-12 paired. Tepals 9, in 3 whorls, obovate; filaments red. Fruit 4-5 cm long, mature carpels apex shortly beaked. Fl. Mar-May. Fr. Jun-Aug. Evergreen broad-leaved forests, ravines or beside trails at 200-1500 m. Distributed in Hainan.

雅致含笑

Michelia elegans Y. W. Law et Y. F. Wu

常绿小乔木，被平伏毛。托叶长 0.5-1厘米，具柔毛，无托叶痕；叶革质，侧脉每边11-13。花单生叶腋，芳香；花被片9，3轮，白色；雌蕊群圆柱形。聚合果穗状；蓇葖果卵圆形或近球形。花期4月，果期10月。生海拔500-800米的林中。产浙江(庆元县)。

Small evergreen trees, with appressed indumenta. Petioles 0.5-1 cm long, pubescent, without a stipular scar; leaf blades leathery, lateral veins 11-13 per side. Flowers solitary and axillary, fragrant; tepals 9, 3-whorled, white; gynoecia cylindric. Aggregate fruits spicate; follicles ovoid or subglobose. Fl. Apr. Fr. Oct. Forests at 500-800 m. Distributed in Zhejiang (Qingyuan County).

金叶含笑

Michelia foveolata Merr. ex Dandy

常绿乔木。芽、幼枝、叶柄、叶背、花梗密被红褐色短绒毛。叶柄长1.5-3厘米，无托叶痕；叶厚革质，侧脉每边16-26条。

雅致含笑 *Michelia elegans*

石碌含笑 *Michelia shiluensis*

花单生叶腋，芳香；花被片9-12，淡黄色，基部带紫色；雄蕊50；雌蕊群长1.7-2厘米，被银灰色短绒毛。聚合果长7-20厘米。花期3-5月，果期9-10月。生海拔500-1800米阴湿林中。产中国西南、华南至华中。越南亦有。

金叶含笑 *Michelia foveolata*

Evergreen trees. Buds, initial branchlets, petioles and leaves abaxially densely rusty-velutinous. Petioles 1.5-3 cm long, without a stipular scar; leaf blades thickly leathery, lateral veins 16-26 per side. Flowers solitary and axillary, fragrant; tepals 9-12, pale yellow, base purplish; stamens 50; gynoecia 1.7-2 cm long, gray-velutinous. Aggregate fruits 7-20 cm long. Fl. Mar-May. Fr. Sep-Oct. Moist and shady forests at 500-1800 m. Distributed in SW, S to C China. Also in Vietnam.

素黄含笑 *Michelia flaviflora*

球花含笑

Michelia sphaerantha C. Y. Wu ex Y. W. Law et Y. F. Wu

常绿乔木。托叶痕长3-4毫米；叶柄长2-2.5厘米；叶革质，倒卵状长圆形或长圆形，基部圆形或钝，下面被毛，托叶离生。花大，芳香；花被片12，白色。聚合果长19-24厘米；蓇葖卵球形，木质。花期3月，果期7月。生海拔1800-2000米的林中。产云南(景东县)。

Evergreen trees. Stipular scars 3-4 mm long; petioles 2-2.5 cm long, pubescent; leaf blades leathery, obovate-oblong or oblong, base rounded or obtuse, abaxially hairy; stipules free. Flowers large, fragrant; tepals 12, white. Aggregate fruits 19-24 cm long; follicles ovoid, woody. Fl. Mar. Fr. Jul. Forests at 1800-2000 m. Distributed in Yunnan (Jingdong County).

素黄含笑

Michelia flaviflora Y. W. Law et Y. F. Wu

常绿乔木。幼枝具褐色绒毛。托叶与叶柄离生；叶柄长0.5-1.2厘米，无托叶痕；叶大，纸质，下面苍白色，被褐色绢毛。花淡黄色，有香味；花被片15，倒披针形；雄蕊90；心皮多数，离生，密被长柔毛。花期2月。生海拔在1400-1500米常绿阔叶林中。产云南(大围山，屏边县)。越南亦有。

Evergreen trees. Young twigs brown tomentulose. Stipules free from petioles; petioles 0.5-1.2 cm long, without a stipular scar; leaf blades large, papery, abaxially glaucous, brown-sericeous. Flowers pale yellow, fragrant; tepals 15, oblanceolate; stamens 90; carpels numerous, free, densely villose. Fl. Feb. Evergreen broad-leaved forests at 1400-1500 m. Distributed in Yunnan (Dawei Mountain, Pingbian County). Also in Vietnam.

球花含笑 *Michelia sphaerantha*

福建含笑 *Michelia fujianensis*

合果木 *Michelia baillonii*

福建含笑

Michelia fujianensis Q. F. Zheng

常绿乔木。叶柄长0.6-1.5厘米，无托叶痕；叶薄革质，叶下面被贴生灰白色或褐色长柔毛，侧脉每边9-15条。花芳香，花梗粗短；花被片12-17，白色，4轮；心皮圆球形，密被短绒毛。聚合果长2-3厘米；蓇葖果黑色。花期4-5，果期8-9月。生海拔300-700米的山坡林中。产福建(永安)和江西。

Evergreen trees. Petioles 0.6-1.5 cm long, without a stipular scar; leaf blades thinly leathery, abaxially grayish-white-appressed or rusty-villose, lateral veins 9-15 per side. Flowers fragrant, peduncles stout; tepals 12-17, white, 4-whorled; carpels globose, densely tomentose. Aggregate fruits 2-3 cm long; follicles black. Fl. Apr-May. Fr. Aug-Sep. Forests on slopes at 300-700 m. Distributed in Fujian (Yong'an) and Jiangxi.

合果木

Michelia baillonii (Pierre) Finet et Gagnep.

常绿大乔木。叶纸质，椭圆形，卵状椭圆形或狭卵形，长6-22(-25)厘米。花芳香；花被片18-21，每轮6枚，白色，花芽期乳黄色。聚合果肉质；蓇葖完全合生。花期3-5月，果期8-10月。生海拔500-1500米的山林中。产云南。印度(阿萨姆邦)、缅甸、泰国、越南和柬埔寨亦有。

Large evergreen trees. Leaves papery, elliptic, ovate-elliptic or narrowly ovate, 6-22(-25) cm long. Flowers fragrant; tepals 18-21, 6 in a whorl, white but cream-colored in bud. Aggregate fruits fleshy; mature carpels completely connate. Fl. Mar-May. Fr. Aug-Oct. Montane forests at 500-1500 m. Distributed in Yunnan. Also in India (Assam), Myanmar, Thailand, Vietnam and Cambodia.

观光木

Michelia odora (Chun) Noot. et B. L. Chen

常绿大乔木。叶厚纸质，倒卵状椭圆形，长8-17厘米。花单生叶腋，芳香，象牙黄色；花被片9-10；雌蕊柄粗壮，长2毫

观光木 *Michelia odora*

鹅掌楸 *Liriodendron chinense*

米；心皮9-13枚。果实长椭圆体形，挂于老枝上；果瓣厚，长1-2厘米。花期3月，果期10-12月。生海拔300-1100米的林中。产云南、广西、广东、海南、福建、湖南和江西。越南北部亦有。

Large evergreen trees. Leaves thickly papery, obovate-elliptic, 8-17 cm long. Flowers solitary and axillary, fragrant, ivory-yellow; tepals 9-10; gynophore stout, ca. 2 mm long; carpels 9-13. Fruits long ellipsoid, hanging on old twigs; valves thick, 1-2 cm long. Fl. Mar. Fr. Oct-Dec. Forests at 300-1100 m. Distributed in Yunnan, Guangxi, Guangdong, Hainan, Fujian, Hunan and Jiangxi. Also in N Vietnam.

鹅掌楸
Liriodendron chinense (Hemsl.) Sarg.

落叶大乔木。小枝灰色至灰褐色。叶膜质至纸质，每侧近基部具一侧生裂片，先端2裂。花杯形；花被片9，淡绿色，内面具黄色纵条纹；雌蕊群于花期超出花被；心皮黄绿色。果实长7-9厘米；小坚果具翅，具1或2粒种子。花期5月，果期9-10月。生海拔900-1000米的林中。产中国西南、东南、华中、华西和华东。越南北部亦有。

Large deciduous trees. Twigs gray to grayish brown. Leaves membranous to papery, with 1 lateral lobe near base of each side, apex 2-lobed. Flowers cupular; tepals 9, pale green, adaxially yellow longitudinal veined; gynoecium exceeding tepals at anthesis; carpels yellowish green. Fruits 7-9 cm long; nutlets winged, 1- or 2-seeded. Fl. May. Fr. Sep-Oct. Forests at 900-1000 m. Distributed in SW, SE, C, W and E China. Also in N Vietnam.

八角科 Illiciaceae

大花八角
Illicium macranthum A. C. Sm.

灌木或乔木。叶2-4枚簇生于上部节上。花梗长6-10毫米；花被片27-32，白色或淡绿色，长圆形(外轮)至狭舌状(最大者)，最大的花被片长15-25毫米；雄蕊21-26枚；花粉粒具3沟。蓇葖果11。花期1-3月，果期7-11月。生海拔1600-2800米的林中或山坡。产云南。

Shrubs or trees. Leaves in clusters of 2-4 at distal nodes. Flower peduncle 6-10 mm long; tepals 27-32, white to slightly greenish, oblong (outer) to narrowly tonguelike (largest), 15-25 mm long(largest); stamens 21-26; pollen grains tricolpate. Follicles 11. Fl. Jan-Mar. Fr. Jul-Nov. Forests or slopes at 1600-2800 m. Distributed in Yunnan.

大花八角 *Illicium macranthum*

大八角 *Illicium majus*

大八角
Illicium majus Hook. f. et Thoms.

灌木或乔木。叶3-6枚簇生于节上；叶长圆状披针形至倒披针形。花梗长1.8-4.5(-6)厘米；花被片15-21，肉质；雄蕊12-21；花粉粒具3合沟。蓇葖果10-14，直径4-4.5厘米。花期4-6月，果期7-10月。生海拔300-2500米的密林中或灌丛中。产云南、贵州、广西、广东和湖南。越南和缅甸亦有。

Shrubs or trees. Leaves in clusters of 3-6 at distal nodes, oblong-lanceolate to oblanceolate. Flower peduncles 1.8-4.5 (-6) cm long; tepals 15-21, fleshy; stamens 12-21; pollen grains trisyncolpate. Follicles 10-14, 4-4.5 cm diam. Fl. Apr-Jun. Fr. Jul-Oct. Dense forests or thickets at 300-2500 m. Distributed in Yunnan, Guizhou, Guangxi, Guangdong and Hunan. Also in Vietnam and Myanmar.

野八角
Illicium simonsii Maxim.

乔木。叶3-5枚簇生于上部节上；叶披针形至椭圆形，中脉在上面凹陷。花被片18-23(-26)，淡黄色，有时乳白色，稀为粉红色，最长的花被片长9-15(-18)毫米；雄蕊16-28；花粉粒具3沟。果实具8-13蓇葖果。花期2-5月，果期6-10月。生海拔1700-3200(-4000)米的林中。产云南、贵州和四川。缅甸北部和印度东北部亦有。

Trees. Leaves in clusters of 3-5 at distal nodes, lanceolate to elliptic, midvein adaxially impressed, Tepals 18-23(-26), pale yellow, sometimes creamy white, rarely pink, longest ones 9-15(-18) mm long; stamens 16-28; pollen grains tricolpate. Fruit with 8-13 follicles. Fl. Feb-May. Fr. Jun-Oct. Forests at 1700-3200(-4000) m. Distributed in Yunnan, Guizhou and Sichuan. Also in N Myanmar and NE India.

莽草 (红毒茴)
Illicium lanceolatum A. C. Sm.

灌木或乔木。叶簇生于上部节；叶披针形、倒披针形或倒卵状椭圆形，中脉在上面稍凹陷。花被片10-15，红色至深红色；雄蕊6-11；花粉粒具3合沟；心皮多数，螺旋状排列。果实具(9或)10-14蓇葖果。花期4-6月，果期8-10月。生海拔300-1500米的混交林、疏林或灌丛中。产华中至中南。

Shrubs or trees. Leaves in clusters at distal nodes; lanceolate, oblanceolate, or obovate-elliptic, midvein adaxially slightly impressed. Tepals 10-15, red to dark red; stamens 6-11; pollen grains trisyncolpate; carpels numerous, spirally arranged. Fruit with (9 or)10-14 follicles. Fl. Apr-Jun. Fr. Aug-Oct. Mixed forests, sparse forests or thickets at 300-1500 m. Distributed in C to SC China.

野八角 *Illicium simonsii*

莽草（红毒茴） *Illicium lanceolatum*

小花八角
Illicium micranthum Dunn

灌木或乔木。叶3-5簇生于上部节处，中脉在上部稍凹陷。花很小，芳香，幼时带白绿色；花被片14-17(-21)，红色或橙红色，最大者5-8 × 3.5-8毫米。蓇葖果6-8，呈星状排列，直径 1.7-2.1厘米。花期4-6月，果期7-9月。生海拔500-2600米灌丛或混交林中、山谷林中或峡谷溪边。产中国西南、华南和华中。

Shrubs or trees. Leaves in clusters of 3-5 at distal nodes, midvein adaxially impressed. Flowers very small, fragrant, initially whitish-green; tepals 14-17(-21), red or orange-red, 5-8 × 3.5-8 mm (largest). Follicles 6-8, stellate-arranged, 1.7-2.1 cm diam. Fl. Apr-Jun. Fr. Jul-Sep. Thickets or mixed forests, forests in valleys, or by streams in valleys at 500-2600 m. Distributed in SW, S and C China.

小花八角 *Illicium micranthum*

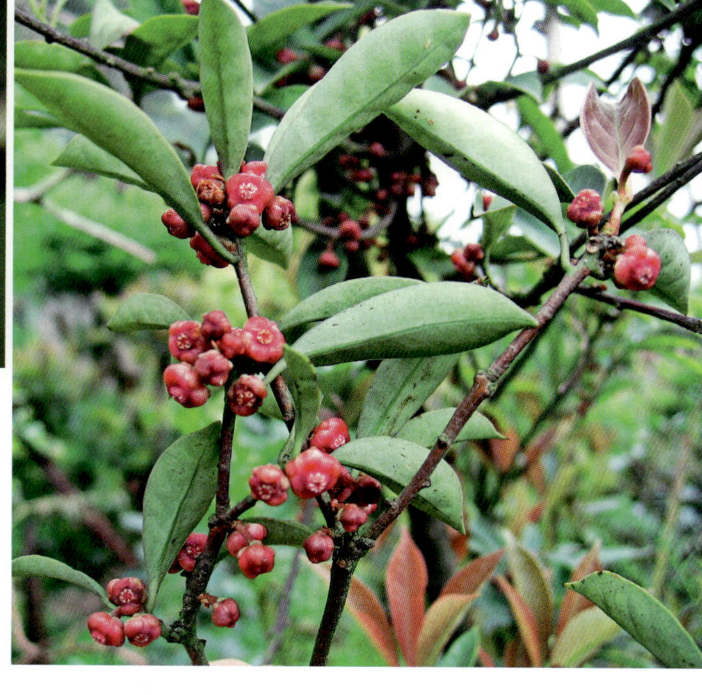

红茴香
Illicium henryi Diels

灌木或乔木。叶2-5簇生于节上部。花梗长1.5-5厘米；花红色；花被片10-15，粉红色至深红色，最大者7-10 × 4-8.5毫米；雄蕊10-14；花粉粒具3合沟。蓇葖果7-9。花期4-6月，果期8-10月。生海拔300-2500米的山地疏林、湿润处或灌丛中。产中国西南、华中和华北。

Shrubs or trees. Leaves in clusters of 2-5 at distal nodes. Flower peduncle 1.5-5 cm long; flowers red; tepals 10-15, pink to dark red, 7-10 × 4-8.5 mm (largest); stamens 10-14; pollen grains tri-syncolpate. Follicles 7-9. Fl. Apr-Jun. Fr. Aug-Oct. Open forests, wet places or thickets at 300-2500 m. Distributed in SW, C and N China.

滇西八角
Illicium merrillianum A. C. Sm.

小乔木。叶3-5枚簇生上部节处；叶椭圆形至披针形。花梗长2-4厘米；花被片15-20，樱桃红色，最大者7-10 × 5-7.5毫米；雄蕊14-19；花粉粒具3合沟。果实具(5-)8蓇葖果。花期12月至翌年1月或3-5月，果期翌年7-9月。生海拔1500-2900米的林中湿润处或疏林下。产云南西部。缅甸亦有。

Small trees. Leaves in clusters of 3-5 at distal nodes; elliptic to lanceolate. Flower peduncle 2-4 cm long; tepals 15-20, cherry red, 7-10 × 5-7.5 mm (largest); stamens 14-19; pollen grains tri-syncolpate. Fruit with (5-)8 follicles. Fl. Dec to next Jan or Mar-May. Fr. next Jul-Sep. Damp sites in forests or in sparse forests at 1500-2900 m. Distributed in W Yunnan. Also in Myanmar.

红茴香 *Illicium henryi*

滇西八角 *Illicium merrillianum*

八角
Illicium verum Hook. f.

乔木。叶3-6枚簇生于上部节处。花梗长1.5-4厘米；花被片(7-)10-11(-12)，粉红至深红色，外层纸质，内层肉质；雄蕊(11-)13或14(-20)；花粉粒具3合沟。果实具约8蓇葖果。第一次花期3-5月，果期9-10月；第二次花期8-10月，果期翌年3-4月。生海拔200-1600米的山地湿润常绿阔叶林中。原产广西，栽培于中国西南和华南。

Trees. Leaves in clusters of 3-6 at distal nodes. Flower peduncle 1.5-4 cm long; tepals (7-)10-11 (-12), pink to dark red, outer ones papery, inner ones fleshy ; stamens (11-)13 or 14(-20); pollen grains trisyncolpate. Fruit with ca. 8 follicles. First Fl. Mar-May. Fr. Sep-Oct. Second Fl. Aug-Oct. Fr. next Mar-Apr. Moist evergreen broad-leaved forests in mountains at 200-1600 m. Native to Guangxi, cultivated in SW and S China.

八角 *Illicium verum*

五味子科 Schisandraceae

黑老虎
Kadsura coccinea (Lem.) A. C. Sm.

木质藤本，全株无毛。叶革质，长圆形或卵状披针形，全缘。花被片10-16，白色、红色或紫红色；雄花：雄蕊10-50；雄花常生于花托顶部；雌花：心皮20-28。离心皮果红色至紫红色。花期5-7月，果期10-12月。生海拔(200-)400-1400(-1900)米的半开放灌丛或林中。产中国西南、华南和华中。缅甸和越南亦有。

Woody vines, glabrous throughout. Leaves coriaceous, oblong or ovate-lanceolate, margin entire. Tepals 10-16, white, red or purplish red; staminate flower: stamens 10-50; staminodes generally present at apex of torus; pistillate flower: carpels 20-28. Apocarps red to purplish red. Fl. May-Jul. Fr. Oct-Dec. Semi-open shrublands or forests at (200-)400-1400(-1900) m. Distributed in SW, S and C China. Also in Myanmar and Vietnam.

黑老虎 *Kadsura coccinea*

异形南五味子 *Kadsura heteroclita*

异形南五味子
Kadsura heteroclita (Roxb.) Craib

木质藤本。叶卵状椭圆形，全

缘或先端稀有疏齿。花单生；花被片10-17(-25)，白色、乳白色或黄色；雄花：雄蕊40-74；无退化雄蕊；雌花：心皮28-72。离心皮果红色；聚合果近球形。花期6-10月，果期10-12月。生海拔800-2000米的山地沟谷或山坡杂木林中。产中国西南、华南和东南。南亚亦有。

Woody vines. Leaves ovate-elliptic, margin entire or apical half sparsely serrulate. Flowers solitary; tepals 10-17(-25), white, cream, or yellow; staminate flower: stamens 40-74; staminodes absent; pistillate flower: carpels 28-72. Apocarps red, subglobose. Fl. Jun-Oct. Fr. Oct-Dec. Mountain valleys or mixed forests on slopes at 800-2000 m. Distributed in SW, S and SE China. Also in S Asia.

大花五味子

Schisandra grandiflora (Wall.) Hook. f. et Thoms.

木质藤本。叶无毛，稍椭圆形，长6.5-13厘米，纸质，不具白粉。单花腋生；花被片6-9，白色，乳白色或有时带粉色，最大者长0.7-1.5(-2.3)厘米；雄蕊30-50，稍离生；花粉具3孔沟；心皮67-120。果梗长4-7厘米。花期4-6月，果期6-10月。生海拔(1800-)2100-3300(-4000)米的阔叶林、针叶林、混合林和丛林中。产中国西南和华中。印度、尼泊尔、不丹、缅甸和泰国亦有。

Woody vines. Leaves glabrous, ±elliptic, 6.5-13 cm long, papery, not glaucous. Flowers axillary, solitary; tepals 6-9, white, cream-white, or sometimes pink-tinged, largest 0.7-1.5(-2.3) cm long; stamens 30-50, ± distinct; pollen 3-colpate; carpels 67-120. Fruit peduncles 4-7 cm long. Fl. Apr-Jun. Fr. Jun-Oct. Broad-leaved forests, coniferous forests, mixed forests, thickets at (1800-)2100-3300(-4000) m. Distributed in SW and C China. Also in India, Nepal, Bhutan, Myanmar and Thailand.

翼梗五味子

Schisandra henryi C. B. Clarke

木质藤本，无毛。叶椭圆形至卵形，纸质，基部常稍下延至叶柄。单花腋生；花被片6-10，黄色至橘黄色，内轮常红色，最大者长5.5-13毫米；雄蕊12-46；花粉具6孔沟；心皮28-65。果梗长3.5-14.5厘米。花期4-8月，果期7-10月。生海拔500-2100(-2300)米的林中或灌丛中。产长江以南地区。泰国和越南北部亦有。

Woody vines, glabrous. Leaves elliptic to ovate, papery, base often ± decurrent on petioles. Flowers axillary, solitary; tepals 6-10, yellow to orange but inner ones often red, largest 5.5-13 mm long; stamens 12-46; pollens 6-colpate; carpels 28-65. Fruit peduncles 3.5-14.5 cm long. Fl. Apr-Aug. Fr. Jul-Oct. Forests or thickets at 500-2100(-2300) m. Distributed in the south of Yangtze River. Also in Thailand and N Vietnam.

大花五味子 *Schisandra grandiflora*

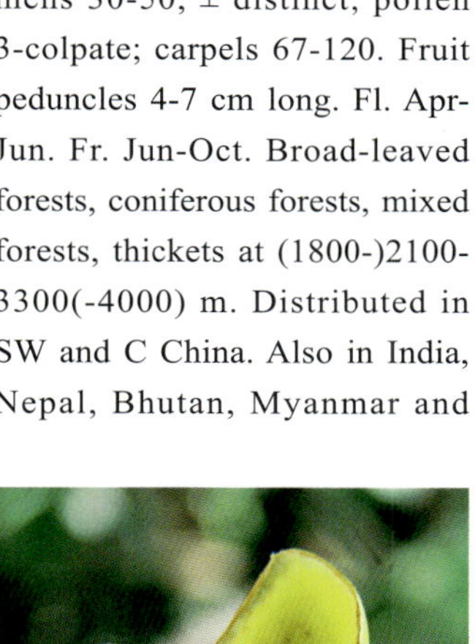

翼梗五味子 *Schisandra henryi*

五味子 *Schisandra chinensis*

五味子
Schisandra chinensis (Turcz.) Baill.

木质藤本。幼枝无翅。叶膜质，不具白粉，侧脉每边3-7条。花被片5-9，白色至黄色，最大者长 6.5-11毫米；雄蕊(4或)5(-7)；花粉具6孔沟；心皮14-40枚。果梗长2-7.5厘米。花期5-7月，果期7-9月。生海拔1200-1700米沟谷、溪边或山坡。产河北、山西、内蒙古、黑龙江、吉林和辽宁。俄罗斯(远东地区)、朝鲜半岛和日本北部亦有。

Woody vines. Young branches lacking wings. Leaves membranous, not glaucous, lateral veins 3-7 per side. Tepals 5-9, white to yellow, largest 6.5-11 mm long; stamens (4 or)5(-7); pollens 6-colpate; carpels 14-40. Fruit peduncles 2-7.5 cm long. Fl. May-Jul. Fr. Jul-Sep. Valleys, by streams, or slopes at 1200-1700 m. Distributed in Hebei, Shanxi, Neimenggu, Heilongjiang, Jilin and Liaoning. Also in Russia (Far East), Korean Peninsula and N Japan.

东亚五味子
Schisandra elongata (Blume) Baill.

木质藤本，雌雄异株。叶无毛，长5-11厘米，下面灰绿色。花被片5-9，黄色或橙黄色；雄蕊11-19，花药内向着生。穗状聚合果长6-17厘米，成熟时红色。花期4-6月，果期7-10月。生海拔100-2600米的山地沟谷、山坡林间或灌丛中。产中国西南、华南、华中和华东。南亚和东南亚亦有。

Woody vines, dioecious. Leaves glabrous, 5-11 cm long, adaxially pale green. Petals 5-9, yellow or orange-yellow; stamens 11-19, anthers introrse. Aggregate fruits spicate, 6-17 cm long, red when matured. Fl. Apr-Jun. Fr. Jul-Oct. Valleys of mountains, forests of slopes, or thickets at 100-2600 m. Distributed in SW, S, C and E China. Also in S and SE Asia.

东亚五味子 *Schisandra elongata*

水青树科
Tetracentraceae

水青树
Tetracentron sinense Oliv.

乔木。树皮灰褐色或灰红色，片状脱落。叶阔卵形，纸质，基部心形，边缘具细齿。花序具短柄，80-125花；花黄绿色，无柄。蓇葖果沿背缝线开裂；每果具种子4-6，纺锤形，两侧具短翅。花期4-7月，果期7-10月。生海拔1100-3500米的常绿阔叶林的溪边或林缘和常绿－落叶混合林中。产中国西南、华中和华西。印度东北部、尼泊尔、不丹、缅甸北部和越南北部亦有。

Trees. Bark gray-brown or gray-red, exfoliate. Leaves broadly ovate, papery, base cordate, margin serrulate. Inflorescences short pedunculate, 80-125-flowered; sepals 4; stamens 4; carpels 4. Follicles dehiscence loculicidal; seeds 4-6 per follicle, spindle-shaped, short winged at each end. Fl. Apr-Jul. Fr. Jul-Oct. Along streams or forest edges in broad-leaved evergreen forests and mixed evergreen-deciduous forests at 1100-3500 m. Distributed in SW, C and W China. Also in NE India, Nepal, Bhutan, N Myanmar and N Vietnam.

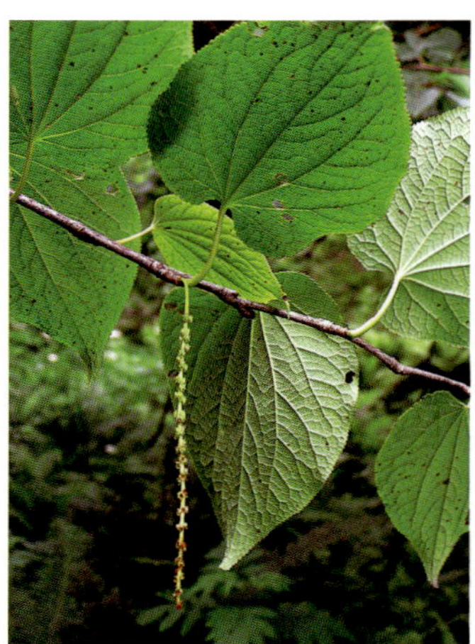

水青树 *Tetracentron sinense*

蜡梅科 Calycanthaceae

夏蜡梅

Calycanthus chinensis (W. C. Cheng et S. Y. Chang) W. C. Cheng et S. Y. Chang ex P. T. Li

灌木。树皮灰白色或灰褐色，皮孔凸。花单个顶生，直径4.5-7厘米；花被片明显二型：外轮花被片10-14，白色，顶端稍带粉色；内轮7-16，浅黄色，基部渐变为白色；雄蕊16-19。瘦果矩圆状。花期5月，果期10月。生海拔600-1000米的山地溪边树下。产浙江北部。

Shrubs. Bark grayish white or grayish brown, with convex lenticels. Flowers terminal, solitary, 4.5-7 cm diam; tepals distinctly dimorphic: outer tepals 10-14, white flushed slightly pink toward margin, inner tepals 7-16, pale yellow becoming white toward base; stamens 16-19. Achenes oblong. Fl. May. Fr. Oct. Under trees near streams in mountainous areas at 600-1000 m. Distributed in N Zhejiang.

蜡梅

Chimonanthus praecox (L.) Link.

落叶灌木。叶卵形、椭圆形至阔椭圆形，或有时长圆状披针形。花单生或成对，先叶开放，直径2-4厘米；花被片15-21，黄色，内轮则常具紫红色斑块，外轮花被片具柔毛，内轮基部明显具爪。瘦果椭圆体形至肾形。花期10月至翌年3月，果期翌年4-11月。生海拔500-1100米的山地林中。产中国西南、华中、华北和华东。

Deciduous shrubs. Leaves ovate, elliptic to broadly elliptic, or sometimes oblong-lanceolate. Flowers solitary or paired, appearing before leaves, 2-4 cm diam; tepals 15-21, yellow but inner ones usually with purplish red pigment, outer tepals puberulent, inner tepals base distinctly clawed. Achenes ellipsoid to reniform. Fl. Oct to next Mar. Fr. next Apr-Nov. Montane forests at 500-1100 m. Distributed in SW, C, N and E China.

蜡梅 *Chimonanthus praecox*

山蜡梅

Chimonanthus nitens Oliv.

常绿灌木或小乔木。叶纸质至近革质，椭圆形至卵状披针形。花单生；花被片20-24，黄色至黄白色，直径3-15毫米，外面具柔毛，内轮花被片基部不具爪。瘦果椭圆体形。花期10月至翌年1月，果期翌年4-8月。生海拔200-2500米的山地疏林中或石灰岩山地。产中国长江以南。

Shrubs or small trees, evergreen. Leaves papery to subcoriaceous, elliptic to ovate-lancelate. Flowers solitary; tepals 20-24, yellow to yellowish white, 3-15 mm diam, outside pubescent, inner tepals base not clawed. Achenes ellipsoid. Fl. Oct to next Jan. Fr. next Apr-Aug. Montane open forests or limestone rocky areas at 200-2500 m. Distributed throughout the south of Yangtze River in China.

夏蜡梅 *Calycanthus chinensis*

山蜡梅 *Chimonanthus nitens*

浙江蜡梅
Chimonanthus zhejiangensis M. C. Liu

常绿灌木。叶椭圆形、卵状椭圆形、长圆状椭圆形或阔卵形。花腋生，芳香；花被片16-20，淡黄色，外被柔毛，内轮花被片基部具爪；心皮6-9。果托 2.5-3.5 × 1.4-1.8厘米。瘦果椭圆体形。花期10-11月，果期翌年6月。生海拔200-900米的山地疏林。产浙江。

Evergreen shrubs. Leaves elliptic, ovate-elliptic, oblong-elliptic or broadly ovate. Flowers axillary, fragrant; tepals 16-20, yellowish, outside pubescent, inner tepals basally clawed; carpels 6-9; fruiting receptacles 2.5-3.5 × 1.4-1.8 cm. Achenes ellipsoid, Fl. Oct-Nov. Fr. next Jun. Montane sparse forests at 200-900 m. Distributed in Zhejiang.

浙江蜡梅 *Chimonanthus zhejiangensis*

番荔枝科 Annonaceae

毛叶藤春
Alphonsea mollis Dunn

常绿乔木。枝幼时密具绒毛，几无毛。叶纸质，椭圆形或卵状长圆形，下面被长柔毛。花序具1或2花；花瓣淡黄白色；心皮3，被绒毛。果1或2，成熟时黄色，卵球形至椭圆体形。花期初春，果期6-8月。生海拔640-1000米的山地常绿林中。产云南南部、广西西南部、广东和海南。

Trees, evergreen. Branches densely tomentose when young, glabrescent. Leaves papery, elliptic or ovate-oblong, abaxially villose. Inflorescences 1- or 2-flowered; petals yellowish white; carpels 3, tomentose. Fruits 1 or 2, yellow when ripe, ovoid to ellipsoid. Fl. early spring. Fr. Jun-Aug. Montane evergreen forests at 640-1000 m. Distributed in S Yunnan, SW Guangxi, Guangdong and Hainan.

鹰爪花
Artabotrys hexapetalus (L. f.) Bhandari

灌木。叶长圆形或披针形，纸质，侧脉于中脉每侧8-16对，上面凸起。花序具1-2花；花芳香；萼片绿色；花瓣淡绿色至淡黄色，长圆状披针形。果实卵球形。花期5-8月，果期5-12月。生海拔1300-1500米的林中。产中国西南、华南和东南。原产印度南部和斯里兰卡。

Shrubs. Leaves oblong or lanceolate, papery, secondary veins 8-16 on each side of midvein and adaxially prominent. Inflorescences 1-2-flowered; flowers fragrant; sepals green; petals greenish to yellowish, oblong-lanceolate. Fruits ovoid. Fl. May-Aug. Fr. May-Dec. Forests at 1300-1500 m. Distributed in SW, S and SE China. Native to S India and Sri Lanka.

香港鹰爪花
Artabotrys hongkongensis Hance

攀援灌木，长达6米。叶革质，无毛或仅下面中脉被微柔毛，侧脉每边8-10条。花序具单花；花梗稍长于钩状花序梗，被微柔毛；外轮花瓣卵状披针形，长1-1.8厘米，加厚。果干后黑色，椭圆体形。花期3-8月，果期5-12月。生海拔300-1500米的山地密林中或山谷阴

毛叶藤春 *Alphonsea mollis*

鹰爪花 *Artabotrys hexapetalus*

香港鹰爪花 *Artabotrys hongkongensis*

依兰 *Cananga odorata*

湿处。产云南、贵州、广西、广东、海南和湖南。越南亦有。

Climbing Shrubs, up to 6 m long. Leaves coriaceous, glabrous or only midvein abaxially puberulent, lateral veins 8-10 per side. Inflorescences 1-flowered; pedicels slightly longer than hooked peduncles, puberulent; outer petals ovate-lanceolate, 1-1.8 cm long, thickened. Fruits black when dry, ellipsoid. Fl. Mar-Aug. Fr. May-Dec. Dense forests on mountains, or shady and moist places in valleys at 300-1500 m. Distributed in Yunnan, Guizhou, Guangxi, Guangdong, Hainan and Hunan. Also in Vietnam.

依兰

Cananga odorata (Lam.) Hook. f. et Thomson

常绿乔木。叶2列，卵形、长圆形或阔椭圆形，膜质至薄纸质。花大，芳香，淡黄绿色，下垂；花瓣条形或条状披针形，5-8 × 0.5-1.8厘米；柱头羽毛状。果实近黑色，卵球形、球形或长圆体形。花期4-5月，果期10月至翌年3月。栽培于海拔120-600米庭院。栽培于云南、广西、福建和广东。产热带亚洲和大洋洲。

Evergreen trees. Leaves in 2 ranks, ovate, oblong, or broadly elliptic, membranous to thinly papery. Flowers large, fragrant, yellowish green, pendulous; petals linear or linear-lanceolate, 5-8 × 0.5-1.8 cm; stigmas plumose. Fruits nearly black, ovoid, globose, or oblong. Fl. Apr-May. Fr. Oct to next Mar. Cultivated in gardens at 120-600 m. Cultivated in Yunnan, Guangxi, Fujian and Guangdong. Distributed in tropical Asia and Oceania.

小依兰

Cananga odorata (Lam.) Hook. f. et Thomson var. **fruticosa** (Craib) J. Sincl.

本变种与依兰的区别在于本变种为灌木，高1-2米。雄蕊和花瓣均条形或条状披针形。花期4-6月，果期10月至翌年3月。产云南和广东。原产印度尼西亚，泰国和马来西亚亦有。

This variety differs from the typical variety in its frutiose habit. Stames and petals linear or linear-lanceolate. Fl. Apr-Jun. Fr. Oct to next Mar. Distributed in Yunnan and Guangdong. Native to Indonesia, also in Thailand and Malaysia.

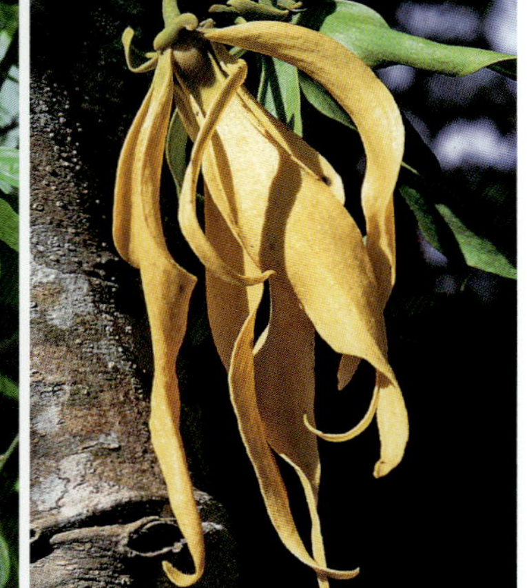
小依兰 *Cananga odorata* var. *fruticosa*

黄花皂帽花 *Dasymaschalon sootepense*

黄花皂帽花

Dasymaschalon sootepense Craib

乔木。叶椭圆形，纸质。花生于新枝上，黄色；花瓣长卵形，长3-4厘米，不扭曲，外面疏被柔毛；花帽长3.7毫米，约占花瓣长度之90%。果实红色，长3-6厘米。种子长17-24毫米。花期4-7月，果期6-9月。生海拔600-1300米的多石或沙冲积层上的常绿阔叶林中。产云南南部。泰国北部亦有。

Trees. Leaves elliptic, papery. Flowers on young growth, yellow; petals long-ovate, 3-4 cm long, not twisted, outside sparsely hairy; floral chamber to 3.7 mm long, ca. 90% of petal length. Fruits red, 3-6 cm long. Seeds 17-24 mm long. Fl. Apr-Jul. Fr. Jun-Sep. Evergreen broad-leaved forests on rocky or sandy alluvium at 600-1300 m. Distributed in S Yunnan. Also in N Thailand.

假鹰爪

Desmos chinensis Lour.

木质藤本。叶6-14 × 2-6.5厘米，薄纸质或膜质，中脉两侧各具7-12条侧脉。花序腋上生或与叶对生，具1花；花黄色，单生，下垂，外轮花瓣长圆形至长圆状披针形，长3-6.5厘米。果念珠状。花期4-10月，果期6-12月。生海拔100-1500米的沟谷荒地和灌丛。产云南东南部、贵州南部、广西、广东和海南。南亚和东南亚亦有。

Climbers woody. Leaves 6-14 × 2-6.5 cm, thinly papery or membranous, secondary veins 7-12 on each side of midvein. Inflorescences superaxillary or leaf-opposed, 1-flowered; flowers yellow, solitary, pendulous; outer petals oblong to oblong-lanceolate, 3-6.5 cm long. Fruits moniliform. Fl. Apr-Oct. Fr. Jun-Dec. Wastelands and thickets in valleys at 100-1500 m. Distributed in SE Yunnan, S Guizhou, Guangxi, Guangdong and Hainan. Also in S and SE Asia.

毛叶假鹰爪

Desmos dumosus (Roxb.) Saff.

木质藤本。幼枝密被毛。叶倒卵状椭圆形至长圆形，背密被柔毛。花序腋外生或与叶对生，具

假鹰爪 *Desmos chinensis*

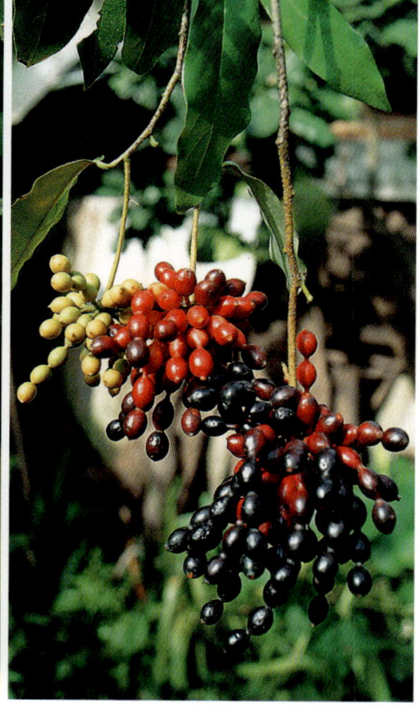

毛叶假鹰爪 *Desmos dumosus*

1朵花；花梗长15-20毫米；花瓣披针形，外轮花瓣长约7厘米；内轮花瓣长3-4(-6.5)厘米。花期4-8月，果期7月至翌年4月。生海拔500-1700米的灌丛或山地疏林。产云南南部、贵州和广西。印度、不丹、老挝、泰国、越南和新加坡亦有。

Climbers woody. Branches densely hairy when young. Leaves obovate-elliptic to oblong, densely pillose. Inflorescences extra-axillary or leaf-opposed, 1-flowered; pedicels 15-20 mm long; petals lanceolate, external petals ca. 7 cm long; inner petals 3-4(-6.5) cm long. Fl. Apr-Aug. Fr. Jul to next Apr. Thickets or sparse forests on mountains at 500-1700 m. Distributed in S Yunnan, Guizhou and Guangxi. Also in India, Bhutan, Laos, Thailand, Vietnam and Singapore.

小萼瓜馥木

Fissistigma polyanthoides (A. DC.) Merr.

木质藤本。叶长圆形、长圆状披针形或有时倒卵状长圆形，下面被黄褐色绒毛。花序与叶对生，近对生或有时顶生，假聚伞状，密被红褐色微绒毛；萼片阔三角形，基部合生。花期5-11月，果期8月至翌年3月。生海拔500-1600米的密林山坡中。产云南中部至南部和贵州。缅甸、老挝、泰国北部和越南亦有。

Climbers woody. Leaves oblong, oblong-lanceolate, or sometimes obovate-oblong, abaxially fulvous tomentose. Inflorescences leaf-opposed, subopposed, or sometimes terminal, pseudo-cymose, densely red fulvous tomentulose; sepals broadly triangular, basally connate. Fl. May-Nov. Fr. Aug to next Mar. Dense forested slopes at 500-1600 m. Distributed in C to S Yunnan and Guizhou. Also in Myanmar, Laos, N Thailand and Vietnam.

白叶瓜馥木

Fissistigma glaucescens (Hance) Merr.

木质藤本。叶长圆形至倒卵状长圆形，背面灰白色，宽1.2-6厘米。花序为顶生聚伞圆锥花序；花瓣卵形，外轮花瓣长约6毫米，密被黄色短绒毛；子房1室；柱头顶端2裂。小果球形，直径约9毫米，是瓜馥木属最小的小果。花期1-9月，果期3-12月。生海拔100-1000米的山坡疏林、灌丛或沟谷。产广西、广东、海南、台湾和福建。越南亦有。

Climbers woody. Leaves oblong to obovate-oblong, gray-white abaxially, 1.2-6 cm wide. Inflorescences terminal, thyrsoid; petals ovate, external petals ca. 6 mm long, with densely yellow short tomenta; ovaries 1-ovuled; stigmas apically 2-cleft, Fruitlets globose, ca. 9 mm diam. Fl. Jan-Sep. Fr. Mar-Dec. Sparsely forested slopes, scrubs, often in ravines at 100-1000 m. Distributed in Guangxi, Guangdong, Hainan, Taiwan and Fujian. Also in Vietnam.

小萼瓜馥木 *Fissistigma polyanthoides*

白叶瓜馥木 *Fissistigma glaucescens*

瓜馥木 *Fissistigma oldhamii*

瓜馥木

Fissistigma oldhamii (Hemsl.) Merr.

木质藤本。小枝被黄褐色短柔毛。叶倒卵状椭圆形至长圆形，革质，下面被短柔毛至渐无毛。花1-3朵集成伞形花序；花瓣长2-2.1厘米；每子房具10胚珠；柱头顶端2裂。果球形。花期4-9月，果期7月至翌年2月。生海拔500-1500米的山谷水旁灌木丛中。产中国西南、华南和东南。

Climbers woody. Branchlets fulvous pubescent. Leaves obovate-elliptic to oblong, coriacous, abaxially pubescent to glabrescent. Flowers 1-3 in umbels; petals 2-2.1 cm long; ovules 10 per ovary; stigmas apically 2-cleft. Fruits globose. Fl. Apr-Sep. Fr. Jul to next Feb. Thickets by streams in valleys at 500-1500 m. Distributed in SW, S and SE China.

多花瓜馥木

Fissistigma polyanthum (Hook. f. et Thomson) Merr.

木质藤本。枝具深灰色至褐色短柔毛或几无毛。叶长圆形、倒卵状长圆形或有时椭圆形，薄革质，下面被微柔毛。花序为团伞花序，常具3-7花；花较小，长约1厘米；花瓣长0.9-1.2厘米；每子房具4-6胚珠；柱头顶端全缘。花期1-10月，果期3-12月。生海拔120-1200米的林中。产中国西南和华南。印度、不丹、越南和缅甸亦有。

Climbers. Branches dark gray to brown pubescent or glabrescent. Leaves oblong, obovate-oblong, or sometimes elliptic, thinly coriaceous, abaxially puberulent. Inflorescences glomerulate, usually 3-7-flowered; flowers small, ca. 1 cm long; petals 0.9-1.2 cm long; ovules 4-6 per ovary; stigmas entire at apex. Fl. Jan-Oct. Fr. Mar-Dec. Forests at 120-1200 m. Distributed in SW and S China. Also in India, Bhutan, Vietnam and Myanmar.

景洪哥纳香

Goniothalamus cheliensis Hu

乔木。枝条被灰黑色硬毛。叶倒卵形，56-76 × 13-19厘米，纸质，下面被疏糙硬毛，中脉密具锈色粗毛。果长圆状椭圆体形，6-9 × 1.5-2厘米，密被锈色粗毛，具皮孔，两头渐尖。果期9月。生海拔约1500米的坡地林中。产云南南部。

Trees. Branches gray-black hirsute.

多花瓜馥木 *Fissistigma polyanthum*

景洪哥纳香 *Goniothalamus cheliensis*

Leaves obovate, 56-76 × 13-19 cm, papery, abaxially sparsely hispid, midvein densely rust-colored hirsute. Fruits oblong-elliptic, 6-9 × 1.5-2 cm, densely rust-colored hirsute, lenticellate, acuminate on both ends. Fr. Sep. Forests of slopes at ca. 1500 m. Distributed in S Yunnan.

大花哥纳香

Goniothalamus calvicarpus Craib

乔木。叶长圆形，17-35 × 5.5-9厘米，纸质，无毛，侧脉在叶面稍凸起。花序腋生或腋外生，具1花；花大；外轮花瓣长圆状披针形，长5-6.5厘米；每子房具2胚珠。果实束生，约1.5 × 0.8厘米。花期4-8月，果期8-11月。生海拔800-1500米的山谷林下或水边。产云南。泰国北部亦有。

Trees. Leaves oblong, 17-35 × 5.5-9 cm, papery, glabrous, lateral nerves adaxially slightly prominent. Inflorescences axillary or extra-axillary, 1-flowered; flowers large; outer petals oblong-lanceolate, 5-6.5 cm long; ovules 2 per ovary. Fruits fascicled, ca. 1.5 × 0.8 cm. Fl. Apr-Aug. Fr. Aug-Nov. Forests in valleys or by waters at 800-1500 m. Distributed in Yunnan. Also in N Thailand.

蕉木

Chieniodendrm hainanense (Merr.) Tsiang et P. T. Li

常绿乔木。小枝被锈色短柔毛。叶长圆形至长圆状披针形，薄纸质，侧脉每边6-10条，斜出。花瓣绿色，勺状；小苞片和花梗均被锈色短柔毛；柱头棍棒状，直立。果实多达8个，具短柄。花期4-12月，果期8月至翌年3月。生海拔300-600米的沟谷密林中。产广西南部和海南。

Trees, evergreen. Branchlets rusty-pubescent. Leaves oblong to oblong-lanceolate, thinly papery, lateral veins 6-10 per side, oblique. Flowers green, cucullate; bracteoles and pedicels rusty-pubescent; stigmas clavate, erect. Fruits to 8, fascicled, shortly stipitate. Fl. Apr-Dec. Fr. Aug to next Mar. Dense forest in valleys at 300-600 m. Distributed in S Guangxi and Hainan.

楔叶野独活 *Miliusa cuneata*

楔叶野独活

Miliusa cuneata Craib

乔木。叶薄纸质，无毛，长14-22厘米，中脉两侧各具12-15条侧脉，基部偏斜，先端钝至钝渐尖。花序腋生于无叶节处，具2-6朵花；花梗长(1-)2-3厘米，下垂；花瓣淡紫色。果实8-14，干后黑色，近球形至椭圆体状卵球形。花期4-9月，果期6月至翌年2月。生海拔500-1500米的山地疏林。产云南南部。泰国北部亦有。

Trees. Leaves thinly papery, glabrous, 14-22 cm long, secondary veins 12-15 on each side of midvein, oblique at base, apex obtuse to bluntly acuminate. Inflorescences axillary at leafless nodes, 2-6-flowered; pedicels (1-)2-3 cm long, pendulous; petals pale purple. Fruits 8-14, drying blackish, subglobose to ellipsoid-ovoid. Fl. Apr-Sep. Fr. Jun to next Feb. Sparse forests in mountains at 500-1500 m. Distributed in S Yunnan. Also in N Thailand.

大花哥纳香 *Goniothalamus calvicarpus*

蕉木 *Chieniodendrm hainanense*

囊瓣木 *Miliusa horsfieldii*

银钩花 *Mitrephora tomentosa*

囊瓣木

Miliusa horsfieldii (Benn.) Baill. ex Pierre

常绿乔木。叶纸质，侧脉每边10-14条。花序为聚伞花序，腋生，具1-30花；花暗红色，单生，腋生；心皮多数，密被绢毛；每子房具8胚珠。果15-30，被短柔毛，初时黄绿色，成熟时暗红色。花期3-4月，果期7-8月。生海拔300-1000米的山谷密林中。产广东和海南。南亚、东南亚和澳大利亚亦有。

Trees, evergreen. Leaves papery, lateral veins 10-14 per side. Inflorescences axillary, cymose, 1-30-flowered; flowers dark-red, solitary, axillary; carpels numerous, densely sericeous; ovules 8 per carpel. Fruits 15-30, pubescent, initially yellow green, dark-red when mature. Fl. Mar-Apr. Fr. Jul-Aug. Dense forests in valleys at 300-1000 m. Distributed in Guangdong and Hainan. Also in S and SE Asia, and Australia.

山蕉 *Mitrephora macclurei*

山蕉

Mitrephora macclurei Weeras. et R. M. K. Saunders

乔木。单叶，披针形，革质，侧脉在中脉每侧7-9条。花序轴不分枝；外轮花瓣白色变黄色，椭圆形至卵形，边缘从不波状；内轮花瓣紫色；每心皮具8-10胚珠。果实倒卵球形，疏被毛，光滑，具纵脊。花期3-5月，果期9-10月。生海拔约800米的林中河畔。产云南南部、贵州南部、广西和海南。老挝、越南和马来西亚亦有。

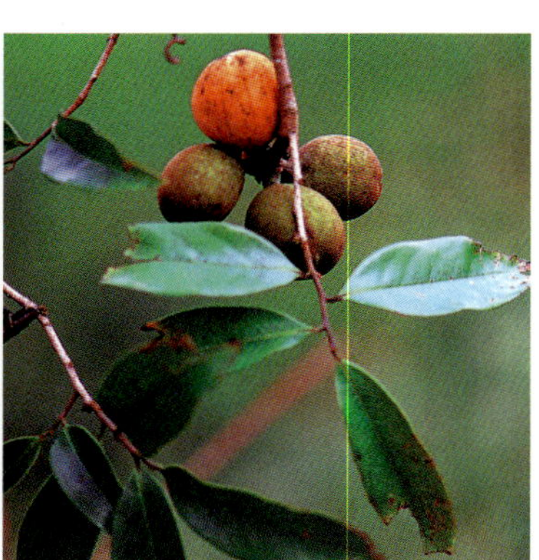

Trees. Leaves simple, lanceolate, coriaceous, secondary veins 7-9 on each side of midvein. Inflorescence rachides unbranched; outer petals white turning yellow, elliptic to ovate, margin never undulate; inner petals purple; ovules 8-10 per ovary. Fruits obovoid, sparsely hairy, smooth, with longitudinal ridge. Fl. Mar-May. Fr. Sep-Oct. Riverine forests at ca. 800 m. Distributed in S Yunnan, S Guizhou, Guangxi and Hainan. Also in Laos, Vietnam and Malaysia.

银钩花

Mitrephora tomentosa Hook. f. et Thomson

乔木。叶革质，下面被锈色长柔毛。花序轴不分枝；外轮花瓣色，灰黄色后为深黄色，老花边缘波状；内轮花瓣乳白色至黄色，顶端具紫条纹；小苞片4.5-7.5 × 3.5-9毫米；萼片5-9 × 5-9毫米；心皮12-17。果近球形，无纵脊。花期1-4月，果期5-9月。生海拔100-1200米的常绿阔叶林中。产云南南部、贵州南部、广西西部和海南。印度、老挝、泰国、越南和柬埔寨亦有。

Trees. Leaves coriaceous, abaxially ferruginous-villose. Inflorescence rachides unbranched; outer petals pale yellow turning

云南银钩花 *Mitrephora wangii*

蚁花 *Orophea laui*

dark yellow, margin undulate on older flowers; inner petals cream to yellow with apical purple streaks; bracteoles 4.5-7.5 × 3.5-9 mm; sepals 5-9 × 5-9 mm; carpels 12-17. Fruits subglobose, without a longitudinal ridge. Fl. Jan-Apr. Fr. May-Sep. Evergreen broad-leaved forests at 100-1200 m. Distributed in S Yunnan, S Guizhou, W Guangxi and Hainan. Also in India, Laos, Thailand, Vietnam and Cambodia.

云南银钩花

Mitrephora wangii Hu

乔木。叶长圆状披针形，革质，侧脉在中脉每边10-14条。花单性，少花簇生；外轮花瓣白色后变为亮黄色，卵形，老花边缘稍波状；内轮花瓣淡紫色。果卵球形或短圆柱状。花期1-5月，果期6-10月。生海拔600-1600米的山地密林中。产云南南部。泰国北部亦有。

Trees. Leaves oblong-lanceolate, coriaceous, secondary veins 10-14 on each side of midvein. Flowers unisexual, few clustered; outer petals white turning bright yellow, ovate, margin ± undulate on older flowers; inner petals purplish. Fruits ovoid or short cylindric. Fl. Jan-May. Fr. Jun-Oct. Montane dense forests at 600-1600 m. Distributed in S Yunnan. Also in N Thailand.

蚁花

Orophea laui Leonardía et P. J. A. Kessler

灌木或小乔木。叶革质，椭圆形至长圆形，两面无毛，中脉每侧具侧7-11条。花直径3-5毫米；心皮9-18；内轮花瓣显著长于外轮花瓣，先端明显反折及加厚。果约10，球形，聚生。花期4月，果期9-10月。生海拔400-1200米的山地林中。产云南南部和海南。

Shrubs or small trees. Leaves coriaceous, elliptic to oblong, both surfaces glabrous, secondary veins of leaves 7-11 on each side of midvein. Flowers 3-5 mm diam; carpels 9-18; inner petals much longer than outer petals, tips conspicuously recurved and thickened. Fruits ca. 10, globose, fascicled. Fl. Apr. Fr. Sep-Oct. Montane forests at 400-1200 m. Distributed in S Yunnan and Hainan.

细基丸

Polyalthia cerasoides (Roxb.) Benth. et Hook. f. ex Bedd.

乔木。叶中脉每边侧脉7或8条。花序腋生，具单花；萼片长8-9毫米；花瓣绿色，干后黑色，内轮花瓣与外轮花瓣近相等或内轮花瓣短于外轮花瓣。果柄长1.5-2厘米，细弱；果实红色，干后黑色，卵球形至近球形，直径约6毫米。花期3-5月，果期4-11月。生海拔100-1100米的山地疏林中。产云南南部、广西南部、广东南部和海南。印度、缅甸、老挝、泰国、越南和柬埔寨亦有。

Trees. Leaves with secondary veins 7 or 8 on each side of midvein. Inflorescences axillary, 1-flowered; sepals 8-9 mm long; petals green but black when dry, subequal or inner petals shorter than outer petals. Fruit stalk 1.5-2 cm long, weak; fruits red but black when dry, ovoid to subglobose, ca. 6 mm diam. Fl. Mar-May. Fr. Apr-Nov. Sparsely forested slopes at 100-1100 m. Distributed in S Yunnan, S Guangxi, S Guangdong and Hainan. Also in India, Myanmar, Laos, Thailand, Vietnam and Cambodia.

细基丸 *Polyalthia cerasoides*

陵水暗罗 *Polyaltia nemoralis*

腺叶暗罗 *Polyalthia simiarum*

陵水暗罗
Polyaltia nemoralis A. DC.

灌木或小乔木，高达5米。叶革质，具短柄，长圆形，长9-18厘米，宽2-6厘米，顶端渐尖，无毛，侧脉8-10对。花白色，直径1-2厘米，与叶对生；萼片3，三角形，长2毫米；花瓣6，成2轮，狭卵形，长6-8毫米；雄蕊小，多数，楔形；心皮7-11。花期4-7月。生林中。产海南和广东南部。越南亦有。

Shrubs or small trees, up to 5 m tall. Leaves shortly petiolate, oblong, 9-18 cm long, 2-6 cm wide at apex acuminate, glabrous, lateral nerves 8-10 pairs. Flower white, 1-2 cm diam, opposite to leaf; sepals 3, triangular, 2 mm long; petals 6, 2-seriate, narrowly ovate, 6-8 mm long; stamens small, numerous, cuneiform; carpels 7-11. Fl. Apr-Jul. Forests. Distributed in Hainan and S Guangdong. Also in Vietnam.

腺叶暗罗
Polyalthia simiarum (Buch.-Ham. ex Hook. f. et Thomson) Hook. f. et Thomson

乔木。枝幼时被微柔毛，老后无毛且具疏皮孔。叶薄纸质，侧脉每边13-17条，膜质至纸质，具透明腺点。花序具1至几朵花；花瓣条状披针形，长25-35毫米。果聚生；总果柄长2-3厘米；果柄长3-3.5厘米。花期4-9月，果期7-12月。生海拔500-1200米的山地密林。产云南南部至西南部。南亚和东南亚亦有。

Trees. Branches puberulent when young, glabrous and sparsely lenticellate with age. Leaves thinly papery, lateral veins 13-17 per side, membranous to papery, hyalopunctate. Inflorescences 1- to several flowered; petals linear-lanceolate, 25-35 mm long. Fruits fascicled; fruit stalk 2-3 cm long; monocarp stipes 3-3.5 cm long. Fl. Apr-Sep. Fr. Jul-Dec. Dense forests on mountains at 500-1200 m. Distributed in S to SW Yunnan. Also in S and SE Asia.

暗罗
Polyalthia suberosa (Roxb.) Thwaites

灌木或小乔木。叶5-11 × 2-4厘米，膜质至纸质，侧脉每边8-10条。花序与叶对生，具1-2朵花；花瓣亮黄色；心皮多数。果柄长8-10毫米，细弱；果多达18，聚生，红色，近球形。花期几乎全年，果期6月至翌年春季。生海拔100-700米的山地或山坡疏林中。产广西南部、广东南部和海南。南亚和东南亚亦有。

Shrubs or small trees. Leaves 5-11 × 2-4 cm, membranous to papery, lateral veins 8-10 per side. Inflorescences leaf-opposed, 1-2-flowered; flowers light yellow; carpels numerous. Monocarp stipes 8-10 mm long, slender; monocarps to 18, fascicled, red, subglobose. Fl. almost throughout year. Fr. Jun to next spring. Mountains or sparse forests on slopes at 100-700 m. Distributed in S Guangxi, S Guangdong and Hainan. Also in S and SE Asia.

暗罗 *Polyalthia suberosa*

嘉陵花 *Popowia pisocarpa*

嘉陵花

Popowia pisocarpa (Blume) Endl.

灌木或小乔木。叶膜质，侧脉每边6-8条。花序具1花或2或3花束生；花白色或黄色，内轮花瓣凹陷且内覆盖雄蕊；心皮5-6，果期成熟心皮离生，但常靠合。果球形，直径6-8毫米。花期1-7月，果期9-11月。生海拔200-300米的山坡林地。产广东和海南。缅甸、泰国、越南、马来西亚、印度尼西亚和菲律宾亦有。

Shrubs or small trees. Leaves membranous, lateral veins 6-8 per side. Inflorescences 1-flowered or flowers 2 or 3 fasciculate; flowers white or yellow, inner petals concave and incurved to cover stamens; carpels 5-6, fruiting carpels free when mature, but usually connivent. Fruits globose, 6-8 mm diam. Fl. Jan-Jul. Fr. Sep-Nov. Forested slopes at 200-300 m. Distributed in Guangdong and Hainan. Also in Myanmar, Thailand, Vietnam, Malaysia, Indonesia and the Philippines.

金钩花

Pseuduvaria trimera (Craib) Y. C. F. Su et R. M. K. Saunders

乔木，雌雄同株。叶薄革质，侧脉在中脉每边10-12(-18)条。花序生幼枝叶腋，每簇3-6，具1或2花；花瓣黄色；外轮花瓣卵形，长2-3毫米；内轮花瓣爪状三角形；心皮7-14；每子房具5或6胚珠。果绿色，球形。花期2-4月，果期4-7月。生海拔200-700(-1500)米的石灰山脚的常绿及落叶阔叶林中。产云南南部。缅甸、泰国和越南亦有。

Trees, monoecious. Leaves thinly coriaceous, secondary veins 10-12(-18) on each side of midvein. Inflorescences on young branches, in clusters of 3-6, axillary, each 1- or 2-flowered; petals yellow; outer petals ovate, 2-3 mm long; inner petals clawed-triangular; carpels 7-14; ovules 5 or 6 per carpel. Fruits green, globose. Fl. Feb-Apr. Fr. Apr-Jul. Evergreen and deciduous broad-leaved forests at base of limestone mountains at 200-700 (-1500) m. Distributed in S Yunnan. Also in Myanmar, Thailand and Vietnam.

金钩花 *Pseuduvaria trimera*

黄花紫玉盘

Uvaria kurzii (King) P. T. Li

攀援灌木。叶膜质，侧脉13-18对。花黄色或淡黄色，直径约3.5厘米；花瓣6片，卵形；雌蕊比雄蕊长；每心皮具10胚珠。果实卵球形至近球形，外果皮密具茶褐色绒毛，无刺。花期5月，果期7-8月。生海拔400-1300米的密林中。产云南和广西。印度亦有。

Shrubs, climbing. Leaves membranous, secondary veins 13-18 on each side of midvein. Flowers yellow or yellowish, ca. 3.5 cm diam; petals 6, ovate; pistils longer than stamens; ovules 10 per carpel. Fruits ovoid to subglobose, epicarps densely tawny brown pubescent, not spiny. Fl. May. Fr. Jul-Aug. Dense forests at 400-1300 m. Distributed in Yunnan and Guangxi. Also in India.

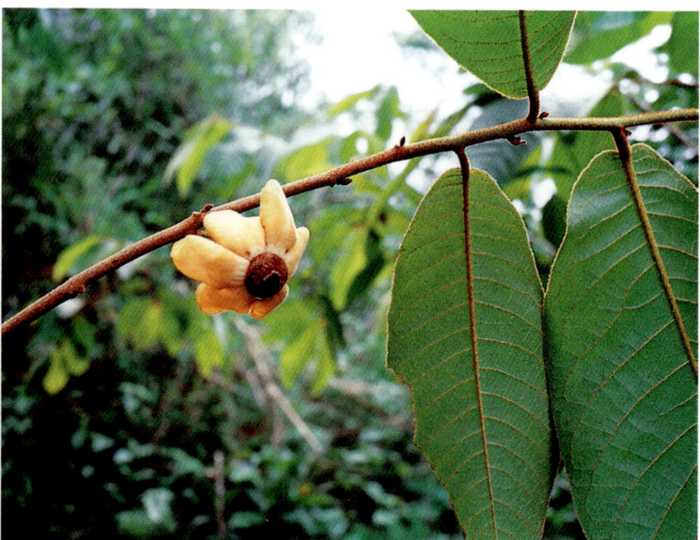
黄花紫玉盘 *Uvaria kurzii*

紫玉盘
Uvaria macrophylla Roxb.

攀援灌木。叶革质，侧脉每边约13条，先端锐尖、钝或圆形，侧脉在上面凹陷。花1-3朵，暗紫色或淡红褐色，直径2.5-4厘米。果橙色，卵球形至近圆柱状，外果皮不具刺。花期3-9月，果期7月至翌年3月。生海拔400-1400米的灌木丛中、丘陵或山地疏林中。产云南东南部、广西、广东、海南、台湾和福建。南亚、东南亚和巴布亚新几内亚亦有。

Shrubs, climbing. Leaves coriaceous, lateral veins ca. 13 per side, apex acute, obtuse, or rounded, secondary veins adaxially impressed. Flowers 1-3, dark-purple or pale-red-brown, 2.5-4 cm diam. Fruits orange, ovoid to subterete, epicarps not spiny. Fl. Mar-Sep. Fr. Jul to next Mar. Thickets, hills, or sparse forests on mountains at 400-1400 m. Distributed in SE Yunnan, Guangxi, Guangdong, Hainan, Taiwan and Fujian. Also in S and SE Asia, and Papua New Guinea.

紫玉盘 *Uvaria macrophylla*

肉豆蔻科 Myristicaceae

红光树
Knema tenuinervia W. J. de Wilde

乔木。叶阔披针形或长圆状披针形，(15-)30-55(-70) × (7-)8-15厘米，近革质，基部心形或圆形，侧脉24-35对。雄花序长1-1.5厘米；雄花大；果序短，常具1或2个果实。果实椭圆体形或卵球形，密具1-3毫米长锈色枝状毛。花期11月至翌年2月，果期7-9月。生海拔500-1000米的湿润密林、山区或峡谷。产云南。印度东北部、尼泊尔、老挝和泰国亦有。

Trees. Leaves broadly lanceolate or oblong-lanceolate, (15-)30-55(-70) × (7-)8-15 cm, subcoriaceous, base cordate or rounded, lateral veins 24-35 pairs. Staminate inflorescences 1-1.5 cm long; staminate flowers large; infructescences short, often with 1 or 2 fruits. Fruits ellipsoid or ovoid, with dense rusty dendritic hairs 1-3 mm long. Fl. Nov to next Feb . Fr. Jul-Sep. Moist dense forests, mountains or ravines at 500-1000 m. Distributed in Yunnan. Also in NE India, Nepal, Laos and Thailand.

假广子
Knema elegans Warburg

乔木。叶纸质至近革质，长圆状披针形至条状披针形，边缘近平行，下面密具锈色或灰褐色具柄星状柔毛，侧脉15-32(-36)对。雄花序具4-8花；果序具1或2果实。果实卵球形或椭圆体形，长2.5-3.2厘米。花期8-9月，果期翌年4-5月。生海拔500-1700米的山坡、低丘陵

红光树 *Knema tenuinervia*

假广子 *Knema elegans*

肉豆蔻 *Myristica fragrans*

或峡谷。产云南。缅甸、泰国、越南和柬埔寨亦有。

Trees. Leaves papery to subcoriaceous, oblong-lanceolate to linear-lanceolate, margins nearly parallel, abaxially with dense rusty or grayish brown stalked stellate pubescence, lateral veins 15-32(-36) pairs. Staminate inflorescences with 4-8 flowers; infructescences of 1 or 2 fruits. Fruit ovoid or ellipsoid, 2.5-3.2 cm long. Fl. Aug-Sep. Fr. next Apr-May. Mountain slopes, low hills or ravines at 500-1700 m. Distributed in Yunnan. Also in Myanmar, Thailand, Vietnam and Cambodia.

云南肉豆蔻
Myristica yunnanensis Y. H. Li

乔木。叶长圆状披针形或长圆状倒披针形，(24-)30-38(-45) × 8-14(-18)厘米，纸质，侧脉20-32对。雄花序，二歧或三歧，为顶生的伞形花序。果实椭圆体形，密具毡毛。花期9-12月，果期翌年3-6月。生海拔500-600米的山坡密林或峡谷。产云南南部。泰国北部亦有。

Trees. Leaves oblong-lanceolate or oblong-oblanceolate, (24-)30-38(-45) × 8-14(-18) cm, papery, lateral veins 20-32 pairs. Staminate inflorescences, dichotomous or trichotomous, flowers in terminal umbels. Fruits ellipsoid, densely lanose hairy. Fl. Sep-Dec. Fr. next Mar-Jun. Dense forests on mountain slopes or ravines at 500-600 m. Distributed in S Yunnan. Also in N Thailand.

肉豆蔻
Myristica fragrans Houtt.

小乔木。叶椭圆形或椭圆状披针形，长4-8厘米，近革质，基部阔楔形或近圆形，侧脉6-10对。雄花序具短梗，不分叉或分叉。果实1或2，橙色或黄色，梨形或近球形。云南、广东和台湾栽培。原产印度尼西亚；热带广泛栽培。

Small trees. Leaves elliptic or elliptic-lanceolate, 4-8 cm long, subcoriceous, base broadly cuneate or nearly rounded, lateral veins 6-10 pairs. Staminate inflorescences shortly peduncled, simple or forked. Fruits 1 or 2, orange or yellow, pyriform or subglobose. Cultivated at Yunnan, Guangdong and Taiwan. Native to Indonesia; widely cultivated in the tropics.

云南肉豆蔻 *Myristica yunnanensis*

云南风吹楠 *Horsfieldia prainii*

云南风吹楠
Horsfieldia prainii (King) Warburg

乔木，单性异株。小枝干后灰色。叶倒卵状长圆形，中部之上最宽，脉9-20对。雄花花被裂片3或4(或5)，聚药棍棒状、近球形，花药4-6。果实椭圆体形至卵球形。外种皮具斑纹。生海拔500-1100米的原始森林、沟谷林中或峡谷。产云南南部。印度、泰国、印度尼西亚、菲律宾和巴布亚新几内亚亦有。

Trees, monoecious. Twigs pale when dry. Leaves obovate-oblong, widest at or above middle, nerves 9-20 pairs. Staminate flower with perianth lobes 3 or 4 (or 5), synandrium stipitate, subglobose, anthers 4-6. Fruits ellipsoid to ovoid. Testa variegated. Primary forests, forests in ravines or valleys at 500-1100 m. Distributed in S Yunnan. Also in India, Thailand, Indonesia, the Philippines and Papua New Guinea.

海南风吹楠
Horsfieldia hainanensis Merr.

乔木，高9-15米。小枝密被锈色星状毛。叶具短柄，近革质，长圆形，长(12-)15-30厘米，宽5-9(-12)厘米，顶端短渐尖，变无毛，侧脉12-16对。雄花序总状圆锥状，长达7厘米；雄花小，密集，球形，长1-1.5毫米；花被3裂；雄蕊合生，花药20；雌花序长3-5厘米；雌花被近球形，厚革质，长约3毫米，3-5裂；雄蕊无花柱，子房近球形，密被星状毛。花期5-8月。生海拔400-450米的山谷密林中。产海南和广西南部。

Tree, 9-15 m tall. Branchlets covered with dense rusty stellate hairs. Leaves shortly petiolate, subcoriaceous, oblong, (12-)15-30 cm long, 5-9(-12) cm broad, at apex shortly acuminate, glabrescent, lateral nerves 12-16 pairs. Staminate inflorescences racemose-paniculate, up to 7 cm long; staminate flowers small, dense, globose, 1-1.5 mm long; perianth 3-parted; stamens connate, with 20 anthers; pistillate inflorescences 3-5 cm long; pistillate perianth subglobose, thickly coriaceous, ca. 3 mm long, 3-5-lobed; pistil without style, ovary subglobose, covered with dense stellate hairs. Fl. May-Aug. Dense forests at 400-450 m. Distributed in Hainan and S Guangxi.

大叶风吹楠
Horsfieldia kingii (Hook. f.) Warb.

乔木，雌雄异株。小枝黑褐色。叶(3-)5列，叶倒卵形或倒披针形，(12-)28-55 × 5-15厘米。雄花花药12-20个合生成为球形聚药雄蕊；雌花花被裂片2或3；子房具柔毛。果实卵球形至长圆状椭圆体形。花期4-8月，果期10-12月。生海拔800-1200米的沟

海南风吹楠 *Horsfieldia hainanensis*

大叶风吹楠 *Horsfieldia kingii*

谷密林中。产云南、广西和海南。印度和泰国北部亦有。

Trees, dioecious. Twigs dark brown. Leaves in (3-)5 rows, leaves obovate or oblanceolate, (12-)28-55 × 5-15 cm. Staminate flower with anthers 12-20, connate into globose synandrium; pistillate flower with perianth lobes 2 or 3; ovary pubescent. Fruit ovoid to oblong-ellipsoid. Fl. Apr-Aug. Fr. Oct-Dec. Dense forests in ravines at 800-1200 m. Distributed in Yunnan, Guangxi and Hainan. Also in India and N Thailand.

风吹楠

Horsfieldia amygdalina (Wall. ex Hook. f. et Thomson) Warb.

乔木，雌雄异株。叶两列或3-5行，叶狭椭圆形或长圆状披针形，纸质。雄花被片2或3(或4)裂；聚药球形或扁球形，无柄；雄蕊8-15，离生；雌花子房无毛。果实橙色，卵球形或椭圆体形。花期8-10月，果期翌年3月至5月。生海拔100-1200米的山坡沟谷密林或稀疏丘林。产云南、广西、广东和海南。印度、孟加拉国、缅甸、老挝、泰国和越南亦有。

Trees, dioecious. Leaves distichous or in 3-5 rows, narrowly elliptic or oblong-lanceolate, papery. Staminate flower perianth 2- or 3(or 4)-lobed; synandrium globose or depressed globose, sessile; staminate 8-15, free; pistillate flower with ovary glabrous. Fruits orange, ovoid or ellipsoid. Fl. Aug-Oct. Fr. next Mar to May. Dense forests on mountain slopes and in ravines, or sparse hilly forests at 100-1200 m. Distributed in Yunnan, Guangxi, Guangdong and Hainan. Also in India, Bangladesh, Myanmar, Laos, Thailand and Vietnam.

风吹楠 *Horsfieldia amygdalina*

樟科 Lauraceae

滇润楠

Machilus yunnanensis Lecomte

乔木。叶倒卵形、倒卵状椭圆形或长圆形至椭圆形，革质，两面无毛。圆锥花序生稍下部短枝，无毛。花浅绿色、淡黄绿色或淡黄白色；花被裂片外面无毛；第三轮花丝具带柄腺体。果实深蓝色，成熟时灰色，椭圆体形或长圆形。花期4-5月，果期5-10月。生海拔1500-2100米的常绿阔叶林中或山坡湿润肥沃处。产云南中部、西北部和西部、四川西部、西藏南部和广西西北部。

Trees. Leaves obovate, obovate-elliptic, or oblong to elliptic, leathery, glabrous on both surfaces. Panicles at lower part of short branchlet, glabrous throughout. Flowers pale green, pale yellowish green, or pale yellowish white; perianth lobes glabrous outside; filaments of 3rd series with stipitate glands. Fruit dark blue, glaucescent when mature, ellipsoid or oblong. Fl. Apr-May. Fr. May-Oct. Evergreen broad-leaved forests, or moist and fertile places on mountain slopes at 1500-2100 m. Distributed in C, NW and W Yunnan, W Sichuan, S Xizang and NW Guangxi.

滇润楠 *Machilus yunnanensis*

凤凰润楠 *Machilus phoenicis*

红楠 *Machilus thunbergii*

凤凰润楠
Machilus phoenicis Dunn

中等乔木。叶椭圆形或长圆形至狭长圆形，厚革质，侧脉每边约12条，基部钝至圆形。圆锥花序多数生于枝顶；花被裂片长6-10毫米，绿色，宿存；子房无毛。果球形，直径约0.9毫米，果柄膨大。生海拔500-900米混交林中。产广东、福建、浙江、湖南和江西。

Medium-sized trees. Leaves elliptic or oblong to narrowly oblong, thickly leathery, lateral nerves ca. 12 on each side, base obtuse to rounded. Panicles numerous clustered on branch apices; perianth-lobes 6-10 mm long, green, persistent; ovary glabrous. Fruits globose, ca. 0.9 mm diam, fruiting pedicels enlarged. Mixed forests at 500-900 m. Distributed in Guangdong, Fujian, Zhejiang, Hunan and Jiangxi.

红楠
Machilus thunbergii Sieb. et Zucc.

常绿乔木。叶革质，侧脉7-12对，两面无毛，中脉在背面明显凸起，先端钝或突渐尖；叶柄淡红色。花序自幼枝基部伸出，多花；子房球形，无毛。果扁球形，初绿色后变黑紫色，花期2月，果期7-8月。生海拔800米以下山坡、沟谷或林中。产华南、东南和华东。朝鲜半岛和日本亦有。

Evergreen trees. Leaves leathery, lateral nerves 7-12 on each side, glabrous on both surfaces, midrib distinctly elevated abaxially, apex obtuse or abruptly cuspidate; petioles reddish. Inflorescences arising from base of young shoots, many flowered; ovary globose, glabrous. Fruits oblate, green when young, black-purple when mature. Fl. Feb. Fr. Jul-Aug. Mountain slopes, valleys or forests below 800 m. Distributed in S, SE and E China. Also in Korean Peninsula and Japan.

绒毛润楠
Machilus velutina Champ. ex Benth.

乔木，所有部位均具锈色绒毛。叶革质，中脉在下面突起，侧脉每边8-11条，基部楔形，先端渐狭或短渐尖。花序顶生，花梗极短；花浅黄色，有香味；子房新鲜时淡红色。果紫红色，球形。花期10-12月，果期翌年2-3月。生海拔500-900米的山谷、阔叶林或林缘。产中国西南、华南和东南。老挝、越南和柬埔寨亦有。

Trees, all parts densely ferruginous tomentose. Leaves leathery, midrib abaxially raised, lateral veins 8-11 pairs, base cuneate, apex attenuate or shortly acuminate. Inflorescences terminal, peduncles very short; flowers pale yellow, fragrant; ovary pale red when alive. Fruits purplish red, globose. Fl. Oct-Dec. Fr. next Feb-Mar. Valleys, broad-leaved forests or forest edges at 500-900 m. Distributed in SW, S and SE China. Also in Laos, Vietnam and Cambodia.

黄绒润楠
Machilus grijsii Hance

乔木。芽、小枝、叶柄和叶片下面被黄褐色短绒毛。叶倒卵状长圆形，革质，基部稍圆形，先端渐狭。花序短，簇生于小枝顶端；花被裂片两面被

绒毛润楠 *Machilus velutina*

黄绒润楠 *Machilus grijsii*

绒毛。果球形。花期3月，果期4月。生海拔500-900米灌丛中或密林中。产广东、海南、福建、浙江和江西。

Trees. Buds, branchlets, petioles and leaves abaxially rusty-velutinous. Leaves obovate-oblong, leathery, base ± rounded, apex attenuate. Inflorescences short, fasciculate at apex of branchlet; perianth lobes tomentose on both surfaces. Fruits globose. Fl. Mar. Fr. Apr. Thickets or dense forests at 500-900 m. Distributed in Guangdong, Hainan, Fujian, Zhejiang and Jiangxi.

黄心树

Machilus gamblei King ex Hook. f.

乔木。叶倒卵形、倒披针形或长圆形，薄革质，叶两面幼时具锈色平伏绢毛。数个圆锥花序自幼枝伸出，长4-13厘米；花白色或淡黄色。果球形，成熟时紫黑色。花期3-4月，果期4-6月。生海拔160-1300米的林中或山谷。产云南、西藏、贵州、广西、广东和海南。南亚和东南亚亦有。

Trees. Leaves obovate, oblanceolate or oblong, thinly leathery, ferruginous appressed sericeous on both surfaces when young. Numerous panicles arising from lower part of young shoots, 4-13 cm long; flowers white or pale yellow. Fruits globose, purple-black when mature. Fl. Mar-Apr. Fr. Apr-Jun. Forests or valleys at 160-1300 m. Distributed in Yunnan, Xizang, Guizhou, Guangxi, Guangdong and Hainan. Also in S and SE Asia.

刨花润楠

Machilus pauhoi Kaneh.

乔木。叶常集生枝顶端，椭圆形或狭椭圆形，革质，下面具平伏绢毛，基部楔形，先端渐尖或长渐尖。聚伞圆锥花序与叶近等长，具疏花。果球形，成熟时黑色。生山坡灌丛或山谷疏林中。产广西、广东、福建、浙江、湖南和江西。

Trees. Leaves usually clustered at branch apices, elliptic or narrowly elliptic, leathery, abaxially appressed sericeous, base cuneate, apex acuminate or long acuminate. Racemose cymes subequal to leaves, with sparse flowers. Fruits globose, black when mature. Bushes on mountain slopes or sparse forests in valleys. Distributed in Guangxi, Guangdong, Fujian, Zhejiang, Hunan and Jiangxi.

黄心树 *Machilus gamblei*

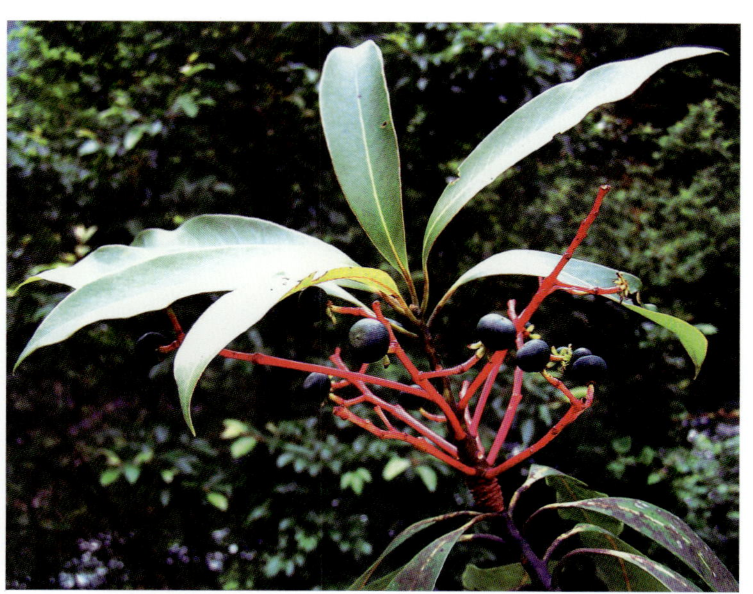
刨花润楠 *Machilus pauhoi*

薄叶润楠
Machilus leptophylla Hand.-Mazz.

大乔木。叶坚纸质，下面具白粉，倒卵状长圆形，侧脉14-20(-24)对，淡红色。圆锥花序多花，密具灰色柔毛；花白色；花被裂片6，排成2轮。果球形，直径约1厘米；果梗长5-10毫米。生海拔400-1200米的阴坡谷地混交林中。产中国西南、华南和东南。

Large trees. Leaves chartaceous, abaxially glaucous, obovate-oblong, lateral nerves 14-20(-24) on each side, pale red. Panicles many-flowered, densely grayish puberulent; flowers white; perianth-lobes 6, 2-whorled. Fruits globose, ca. 1 cm diam; fruiting pedicels 5-10 mm long. Mixed forests of shady valleys at 400-1200 m. Distributed in SW, S and SE China.

薄叶润楠 *Machilus leptophylla*

柳叶润楠 *Machilus salicina*

柳叶润楠
Machilus salicina Hance

灌木。叶狭披针形，薄革质，基部渐狭，先端渐尖。聚伞圆锥花序常长3厘米；花淡黄色，花被裂片较薄，两面都有绢毛或灰白色微毛。果球形，成熟时紫黑色；果梗鲜时红色。花期2-3月，果期4-6月。生海拔490-550米的河岸林中。产云南、贵州、广西、广东和海南。柬埔寨、老挝和越南亦有。

Shrubs. Leaves narrowly lanceolate, thinly leathery, base attenuate, apex acuminate. Cymose panicles usually 3 cm long; flowers yellowish; perianth lobes thin, sericeous or minutely gray hairy on both sides. Fruits globose, purple-black when mature; fruiting pedicels red when fresh. Fl. Feb-Mar. Fr. Apr-Jun. Forests by rivers at 490-550 m. Distributed in Yunnan, Guizhou, Guangxi, Guangdong and Hainan. Also in Cambodia, Laos and Vietnam.

细毛润楠
Machilus tenuipila H. W. Li

乔木。花序轴、总梗、花梗、花被裂片两面和小枝均被淡黄色微柔毛。叶椭圆形至长圆形，坚纸质，两面无毛，基部

细毛润楠 *Machilus tenuipila*

粗壮润楠 *Machilus robusta*

红梗润楠 *Machilus rufipes*

楔形，边缘反卷，先端短渐尖。圆锥花序近顶生，具少花；花浅绿色。果实球形，成熟后蓝黑色；宿存花被裂片膜质。花期3-4月，果期8-9月。生海拔1400-2400米的林中或灌丛。产云南西南部。

Trees. Inflorescence-rhachis, peduncles, pedicels, both surfaces of perianth-lobes and branchlets all pale yellow pubescent. Leaves elliptic to oblong, firmly papery, glabrous on both surfaces, base cuneate, margin revolute, apex shortly acuminate. Panicles subterminal, few flowered; flowers greenish. Fruits globose, blue-black when matured; persistent perianth lobes membranous. Fl. Mar-Apr. Fr. Aug-Sep. Forests or thickets at 1400-2400 m. Distributed in SW Yunnan.

粗壮润楠

Machilus robusta W. W. Sm.

乔木。叶较大，长10-26厘米，侧脉每边(5-)7-9条，革质，两面无毛，中脉淡红色，先端近锐尖，有时短渐尖。圆锥花序顶生或近顶生，多分枝，多花；花大，灰绿、黄绿或黄色。果球形，直径2.5-3厘米，成熟时蓝黑色。花期1-4月，果期4-6月。生海拔600-2100米的常绿阔叶林或开阔灌丛中。产云南、西藏、贵州、广西、广东和海南。缅甸亦有。

Trees. Leaves large, 10-26 cm long, lateral nerves (5-)7-9 on each side, leathery, glabrous on both surfaces, midrib pale red, apex subacute, sometimes shortly acuminate. Panicles terminal or subterminal, much branched, many-flowered; flowers large, gray-green, yellow-green or yellow. Fruits globose, 2.5-3 cm diam, blue-black when mature. Fl. Jan-Apr. Fr. Apr-Jun. Evergreen borad-leaved forests or open thickets at 600-2100 m. Distributed in Yunnan, Xizang, Guizhou, Guangxi, Guangdong and Hainan. Also in Myanmar.

红梗润楠

Machilus rufipes H. W. Li

乔木。叶长圆形，近革质，下面密具金黄色柔毛，先端短或近长渐尖。圆锥花序由1-3花的聚伞花序组成，生于幼枝下部；花大，长达9毫米；总梗、序轴和花梗扁平，红紫色。果球形，成熟时紫黑色。花期3-4月，果期5-9月。生海拔1500-2000米的山脊或多苔藓林中。产云南和西藏。

Trees. Leaves oblong, subleathery, abaxially densely golden yellow villous, apex shortly or nearly long acuminate. Panicles composed from 1-3-flowered cymes, inserted on the lower part of branchlets; flowers large, up to 9 mm long; peduncles, rhachis and pedicels complanate, red-purple. Fruits globose, purple-black when mature. Fl. Mar-Apr. Fr. May-Sep. Mountain ridges or mossy forests at 1500-2000 m. Distributed in Yunnan and Xizang.

皱皮油丹

Alseodaphne rugosa Merr. et Chun

乔木。小枝粗壮。叶密集，近轮生，长圆状倒卵形或长圆状倒披针形，长15-36厘米。果序近顶生，长约12.5厘米，粗壮；果扁球形，直径约3厘米。果梗粗，长1-1.5厘米，幼时肉质，多疣。果期7-12月。生海拔1200-1300米山谷混交林中。产海南。

Trees. Branchlets stout. Leaves densely aggregated and nearly verticillate, oblong-obovate or oblong-oblanceolate, 15-36 cm long. Infructescences subterminal, ca. 12.5 cm long, robust; fruits oblate, ca. 3 cm diam. Fruiting pedicels stout, 1-1.5 cm long, fleshy when young, with many warts. Fr. Jul-Dec. Mixed forests in valleys at 1200-1300 m. Distributed in Hainan.

皱皮油丹 *Alseodaphne rugosa*

西畴油丹 *Alseodaphe sichourensis*

油丹 *Alseodaphne hainanensis*

西畴油丹

Alseodaphe sichourensis H. W. Li

常绿乔木。叶对生，疏离，长圆形，两面无毛；叶柄淡红色。果序圆锥状，长5-8.5厘米，仅具1枚发育良好的果实；果序轴淡红色，无毛。果实红色，椭圆体形，达5 × 3厘米，无毛，果梗粗壮，肉质。生海拔1300-1500米的石灰山常绿阔叶林中。产云南东南部。

Evergreen trees. Leaves opposite, remote, oblong, glabrous on both surfaces; petioles reddish. Infructescences paniculate, 5-8.5 cm long, only with 1 well-developed fruit; rachis pale red, glabrous. Fruits red, ellipsoid, up to 5 × 3 cm, glabrous; fruit stalks robust, fleshy. Evergreen broad-leaved forests of limestone mountains at 1300-1500 m. Distributed in SE Yunnan.

油丹

Alseodaphne hainanensis Merr.

乔木。叶多数，簇生于枝顶，狭椭圆形，上面具蜂窝状小窝穴。圆锥花序生上部小枝腋处，少分枝，无毛；花梗细弱，果期膨大，不具疣。果球形或卵球形，干时黑色。花期7月，果期10月至翌年2月。生海拔1400-1700米的山谷或密林中。产海南。越南北部亦有。

Trees. Leaves numerous, clustered at top of branches, narrowly elliptic, shallowly foveolate adaxially. Panicles axillary on upper part of branchlet, few branched, glabrous; pedicels slender, dilated in fruit, not verrucose. Fruits globose or ovoid, black when dry. Fl. Jul. Fr. Oct to next Feb. Valleys or dense forests at 1400-1700 m. Distributed in Hainan. Also in N Vietnam.

毛叶油丹

Alseodaphne andersonii (King ex Hook. f.) Kosterm.

乔木。叶椭圆形，两面有明显的蜂巢状小窝穴，下面被锈色微柔毛。圆锥花序腋生于小枝上部，多分枝；花梗细弱，长约2毫米，果时膨大。果长圆形，长达5厘米，鲜时绿色，熟时紫黑色，果柄肉质，紫红色。花期7-8月，果期10月至翌年3月。生海拔(1000-)1200-1500(-1900)米潮湿山谷或常绿阔叶林中。产云南和西藏。印度东北部、缅甸、老挝、越南和泰国亦有。

Trees. Leaves elliptic, conspicuously foveolate on both surfaces,

毛叶油丹 *Alseodaphne andersonii*

长柄油丹 *Alseodaphne petiolaris*

abaxially rusty-pubescent. Panicles axillary on upper part of branchlet, many-branched; pedicels slender, ca. 2 mm long, dilated in fruit. Fruits oblong, up to 5 cm long, green when young, purple-black when mature; fruit stalks fleshy, purple-red. Fl. Jul-Aug. Fr. Oct to next Mar. Moist valleys or evergreen broad-leaved forests at (1000-)1200-1500(-1900) m. Distributed in Yunnan and Xizang. Also in NE India, Myanmar, Laos, Vietnam and Thailand.

长柄油丹

Alseodaphne petiolaris (Meissn.) Hook. f.

乔木。叶大，长14-26厘米，倒卵状长圆形或长圆形，厚革质。圆锥花序近顶生，簇生于小枝茎顶，多花；花梗长约2毫米，具锈色柔毛；花小；花被片6，两面被柔毛；柱头3裂。果实长圆状卵球形，肉质；果梗强壮。花期10-11月，果期12月至翌年4-5月。生海拔600-900米干燥疏林或常绿阔叶林中。产云南南部。印度和缅甸亦有。

Trees. Leaves, large, 14-26 cm long, obovate-oblong or oblong, thickly leathery. Panicles subterminal, clustered at apex of branchlet, many-flowered; pedicels ca. 2 mm long, rusty pubescent; flowers small; perianth-lobes 6, rusty-pubescent on both surfaces; stigmas 3-lobed. Fruits oblong-ovoid, fleshy; fruit stalks robust. Fl. Oct-Nov. Fr. Dec to next Apr-May. Sparse forests at dry places or evergreen broad-leaved forests at 600-900 m. Distributed in S Yunnan. Also in India and Myanmar.

檬果樟

Caryodaphnopsis tonkinensis (Lecomte) Airy Shaw

小或中等大小的乔木。叶对生或近对生，卵状长圆形、椭圆形或圆椭圆形，纸质，两面无毛。圆锥花序狭且细弱，腋生或近顶生，具短分枝；花白色或浅绿色。果实狭椭球状球形或椭圆体形，约7 × 5厘米。花期3-6月，果期6-8月。生海拔100-1200米的山谷疏林中或林缘路旁。产云南。越南、马来西亚和菲律宾亦有。

Small or middle-sized trees. Leaves opposite or subopposite, ovate-oblong, elliptic or orbicular-elliptic, papery, glabrous on both surfaces. Panicles narrow and slender, axillary or subterminal, shortly branched; flowers white or pale green. Fruits narrowly ellipsoid-globose or ellipsoid, ca. 7 × 5 cm. Fl. Mar-Jun. Fr. Jun-Aug. Sparse forests in valleys or roadsides of forest edges at 100-1200 m. Distributed in Yunnan. Also in Vietnam, Malaysia and the Philippines.

檬果樟 *Caryodaphnopsis tonkinensis*

披针叶楠

Phoebe lanceolata (Wall. ex Nees) Nees

乔木。叶对生，革质，披针形，长13-22厘米。圆锥花序腋生或顶生，长12-15厘米。果卵形，长7-12毫米。花期4-5月，果期7-9月。生海拔300-1300米的混交林中。产云南。

Trees. Leaves opposite, leathery, lanceolate, 13-22 cm long. Panicles axillary or terminal, 12-15 cm long. Fruits ovate, 7-12 mm long. Fl. Apr-May. Fr. Jul-Sep. Mixed forests at 300-1300 m. Distributed in Yunnan.

披针叶楠 *Phoebe lanceolata*

雅砻江楠
Phoebe legendrei Lec.

乔木。叶革质，椭圆形，长9-12厘米。圆锥花序腋生或顶生，长6-10厘米。果卵形，长7-9毫米。花期6月，果期10-11月。生海拔1500-1900米的密林中。产四川和云南。

Trees. Leaves leathery, elliptic, 9-12 cm long. Panicles axillary or terminal, 6-10 cm long. Fruits ovate, 7-9 mm long. Fl. Jul. Fr. Oct-Nov. Dense forests at 1500-1900 m. Distributed in Sichuan and Yunnan.

雅砻江楠 *Phoebe legendrei*

大果楠
Phoebe macrocarpa C. Y. Wu

大乔木。叶椭圆状倒披针形或倒披针形，薄革质，下面疏具黄褐色柔毛。圆锥花序长10-21厘米，密具黄褐色硬毛，顶部分枝。花淡黄绿色。果序近木质。果实椭圆体形或近长圆形，长3.5-3.8(-4.2)厘米，直径1.9-2.2厘米，先端无毛。花期4-5月，果期10-12月。生海拔1200-1800米的混交阔叶林中。产云南东南部。越南北部亦有。

Large trees. Leaves elliptic-oblanceolate or oblanceolate, thinly leathery, abaxially sparsely yellowish brown pubescent. Panicles 10-21 cm long, densely yellowish brown strigose, branched at top. Flowers pale yellow green. Infructescences subligneous. Fruits ellipsoid or suboblong, 3.5-3.8(-4.2) cm long, 1.9-2.2 cm diam, apex glabrous. Fl. Apr-May. Fr. Oct-Dec. Mixed broad-leaved forests at 1200-1800 m. Distributed in SE Yunnan. Also in N Vietnam.

紫楠
Phoebe sheareri (Hemsl.) Gamble

大到小乔木。叶革质，倒卵形、椭圆状倒卵形、阔倒披针形或倒披针形，长12-18厘米，下面密被长柔毛，侧脉每边8-13条。圆锥花序长7-15(-18)厘米，在花梗顶端分枝；花被裂片两面被毛。果实卵球形。花

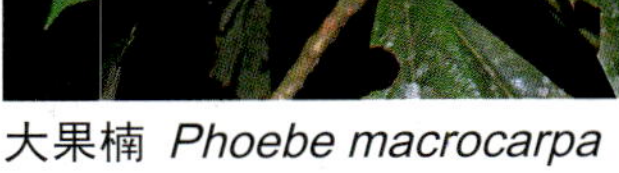
大果楠 *Phoebe macrocarpa*

紫楠 *Phoebe sheareri*

普文楠 *Phoebe puwenensis*

闽楠 *Phoebe bournei*

期4-5月，果期9-10月。生海拔1000米以下的混交阔叶林中。产华西、东南，华中和华东。越南亦有。

Large to small trees. Leaves leathery, obovate, elliptic-obovate, broadly oblanceolate or oblanceolate, 12-18 cm long, abaxially densely villose, lateral nerves 8-13 on each side. Panicles 7-15(-18) cm long, branched at top of peduncle; perianth-lobes hairy on both surfaces. Fruits ovoid. Fl. Apr-May. Fr. Sep-Oct. Mixed broad-leaved forests below 1000 m. Distributed W, SE, C and E China. Also in Vietnam.

普文楠

Phoebe puwenensis W. C. Cheng

大乔木。小枝和叶柄密被黄褐色绒毛。叶倒卵状椭圆形或倒卵状披针形，基部狭楔形。圆锥花序长4.5-22(-25)厘米，于花梗近顶端分枝，具黄褐色绒毛；花淡黄色。果实卵球形，果梗不膨大。花期3-4月，果期6-7月。生海拔800-1500米的常绿阔叶林中。产云南南部。

Large trees. Branchlets and petioles densely brown-tomentose. Leaves obovate-elliptic or obovate-lanceolate, base narrowly cuneate. Panicles 4.5-22(-25) cm long, branched near top of peduncle, yellowish brown tomentose; flowers pale yellow. Fruits ovoid, fruiting pedicels not enlarged. Fl. Mar-Apr. Fr. Jun-Jul. Evergreen broad-leaved forests at 800-1500 m. Distributed in S Yunnan.

浙江楠

Phoebe chekiangensis P. T. Li

大乔木。叶倒卵状椭圆形或倒卵状披针形，革质，侧脉每边8-10条，基部楔形或近圆形。圆锥花序长5-6厘米，密被黄褐色绒毛；柱头盘状。果熟时外被白粉，椭圆体状卵球形。花期4-5月，果期9-10月。生海拔1000米以下山地阔叶林中。产浙江、福建和江西；亦有栽培。

Large trees. Leaves obovate-elliptic or obovate-lanceolate, leathery, lateral nerves 8-10 on each side, base cuneate or subrounded. Panicles 5-6 cm long, densely yellowish-brown tomentose; stigmas dish-shaped. Fruits white-farinose when mature, ellipsoid-ovoid. Fl. Apr-May. Fr. Sep-Oct. Broad-leaved forests of mountains below 1000 m. Distributed in Zhejiang, Fujian and Jiangxi; also cultivated.

闽楠

Phoebe bournei (Hemsl.) Yen C. Yang

大乔木。叶片披针形或倒披针形，革质，侧脉10-14对，基部渐狭或楔形。圆锥花序长3-7(-10)厘米；子房近球形；柱头帽状。果实椭圆体形或长圆形。花期4月，果期10-11月。生海拔1000米以下沟谷或阔叶林中。产华南、东南和华西；也常栽培。

Large trees. Leaves lanceolate or oblanceolate, leathery, lateral nerves 10-14 on each side, base attenuate or cuneate. Panicles 3-7(-10) cm long; ovary subglobose; stigma capitate. Fruits ellipsoid or oblong. Fl. Apr. Fr. Oct-Nov. Valleys or broad-leaved forests below 1000 m. Distributed in S, SE and W China; also cultivated.

浙江楠 *Phoebe chekiangensis*

滇琼楠 *Beilschmiedia yunnanensis*

滇琼楠
Beilschmiedia yunnanensis Hu

乔木。顶芽小，被毛。叶互生，革质，椭圆形或披针形，长8-16厘米。圆锥花序腋生或顶生，长2-8厘米。果椭球体形，长2-4厘米。花期10-12月，果期翌年5-7月。生海拔900-2000米的常绿阔叶林中。产云南、广东和广西。

Trees. Terminal buds small, pubescent. Leaves alternate, leathery, elliptic or lanceolate, 8-16 cm long. Panicles axillary or terminal, 2-8 cm long. Fruits elliptic-globose, 2-4 cm long. Fl. Oct-Dec. Fr. next May-Jul. Evergreen broad-leaved forests at 900-2000 m. Distributed in Yunnan, Guangdong and Guangxi.

美脉琼楠
Beilschmiedia delicata S. Lee et Y. T. Wei

乔木或灌木。顶芽小，被毛。叶对生，革质，椭圆形或披针形，长7-12厘米。圆锥花序腋生或顶生，长3-6厘米。果椭球体形，长2-3厘米。花期5-10月，果期9-12月。生海拔600-1700米的山谷、溪边和密林中。产广东、广西、贵州和云南。

Trees or shrubs. Terminal buds small, pubescent. Leaves opposite, leathery, elliptic or lanceolate, 7-12 cm long. Panicles axillary or terminal, 3-6 cm long. Fruits elliptic-globose, 2-3 cm long. Fl. May-Oct. Fr. Sep-Dec. Valley, streamside and dense forests at 600-1700 m. Distributed in Guangdong, Guangxi, Guizhou and Yunnan.

美脉琼楠 *Beilschmiedia delicata*

粗壮琼楠
Beilschmiedia robusta Allen

乔木。顶芽大，卵球形。叶对生或互生，革质，披针形或长椭圆形，下面干后常密具腺点。圆锥花序腋生或近顶生，长至6厘米，具少花。果倒卵珠形或近陀螺形，直径约3厘米。花期4-5月，果期6-11月。生海拔1000-2400米的湿润山地沟谷或林中。产云南、西藏、贵州

粗壮琼楠 *Beilschmiedia robusta*

厚叶琼楠 *Beilschmiedia percoriacea*

和广西。

Trees. Terminal buds large, ovoid. Leaves opposite or alternate, leathery, lanceolate or elliptic, abaxially always densely glandular-punctate when dry. Panicles axillary or subterminal, up to 6 cm long, few-flowered. Fruits obovoid or subturbinate, ca. 3 cm diam. Fl. Apr-May. Fr. Jun-Nov. Humid montane valleys or forests at 1000-2400 m. Distributed in Yunnan, Xizang, Guizhou and Guangxi.

厚叶琼楠

Beilschmiedia percoriacea Allen

乔木，全株无毛。顶芽卵球形。叶对生或近对生，狭椭圆形或椭圆形。圆锥花序或总状花序长1.5-5厘米，少量簇生于小枝顶端；花被片边缘具缘毛。果实幼时绿色，成熟后深红色或黑褐色，狭椭圆体形，光滑；果梗两面不加厚。花期5月，果期6-12月。生海拔约1000米的密林湿润处。产云南、广西和海南。

Trees, whole plant glabrous. Terminal buds ovoid. Leaves opposite or subopposite, narrowly elliptic or elliptic. Panicles or racemes 1.5-5 cm long, a few clustered at apex of branchlet; perianth-lobes ciliolate on margin. Fruits green when young, dark red or blackish brown when mature, narrowly ellipsoid, smooth; fruiting pedicels not thickened at both ends. Fl. May. Fr. Jun-Dec. Moist places in dense forests at ca. 1000 m. Distributed in Yunnan, Guangxi and Hainan.

李榄琼楠 *Beilschmiedia linocieroides*

李榄琼楠

Beilschmiedia linocieroides H. W. Li

乔木。顶芽大型，无毛。叶对生，革质，椭圆形，长9-21厘米。圆锥花序腋生或顶生，长3-8厘米。果椭球体形，长2-4厘米。花期4-6月，果期8-11月。生海拔680-1350米的混交林中。产云南。

Trees. Terminal buds large, glabrous. Leaves opposite, leathery, elliptic, 9-21 cm long. Panicles axillary or terminal, 3-8 cm long. Fruits elliptic-globose, 2-4 cm long. Fl. Apr-Jun. Fr. Aug-Nov. Mixed forests at 680-1350 m. Distributed in Yunnan.

香港油果樟

Sinopora hongkongensis (N. H. Xia et al.) J. Li et al.

乔木。叶互生，革质，长5-10厘米，无毛，侧脉每边3-5条。果球形，直径3.5-4厘米；果梗粗壮，圆柱形，顶端增粗。果期10月。生沟谷林中。产香港。

Trees. Leaves alternate, coriaceous, 5-10 cm long, glabrous; lateral nerves 3-5 on each side. Fruits globose, 3.5-4 cm diam; fruiting pedicels stout, teret, apex inflated. Fr. Oct. Valley forests. Distributed in Hong Kong.

香港油果樟 *Sinopora hongkongensis*

细毛樟 *Cinnamomum tenuipilum*

黄樟 *Cinnamomum parthenoxylon*

细毛樟

Cinnamomum tenuipilum Kosterm.

小至大乔木。叶倒卵形或椭圆形，纸质，下面密被灰色绒毛。圆锥花序腋生或近顶生，具12-20花，密被灰色绒毛；花淡黄色，小。果近球形，紫红色。花期2-4月，果期6-10月。生海拔500-2100米的山谷，谷地的林中或灌丛中。产云南。

Small to large trees. Leaves obovate or elliptic, papery, abaxially densely gray-tomentose. Panicles axillary or subterminal, 12-20-flowered, densely gray-tomentose; flowers pale yellow, small. Fruits subglobose, red-purple. Fl. Feb-Apr. Fr. Jun-Oct. Valleys, forests or thickets on hills at 500-2100 m. Distributed in Yunnan.

黄樟

Cinnamomum parthenoxylon (Jack) Meisn.

常绿乔木。叶上面浅绿色或粉绿色，羽状脉，常椭圆状卵形或狭椭圆状卵形，下面侧脉脉窝不明显，上面相应处也不明显呈泡状隆起。花黄绿色，小。果球形，黑色。花期3-5月，果期4-10月。生海拔1500米以下的常绿阔叶林中或灌丛。产中国西南、华南和东南。南亚和东南亚亦有。

Evergreen trees. Leaves pale green or glaucous green abaxially, pinnatinervate, usually elliptic-ovate or narrowly elliptic-ovate, the nerves axils adaxially inconspicuously bullate and abaxially with inconspicuously dome-shaped protuberance. Flowers green-yellow, small. Fruits globose, black. Fl. Mar-May. Fr. Apr-Oct. Evergreen broad-leaved forests or thickets below 1500 m. Distributed in SW, S and SE China. Also in S and SE Asia.

沉水樟

Cinnamomum micranthum (Hayata) Hayata

乔木。叶坚纸质或近革质，干后下面黄褐色，上面黄绿色，长圆状椭圆形或卵状椭圆形，两面无毛，侧脉4-5对。圆锥花序顶生或腋生；花白色或紫红色，有香气；花被裂片6。果实椭圆体形。花期7-9月，果期10月。生海拔300-700(-1800)米的山坡、山谷密林中或河边。产中国西南、东南和华南。越南北部亦有。

Trees. Leaves chartaceous or nearly leathery, yellow-brown abaxially and yellow-green adaxially when dry, oblong-elliptic or ovate-elliptic, glabrous on both surfaces, lateral nerves 4-5-paired. Panicles terminal or axillary; flowers white or purple-red, fragrant; perianth-lobes 6. Fruits ellipsoid. Fl. Jul-Sep. Fr. Oct. Slopes, dense forests in valleys or riversides at 300-700(-1800) m. Distributed in SW, SE and S China. Also in N Vietnam.

樟

Cinnamomum camphora (L.) Presl

常绿大乔木。叶卵状椭圆形，近革质，两面无毛，离基三出脉，下面黄绿色或灰绿色。圆锥花序腋生；花浅绿色或浅黄色。果实黑紫色，卵球形或近球形。花期4-5月，果期8-11月。生海拔100-1800米山坡或沟谷。长江以南地区广泛栽培。世界多地广泛引入或栽培。越南、朝鲜半岛和日本亦有。

Evergreen large trees. Leaves ovate-elliptic, subleathery, glabrous on both surfaces, triplinerved, yellow-green or gray-green abaxially. Panicles axillary; flowers green-white or pale yellow. Fruits purple-black, ovoid or subglobose. Fl. Apr-May. Fr. Aug-Nov. Slopes or valleys at 100-1800 m. Commonly cultivated

沉水樟 *Cinnamomum micranthum*

樟 *Cinnamomum camphora*

云南樟 *Cinnamomum glanduliferum*

in the south of Yangtze River. Introduced or cultivated in many countries around the world. Also in Vietnam, Korean Peninsula and Japan.

云南樟

Cinnamomum glanduliferum
(Wall.) Meisn.

常绿乔木。叶互生，下面常灰色和浅绿色，椭圆形至卵状椭圆形或披针形，具羽状脉。圆锥花序腋生，较叶短；花小，淡黄色。果实黑色，球形；果托红色。花期3-5月，果期7-9月。生海拔1500-2500(-3000)米的山地常绿阔叶林中。产云南、四川、西藏和贵州。印度、尼泊尔、不丹、缅甸和马来西亚亦有。

Evergreen trees. Leaves alternate, usually glaucous and pale green abaxially, elliptic to ovate-elliptic or lanceolate, pinninerved. Panicles axillary, shorter than leaves; flowers small, pale yellow. Fruits black, globose; perianth cups in fruit red. Fl. Mar-May. Fr. Jul-Sep. Evergreen broad-leaved forests on hilly land at 1500-2500(-3000) m. Distributed in Yunnan, Sichuan, Xizang and Guizhou. Also in India, Nepal, Bhutan, Myanmar and Malaysia.

少花桂

Cinnamomum pauciflorum
Ness

乔木。叶厚革质，下面粉绿色且晦暗，三出脉或离基三出脉。具3-5(-7)花，常呈聚伞状，黄白色；花被裂片6；柱头盘状。果实椭圆体形，成熟后紫黑色，具鸡冠状斑点。花期3-8月，果期9-10月。生海拔400-1800(-2200)米石灰岩或沙岩的山地或山谷林中。产中国西南、华南和华中。印度和尼泊尔亦有。

Trees. Leaves thickly leathery, glaucous green and opaque abaxially, trinervious or triplinerved. Flowers 3-5(-7), in cymes, yellowish white; perianth-lobes 6; stigmas disk-shaped. Fruits ellipsoid, purple-black when mature, corky-maculate. Fl. Mar-Aug. Fr. Sep-Oct. Limestone areas or sandy-gravel mountains or forests at 400-1800(-2200) m. Distributed in SW, S and C China. Also in India and Nepal.

少花桂 *Cinnamomum pauciflorum*

阴香 *Cinnamomum burmannii*

钝叶桂 *Cinnamomum bejolghota*

阴香
Cinnamomum burmannii
(Nees et T. Nees) Blume

乔木。叶互生或近对生，叶下灰绿色，晦暗。圆锥花序腋生或近顶生，短于叶，具少花，疏松，密具灰色柔毛，具分枝，分枝顶端具1个3花组成的聚伞花序；花绿白色。果实卵球形。花期3-4月，果期10-11月。生海拔100-1400(-2100)米的林中、灌丛或溪畔。产云南、广西、广东、海南和福建。印度、缅甸、越南、菲律宾和印度尼西亚亦有。

Trees. Leaves alternate or subopposite, greyish green and opaque abaxially. Panicles axillary or subterminal, shorter than leaf, few flowered, lax, densely gray puberulent, branched, apex of branch bearing a 3-flowered cyme; flowers green-white. Fruits ovoid. Fl. Mar-Apr. Fr. Oct-Nov. Forests, thickets or roadsides along streams at 100-1400 (-2100) m. Distributed in Yunnan, Guangxi, Guangdong, Hainan and Fujian. Also in India, Myanmar, Vietnam, the Philippines and Indonesia.

钝叶桂
Cinnamomum bejolghota
(Buch.-Ham.) Sweet

小至大乔木。叶浅绿色或黄绿色，下面稍苍白色，椭圆状长圆形，厚革质，两面无毛，三出脉或离基三出脉，先端钝、锐或渐尖。圆锥花序长13-16厘米，密具花，多分枝；花黄色。果实椭圆体形；果梗紫色，稍膨大。花期3-4月，果期5-7月。生海拔600-1800米的山坡沟谷林中。产云南、广东和海南。印度、尼泊尔、不丹、孟加拉国、缅甸、老挝、泰国和越南亦有。

Small to large trees. Leaves pale green or yellow-green and ± glaucous abaxially, elliptic-oblong, thickly leathery, glabrous on both surfaces, trinerved or tri-plinerved, apex obtuse, acute or acuminate. Panicles 13-16 cm long, densely many flowered, much branched; flowers yellow. Fruits ellipsoid; fruit stalks purple, somewhat dilated. Fl. Mar-Apr. Fr. May-Jul. Forests in valleys at 600-1800 m. Distributed in Yunnan, Guangdong and Hainan. Also in India, Nepal, Bhutan, Bangladesh, Myanmar, Laos, Thailand and Vietnam.

狭叶阴香
Cinnamomum heyneanum
Nees

小乔木。叶互生或近对生，下面苍绿色且晦暗，条形、条状披针形或披针形，纸质，离基三出脉，先端渐尖。圆锥花序腋生，具少花，短于叶；花绿

狭叶阴香 *Cinnamomum heyneanum*

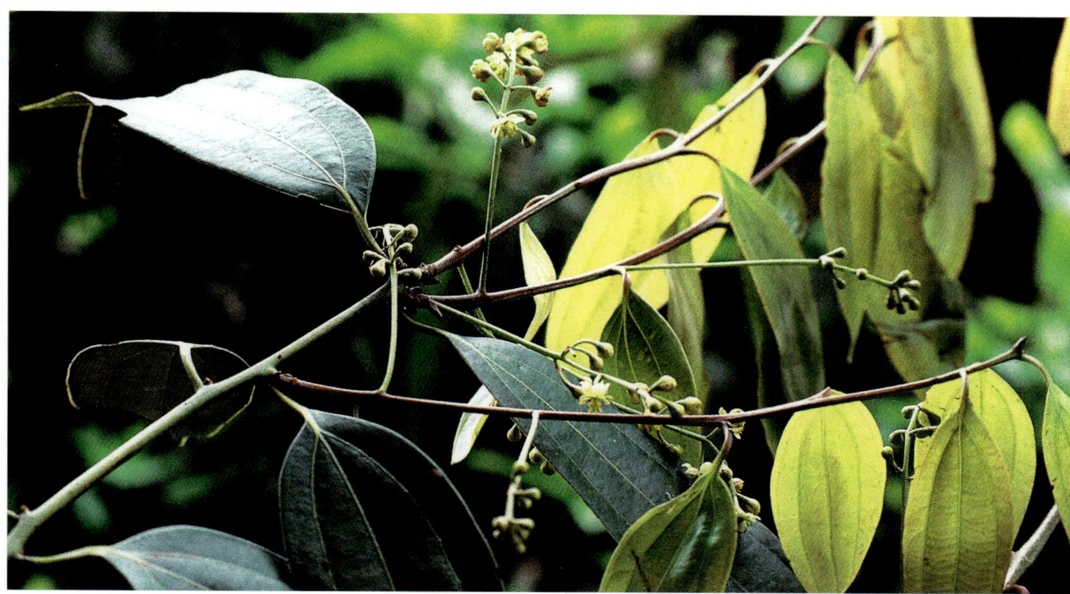
柴桂 *Cinnamomum tamala*

白色。花期4月，果期9-11月。生海拔100-500米的山坡灌丛或河边。产云南、四川、贵州、广西和湖北。印度至印度尼西亚亦有。

Small trees. Leaves alternate or subopposite, glaucous green and opaque abaxially, linear, linear-lanceolate or lanceolate, papery, triplinerved, apex acuminate. Panicles axillary, few-flowered, shorter than leaf; flowers green-white. Fl. Apr. Fr. Sep-Nov. Bushes on slopes or along rivers at 100-500 m. Distributed in Yunnan, Sichuan, Guizhou, Guangxi and Hubei. Also in India to Indonesia.

柴桂

Cinnamomum tamala (Buch.-Ham.) Nees et Eberm.

乔木。叶互生或幼枝上有时近对生，叶下绿白色且晦暗，卵形、长圆形或披针形，薄革质，两面无毛，离基三出脉，先端长渐尖。圆锥花序腋生或顶生，多花；花绿白色；花被筒倒圆锥形，短小。果实倒卵球形或椭圆体形。花期4-5月，果期6-11月。生海拔1100-2000米山坡、山谷常绿阔叶林中或水边。产云南西部。印度、尼泊尔和不丹亦有。

Trees. Leaves alternate or those on young branchlets sometimes subopposite, green-white and opaque abaxially, ovate, oblong or lanceolate, thinly leathery, glabrous on both surfaces, triplinerved, apex long acuminate. Panicles axillary or terminal, many-flowered; flowers white-green; perianth-tube obconical, short. Fruits obovoid or ellipsoid. Fl. Apr-May. Fr. Jun-Nov. Slopes, evergreen broad-leaved forests in valleys or by streams at 1100-2000 m. Distributed in W Yunnan. Also in India, Nepal and Bhutan.

肉桂

Cinnamomum cassia (L.) J. Presl

中等乔木。叶片长圆形至近披针形，三出脉，下面花被灰黄色短绒毛。圆锥花序腋生或近顶生，三回分枝，与叶等长；花白色。果实椭圆体形，成熟后紫黑色。花期6-8月，果期10-12月。在海拔1000米以下的热带栽培。中国西南和华南有栽培。印度、老挝、泰国、越南、马来西亚和印度尼西亚亦有。原产中国。

Medium-sized trees. Leaves elliptic to sublanceolate, trinerved, abaxially densely gray yellow-tomentellate. Panicles axillary or subterminal, triplicate-branched, equal to leaves; flowers white. Fruits ellipsoid, black-purple when mature. Fl. Jun-Aug. Fr. Oct-Dec. Cultivated in tropical regions below 1000 m. Cultivated in SW and S China. Also in India, Laos, Thailand, Vietnam, Malaysia and Indonesia. Originated from China.

肉桂 *Cinnamomum cassia*

刀把木 *Cinnamomum pittosporoides*

香桂 *Cinnamomum subavenium*

刀把木
Cinnamomum pittosporoides Hand.-Mazz.

乔木。叶互生，叶下紫灰色，椭圆形或披针状椭圆形，薄革质，侧脉每边3-4条。圆锥花序腋生于小枝近顶端且常密簇生，短，具1-7花；花金黄色，外面密被污黄色短绒毛，内面被丝毛；花柱被毛。果实卵球形。花期2-5月，果期6-10月。生海拔1800-2500米的常绿阔叶林中。产云南和四川。

Trees. Leaves alternate, purplish glaucous abaxially, elliptic or lanceolate-elliptic, thinly leathery, lateral veins 3-4 on each side. Panicles axillary in leaf axils nearly at apex of branchlet and always densely clustered, short, 1-7-flowered; flowers golden yellow, abaxially densely dirty yellow-tomentellate, adaxially sericeous; styles hairy. Fruits ovoid. Fl. Feb-May. Fr. Jun-Oct. Evergreen broad-leaved forests at 1800-2500 m. Distributed in Yunnan and Sichuan.

华南桂
Cinnamomum austrosinensis H. T. Chang

乔木。叶对生或互生，叶下浅绿色且晦暗，椭圆形，薄革质或革质，下面密具灰褐色短的平伏柔毛，离基三出脉或三出脉。圆锥花序三回分枝；花黄绿色。果椭圆体形。花期6-8月，果期8-10月。生海拔600-700米的山坡、溪边常绿阔叶林中或灌丛中。产中国西南和东南。

Trees. Leaves opposite or alternate, pale green and opaque abaxially, elliptic, thinly leathery or leathery, densely covered with gray-brown short and appressed puberulent hairs abaxially, trinerved or triplinerved. Panicles triplicate-branched; flowers yellow-green. Fruits ellipsoid. Fl. Jun-Aug. Fr. Aug-Oct. Slopes, evergreen broad-leaved forests by streams or thickets at 600-700 m. Distributed in SW and SE China.

香桂
Cinnamomum subavenium Miq.

乔木。叶在幼枝上近对生，老枝上互生，叶下黄绿色且晦暗，革质，密具平伏黄色绢状柔毛，成熟后渐稀疏，三出脉或离基三出脉。花淡黄色。果实椭圆体形，成熟后蓝黑色，果托杯状。花期6-7月，果期8-10月。生海拔400-1100(-2500)米的山坡或山谷常绿阔叶林中。产中国西南、华南、东南和华东。南亚和东南亚亦有。

华南桂 *Cinnamomum austrosinensis*

滇新樟 *Neocinnamomum caudatum*

Trees. Leaves opposite at branchlets, alternate at branches, yellow-green and opaque abaxially, leathery, densely appressed yellow sericeous-pubescent but sparsely so when mature abaxially, trinerved or triplinerved. Flowers pale yellow. Fruits ellipsoid, blue-black when mature, perianth cups in fruit cupuliform. Fl. Jun-Jul. Fr. Aug-Oct. Evergreen broad-leaved forests on mountain slopes or in valleys at 400-1100(-2500) m. Distributed in SW, S, SE and E China. Also in S and SE Asia.

滇新樟

Neocinnamomum caudatum
(Nees) Merr.

乔木。叶互生，卵形或卵状长圆形，纸质，两面无毛，三出脉。团伞花序常具5至6花，疏离并组成圆锥花序；花小，黄绿色。果实狭椭圆体形，成熟时红色，果杯漏斗形。花期6-10月，果期10月至翌年2月。生海拔500-1800米的林中或山谷中。产云南、贵州和广西。印度、尼泊尔、不丹、缅甸、泰国和越南亦有。

Trees. Leaves alternate, ovate or ovate-oblong, papery, glabrous on both surfaces, trinerved. Glomerules usually 5-6-flowered, remotely arranged into a panicle; flowers small, yellow-green. Fruits narrowly ellipsoid, red when mature; perianth cups in fruit crateriform. Fl. Jun-Oct. Fr. Oct to next Feb. Forests or valleys at 500-1800 m. Distributed in Yunnan, Guizhou and Guangxi. Also in India, Nepal, Bhutan, Myanmar, Thailand and Vietnam.

新樟

Neocinnamomum delavayi
(Lecomte) Liou

乔木或灌木。枝条及叶下面密被锈色或白色细绢状微柔毛。叶互生，椭圆状披针形至卵形或阔卵形，近革质。腋生团伞花序具(1-)4-6(-10)花；花小，黄绿色。成熟果实红色，卵球形，果杯漏斗形。花期4-9月，果期9月至翌年1月。生海拔1100-2300米的灌丛中、林中、石灰山或河岸。产云南、四川和西藏。

Trees or shrubs. Branches and leaves abaxially densely rusty- or white-sericeous. Leaves alternate, elliptic-lanceolate to ovate or broadly ovate, subleathery. Glomerules axillary, (1-)4-6 (-10)-flowered; flowers small, yellow-green. Mature fruit red, ovoid; perianth cups in fruit crateriform. Fl. Apr-Sep. Fr. Sep to next Jan. Thickets, forests, limestone hills or river banks at 1100-2300 m. Distributed in Yunnan, Sichuan and Xizang.

新樟 *Neocinnamomum delavayi*

檫树 *Sassafras tzumu*

檫树

Sassafras tzumu (Hemsl.) Hemsl.

落叶乔木。叶互生，集生枝顶，全缘或2-3浅裂，羽状脉或离基三出脉。总状花序顶生，先叶开放，多花，具花序梗；花黄色。果球形，成熟时蓝黑色，带白蜡粉。花期3-4月，果期5-9月。生海拔100-1900米的疏林或密林中。产中国西南、东南、华中和华东。

Deciduous trees. Leaves alternate, aggregated at branchlet apices, margin entire or 2-3-lobed, pinnate or triplinerved. Racemes terminal, appearing before leaves, many-flowered, pedunculate; flowers yellow. Fruits globose, blue-black when mature. Fl. Mar-Apr. Fr. May-Sep. Sparse or dense forests at 100-1900 m. Distributed in SW, SE, C and E China.

倒卵叶黄肉楠

Actinodaphne obovata (Nees) Blume

乔木，雌雄异株。小枝密具锈色柔毛。叶3-5，簇生于枝顶，近轮生，倒卵形、倒卵状长圆形或椭圆状长圆形，离基三出脉。花单性；总状花序由伞形花序组成，具5花；花被裂片6，黄色，卵形。果实长圆状或椭圆体形。花期4-5月，果期7-11月，偶有延至翌年3月。生海拔1000-2700米的山谷、溪旁或湿润的混交林中。产云南和西藏。印度、尼泊尔和不丹亦有。

Trees, dioecious. Branchlets densely ferruginous pubescent. Leaves 3-5, clustered at branch apices, subverticillate, obovate, obovate-oblong or elliptic-oblong, triplinerved. Flowers unisexual; racemes composed of umbels, 5-flowered; perianth segments 6, yellow, ovate. Fruits oblong or ellipsoid. Fl. Apr-May. Fr. Jul-Nov, occasionally to next Mar. Valleys, by streams, or moist mixed forests at 1000-2700 m. Distributed in Yunnan and Xizang. Also in India, Nepal and Bhutan.

毛果黄肉楠

Actinodaphne trichocarpa Allen

灌木或小乔木。叶近轮生，常3-5枚聚生枝顶，叶片长圆形，长5-14厘米。伞形花序单生或簇生于枝侧；花黄色。果球形，径1.2-1.6厘米，密被贴伏黄褐色短绒毛，生于果托中。

倒卵叶黄肉楠 *Actinodaphne obovata*

毛果黄肉楠 *Actinodaphne trichocarpa*

花期3-4月，果期6-8月。生海拔1000-2600米的山坡、灌丛中。产贵州、四川和云南。

Shrubs or small trees. Leaves subverticillate, usually 3-5-crowded on top branchlets, leaf blade oblong, 5-14 cm long. Umbel solitary or clustered on lateral side of branchlet; flowers yellow. Fruit globose, 1.2-1.6 cm diam, densely yellowish brown appressed tomentose, seated on cupule. Fl. Mar-Apr. Fr. Jul-Aug. Mountain slopes, thickets at 1000-2600 m. Distributed in Guizhou, Sichuan and Yunnan.

峨眉黄肉楠

Actinodaphne omeiensis
(Liou) C. K. Allen

灌木或小乔木。幼枝具绒毛，渐无毛。叶常4-6簇生于小枝顶端，近轮生，革质，两面无毛，侧脉每边12-15。假伞形花序，单生或2簇生；雌花花柱膨大；柱头头状，2裂。果近球形。花期2-3月，果期8-9月。生海拔500-1700米的山谷、路边灌丛中或杂木林中。产四川和贵州(梵净山)。

Shrubs or small trees. Young branchlets villous and becoming glabrous. Leaves often 4-6 clustered at apex of branchlet, subverticillate, leathery, glabrous on both surfaces, lateral nerves 12-15 per side. Pseudo-umbels solitary or 2 fascicled; pistillate flowers with inflated styles; stigmas capitate, 2-lobed. Fruits subglobose. Fl. Feb-Mar. Fr. Aug-Sep. Valleys, thickets by roads or mixed forests at 500-1700 m. Distributed in Sichuan and Guizhou (Fanjing Mountain).

黔桂黄肉楠

Actinodaphne kweichowensis
Yang et P. H. Huang

乔木。叶近轮生，叶片长圆状椭圆形，长11-27厘米。伞形花序单生或簇生于枝侧；花黄色。果长圆形，长1.5-1.7厘米，生于果托中。花期3-6月，果期10月。生海拔1000-1300米的山地混交林中。产广西、贵州和云南。

Trees. Leaves subverticillate, leaf blade oblong-elliptic, 11-27 cm long. Umbel solitary or clustered on lateral side of branchlet; flowers yellow. Fruit oblong, 1.5-1.7 cm long, seated on cupule. Fl. Mar-Jun. Fr. Oct. Mixed montane forests at 1000-1300 m. Distributed in Guangxi and Guizhou and Yunnan.

峨眉黄肉楠 *Actinodaphne omeiensis*

黔桂黄肉楠 *Actinodaphne kweichowensis*

毛尖树 *Actinodaphne forrestii*

毛尖树

Actinodaphne forrestii (Allen) Kosterm.

乔木。叶近轮生，常6-7枚聚生枝顶，椭圆状披针形，长9-27厘米。伞形花序簇生于枝侧；花黄色。果长圆形，长1.4-1.6厘米，熟时黑色，生于果托中。花期11月至翌年3月，果期翌年8-9月。生海拔1000-2700米的石灰岩灌丛或山地混交林中。产广西、贵州和云南。

Trees. Leaves subverticillate, usually 6-7 crowded on top branchlets, elliptic-lanceolate, 9-27 cm long. Umbel clustered on lateral side of branchlet; flowers yellow. Fruit oblong, 1.4-1.6 cm long, black at maturity, seated on cupule. Fl. Nov to next Mar. Fr. next Aug-Sep. Thickets on calcareous rocks, or mixed montane forests at 1000-2700 m. Distributed in Guangxi, Guizhou and Yunnan.

红果黄肉楠

Actinodaphne cupularis (Hemsl.) Gamble

灌木或小乔木。叶近轮生，常5-6枚聚生枝顶，长圆形至长圆状披针形，长5.5-13.5厘米。伞形花序单生或数个簇生于枝侧；花黄色。果卵形，长1.2-1.4厘米，熟时红色，生于果托中。花期10-11月，果期翌年8-9月。生海拔300-1300米的山坡、密林、溪旁和灌丛中。产广西、贵州、湖北、湖南、四川和云南。

Shrubs or small trees. Leaves subverticillate, usually 5-6 crowded on top of branchlets, oblong to oblong-lanceolate, 5.5-13.5 cm long. Umbel solitary or numerous in lateral side of branchlet; flowers yellow. Fruit ovoid, 1.2-1.4 cm long, red at maturity, seated on cupule. Fl. Oct-Nov. Fr. next Aug-Sep. Mountain slopes, dense forests, streamside and thickets at 300-1300 m. Distributed in Guangxi, Guizhou, Hubei, Hunan, Sichuan and Yunnan.

毛黄肉楠

Actinodaphne pilosa (Lour.) Merr.

乔木或灌木。幼枝、芽鳞、幼叶两面和叶柄密具锈色绒毛。叶互生或3-5枚簇生，叶革质，羽状脉，侧脉5-7(-10)对。伞形花序簇生成圆锥状，具5花；花被片6。果实球形。花期8-12月，果期翌年2-3月。生海拔500米以下旷野丛林或混交林中。产广西和广东。老挝和越南亦有。

Trees or shrubs. Young branchlets, bud scales, young leaves on

红果黄肉楠 *Actinodaphne cupularis*

毛黄肉楠 *Actinodaphne pilosa*

杨叶木姜子 *Litsea populifolia*

both surfaces, and petioles densely ferruginous tomentose. Leaves alternate or in clusters of 3-5; leathery, penninerved, lateral veins 5-7(-10) pairs. Umbels clustered in a panicle, 5-flowered; perianth-lobes 6. Fruits globose. Fl. Aug-Dec. Fr. next Feb-Mar. Wildernesss or mixed forests below 500 m. Distributed in Guangxi and Guangdong. Also in Laos and Vietnam.

杨叶木姜子

Litsea populifolia (Hemsl.) Gamble

落叶乔木。叶互生，圆形，长6-8厘米。伞形花序生于枝梢。果球形，径5-6毫米，生于杯状果托上。花期4-5月，果期8-9月。生海拔700-2000米的山地阳坡、河谷两岸、阴坡灌丛或干瘠土层的次生林中。产四川、西藏和云南。

Deciduous trees. Leaves alternative, orbicular, 6-8 cm long. Umbels clustered at apex of branchlet. Fruits globose, 5-6 mm diam. Fl. Apr-May. Fr. Aug-Sep. Sunny slopes on mountains, along banks in river valleys, thickets on shady slopes, arid and barren secondary forests at 700-2000 m. Distributed in Sichuan, Xizang and Yunnan.

山鸡椒

Litsea cubeba (Lour.) Pers.

落叶灌木或小乔木。叶纸质，披针形、长圆形或椭圆形，侧脉每边6-10条。假伞形花序单生或簇生，具4-6花；花先叶开放或与叶同期。果球形，直径约5毫米。花期2-3月，果期7-8月。生海拔300-3200米的向阳山地、灌丛、疏林或水边。产中国西南、华南、东南、华中和华东。南亚和东南亚亦有。

Deciduous shrubs or small trees. Leaves papery, lanceolate, oblong or elliptic, lateral nerves 6-10 on each side. Pseudo-umbels solitary or clustered, 4-6-flowered, flowering before leaves or with leaves. Fruits globose, ca. 5 mm diam. Fl. Feb-Mar. Fr. Jul-Aug. Open mountain regions, thickets, sparse forests or by streams at 300-3200 m. Distributed in SW, S, SE, C and E China. Also in S and SE Asia.

山鸡椒 *Litsea cubeba*

红叶木姜子 *Litsea rubescens*

红叶木姜子
Litsea rubescens Lecomte

落叶灌木或小乔木。叶互生椭圆形、披针状椭圆形或圆椭圆形，两面无毛，羽状脉；叶柄幼时红色，无毛。伞形花序腋生；雄性伞形花序具10-12花；花被裂片6，黄色。果实球形。花期3-4月，果期9-10月。生海拔700-3800米的山谷常绿阔叶林中或林缘。产中国西南和华中。

Deciduous shrubs or small trees. Leaves alternate; elliptic, lanceolate-elliptic or rounded-elliptic, glabrous on both surfaces, penninerved; petioles red when young, glabrous. Umbels axillary; male umbels 10-12-flowered; perianth segments 6, yellow. Fruits globose. Fl. Mar-Apr. Fr. Sep-Oct. Evergreen broad-leaved forests of valleys or forest edges at 700-3800 m. Distributed in SW and C China.

高山木姜子
Litsea chunii W. C. Cheng

落叶灌木。幼枝无毛。叶互生，椭圆形、椭圆状披针形或椭圆状倒卵形，无毛或近无毛，羽状脉。伞形花序单生；雄伞形花序具8-12花；花被裂片6。果实卵球形。花期3-4月，果期7-8月。生海拔1500-3400米向阳山坡，溪边或灌丛中。产云南西北部、四川西部和甘肃南部。

Deciduous shrubs. Young branchlets glabrous. Leaves alternate, elliptic, elliptic-lanceolate, or elliptic-obovate, glabrous or subglabrous, penninerved. Umbels solitary; male umbels 8-12-flowered; perianth segments 6. Fruits ovoid. Fl. Mar-Apr. Fr. Jul-Aug. Open slopes, by streams or thickets at 1500-3400 m. Distributed in NW Yunnan, W Sichuan and S Gansu.

独龙木姜子
Litsea taronensis H. W. Li

落叶乔木。小枝具灰黄色柔毛。叶互生，椭圆形或长圆形，密具短绒毛，下面沿脉疏具柔毛，羽状脉。伞形花序成对生短小枝上；雄性伞形花序具12-14花；雄蕊9。花期11月。生海拔约2200米的山坡阔叶林中。产云南西北部。

Deciduous trees. Branchlets gray-yellow pubescent. Leaves alternate, elliptic or oblong, densely shortly tomentose and along veins sparsely pubescent abaxially, penninerved. Umbels in pairs seated on short branchlets; male umbels 12-14-flowered; stamens 9. Fl. Nov. Broad-leaved forests on slopes at ca. 2200 m. Distributed in NW Yunnan.

毛叶木姜子
Litsea mollifolia Chun

落叶灌木或小乔木。幼枝具柔毛。叶互生或聚生枝顶，长圆形或椭圆形，下面苍白色，密具白色柔毛。伞形花序腋生，常2或3个簇生于短枝；雄性伞形花序具4-6花；花先叶开放或与叶同期。果球形，成熟时蓝黑色。花期3-4月，果期9-10月。生海拔600-2800米的山坡灌丛中或阔叶林中。产中国西南、华南和华中。泰国北部亦有。

Deciduous shrubs or small trees. Young branchlets covered with

高山木姜子 *Litsea chunii*

独龙木姜子 *Litsea taronensis*

毛叶木姜子 *Litsea mollifolia*

pubescence. Leaves alternate or clustered at apex of branchlet, oblong or elliptic, glaucous and with dense white pubescence abaxially. Umbels axillary, often 2 or 3 fascicled on short branchlets; male umbels 4-6-flowered, flowering before leaves or with leaves. Fruits globose, blue-black at maturity. Fl. Mar-Apr. Fr. Sep-Oct. Thickets on slopes or broad-leaved forests at 600-2800 m. Distributed in SW, S and C China. Also in N Thailand.

木姜子
Litsea pungens Hemsl.

落叶小乔木。幼枝具柔毛。叶互生，常聚集在枝顶，披针形或倒卵状披针形，下面幼时具绢状柔毛，后无毛或沿中脉具毛，羽状脉。伞形花序腋生，具8-12花。果球形。花期3-5月，果期7-9月。生海拔800-2300米的溪边、林中或林缘。产中国西南、华南、华中、华北和华西。

Deciduous small trees. Young branchlets pubescent. Leaves alternate, often clustered at apex of branchlet, lanceolate or obovate-lanceolate, sericeous-pubescent abaxially when young and becoming glabrous or pilose along midrib, penninerved. Umbels axillary, 8-12-flowered. Fruits globose. Fl. Mar-May. Fr. Jul-Sep. Streamsides, forests or forest edges at 800-2300 m. Distributed in SW, S, C, N and W China.

木姜子 *Litsea pungens*

潺槁木姜子
Litsea glutinosa (Lour.) C. B. Rob.

常绿或落叶乔木。小枝具灰黄色绒毛。叶互生，倒卵形、倒卵状长圆形或椭圆状披针形，两面幼时具绒毛，羽状脉。伞形花序单生或数个生短枝上，具少花。果球形。花期5-6月，果期9-10月。生海拔500-1900米的林中、灌丛或溪边。产广东、广西、海南、福建和云南南部。印度、缅甸、泰国、越南和菲律宾亦有。

Evergreen or deciduous trees. Young branchlets gray-yellow tomentose. Leaves alternate, obovate, obovate-oblong or elliptic-lanceolate, tomentose on both surfaces when young, penninerved. Umbels solitary or several on short branchlets, few flowered. Fruits globose. Fl. May-Jun. Fr. Sep-Oct. Forests, thickets or streamsides at 500-1900 m. Distributed in Guangdong, Guangxi, Hainan, Fujian and S Yunnan. Also in India, Myanmar, Thailand, Vietnam and the Philippines.

潺槁木姜子 *Litsea glutinosa*

轮叶木姜子 *Litsea verticillata*

剑叶木姜子 *Litsea lancifolia*

轮叶木姜子
Litsea verticillata Hance

常绿灌木或小乔木。幼枝密具黄色毛被，后无毛。叶4-5片轮生，羽状脉，每边12-14条。伞形花序2-10个簇生于小枝顶端，具5-8花，浅黄色，近无柄；总苞片4-7，披针形。果实卵球形或椭圆体形。花期4-11月，果期11月至翌年1月。生海拔1300米以下的山谷、溪边、密林或杂木林中。产云南南部、广西、广东和海南。泰国东北、越南和柬埔寨亦有。

Evergreen shrubs or small trees. Young branchlets densely yellow hirsute and becoming glabrous. Leaves 4-5 verticillate, penni nerved, lateral nerves 12-14 on each side. Umbels in cluster of 2-10 at apex of branchlet, 5-8-flowered, pale yellow, sub-sessile; involucral bracts 4-7, lanceolate. Fruits ovoid or ellipsoid. Fl. Apr-Nov. Fr. Nov to next Jan. Valleys, by streams, thickets or mixed forests below 1300 m. Distributed in S Yunnan, Guangxi, Guangdong and Hainan. Also in NE Thailand, Vietnam and Cambodia.

剑叶木姜子
Litsea lancifolia (Roxb. ex Nees) Benth. et Hook. f.

乔木。叶对生，椭圆形，长5-10厘米。伞形花序簇生于叶腋。果球形，径1厘米。花期5-6月，果期7-8月。生海拔100-2000米的林中。产广西、海南和云南。

Trees. Leaves opposite, oblong, 5-10 cm long. Umbels clustered, axillary. Fruits globose, 1 cm diam. Fl. May-Jun. Fr. Jul-Aug. Forest at 100-2000 m. Distributed in Guangxi, Hainan and Yunnan.

豺皮樟
Litsea rotundifolia var. **oblongifolia** (Nees) Allen

灌木或小乔木。叶散生，卵状长圆形，长2.5-5.5厘米。伞形花序簇生于叶腋。果长圆形，径6毫米。花期8-9月，果期9-10月。生海拔20-800米的疏林中。产福建、广东、广西、

豺皮樟 *Litsea rotundifolia* var. *oblongifolia*

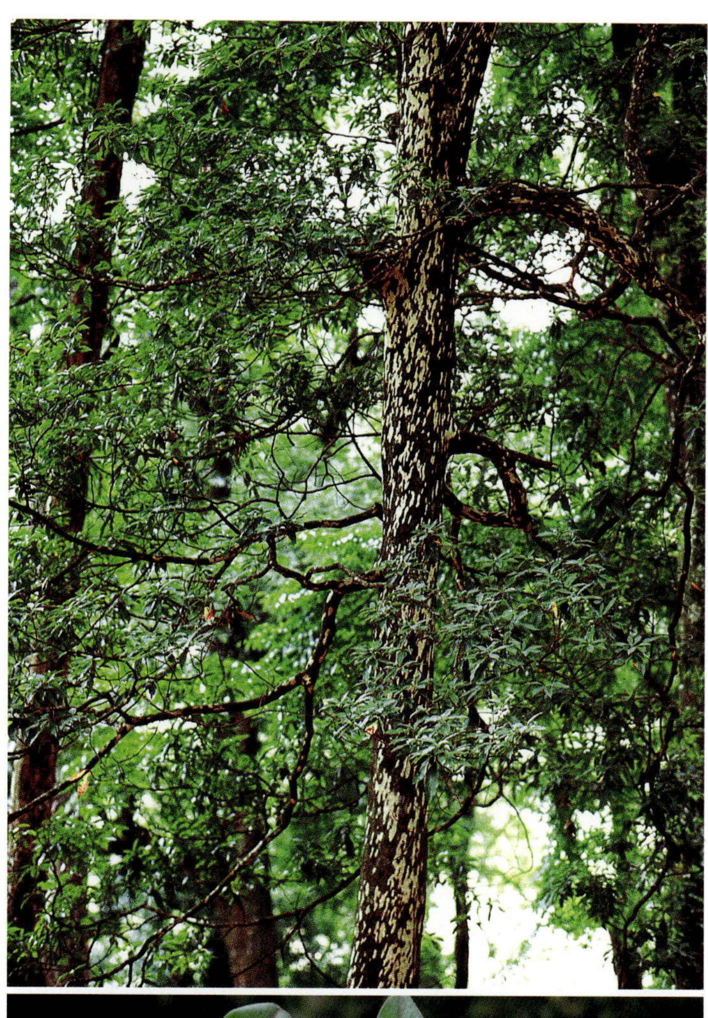

海南、湖南、江西、台湾和浙江。

Shrubs or small trees. Leaves scattered, ovate-oblong, 2.5-5.5 cm long. Umbels clustered, axillary. Fruits globose, 6 mm diam. Fl. Aug-Sep. Fr. Sep-Nov. Sparse forests at 20-800 m. Distributed in Fujian, Guangdong, Guangxi, Hainan, Hunan, Jiangxi, Taiwan and Zhejiang.

毛豹皮樟

Litsea coreana Lévl. var. **lanuginosa** (Migo) Yang et P. H. Huang

常绿乔木。树皮灰色，呈小鳞片状剥落。嫩枝密被灰黄色长柔毛。幼叶两面被黄色长柔毛；叶柄被灰黄色长柔毛。假伞形花序腋生；总苞片4。果近球形。花期8-9月，果期翌年夏季。生海拔300-2300米山谷杂木林中。产中国西南、东南、华中和华东。

Evergreen trees. Bark gray-colored, scaly peeled off. Branchlets initially densely gray-yellow-villose. Young leaves on both surfaces gray-yellow-villose; petioles gray-yellow-villose. Pseudo-umbels axillary; involucral bracts 4. Fruits subglobose. Fl. Aug-Sep. Fr. next summer. Mixed forests in valleys at 300-2300 m. Distributed in SW, SE, C and E China.

红河木姜子

Litsea honghoensis Liou

乔木。叶互生，长椭圆形，长10-19厘米。伞形花序簇生。果球形，径2-3厘米。花期2-3月，果期8-9月。生海拔1300-2200米的密林中。产云南。

Trees. Leaves alternate, oblong-elliptic, 10-19 cm long. Umbels clustered. Fruits globose, 2-3 cm diam. Fl. Feb-Mar. Fr. Aug-Sep. Dense forest at 1300-2200 m. Distributed in Yunnan.

毛豹皮樟 *Litsea coreana* var. *lanuginosa*

红河木姜子 *Litsea honghoensis*

琼楠叶木姜子 *Litsea beilschmiediifolia*

琼楠叶木姜子

Litsea beilschmiediifolia H. W. Li

常绿乔木。幼枝具黄褐色绢状柔毛。叶互生，椭圆形，无毛，两面明显具凹穴，具羽状脉，基部镰刀状，先端长尾尖。伞形花序2-4个生短枝，具5-6花。果球形，成熟时黑色。花期4-5月，果期9-10月。生海拔1700-1900米的山地湿润疏林区域。产云南东南部。

Evergreen trees. Young branchlets yellow-brown sericeous-pubescent. Leaves alternate, elliptic, glabrous, conspicuously foveolate on both surfaces, penninerved, base falcate-curved, apex long caudate-acute. Umbels 2-4 clustered on short branch, 5-6-flowered. Fruits globose, black when mature. Fl. Apr-May. Fr. Sep-Oct. Moist areas of sparse forests on mountains at 1700-1900 m. Distributed in SE Yunnan.

黑木姜子

Litsea salicifolia (Roxb. ex Wall.) Hook. f.

乔木。叶互生，长椭圆形，长9-19厘米。伞形花序簇生于叶腋。果矩圆形，长1-1.1厘米。花期4-5月，果期6-9月。生海拔300-1200米的疏林中。产广东、广西、贵州、海南和云南。

Trees. Leaves alternative, long elliptic, 9-19 cm long. Umbels clustered, axillary. Fruits oblong, 1-1.1 cm long. Fl. Apr-May. Fr. Jun-Sep. Sparse forests at 300-1200 m. Distributed in Guangdong, Guangxi, Guizhou, Hainan and Yunnan.

假柿木姜子

Litsea monopetala (Roxb.) Pers.

常绿乔木。小枝、叶柄和叶片下面均被锈色短柔毛。叶互生，阔卵形或倒卵形至卵状长圆形；叶柄密具毛状小枝。伞形花序簇生于最短小枝上，具4-6花或更多；花被裂片黄白色。果实长卵球形。花期11月至翌年6月，果期翌年6-7月。生海拔1500米以下的林中、灌丛或阳坡。产云南南部、贵州、广西、广东和海南。东亚和东南亚亦有。

黑木姜子 *Litsea salicifolia*

假柿木姜子 *Litsea monopetala*

Evergreen trees. Branchlets, petioles and leaves abaxially rust-pubescent. Leaves alternate, broadly ovate or obovate to ovate-oblong; petioles densely hairy like branchlets. Umbels clustered on shortest branchlets, 4-6-flowered or more; perianth segments yellow-white. Fruits long ovoid. Fl. Nov to next Jun. Fr. next Jun-Jul. Forests, thickets or sunny slopes below 1500 m. Distributed in S Yunnan, Guizhou, Guangxi, Guangdong and Hainan. Also in E and SE Asia.

五桠果叶木姜子

Litsea dilleniifolia P. Y. Pai et P. H. Huang

乔木。叶互生，长圆形，长21-50厘米。伞形花序簇生于短枝上。果扁球形，径2-2.3厘米，生于杯状果托上。花期5-6月，果期12月至翌年3月。生海拔500米的雨林中。产云南。

Trees. Leaves alternate, oblong, 21-50 cm long. Umbels clustered on short branchlets. Fruits compressed globose, 2-2.3 cm diam, seated on cup-shaped perianth tube. Fl. May-Jun. Fr. Dec to next Mar. Rain forest at 500 m. Distributed in Yunnan.

思茅木姜子

Litsea szemaois (H. Liou) J. Li et H. W. Li

常绿乔木。小枝无毛。叶互生，椭圆形或长圆状椭圆形，两面无毛，基部和顶端渐狭。伞形花序3-5个在短小枝上组成总状，具4或5花。果实球形或扁球形。花期6-7月，果期10-11月。生海拔800-1500米的阔叶混交林中。产云南南部。

Evergreen trees. Branchlets glabrous. Leaves alternate, elliptic or oblong-elliptic, glabrous on both surfaces, base and apex attenuate. Umbels 3-5 in racemes on short branchlets, 4- or 5-flowered. Fruits globose or compressed globose. Fl. Jun-Jul. Fr. Oct-Nov. Broad-leaved mixed forests at 800-1500 m. Distributed in S Yunnan.

五桠果叶木姜子 *Litsea dilleniifolia*

思茅木姜子 *Litsea szemaois*

滇南木姜子 *Litsea martabanica*

滇南木姜子

Litsea martabanica (Kurz) Hook. f.

乔木。叶互生，长椭圆形，长8-16厘米。伞形花序簇生于短枝上。果长圆形，长1-1.5厘米，生于杯状果托上。花期10-11月，果期翌年6-7月。生海拔500-2000米的阔叶林中。产云南。

Trees. Leaves alternate, oblong-elliptic, 8-16 cm long. Umbels clustered on short branchlets. Fruits oblong, 1-1.5 cm long, seated on cup-shaped perianth tube. Fl. Oct-Nov. Fr. next Jun-Jul. Broad-leaved forest at 500-2000 m. Distributed in Yunnan.

干香柴

Litsea viridis H. Liou

常绿小乔木。幼枝具柔毛，后无毛。叶互生，椭圆形，下面具灰黄色柔毛，羽状脉。伞形花序常2-5个簇生于短枝；雄花花丝无毛；无退化雌雄蕊。果实椭圆体形。花期8月，果期11-12月。生海拔400-1100米的沟谷溪岸边疏林中。产云南东南部。越南亦有。

Evergreen small trees. Young branchlets puberulent and becoming glabrous. Leaves alternate, elliptic, gray-yellow puberulent abaxially, penninerved. Umbels often in cluster of 2-5 on short branchlets; filaments of staminate flowers glabrous; pistillodes lacking. Fruits ellipsoid. Fl. Aug. Fr. Nov-Dec. Sparse forests on riverbanks in valleys at 400-1100 m. Distributed in SE Yunnan. Also in Vietnam.

云南木姜子

Litsea yunnanensis Y. C. Yang et P. H. Huang

常绿乔木。幼枝具黄褐色柔毛，后无毛。叶互生，具羽状脉。伞形花序2-5个簇生于短枝上，腋生；雄性伞形花序具5或

干香柴 *Litsea viridis*

云南木姜子 *Litsea yunnanensis*

6花；总苞片4，外被黄褐色短柔毛。花期5月，果期10-11月。生海拔800-1900米的山坡、路旁、沟边疏林中或混交林中。产云南东南部和广西西南部。越南亦有。

Evergreen trees. Young branchlets gray-yellow pubescent and becoming glabrous. Leaves alternate, penninerved. Umbels in cluster of 2-5 on short branchlets, axillary; male umbels 5- or 6-flowered; involucral bracts 4, abaxially yellowish-brown-pubescent. Fl. May. Fr. Oct-Nov. Slopes, roadsides, sparse forests by streams, or mixed forests at 800-1900 m. Distributed in SE Yunnan and SW Guangxi. Also in Vietnam.

大萼木姜子 *Litsea baviensis*

大萼木姜子
Litsea baviensis Lec.

乔木。叶互生，长椭圆形，长11-20厘米。伞形花序数个簇生于短枝上。果椭球形，长2.5-3厘米，生于厚木革质的杯状果托上。花期5-6月，果期12月至翌年3月。生海拔400-2000米的密林中。产广西、海南和云南。

Trees. Leaves alternate, oblong-elliptic, 11-20 cm long. Umbels few in cluster on short branchlets. Fruits ellipsoid, 2.5-3 cm long, seated on thickly woody cupular perianth tube. Fl. May-Jun. Fr. Dec to next Mar. Dense forest at 400-2000 m. Distributed in Guangxi, Hainan and Yunnan.

黄丹木姜子
Litsea elongata (Wall. ex Nees) Benth. et Hook. f.

常绿小或中乔木。小枝密具褐色绒毛。叶互生或近轮生，革质，侧脉10-20对。伞形花序单生，每个具4或5花；花被裂片6。果长圆形，成熟时黑紫色。花期5-11月，果期翌年2-6月。生海拔500-2300米的河边或杂木林下。产中国西南、华南、东南、华中和华东。印度和尼泊尔亦有。

Evergreen small or medium-sized trees. Branchlets densely brown tomentose. Leaves alternate or subverticillate, leathery, lateral nerves 10-20 pairs. Umbels solitary, 4- or 5-flowered per umbel; perianth-lobes 6. Fruits oblong, black-purple when mature. Fl. May-Nov. Fr. next Feb-Jun. By streams or in mixed forests at 500-2300 m. Distributed in SW, S, SE, C and E China. Also in India and Nepal.

黄丹木姜子 *Litsea elongata*

簇叶新木姜子 *Neolitsea confertifolia*

簇叶新木姜子
Neolitsea confertifolia (Hemsl.) Merr.

乔木。叶密集呈轮生状，披针形，长5-12厘米。伞形花序簇生于叶腋或节间。果椭球形，长0.8-1.2厘米。花期4-5月，果期9-10月。生海拔400-2000米的灌丛和密林中。产广东、广西、贵州、河南、湖北、湖南、江西、陕西和四川。

Trees. Leaves conferted, subverticillate, lanceolate, 5-12 cm long. Umbels axillary or lateral, clustered. Fruits ellipsoid, 0.8-1.2 cm long. Fl. Apr-May. Fr. Sep-Oct. Thickets and dense forests at 400-2000 m. Distributed in Guangdong, Guangxi, Guizhou, Henan, Hubei, Hunan, Jiangxi, Shaanxi and Sichuan.

新木姜子
Neolitsea aurata (Hayata) Koidz.

乔木。叶革质，离基三出脉，侧脉每边3-4条，下面密被黄色绢毛。伞形花序3-5簇生于小枝顶端或节间，具5花；花被裂片4；花柱无毛。果实椭圆体形。花期2-3月，果期9-10月。生海拔500-1900米的阔叶林和山坡林缘或杂林中。产中国西南、华南、东南、华中和华东。日本亦有。

Trees. Leaves leathery, triplinerved, lateral nerves 3-4 on each side, abaxially densely yellow-sericeous. Umbels 3-5 fascicled toward apex of branchlet or internode, 5-flowered; perianth-lobes 4; styles glabrous. Fruits ellipsoid. Fl. Feb-Mar. Fr. Sep-Oct. Broad-leaved forests and forest edges on slopes or mixed forests at 500-1900 m. Distributed in SW, S, SE, C and E China. Also in Japan.

毛叶新木姜子
Neolitsea velutina W. T. Wang

小乔木。叶密集呈轮生状，宽倒卵形，长5-7.5厘米。伞形花序簇生。果椭球形，长0.8-1厘米。花期11-12月，果期翌年5-6月。生海拔600-1400米的混交林中。产广东、广西和云南。

Small trees. Leaves conferted, subverticillate, broadly obovate, 5-7.5 cm long. Umbels clustered. Fruits ellipsoid, 0.8-1 cm long. Fl. Nov-Dec. Fr. next May-Jun. Mixed forests at 600-1400 m. Distributed in Guangdong, Guangxi and Yunnan.

新木姜子 *Neolitsea aurata*

毛叶新木姜子 *Neolitsea velutina*

大叶新木姜子 *Neolitsea levinei*

鸭公树 *Neolitsea chuii*

大叶新木姜子
Neolitsea levinei Merr.

乔木。叶4-5片轮生，下面密被黄褐色长柔毛，离基三出脉，侧脉每边3-4条，15-31 × 4.5-9厘米。花被裂片4；子房无毛。花期3-4月，果期8-10月。生海拔300-1300米的山坡路边、水旁或山谷密林中。产中国西南、华南、东南和华中。

Trees. Leaves 4-5 whorled, abaxially densely rusty-villose, triplinerved, lateral nerves 3-4 on each side, 15-31 × 4.5-9 cm. Perianth-lobes 4; ovary glabrous. Fl. Mar-Apr. Fr. Aug-Oct. Roadsides on slopes, by waters or dense forests in valleys at 300-1300 m. Distributed in SW, S, SE and C China.

鸭公树
Neolitsea chuii Merr.

乔木，除花序具柔毛外无毛。叶互生或聚生枝顶呈轮生状，8-16 × 2.7-9厘米，离基三出脉，侧脉3-5对。伞形花序腋生或侧生，多个密集。果实椭圆体形或近球形。花期9-10月，果期12月。生海拔500-1400米的山谷或丘陵地疏林。产云南、广西、广东、福建、湖南和江西。

Trees, glabrous, except for pubescent inflorescence. Leaves alternate or clustered at branch apices to verticillate, 8-16 × 2.7-9 cm, triplinerved, lateral veins 3-5 pairs. Umbels axillary or lateral, many clustered. Fruits ellipsoid or subglobose. Fl. Sep-Oct. Fr. Dec. Valleys or sparse forests on hills at 500-1400 m. Distributed in Yunnan, Guangxi, Guangdong, Fujian, Hunan and Jiangxi.

黑壳楠
Lindera megaphylla Hemsl.

常绿乔木。小枝圆柱形，粗壮，紫黑色。叶互生，倒披针形至倒卵状长圆形，革质，两面无毛，羽状脉。伞形花序多花，具16雄花和12雌花；花被片6。果实椭圆体形或卵球形。花期2-4月，果期9-12月。生海拔1600-2000米的山坡、谷地、湿润常绿阔叶林中或灌丛中。产中国西南、华南、华中、华西和华东。

Evergreen trees. Branchlets terete, robust, purple-black. Leaves alternate, oblanceolate to obovate-oblong, leathery, glabrous on both surfaces, penninerved. Umbels many-flowered, with 16 staminate flowers and 12 pistillate flowers; tepals 6. Fruits ellipsoid or ovoid. Fl. Feb-Apr. Fr. Sep-Dec. Slopes, valleys, moist evergreen broad-leaved forests or thickets at 1600-2000 m. Distributed in SW, S, C, W and E China.

黑壳楠 *Lindera megaphylla*

红果山胡椒 *Lindera erythrocarpa*

红果山胡椒
Lindera erythrocarpa Makino

灌木或乔木。叶纸质，羽状脉，侧脉每边4-5条。假伞形花序具15-17花；总苞片4；雄花花被片6；雄蕊9；雌花较小；退化雄蕊9，条形；柱头盘状。果球形，成熟后红色。花期4月，果期9-10月。生海拔1000米以下的山坡、山谷、溪边或林下。产中国西南、华南、华中和华东。朝鲜半岛和日本亦有。

Shrubs or trees. Leaves papery, penninervate, lateral nerves 4-5 on each side. Pseudo-umbels 15-17-flowered; involucral bracts 4; staminate flower: perianth-lobes 6; stamens 9; pistillate flower smaller; staminodes 9, linear; stigmas dish-shaped. Fruits globose, red at maturity. Fl. Apr. Fr. Sep-Oct. Mountain slopes, valleys, by streams or forests below 1000 m. Distributed in SW, S, C and E China. Also in Korean Peninsula and Japan.

山橿
Lindera reflexa Hemsl.

灌木或乔木。叶互生，常卵形或倒卵状椭圆形，羽状脉，侧脉每边6-8(-10)条，下面被白色柔毛。假伞形花序：总苞片4。果球形，直径约7毫米，熟时红色。花期4月，果期8月。生海拔1000米以下的山谷，山坡林中或灌丛中。产中国西南、东南、华中和华东。

Shrubs or trees. Leaves alternate, usually ovate or obovate-elliptic, penninerved, lateral nerves 6-8 (-10) on each side, abaxially white-pubescent. Pseudo-umbel: involucral bracts 4. Fruits globose, ca. 7 mm diam, red when mature. Fl. Apr. Fr. Aug. Valleys, forests on slopes or thickets below 1000 m. Distributed in SW, SE, C and E China.

山胡椒
Lindera glauca (Sieb. et Zucc.) Blume

落叶灌木或乔木。叶互生，阔椭圆形、椭圆形、倒卵形或狭倒卵形，纸质，侧脉(4或)5或6对。假伞形花序腋生，3-8花；花被片黄色；子房椭圆体形；柱头盘状。花期3-4月，果期7-8月。生海拔900米以下的山坡、林缘或路旁。产中国除西北和东北外的大部分地区。缅甸、越南、朝鲜半岛和日本亦有。

Deciduous shrubs or trees. Leaves alternate, broadly elliptic,

山橿 *Lindera reflexa*

山胡椒 *Lindera glauca*

狭叶山胡椒 *Lindera angustifolia*

elliptic, obovate or narrowly obovate, papery, lateral veins (4 or)5 or 6 pairs. Pseudo-umbels axillary, 3-8-flowered; tepals yellow; ovary ellipsoid; stigmas dish-shaped. Fl. Mar-Apr. Fr. Jul-Aug. Slopes, forest edges or roadsides below 900 m. Distributed in most parts except NW and NE China. Also in Myanmar, Vietnam, Korean Peninsula and Japan.

狭叶山胡椒

Lindera angustifolia Cheng

落叶灌木或小乔木。叶互生，披针形，长6-14厘米。伞形花序簇生。果球形，径约8毫米。花期3-4月，果期9-10月。生海拔300-1500米的山坡灌丛或疏林中。产安徽、福建、广东、广西、河南、湖北、江苏、江西、陕西、山东和浙江。

Deciduous shrubs or small trees. Leaves alternate, lanceolate, 6-14 cm long. Umbel clustered. Fruits globose, ca. 8 mm diam. Fl. Mar-Apr. Fr. Sep-Oct. Thickets or sparse forests on mountain slopes at 300-1500 m. Distributed in Anhui, Fujian, Guangdong, Guangxi, Henan, Hubei, Jiangsu, Jiangxi, Shaanxi, Shandong and Zhejiang.

更里山胡椒

Lindera kariensis W. W. Sm.

落叶灌木或乔木。叶互生，椭圆形、倒卵形或倒披针形，膜质，成熟后纸质，具羽状脉，侧脉4或5对。伞形花序具(2或)3-6花；雄蕊9。果实卵球形至近球形。花期3-6月，果期7-10月。生海拔(2700-)3000-3700米的杂木林、高山竹林、灌丛、林缘或河边。产云南西北部。

Deciduous shrubs or trees. Leaves alternate, elliptic, obovate or oblanceolate, membranous, papery at maturity, penninerved, lateral veins 4 or 5 pairs. Umbels (2 or)3-6-flowered; stamens 9. Fruits ovoid to subglobose. Fl. Mar-Jun. Fr. Jul-Oct. Mixed forests, alpine bamboo thickets, thickets, forest edges or by streams at (2700-)3000-3700 m. Distributed in NW Yunnan.

纤梗山胡椒

Lindera gracilipes H. W. Li

小乔木。叶互生，长圆形，长12-20厘米。伞形花序簇生，花序梗细长。果卵形，长1.3厘米。花期8-10月，果期10-12月。生海拔700-1900米的沟谷密林中。产西藏和云南。

Small trees. Leaves alternate, oblong, 12-20 cm long. Umbel clustered; peduncles slender. Fruits ovoid, 1.3 cm long. Fl. Aug-Oct. Fr. Oct-Dec. Thickets in valleys at 700-1900 m. Distributed in Xizang and Yunnan.

更里山胡椒 *Lindera kariensis*

纤梗山胡椒 *Lindera gracilipes*

团香果 *Lindera latifolia*

团香果
Lindera latifolia Hook. f.

常绿乔木。叶互生，倒卵形或长圆形，纸质，下面密被灰白或灰黄色长硬毛，羽状脉，侧脉6-8对。伞形花序1-3，具10-12(或13)花；总苞4；雄蕊8-10。果实球形，成熟后紫红色。花期2-4月，果期5-11月。生海拔1500-2300(-2900)米的林中、灌丛、山坡路边或沟谷中。产云南和西藏。印度、孟加拉国和越南北部亦有。

Evergreen trees. Leaves alternate. obovate or oblong, papery, densely gray-white or gray-yellow hirsute abaxially, penninerved, lateral veins 6-8 pairs. Umbels 1-3, 10-12 (or 13)-flowered; involucral bracts 4; stamens 8-10. Fruits globose, purple-red at maturity. Fl. Feb-Apr. Fr. May-Nov. Forests, shrubs, roadsides on mountain slopes or in ravines at 1500-2300(-2900) m. Distributed in Yunnan and Xizang. Also in India, Bangladesh, and N Vietnam.

绒毛山胡椒
Lindera nacusua (D. Don) Merr.

常绿灌木或乔木。叶互生，阔卵形、椭圆形、狭椭圆形、椭圆状披针形或披针形，下面具黄褐色绒毛，侧脉6-8对，极厚。伞形花序单生或2-4簇生于叶腋。果近球形，成熟时红色。花期5-6月，果期7-10月。生海拔700-2500米的谷地或山坡常绿阔叶林中。产中国西南、华南和东南。印度、尼泊尔、不丹、缅甸和越南亦有。

Evergreen shrubs or trees. Leaves alternate, broadly ovate, elliptic, narrowly elliptic, elliptic-lanceolate or lanceolate, yellow-brown villous abaxially, lateral veins 6-8 pairs, rather thick. Umbels solitary or 2-4 fascicled in leaf axil. Fruits subglobose, red at maturity. Fl. May-Jun. Fr. Jul-Oct. Valleys or evergreen broad-leaved forests of slopes at 700-2500 m. Distributed in SW, S and SE China. Also in India, Nepal, Bhutan, Myanmar and Vietnam.

香叶树
Lindera communis Hemsl.

常绿灌木或小乔木。叶互生，披针形、卵形或椭圆形，薄革质或厚革质，具黄褐色柔毛，后疏具柔毛或下面无毛，羽状脉，侧脉5-7对。伞形花序单生或双生。果实卵球形，成熟后红色。花期3-4月，果期9-10月。生海拔2000米以下的常绿阔叶林下或干旱沙地。产中国西南、华南、东南、华中和华西。印度、缅甸、老挝、泰国、柬埔寨和越南亦有。

Evergreen shrubs or trees. Leaves alternate, lanceolate, ovate or elliptic, thinly leathery or thickly leathery, yellow-brown pubescent, later laxly pubescent or glabrous abaxially, penninerved, lateral veins 5-7 pairs. Umbels solitary or 2. Fruits ovoid, red at maturity. Fl. Mar-Apr. Fr. Sep-Oct. Evergreen broad-leaved forests or dry sandy places below 2000 m. Distributed in SW, S, SE, C and W China. Also in India, Myanmar, Laos, Thailand, Cambodia and Vietnam.

三桠乌药
Lindera obtusiloba Blume

落叶乔木或灌木。叶互生，全缘或3浅裂，近圆形、扁圆形或椭圆形，下面具黄褐色柔毛或无毛，离基三出脉。花芽5或6，伞形花序内藏；总苞片4。果成熟时红色，之后变紫黑色。花期3-4月，果期8-9月。生海拔3000米以下的山谷，密林或灌木中。产中国西南、华南、华中、华北、华西和华东。印度、不丹、尼泊尔、朝

绒毛山胡椒 *Lindera nacusua*

香叶树 *Lindera communis*

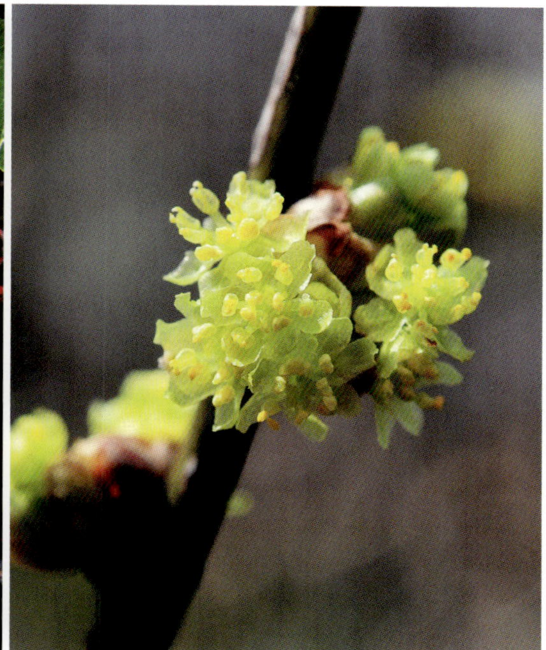

三桠乌药 *Lindera obtusiloba*

鲜半岛和日本亦有。

Deciduous trees or shrubs. Leaves alternate, entire or 3-lobed, suborbicular, compressed-rounded or elliptic, brown-yellow pubescent or glabrate abaxially, triplinerved. Floral buds 5 or 6, umbels included; involucral bracts 4. Fruits red when mature, then purple-black. Fl. Mar-Apr. Fr. Aug-Sep. Valleys, dense forests, or thickets below 3000 m. Distributed in SW, S, C, N, W and E China. Also in India, Bhutan, Nepal, Korean Peninsula and Japan.

香叶子

Lindera fragrans Oliv.

乔木。叶互生，披针形或狭卵形，三出脉，长2-5厘米。伞形花序簇生。果卵形，长1厘米。花期3-4月，果期9-10月。生海拔700-2100米的沟边或山坡灌丛中。产广西、贵州、湖北、陕西和四川。

Trees. Leaves alternate, lanceolate or narrowly ovate, trinerved, 2-5 cm long. Umbel clustered. Fruits ovate, 1 cm long. Fl. Mar-Apr. Fr. Sep-Oct. Ditch sides or thickets on mountain slopes at 700-2100 m. Distributed in Guangxi, Guizhou, Hubei, Shaanxi and Sichuan.

三股筋香

Lindera thomsonii C. K. Allen

常绿乔木。叶互生，卵形、狭卵形或披针形，纸质，幼时两面密具贴伏的白色或黄色绢状柔毛，三出脉或离基三出脉。花单性；总苞片4；雄花黄色；雌花白色、黄色或绿色。果实椭圆体形，成熟后由红色变为黑色。花期2-3月，果期6-9月。生海拔1100-2500(-3000)米的山坡疏林中。产云南、贵州和广西。印度、缅甸和越南北部亦有。

Evergreen trees. Leaves alternate, ovate, narrowly ovate or lanceolate, papery, densely appressed white or yellow sericeous-pubescent on both surfaces when young, trinerved or triplinerved. Flowers unisexual; involucral bracts 4; staminate flowers yellow; pistillate flowers white, yellow or green. Fruits ellipsoid, red becoming black at maturity. Fl. Feb-Mar. Fr. Jun-Sep. Sparse forests on slopes at 1100-2500 (-3000) m. Distributed in Yunnan, Guizhou and Guangxi. Also in India, Myanmar and N Vietnam.

香叶子 *Lindera fragrans*

三股筋香 *Lindera thomsonii*

川钓樟 *Lindera pulcherrima* var. *hemsleyana*

绒毛钓樟 *Lindera floribunda*

川钓樟

Lindera pulcherrima (Wall.) Benth. var. **hemsleyana** (Diels) H. P. Tsui

常绿乔木。叶互生，椭圆形、长圆形或倒卵形，先端渐尖或短尾状渐尖，三出脉。假伞形花序，无总梗或具极短总梗；花被片6；子房无毛；花柱疏或密具柔毛。幼果无毛。果期6-8月。生海拔约2000米的山坡、灌丛或林缘。产中国西南和华中。

Evergreen trees. Leaves alternate, elliptic, oblong or obovate, apex acuminate or shortly caudate-acuminate, trinervious. Pseudo-umbels expedunculate or shortly pedunculate; perianth-lobes 6; ovary glabrous; styles laxly or densely pubescent. Fruits glabrous when young. Fr. Jun-Aug. Slopes, thickets or forest edges at ca. 2000 m. Distributed in SW and C China.

绒毛钓樟

Lindera floribunda (Allen) H. P. Tsui

小乔木。叶互生，倒卵形或椭圆形，长7-10厘米。伞形花序簇生。果椭球形，长0.8厘米。花期3-4月，果期4-8月。生海拔300-1300米的山坡或河边。产甘肃、广东、贵州、湖北、湖南、陕西和四川。

Small trees. Leaves alternate, obovate or elliptic, 7-10 cm long. Umbel clustered. Fruits ellipsoid, 0.8 cm long. Fl. Mar-Apr. Fr. Apr-Aug. Mountain slopes or riversides at 300-1300 m. Distributed in Gansu, Guangdong, Guizhou, Hubei, Hunan, Shaanxi and Sichuan.

乌药

Lindera aggregata (Sims.) Kosterm.

常绿灌木或小乔木。叶革质或近革质，下面密具褐色柔毛，后渐脱落，狭卵形、阔椭圆形、近圆形或披针形，三出脉。假伞形花序有6-8花；花被裂片6。果实卵球形或有时近圆形。花期3-4月，果期5-11月。生海拔200-1000米的向阳坡地、山谷、疏林或灌丛中。产中国西南、华南、东南和华东。越南和菲律宾亦有。

Evergreen shrubs or small trees. Leaves leathery or subleathery, densely brown pubescent abaxially, later gradually deciduous,

乌药 *Lindera aggregata*

香面叶 *Iteadaphne caudata*

narrowly ovate, broadly elliptic, subrounded, or lanceolate, trinervious. Pseudo-umbels 6-8-flowered; perianth lobes 6. Fruits ovoid or sometimes subrounded. Fl. Mar-Apr. Fr. May-Nov. Open mountain slopes, valleys, sparse forests or thickets at 200-1000 m. Distributed in SW, S, SE and E China. Also in Vietnam and the Philippines.

香面叶

Iteadaphne caudata (Nees) H. W. Li

灌木或小乔木。叶互生，狭卵形或长圆状披针形，薄革质，两面密具黄褐色柔毛，离基三出脉。假伞形花序只有1花，无总花梗，每个具1苞片和2个总苞。果实球形。花期10月至翌年4月，果期翌年3-10月。生海拔700-2300米的灌丛、疏林、林缘或路旁。产云南和广西。印度、缅甸、老挝、泰国和越南亦有。

Shrubs or small trees. Leaves alternate, narrowly ovate or oblong-lanceolate, thinly leathery, densely yellow-brown pubescent on both surfaces, triplinerved. Pseudoumbels 1-flowered, expedunculate, each with 1 bract and 2 involucral bracts. Fruits globose. Fl. Oct to next Apr. Fr. next Mar-Oct. Bushes, sparse forests, forest edges or roadsides at 700-2300 m. Distributed in Yunnan and Guangxi. Also in India, Myanmar, Laos, Thailand and Vietnam.

厚壳桂

Cryptocarya chinensis (Hance) Hemsl.

乔木。叶互生或对生，狭椭圆形，革质，幼时具灰褐色绒毛，两面渐无毛，离基三出脉。圆锥花序腋生或顶生；花丝基部有1对腺体；子房棍棒状。果实球形或扁球形，成熟后紫黑色，具12-15棱。花期4-5月，果期8-12月。生海拔300-1100米的山谷阴湿的阔叶林中。产中国西南至华南。

Trees. Leaves alternate or opposite, narrowly elliptic, leathery, grayish brown tomentulose when young, gradually glabrate on both surfaces, triplinerved. Panicles axillary or terminal; filaments 2-glandular at base; ovary clavate. Fruits globose or oblate, purple-black when mature, 12-15-angulate. Fl. Apr-May. Fr. Aug-Dec. Moist broad-leaved forests in valleys at 300-1100 m. Distributed in SW to S China.

厚壳桂 *Cryptocarya chinensis*

岩生厚壳桂 *Cryptocarya calcicola*

尖叶厚壳桂 *Cryptocarya acutifolia*

岩生厚壳桂

Cryptocarya calcicola H. W. Li

乔木。叶长圆形或椭圆形至卵圆形，羽状脉，下面疏具黄褐色柔毛。圆锥花序腋生与顶生，腋生生少数分枝，近穗状，顶生或近顶生者多分枝，疏松；花淡绿色。果紫黑色，近球形，具不明显12棱。花期4-5月，果期5-10月。生海拔(500-)700-1000米的石灰山山坡、溪边或山谷林中。产云南、贵州和广西。

Trees. Leaves oblong or elliptic to ovate, penninerved, abaxially sparsely yellowish brown pubescent. Panicles axillary and terminal, axillary ones generally few branched and nearly spikelike, terminal or subterminal ones many branched and lax; flowers pale green. Fruits purple-black, subglobose, inconspicuously 12-angulate. Fl. Apr-May. Fr. May-Oct. Slopes of limestone hills, streamsides or forests in valleys at (500-)700-1000 m. Distributed in Yunnan, Guizhou and Guangxi.

尖叶厚壳桂

Cryptocarya acutifolia H. W. Li

乔木。叶互生，狭椭圆形，革质，下面具柔毛。圆锥花序腋生及顶生，腋生者短，顶生者长达19厘米，塔形，具分枝，密具锈色柔毛；花淡黄色。成熟果实黑紫色，椭圆体形，具不明显12棱。花期3-5月，果期6-12月。生海拔500-700米的林中。产云南。

Trees. Leaves alternate, narrowly elliptic, leathery, abaxially pubescent. Panicles axillary and terminal, those arising from upper leaf axils shorter, terminal ones longer, up to 19 cm long, all tower-like, branched, densely rusty pubescent; flowers pale yellow. Mature fruit black-purple, ellipsoid, inconspicuously 12-angulate. Fl. Mar-May. Fr. Jun-Dec. Forests at 500-700 m. Distributed in Yunnan.

黄果厚壳桂

Cryptocarya concinna Hance

乔木。叶互生，椭圆状长圆形或长圆形，纸质，稍具柔毛，下面迅速脱落，侧脉4-7对。圆锥花序顶生及腋生；第三轮花丝有1对具柄腺体，其他无腺；子房狭倒卵球形。果实幼时深绿色，成熟后黑色或蓝黑色，狭卵球形，幼时具不明显12棱。花期3-5月，果期6-12月。生海拔600米以下的山坡常绿阔叶林或沟谷。产广西、贵州东南部、广东、台湾、海南和江西。越南北部亦有。

Trees. Leaves alternate, elliptic-oblong or oblong, papery, slightly pubescent but soon glabrate abaxially, lateral veins 4-7 pairs. Panicles axillary and terminal; filaments of 3rd whorl each with 2 stalked glands, others glandless; ovary narrowly obovoid. Fruit dark green when young, black or blue-black when mature, narrowly ellipsoid, inconspicu-

黄果厚壳桂 *Cryptocarya concinna*

硬壳桂 *Cryptocarya chingii*

丛花厚壳桂 *Cryptocarya densiflora*

ously 12-angulate when young. Fl. Mar-May. Fr. Jun-Dec. Evergreen broad-leaved forests on slopes or in valleys below 600 m. Distributed in Guangxi, SE Guizhou, Guangdong, Taiwan, Hainan and Jiangxi. Also in N Vietnam.

硬壳桂

Cryptocarya chingii Cheng

小乔木。叶互生，长圆形或椭圆状长圆形，革质，两面具灰黄色平伏绢状柔毛。圆锥花序腋生和顶生，被毛；花被裂片6；子房棍棒状。果幼时淡绿色，成熟时污红色，椭圆体形，具12棱。花期6-10月，果期9月至翌年3月。生海拔300-800(-2400)米的常绿阔叶林中。产华南和东南。越南北部亦有。

Small trees. Leaves alternate, oblong or elliptic-oblong, leathery, gray-yellow appressed sericeous-pubescent on both surfaces. Panicles axillary and terminal, hairy; perianth-lobes 6; ovary clavate. Fruits pale green when young and dirty red when mature, ellipsoid, 12-angulate. Fl. Jun-Oct. Fr. Sep to next Mar. Evergreen broad-leaved forests at 300-800(-2400) m. Distributed in S and SE China. Also in N Vietnam.

丛花厚壳桂

Cryptocarya densiflora Blume

乔木。叶长椭圆形，革质，离基三出脉。圆锥花序腋生兼顶生，长2.5-8厘米，花密集；白色。果扁球形，直径1.5-2.5厘米，具不明显的纵棱，熟时黑色。花期4-6月，果期7-11月。生海拔650-1700米的常绿阔叶林中。产广东、广西、福建和云南。

Trees. Leaves narrowly elliptic, leathery, triplinerved. Panicles axillary and terminal, 2.5-8 cm long, densely many flowered; flowers white. Fruits compressed globose, 1.5-2.5 cm diam. inconspicuously angled, black when mature. Fl. Apr-Jun. Fr. Jul-Nov. Evergreen broad-leaved forests at 650-1700 m. Distributed in Guangdong, Guangxi, Fujian and Yunnan.

无根藤

Cassytha filiformis L.

寄生缠绕草本，借吸根攀附寄主植物上。茎线形。叶退化为微小鳞片。穗状花序长2.5厘米，密具锈色柔毛；花白色，小；花被片6；可育雄蕊9，第三轮外向。果实小，卵球形，花后包藏于膨大肉质花被筒内。花果期5-12月。生海拔1600米以下的山坡灌丛或疏林中。产中国西南、华南和东南。热带亚洲、澳大利亚和非洲亦有。

Parasitic twining herbs, adhering to the host plants by haustoria. Stems filiform. Leaves small, scaly. Spikes ca. 2.5 cm long, densely rusty pubescent; flowers white, small; perianth lobes 6; fertile stamens 9, those of 3rd whorl extrorse. Fruits small, ovoid, included in dilated and fleshy perianth tube after anthesis. Fl. and fr. May-Dec. Bushes or sparse forests on slopes below 1600 m. Distributed in SW, S and SE China. Also in tropical Asia, Australia and Africa.

无根藤 *Cassytha filiformis*

莲叶桐科 Hernandiaceae

莲叶桐

Hernandia nymphifolia (Presl) Kubizki

乔木，高达20米。叶互生，具长柄，无毛；叶片正三角形，宽15-30厘米，全缘，具掌状脉，基部盾状。聚伞圆锥花序腋生，具长梗；小聚伞花序具3花，基部有4轮生小苞片；2侧生花为雄花：花被片6，成2层；雄蕊3，退化雄蕊6；1中央花为雌花，基部具1柄状总苞：花被片8，成2层；子房下位，具1室和1胚珠；花柱短，柱头扁球形。果实为核果。生海岩上。产海南东部和台湾南部。广布于旧世界热带地区。

Trees, up to 20 m tall. Leaves alternate, long petiolate, glabrous; blades deltoid, 15-30 cm broad, entire, palmately nerved, base peltate. Thyrses axillary, long pedunculate; cymule 3-flowered, subtended by 4 bracteoles; 2 lateral flowers staminate: tepals 6, 2-seriate; stamens 3, and staminodes 6; 1 central flower pistillate: tepals 8, 2-seriate; ovary inferior, 1-locular and with 1 ovule; style short, stigma depressed-globose. Fruit a drupe. Along shores. Distributed in E Hainan and S Taiwan. Widespread in the tropics of the Old World.

多毛青藤

Illigera cordata var. **mollissima** (W. W. Sm.) Kubitzki

藤本。叶为指状3小叶；小叶被黄色长柔毛，卵形至椭圆形，长8-12厘米；叶柄长4-12厘米。聚伞花序腋生；花黄色。果实有4翅，2大2小，较大的宽1.8-2.2厘米。花期5-6月，果期8-9月。生海拔1000-1900米的山坡密林或灌丛中。产四川和云南。

Lianas. Leaves 3-foliolate; leaflets yellowish villous, blade ovate to elliptic, 8-12 cm long; petiole 4-12 cm; Cymes axillary; flowers yellow. Fruit 4-winged, wings 2 large and 2 small, large ones 1.8-2.2 cm wide. Fl. May-Jun. Fr. Aug-Sep. Dense forests or thickets at 1000-1900 m. Distributed in Sichuan and Yunnan.

莲叶桐 *Hernandia nymphifolia*

多毛青藤 *Illigera cordata* var. *mollissima*

红花青藤
Illigera rhodantha Hance

藤本。叶为指状3小叶；小叶被金黄褐色绒毛，卵形至倒卵状椭圆形，长6-11厘米。聚伞圆锥花序腋生，花红色。果实有4翅，较大的宽2.5-3.5厘米。花期9-11月，果期翌年4-5月生海拔300-2100米的山坡密林或灌丛中。产广东、广西、贵州、海南和云南。

Lianas. Leaves 3-foliolate; leaflets densely golden brown tomentose, blade ovate or obovate-elliptic, 6-11 cm long. Cymose panicles axillary, flowers red. Fruit 4-winged, large ones 2.5-3.5 cm wide. Fl. Sep-Nov. Fr. next Apr-May. Dense forests or thickets at 300-2100 m. Distributed in Guangdong, Guangxi, Guizhou, Hainan and Yunnan.

红花青藤 *Illigera rhodantha*

宽药青藤
Illigera celebica Miq.

藤本。叶为3小叶；小叶上方光亮。聚伞圆锥花序腋生，疏松，长约20厘米；花绿白色；雄蕊长过花瓣2倍；花丝蕾期在花药周围弯折，花被状，花丝基部宽达1.5-2.5毫米。花期4-10月，果期6-11月。生海拔(160-)320-1300米的林中。产云南、广西、广东和海南。泰国、柬埔寨、越南、马来西亚、印度尼西亚和菲律宾亦有。

Lianas. Leaves 3-foliolate; leaflets adaxially nitid. Cymose panicles axillary, lax, ca. 20 cm long; flowers green-white; stamens 2 times longer than petals; filaments curved around anthers in bud, tepal-like, base 1.5-2.5 mm broad. Fl. Apr-Oct. Fr. Jun-Nov. Forests at (160-)320-1300 m. Distributed in Yunnan, Guangxi, Guangdong and Hainan. Also in Thailand, Cambodia, Vietnam, Malaysia, Indonesia and the Philippines.

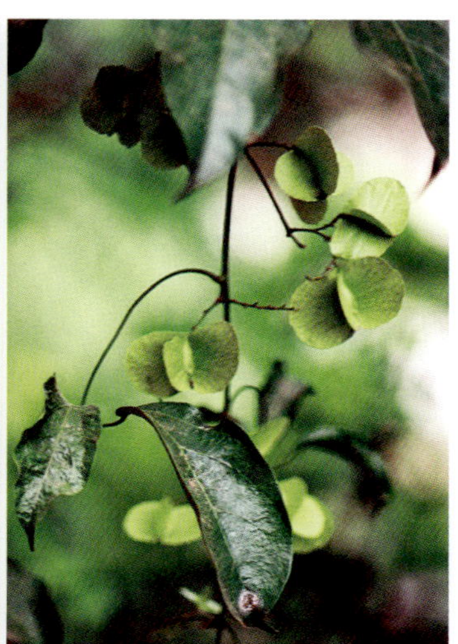

宽药青藤 *Illigera celebica*

中文名索引

D

J

K

Y

Z

拉丁学名索引

D

E

N

O

P

T